Cerebral Ischemia and Calcium

Edited by
Alexander Hartmann and Wolfgang Kuschinsky

With 187 Figures

Springer-Verlag Berlin Heidelberg New York
London Paris Tokyo Hong Kong

Professor Dr. ALEXANDER HARTMANN
Neurologische Universitätsklinik
Sigmund-Freud-Str. 25
D-5300 Bonn 1, FRG

Professor Dr. WOLFGANG KUSCHINSKY
Physiologisches Institut der Universität
Im Neuenheimer Feld 326
D-6900 Heidelberg, FRG

Springer-Verlag Berlin Heidelberg New York
Springer-Verlag New York Berlin Heidelberg

Library of Congress Cataloging-in-Publication Data. Cerebral Ischemia and Calcium / edited by Alexander Hartmann and Wolfgang Kuschinsky. "Proceedings of the International Symposium, Cerebral Ischemia and Calcium, held in Prien/Chiemsee, Federal Republic of Germany, in June 1988" – Pref. 1. Cerebral Ischemia – Pathophysiology – Congresses. 2. Cerebral Ischemia – Chemotherapy – Congresses. 3. Calcium – Antagonists – Congresses. I. Hartmann, A. (Alexander), 1943–. II. Kuschinsky, Wolfgang, 1944–. III. International Symposium, Cerebral Ischemia and Calcium, (1988:Prien am Chiemsee, Germany) [DNLM: 1. Calcium – Pharmacology – Congresses. 2. Calcium Channel Blockers – Therapeutic Use – Congresses. 3. Cerebral Ischemia – Congresses. WL 355 C41138 1988]
RC388.5.C3964 1989 616.8'1 – dc20 89-19733 CIP

Softcover reprint of the hardcover 1st edition 1989

ISBN 978-3-642-85865-9 ISBN 978-3-642-85863-5 (eBook)
DOI 10.1007/978-3-642-85863-5

List of Contributors

You will find the addresses at the beginning of the respective contribution

Preface

In this volume we are pleased to present the proceedings of the International Symposium *Cerebral Ischemia and Calcium,* held in Prien/Chiemsee, Federal Republic of Germany, in June 1988.

Knowledge of the pathophysiology of cerebral ischemia is limited although an abundance of studies have been performed on this subject. Recently, a major advance has resulted from focusing on the role of calcium in the development of ischemic lesions in the brain. Although it may at present not be justified to ascribe to calcium an exclusive or even causal role in the generation of an ischemic lesion, it appears mandatory to devote due attention to this topic.

This book contains the results of a broad spectrum of experiments and measurements. Measurements performed in patients with cerebral ischemia are presented as well as studies on the pathophysiological role of calcium in experimental ischemia, including investigations on the cellular level and in vitro data. Therapeutic strategies using calcium antagonists are also pertinent in this context. In addition recent studies on calcium antagonists in migraine, epilepsy, and subarachnoid hemorrhage are included.

We acknowledge the help of the advisory board in the selection of papers. We would like to thank the members of the advisory board for their critical analysis and helpful suggestions. They are: K.-A. Hossmann, Cologne; J. Olesen, Copenhagen; K. J. Reulen, Bern; L. Symon, London; P. A. van Zwieten, Amsterdam; and F. Yatsu, Houston.

Bonn, Heidelberg, June 1989

ALEXANDER HARTMANN
WOLFGANG KUSCHINSKY

Contents

Pharmacology of Calcium and Cerebral Ischemia

Cerebral Ischemia: Pathophysiology and Biochemistry

Experimental Cerebral Ischemia and Calcium

Clinical Cerebral Ischemia and Calcium

Migraine

Epilepsy

Brain Protection

Subarachnoid Hemorrhage

Pharmacology of Calcium and Cerebral Ischemia

Academic Hemianopia and the Treatment of Cerebrovascular Disease

M. D. O'Brien

The term academic scotomata has often been used to describe defects in our perception of problems and the analogy can be extended to a variety of field defects, including tunnel vision. Hemianopic defects can be of several sorts, of which the commonest is probably the extrapolation of graphs beyond the data. This paper considers the often erroneous assumption that if a problem is caused by too little of something, then more of it must be beneficial.

The history of treatment of cerebrovascular disease is full of examples and it would be tedious to enumerate all of them, though some examples illustrate the point. These may be divided into, firstly, alterations of physiological variables including carbon dioxide, blood pressure, glucose, oxygen, blood viscosity and surgical methods of improving perfusion and, secondly, drug therapy.

The effect of carbon dioxide on the cerebral circulation was recognised more than half a century ago and led to the obvious thought that hypercapnia might help patients with stroke. No obvious benefit resulted and the practice was abandoned; but in the early 1970s blood flow measurement showed that although hypercapnia markedly increased blood flow in normal brain, in some patients there was a reduction in flow in or near an infarct, the so-called intracerebral steal phenomenon. This phenomenon was not invariable and actually depended on a number of other features, including a limitation of input and the presence of suitable collaterals. Although a relative hyperaemia in association with an infarct had already been well recognised, the so-called luxury perfusion syndrome, it was not clear whether the steal phenomenon only occurred when the flow was already inappropriately high or whether it could actually reduce the flow to metabolically important levels, considerably lower than normal values. Despite this, the suggestion was made that if vasodilation by carbon dioxide inhalation could cause a reduction in flow in an infarct, then vasoconstriction by hyperventilation might cause a rise in flow in the ischaemic area and, furthermore, that this must be beneficial. Several large studies involving considerable doctor-patient time and effort were needed to show that this fallacious argument had no merit. Indeed, there are even circumstances in which a reduction in blood flow in normal brain by hyperventilation might reduce the blood supply to an infarct if the collateral circulation is affected by the vasoconstriction.

A similar problem exists over the control of blood pressure. The association of stroke with hypertension has been known for many years and this has often led to overtreatment because, once an infarct has occurred, hypotension is prob-

Department of Neurology, Guy's Hospital, St. Thomas Street, London SE1 9RT, U.K.

Cerebral Ischemia and Calcium
Ed. by A. Hartmann and W. Kuschinsky

ably more harmful than moderate hypertension, particularly if the collateral circulation is pressure dependent. The critical point is to what extent the systemic pressure is transmitted to capillaries. If the main supply to an infarct is patent, which is probably not the case in most patients with stroke, then the development of oedema and blood-brain barrier damage is directly proportional to the blood pressure. If the only supply to an infarct is via a tenuous collateral circulation, some head of pressure may be necessary. If there is ischaemia but no infarct and patent vessels, as may occur after subarachnoid haemorrhage, then relative hypertension may reverse a neurological deficit caused by vasospasm. It is, therefore, impossible to determine the optimal blood pressure in cerebrovascular disease, as the optimum will vary considerably according to circumstances, most of whose parameters are as yet unmeasurable.

The effect of an elevated haematocrit has recently caused some confusion. There is little argument about how to manage patients with true polycythaemia, but the rise in flow which follows venesection in patients with moderately elevated haematocrit was thought to be beneficial, until it became clear that this was a response to reduced tissue oxygen delivery because of haemodilution, rather than being due to a change in viscosity. Two large recent multicentre trials of haemodilution in stroke from Italy [1] and Scandinavia [2] have shown no benefit.

The effect of changes in blood sugar provides another interesting example. Hyperglycaemia is usually considered to be harmful, but this is based on the outcome in animal experiments with pre-existing hyperglycaemia, which is of course applicable to diabetics, and in reperfusion experiments. However, the high blood sugar has to be delivered to the area of ischaemia, and Ginsberg et al. [3] showed that where end-arteries were occluded, this deleterious effect was not observed; indeed, the infarcts in these hyperglycaemic animals were actually smaller than in controls, possibly due to improved local energy metabolism at the periphery of a lesion. Once again, it might be difficult to determine the optimum blood sugar level in patients with stroke.

Even the delivery of oxygen is not as obvious as might first appear. Clearly hypoxia should be avoided but there is some evidence that 100% oxygen may be harmful. Mickel et al. [4] showed that after 15 min of carotid occlusion in the ubiquitous gerbil global ischaemia model, there was much more extensive white matter damage in the animals that breathed 100% oxygen for 3 h than in those that breathed normal air, and they postulated that this was because of increased lipid peroxidation due to free radical formation.

Many of these measures are designed to supply more blood or metabolic substrate to an infarct; the argument is that more blood must be an advantage, but other than shortly after an infarct this may not be correct. In most cases the blood supply, even if reduced, is sufficient for the metabolic needs of the tissue and in stroke this is because the metabolic demand has been reduced by infarction to match the blood supply available. There is, of course, a therapeutic window; but this is measured in minutes or hours and not days and weeks. In the early days of carotid endarterectomy, surgeons were confident that it must be beneficial to remove stenoses significant in terms of local vessel flow, though measurement of cerebral blood flow, after the reactive hyperaemia had settled, showed no improvement [5]. The more recent EC-IC bypass study clearly demonstrated that even in those patients with haemodynamically determined events, there was no benefit in improving the blood supply by shunting, nor did this study lend support to the argument that these surgical procedures reduce the risk of strokes in the future by increasing the cerebrovascular reserve [6].

In most of these circumstances "not too little, not too much, but just right" would appear to be an apposite motto and the problem is how to determine what is just right, particularly as this changes during the time of the "therapeutic window" when events might be influenced.

The drug treatment of cerebrovascular disease also presents considerable problems. Perhaps one of the major difficulties is that delivery of the drug to an infarct cannot be determined. This may be particularly important if the drug requires a critical concentration to produce its effect. It is not possible to extrapolate the effect of a drug tested in vitro, or in animal experiments, directly to man. It is not just a question of calculating a comparable dose, for there is often considerable interspecies variance in drug response. Animal experiments are often done following intraperitoneal or intravenous injection, which may have little relevance to an oral preparation in man, particularly those drugs, including many of the calcium channel blockers, which have a considerable hepatic first pass effect. The drug must cross the blood-brain barrier, which very effectively excludes many preparations in normal brain, though the permeability may be altered by ischaemia. It is also important to determine the effect of the drug on damaged tissue, which is unlikely to be the same as the effect on normal brain, and the degree of damage may well be relevant. The timing of administration must also be considered and may be critical. In most patients with stroke it is only possible to treat after the event and logistical problems usually entail that this is after some hours. However, studies involving pretreatment are not entirely irrelevant since they may be of great benefit to neurosurgeons, who can accurately time their vascular insult.

A large number of drugs have been tried in the treatment of cerebrovascular disease and the results are often conflicting. It is not difficult to see why. Steroids, which have such a striking effect on brain tumours, seem to be almost completely ineffective in stroke; this is probably because the drug is not delivered to the site in adequate concentration, and they appear to need intact cell membranes to produce an effect. In stroke, the damage to cell membranes, and the rate at which this occurs, precludes any real benefit. Osmotic agents have also been widely used, but they too need to be delivered to the ischaemic area and then only dialyse across intact membranes and so effectively dehydrate normal brain while having little or no effect on ischaemic brain.

Finally to the question of calcium channel antagonists, the topic of this symposium. Experimental evidence available to date suggests that calcium enters the cells in large quantities at the same time as the potassium flux, which is usually considered to be a marker of cell membrane failure. If calcium channel antagonists are to protect brain cells, it will be necessary to demonstrate that there is significant calcium overload before membrane failure and that prevention of this overload protects the cell. Of course, these drugs must reach the infarct in sufficient concentration and early enough to be effective and that too may be a problem. Many of the effects of calcium channel blockers so far reported can be ascribed to their effect on the vascular smooth muscle, causing vasodilation, and perhaps those who claim a protective effect on ischaemic neurones, but who are really observing an effect on vascular smooth muscle, are also demonstrating a form of academic hemianopia.

References

1. Italian Acute Stroke Study Group (1988) The Italian hemodilution trial in acute stroke. Stroke 19:145
2. Scandinavian Stroke Study Group (1987) Multicentre trial of hemodilution in ischaemic stroke. Stroke 18:691–699
3. Ginsberg MD, Prado R, Deietrich D, Busto R, Watson BD (1987) Hypoglycaemia reduces the extent of cerebral infarction in rats. Stroke 18:570–574
4. Mickel HS, Vaishnav YN, Kempski O, Lubitz D, von Weiss JF, Feuerstein G (1987) Breathing 100% oxygen after global brain ischaemia in mongolian gerbils results in increased lipid peroxidation and increased mortality. Stroke 18:426–430
5. O'Brien MD, Veall N, Luck RJ, Irvine WT (1967) Cerebral cortex perfusion rates in extracranial cerebrovascular disease and the effects of operation. Lancet II:392–395
6. ECIC Bypass Study Group (1985) Failure of extracranial intracranial arterial bypass to reduce the risk of ischaemic stroke. N Engl J Med 313:1191–1250

The Classification of Calcium Antagonists

T. GODFRAIND

Classifications in pharmacology are usually established in order to identify chemical compounds which share in common a major biological property. Recognition of a class of drugs does not imply that drugs belonging to a given class have an identical pharmacological profile. Differences may be the basis for recognition of subclasses. The major common properties as well as the subtle differences need to be characterized using well established quantitative pharmacological criteria. Drug classification is not an academic game. Pharmacological classifications have not only interest per se; their purpose is to establish the criteria for a rational choice between various drugs in order to optimalize their clinical use. This will be possible if the pharmacological tests chosen for the assay of the drugs present a sufficient predictive value as far as therapy is concerned.

Objective criteria are most easily obtained when experimental studies are performed at the first levels of the pharmacotherapeutic cascade (Table 1). Indeed, at molecular, cellular, and tissue levels, accurate estimates of quantitative parameters of drug action are usually possible. Estimates of such parameters in vivo, in either physiological or pathological conditions, are often altered by factors unrelated to the chosen criteria.

The purpose of this paper is to illustrate the experimental protocol allowing recognition of a drug as a calcium antagonist. We will also examine on which pharmacological grounds, differences and similarities may be found in this group of agents.

Table 1. The pharmacotherapeutic cascade of calcium antagonists (Godfraind et al. [14])

Level of pharmacological action	Qualitative effect	Quantitative parameter
1. Molecule	Binding to Ca^{2+} channel	K_d, K_i
2. Cell	Changes in Ca fluxes	IC_{50}
3. Tissue (in vitro)	Cardiac negative inotropic and chronotropic effects, smooth muscle relaxation	IC_{50}, PA_2
4. Organ (in vivo)	Hemodynamic modifications (ECG, cardiac output, blood pressure)	Plasma levels
5. Clinical disorder	Antianginal and antihypertensive effects	Plasma levels

Laboratoire de Pharmacodynamie Générale et de Pharmacologie, Université Catholique de Louvain, UCL 7350, Avenue Emmanuel Mounier 73, B-1200 Bruxelles, Belgium.

Cerebral Ischemia and Calcium
Ed. by A. Hartmann and W. Kuschinsky

Major Pharmacological Properties and Terminology

The term calcium antagonist was used coincidentally by Fleckenstein and Godfraind in the 1960s as a conclusion of independent studies in arterial preparations [13] and cardiac tissues [5].

A first indication that a drug could be a calcium antagonist may be given by the observation that the agent concerned is acting as a polyvalent antagonist. For instance, in some vascular smooth muscles, dose-dependent inhibition of the contractile response evoked by a large variety of vasoconstrictors may be observed within the same range of concentrations of the antagonist, independently of the agonist utilized. This indicates that the antagonist interferes with a process common to the various agonists.

The common process sensitive to such polyvalent antagonists is the increase in calcium permeability evoked either by KCl-evoked depolarization of smooth muscle cells, or by specific interaction of agonists with their receptors [11]. Figure 1 shows the effect of cinnarizine on contractions evoked by Ca^{2+} in K^+ -depolarized rabbit mesenteric arteries. Arterial preparations were preincubated in Ca^{2+} free physiological solution, depolarized in Ca^{2+} free KCl-rich solution, and then further incubated with increasing Ca^{2+} concentrations. Cumulative concentration-effect curves were obtained before and after addition of various concentrations of cinnarizine. The inhibitory effect of cinnarizine is observed at

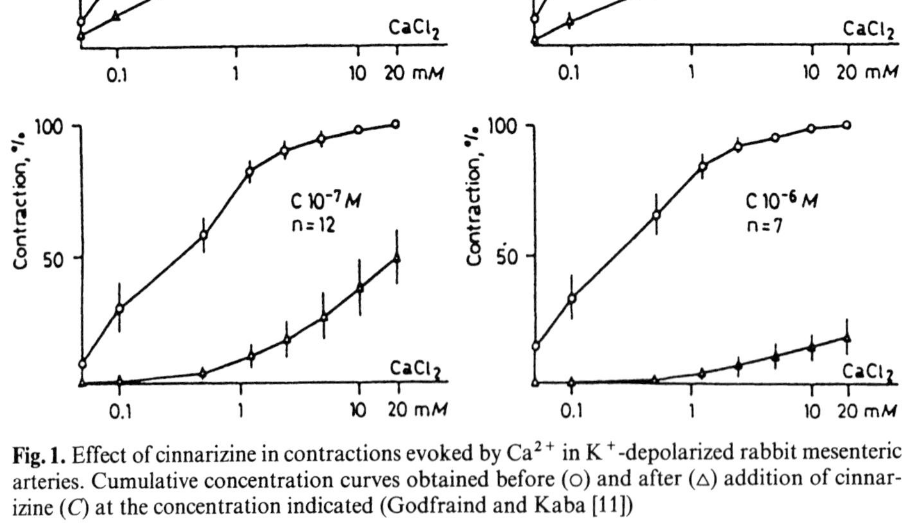

Fig. 1. Effect of cinnarizine in contractions evoked by Ca^{2+} in K^+-depolarized rabbit mesenteric arteries. Cumulative concentration curves obtained before (○) and after (△) addition of cinnarizine (*C*) at the concentration indicated (Godfraind and Kaba [11])

concentrations as low as 1 nM and resembles the action of antagonists in receptor studies. In view of this similarity, the term "calcium antagonist" was suggested to describe the action of cinnarizine. Since this first description, a large variety of drugs have been shown to possess this property of cinnarizine. As shown in Fig. 2, they belong to various chemical families which show a similar effect on calcium contraction in vascular smooth muscle. It has recently been shown that modifications in the group of dihydropyridines result in compounds which *facilitate* Ca entry; such agents have been termed "calcium agonists" [3].

Several experimental protocols have been designed in order to estimate unidirectional Ca fluxes in vascular (and other) smooth muscles [6]. In resting conditions, when a muscle is bathed in physiological solution containing ^{45}Ca, the specific activity of intracellular calcium increases slowly with the duration of incuba-

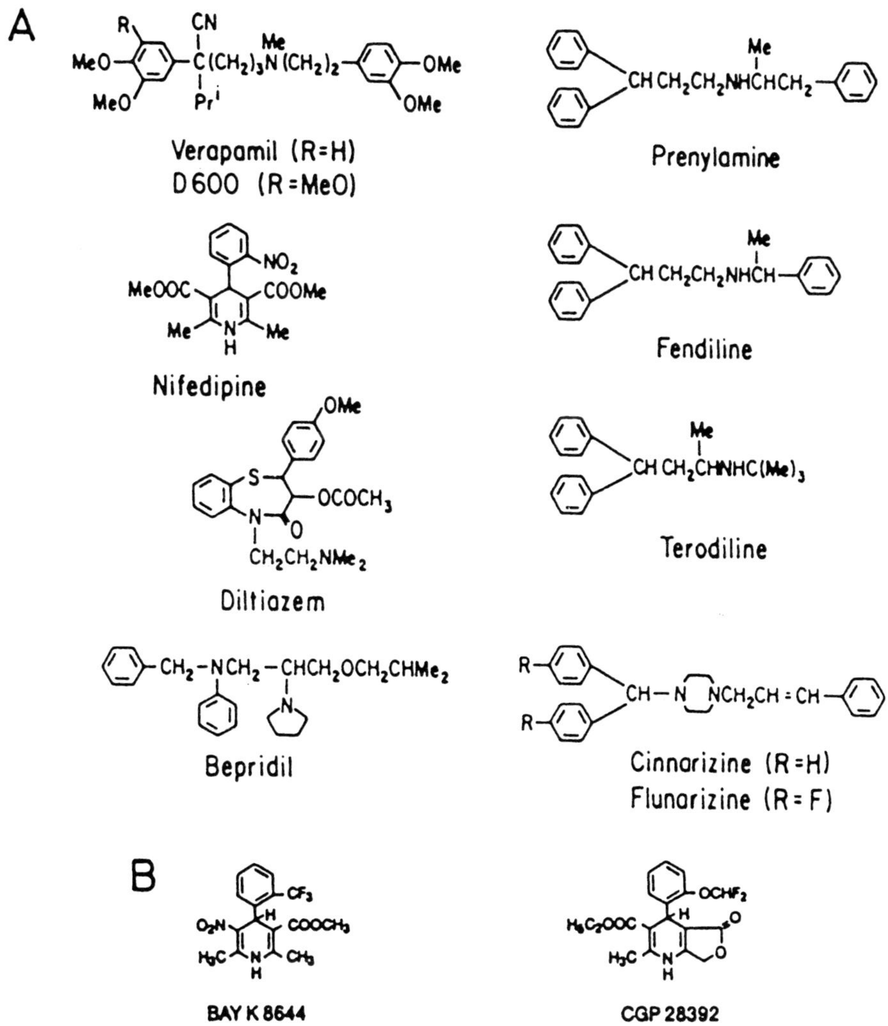

Fig. 2. Structure of some calcium entry blockers (**A**) and of calcium agonists (**B**)

tion and reaches a plateau after 1 or 2 h, when inward and outward movements of ^{45}Ca become equal. The initial rate of uptake of ^{45}Ca may be taken as reflecting an inward influx. When the intracellular fraction is completely loaded, the initial rate of ^{45}Ca loss after transfer to nonradioactive solution depends mainly on outward movements.

Calcium influx rate under resting conditions is about 0.014 and 0.03 pmol $cm^{-2} \cdot s^{-1}$ respectively in rat aorta and in rabbit aorta [10]. This calcium influx has been termed passive Ca leak, because it is thought to occur through pathways that differ from those involved in calcium influx evoked by stimulating agents. This calcium influx through leak channels is resistant to calcium antagonists. As shown by us some years ago [6, 12], norepinephrine and KCl stimulations evoke a 20-fold increase in the rate of calcium influx through channels sensitive to calcium antagonists. This stimulated influx may be completely blocked by those agents at concentrations blocking the contractile response. Furthermore, the dose-effect curves relating the degree of inhibition of Ca influx and of contraction to the concentration of any given calcium antagonist are superimposed. This indicates that inhibition of contraction may be related to blockade of receptor- and voltage-operated Ca channels [14].

An interesting observation was made at the earlier stage of the study of calcium antagonists; it concerns their selectivity of action according to the vascular bed examined [13]. It was observed that inhibition of the contractile response to norepinephrine could be completely blocked in some vascular beds, but that in others a resistant part of the contraction could be identified. The ratio between the contractions resistant and sensitive to calcium antagonists may vary from one vessel to another. The magnitude of the contraction resistant to calcium antagonists in one given vessel is similar to the contraction evoked by the same agonist after preincubation of the vessel in Ca free-EGTA-physiological solution for a time sufficient to displace all extracellular free Ca^{2+}. Therefore, it has been proposed that this contraction is supported by Ca^{2+} released from intracellular stores during receptor activation.

It has recently been shown that agonist-receptor interactions result in the activation of the phosphatidylinositol cascade with the formation of IP_3, which is responsible for the release of calcium from the sarcoplasmic reticulum [1]; this process is not affected by calcium antagonists. Because calcium antagonists inhibit stimuli-evoked Ca entry, they have also been termed calcium entry blockers. Electrophysiological studies have shown that in cardiac tissue they block the slow Ca current; therefore the term slow channels blockers has also been proposed.

Tissue Selectivity of Calcium Antagonists

Membranes isolated from various tissues possess specific binding sites for at least three chemical classes of calcium entry blockers: dihydropyridines, phenylalkylamines, and benzothiazepines. These sites are distinct but interact with each other. Overall, the properties of the binding sites are remarkably similar in isolated membranes from different tissues, including nervous and endocrine tissues.

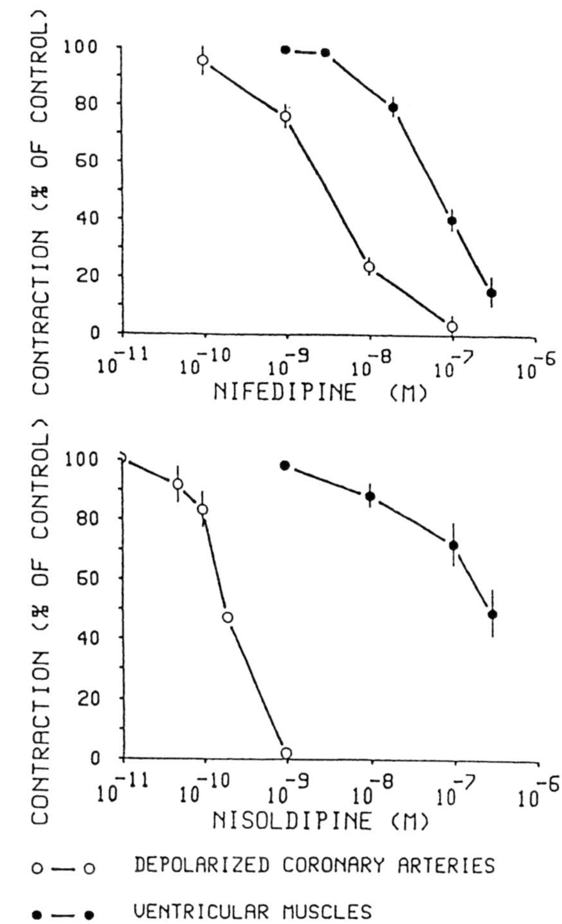

Fig. 3. In vitro study of K^+-depolarization-induced contraction of human coronary arteries and of the contraction of electrically paced human ventricular muscles by nifedipine and nisoldipine [15]

Furthermore, the size and immunochemical features of the polypeptide bearing the three binding sites are very similar in skeletal, cardiac, and smooth muscle [19]. Thus, at the *molecular level*, little selectivity may be detected in the interaction of calcium entry blockers with various tissues (for recent reviews, see refs. 14 and 20).

In contrast, at the *tissular level*, the selectivity of calcium entry blockers is obvious [14, 22]. It is well known, for instance, that neurotransmitter release from brain tissue is little affected by dihydropyridines, and that cardiac contractility is much less sensitive to dihydropyridines than is smooth muscle contraction. Moreover, as illustrated in Fig. 3, the preferential action on smooth muscle is more marked with nisoldipine than with nifedipine. Some agents may even show a selectivity within the arterial tree [4]. For example, whereas nifedipine appears equipotent on rat aorta and mesenteric artery (contracted by potassium depolarization or receptor stimulation), flunarizine is distinctly more potent on the mesenteric artery [14]. When comparing different tissues as regards the sensitivity of calcium-dependent processes to calcium entry blockade, a first source of variation to be considered is the contribution of intracellular calcium stores to activation. For instance, in rat aorta, some 50% of the contraction evoked by norepinephrine is resistant to calcium entry blockers, whereas in rat mesenteric artery, this proportion is only 10%. This difference is related to the greater role of intracellular calcium mobilization upon α-adrenoceptor stimulation in aorta.

Over the last few years, two fundamental causes of tissular selectivity have emerged: multiplicity of calcium channel subtypes and of calcium channel states. These factors have been revealed essentially thanks to the use of the powerful patch-clamp technique, and they are being confirmed in ligand binding studies on intact cells.

Numerous electrophysiological studies show the existence of at least two subtypes (L and T) of voltage-dependent calcium channels in several cell types, including neurons and cardiac and smooth muscle cells. Only one of them, the L type, is blocked by "classical" calcium entry blockers. Neurotransmitter release may depend primarily on calcium influx via N channels, which are insensitive to dihydropyridines but are blocked by ω-conotoxin [17].

Three main states of voltage-dependent calcium channels have been postulated: resting, open, and inactivated. The state of the L channel dramatically influences its affinity for calcium channel blockers. In particular, in both cardiac and smooth muscle cells, dihydropyridines may show an affinity three orders of magnitude higher for the inactivated state as compared with the resting state [2, 16, 18]. As the inactivated state is promoted by membrane depolarization, these drugs will be more effective in smooth muscle tissues, which are characterized by less negative resting potentials and long-lasting depolarizing stimuli, than in myo-

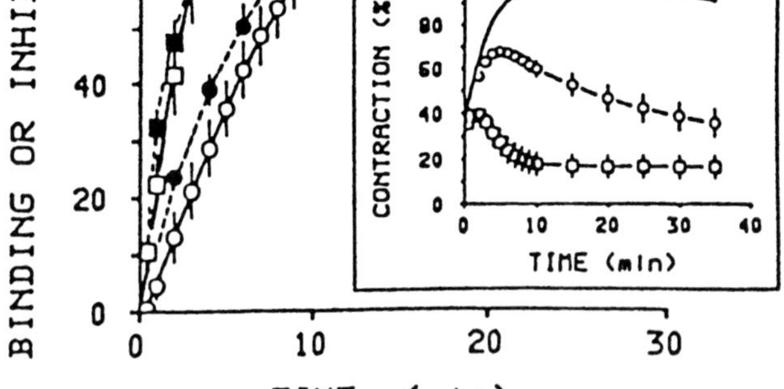

Fig. 4. Comparison between kinetics of ^{3}H-(+)PN 200-110 association to its specific binding site (*broken line* – ●: 0.1 nM; ■: 0.38 nM) and (+)PN 200-110 inhibition of KCl-evoked contraction of rat aorta (*solid line* – ○: 0.1 nM; □: 0.3 nM). *Inset*: Mechanical responses to KCl depolarization of rat aortic rings; control (*upper curve*), rings treated with 0.1 nM (○) or 0.3 nM (□) (+)PN 200-110 [23]

cardial tissue, which shows a more negative resting potential and is usually submitted to short repetitive depolarizing stimuli.

Using ^{3}H-(+)PN 200-110, we have recently performed an analysis of dihydropyridine receptor sites in intact mesenteric arteries [16]. Arterial segments incubated in depolarizing medium bind this ligand with high affinity (K_D 44 pM). The affinity is lower (K_D 200 pM) when segments are incubated in physiological medium. The B_{max} is not significantly influenced by depolarization.

These results are compatible with the view that dihydropyridines bind preferentially to the inactivated state of the channel. The effect of depolarization on affinity allows us to understand why the inhibitory effect of (+)PN 200-110 on arterial contraction increases with the time after depolarization (Fig. 4). Upon depolarization, the inhibition of contraction of rat aorta by (+)PN 200-110 develops with a time course similar to that of the specific binding to the high affinity state of the channel in isolated membranes (Fig. 4) [23]. Thus, binding to this high affinity state, which is induced by depolarization in intact arteries, appears to be the rate-limiting step of the inhibition of the contraction.

In addition to membrane potential, a variety of factors (protein kinases, G proteins) may influence the functioning of calcium channels. These regulations may differ according to the tissue and contribute to the tissular selectivity of calcium entry blockers. It has been shown that, in arteries stimulated by norepinephrine, calcium entry into smooth muscle cells is reduced by an endothelial factor (EDRF) acting via cyclic GMP [9].

Pharmacological Classification of Calcium Antagonists

Studies of tissue selectivity indicate that a pharmacological classification of calcium antagonists may actually provide a better understanding of this group than a molecular classification. This view has been agreed upon by several authors [21]. The classification summarized in Table 2 is based on such an analysis. Furthermore, on the basis of functional criteria, differentiation into four subclasses of selective calcium entry blockers is possible. These criteria are as follows:

1. Cardiac effects resulting mainly from inhibition of slow inward current: antiarrythmic action, slowing of conduction, and negative inotropic effects
2. Vascular effects: inhibition of myogenic contraction of rat portal vein and of contractions of isolated arteries
3. Tissue protection: against myocardial necrosis due to intracellular calcium overload resulting from anoxia, ischemia, or isoproterenol intoxication; against ischemic and anoxic alterations of brain tissue
4. Hemorrheological effects: restoration of red blood cell deformability altered by calcium-induced stiffness
5. Specific interaction with [^{3}H] dihydropyridine binding sites

The four subclasses obtained in this manner correspond to the four subgroups of specific calcium entry blockers listed in Table 2, which also indicates the WHO classification (types I–IV). The different clinical indications for diltiazem, verapa-

Table 2. Classification of calcium antagonists (calcium entry blockers)

Group I:	Specific calcium entry blockers
	Subgroup I. A: Agents selective for slow calcium channels in myocardium (slow channel blockers)
	Phenylalkylamines: lead drug: verapamil (Type I)
	Dihydropyridines: lead drug: nifedipine (Type II)
	Benzothiazepines: lead drug: diltiazem (Type III)
	Subgroup I. B: Agents not selective for slow calcium channels in myocardium
	Diphenylpiperazines: lead drug flunarizine (Type IV)
Group II:	Nonspecific calcium entry blockers
	Subgroup II. A: Agents acting at similar concentrations on calcium channels and fast sodium channels
	Lead drugs: bepridil, lidoflazine, tiapamil
	Subgroup II. B: Agents interacting with calcium channels while having another primary site of action
	They include, among others: agents acting on sodium channels (local anesthetics, phenytoin); on catecholamine receptors (examples: phenothiazines, yohimbine derivatives); on benzodiazepine receptors (diazepam, flurazepam); on opiate receptors (examples: loperamide, fluperamide); on cyclic nucleotide phosphodiesterases (examples: amrinone, cromoglycate, papaverine); barbiturates; cyproheptadine; indomethacin; reserpine

mil, nifedipine, and flunarizine are consistent with this classification, which emphasizes the importance of tissue selectivity in the pharmacological and therapeutic actions of calcium entry blockers [14].

Concluding Remarks

As pointed out in the introduction to this paper, the choice of the most appropriate drug for a given pathological situation is peremptory in therapy. We have provided experimental data which illustrate the existence of different pharmacological profiles among specific calcium antagonists. Such differences may be related to the mode of interaction of these drugs with calcium channels depending upon the state of the channels and upon the localization of their receptor sites within the channels. Analysis of the differences provides a rational explanation for the selective indications for these drugs either in cardiovascular disorders or in neurological disorders [14].

Acknowledgments. This work has been supported by F.R.S.M. Grant no 3.9006.87.

References

1. Abdel-Latif AA (1986) Calcium-mobilizing receptors, polyphosphoinositides, and the generation of second messengers. Pharmacol Rev 38:227–272
2. Bean B, Sturek M, Puga A, Hermsmeyer K (1986) Calcium channels in muscle cells isolated from rat mesenteric arteries: modulation by dihydropyridine drugs. Circ Res 59:229–235
3. Bechem M, Hebisch S, Schramm M (1988) Ca^{2+} agonists: new, sensitive probes for Ca^{2+} channels. Trends Pharmacol Sci 9:257–261
4. Cauvin C, van Breemen C (1987) Vascular effects of organic calcium antagonists. ISI Atlas of Science Pharmacology 1:13–19
5. Fleckenstein A (1977) Specific pharmacology of calcium in myocardium, cardiac pacemakers, and vascular smooth muscle. Annu Rev Pharmacol Toxicol 17:149–166
6. Godfraind T (1976) Calcium exchange in vascular smooth muscle, action of noradrenaline and lanthanum. J Physiol 260:21–35
7. Godfraind T (1983) Actions of nifedipine on calcium fluxes and contraction in isolated rat arteries. J Pharmacol Exp Ther 224:443–450
8. Godfraind T (1985) Cellular and subcellular approaches to the mechanism of action of calcium antagonists. In: Rubin RP, Weiss GB, Putney JW (eds) Calcium in biological systems. Plenum, New York, pp 411–421
9. Godfraind T (1986) EDRF and cyclic GMP control gating of receptor-operated calcium channels in vascular smooth muscle. Eur J Pharmacol 126:341–343
10. Godfraind T (1988) Pharmacology of calcium metabolism in smooth muscle. In: Baker PF (ed) Calcium in drug action. Springer, Berlin Heidelberg New York (Handbook of experimental pharmacology, vol 83, pp 361–382)
11. Godfraind T, Kaba A (1969) Blockade or reversal of contraction induced by calcium and adrenaline in depolarized arterial smooth muscle. Br J Pharmacol 36:549–560
12. Godfraind T, Miller RC (1983) Specificity of action of Ca++ entry blockers. A comparison of their actions in rat arteries and in human coronary arteries. Circ Res 52:181–191
13. Godfraind T, Kaba A, Polster P (1968) Differences in sensitivity of arterial smooth muscles to inhibition of their contractile response to depolarization by potassium. Arch Int Pharmacodyn Ther 172:235–239
14. Godfraind T, Miller RC, Wibo M (1986) Calcium antagonism and calcium entry blockers. Pharmacol Rev 38:321–416
15. Godfraind T, Eglème C, Finet M, Jaumin P (1987) The actions of nifedipine and nisoldipine on the contractile activity of human coronary arteries and human cardiac tissue in vitro. Pharmacol Toxicol 61:79–84
16. Morel N, Godfraind T (1987) Prolonged depolarization increases the pharmacological effect of dihydropyridines and their binding affinity for calcium channels of vascular smooth muscle. J Pharmacol Exp Ther 243:711–715
17. Rivier J, Galyean R, Gray MR, Azimi-Zonooz A, McIntosh JM, Cruz LJ, Olivera BM (1987) Neuronal calcium channel inhibitors. Synthesis of W-conotoxin GVIA and effects on 45 Ca uptake by synaptosomes. J Biol Chem 262:1194–1198
18. Sanguinetti MC, Kass RS (1984) Voltage-dependent block of calcium channel current in the calf cardiac Purkinje fiber by dihydropyridine calcium channel antagonists. Circ Res 55:336–348
19. Schmid A, Barhanin J, Coppola T, Borsotto M, Lazdunski M (1986) Immunochemical analysis of subunit structures of 1,4-dihydropyridine receptors associated with voltage-dependent Ca^{2+} channels in skeletal, cardiac, and smooth muscles. Biochemistry 25:3492–3495
20. Triggle DJ, Janis RA (1987) Calcium channel ligands. Ann Rev Pharmacol Toxicol 27:347–369
21. Vanhoutte PM, Paoletti R (1987) The WHO classification of calcium antagonists. Trends Pharmacol Sci 8:4–5
22. Van Nueten JM (1969) Comparative bioassay of vasoactive drugs using isolated perfused rabbit arteries. Eur J Pharmacol 6:286–293
23. Wibo M, DeRoth L, Godfraind T (1988) Pharmacological relevance of dihydropyridine binding sites in membranes from rat aorta: kinetic and equilibrium studies. Circ Res 62:91–96

Pharmacological Activities of Calcium Antagonists: Therapeutic Applications and Adverse Reactions

P. A. van Zwieten

Introduction

Calcium antagonists or calcium entry blockers are considered to be a breakthrough in the drug treatment of cardiovascular disease, while their importance as tools in fundamental research is generally recognized as well. More recently, certain types of calcium antagonists have been studied as potentially useful drugs in the treatment of neurological disorders, such as migraine, cerebral ischemia and hypoxia, certain forms of vertigo, epilepsy, and subarachnoid hemorrhage.

It is the aim of the present communication to review the basic properties and potential application of these drugs in various disorders, with an emphasis on the pharmacological background of the therapeutic activity and adverse effects. Organ specificity, an obvious characteristic of the calcium antagonists, will be discussed as well.

Nomenclature and Definitions

According to Fleckenstein's classical and widely accepted definition [7], a calcium antagonist is a drug which inhibits the slow inward calcium current (cf. Fig. 1) in a dose-dependent manner, while a second requirement is that this effect can be reversed by calcium ions. Accordingly, the drug-induced inhibition of the slow inward current, which is known to be carried by calcium ions, obeys the laws of competitive inhibition. In this connection the potency of the calcium antagonists catches the attention: a single molecule of such a drug is able to prevent the influx of several thousands of calcium ions. From a pharmacological point of view the term "calcium antagonism" may be subject to fundamental criticism. In the pharmacological sense of the word an antagonist is a drug which inhibits the effect of another drug, e.g., atropine vs acetylcholine, or propranolol vs isoproterenol. Verapamil and other "calcium antagonists," however, do not really inhibit the *effects* of calcium ions; rather they inhibit the passage of calcium ions through specific channels in cell membranes. For this reason the term "calcium entry blocker," proposed by Vanhoutte and Bohr [42], would seem more appropriate.

Division of Pharmacotherapy, Academic Medical Centre, University of Amsterdam, Meibergdreef 15, NL-1105 AZ Amsterdam, The Netherlands.

Cerebral Ischemia and Calcium
Ed. by A. Hartmann and W. Kuschinsky

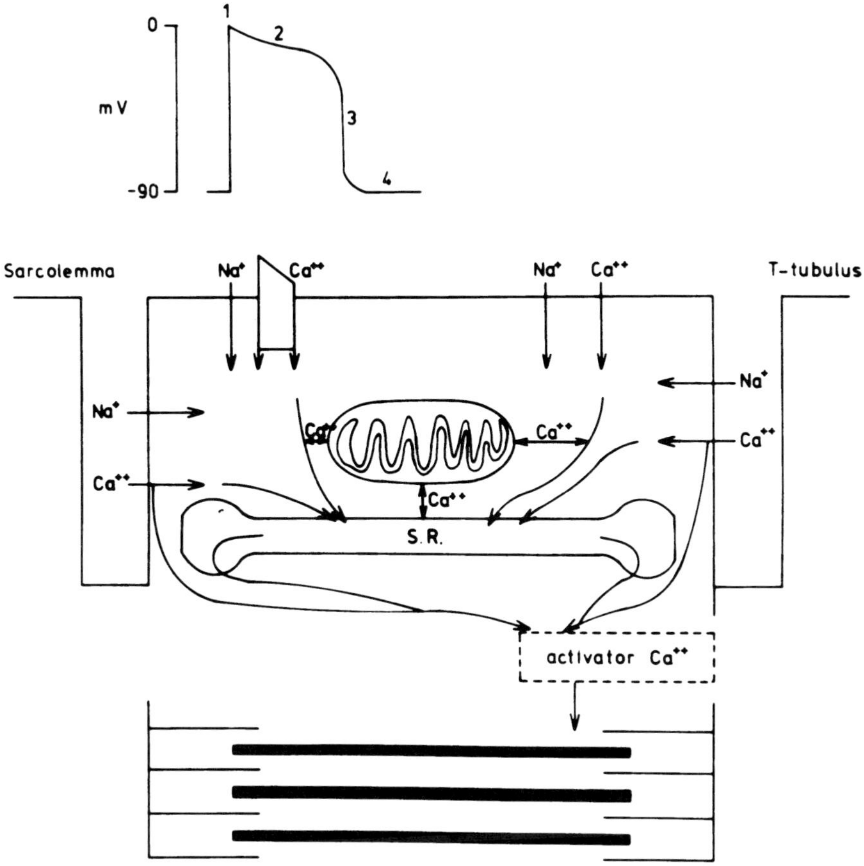

Fig. 1. Effect of CAs on a cardiac cell. *Top:* typical cardiac action potential. The calcium (slow) inward current flows during the characteristic plateau phase (phase 2) of the action potential. This calcium influx is selectively inhibited by CAs. Activation of the sarcoplasmatic reticulum (*S.R.*) and other cellular calcium pools occurs via Ca^{2+} and Na^{+} ions which flow into the cell. The SR and other pools donate activator Ca^{2+} ions which stimulate the contractile proteins. The presence of tubular systems (invaginations), which are characteristic of cardiac tissues, results in a considerable enlargement of the cellular surface, thus enabling an effective influx of Na^{+} and Ca^{2+} ions. Inhibition of the calcium inward flux by a CA causes diminished activation of the contractile proteins

An equivalent, alternative expression would be "slow channel inhibitors" or "slow channel blockers" or "calcium channel blockers," as proposed by Nayler and Poole-Wilson [24, 25]. The term "calcium overload blockade" has recently been submitted by Borgers [4] and by ourselves [44, 45] for the drug-induced inhibition of the overloading of the cell under pathological, ischemic conditions. Since, however, the nomenclature *calcium antagonists* has been maintained in the recently issued WHO classification [32], we shall henceforth apply this term throughout and use its abbreviation, *CA*.

Chemistry

As is obvious from Fig. 2, the CAs are a most heterogeneous series of compounds. Accordingly, several different subgroups should be distinguished.

The oldest compounds, that is verapamil and gallopamil (D 600), are at least chemically related to the β-blockers, although these compounds do not possess significant β-adrenoceptor-antagonistic activity. Chemically speaking, diltiazem as a benzothiazepine derivative shows some resemblance to the benzodiazepines, although there is no pharmacological similarity. The dihydropyridines form a rather extensive group of CAs for which a quantitative structure-activity relationship has been proposed [20, 35, 36]. Flunarizine and cinnarizine are diphenylalkylamines (Fig. 2). They possess a voluminous lipophilic group (the diphenyl moiety) separated by a 2 or 3 carbon atom chain or by a piperazine ring from an alkaline-substituted nitrogen atom. It seems of interest to note that the structure-activity requirements within the diphenylalkylamines follow those of calmodulin inhibitors [38].

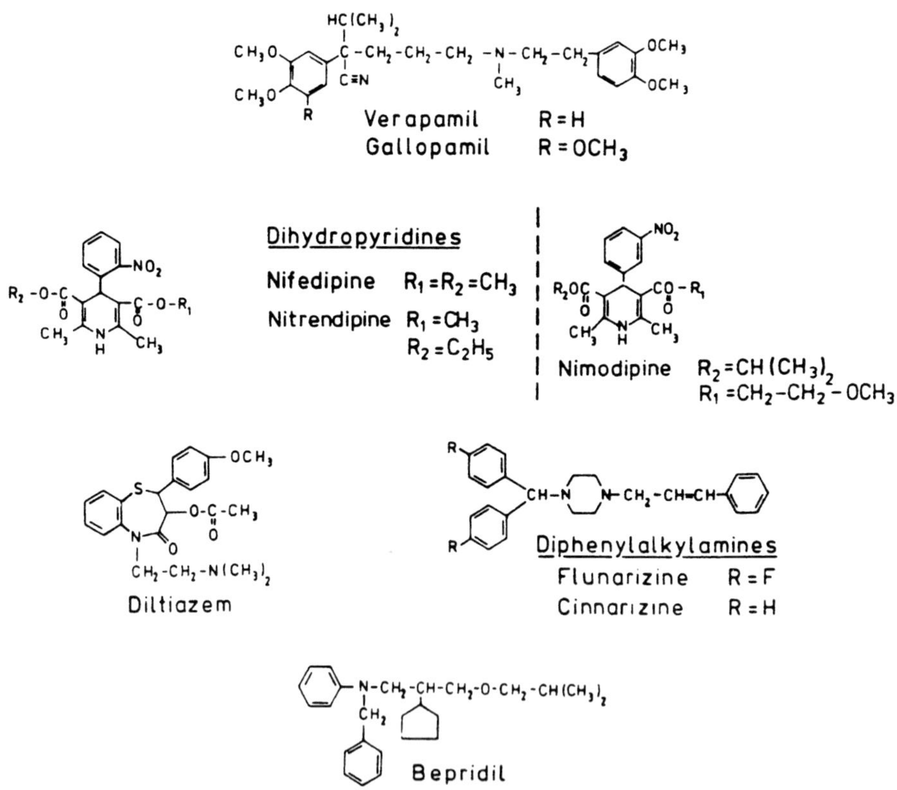

Fig. 2. Structural formulae of CAs. Note heterogeneity of structures. A relationship between chemical structure and pharmacological activity has so far been discovered only within the 1,4-dihydropyridines (nifedipine and related compounds) and within the diphenylalkylamine series (flunarizine and related drugs). Note the deviating structures of diltiazem and bepridil. Flunarizine and nimodipine are the CAs used most frequently in neurological disorders

With respect to lipophilicity, the following order of potency holds: flunarizine $\gg$ verapamil > diltiazem > dihydropyridines.

In particular flunarizine and nimodipine readily penetrate into the brain and are known to achieve high concentrations in brain tissues and in the cerebrospinal fluid. Most CAs also achieve high concentrations in the cell membranes of various central and peripheral tissues.

Radioligand Binding Studies

Radioligand binding studies have mainly been carried out with radioactively labeled dihydropyridine derivatives, such as [^{3}H]-nitrendipine, in rat brain membranes [10], and in a later stage also with cardiac and skeletal muscle preparations. For the dihydropyridines three different binding sites have been identified, which might lead to the subdivision of the calcium channel blockers into three subgroups with different affinities for these binding sites [10]. However, this subdivision may be different from that based upon pharmacological techniques and it remains an open question whether the binding sites found in radioligand binding studies are identical to those associated with calcium channels. Nevertheless, it is of interest to note that the dihydropyridine calcium channel blockers can effectively displace the dihydropyridine radioligands from their specific binding sites in a specific manner, suggesting that these binding sites and calcium channels may be somehow associated [15, 43].

Calcium Channels

Through the years the concept of specific calcium channels has become more sophisticated and it is now generally assumed that these channels are by no means homogeneous, at least three different subtypes being distinguished at present. Originally, a clear distinction was made between potential (or voltage) operated channels (POCs) and receptor operated channels (ROCs). The major effect of the CAs was assumed to occur at the level of the POCs, whereas the influence of CAs on ROCs is still subject to some debate. More recently, the nomenclature L-, T-, and N-channels has been introduced [16] for the distinction of the three different calcium channels which are demonstrable in various tissues. This field is advancing at a rapid pace. For reviews see refs. [10, 15, 32, and 44].

With respect to the influence of CAs on the ion channels, it is assumed that a drug/receptor (binding site) occurs at a contact domain of distinct elements which forms and decays, for instance as a function of state, i.e., opened/closed.

Pharmacology

The blockade of the aforementioned specific calcium channels remains the one and common mechanism of all CAs, notwithstanding their obviously different chemical and other characteristics. Events at the cellular membrane and their impairment by CAs have been extensively reviewed [22, 23, 47–49, 33]. Our group [22, 23, 47, 48] has shown that the vasoconstriction induced by the stimulation of postsynaptic α_2-adrenoceptors in resistance vessels by appropriate agonists is selectively impaired by calcium channel blockers. This effect is strictly correlated to calcium channel blockade, reflecting the initiation of a calcium inward flux sensitive to CAs upon formation of the α_2-adrenoceptor-agonist complex. This finding suggests that the receptor involved in triggering the ROCs is probably an α_2-adrenoceptor, although in certain tissues and species α_1-adrenoceptors may play a role as well.

The role of α-adrenoceptors in triggering the opening of ROCs is probably of fundamental importance. It may also contribute to the vasodilator activity which is characteristic for virtually all CAs. Accordingly, the calcium influx triggered by α-adrenoceptor stimulation is inhibited by the CAs, hence reducing the activation of contractile proteins. It should be recalled that both α_1- and α_2-adrenoceptors are stimulated by endogenous catecholamines; the attenuation of the sequelae of this stimulation by CAs leads to vasodilation. Accordingly, it seems justified to assume that there exists a refined interaction between CAs and the peripheral sympathetic nervous system. Very recently, we have presented experimental findings favoring the view that besides α-adrenoceptors the angiotensin II receptor also plays a mediator role in triggering the opening of ROCs [50]. Accordingly, angiotensin II receptors may also be involved, though indirectly, in the vasodilator potency of the CAs.

Organ Specificity

The heterogeneity of the various CAs is also reflected by considerable differences in organ specificity. This is a very important issue which may lead to potentially new therapeutic approaches. With respect to the heart the newly introduced antianginal compound bepridil has probably the highest specificity for this organ [18, 21, 48].

Verapamil and gallopamil (D 600) act predominantly on the heart, with a special affinity for the A-V node or the pacemaker cells, which is the basis for their antiarrythmic potency. Apart from this well-known activity on cardiac tissues verapamil is also a potent vasodilator, which is the basis of its potential use in the treatment of arterial hypertension and its antianginal activity. Diltiazem shows a similar profile although its vascular component may be somewhat stronger than that of verapamil. Its impairing influence on A-V conduction is possibly less than that of verapamil. As such it is intermediate between verapamil and nifedipine.

The *dihydropyridines*, of which nifedipine is the prototype, mainly cause relaxation of blood vessels, in particular of the coronary and peripheral arteries. Given by mouth in therapeutic doses nifedipine does not display direct activity on cardiac tissues and the compound is not used as an antiarrythmic agent. The mild and transient tachycardia caused by nifedipine reflects a reflex event mediated by the baroreceptor system [19]. Within the series of dihydropyridines compounds with special effects on distinct vascular beds may be anticipated. Nimodipine, for instance, is put forward as a potential drug used for cerebral vasodilatation and it appears possibly useful in conditions of cerebral ischemia [9, 30]. Apart from cerebral vasodilatation nimodipine is assumed to display anti-ischemic activity at a cellular level. Under physiological conditions the diphenylalkylamines flunarizine or cinnarizine have little or no influence on the peripheral cardiovascular system [28] or on the brain. However, their protective activity on the *ischemic* brain is well established and therapeutically appreciated [43]. A protective effect of these compounds in conditions of myocardial ischemia is the subject of present studies by our group. It should be realized that flunarizine and cinnarizine display little or no vasodilator activity, although they slightly and probably nonspecifically inhibit vasopressor effects of a variety of vasoconstrictors, including catecholamines, serotonin, and angiotensin II [28]. Flunarizine can be shown to depress myocardial contractility under particular experimental conditions [2], although this effect is very weak and irrelevant during oral treatment with flunarizine.

It may be of interest to add that flunarizine is an effective antidote to digitalis poisoning in guinea pigs [16], probably because of its potency to impair the overloading of cardiac cells with calcium ions, as observed in the presence of toxic concentrations of a cardiac glycoside.

Pharmacological Background of Therapeutic Applications

Cardiovascular Disease

Cardiac Dysrhythmia

Both the automaticity of the cardiac pacemaker cells and the process of A-V conduction are dependent upon the influx of calcium ions through specific channels. Conversely, specific inhibition of the calcium inward flux by CAs is the basis of the antidysrhythmic effect of these drugs. In particular verapamil and gallopamil are used as antidysrhythmic agents in cardiology in order to counteract supraventricular tachycardia. In view of the foregoing, A-V blockade is a logical side-effect of these CAs, reflecting the calcium dependency of A-V conduction. Diltiazem displays antidysrhythmic activity but it is not currently used as such. For review of the antidysrhythmic activity of CAs, see refs. 13 and 31.

Angina Pectoris

Angina pectoris is a well-established indication for verapamil, the dihydropyridines, and diltiazem. The mechanisms underlying this therapeutic benefit are manifold: relief of coronary spasm (when present); coronary vasodilatation *not* accompanied by a steal effect; a reduction in total peripheral resistance and, hence, cardiac afterload; and, in the case of verapamil and diltiazem (thus *not* for the dihydropyridines), a decrease in cardiac frequency. An anti-ischemic mechanism at the cellular level may contribute to the antianginal effect of the CA, although this so far remains uncertain. This combination of mechanisms leads to an improvement in the unfavorable balance between O_2 supply to the heart and O_2 consumption which is characteristic for angina pectoris. For reviews of this subject, see refs. 5 and 46.

Arterial Hypertension

Vasodilatation, predominantly in the arterial (arteriolar) vascular bed, is the main mechanism which satisfactorily explains the antihypertensive activity and hemodynamic properties of most of the dihydropyridines (nifedipine, nitrendipine) and also of verapamil. In modest doses the vasodilatation is limited to the arteriolar vascular bed, leaving the capacitance vessels unaffected. Accordingly, total peripheral resistance is reduced. The dihydropyridines cause modest and transient reflex tachycardia as a response to peripheral vasodilatation. Verapamil, however, does not induce tachycardia but rather a tendency to bradycardia as a result of impaired A-V conduction. For reviews see refs. 5 and 48.

Other Applications of CAs in Cardiovascular Disease

Transient relief of Raynaud's phenomenon by nifedipine is without any doubt based upon vasodilatation in the digital arteries. Hypertrophic obstructive cardiomyopathy, pulmonary hypertension, congestive heart failure, the sick sinus syndrome, and acute digitalis intoxication are other potential applications of the CAs in cardiovascular disease. For reviews see refs. 6 and 14.

Calcium Antagonists in the Treatment of Neurological Disease

In contrast to the now well-established position of the CAs in the management of several cardiovascular diseases, the attempted treatment of certain neurological disorders with these drugs represents a new field of investigation. This field, although most promising, is still full of uncertainty, both from the therapeutic and from the fundamental point of view. The results so far obtained strongly suggest that a thorough further exploration of the application of CAs in neurological disorders will be well worthwhile.

Compounds

So far very few CAs have been studied appropriately as potential therapeutic agents in neurological disease. The diphenylalkylamine (see Fig. 2) *flunarizine* has been studied widely and has become one of the prototypes of CAs applied in neurological disorders. The efficacy of flunarizine in migraine prophylaxis and in the treatment of certain forms of vertigo appears to be generally accepted, although its usefulness in a variety of other disorders remains to be demonstrated. For a variety of reasons flunarizine appears to be preferable to cinnarizine. For reviews see refs. 8, 15, and 26.

Nimodipine is a dihydropyridine and thus related to nifedipine. This CA is claimed to be a vasodilator with a certain degree of specificity towards the cerebral vascular bed over the peripheral arterial system, although its specificity has been contested [1, 17, 40]. The vascular activity of nimodipine is much stronger than that of flunarizine, which as such is not a vasodilator but rather causes a modest attenuation of the vasoconstrictor effects of endogenous factors like norepinephrine, angiotensin II, or serotonin. Apart from its vasodilator potency nimodipine, like flunarizine, should be expected to display cytoprotective activity, especially in a condition of ischemia when the cellular calcium concentration becomes excessively high. Several therapeutic claims have been issued for nimodipine, e.g., in subarachnoid hemorrhage and cerebral ischemia. Nimodipine's potential activity in migraine is the subject of investigation.

Mode of Action

In contrast to the application of CAs in cardiovascular disease, where the mode of action of the drugs is understood reasonably well, the mechanisms underlying the therapeutic application of CAs in neurology remain rather uncertain and subject to debate.

Improved perfusion of ischemic neuronal tissues appears to be a prominent effect of nimodipine, a CA with obvious vasodilator potency in the cerebral region. However, it should be realized that this vascular activity is probably not

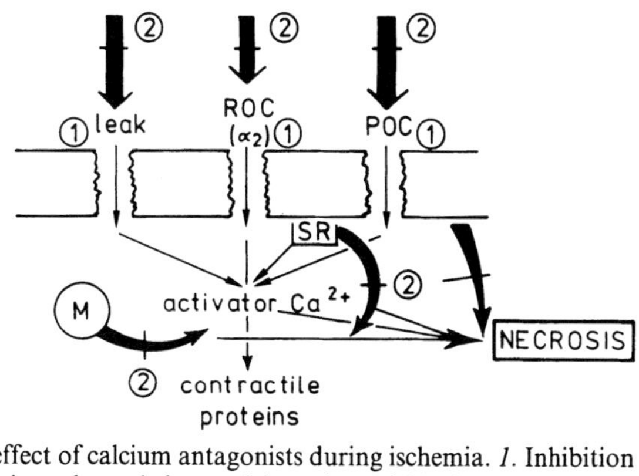

Fig. 3. Schematic presentation of the effect of calcium antagonists during ischemia. *1.* Inhibition of the calcium influx via different calcium channels by a calcium channel blocker under physiological conditions: leak channel; *ROC*, receptor operated channel, operated via α_2-adrenoceptors; *POC*, potential operated channel, triggered by depolarization. *2.* Inhibition of the pathologically increased calcium influx under ischemic conditions, as indicated by the *large black arrow*. Apart from an excessive influx of calcium through the physiological channels there occurs an enhanced release of calcium from the inner side of the cell membrane (*far right*), as well as from subcellular pools like the sarcoplasmic reticulum (*SR*), mitochondria (*M*), or microsomes (*M*), resulting in a grossly increased cystolic calcium concentration which significantly contributes to necrosis

limited to the cerebral vessels; rather peripheral vasodilatation and concomitant hypotension should be anticipated as well. Furthermore, vasodilatation in the cerebral region may be the basis of steal effects which have indeed been claimed to occur.

Flunarizine, when given orally in low doses (5–10 mg daily), hardly displays any cardiovascular activity. Only in vitro has it been demonstrated that the vasoconstrictor activity of endogenous agonists, like serotonin, catecholamines, KCl, and $PGF_{2\alpha}$, is blunted upon pretreatment with flunarizine. It therefore seems likely that the therapeutic activity of flunarizine is hardly caused to a substantial degree by vasodilatation (or antivasoconstriction, as it is sometimes called) but rather is to be explained on the basis of cytoprotection in a condition of ischemia, possibly by reducing the cellular overload with calcium, which is characteristic for the ischemic situation. A schematic representation of calcium overload blockade is shown in Fig. 3. Such a cytoprotective mechanism should also be claimed for nimodipine, since its vasodilator potency is probably not sufficient to explain all of its therapeutic effects. For reviews see refs. 47–49. An excessive rise in the concentration of extracellular potassium ions during cerebral ischemia is a generally recognized pathogenetic process. This phenomenon is usually accompanied by a reduction in extracellular space which impairs the diffusion of oxygen and hence accelerates the development of ischemia. The elevated concentration of potassium in the extracellular space reduces the transmembrane resting potentials of several types of cell. Owing to the potassium-induced depolarization the influx of calcium ions into the cells is enhanced, causing vasoconstriction, release of neurotransmitters, and calcium overload, leading to necrosis. Excessive calcium concentrations also cause an increased stiffness of the erythrocyte which is counteracted by flunarizine. Accordingly, flunarizine-induced changes in the rheology of the blood may contribute to the drug's therapeutic activity in cerebral ischemia.

The CAs of the flunarizine and nimodipine types are assumed to interfere with these extremely complex series of events, predominantly by inhibiting the cellular influx of calcium ions. It goes without saying that many details in this mechanism remain to be clarified.

Flunarizine has been claimed to be of therapeutic benefit as add-on medication in the treatment of certain forms of epilepsy [29], although this activity has been denied by others. The theoretical background is based upon the existence of a particular potential, derived from the brain's surface during an epileptic discharge, called the paroxysmal depolarizing shift (PDS). The PDS spreads to deeper brain structures during more prolonged epileptic discharges. In experimental animals it has been shown that the PDS is frequently preceded by a pathological influx of calcium ions, which may be sensitive to calcium antagonistic drugs. Experimental studies have shown that the extension of the PDS is indeed strongly inhibited by verapamil and flunarizine. The relevance of these observations in human epilepsy remains to be demonstrated.

Finally, it should be emphasized that there is no conclusive evidence that the various therapeutic effects and claimed effects of CAs in neurological disorders are indeed directly related to calcium entry blockade, although this seems very likely.

Pharmacological Background of Adverse Reactions

The heterogeneity of the various CAs implicates that the adverse reactions are also rather different for the various individual compounds. However, vasodilatation, with its sequelae, is a very common side-effect for all dihydropyridines, for verapamil, and to some degree also for diltiazem. The background of the various side-effects will be discussed here for the individual CA or groups of CAs.

Dihydropyridines (Nifedipine, Nitrendipine, Nimodipine, and Related Dihydropyridines)

Headache, flush, hypotension, and reflex tachycardia are well-known side-effects to the dihydropyridines. These adverse reactions all reflect vasodilatation, based upon the inhibition of calcium influx through the specific channels. For nifedipine peripheral edema, in particular in the legs, is a well-known side-effect, which is *not* based upon the retention of sodium and fluid. This type of edema does not respond adequately to treatment with diuretics. It has been presumed that nifedipine would promote the opening of the sphincters of the capillary vessels and hence provoke the aforementioned edema.

Verapamil and Gallopamil

Similar to the dihydropyridines (see above), verapamil causes the side-effects which are characteristic for vasodilatation, with the exception of reflex tachycardia. Verapamil rather causes bradycardia and impairment of A-V conduction, as a result of its depressant influence on calcium influx dependent processes in the sinus node and in the A-V node. Obstipation, a frequently reported side-effect of verapamil, is caused by the relaxant effect of verapamil on colon movements, probably as a result of the inhibition of calcium fluxes in intestinal smooth muscle. As far as can be judged at present the side-effects of gallopamil are very similar to those of verapamil.

Diltiazem

Headache and hypotension caused by diltiazem again reflect vasodilatation. Bradycardia, although occurring infrequently, is probably of the same origin as that described for verapamil. As a whole the impression is obtained that diltiazem is subjectively better tolerated than the dihydropyridines or verapamil, although hard data substantiating this impression are not available at present.

Flunarizine

Flunarizine, unlike the other CAs, displays but little or no influence on the cardiovascular system and does not show the adverse reactions as described above for other CAs, like vasodilatation or cardiac depression. The most relevant adverse reactions to flunarizine are all based upon central nervous mechanisms.

Sedation is a rather common side-effect of flunarizine, probably resulting from a rather unspecific mechanism. Depressions caused by flunarizine have sometimes been observed, although usually in patients suffering from migraine. The association migraine/depression is not uncommon and it is therefore difficult to judge the role of flunarizine as such in this matter.

Extrapyramidal side-effects attributed to flunarizine have been reported in rare cases and virtually only in the elderly. The mechanism of this adverse reaction so far remains unknown and is difficult to explain on a molecular basis, since flunarizine has no relevant affinity for dopaminergic receptors.

Finally, flunarizine when applied for prolonged periods sometimes causes a slight increase in body weight, which is usually transient. The mechanism of this effect is unknown. It should be recalled that several drugs with obvious central nervous activity, like benzodiazepines, neuroleptics, and antidepressants cause a similar gain in body weight. It seems likely that the effect of flunarizine is not particularly specific.

Conclusions

The CAs have conquered a generally recognized position in the treatment of cardiovascular disease, in particular of angina pectoris, arterial hypertension, and certain cardiac dysrhythmias. Most of their therapeutic activities, including their adverse reactions in the cardiovascular field, are understood reasonably well and as such an extrapolation of the intellectually satisfactory picture of these drugs in basic pharmacology.

The application of certain CAs in neurology has opened a new field of enormous potential interest, still characterized by many uncertainties and speculations, both from the fundamental/mechanistic and from the practical/therapeutic point of view.

Migraine and vertigo have become routine indications for flunarizine, whereas the therapeutic claims of nimodipine in subarachnoid hemorrhage and in acute cerebral ischemia appear to be quite firm and well substantiated. Flunarizine might develop as an antiepileptic drug of a novel type and with a relatively low incidence of adverse reactions. Owing to the availability of flunarizine for *intravenous* application it is now possible to study its potential benefit in acute stroke. Trials designed to study this issue are proceeding at present. Most of the results strongly indicate that a field of new therapeutic possibilities has been opened. The enormous investment in further clinical and fundamental research required to elucidate the many uncertainties appears to be well worthwhile. It will be the task of basic pharmacologists to unravel the mechanisms underlying the therapeutic efficacy and side-effects of the CAs in neurological disorders.

References

1. Betz E, Deck K, Hoffmeister F (eds) (1985) Nimodipine. Pharmacological and clinical properties. Proceedings of the 1st International Nimotop-Symposium, Munich 1984. Schattauer, Stuttgart
2. Boddeke HWGM, Heynis JB, Jonkman FAM, Van Zwieten PA (1986) Characterization of calcium entry blockers in isolated guinea-pig hearts; cardiodepressant activity of flunarizine. Br J Pharmacol 89:615
3. Bou J, Llenas J, Massingham R (1983) Calcium entry blocking drugs, "calcium antagonists" and vascular smooth muscle function. J Auton Pharmacol 3:219–232
4. Borgers M. Personal communication
5. Braunwald E (1982) Mechanism of action of calcium channel blocking agents. N Engl J Med 307:1618–1627
6. Bussey HF, Talbert RL (1984) Promising uses of calcium channel blocking agents. Pharmacotherapy 4:137–143
7. Fleckenstein A (1977) Specific pharmacology of calcium in the myocardium cardiac pacemaker and vascular smooth muscle. Ann Rev Pharmacol Toxicol 17:149–166
8. Fleckenstein A (1983) Calcium antagonism in heart and smooth muscle. Experimental facts and therapeutic prospects. Wiley, New York
9. Gelmers HJ (1984) The effects of nimodipine on the clinical course of patients with acute ischemic stroke. Acta Neurol Scand 69:232–239
10. Glossmann H, Ferry DR, Lubbecke F, Mewes R, Hofmann F (1982) Calcium channels: direct identification with radioligand binding studies. Trends Pharmacol Sci 3:431–437
11. Glossmann H, Goll A, Rombusch M, Ferry DR (1985) Molecular pharmacology of Ca^{2+}-channels: receptor binding studies. In: Betz A, Deck K, Hoffmeister A (eds) Nimodipine; pharmacological and clinical properties. Schattauer, Stuttgart, pp 57–76
12. Godfraind T, Miller RC, Socrates Lima J (1982) Selective α_1- and α_2-adrenoceptor agonist-induced contractions and ^{45}Ca-fluxes in the rat isolated aorta. Br J Pharmacol 77:597–604
13. Godfraind T, Herman A, Wellens H (1984) Calcium entry blockers in cardiovascular and cerebral dysfunctions. Nijhoff, Den Haag
14. Gray Ellrodt A, Singgh BN (1983) Clinical applications of slow channel blocking compounds. Clin Pharmacol Ther 23:1–43
15. Hartmann A, Van Zwieten PA (1986) Pharmacological aspects and neurological potential of calcium entry blockers. XIIIth World Congress of Neurology, Hamburg 1985. Eur Neurol 25 (Suppl 1):1–126
16. Jonkman FAM, Boddeke HWGM, Van Zwieten PA (1986) Protective activity of calcium entry blockers against ouabain intoxication in anaesthetized guinea pigs. J Cardiovasc Pharmacol 8:1009–1013
17. Kazda SR (1982) Nimodipine: a new calcium antagonistic drug with a preferential cerebrovascular action. Acta Neurochir 63:259–265
18. Labrid C, Grosset A, Dureng G, Mironneau J, Duchêne-Marillaz P (1979) Some membrane interactions with bepridil, a new antianginal agent. J Pharmacol Exp Ther 211:546–554
19. Lederballe Pedersen O, Christensen NJ, Ramsch KD (1980) Comparison of acute effects of nifedipine in normotensive and hypertensive man. J Cardiovasc Pharmacol 2:357–361
20. Mannhold R, Rodenkirchen R, Bayer R (1982) Qualitative and quantitative structure-activity relationships of specific Ca-antagonists. Prog Pharmacol 5:25–52
21. Marshall RJ, Winslow E, Lamar JC, Apoil E (1984) Bepridil. In: Scriabine A (ed) New drugs annuals: cardiovascular drugs, vol 2. Raven, New York, pp 157–176
22. Van Meel JCA, De Jonge A, Wilffert B, Kalkman HO, Timmermans PBMWM, Van Zwieten PA (1981) Organic and inorganic calcium antagonists reduce vasoconstriction in vivo mediated by postsynaptic α_2-adrenoceptors. Naunyn Schmiedebergs Arch Pharmacol 316:288–293
23. Van Meel JCA, Towart R, Timmermans PBMWM, Van Zwieten PA (1983) Correlation between the inhibitory activities of calcium antagonists on vascular smooth muscle constriction in vitro after depolarisation and in vivo after α_2-adrenoceptor stimulation. Naunyn Schmiedebergs Arch Pharmacol 322:34–37

24. Nayler WG, Poole-Wilson P (1981) Calcium antagonists: definition and mode of action. Basic Res Cardiol 76:1–15
25. Nayler WG (1980) Calcium antagonists. Eur Heart J 1:225–237
26. Van Nueten JM, Janssen PAJ (1973) Comparative study of the effects of flunarizine and cinnarizine on smooth muscles and cardiac tissues. Arch Int Pharmacodyn Ther 204:37–55
27. Van Nueten JM, Van Beek J, Janssen PAJ (1978) Effect of flunarizine on calcium-induced responses of peripheral vascular smooth muscle. Arch Int Pharmacodyn Ther 232:42–52
28. Van Nueten JM, Vanhoutte PM (1982) Calcium entry blockers and vasospasm. In: Vanhoutte PM, Leusen I (eds) Vasodilatation. Raven, New York, pp 459–468
29. Overweg J, Ashton D, De Beukelaar F, Binnie CD, Wauquier A, Van Wieringen A (1986) Add-on therapy in epilepsy with calcium entry blockers. Eur J Neurol 25:93–101
30. Paci A, Ottaviono B, Trenta A, Iannone G, Acuto G, Carosi M. Treatment of acute ischemic cerebral infarction with nimodipine (BAY e 9736). Cited in ref. 47
31. Packer M, Frishman WH (1984) Calcium channel antagonists in cardiovascular disease. Appleton-Century-Crofts, Norwalk (Conn), pp 225–269
32. Paoletti R, Vanhoutte PM (1987) International symposium on calcium antagonists. Pharmacology and clinical research. New York City
33. Peters T (1983) „Calcium-Antagonismus". Ein einheitlicher Begriff für uneinheitliche Wirkmechanismen. Med Klinik 78:368–375
34. Peters T (1986) Calcium in physiological and pathophysiological cell function. Eur Neurol 25 (Suppl 1):27–44
35. Rodenkirchen R, Bayer R, Mannhold R (1982) Specific and nonspecific analysis of cardiodepressive drugs. Prog Pharmacol 5:9–23
36. Rodenkirchen R, Bayer R, Steiner R, Bossert F, Meyer H, Möller E (1979) Structure-activity studies on nifedipine in isolated cardiac muscle. Naunyn Schmiedebergs Arch Pharmacol 310:69–78
37. Schwartz A, Triggle DJ (1984) Cellular action of calcium channel blocking drugs. Ann Rev Med 35:325–339
38. Spedding M (1985) Calcium antagonist subgroups. Trends Pharmacol Sci 6:109–114
39. Spedding M (1987) Three types of Ca^{2+}-channels explain discrepancies. Trends Pharmacol Sci 8:115–117
40. Tanaka K, Gotoh F, Muranratsu F et al. (1980) Effects of nimodipine (BAY e 9736) on cerebral circulation in cats. Arzneimittelforschung 26:2172–2176
41. Triggle DJ, Swamy VC (1983) Calcium antagonists. Some chemical-pharmacologic aspects. Circ Res 52 (Suppl I):17–28
42. Vanhoutte PM, Bohr DF (1981) Calcium entry blockers and the cardiovascular system. Fed Proc 40:2851
43. Wauquier A, Edmonds HC, Van den Broek WAE, Melis W, Van Loon J (1985) Effects of flunarizine on recovery of cerebral function following cardiac arrest. Neuropsychobiology 13:147–151
44. Van Zwieten PA (1985) Calcium antagonists – terminology, classification and comparison. Arzneimittelforschung 35:298–301
45. Van Zwieten PA (1986) Differentiation of calcium entry blockers into calcium channel blockers and calcium overload blockers. Eur Neurol 25 (Suppl 1):57–67
46. Van Zwieten PA (1985) Drug targets in unstable angina. In: Hugenholtz PG, Goldman BS (eds) Unstable angina, current concepts and management. Schattauer, Stuttgart, pp 151–157
47. Van Zwieten PA, Timmermans PBMWM (1985) Pharmacological basis of the antihypertensive action of calcium entry blockers. J Cardiovasc Pharmacol 7:S11–S17
48. Van Zwieten PA, Schönbaum E (1983) Calcium antagonists and blood vessels. Prog Pharmacol 5:1–142
49. Van Zwieten PA, Van Meel JCA, Timmermans PBMWM (1983) Pharmacological basis of the therapeutic effect of calcium entry blockers. Interaction with vascular α-adrenoceptors. Hypertension 5 (Suppl II):II-8–II-17
50. Van Zwieten PA, Timmermans PBMWM, Van Heiningen PNM (1987) Receptor subtypes involved in the action of calcium antagonists. J Cardiovasc Pharmacol 5 (Suppl 4):S21–S28

Cellular Movements and Distribution of Calcium

T. Peters

Introduction

According to present knowledge any physiological or pharmacological stimulus somewhere in the chain of events occurring between fixation of an agent to its binding sites or receptor and final response requires ionized calcium (Ca^{2+}) as a mediator (ionic messenger). This concerns a variety of physiological events like smooth, skeletal, and heart muscle contraction, exocytosis, phagocytosis, axonal transport, cell shape changes, cilial movement, cell division, sperm motility, conductance changes for other ions and for calcium itself, and regulation of metabolic activities (e.g., lipase and protein kinase activation), to give only a few examples. Proper physiological activation is, however, bound to a graded cytosolic increase of Ca^{2+} of well-controlled duration.

The Ca^{2+} Gradients

In non-stimulated cells the Ca^{2+} concentration of the cytosol ranges below 10^{-7} mol/liter, whereas upon physiological stimulation this concentration is increased to any value up to about 5×10^{-6} mol/liter (Portzehl et al. 1964; Ashley and Ridgway 1970; Blinks et al. 1978; Rudel 1979; Fay et al. 1979; Allen and Blinks 1978; Lee et al. 1980; Lee et al. 1987).

These very low cytosolic Ca^{2+} concentrations are in contrast to the extracellular Ca^{2+} concentration of 1.3 mmol/liter and the total cell calcium (nonionized) of up to about 2.5 mmol/kg w.w. in eukaryotic cells containing mitochondria (Manery 1969; Carafoli and Crompton 1976; Bygrave 1977), sarcoplasmic reticulum (Ebashi 1960, 1961; Endo 1977), nucleus, and plasma membrane located calcium stores (Lüllmann et al. 1983). From this it becomes obvious that under any conditions – rest or excitation – gradients exist from both extracellular space and intracellular binding or storage sites to the cytosol in the order of 10000 to 1 down to 1000 to 1. Consequently, both small increases in permeability and minimal release from binding or storage sites will enhance the cytosolic Ca^{2+} concentration substantially (Fig. 1).

Department of Experimental Medicine, Janssen Research Foundation, Raiffeisenstr. 8, D-4040 Neuss 21, FRG.

Cerebral Ischemia and Calcium
Ed. by A. Hartmann and W. Kuschinsky

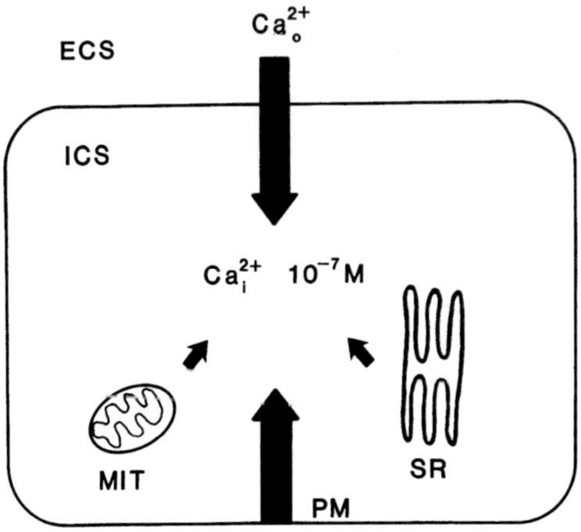

Fig. 1. Calcium gradients directed into the cytosol. *ECS*, extracellular space; *ICS*, intracellular space; *PM*, plasma membrane; *MIT*, mitochondrion; *SR*, sarcoplasmic reticulum. Under nonstimulated conditions the cytosolic Ca^{2+} concentration is maintained below 10^{-7} mol/liter. This is less than 0.01% of total cell calcium and creates a gradient from the outside to the inside of about 10000 to 1. Schematically, the *thickness of arrows* indicates the size of the respective gradients

Plasmalemmal Mechanisms Controlling Ca^{2+} Entry

The plasmalemma consists of a phospholipid bilayer with interspersed proteins and with water molecules forming polymers predominantly in the proximity of the outer and inner charged lipid head groups (for details see Benga 1985; Yeagle 1987).

Phospholipids

Different classes of phospholipid are asymmetrically distributed within cell membranes (Zwaal et al. 1973; Post et al. in press), and the lipid bilayer without participation of proteins may display selective permeability for ions, depending upon certain conditions like transmembrane potential and fatty acid composition (Cullis et al. 1980; Yoshikawa et al. 1988).

Under the influence of a transmembrane potential in some cell types calcium appears to be bound to acidic lipids located at the internal surface of the plasmalemma, from where it is released in the form of Ca^{2+} upon depolarization (Lüllmann and Peters 1977). Additionally, after having released Ca^{2+} these acidic phospholipids appear to rearrange in a way that allows Ca^{2+} to enter the inside of cells from the outside (Campbell 1985). As stated above, the permeability function of the lipid bilayer of mammalian cells is strongly influenced by the size of the transmembrane potential, which in turn depends upon proper gradients for potassium and sodium. Basic Ca^{2+} permeability is, therefore, a function of Na^+ -K^+ distribution and thus depends upon the energy supply for and activity of the Na-K pump. Depending upon the membrane potential lipid bilayers not only control permeability for Ca^{2+} and other ions but also govern, together with the transmembrane potential, the molecular arrangement of voltage-dependent gates or "channels" for cations. In particular, by means of internally located acidic lipids bilayers offer the possibility to participate in physiological regulations by binding and releasing Ca^{2+} (Seimiya and Ohki 1973).

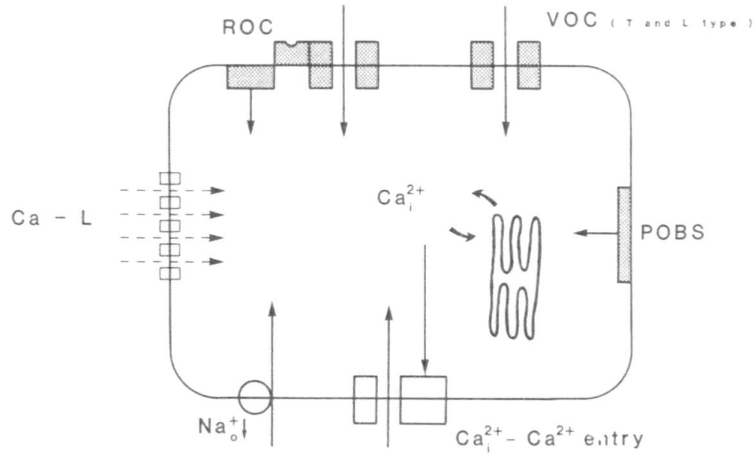

Fig. 2. Schematic representation of mechanisms to increase cytosolic Ca^{2+} concentrations. *ROC*, receptor-operated channel; *VOC*, voltage-operated channel; *Ca-L*, calcium leaks through the bilayer of the plasmalemma; *POBS*, potential operated plasmalemmal binding sites; $Na_o^+\downarrow$, Na-Ca exchange system; Ca_i^{2+}-Ca^{2+} *entry*, calcium entry through a cation channel activated by Ca^{2+} in the cytosol (Ca_i^{2+}); *thick bent arrows* symbolize Ca-induced Ca release from endoplasmic reticulum. For details see text

Ca^{2+} Channels

Biomembranes and particularly plasma membranes contain an abundance of protein molecules. Among them receptor proteins, channel proteins, and pump proteins are of particular interest for passive ion fluxes in the direction of and transport against their concentration gradients. For many of those proteins the amino acid sequence and the respective tertiary structure have been disclosed [the interested reader is referred to Benga (1985) and Yeagle (1987)]. However, the structure of Ca^{2+} channels is as yet unknown. Nevertheless, according to "receptor binding" studies and electrophysiological investigations different types of calcium channel could be classified in different tissues (Hess and Tsien 1984; Rosenberg et al. 1986; Narahashi et al. 1987; Hescheler et al. 1987).

Principally one has to distinguish between receptor- and voltage-operated Ca^{2+} channels (Fig. 2). Of the latter class subtypes named L, T, and N channels have been identified in different tissues, which permit entry of Ca^{2+} in different segments of the transmembrane potential (Reuter 1986).

$Na^+{}_i$-Mediated Increases in Intracellular Ca^{2+}

In principle, three ways have been described in different cell types which allow Na^+ to interfere with cytosolic ionization of calcium: (1) inhibition of the Na-K pump by cardiotonic steroids, (2) prolongation of the open state of the fast sodium channel (e.g., by veratridine), and (3) stimulation of the Na^+-H^+ exchange mechanism by increased intracellular proton production.

In each case it appears that an increase of intracellular Na^+ concentration induces an Na-Ca exchange mechanism by which three Na ions are extruded per Ca ion entering the cell (Catteral et al. 1982; Marban and Tsien 1982; Cavero and Spedding 1983; Eisner et al. 1984; Lee 1985). It is at present, however, not fully understood to what extent physiological regulation of intracellular calcium ionization involves one or the other type of Na-mediated mechanisms.

Intracellular Distribution of Calcium

In living cells the contribution of different binding and storage systems is extremely difficult to estimate because of their variable capacities and affinities which may mutually influence each other and which appear to be controlled by still unknown cytosolic factors (see, e.g., Baker and Umbach 1987). Moreover, most of the pharmacological tools used to block calcium binding or uptake into specific subcellular organelles lack specificity and under in vitro conditions mitochondria or preparations containing sarcoplasmic reticular fragments may behave differently from the in vivo situation (Baker and Umbach 1987). It is, therefore, at present impossible to describe exactly the interplay of intracellular systems participating in the regulation of Ca^{2+}. It is, however, well accepted that intracellular systems display a tremendous capacity to inactivate or buffer Ca^{2+} since the free Ca^{2+} concentration is maintained well below 0.1% of the total cellular amount of calcium (Baker 1976, 1978). It is also evident that binding of Ca^{2+} can be fractionated into energy-dependent and energy-independent components (Baker and Schlaepfer 1975, 1978).

While the energy-independent components mainly comprise anionic anorganic and organic binding sites (e.g., the protein calmodulin or the bulk of negatively charged lipids), the energy-dependent component consists of mitochondria (Moore 1971) and the endoplasmic reticulum (Hasselbach 1963) and some special acidic lipids at the internal surface of the plasmalemma (see above) which bind or release Ca^{2+} according to the actual electrical potential across the plasmalemma (Lüllmann and Peters 1977; Lüllmann et al. 1983). The role of mitochondria becomes, however, particularly prominent under pathological conditions, e.g., ischemia, when cellular Ca^{2+} overload is counteracted by mitochondrial Ca^{2+} uptake. This mechanism represents a major factor in ischemic pathology (Siesjö 1985).

Extrusion of Ca^{2+} from Cells

In order to maintain calcium homeostasis inward directed Ca^{2+} fluxes and leaks have to be balanced by equivalent outward movements of this ion. In principle, two mechanisms are known to serve this purpose: (1) the ATP-driven Ca-Mg-ATPase of the sarcolemma, and (2) the Na-Ca exchange process (Fig. 3).

Fig. 3. Schematic presentation of mechanisms responsible for the extrusion of calcium from cells. The Ca^{2+} concentration in the cytosol (Ca_i^{2+}) is maintained below 10^{-7} mol/liter (the threshold for activation of physiological processes) by the Ca-Mg-ATPase which transports Ca into the extracellular space and countertransports Mg. For this activity ATP is needed. A second transport system is driven by the gradient of Na from the outside to the inside. Upon entry of three molecules of sodium this exchange system extrudes one molecule of calcium. For details see text

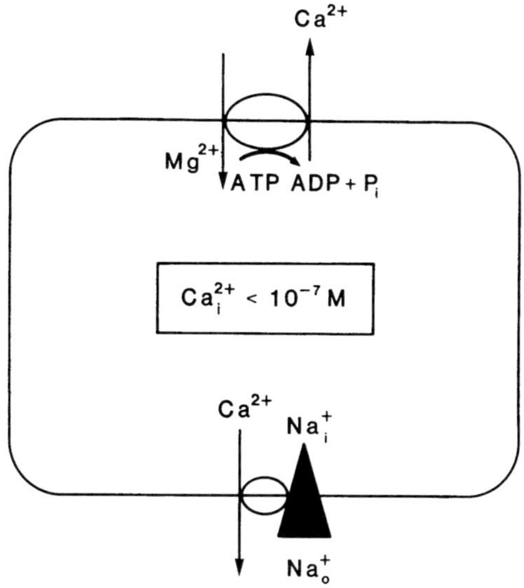

The sarcolemmal Ca-Mg-ATPase displays high affinity for Ca^{2+}, generally ranging below 10^{-7} mol/liter, yet its capacity appears variable and seems to depend upon both the cell type under investigation and the experimental conditions (Schatzmann 1966; Vincenzi and Schatzmann 1967; Lüllmann and Peters 1976; Di Polo 1978; Preuner 1981; Caroni et al. 1983). Nevertheless, this integral plasmalemmal protein seems to fulfill all the requirements necessary to maintain the intracellular Ca^{2+} concentration below the activation threshold of 10^{-7} mol/liter. A proper function of the Na-Ca exchange mechanism in the desired direction requires both a well-maintained Na gradient, which determines the driving force, and a polarized state of the plasmalemma (Glitsch et al. 1969; Stahl and Swanson 1972; Reuter et al. 1973; Blaustein 1974; Baker 1976; Kalix 1977).

From the foregoing it becomes obvious that the Ca-Mg pump and the Na-Ca exchange mechanism are both energy dependent and will fail under conditions of energy lack. It is not clear as to how both extrusion mechanisms interplay under physiological or pathological conditions.

Nonselective Ca^{2+} Entry

In a large variety of cells it could be demonstrated that the cytosolic Ca^{2+} concentration under certain conditions induces an increased K^+ permeability (for review see Latorre et al. 1984). This mechanism primarily causes hyperpolarization, which would counteract voltage-operated Ca^{2+} entry. If, however, increased K^+ permeability persists for a longer period, depolarization has to occur due to critical cellular K^+ loss. Recently it could be demonstrated that upon prolonged increases of intracellular Ca^{2+} concentrations an entry of Na^+ and Ca^{2+} through nonselective cation channels was favored. This observation is of particular interest because it would explain the inhibition of cellular Ca^{2+} and Na^{+} overload by pharmacological agents which interfere neither with slow Ca^{2+} channels nor with fast Na^{+} channels nor with the repolarizing K^{+} channel (Vollmer et al. 1987; Heers et al. 1988; Schneider et al. 1988).

Further experiments are needed to define whether such nonselective cation channels represent converted selective channels or whether they exist as a consequence of altered lipid–lipid or lipid–protein interactions. Independent of the respective underlying events, it seems of interest that the cytosolic Ca^{2+} concentration may change the permeability function of the plasmalemma not only for other cations but even for Ca^{2+} itself.

Summary

The essential prerequisite for the vitality and proper function of eukaryotic cells is an asymmetric distribution of ions between the intra- and the extracellular space. The maintenance of this asymmetric distribution, which is far removed from an equilibrium state, requires a selective permeability of the plasmalemma as well as intracellular binding and buffering systems for the ions. In addition, pump systems have to be active in order to transport ions against their chemical and electrical gradients. Moreover, exchange mechanisms have to be involved to allow countertransport of ions across both external and internal biomembranes.

The mechanisms involved in the maintenance of calcium homeostasis of cells in general have been briefly reviewed. Since calcium homeostatic mechanisms are inseparably linked to sodium and potassium homeostasis, it is necessary to discuss relevant aspects in parallel.

References

Allen DG, Blinks JR (1978) Calcium transients in aequorin-injected frog cardiac muscle. Nature 273:509–513

Ashley CC, Ridgway EB (1970) On the relationship between membrane potential, calcium transient and tension in single barnacle muscle fibres. J Physiol (Lond) 209:105–130

Baker PF (1976) The regulation of intracellular calcium. Symp Soc Exp Biol 30:67–88

Baker PF (1978) The regulation of intracellular calcium in giant axons of loligo and myxicola. Ann NY Acad Sci 307:250–268

Baker PF, Schlaepfer WW (1975) Calcium uptake by axoplasm extruded from giant axons of loligo. J Physiol 249:37–38

Baker PF, Schlaepfer WW (1978) Uptake and binding of calcium by axoplasm isolated from giant axons of loligo and myxicola. J Physiol 276:103–125

Baker PF, Umbach JA (1987) Calcium buffering in axons and axoplasm of loligo. J Physiol 383:369–394

Benga MP (1985) Structure and properties of cell membranes, vol. I, II, III. Benga G (ed). CRC Press, Boca Raton

Blaustein MP (1974) Interrelationships between sodium and calcium fluxes across cell membranes. Rev Physiol Biochem Pharmacol 70:33–82

Blinks JR, Rüdel R, Taylor SR (1978) Calcium transients in isolated amphibian skeletal muscles. J Physiol (Lond) 277:291–323
Bygrave FL (1977) Mitochondrial calcium transport. Curr Top Bioenerg 6:260–318
Campbell AK (1985) Intracellular calcium, its universal role as regulator. Wiley, Chichester
Carafoli E, Crompton M (1976) Calcium ion and mitochondria. Symp Soc Exp Biol 30:89–115
Caroni P, Zurini M, Clark A, Carafoli E (1983) Further characterization and reconstitution of the purified Ca^{2+}-pumping ATPase of heart sarcolemma. J Biol Chem 258:7305
Catteral WA, Hartshorne RP, Beneski DA (1982) Molecular properties of neurotoxin receptor sites associated with sodium channels from mammalian brain. Toxicon 20:27–40
Cavero I, Spedding M (1983) "Calcium antagonists": a class of drugs with a bright future. Part I. Cellular calcium homeostasis and calcium as a coupling messenger. Life Sci 33:2571–2581
Cullis PR, Kruijff B de, Hope MJ, Nayar R, Schmid SL (1980) Phospholipids and membrane transport. Can J Biochem 58:1091–1100
Di Polo R (1978) Ca pump driven by ATP in squid axons? Nature 274:340
Ebashi S (1960) Calcium binding and relaxation in actomysin system. J Biochem 48:150–151
Ebashi S (1961) Calcium binding activity of vesicular relaxing factor. J Biochem 50:236–244
Eisner DA, Lederer WJ, Vaughn-Jones RD (1984) The quantitative relationship between twitch tension and intracellular sodium activity. J Physiol 355:251–266
Endo (1977) Calcium release from the sarcoplasmic reticulum. Physiol Rev 57:71–108
Fay FS, Shelvin HH, Granger WC, Taylor SR (1979) Aequorin luminescense during activation of single isolated smooth muscle cells. Nature 280:506–508
Glitsch HG, Reuter H, Scholz H (1969) Influence of intracellular sodium concentrations on calcium influx of isolated guinea-pig auricles. Arch Pharmakol 264:236–237
Hasselbach W (1963) Relaxing factor and the relaxation of muscle. Prog Biophys Mol Biol 14:167–222
Heers C, Scheufler E, Wilhelm D, Wermelskirchen D, Wilffert B, Peters T (1988) The antiarrythmic effects of R 56865 in cardiac glycoside toxicity are not caused by inhibition of receptor binding. Br J Pharmacol 93:273P
Hescheler J, Rosenthal W, Trautwein W, Schultz G (1987) The GTP-binding protein, G_o, regulates neuronal calcium channels. Nature 325:445–447
Hess P, Tsien RW (1984) Mechanism of ion permeation through calcium channels. Nature 309:453–456
Kalix P (1977) Uptake and release of calcium in rabbit vagus nerve. Pflügers Arch 326:1–14
Latorre R, Coronado R, Vergara C (1984) K^+-channels gated by voltage and ions. Ann Rev Physiol 46:485–495
Lee CO (1985) 200 years of digitalis: the emerging central role of the sodium ion in the control of cardiac force. Am J Physiol 249:C367–C378
Lee CO, Taylor A, Windhager EC (1980) Cytosolic calcium ion activity in epithelial cells of necturus kidney. Nature 287:859–861
Lee H-C, Smith N, Mohabir R, Clusin WT (1987) Cytosolic calcium transients from the beating mammalian heart. Proc Natl Acad Sci USA 84:7793–7797
Lüllmann H, Peters T (1976) On the sarcolemmal site of action of cardiac glycosides. In Roy, Dhalla (eds) Recent advances in studies on cardiac structure and metabolism, vol 9. The sarcolemma. University Park Press, Baltimore, pp 311–328
Lüllmann H, Peters T (1977) Plasmalemmal calcium in cardiac excitation-contraction coupling. Clin Exp Pharmacol Physiol 4:49–57
Lüllmann H, Peters T, Preuner J (1983) Role of the plasmalemma for calcium homeostasis and for excitation-contraction coupling in cardiac muscle. In: Drake, Holland, Noble (eds) Cardiac metabolism. Wiley, Chichester, pp 1–18
Manery JF (1969) Calcium and membranes. In: Comer, Bronner (eds) Mineral metabolism, vol 3. Academic Press, New York, pp 405–452
Marban E, Tsien RW (1982) Enhancement of Ca-current during digitalis inotropy. J Physiol 329:589–614
Moore CL (1971) Specific inhibition of mitochondrial calcium transport by ruthenium red. Biochem Biophys Res Commun 42:298–305

Narahashi T, Tsunoo A, Yoshii M (1987) Characterization of two types of calcium channels in mouse neuroblastoma cells. J Physiol 383:231–249
Portzehl H, Caldwell PC, Rüegg JC (1964) The dependence of contraction and relaxation of muscle fibres from the crab *Maia squinado* on the internal concentration of free calcium ions. Biochim Biophys Acta 79:581–591
Post JA, Langer GA, Op den Kamp JAF, Verkleij AJ (in press) Phospholipid asymmetry in cardiac sarcolemma. Biochim Biophys Acta
Preuner J (1981) Ca-homeostasis in cardiac muscle cell: an active Ca-pump and its functional dependence on plasma membrane bound Ca. Arch Pharmacol 316:30
Reuter H (1986) Voltage-dependent mechanisms for raising intracellular free calcium concentration: calcium channels. In: Calcium and the cell (Ciba Foundation Symposium 122). Wiley, Chichester, pp 5–22
Reuter H, Blaustein MP, Haeusler G (1973) Na-Ca-exchange and tension development in arterial smooth muscle. Phil Trans R Soc 265:87–94
Rosenberg RL, Hess P, Reeves JP, Smilowitz H, Tsien RW (1986) Calcium channels in planar lipid bilayers: insights into mechanisms of ion permeation and gating. Science 231:1564–1566
Rudel R (1979) In: Ashley, Campbell (eds) Detection and measurement of free Ca^{++} in cells. Elsevier/North-Holland, Amsterdam, pp 153–158
Schatzmann HJ (1966) ATP-dependent Ca^{++}-extrusion from human red cells. Experientia 22:364
Schneider J, Beck E, Wilffert B, Peters T (1988) A dose-dependent inhibition of digitalis-induced toxicity by R 56865 in the guinea-pig heart-lung preparation. Br J Pharmacol 93:272P
Seimiya T, Ohki S (1973) Ionic structure of phospholipid membranes, and binding of calcium ions. Biochim Biophys Acta 298:546–561
Siesjö BK (1985) Oxygen deficiency and brain damage: Localization, evolution in time, and mechanism of damage. Clin Tox 23:4–6
Stahl WL, Swanson PD (1972) Calcium movements in brain slices in low sodium or calcium media. J Neurochem 19:2395–2407
Vincenzi FF, Schatzmann HJ (1967) Some properties of Ca-activated ATPase in human red cells. Helv Physiol Pharmacol Acta 25:233
Vollmer B, Meuter C, Janssen PAJ (1987) R 56865 prevents electrical and mechanical signs of ouabain intoxication in guinea-pig papillary muscle. Eur J Pharmacol 142:137–140
Yeagle P (1987) The membranes of cells. Academic Press, New York
Yoshikawa K, Fujimoto T, Shimooka T, Terada H, Kumazawa N, Ishii T (1988) Electrical oscillation and fluctuation in phospholipid membranes. Phospholipids can form a channel without protein. Biophys Chem 29:293–299
Zwaal RFA, Roelofsen B, Colley CM (1973) Localization of red cell membrane constituents. Biochim Biophys Acta 300:159–183

Cerebral Ischemia: Pathophysiology and Biochemistry

New Methods for Topographical Imaging of Physiologically Defined Boundary Zones in Focal Cerebral Ischemia

A. J. STRONG, A. R. HOGG, and E. MONTEIRO

Introduction

The concept of a penumbra surrounding areas of focal cortical ischaemia was originally developed [1] on the basis of differences in the sensitivity of evoked potential amplitude and extracellular potassium homeostasis to reductions in local blood flow. Where flow was reduced, but remained above a critical threshold for evoked potential amplitude suppression, such suppression could nevertheless be induced by haemorrhagic hypotension, and was followed, but only with further hypotension and at significantly lower flows, by loss of extracellular potassium homeostasis [3]. One may therefore envisage a cortical boundary zone with intermediate degrees of flow reduction resulting from partial but incomplete collateralisation: the zone could be relocated away from the occluded input by systemic arterial hypotension, and towards it by hypertension. Cortex in this zone is regarded as subject to functional suppression, but structure is largely but not completely preserved.

The conclusion that neural function and structure are differentially sensitive to ischaemia is of considerable importance, but the nett value of any potential therapeutic regimen targeted at the penumbra depends to some extent on a quantitative assessment of the relative extents of core and penumbral territories. The need therefore arises for a topographical approach to the delineation of ischaemic boundary zones, using criteria which correspond as closely as possible to the original electrophysiological methods.

In these studies we have used pH reduction as a marker of early ischaemic change, and depletion of the total potassium pool as a marker of core ischaemia. The reasons for choice of these variables as markers are considered in the discussion below.

Methods

Experiments were carried out in cats of either sex of a minimum weight of 3 kg, in which anaesthesia was induced with halothane (4%) and maintained after intubation with chloralose. The animals were paralysed and ventilated to physiological $PaCO_2$. The right hemisphere was exposed by craniectomy and protected

Neurosurgical Research Group, Institute of Psychiatry, London SE5 8AF, U.K.

Cerebral Ischemia and Calcium
Ed. by A. Hartmann and W. Kuschinsky

with a pool of warm mineral oil, the temperature of which was monitored, and maintained at 37° C by an overtable lamp of variable height. The right middle cerebral artery was exposed by the transorbital route and occluded by clip for 2 h. The mineral oil was removed, and the brain then rinsed with warm mock CSF, and freeze trapped with isopentane chilled in liquid nitrogen, followed by liquid nitrogen. Brains were preserved at −37° or −70° C, and were subsequently dissected in a cold chamber. Cryostat sections (20 μm) were cut and adjacent sections taken for either pH or K^+ imaging.

pH Imaging (Csiba et al. [4])

Sections were laid on electrophoresis paper impregnated with umbelliferone and then allowed to thaw to 0° C on a Peltier plate. The intensity of 450 nm fluorescence of umbelliferone is proportional to pH, if an excitation wavelength of 370 nm is used. Fluorescence was photographed, and measured subsequently by manual or computerised densitometry.

K^+ Imaging (Mies et al. [9])

Cryostat sections were laid on glass slides, dried on a hotplate, and stained for potassium content using the sodium cobaltinitrite reaction, counterstaining with ammonium polysulphide in alcohol. This technique yields a stained section in which the intensity of staining is approximately proportional to potassium content.

Analysis of Images

Section and lesion areas were defined manually by planimetry and calculated by microcomputer.

Results

In some experiments, pH and K^+ lesion areas were confined to the lenticulostriate territory and coincided, confirming the absence of any boundary zone in this end-vessel territory. However, since the calculated lesion areas lay within the measurement error of the technique, this group of results will not be further considered.

In the case of larger lesions involving cortex, centrum semiovale, and head of caudate, there was a strong correlation between the potassium and pH lesion areas expressed as percent of hemisphere (52 sections in 11 experiments) (Fig. 1). The equation of linear regression was $Y = 0.642X - 0.124$, where Y = potassium

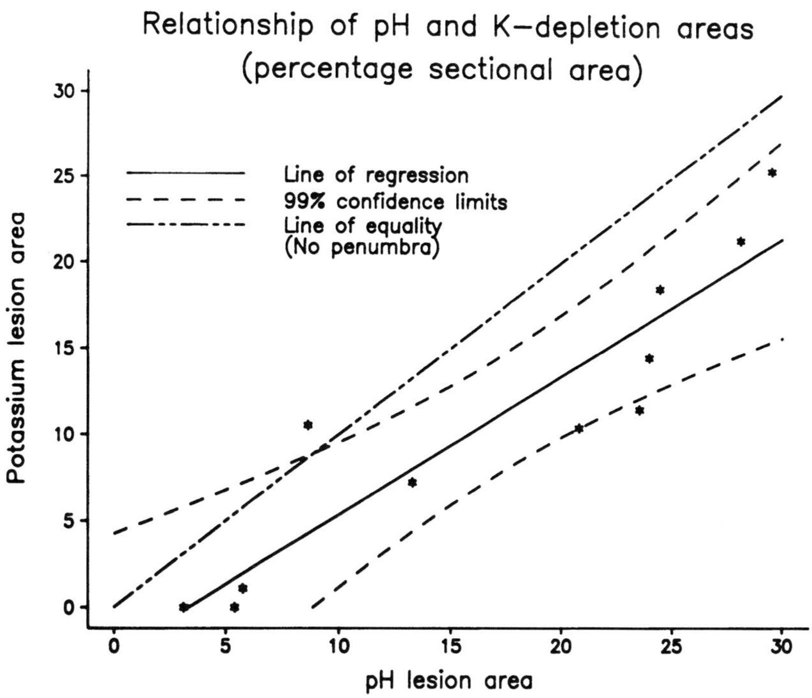

Fig. 1. Relationship of pH and potassium lesion sizes 2 h following onset of ischaemia. Each *point* represents the results of one experiment, and is the mean of data from 3, 5 or 6 planes (mean = 4.73 planes). The calculated line of regression is shown, together with the 99% confidence limits and the line of equality, which represents a hypothesis of no penumbra

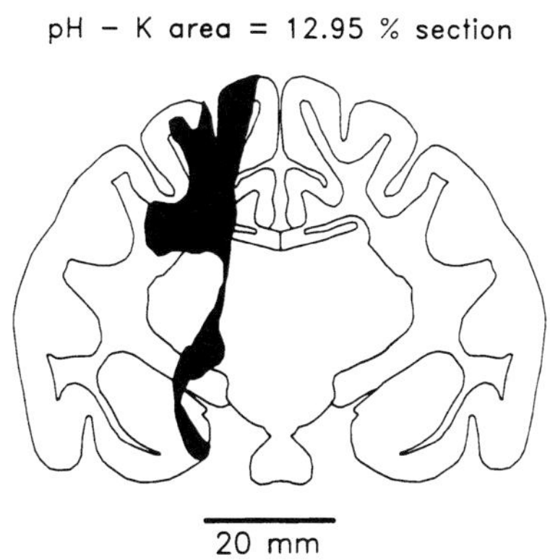

Fig. 2. Illustration of results of comparison of pH and K lesion sizes at one plane in a single experiment. The *shaded area* is the territory in which pH but not potassium content was reduced, thus meeting the criteria for penumbra

lesion area, and X = pH lesion area, expressed as percent of whole section area. For pH lesion sizes above 10% of the section, the line of equality (representing a hypothesis of no penumbra) lay above the upper 99% confidence limit of the regression equation, confirming the existence of a significant boundary zone. For larger lesions this represents some 13% of the whole section (Fig. 2).

Discussion

These studies represent an initial attempt to establish the size and location of ischaemic boundary zones associated with occlusion of a single major territory artery, using topographical techniques, and applying criteria for penumbra resembling the physiological indices on which the concept was originally based. We have substituted the pooled cortical pH value for evoked potential amplitude, and depletion of the total K^+ pool for a sustained increase in extracellular K^+. The first substitution is based on the fact that the relationship of both variables with ischaemic CBF is very similar: in both cases a threshold exists, at an intermediate degree of ischaemia [2, 6]. Our topographical data suggest that there is frequently a relatively sharp interface between normal and acidotic brain, often near the convexity of the marginal gyrus and perpendicular to it. The area of acidosis most peripheral in the middle cerebral territory therefore corresponds well with the site on the surface of the marginal gyrus where we have observed pH reductions using an electrode technique [5]. However, it is not clear from the present data whether a decrease in pial surface or extracellular pH would be likely to precede a decrease in total cortical pH (dominated by the intracellular compartment), and the possibility that the glia may become selectively acidotic is now also recognised [8].

Our use of total K^+ depletion as a marker for core ischaemia derives from the knowledge that the total cortical K^+ pool is depleted in core ischaemia [7], whereas there is no evidence of K^+ clearance from the (rat) brain under conditions resembling penumbral ischaemia [10].

The data provide clear evidence of a peripheral zone in which brain pH is reduced but K^+ content is preserved. Although apparently only some 13% of the whole section area in the case of larger lesions, or 26% of the hemisphere (on the arguable assumption of no swelling), the pH lesion area is nevertheless greater than that of the K^+ lesion by some 50%. The location of the penumbral area (Fig. 2) corresponds closely with the area previously identified as penumbra in the cat MCAO model using electrode techniques both for flow topography (hydrogen clearance) and ion homeostasis (pial surface valinomycin electrodes) [11]. We have also demonstrated impaired vascular and NAD reactivity to local cortical stimulation in this territory, and some elements of these latter data suggest the possibility that functional suppression or deafferentation might contribute to flow reduction in the penumbra [12].

Our present data relate only to a single time point 2 h following occlusion, and it would clearly be of interest to examine shorter and longer durations. Application of this and similar differential imaging techniques may help considerably in developing our understanding of how an ischaemic lesion evolves or may be protected.

Acknowledgement. We thank the MRC for financial support for this study (G8313957).

References

1. Astrup J, Symon L, Branston NM, Lassen NA (1977) Cortical evoked potential and extracellular K^+ and H^+ at critical levels of brain ischaemia. Stroke 8:51–57
2. Branston NM, Symon L, Crockard HA, Pasztor E (1974) Relationship between the cortical evoked potential and local cortical blood flow following acute middle cerebral artery occlusion in the baboon. Exp Neurol 45:195–208
3. Branston NM, Strong AJ, Symon L (1977) Extracellular potassium activity, evoked potential and tissue blood flow: relationships during progressive ischaemia in baboon cerebral cortex. J Neurol Sci 32:305–321
4. Csiba L, Paschen W, Hossmann K-A (1983) A topographic quantitative method for measuring brain tissue pH under physiological and pathophysiological conditions. Brain Res 289:334–337
5. Gibson G, Miller SA, Venables GS, Strong AJ (1983) Evidence of acidosis in the ischaemic penumbra. J Cereb Blood Flow Metab 3,1:401–402
6. Harris RJ, Richards PG, Symon L, Habib A-HA, Rosenstein J (1987) pH, K^+, and PO_2 of the extracellular space during ischaemia of the primate cerebral cortex. J Cereb Blood Flow Metab 7:599–604
7. Hossmann K-A, Schuier FJ (1980) Experimental brain infarcts in cats. 2. Ischemic brain edema. Stroke 11:583–592
8. Kraig RP, Petito CP, Plum F, Pulsinelli WF (1987) Hydrogen ions kill brain at concentrations reached in ischemia. J Cereb Blood Flow Metab 7:379–386
9. Mies G, Kloiber O, Drewes LR, Hossmann K-A (1984) Cerebral blood flow and regional potassium distribution during focal ischemia of gerbil brain. Ann Neurol 16:232–237
10. Strong AJ, Morii S, Ngai A, Winn HR (1983) Early impairment of the blood brain barrier to potassium in systemic hypoxia and mild hypotension. J Cereb Blood Flow Metab 3 (Suppl 1):S638–639
11. Strong AJ, Venables GS, Gibson G (1983) The cortical ischaemic penumbra associated with occlusion of the middle cerebral artery in the cat. 1. Topography of changes in blood flow, potassium ion activity and EEG. J Cereb Blood Flow Metab 3:86–96
12. Strong AJ, Gibson G, Miller SA, Venables GS (1988) Changes in vascular and metabolic reactivity as indices of ischaemia in the penumbra. J Cereb Blood Flow Metab 8:79–88

Experimental Hypoglycemia: Correlation Between EEG Abnormalities and Structural Alterations

J. H. Garcia, M. L. Anderson, and J. H. Halsey, Jr.

Introduction

Ischemic injury is characterized by a combination of decreased arterial flow, decreased tissue oxygen, limited availability of energy substrates, anaerobic glycolysis, retention of metabolic end-products (such as lactic acid), and depressed function [5]. The latter, in the case of the brain, is usually reflected in disappearance of delta waves in the electroencephalogram [1].

Under special circumstances the cellular injury initiated by ischemia is either continued or, perhaps, aggravated *after* the ischemic conditions have ceased, that is: during reperfusion [5]. Numerous factors have been held responsible for the structural and functional abnormalities characteristic of ischemia, including tissue acidosis, energy failure, and calcium entry into cells with injured plasma membranes [4, 9, 12].

Preliminary observations by von Lubitz and Diemer [14] and by Auer et al. [2] suggest that the electrical silence typical of a brain area injured by either ischemia or hypoglycemia may reflect abnormalities involving the dendritic tree of the respective neurons. We have explored the possible relationship between EEG tracings and neuronal abnormalities in an experimental model of hypoglycemia, a condition that shares with ischemia the absence of energy substrate. Furthermore the method to induce hypoglycemia is more reproducible, and more compatible with the maintenance of "steady-state" conditions than is ischemia.

Materials and Methods

Twenty-two adult Wistar rats (mean body weight 300 g) were anesthetized with a mixture of halothane (0.5%) and nitrous oxide (50%–60%). EEG was recorded from needle electrodes placed in the frontal and occipital scalp at the beginning of the experiment after anesthesia had been induced. EEG tracings were displayed on a channel of a Grass model 7 polygraph using an AC preamplifier. The amplification was adjusted so that the maximum amplitude of the EEG potentials was about 50% of the full pen deflection. Recordings were obtained before, during, and after the end of the "injury period" (Fig. 1).

Departments of Pathology and Neurology, The University of Alabama at Birmingham, UAB Station, Birmingham, Alabama 35294, USA.

Cerebral Ischemia and Calcium
Ed. by A. Hartmann and W. Kuschinsky

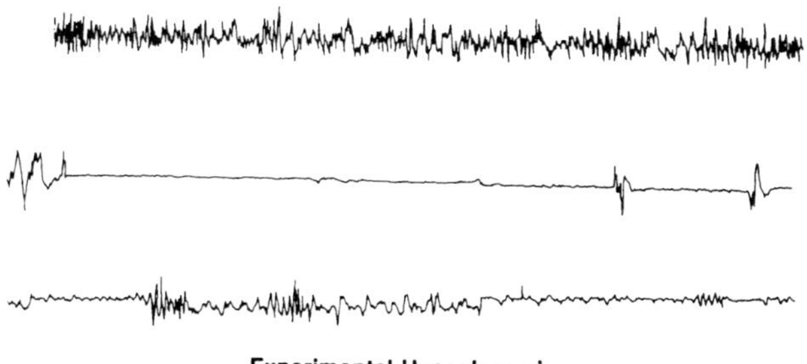

Fig. 1. Typical sample of EEG tracing, as obtained before (*top*), during (*middle*), and after (*bottom*) the period of injury, as defined in the text

Insulin was injected intravenously at a rate of 100 U/h, until the EEG became isoelectric (Fig. 1). The beginning of the "injury" was calculated as the time when the EEG tracing became flat, in accordance with criteria spelled out by other investigators [1], and in most instances such "injury" ranged between 4 and 35 min. Hypoglycemia was terminated by manual intravenous injection of 1.0 cc 10% glucose over an interval of 30 s. Thereafter 50% glucose was infused by syringe pump at an average rate of 500 mg/h. Blood glucose was measured at intervals of 10–30 min and the infusion rate adjusted to maintain a blood glucose level between 70 and 150 mg%.

The EEG was also recorded through two bandpass filters to obtain selectively the relative proportions of "high" and "low" frequency activity. The low frequency filter passed 100% amplitude at 2 Hz, 20% at 7 Hz, and zero at 20 Hz. The high frequency filter passed zero at 2 Hz, 100% at 7 Hz, and 80% at 20 Hz. The potentials from the filter were rectified and their amplitude continuously averaged over approximately 2 s with the aid of the low linear frequency switch on the polygraph. The display was adjusted so that the full-scale pen deflection represented 100% of the 2-Hz response for the low frequency output and of the 7-Hz response for the high frequency output.

A separate group of 20 rats was made hypoglycemic as above while ^{31}P nuclear magnetic resonance spectra were being obtained from an 8.5-Tesla MRS spectrometer. Spectra were obtained every 4 min during a total period of 3 h that included an "injury period" ranging between 10 and 40 min. The beginning of this interval was indicated by a rapid fall in the phosphocreatine/inorganic phosphorus (PCr/Pi) ratio and the end, by the intravenous glucose injection.

About 3 h after the "injury period" was terminated, the animals were killed and injected (via the vascular system) with a buffered aldehyde solution. Histological samples were prepared from coronal slabs of both cerebral hemispheres (including the hippocampal formation). Staining with H&E was completed in samples from all specimens and histological quantitation of the tissue injury was carried out (first separately and then jointly) by two blinded observers. A minimum of two preparations from each animal's brain were evaluated for extent and severity of "neuronal changes." Each histological sample was graded as being: (A) normal (0), (B) mildly injured (+) or (C) markedly injured (++) based on the extent of changes in volume and stainability of neuronal perikarya located in the cerebral cortex and subcortical structures of the cerebral hemispheres.

Brain samples from five animals were additionally examined after being stained by a modified Golgi-Kopsch silver impregnation method, as follows:

coronal slices of formalin-fixed rat brains, trimmed to a thickness of 3–5 mm, were placed over glass wool in a chromating solution composed of 3% potassium dichromate (100 ml), 10% unbuffered formalin (5 ml), and sucrose (12.5 g). Three days later the tissues were transferred to a second chromating solution identical to the first. Following immersion for an additional week, the sections were blotted dry of excess chromating solution and rinsed in 0.75% silver nitrate. Then the sections were placed over glass wool once more in a fresh 0.75% silver nitrate solution for 3 days. Impregnated tissue blocks were dehydrated in a series of ethanol solutions of ascending concentration, and subsequently embedded in celloidin. Sections wetted with 70% ethanol were cut at 100 μm on a sliding microtome, cleared with α-terpinol, and mounted in a neutral synthetic resin (Permount) under #1 coverslips.

A correlative analysis was performed to look for a possible association between severity of EEG abnormalities and the extent/degree of the structural alterations detected in some of the dendritic spines of cortical and subcortical neurons.

Results

Pilot phantom experiments demonstrated that about 90% of the NMR signal was being recorded from the outer 2 mm of the cerebral convexity, roughly including the frontoparietal cortex and the dorsal portion of the hippocampus.

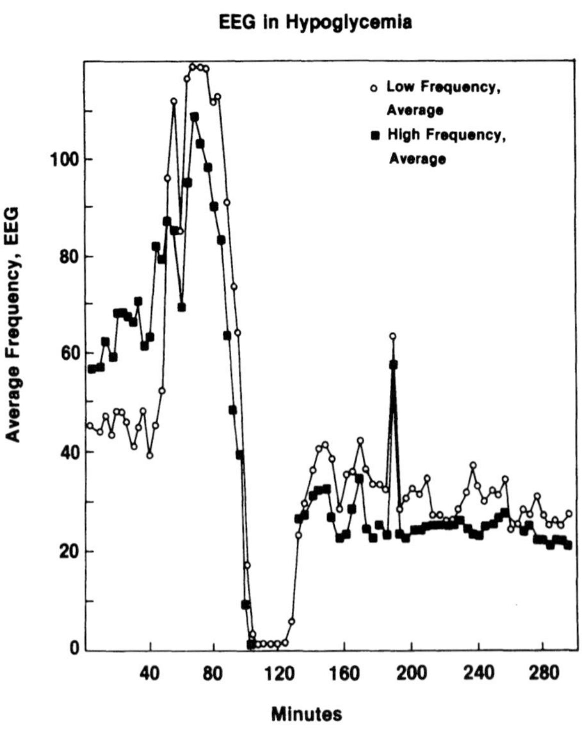

Fig. 2. Composite demonstration of EEG abnormalities in rats made hypoglycemic by insulin injection; after a period of injury, ranging between 5 and 40 min, there is incomplete "recovery" of EEG activity despite the full recovery of ATP and PCr demonstrated in NMR spectra

Energy failure was manifested by rapid depletion of β-ATP and PCr and rise in the Pi. The ratio PCr/Pi fell to 1 and was followed within about 5 min by virtually total depletion of ATP. We succeeded in recording EEG from within the NMR spectrometer in five animals in which the EEG became isoelectric almost coincidental with the fall of the PCr/Pi to 1–1.5, 5–10 min before total depletion of PCr/Pi and 10–15 min before total depletion of β-ATP; restitution of glucose was associated with a prompt, brisk rise in the PCr/Pi ratio to 60%–80% of the preinjury level and recovery of β-ATP to about 80%. This recovery occurred within less than 10 min of the glucose restitution, and was associated with an average fall in brain intracellular pH of about 0.2 units.

EEG recovery was considerably delayed, sometimes as long as 30 min after glucose restitution, and generally was incomplete (Fig. 2). After as short a time as 5 min of "injury," there was persistent reduction of high frequency activity averaging about 30%–50% of the pre-energy failure levels (Fig. 3). This degree of high frequency attenuation was not worse after up to 40 min of energy failure. In general, the maximum high frequency recovery was complete within about an hour after glucose restitution and did not change further after 3 h survival (Fig. 3). Low frequency recovery was generally more nearly complete and by 3 h was increased above pre-energy failure levels by up to 100%.

Detailed histological evaluation of all samples stained with H&E disclosed a minimal degree of neuronal shrinkage and hyperchromatism ("dark neurons") that could be seen in scattered cells of both the cerebral cortex and the subcortical nuclei. These changes were more pronounced in animals having the longest injury period, i.e., 30 min. Dendritic spines, as revealed by the Golgi method, were less numerous in animals that sustained longer periods of EEG electrical silence.

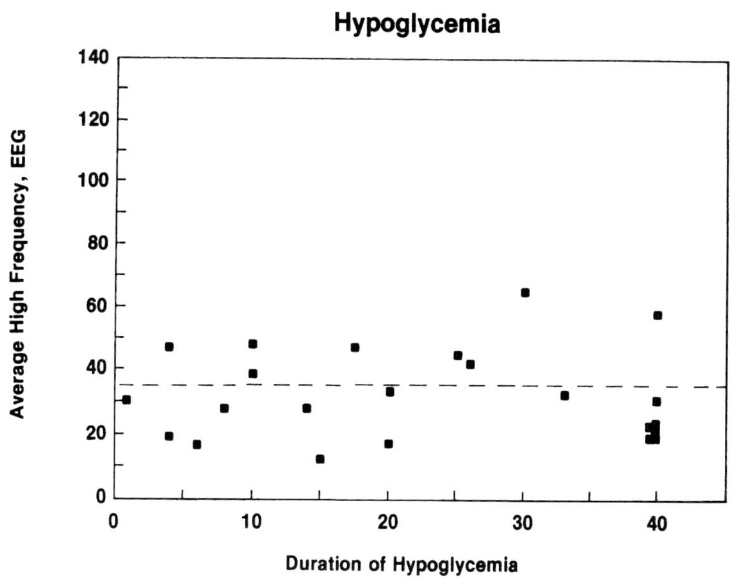

Fig. 3. The "recovery" of the high frequency EEG in animal groups exposed to an injury lasting 5–40 min was similar in all groups and averaged 30%–50% of the preinjury levels

However, at the moment, the evaluation of the dendritic spines has been completed in only five of the experimental rats.

Discussion

The ionic and chemical changes in hypoglycemia take place abruptly at the onset of EEG "isoelectricity." Neuronal necrosis is absent in hypoglycemia unless the EEG becomes isoelectric, regardless of the blood sugar levels. The duration of the electric cerebral silence roughly determines the degree of the resultant brain damage [1, 2].

The neuropathology of experimental hypoglycemic injury has been described in rats maintained at isoelectric EEG conditions for 60 min and allowed to survive for 1 week. In these animals, progressively longer periods of isoelectricity induced neuronal necrosis that appeared first in susceptible brain regions exposed to the subarachnoid spaces, second in portions of the brain apposed to other brain surfaces, and last in the deeper regions [1].

The minimal degree of structural alterations observed in our experiments is attributed to a short "injury period" (30 min) and a short postinjury survival period (2–3 h). In contrast to the minimal changes disclosed by the histological preparations (stained with H&E), our Golgi-impregnated preparations suggest that dendritic spine alterations occur early in the course of systemic hypoglycemia of sufficient magnitude to cause electrical failure. Furthermore, the temporal evolution of spine fragmentation and spine loss is quite consistent with the sequential dendritic changes described by Auer et al. [1, 2].

The earliest ultrastructural changes induced by hypoglycemia occur within the dendritic arbor; dendritic swelling is evident as early as 10 min after the onset of insulin-induced cerebral isoelectricity, with a steadily progressive deterioration of dendritic membrane integrity occurring thereafter. In contrast, interspersed axons-en-passant, terminal boutons, and neuronal perikarya are initially spared; neuronal perikarya are intact up to approximately 30 min of EEG isoelectricity [1]. "Dark" neurons of the type observed in the specimens obtained from our experimental animals have been thought to represent early and reversible signs of injury of a nonspecific nature [2].

In further support of the hypothesis that there is a selective synaptic (dendritic) insult in hypoglycemic animals that is responsible for the electrical silence we offer the following arguments:

1. The EEG and the NMR are recorded from roughly the same tissue regions, yet the EEG fails well in advance of total energy depletion.
2. After glucose restitution the recovery of high energy phosphorus metabolites is prompt and up to 80% complete while the EEG recovery is greatly delayed.
3. The EEG recovery after glucose restitution is poorer for the high frequency activity than the low frequency.
4. Impairment of high frequency EEG is disproportionate to the minimal cell body changes seen by ordinary (H&E) histology.

The relationship between dendritic spine loss and the impaired recovery of the high frequency components of the EEG remains speculative. However, anatomical and biophysical lines of evidence make a causal relationship at least plausible. Dendritic spines are the preferential sites for synaptic interactions between cortical neurons and their afferents [11]. Also, the EEG signal depends heavily upon the passive electrical (or "cable") properties of cortical neurons [7]. Specifically, the electrical space constant of the pyramidal cell soma is so large that the cell body essentially behaves as an isopotential unit, lacking any dipole properties. On the other hand, a dendritic process, possessing high internal resistance, is generally characterized by a space constant much less than its overall length. Consequently, electrotonic conduction or temporospatial spread of local postsynaptic potential change is diminished, and the pyramidal dendritic tree processes function collectively as an electrical dipole. For this reason, the EEG is best interpreted as the summation of dendritic (essentially spine) activities as opposed to postsynaptic somal potential changes [13]. Quick transient membrane potentials presumably are responsible for the high frequency range of the EEG filtered spectrum. Because of the long time constant of pyramidal neurons [3], as well as the small space constant of dendrites, considerable delay of electrotonic spread and distortion of traveling *axosomatic* potential waves will occur at recording surface electrodes. It seems reasonable then that most of the high frequency EEG spectrum reflects *axodendritic* interactions, and that the contributions of axosomatic synapses are negligible. For these reasons pathological changes in spine morphology of the type we found should correlate well with suppressed high frequency EEG activity.

In addition to the intrinsic value of these studies on the effects of hypoglycemia, the experimental model may be useful, because it is more reproducible than ischemia, in the evaluation of some of the factors that initiate and aggravate ischemic neuronal injury.

Auer [1] has written that the conditions created by hypoglycemia and ischemia differ from one another because (a) energy loss is incomplete in hypoglycemia (compared to ischemia) and (b) tissue acidosis, which is an important component of ischemia [6, 9], does not occur in hypoglycemic animals.

In hypoglycemia with "isoelectricity," brain ATP levels are said to be over one-third of the normal values [1], yet our NMR spectra clearly show total but reversible depletion of ATP that preceded by several minutes the disappearance of EEG activity. As for the brain tissue pH, Auer [1] reports brain tissue alkalosis in hypoglycemic rats which he attributes, among other factors, to the formation of ammonia and absence of lactic acid. However, pH calculations from the NMR spectra in our animals show a consistent drop of about 0.2 units in the intracerebral pH of hypoglycemic animals with isoelectric EEG. Acidosis in experimental ischemia has been held responsible for injury to the astroglia and, secondarily, necrosis of some neurons [8]. Kraig et al. [6] have recently reported induction of brain tissue necrosis by intracerebral injections of lactic acid at concentrations known to be attained in areas of brain ischemia. The slight drop in pH observed in our experiments seemingly was insufficient to cause either astrocytic nuclear swelling or changes in volume/stainability of neuronal perikarya.

Considerable emphasis has been placed on the possible role that calcium ions may play in the process that is initiated by ischemia and eventually leads to coagulative necrosis. The calcium overload of cells whose plasma membranes had been altered by the energy-deficient status created by ischemia has been well demonstrated in reperfused myocardium [5]. Moreover, the deleterious effects of intracellular calcium overload can be traced directly to the cellular respiratory failure secondary to the presence of calcium salts in the mitochondrial matrix. Demonstration, by electron microscopy, of such calcium precipitates has been used by some as an early indicator or irreversible cell injury [5].

Excessive intracellular calcium is also injurious to the cell because calcium activates lipases and proteases, causes disaggregation of neurotubuli, and enhances protein phosphorylation [12]. In the case of the neuron, calcium entry is governed not only by the effect of energy metabolites on the plasma membrane but also by the mechanisms involved in synaptic transmission [9, 10]; thus excitatory neurotransmitters such as glutamate are known to be closely linked to the generation of action potentials and ionic calcium exchanges between the extra- and intracellular compartments [10].

The initiating event in both ischemia and hypoglycemia seems to injure a selected group of neurons. The selectivity, in the case of experimental hypoglycemia, is said to be traceable to their proximity to the spinal fluid compartment [1]. In terms of individual neurons, the selective injury in both ischemia and hypoglycemia may occur at the dendritic spines where it could be mediated by the effect of "excitatory neurotransmitters" on the movement of Ca^{2+} across the corresponding membrane channels. The potential effect that hypoglycemia may exert on dendritic spines awaits verification by appropriate quantitation of Golgi-impregnated preparations.

Acknowledgements. The authors gratefully acknowledge the secretarial support of Ms. Jeannie Cornelius and the photographic skills of Mr. Ralph Roseman. This work was financially supported by a grant from the USPHS #NSO8802 and the Department of Pathology at UAB.

References

1. Auer RN (1986) Progress review: hypoglycemic brain damage. Stroke 17:699–708
2. Auer R, Kalimo H, Olsson Y, Wieloch T (1985) The dentate gyrus in hypoglycemia: pathology implicating excitotoxin-mediated neuronal necrosis. Acta Neuropathol 67:279–288
3. Creutzfeldt OD, Lux HD, Nacimiento AC (1964) Intracellulare Reizung corticaler Nervenzellen. Pflügers Arch Ges Physiol 281:129–151
4. Garcia JH, Anderson ML (1988) Physiopathology of cerebral ischemia. CRC Critical Reviews in neurobiology (in press)
5. Jennings RB, Reimer KA, Steenbergen C (1988) Myocardial ischemia revisited: the osmolar load, membrane damage, and reperfusion. J Mol Cell Cardiol 18:769–773
6. Kraig RP, Petito CK, Plum F, Pulsinelli WA (1987) Hydrogen ions kill brain at concentrations reached in ischemia. J Cereb Blood Flow Metabol 7:379–386
7. Lopes da Silva F, Van Rotterdam A (1987) Biophysical aspects of EEG and magnetoencephalogram generation. In: Niedermeyer E, Lopes da Silva F (eds) Electroencephalography: basic principles, clinical applications, and related fields. Urban & Schwarzenberg, Baltimore, pp 19–41

8. Plum F (1983) What causes infarction in ischemic brain? The Robert Wartenberg lecture. Neurology 33:222–233
9. Raichle ME (1983) The pathophysiology of brain ischemia. Ann Neurol 13:2–10
10. Rothman SM, Olney JW (1986) Glutamate and the pathophysiology of hypoxic-ischemic brain damage. Ann Neurol 19:105–111
11. Shepherd GM (1979) The synaptic organization of the brain. Oxford University Press, New York, pp 362–364, pp 383–387
12. Siesjö BK (1986) Calcium and ischemic brain damage. Eur Neurol 25 (Suppl 1):45–56
13. Towe AL (1979) Electrophysiology of the cerebral cortex. In: Ruch R, Patton HD (eds) Physiology and biophysics: the brain and neural function. Saunders, Philadelphia, pp 563–588
14. von Lubitz DKJE, Diemer NH (1982) Complete cerebral ischaemia in the rat: an ultrastructural and stereological analysis of the distal stratum radiatum in the hippocampal CA-1 region. Neuropathol Appl Neurol 8:197–215

Blood-Brain Barrier Permeability to Micromolecules After Focal Cerebral Ischemia

S. SAMPAOLO, N. HEYE, and J. CERVOS-NAVARRO

Introduction

During the initial 1–6 h of ischemia the blood-brain barrier (BBB) is grossly impermeable to protein tracers such as Evans blue (EB) [9–11, 19]. Tissue water content in ischemic brain increases within the first 2 h of arterial occlusion [8–10, 19, 21] and a significant decrease of the specific gravity of the ischemic frontal cortex has been reported as early as 30 min after permanent middle cerebral artery (MCA) occlusion in the rat [22]. Fluid accumulation is believed to occur in response to an increase of sodium into the lesion site [27] or reduction in sodium clearance from the tissue [12].

Little is known about BBB permeability to molecules smaller than proteins during the first hour of continuous incomplete ischemia, and systematic ultrastructural investigations with ionic electron dense tracers have never been performed.

Sodium fluorescein (F) in its free form acts as a low molecular weight tracer (MW 376), detectable macro- and microscopically. Ionic lanthanum (La^{3+}) is the smallest electron dense tracer (MW 138.9; diameter 0.5 mm) in use in biological investigations. Its passage through the endothelial lining is restricted in the normal brain by the tight junctions [2].

In the present study we examined the local changes of cerebrovascular permeability to EB, F, and La^{3+} after 15, 30, and 60 min of permanent MCA occlusion in the rat.

Material and Methods

Surgical Procedure

Twenty-seven adult Wistar rats of either sex, weighing 250–320 g, were anesthetized with ketamine (15 mg/100 g body wt. plus 0.5 ml atropine sulphate intraperitoneally). Spontaneous breathing was allowed. Rectal temperature was monitored and normothermia (37 °C) maintained by means of a heating pad. Polyethylene catheters were inserted into one femoral artery and vein in order to monitor

Institute of Neuropathology, Universitätsklinikum Steglitz, Free University of Berlin, Hindenburgdamm 30, D-1000 Berlin 45, FRG.

Cerebral Ischemia and Calcium
Ed. by A. Hartmann and W. Kuschinsky

continuously the mean arterial blood pressure, to determine intermittently the arterial blood gas concentrations and pH, and to administer the tracers. With the aid of an operating microscope the right MCA was exposed transcranially without lesioning the facial nerve and avoiding damage to the zygomatic bone [1]. The MCA was occluded and transected with a microbipolar coagulator using a low power setting and continuous saline irrigation. The artery was occluded at a point 2 mm proximal to the olfactory tract.

Experimental Groups

In a series of 15 rats with occluded MCA, each animal received EB and F (series A). A second series of 12 rats was given La^{3+} (series B). Five, five, and five rats of group A and four, four, and four rats of group B were killed respectively after 15, 30, and 60 min of MCA occlusion.

Administration of the Tracers

EB and F

In a syringe 0.2 ml EB (1% solution in NaCl 0.9%) was mixed with 0.8 ml of autologous blood and injected via the femoral vein catheter. Through the same catheter 0.2 ml F (10% solution in NaCl 0.9%) was injected 10 min before the animals were killed. At the end of the survival time, the animals were perfused through the ascending aorta with Karnovsky fixative solution.

La^{3+}

After administration of 50 ml NaCl 0.9%, perfused through the ascending aorta, a solution containing 2% glutaraldehyde and 0.2% paraformaldehyde in 0.1 M cacodylate buffer was perfused for about 5 min. The same fixative solution containing 20 mmol $LaCl_3$ (Sigma) was then perfused for a further 20 min. A more concentrated fixative (4% glutaraldehyde and paraformaldehyde) was perfused afterwards until the animals appeared to be well fixed. At the end of the perfusion, 3 ml of a solution of 3% methylene blue in the more concentrated fixative was injected through the aortic cannula to detect perfused and nonperfused territories in the brain.

Morphology

EB and F Study

After perfusion fixation the brain was immediately removed and sliced in standard coronal sections 2 mm thick. The sections were postfixed for an additional 3 h in fixative solutions having the same composition as those used for vascular

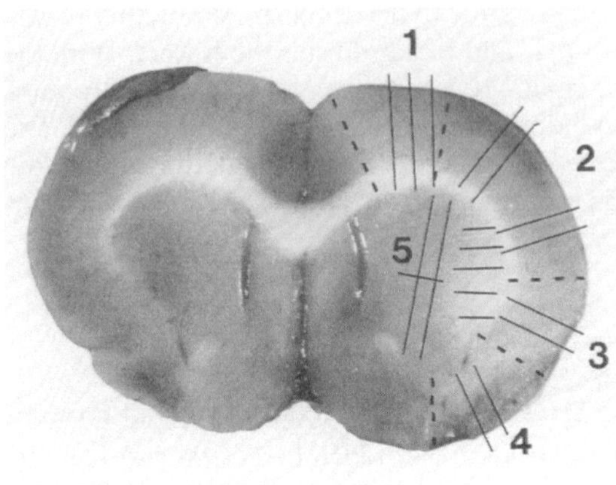

Fig. 1. Tissue samples dissected from the ischemic hemisphere for EM. Samples from the contralateral side served as controls

perfusion. Eventual extravasation of EB was investigated by gross examination. Distribution of sodium fluorescein was studied on gross sections under a Zeiss operating microscope equipped for fluoroscopy.

La^{3+} Study

For the study of La^{3+} distribution the brain was removed immediately after perfusion fixation and a 3 mm thick coronal slice of brain was cut behind the caudal limit of the olfactory bulb. In order to choose specimens for transmission electron microscopy (EM) the distribution of methylene blue staining was studied on the brain slice. In such a slice the whole brain appeared methylene blue stained, except for a pale area in the ischemic hemisphere. This pale area anatomically comprised the parietal cortex, insular cortex, a portion of the piriform cortex, the white matter subjacent to the above structures, and a rim of the adjacent lateral neostriatum. Tissue samples approximately 2 mm^3 in volume were dissected from both the methylene blue stained tissues and the adjacent pale area (Fig. 1). Tissue samples of corresponding areas were dissected from the contralateral hemisphere. All the tissue blocks were postfixed in OsO_4, dehydrated through graded acetone series, and embedded in Araldite. Ultrathin sections were contrasted with or without heavy metal staining and studied with an EM 10 Zeiss electron microscope.

Results

Occlusion of the MCA for 60 min resulted in a slight swelling of the occluded hemisphere. No gross change was detected in the brains of animals with shorter times of occlusion.

Tracer Studies

After 15 min of MCA occlusion, a patch of F staining was seen in the insular cortex adjacent to the operation site. At 30 min faint F fluorescence was seen extending to the piriform cortex, parietal cortex, and subjacent white matter (Fig. 2) in the ischemic site. At 60 min a more intense fluorescence was seen in the abovementioned areas.

None of the periods of MCA occlusion studied by us resulted in EB extravasation in any of the animals.

By studying electron microscopically the La^{3+} distribution a large amount of the tracer was generally demonstrated in the lumen of the vessels. However, only spots of La^{3+} could be demonstrated in the vessels of the parietal cortex, insular cortex, and piriform cortex in the ischemic site (Fig. 2).

At 30 min of MCA occlusion La^{3+} was seen meandering the interendothelial clefts and accumulating in the basal membrane of several vessels in the occluded hemisphere. Small amounts of La^{3+} accumulated in the brain extracellular space (Fig. 3a). MCA occlusion for 60 min resulted in more conspicuous extravasation of La^{3+} (Fig. 3b). In the areas showing tracer extravasation widening of the extracellular space and slight swelling of astrocytic foot processes was noted.

These ultrastructural features were always observed in the frontal cortex and subjacent white matter and in the lateral neostriatum in the ischemic site. In all other tissue specimens sampled from both the ischemic and the contralateral hemisphere, La^{3+} remained in the vascular lumen. After MCA occlusion for 15 min La^{3+} extravasation was never seen.

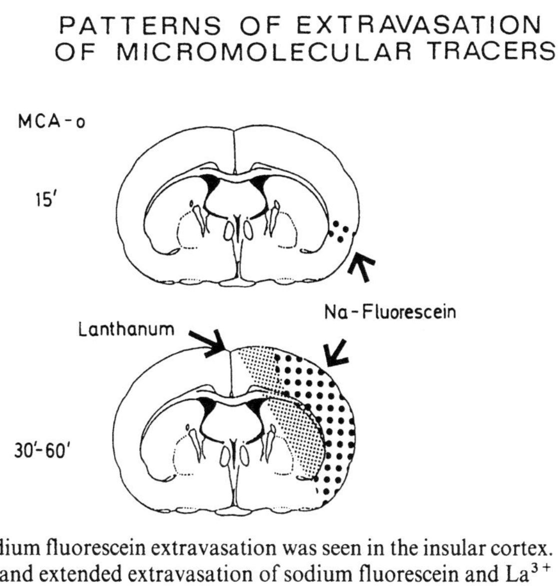

Fig. 2. At 15 min ischemia spotty sodium fluorescein extravasation was seen in the insular cortex. After 30 and 60 min a more intense and extended extravasation of sodium fluorescein and La^{3+} was seen

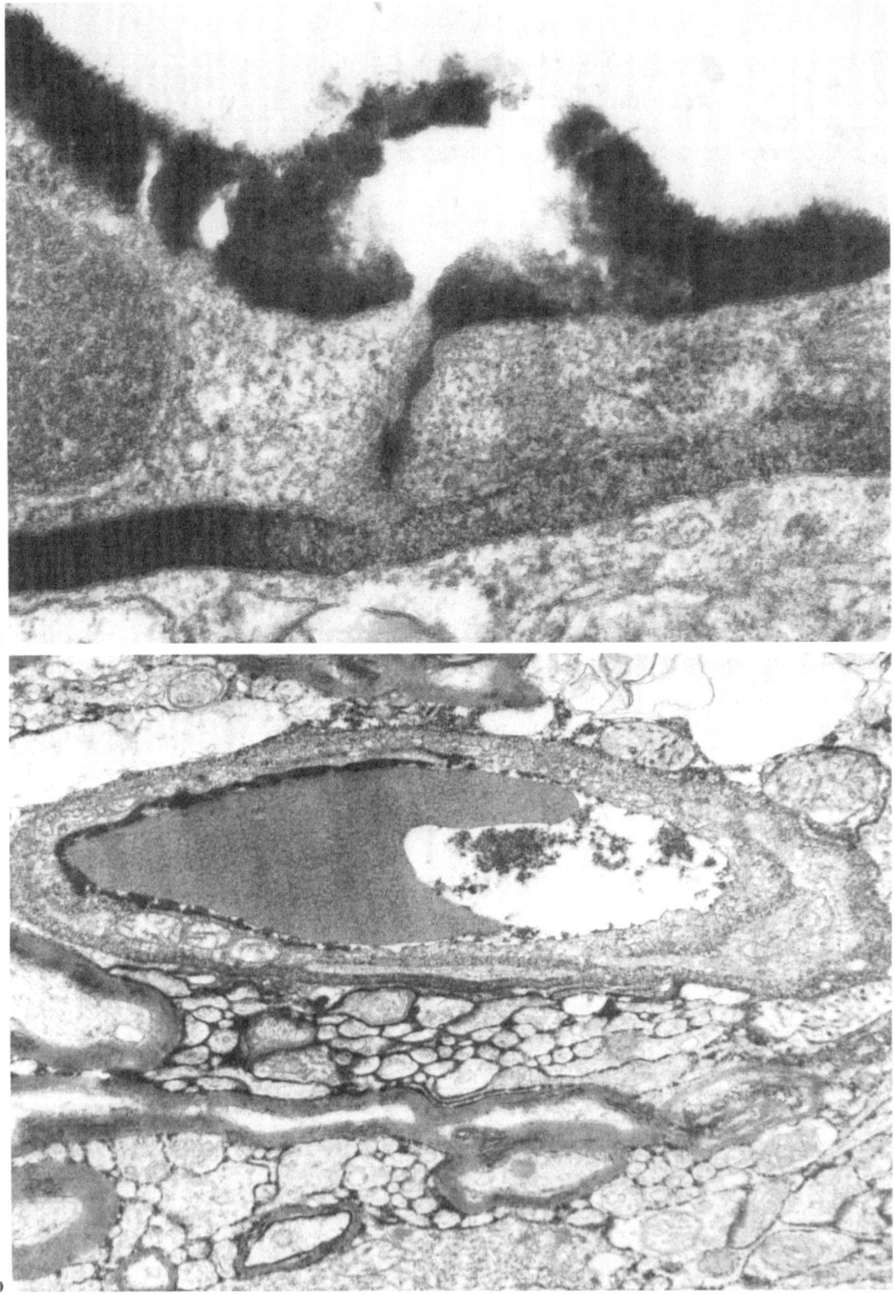

Fig. 3. a La^{3+} crosses the interendothelial clefts of cerebral vessels after 30 min of ischemia. **b** Conspicuous La^{3+} extravasation was seen after 60 min

Discussion

Ischemic brain edema was believed to be initially (for a not too precise time interval) of the so-called cytotoxic type without impairment of the BBB. Cytotoxic edema has been attributed to a derangement of the ionic pumps in the plasma membranes that leads to an uptake of sodium and a loss of potassium. In this condition the cells may swell at the expense of the interstitial fluid even if no net increase in water content occurs. However, net water content of the brain increases during the first 2 h of ischemia in spite of a BBB normal for EB [10, 19] or for radiolabeled albumin [9, 21]. To investigate the function of the BBB in the very early stages of ischemia we used micromolecular tracers rather than the classical protein tracer EB.

In the present investigation F extravasation was evident after 30 and 60 min of permanent MCA occlusion in the rat, and circumscribed leakage of the tracer was seen in the cerebral cortex subjacent to the operative site after 15 min. In other studies, F extravasation was observed at 90 min of permanent MCA occlusion in the cat [5] but bilateral carotid ligation in spontaneously hypertensive rats did not result in F extravasation until 4 h after the occlusion [25]. When F alone is administered in the concentration used in the present study a small aliquot of the tracer circulating in free form acts as a micromolecular tracer, the main aliquot being loosely or stably bound to plasma proteins [3]. We administered EB before F in order to bind to the first tracer as much protein as possible. Our results may have been influenced by administering the tracers in this manner. BBB impairment has been reported to be related to the levels of residual CBF in the brain territory supplied by the occluded artery [7, 23]. In the present model F extravasation began in the insular cortex and extended to the piriform and parietal cortex. These are the same brain regions as remain unstained after methylene blue perfusion in the ischemic hemisphere. Autoradiographic measurement of CBF following 30 min, 2 h, and 4 h of MCA occlusion in the rat showed that these regions have the lowest blood flow in the ischemic hemisphere [23].

One of the ultrastructural substrates for the early increase of BBB permeability in cerebral ischemia is the greatly intensified vesicular transport of capillary endothelium, which has been demonstrated using horseradish peroxidase as a tracer [24]. In the present study La^{3+} penetrated the tight junctions of several vessels and accumulated in the cerebral extracellular space 30 and 60 min after MCA occlusion. This pattern has been observed in brain areas showing no parenchymal damage by either light or electron microscopy. Because of lacking perfusion with La^{3+} an increased permeability of endothelial tight junctions could not be demonstrated in the regions in which F extravasation was seen. However, local changes in perfusion pressure and hydrostatic forces may damage the endothelial junction of brain vessels in areas of main circulatory impairment. Using horseradish peroxidase as a tracer early junctional opening has been demonstrated in an animal model of osmotic shock [15, 16], in a pressure pulse model, and in acute arterial hypertension [14]. However, La^{3+} seems to possess unique characteristics to study alterations of the interendothelial junctions. After "in vivo injection" in the lateral ventricle of hydrocephalic rats, La^{3+} diffuses from the brain to the blood throughout the interendothelial junctions of arterioles and venules, whereas horseradish peroxidase does not [18]. Indirect evidence suggests that endothelial junctions are partially permeable to small cations in the normal cerebral microcirculation [4].

The passage of La^{3+} across the interendothelial clefts, which in this experiment occurred "post mortem," may or may not occur in the living animal. However, the more pronounced extravasation of La^{3+} for 1 h occlusion, compared to that observed after 30 min, suggests that changes in the interendothelial junctions have evolved "in vivo." Enhancement of the endothelial permeability for small cations may be an early consequence of incomplete cerebral ischemia.

Dehiscence of interendothelial junctions has been induced in brain vessels by either perfusion with a polycation [13] or with Ca^{2+} depletion [6]. This suggests that the endothelial cell membranes' electric charge or the active contractile mechanisms in the endothelial cells [17], or both, are important factors in preserving the integrity of the interendothelial junctions. Alterations of both cerebral endothelium charge [26] and Ca^{2+} homeostasis [27] may be early consequences of cerebral ischemia and may cause the early BBB impairment for low molecular weight tracers.

In Conclusion

Extravasation of molecules, smaller than proteins, begins during the first hour of MCA occlusion in the rat. Brain regions having various levels of residual blood flow within the ischemic hemisphere may develop different patterns of BBB impairment. Selective patency of the endothelial junctions is evident by means of electron-dense ionic lanthanum in brain regions neighboring the zone of most severe CBF decrease. A paravascular pathway may account for the very early increase of brain tissue water content observed in the acute stage of incomplete ischemia.

References

1. Albanese V, Tommasino C, Spadaro A, Tomasello F (1980) A transbasisphenoidal approach for selective occlusion of the middle cerebral artery in rats. Experientia 36:1302–1304
2. Bouldin TW, Krigman MR (1975) Differential permeability of cerebral capillary and choroid plexus to lanthanum ions. Brain Res 241:57–65
3. Crockard A, Iannotti F, Hunstock AT, Smith RD, Harris RJ, Symon L (1980) Cerebral blood flow and edema following carotid occlusion in the gerbil. Stroke 11:494–498
4. Crone C (1984) Lack of selectivity to small ions in paracellular pathways in cerebral and muscle capillaries of the frog. J Physiol 353:317–337
5. Hayashi N, Tsubokawa T, Kimura S, Makiyama Y, Toeda M (1985) Simultaneous topographic analysis of brain edema, parenchymal energy metabolism (NAD/NADH redox state, reduced cytochrome aa3, ATP) and intracellular calcium during cerebral ischemia. In: Inaba Y, Klatzo I, Spatz M (eds) Brain edema. Springer, Berlin, pp 258–262
6. Herman IM, Pollard TD, Weng AJ (1982) Contractile proteins in endothelial cells. Ann NY Acad Sci 401:50–60

7. Iannotti F, Hoff JT (1983) Ischemic brain edema with and without reperfusion: an experimental study in gerbils. Stroke 14:562–567
8. Inao S, Kuchiwaki H, Hirai N (1985) Dynamics of tissue water content free and bound components, associated with ischemic brain edema. In: Inaba I, Klatzo I, Spatz M (eds) Brain edema. Springer, Berlin, pp 360–366
9. Ito U, Ohno K, Nakamura R, Suganuma F, Inaba Y (1979) Brain edema during ischemia and after restoration of blood flow. Stroke 10:542–547
10. Kamijyo Y, Garcia JH, Cooper J (1977) Temporary regional cerebral ischemia in the cat. A model of hemorrhagic and subcortical infarction. J Neuropathol Exp Neurol 36:338–350
11. Laursen H, Suzuki R, Kuroiwa T, Fenton I, Klatzo I (1983) Changes of cerebrovascular permeability in middle cerebral artery occlusion. In: Stefanovich V (ed) Stroke animal models. Pergamon, Oxford (Advances in the Biosciences 43:13–25)
12. Lo WD, Betz AL, Schielke GP, Hoff JT (1987) Transport of sodium from blood to brain in ischemic brain edema. Stroke 18:150–157
13. Nagy Z, Göhlert MG, Wolfe LS, Hüttner I (1985) Ca^{2+} depletion induced disconnection of tight junctions in isolated rat brain microvessels. Acta Neuropathol 68:48–52
14. Nagy Z, Mathieson G, Hüttner I (1979) Blood brain barrier opening to horseradish peroxidase in acute arterial hypertension. Acta Neuropathol 48:45–53
15. Nagy Z, Mathieson G, Hüttner I (1979) Opening of tight junctions in cerebral endothelium II: effect of pressure pulse induced acute arterial hypertension. J Comp Neurol 185:579–585
16. Nagy Z, Pappius HM, Mathieson G, Hüttner I (1979) Opening of tight junctions in cerebral endothelium I: effect of hyperosmolar mannitol infused through the internal carotid artery. J Comp Neurol 185:569–578
17. Nagy Z, Peters H, Hüttner I (1983) Charge related alterations of the cerebral endothelium. Lab Invest 49:662–671
18. Nakagawa Y, Cervos-Navarro J, Artigas J (1985) Tracer study on a paracellular route in experimental hydrocephalus. Acta Neuropathol 65:247–254
19. Olsson Y, Crowell RM, Klatzo I (1971) The blood brain barrier to protein tracers in focal cerebral ischemia and infarction caused by occlusion of the middle cerebral artery. Acta Neuropathol 18:89–102
20. Paxinos G, Watson C (1986) The rat brain, 2nd edn. Academic, Sydney
21. Sadoshima S, Fujishima M, Ogata J, Ibayashi S, Shiokawa O, Omae T (1983) Distribution of blood brain barrier following bilateral carotid artery occlusion in spontaneously hypertensive rats. Stroke 14:876–882
22. Shigeno T, Taesdale GM, McCulloch J, Graham DI (1981) Recirculation model following MCA occlusion in rats. J Neurosurg 63:272–277
23. Teasdale G, Tyson G, Tamura A, Graham ID, McCulloch J (1983) Focal cerebral ischemia in the rat: neuropathology, local cerebral blood flow and cerebrovascular permeability. In: Stefanovich V (ed) Stroke animal models. Pergamon, Oxford (Advances in the Biosciences 43:83–97)
24. Westergaard E, Go KG, Klatzo I, Spatz M (1976) Increased permeability of cerebral vessels to horseradish peroxidase induced by ischemia in Mongolian gerbils. Acta Neuropathol 35:307–325
25. Wolman M, Chui E, Wilmes F, Mishimoto K, Fujiwara K, Spatz M (1981) Evolution of the dye-protein tracers in pathology of the blood-brain barrier. Acta Neuropathol 54:55–61
26. Yanagihara T, McCall JR (1982) Ionic shift in cerebral ischemia. Life Sci 30:1921–1925
27. Young W, Rappaport H, Chalif DJ, Flamm ES (1987) Regional brain sodium, potassium and water changes in the rat middle cerebral artery occlusion model of ischemia. Stroke 18:751–759

Mediation of Platelet and Coagulation Events by Blood Flow: Basic Mechanisms of Thrombosis in Cerebral Ischemia

V. T. Turitto[1], H. J. Weiss[2], L. Badimon[1], C. Gemmell[1], and Y. Nemerson[1]

Introduction

Atherosclerosis is the principal underlying disease process in myocardial and cerebral infarction and is the major cause of death in the United States and Western Europe [22]. Recent studies have delineated the importance of thrombosis in the latter stages of atherosclerotic disease [1, 10, 11]. While there is limited evidence that thrombotic events contribute to the progression of atherosclerotic narrowing of the vessel lumen [18, 44], there is little doubt that in a majority of cases, final vascular occlusion is associated with thrombotic deposits [11]. The mechanisms by which thrombi form in the vicinity of a stenotic lesion are not fully understood, but clearly involve the local blood flow conditions, the composition of the underlying vascular material, and the participation of certain blood factors [14, 35, 36]. While considerable effort has been expended in uncovering the biochemical mechanisms in the blood and vascular wall which lead progressively to atherosclerotic disease, little emphasis has been focused on the fluid dynamic factors which modulate the disease process. It has long been known that initiation and progression of vascular lesions occur at sites which are distinguished primarily by the local fluid dynamic conditions; that is, lesions are characterized by their focal, rather than ubiquitous nature [19, 21]. Both high and low shear stresses have been implicated in the pathogenesis of arterial disease [12, 15, 18, 45, 46], and early fluid dynamic studies were primarily directed at a delineation of local flow disturbances at vascular bifurcations and sinuses or in geometrical models of atherosclerotic lesions. While controversy still exists with respect to the importance of low versus high shear stresses in the initiation of atherosclerotic disease, it is a general observation that both levels of shear stress exist in close proximity in the presence of advanced vascular lesions, a condition which is not observed in streamline flow through normal arterial vessels (Fig. 1). Such differing levels of shear may influence thrombotic mechanisms through dramatically different pathways. For example, in a previous symposium we presented evidence that low shear was conducive for the deposition of fibrin on injured vessel walls, whereas high shear leads to the enhanced deposition of platelets [33, 39]. Thus, both coagulative and platelet events may be stimulated, perhaps synergistically, in these

[1] Division of Cardiology, Department of Medicine, Mount Sinai Medical Center, New York 10029, USA.
[2] St. Lukes Roosevelt Hospital, Division of Hematology-Oncology, Department of Medicine, Columbia University College of Physicians and Surgeons, New York 10019, USA.

Cerebral Ischemia and Calcium
Ed. by A. Hartmann and W. Kuschinsky

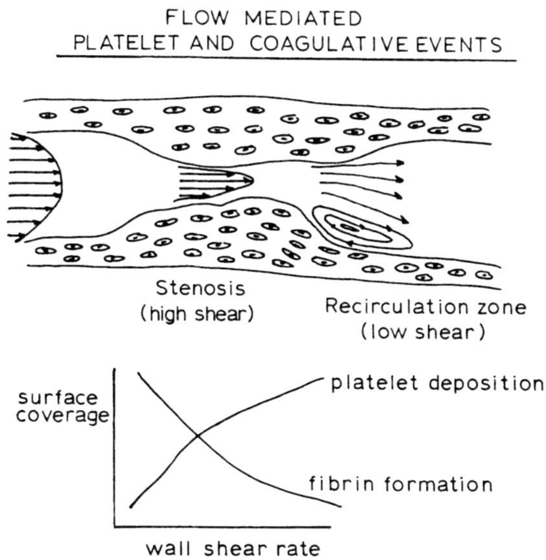

Fig. 1. *Above:* Drawing depicting blood flow through an atherosclerotic vessel. The flow entering the stenosed region becomes accelerated such that shear values are greatly increased at the local constriction. Upon leaving this region, expansion produces an adverse pressure gradient which can result in recirculatory flow and low shear values. *Below:* Typical behavior observed when injured vessel (subendothelium) is exposed to blood in streamline flow. The deposition of platelets increases, whereas the formation of fibrin deposits decreases, with increasing wall shear rates. In stenotic zones both high and low shear values in close proximity could lead to mechanisms whereby the deposition of both platelets and fibrin is greatly enhanced

disturbed flow regions. The extent to which various thrombotic mechanisms may be influenced not only quantitatively, but also qualitatively, by shear forces is the subject of the present manuscript. For such studies we have utilized an annular perfusion chamber in which de-endothelialized aorta is exposed to blood flowing under parallel streamline flow conditions [28]. In a companion paper, we consider the interaction of blood elements with injured vessel wall modified so as to stimulate the disturbed flow in stenotic regions [4].

Platelet Events in Thrombosis: Role of GPIIb-IIIa at High Shear Rates

Two glycoproteins on the surface of the platelet have been identified as being important in platelet-vessel wall interaction. Early studies in patients (Bernard-Soulier) with abnormalities of platelet GPIb revealed the importance of this protein for normal platelet adhesion to subendothelium at high shear rates [27, 37]. This glycoprotein is known to bind von Willebrand factor (vWF) upon stimulation with the antibiotic ristocetin, and original studies suggested that decreased adhesion of platelets from patients with von Willebrand's disease (vWD) at high shear rates (>800 s^{-1}) was due to reduced interaction between GPIb and vWF. Thus, platelet attachment to subendothelium originally appeared to be related to the formation of available sites on subendothelium for vWF which then bound to platelet GPIb [29, 33]. More recently, vWF has been shown to bind to GPIIb-IIIa upon stimulation with adenosine diphosphate [23, 24]. Therefore, a role for this glycoprotein in both adhesion and the platelet–platelet interactions leading to thrombus formation is also possible. Indeed, such a role has been indicated by experiments in which patients with thrombasthenia, whose platelets are deficient in GPIIb-IIIa, or normal subjects treated with an antibody to GPIIb-IIIa showed reduced levels of platelet adhesion on subendothelium at high shear rates [27, 40]. In addition, platelet–platelet interactions in this disorder were absent at all shear conditions studied. Thus, the consensus is that at high shear rate conditions platelet glycoproteins Ib and IIb-IIIa both appear to be involved in the events of platelet adhesion, whereas GPIIb-IIIa may be involved predominantly in platelet–platelet interactions, although a role of GPIb in the latter interactions has not been investigated as yet (Fig. 2).

An additional glycoprotein on platelets, GPIa, has recently been related to the ability of platelets to adhere directly to subendothelium at both low and high shear rates [20] and may be responsible for the adhesion observed at low shear rate in the absence of vWF. Few studies have been conducted to determine the adhesive/aggregative mechanisms involved in platelet–platelet attachment at low shear rates (<300 s^{-1}). Generally, platelet–platelet interactions are significantly reduced in normal subjects at these low shear conditions [39]; however, as indicated previously, in thrombasthenic patients all platelet–platelet interactions are absent [40], a finding which suggests the importance of this glycoprotein complex

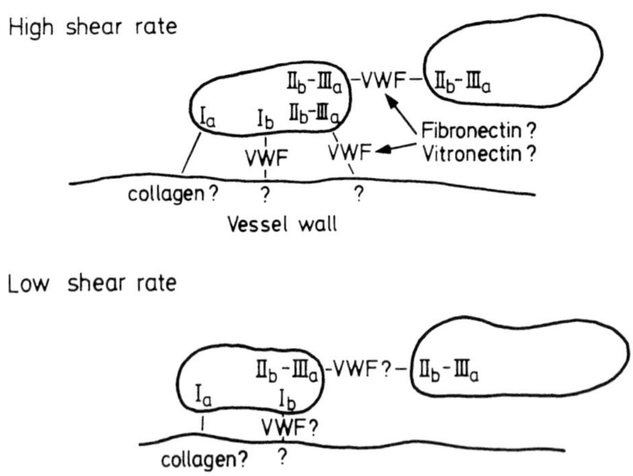

Fig. 2. Platelet attachment to an injured vessel wall (subendothelium). Various glycoproteins (*GP*) on the platelet are shown to be involved in platelet attachment to vessel wall and in platelet–platelet cohesion at high wall shear rates. The vWF protein appears to be particularly important in this interaction, whereas fibrinogen does not appear to be necessary. Other proteins such as fibronectin and vitronectin are of potential importance, but have not been studied as yet. At low shear rates the factors involved in attachment and aggregation are different to some extent, but have not been fully investigated

even at low shear rates. A role for GPIb has also been indicated at low shear rates since an antibody to this glycoprotein produces reduced levels of platelet adhesion under these shear conditions [26]. The vWF protein may also be involved in the limited platelet–platelet interactions, but probably not in the platelet-vessel wall adhesion observed at low shear rates since recent studies have shown an abnormality of platelet deposition under such conditions in vWD pigs [2].

The specific plasma/platelet proteins which are involved in platelet–platelet interactions under various shear conditions remain to be determined. The absence of platelet aggregation (low shear, nonflow system) in thrombasthenia has been generally ascribed to inability of platelets to bind fibrinogen, since platelets are known to require fibrinogen for aggregation in plasma or buffer; however, this requirement is not absolute [6, 30]. Recent studies have indicated that platelet attachment or aggregate build-up (thrombus formation) on subendothelium is normal in afibrinogenemia under a wide variety of shear conditions [32, 37, 38]. Antibodies to fibrinogen, even when added to afibrinogemic blood in order to remove any small trace of residual fibrinogen, did not inhibit the platelet interaction with the vessel wall [41]. Such treatment did, however, reduce aggregation with ADP and collagen when tested in an aggregometer. Treatment of the GPIIb-IIIa site on platelets with either an antibody (LJ-CP8) which blocks the general binding of adhesive proteins or with various peptides which simulate the common sequence (RGD) in a variety of adhesive proteins inhibits both platelet adhesion and thrombus formation in flowing blood at high shear rates. These findings reinforce the importance of the GPIIb-IIIa site in platelet–vessel wall interaction and suggest that fibrinogen is not a necessary component for such interactions. Moreover, the results are also consistent with perfusion studies which demonstrated the importance of vWF in both platelet–platelet and platelet–vessel wall interactions, as described above [32, 37]. Direct support for the ability of adhesive proteins other than fibrinogen to participate in platelet–platelet interactions has been obtained from studies conducted with a monoclonal antibody (LJ-P5) to GPIIb-IIIa which blocks the binding of vWF and other adhesive proteins, but not fibrinogen, to platelets [16]. This antibody produced reduced levels of both platelet–vessel and platelet–platelet interaction on subendothelium, suggesting that vWF or one of the other proteins blocked by the antibody could participate in thrombotic events occurring in flowing blood [42].

The mechanisms by which platelets form thrombi have become more complex in recent years and the simple concept that adhesion and aggregation events, which are distinct morphological observations, are completely independent mechanisms is no longer valid. Both adhesion and aggregate formation appear to be dependent to some extent on GPIb and GPIIb-IIIa receptors and both appear to depend on similar adhesive proteins, especially vWF. What remains to be determined is how the two mechanisms interact in flowing blood to produce a stable thrombus. For example, neither receptor will bind protein unless stimulated by an agonist, GPIb by ristocetin and GPIIb-IIIa by adenosine diphosphate or thrombin. It has often been assumed that, since ristocetin is not available intravascularly, the vessel wall can directly activate this receptor. However, that the vessel wall or a vascular component can induce binding of vWF to GPIb has never been experimentally demonstrated; in fact, little is known about the actual binding sites for initial adhesion in vascular substrates. It is known that vWF will bind to subendothelium and promote adhesion in a concentration-dependent manner [25]. However, the material to which vWF binds or whether a change in the vWF protein or the platelet membrane itself is induced by proximity to the vessel wall is currently unknown.

Coagulation and Thrombosis: Role of Tissue Factor Mediated Pathway

While the presence of procoagulant moieties in the vessel wall has been known for some time [2, 43], it is only recently that such activity has been associated with blood cells and cells of the vascular lining [9, 17]. More recently, we have demonstrated that vascular subendothelium which is completely devoid of endothelial cells is able to clot whole plasma [38] and more specifically activate factor X in the presence of factor VII [42]. This activity can also be blocked by a monoclonal antibody to tissue factor (TF) [42]. Thus, TF appears to be a major procoagulant factor in the vascular space immediately underlying the endothelial lining of arteries, a site which might be readily accessible upon local injury.

Few studies have considered the influence of flow on procoagulant activity, in contrast to the relatively abundant literature which has appeared in the past 15 years with respect to platelet function. The observation that fibrin formation is diminished at increasing shear rates [33] is currently unexplained. While dilution of procoagulant moieties has generally been proposed as the mechanism by which flow reduces coagulative events at surfaces, such a mechanism has never been verified experimentally and, in fact, there are theoretical grounds for suspecting the validity of such a hypothesis [5]. It is quite plausible that flow may have direct effects on certain enzyme or polymerization kinetics involved in thrombosis, in addition to the well-defined effect that flow has in enhancing transport of reactants and products to and from the vessel wall. Such effects of flow on immobilized enzymes have occasionally been observed, but never studied with respect to coagulative processes [7, 8]. Since coagulation primarily initiates and proceeds at the surface of cells or other solid surfaces where enzymes are bound, immobilized enzyme studies seem particularly appropriate. A system has been developed in our laboratory for investigating the influence of wall shear rate on phospholipid-dependent reaction kinetics [13] (Fig. 3). Basically, glass capillary tubes are treated so as to form a phospholipid bilayer in which purified TF can be inserted into the membrane at various ratios ($1/10^5$-10^6, TF/phospholipid molecules, respectively). The planar membrane so formed is quite stable, persisting over days when incubated with buffer, and is resistant to removal by high shear forces (3000 s^{-1} for prolonged time periods). The conversion of factor X to Xa in the presence of factor VII reaches a steady state production level which remains constant for periods of time as long as 45 min (the longest period studied). Production of factor Xa is dependent on the calcium level and the presence of factor VII, but is independent of factor VII concentrations as low as 1% of that normally found in plasma (10 nM). The system is highly reproducible, economical

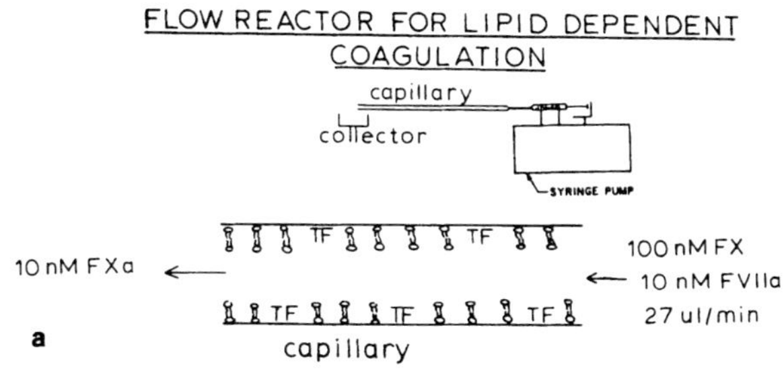

a

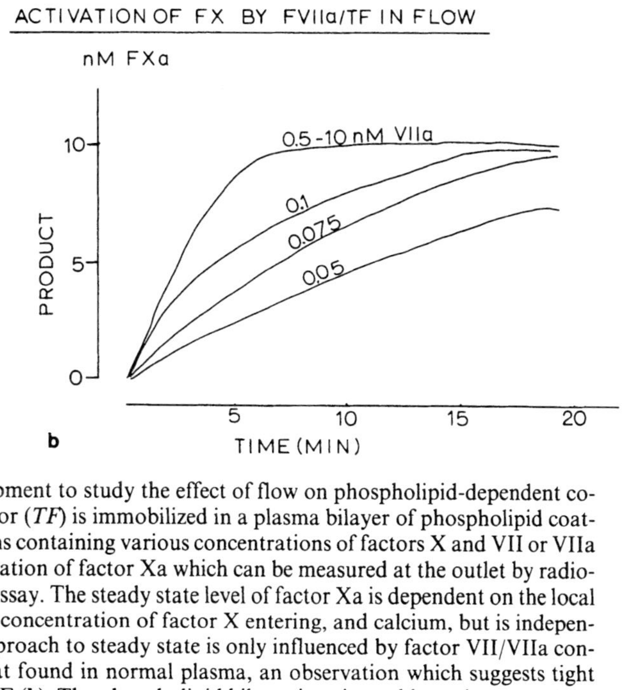

b

Fig. 3 a, b. Flow reactor development to study the effect of flow on phospholipid-dependent coagulation reactions. Tissue factor (*TF*) is immobilized in a plasma bilayer of phospholipid coating a capillary tube (**a**). Solutions containing various concentrations of factors X and VII or VIIa produce a steady state concentration of factor Xa which can be measured at the outlet by radioactive peptide or chromagenic assay. The steady state level of factor Xa is dependent on the local shear condition at the wall, the concentration of factor X entering, and calcium, but is independent of factor VII/VIIa; the approach to steady state is only influenced by factor VII/VIIa concentrations less than 1% of that found in normal plasma, an observation which suggests tight binding of factor VII/VIIa to TF (**b**). The phospholipid bilayer is quite stable, resistant to shear removal, and reusable; the system offers the potential for investigating the influence of flow on a variety of coagulation reactions

of components, and versatile in that in general all proteins which can be inserted into phospholipid vesicles can be studied under flow conditions. Finally, the system offers the advantage that a broad range of physiologically relevant flow conditions can be studied for their effect on the transport kinetics involved in coagulation and thrombosis.

The long range goals envisioned by us for the capillary system are (1) to define basic reaction kinetics of purified coagulation enzymes in flowing fluids and (2) to help determine the coagulation mechanisms which are controlling the overall deposition of fibrin formation on vascular substrates under more complicated experimental conditions which involve plasma and blood cells, as well as vascular components. With such a system the complicated platelet and coagulation reactions which occur in the vicinity of diseased vessels can begin to be understood on a basic level.

Acknowledgement. This work was supported in part by U.S. Public Health Service Grants HL 38933, 27346, 29019, and 39840.

References

1. Adams P, Fuster V, Badimon L, Badimon JJ, Chesebro JH (1987) Platelet/vessel wall interactions, rheologic factors and thrombogenic substrate in acute coronary syndromes: preventive strategies. Am J Cardiol 60:9G–16G
2. Astrup T, Buluk K (1962) Thromboplastic and fibrinolytic activities in vessels of animals. Circ Res 13:253–260
3. Badimon L, Badimon JJ, Rand J, Turitto VT, Fuster V (1987) Platelet deposition on von Willebrand factor-deficient vessels. Extracorporeal perfusion studies in swine with von Willebrand's disease using native and heparinized blood. J Lab Clin Med 110:634–647
4. Badimon L, Badimon JJ, Lassila R, Turitto VT, Fuster V. Thrombosis on stenosis: mechanisms of thrombus growth and embolization on carotid atherosclerotic plaque. (This volume)
5. Basmadjian D, Sefton MV (1986) A model of thrombin inactivation in heparinized and nonheparinized tubes with consequences for thrombus formation. J Biomed Mater Res 20:633–651
6. Cattaneo M, Kinlough-Rathbone RL, Lecchi A, Bevilacqua C, Packham MA, Mustard JF (1987) Fibrinogen-independent aggregation and deaggregation of human platelets: studies in two afibrinogenemic patients. Blood 70:221–226
7. Charm SE, Wong BL (1970) Enzyme inactivation with shearing. Biotechnol Bioeng 12:1103
8. Charm SE, Lai CJ (1971) Comparison of ultrafiltration systems for concentration of biologicals. Biotechnol Bioeng 12:185
9. Colucci M, Balconi G, Lorenzet R, Pietra A, Locati D, Donati MB, Semeraro N (1983) Cultured human endothelial cells generate tissue factor in response to endotoxin. J Clin Invest 71:1893–1896
10. Davies MJ, Thomas AC (1983) Plaque fissuring – the cause of acute myocardial infarction, sudden ischemic death and crescendo angina. Br Heart J 53:363–373
11. Falk E (1985) Unstable angina with fatal outcome: dynamic coronary thrombosis leading to infarction and/or sudden death. Circulation 71:699–708
12. Fry DL (1976) Hemodynamic forces in atherogenesis. In: Schemberg R (ed) Cerebrovascular diseases. Raven, New York, pp 77–95
13. Gemmell C, Turitto VT, Nemerson Y (1988) Flow regulates the initation of blood coagulation by tissue factor. Blood 72:1404–1406
14. Goldsmith HG, Turitto VT (1986) Rheological aspects of thrombosis and hemostasis: basic principles and applications. Thromb Haemost 55:415–435
15. Langille BL, Reidy MA, Kline RL (1986) Injury and repair of endothelium at sites of flow disturbances near abdominal aortic coarctions in rabbits. Arteriosclerosis 6:148
16. Lombardo VT, Hodson E, Roberts JR, Kunicki TJ, Zimmerman TS, Ruggeri ZM (1985) Independent modulation of von Willebrand factor and fibrinogen binding to the platelet membrane glycoprotein IIb/IIIa complex as demonstrated by monoclonal antibody. J Clin Invest 76:1950–1958
17. Maynard JR, Dreyer BE, Stemerman MB, Pitlick FA (1977) Tissue factor coagulant activity of cultured human endothelial cells and smooth muscle cells and fibroblasts. Blood 50:387–396

18. Moore S (1973) Thromboatherosclerosis in normolipemic rabbits: a result of continued endothelial damage. Lab Invest 29:478–487
19. Nerem RM, Levesque M (1987) Hemodynamics and the arterial wall. In: Strandness DE, Didesheim PD, Clowes AW, Watson JT (eds) Vascular diseases. Grune and Stratton, New York, pp 295–317 (Current research and clinical application)
20. Nieuwenhuis HK, Sakariassen KS, Houdijk WP, Nievelstein FE, Sixma J (1986) Deficiency of platelet membrane glycoprotein Ia associated with a decreased platelet adhesion to subendothelium: a defect in platelet spreading. Blood 68:692–695
21. Packham MA, Mustard JF (1986) The role of platelets in the development and complications of atherosclerosis. Semin Hematol 23:1–17
22. Ross R (1986) The pathogenesis of atherosclerosis – an update. N Engl J Med 314:488–500
23. Ruggeri ZM, Bader R, de Marco L (1982) Glanzmann's thrombasthenia: deficient binding of von Willebrand factor to thrombin stimulated platelets. Proc Natl Acad Sci USA 79:6038–6041
24. Ruggeri ZM, de Marco L, Gatto L, Bader R, Montgomery RR (1983) Platelets have more than one binding site for von Willebrand factor. J Clin Invest 72:1–12
25. Sakariassen K, Bolhuis PA, Sixma JJ (1979) Human blood platelet adhesion to artery subendothelium is mediated by factor VIII-von Willebrand factor bound to the subendothelium. Nature 279:636–638
26. Sakariassen KS, Nievelstein PF, Coller BS, Sixma JJ (1986) The role of glycoproteins Ib and IIb-IIIa in platelet adherence to human artery subendothelium. Br J Haematol 63:681–691
27. Sakariassen KS, Nievelstein PF, Coller BS, Sixma JJ (1986) The role of platelet membrane glycoproteins Ib and IIb-IIIa in platelet adherence to human artery subendothelium. Br J Haematol 63:681–691
28. Sakariassen KS, Muggli R, Baumgartner HR (1988) Measurements of platelet interaction with components of the vessel wall in flowing blood. In: Hawiger J (ed) Methods in enzymology. Academic, New York (in press)
29. Sixma JJ (1987) Role of blood platelets, plasma proteins and the vessel wall in hemostasis. In: Bloom AL, Thomas DP (eds) Hemostasis and thrombosis. Churchill Livingstone, New York, pp 283–302
30. Soria J, Soria C, Borg JY, Mirshaki M, Piguet H, Tron P, Fessard C, Caen JP (1985) Platelet aggregation occurs in congenital afibrinogenemia despite the absence of fibrinogen or its fragments in plasma and platelets, as demonstrated by immunoenzymology. Br J Haematol 60:503
31. Turitto VT, Baumgartner HR (1983) Platelet adhesion. In: Harker LA, Zimmerman TS (eds) Methods in haematology: measurements of platelet function. Churchill-Livingstone, New York, pp 46–63
32. Turitto VT, Weiss HJ, Baumgartner HR (1984) Platelet interaction with subendothelium in von Willebrand's disease: altered thrombus formation distinct from defective platelet adhesion. J Clin Invest 74:1730–1741
33. Turitto VT, Baumgartner HR (1987) Platelet-surface interactions. In: Colman R, Hirsh J, Marder V, Salzman E (eds) Hemostasis and thrombosis, basic principles and clinical practice, 2nd edn. J.P. Lippincott, New York, pp 555–571
34. Turitto VT, Weiss HJ, Baumgartner HR, Badimon L, Fuster V (1987) Fluid dynamics, the factor VIII/vWF complex and thrombosis on vessel walls. In: Hartmann A, Kuschinsky W (eds) Cerebral ischemia and hemorheology. Springer, Berlin Heidelberg New York, pp 102–112
35. Turitto VT, Goldsmith HG (in press) Rheology, transport and thrombosis in the circulation. In: Loscalzo J, Creager M, Dzau V (eds) Textbook of vascular medicine. Little, Brown and Co., New York
36. Wechsler LR (1988) Ulceration and carotid artery disease. Stroke 19:650–653
37. Weiss HJ, Turitto VT, Baumgartner HR (1978) Effect of shear rate on platelet interaction with subendothelium in citrated and native blood. I. Shear dependent decrease of adhesion in von Willebrand's disease and the Bernard-Soulier syndrome. J Lab Clin Med 92:750
38. Weiss HJ, Turitto VT, Vicic WJ, Baumgartner HR (1984) Fibrin formation fibrinopeptide A release and platelet thrombus dimensions on subendothelium exposed to flowing native blood: greater in factor XII and XI than in factor VIII and IX deficiency. Blood 63:1004–1014

39. Weiss HJ, Turitto VT, Baumgartner HR (1986) Role of shear rate and platelets in promoting fibrin formation on rabbit subendothelium. Studies utilizing patients with quantitative and qualitative platelet defects. J Clin Invest 78:1072–1082
40. Weiss HJ, Turitto VT, Baumgartner HR (1986) Platelet adhesion and thrombus formation on subendothelium in platelets deficient in glycoproteins IIb-IIIa, I, and storage granules. Blood 67:322–330
41. Weiss HJ, Hawiger J, Ruggeri ZM, Turitto VT, Thiagarajan P, Hoffmann T (1989) Fibrinogen-independent interaction of platelets with subendothelium mediated by glycoprotein II_b-III_a complex at high shear rate. J Clin Invest 83:288–297
42. Weiss HJ, Turitto VT, Baumgartner HR, Nemerson Y, Hoffmann T (1988) Evidence for the presence of tissue-factor activity on de-endothelialized vessel segments. Blood 73:968–976
43. Witte S, Bressel D (1958) Über die Thrombokinase-Activität der Blutgefässe. Folia Haematol (Leipz) 2:236–244
44. Woolf N (1987) Thrombosis and atherosclerosis. In: Bloom AL, Thomas DP (eds) Haemostasis and thrombosis, 2nd edn. Churchill Livingstone, New York, pp 651–678
45. Zarins CK, Giddens DP, Glagov S (1983) Atherosclerotic plaque distribution and flow velocity profiles in the carotid bifurcation. In: Bergan JJ, Yas JS (eds) Cerebrovascular insufficiency. Grune and Stratton, New York, p 19
46. Zarins CK, Giddens DP, Bharadvaj BK (1983) Carotid bifurcation atherosclerosis. Quantitative correlation of plaque localization with flow velocity profiles and wall shear stresses. Circ Res 53:502–514

Altered Cerebral Collaterals and Protection from Infarction

P. COYLE

Introduction

Sudden occlusion of the middle cerebral artery (MCA) above the rhinal fissure reduces blood flow to the territory of the occluded artery and invariably results in infarction in stroke-prone spontaneously hypertensive rats (SHRSPs) [8, 14, 27]. Normotensive rats usually do not have infarction after this occlusion [1, 3, 6, 8, 10, 11, 13]. Following the occlusion in normotensive rats, the anastomosing collaterals enlarge to provide blood supply to the territory of the occluded MCA [9, 12, 15]. Within a month after the occlusion, blood flow to the territory of the occluded MCA returns to normal levels in the normotensive rat [15].

The pathogenesis responsible for ischemia and the subsequent infarction in SHRSPs is not clear. At occlusion time the collaterals are smaller in luminal diameter in SHRSP than in normotensive rats [7, 16]. One possibility is that the collaterals become smaller in luminal diameter as the vessel wall thickens during the development of hypertension [19, 25, 26]. Smaller diameter collaterals or vessels with less dilator reserve could increase the risk of infarction following occlusion of a cerebral artery. Another possibility is that glutamate neurotoxicity [5] is preferentially increased in genetically determined hypertensive rats and that predisposes to infarction during reduced blood flow to cerebrum.

The first goal of the present study was to determine whether the anastomosing collaterals in SHRSPs enlarge after occlusion of a major cerebral artery. The second objective was to examine the possibility that genetically normotensive rats made hypertensive are at increased risk of infarction following an arterial occlusion. Lastly MK-801, the *N*-methyl-D-aspartate (NMDA) receptor antagonist [5, 20], was evaluated for protection against infarction after permanent occlusion of an artery in genetically hypertensive rats.

Methods

Middle cerebral artery occlusion was used to test for changes in the anastomosing collaterals and increased susceptibility to infarction. I studied anastomosing collaterals in 16 SHRSP and 5 normotensive Wistar rats (NWs) including both sexes.

Department of Anatomy and Cell Biology, 5714 Medical Science II, The University of Michigan, Ann Arbor, MI 48109, USA.

Cerebral Ischemia and Calcium
Ed. by A. Hartmann and W. Kuschinsky

The effect of hypertension on susceptibility to infarction was evaluated in male Sprague-Dawley rats. Sprague-Dawley rats are *genetically normotensive rats*, but four rats became hypertensive after constriction of a renal artery and six controls were normotensive. Tail systolic blood pressure measurements were made in awake animals [10]. MK-801 was tested for protective effects against infarction in eight spontaneously hypertensive rats. Twelve control rats, including both sexes, received saline. All rats were 1–3 months old.

MCA Occlusion

In all rats, the MCA was exposed by a temporal approach during anesthesia with ketamine (130 mg/kg body weight) [6]. Occlusion of the MCA was 1.0 ± 0.3 mm dorsal to the rhinal fissure. A monofilament nylon ligature, about 35 μm in diameter, was used for the occlusion. After the skin was sutured, bicillin or penicillin G procaine (30000 units) was injected intramuscularly.

Visualization of the Dorsal Collaterals

One week after MCA occlusion, SHRSPs and NWs were reanesthetized. Papaverine hydrochloride (saturated solution) was infused into the jugular vein to produce maximal vasodilatation of the cerebral vessels. After death undiluted Vultex, a white latex-based compound, was injected into the ascending aorta in order to visualize the luminal course and size of the cerebral arterial tree [7]. Filling of the pial surface collaterals with latex was observed through a closed cranial window [7]. Brains were fixed by immersion fixation in 10% neutral buffered formalin.

All brains were photographed in the standard dorsal perspective. Prints were made at ×35–70 magnification for measurements that utilized a digitizing tablet interfaced to a microcomputer [16]. The anastomosis between each distal branch of the anterior cerebral artery (ACA) and MCA is located between branch angle reversal points [7] (see stars on Fig. 2B). Luminal (internal) diameter measurements were made on the ACA-MCA anastomosis ipsilateral to the MCA occlusion. The anastomosis on the unoccluded contralateral side was a control. Luminal diameter values of the ten largest anastomoses on the MCA occluded side were averaged to yield one experimental value for each rat. One control value for each rat was obtained by averaging diameter values of the ten largest ACA-MCA anastomoses on the unoccluded side.

Two-Kidney One-Clip Rats (2k1c)

Six-week-old male Sprague-Dawley rats were anesthetized with ketamine. An incision, 1–2 cm in length, was made in the flank near the left kidney. A 2 mm wide, 0.5 cm long, U-shaped hardened silver clip with a 0.20-mm aperture was used to constrict the left renal artery. Muscle and skin were sutured. MCA occlusion was undertaken 7 ± 1 (SEM) weeks later. Evans blue dye (1–2 ml, 3% in sodium bicarbonate buffered saline) was injected intraperitoneally during the ketamine anesthetic period. The dye was transported to brain. Lesioned tissue was marked blue [6, 8]. Three days following MCA occlusion, tail systolic blood pressure was measured and the rats were reanesthetized. Brains were perfused with 10% neutral buffered formalin.

MK-801 Treated Rats

MK-801 (1.0 mg/kg body weight) or vehicle (physiological saline) was administered intraperitoneally 1–1.5 h before MCA occlusion. Five days later the rats were anesthetized and papaverine hydrochloride was infused intravenously. Within minutes following death, the brain was perfused with 2,3,5-triphenyltetrazolium chloride (100 ml 1% solution in physiological saline, 40° C) that histochemically stained cortical tissue red and differentiated the infarcted region by noncoloring [2] (see Fig. 2 C). Three minutes after the perfusion, 150 ml 10% neutral buffered formalin was injected for tissue fixation.

Lesion Measurements

Infarct size was computed after digitizing the perimeter of the lesion and adjustments were made for brain curvature [21]. Brain sections were cut and studied microscopically after tissue staining with basic fuchsin to confirm the presence or absence of tissue infarction.

Statistical Tests

All values reported are means $\pm$ one standard error of the mean (SEM). A paired *t*-test was used to compare the mean internal diameter of the anastomosis on occluded and unoccluded sides of the brain. Student's *t*-test and profile analysis [28] were used to compare mean infarct sizes and mean infarct borders, respectively, in MK-801 or saline-treated rats. *P*-values under 0.05 were considered to be significant.

Results

Dorsal Anastomosis Before MCA Occlusion

In rat, the unpaired ACA courses in the interhemispheric fissure. Branches of the ACA distribute bilaterally to the medial cortex on the dorsal aspect of the brain (Fig. 1). The bilateral MCA is positioned laterally to supply the greatest portion of the neocortex (Fig. 1). The dorsal ACA-MCA anastomosis is located at the interface of distal branches of the ACA and MCA (*arrowheads*, Fig. 1). In SHRSPs

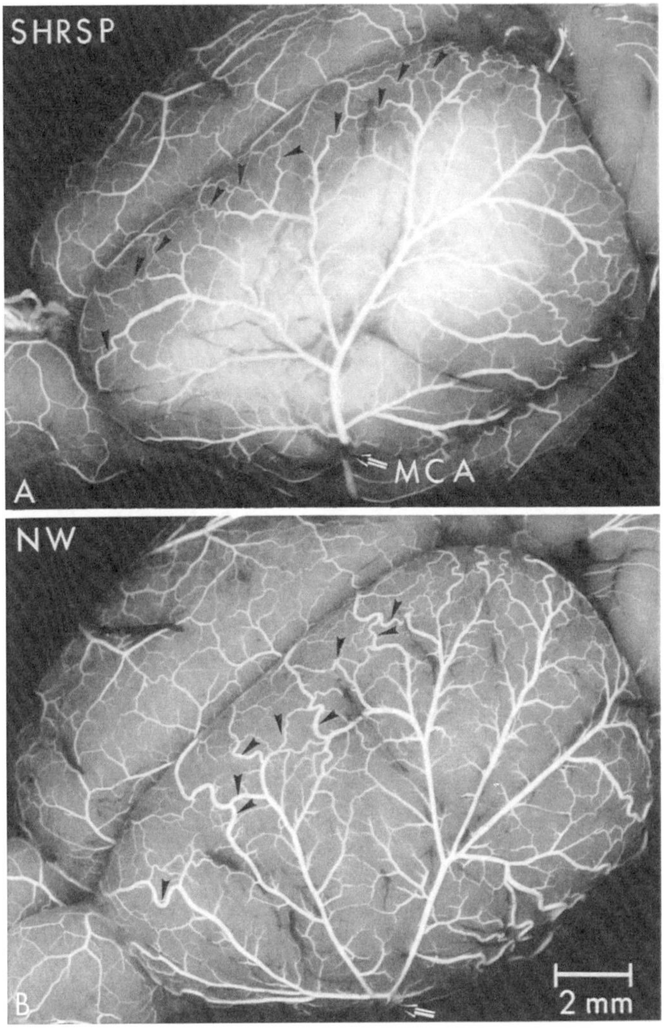

Fig. 1 A, B. Dorsal-lateral view of two rat brains 1 week following occlusion (*arrow*) of the left MCA. After treatment with papaverine, latex was injected into the arterial tree to visualize the luminal size and course of anastomosing branches (*arrowheads*) of the ACA and MCA. **A** Stroke-prone spontaneously hypertensive rat; **B** Normotensive Wistar rat

and NWs without vascular occlusion, the number of ACA-MCA anastomoses averages 27 ± 1 and 26 ± 1, respectively [16]. The anastomosis is positioned nearly parallel to the midline and lateral to it (see Figs. 1, 3 C). The mean internal diameter of the anastomosis is significantly smaller in SHRSPs (32 ± 2 μm) than in the normotensive rat (55 ± 3 μm) [16].

Dorsal Anastomosis After MCA Occlusion

In SHRSPs and NWs 1 week after occlusion of the MCA, the internal diameter of the ACA-MCA anastomosis was significantly ($P<0.05$) larger on the side of MCA occlusion than on the unoperated control side (Figs. 1, 3 A). On the side of MCA occlusion, the anastomosis was significantly ($P<0.05$) smaller in luminal diameter in SHRSPs (62 ± 1 μm) than in NWs (83 ± 5 μm). Upstream to the anastomosis, there were fewer large diameter ACA branches in SHRSPs than in NWs (compare Figs. 1 A and 1 B). In SHRSPs, MCA branches immediately downstream to the anastomosis appeared to be larger in diameter than the upstream ACA counterparts (compare in Fig. 2 A). One week after MCA occlusion vessel taper across the anastomosis was greater in SHRSPs than in NWs, thus producing an anastomosis with a "bottleneck" appearance in SHRSPs (Fig. 2 A).

Outcome to MCA Occlusion in Rats Made Hypertensive

In rats made severely hypertensive (200 ± 7 mmHg) by constriction of a renal artery before MCA occlusion, infarction frequency was 100% after the occlusion. Mean infarct size was 48 ± 4 mm^2 for the hypertensive rats (Fig. 3 B). For the normotensive (122 ± 5 mmHg) control rats, infarction frequency was 17% and mean lesion size was 3 ± 1 mm^2 (Fig. 3 B). Thus after MCA occlusion, frequency of infarction and lesion size were greater in rats made hypertensive than in normotensive control rats.

Effect of MK-801 on Infarction Frequency and Lesion Size

All (100%) of the genetically hypertensive rats had tissue infarction localized within the territory of the occluded MCA (Fig. 2 C). In spontaneously hypertensive rats receiving MK-801 before MCA occlusion, infarct size (71 ± 3 mm^2) was significantly smaller than infarct size (84 ± 4 mm^2) of the saline control rats (Fig. 3 B). In rats receiving MK-801, the infarct border was positioned at a significant (profile analysis, $P<0.05$) distance downstream from the infarct border of saline control rats (Fig. 3 C). Thus, the infarct border was further downstream from the source of collateral supply (dorsal anastomosis) in the rats receiving MK-801 than in the saline controls (Fig. 3 C).

Discussion

The major new findings of this study are first, the anastomosing collaterals were larger in internal diameter on the side of MCA occlusion than on the control side, which suggests that blood flow to the territory of the occluded MCA increases with time after MCA occlusion in SHRSPs. Second, genetically normotensive rats made hypertensive by constriction of a renal artery before MCA occlusion

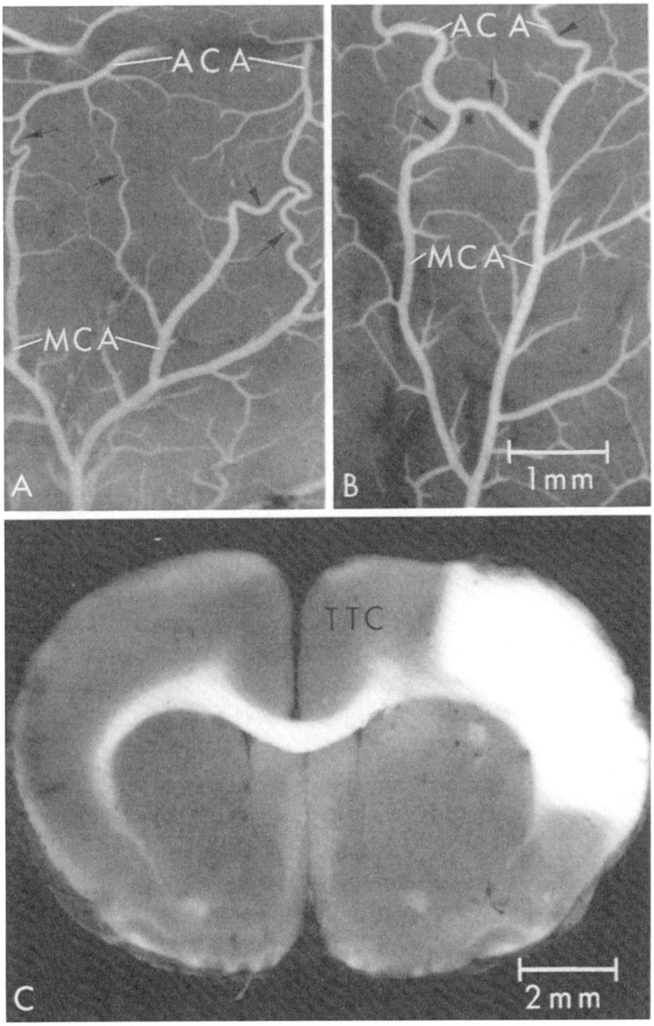

Fig. 2 A–C. Pial surface branches (**A, B**) of the ACA and MCA anastomose (*arrows*) at the anatomical boundary between vascular territories normally supplied by the two arteries. Same type of preparation as in Fig. 1. **A** Stroke-prone spontaneously hypertensive rat; **B** Normotensive Wistar rat; **C** Coronal brain section through the septum of a spontaneously hypertensive rat after MCA occlusion and treatment with triphenyltetrazolium chloride to demarcate the infarct border

had, after MCA occlusion, large infarcts whereas normotensive controls were usually protected by collateral circulation. This finding suggests that the mechanism of hypertension or some factor secondary to hypertension exacerbates the occlusion test outcome. Third, infarct size was smaller in rats receiving MK-801 than in controls receiving saline, which means that MK-801 provides neuroprotection in this hypertensive model with permanent vessel occlusion.

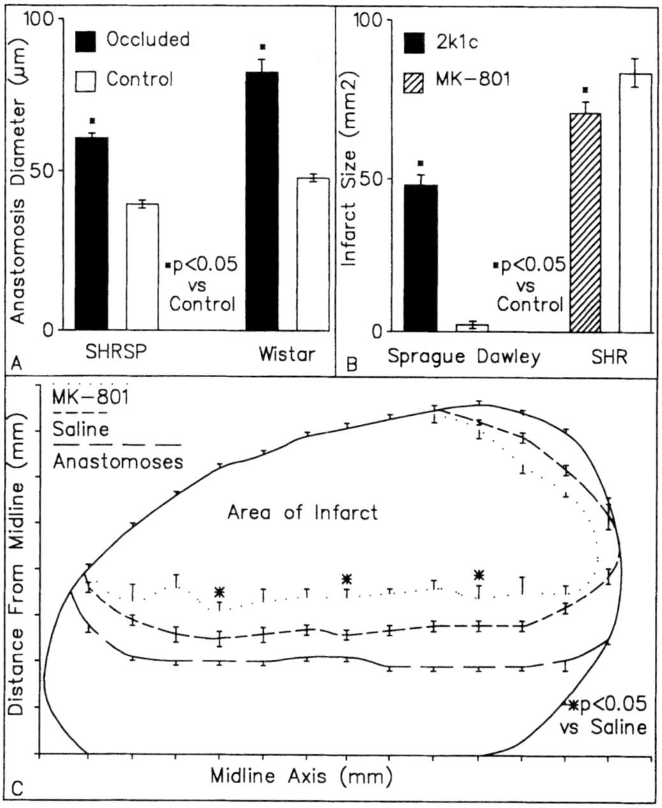

Fig. 3 A–C. A Effect of 1 week of MCA occlusion on the anastomosis diameter on occluded and unoccluded (control) sides in SHRSPs and NWs. **B** Infarct sizes in two-kidney one-clip (2k1c) hypertensive rats and normotensive control rats (*open column*) after MCA occlusion. Infarct sizes in spontaneously hypertensive rats receiving MK-801 or saline (control, *open column*) before MCA occlusion. **C** Relative spatial positions (means) of the infarct border and the anastomosing collaterals in rats receiving MK-801 or saline before MCA occlusion. Infarct borders are statistically parallel but separated spatially (profile analysis) and * means differ ($P<0.05$) from saline control by Student's *t*-test

Earlier studies on collateral circulation in the brain provided evidence that vascular responses are impaired by acute occlusion of the MCA, which impairs autoregulation of blood flow over a large range of blood pressure. After proximal occlusion of the MCA there was progressive dilatation of pial arteries with a time course of at least several hours [23]. A normalization of vascular responses to hypercapnia occurs progressively during days and weeks after MCA occlusion in cats [18]. After 1 month of MCA occlusion in rat, blood flow and vascular coinductance to the territory of the occluded MCA increased significantly during seizure, which means that the collateral vessels were not maximally dilated during control conditions [15]. This finding contrasts strikingly with the finding that vasodilatation 20 min after MCA occlusion produced only minimal increases in blood flow to the territory of the occluded MCA [14]. Thus, vasodilator reserve for the territory of the occluded MCA increased greatly during the month after occlusion of the artery in the normotensive rat [15].

The mechanism by which vasodilator reserve increases after MCA occlusion probably involves structural remodeling of existing pial surface vessels rather than growth of new vessels either on the pial surface or within the cortex [24].

Twenty minutes after MCA occlusion there was significantly less blood flow to the territory of the occluded MCA in SHRSPs than in the normotensive rat [14]. SHRSPs, but not the normotensive rats, developed cerebral infarction [15]. Even in SHRSPs with tissue infarction, the collaterals were significantly larger in internal diameter 1 week after MCA occlusion. Thus, tissue infarction probably was the result of inadequate vasodilatory reserve of the collaterals at the time of MCA occlusion in SHRSPs, not that the collaterals were incapable of structural remodeling with time.

High risk of cerebral infarction after MCA occlusion above the rhinal fissure is not unique to inbred spontaneously hypertensive rats. Genetically normotensive Sprague-Dawley rats made severely hypertensive by constriction of a renal artery had large infarcts after MCA occlusion. Angiotension II, a potent vasoconstrictor, is elevated in the early phase of this form of renal hypertension, but not in mineralocorticoid hypertension. Genetically normotensive rats made hypertensive by deoxycorticosterone acetate and salt are not at high risk for infarction after MCA occlusion [10]. Thus, different outcomes to the MCA occlusion test in different forms of hypertension may depend on a modulatory effect of the renin-angiotensin system on the anastomosing collaterals. In contrast to a possible deleterious action of angiotensin II, others propose that angiotensin II promotes the early development of collateral circulation to brain independent of its pressor property [17].

Protection from large infarcts was obtained by treating spontaneously hypertensive rats with MK-801 prior to MCA occlusion. Brain calcium content is increased 3 days after cerebral infarction is induced in rats [4]. In the presence of calcium, excess exposure to glutamate produces neuronal injury and death [5]. Glutamate exposure most likely leads to calcium influx both through glutamate-activated cation channels and through voltage-dependent calcium channels activated by membrane depolarization [5]. NMDA receptors may play a dominant role in mediating glutamate-induced lethal calcium influx [5]. MK-801 is thought to protect by blocking calcium entry through the glutamate-activated channel [5]. After receiving MK-801, protection from infarction occurred at the lesion border. This finding suggests areas of marginal blood supply receive protection from MK-801 that outlasts any protective effects of ketamine, which also acts on the NMDA receptor.

In summary, after MCA occlusion in spontaneously hypertensive rats, collaterals to the territory of the occluded artery are larger in diameter than are controls and smaller infarcts occur after pretreatment with MK-801, a glutamate receptor antagonist. Genetically normotensive rats made hypertensive by constriction of a renal artery are at high risk of cerebral infarction after MCA occlusion whereas mineralocorticoid hypertensive rats are not at high risk. Thus, the increased susceptibility to infarction after occlusion of a cerebral artery is not unique to spontaneously hypertensive rats nor common to all forms of hypertension, but may, in part, result from an impaired collateral circulation.

Acknowledgments. The author thanks Merck Sharp and Dohme Research Laboratories for generously providing the MK-801 used in this study. This work was supported by NIH Grants HL 18575 and HL 31468.

References

1. Bederson JB, Pitts LH, Tsuji M, Nishimura MC, Davis RL, Bartkowski H (1986) Rat middle cerebral artery occlusion: evaluation of the model and development of a neurologic examination. Stroke 17:472–476
2. Bederson JB, Pitts LH, Germano SM, Nishimura MC, Davis RL, Bartkowski HM (1986) Evaluation of 2,3,5-triphenyltetrazolium chloride as a stain for detection and quantification of experimental cerebral infarction in rats. Stroke 17:1304–1308
3. Chen ST, Hsu CY, Hogan EL, Maricq H, Balentine JD (1986) A model of focal ischemic stroke in the rat: reproducible extensive cortical infarction. Stroke 17:738–743
4. Chen ST, Hsu CY, Hogan EL, Juan HY, Banik NL, Balentine JD (1987) Brain calcium content in ischemic infarction. Neurology 37:1227–1229
5. Choi DW (1987) Ionic dependence of glutamate neurotoxicity. J Neurosci 7:369–379
6. Coyle P (1982) Middle cerebral artery occlusion in the young rat. Stroke 13:855–859
7. Coyle P, Jokelainen PT (1982) Dorsal cerebral arterial collaterals of the rat. Anat Rec 203:397–404
8. Coyle P, Jokelainen PT (1983) Differential outcome to middle cerebral artery occlusion in spontaneously hypertensive stroke-prone rats (SHRSP) and Wistar Kyoto (WKY) rats. Stroke 14:605–611
9. Coyle P (1984) Diameter and length changes in cerebral collaterals after middle cerebral artery occlusion in the young rat. Anat Rec 210:357–364
10. Coyle P (1984) Outcomes to middle cerebral artery occlusion in hypertensive and normotensive rats. Hypertension 6 (Suppl I):I-69–I-74
11. Coyle P, Odenheimer DJ, Sing CF (1984) Cerebral infarction after middle cerebral artery occlusion in progenies of spontaneously stroke-prone and normal rats. Stroke 15:711–716
12. Coyle P (1985) Interruption of the middle cerebral artery in 10-day old rat alters normal development of distal collaterals. Anat Rec 212:179–182
13. Coyle P (1986) Different susceptibilities to cerebral infarction in spontaneously hypertensive (SHR) and normotensive Sprague-Dawley rats. Stroke 17:520–525
14. Coyle P, Heistad DD (1986) Blood flow through cerebral collateral vessels in hypertensive and normotensive rats. Hypertension 8 (Suppl II):II-67–II-71
15. Coyle P, Heistad DD (1987) Blood flow through cerebral collaterals one month after middle cerebral artery occlusion. Stroke 18:407–411
16. Coyle P (1987) Dorsal cerebral collaterals of stroke-prone spontaneously hypertensive rats (SHRSP) and Wistar Kyoto rats (WKY). Anat Rec 218:40–44
17. Fernandez LA, Spencer DD, Kaczmar T Jr (1986) Angiotensin II decreases mortality rate in gerbils with unilateral carotid ligation. Stroke 17:82–85
18. Halsey JH, Clark LC (1970) Some regional circulatory abnormalities following experimental cerebral infarction. Neurology 20:238–246
19. Hart MN, Heistad DD, Brody MJ (1980) Effect of chronic hypertension and sympathetic denervation on wall/lumen ratio of cerebral vessels. Hypertension 2:419–423
20. Iversen LL, Woodruff GN, Kemp JA, Foster AC, Gill R, Wong EHF (1988) Pharmacology and neuroprotective effects of the NMDA antagonist MK-801. In: Domino EF, Kamenka J-M (eds) Sigma and phencyclidine-like compounds as molecular probes in biology. NPP Books, Ann Arbor, pp 757–766
21. Jones PG, Coyle P (1984) Microcomputer assisted lesion size measurements in spontaneously hypertensive stroke-prone rats. J Electrophysiol Tech 11:71–78

22. König JFR, Klippel RA (1963) The rat brain. A stereotaxic atlas of the forebrain and lower parts of the brain stem. Williams and Wilkins, Baltimore
23. Meyer JS, Denny-Brown D (1957) The cerebral collateral circulation. 1. Factors influencing collateral blood flow. Neurology 7:447–458
24. Myers SF, Coyle P (1987) Differential uptake of ^{3}H-thymidine by structures of the pia-arachnoid and cerebral cortex distal to the interrupted middle cerebral artery. Anat Rec 218:162–165
25. Nordborg C, Johansson BB (1980) Morphometric study on cerebral vessels in spontaneously hypertensive rats. Stroke 11:266–270
26. Nordborg C, Fredriksson K, Johansson BB (1985) The morphometry of consecutive segments in cerebral arteries of normotensive and spontaneously hypertensive rats. Stroke 16:313–320
27. Okamoto K, Yamori Y, Nagaoka A (1974) Establishment of the stroke-prone spontaneously hypertensive rat (SHR). Circ Res 34&35:143–153
28. Spence JT, Underwood BJ, Duncan CP, Cotton JW (1968) Elementary statistics, 2nd edn. Appleton-Century-Crofts, New York, pp 195–207

Physical Activity Determines Survival of Mongolian Gerbils from Forebrain Ischemia

K. WEBER, B. TRANMER, L. SCHÜRER, A. BAETHMANN, and O. KEMPSKI

Introduction

The Mongolian gerbil is frequently used in studies on the pathophysiology of cerebral ischemia. Due to the incomplete circle of Willis, simultaneous occlusion of both common carotid arteries reduces blood flow in the forebrain below a critical threshold. The model permits reperfusion of the brain after periods of variable length. With 5 min of forebrain ischemia, most of the animals survive the insult, and brain tissue remains morphologically intact except for the CA1 sector of the hippocampus, where loss of pyramidal cells becomes evident after 2–4 days (Kirino 1982). This phenomenon has been termed "delayed neuronal death" or "maturation phenomenon" (Ito et al. 1975; Kirino 1982). Extensive periods even of unilateral ischemia, e.g., 15–30 min, are followed by tissue damage involving the adjacent CA2/3 sectors of the hippocampus, striatum, and cerebral cortex in addition (Yanagihara et al. 1985). Assessment of survival from ischemia as well as of nerve cell loss has been used to evaluate the effect of pharmacological treatment. Yet, even under tightly controlled conditions, survival and brain tissue damage in individual animals are quite variable. Focal neurological deficits are not induced by this form of global cerebral ischemia. Therefore, behavioral parameters such as locomotor activity were utilized as an additional criterion to evaluate the functional consequences of the ischemic insult on a quantitative basis.

Methods

Male Mongolian gerbils (60–90 g body wt.) were anesthetized with halothane (1.5%). Both common carotid arteries were exposed and carefully isolated from surrounding tissue using an operating microscope. A 6-0 monofilament thread was looped around both arteries, while special care was taken not to damage peripheral nerve tissue. Occlusion of the carotid arteries was induced by attaching a weight of 15 g to both ends of the thread. Halothane was reduced to 0.5% during ischemia. After either 15 or 20 min, the ligature was removed and the skin wound closed. During the surgical preparation, ischemia, and 30 min of recircu-

Institute for Surgical Research, Klinikum Großhadern, Ludwig-Maximilians Universität München, Marchioninistr. 15, D-8000 München 70, FRG.

Cerebral Ischemia and Calcium
Ed. by A. Hartmann and W. Kuschinsky

lation after ischemia, the body temperature was kept at 37 °C by a feedback-controlled heating pad. Two experimental groups were formed:

1. One group of animals ($n=55$) was housed in conventional cages with free access to food and water before and after ischemia.

2. Another group of animals ($n=37$) was placed, 3 weeks before induction of forebrain ischemia, in activity cages (27 × 15 × 13 cm) furnished with running wheels of 34 cm diameter. The animals had access to the running wheels all day and night. The number of revolutions was monitored and stored at 4-min intervals by a computer logging system (Datalog, Fa. Kontron, Eching, FRG). The animals were allowed to get accustomed to the environment for 1 week before recording of their activity. A preischemic control phase of 3 weeks passed to let the animals reach a steady state in running. Thereafter, anesthesia (halothane) was administered again for induction of forebrain ischemia. After regaining consciousness the gerbils were returned to the activity cages for further recording of locomotion during an additional 3 weeks. Another series of experiments using 20 min instead of 15 min of bilateral carotid occlusion was added utilizing the same experimental protocol.

Results

Fifteen Minutes' Forebrain Ischemia

Mean preischemic running activity during the control phase was 855 ± 142 (mean ± SEM) revolutions/24 h. Within 3 days after ischemia, running increased significantly to 400% of the control level ($P<0.05$). Then, 6–10 days after ischemia, locomotor activity gradually decreased to the preischemic level. In control animals ($n=6$) the level of wheel running was significantly depressed for 1 week after sham operation ($P<0.05$).

The locomotor activity proved to predict outcome of the experimental animals from ischemia. The largest increase in voluntary wheel running was seen in those animals in which loss of nerve cells in the CA2/3 sectors of the hippocampus was most severe (data not shown). A most striking result, however, was that wheel running was found to protect the animals against the sequelae of cerebral ischemia. Gerbils which were housed in conventional cages had a mortality of 56% from 15 min of forebrain ischemia within 2 weeks, whereas the animals kept in activity cages had a mortality of only 15% ($P<0.0001$) (Fig. 1). Thus, the physical exercise as currently studied in running wheels, to the best of our knowledge improved outcome from ischemia *better* than any pharmacological method tested so far.

Twenty Minutes' Forebrain Ischemia

As in the above experiments protection of animals from forebrain ischemia by voluntary wheel running was also observed in gerbils after a 20 min period of

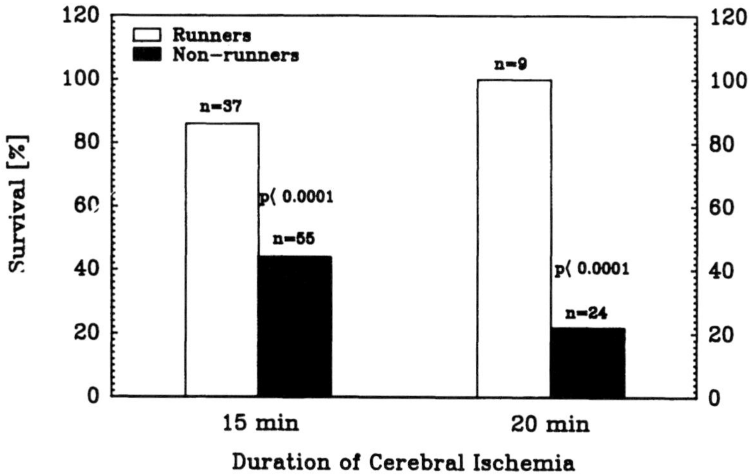

Fig. 1. Survival of Mongolian gerbils after 15 or 20 min of forebrain ischemia induced by bilateral occlusion of the common carotid arteries, according to whether they were maintained in conventional cages or with an opportunity of physical activity in running wheels ("runners"). Reduction of mortality by wheel running was statistically highly significant after 15 or 20 min of forebrain ischemia ($P < 0.0001$)

forebrain ischemia. The rate of running in the preischemic control phase ($n = 9$) was 880 ± 229 (mean $\pm$ SEM) revolutions/24 h. Postischemic running was again significantly increased by ca. 400% during the first 4 days ($P < 0.05$). Survival after 2 weeks in control animals which were not placed in activity cages was only 22% after 20 min of ischemia ($n = 24$), which is lower than what was found in control animals after 15 min of forebrain ischemia. On the other hand, experimental animals maintained in cages with running wheels had a survival of 100% after 20 min of forebrain ischemia.

Discussion

Locomotor hyperactivity is a well-known phenomenon after selective hippocampectomy in both rats and gerbils (Teitelbaum and Milner 1963; Kimble 1963; Douglas and Isaacson 1964; Glickman et al. 1970). However, it is still under discussion whether the phenomenon is due to retardation of habituation in a novel environment eliciting exploratory behavior or due to the loss of inhibition from the hippocampus which, under physiological conditions, attenuates activation from lower structures, such as the reticular formation (Leaton 1965; Jarrard 1968). Many investigations using the model of transient bilateral carotid occlusion in gerbils have observed the phenomenon of selective vulnerability of the CA1 sector in the hippocampus (Ito et al. 1975; Kirino 1982; Kirino and Sano 1984). Thus, the increased locomotor drive as found in the current study might have resulted from ischemic damage of hippocampal structures. This contention is supported by observations on stimulation of locomotor motor activity in animals with severe hippocampal lesions (Teitelbaum and Milner 1963; Kimble 1963; Douglas and Isaacson 1964; Glickman et al. 1970).

The impressive effects of wheel running on survival after 15 or 20 min of forebrain ischemia, however, remain unexplained at the moment. Present research of this laboratory focuses on a role of the general circulatory system, alterations of transmitter metabolism, and of endocrinological mediators. However, irrespective of the as yet unknown mechanisms of protection afforded by pre- and post-ischemic wheel running, the findings provide interesting perspectives as to future research to help improve the poor outcome of patients afflicted with cerebral ischemia.

Acknowledgement. This work was supported by Deutsche Forschungsgemeinschaft DFG Ba 452/6-6.

References

Douglas RJ, Isaacson RL (1964) Hippocampal lesions and activity. Psychon Sci 1:187–188
Glickman SE, Higgins TJ, Isaacson L (1970) Some effects of hippocampal lesions on the behavior of Mongolian gerbils. Physiol Behav 5:931–938
Ito U, Spatz M, Walker JT Jr, Klatzo I (1975) Experimental cerebral ischemia in Mongolian gerbils. I. Light microscopic observations. Acta Neuropathol 32:209–223
Jarrard LE (1968) Behavior of hippocampal lesioned rats in home cage and novel situations. Physiol Behav 3:65–70
Kimble DP (1963) The effects of bilateral hippocampal lesions in rats. J Comp Physiol Psychol 56:273–283
Kirino T (1982) Delayed neuronal death in the gerbil hippocampus following ischemia. Brain Res 239:57–69
Kirino T, Sano K (1984) Selective vulnerability in the gerbil hippocampus following transient ischemia. Acta Neuropathol 62:201–208
Leaton RN (1965) Exploratory behavior in rats with hippocampal lesions. J Comp Physiol Psychol 59:325–330
Teitelbaum H, Milner P (1963) Activity changes following partial hippocampal lesions in rats. J Comp Physiol Psychol 56:284–289
Yanagihara T, Yoshimine T, Morimoto K, Yamamoto K, Homburger H (1985) Immunohistochemical investigation of cerebral ischemia in gerbils. J Neuropath Exp Neurol 44:204–215

Immunocytochemical Studies on Some Brain Regions in a Rat Model of Incomplete Cerebral Ischemia. Relationship to Behavior and Aging

G. NÉMETH[1]*, A. CINTRA[2], K. FUXE[2], L. PERSSON[3], L. F. AGNATI[4], M. GOLDSTEIN[5], and S. HOYER[1]

Introduction

Several pathogenetic mechanisms have been proposed to explain the peculiar histological features of brain ischemia, i.e., the selective neuronal death and the so-called maturation phenomenon [17] – the extension of the lesions during the reperfusion.

One way to assess the problem of selective vulnerability of neurons to ischemia is the study of neurochemically defined populations of neurons as well as the release and turnover of neurotransmitters. The role of monoamines [14, 16], GABA, acetylcholine [11], and somatostatin [18] has been discussed. The involvement of synaptic input to lesioned neurons leading to cell death was postulated, considering that excitatory amino acids such as aspartate and glutamate increased after ischemia [2], deafferentation [30] or blocking of NMDA receptors [32] protected against ischemic lesion, and Mg^{2+}, which eliminates synaptic activity, also prevented death of cultured neurons exposed to anoxia [31]. The promotion of brain lesions similar to those in ischemia by the injection of excitatory amino acids such as glutamate or kainic acid [29] indirectly supports such a mechanism.

Recently it was found that cerebral ischemia in some brain areas led to an accumulation of putrescine, the first product in the pathway of the polyamines [25]. It was suggested that these substances could be a pathogenic factor in ischemia, by a direct effect or by causing Ca^{2+} overloading or release of excitatory amino acids [3]. However, studies on mechanical lesions indicate that putrescine may be a trophic factor, since inhibitors of polyamine synthesis block ganglioside 1 (GM1) promoted regeneration in the nigrostriatal dopamine system [1].

In the present study we used a model of incomplete cerebral ischemia (ICI) in the rat to analyze the behavioral parameters and the histological and immunohistochemical characteristics of the lesions 72 and 144 h following recirculation.

[1] Department of Pathochemistry and General Neurochemistry, University of Heidelberg, Im Neuenheimer Feld 220–221, D-6900 Heidelberg, FRG.
[2] Department of Histology and Neurobiology, Karolinska Institute, Stockholm, Sweden
[3] Department of Physiology, University of Lund, Lund, Sweden.
[4] Department of Human Physiology, University of Modena, Modena, Italy.
[5] Neurochemical Laboratories, Department of Psychiatry, New York, University School of Medicine, New York, NY, USA.
* On leave from Department of Neurology, University of Debrecen, Hungary; stipendary of Alexander von Humboldt Foundation, Bonn, FRG.

Cerebral Ischemia and Calcium
Ed. by A. Hartmann and W. Kuschinsky

The role of polyamines and the effect of the administration of GM1 on the evolution of the brain lesions were also investigated.

Materials and Methods

Surgical Procedure

One-year-old (adult) and 2-year-old (aged) male Wistar rats were used in this study. Experimental groups (Table 1) were defined according to age, ischemic procedure, and GM1 administration (10 mg/kg, once daily, i.p.). The common carotid arteries were exposed and wound around with threads without damaging the vessels or the vagus nerve. Femoral arteries and veins were also exposed. With this the operative procedure was concluded in the sham-operated rats. In the ischemic groups the femoral arteries and veins were catheterized to measure arterial blood pressure and to perform hypovolemic hypotension. Subsequently bilateral ICI was induced for 15 min by looping the threads wound around the common carotid arteries. At the same time the mean arterial blood pressure was reduced to 30–40 mmHg by hypovolemic hypotension. After 15 min ICI, perfusion of the brain was restored. The animals were allowed to recover for 72 or 144 h. Before this endpoint of the recovery periods a series of behavioral tests was performed on both sham-operated ($n=10$/age group) and ICI rats ($n=10$/age group).

Behavioral Tests

Behavioral tests were carried out preoperatively and on the second postoperative day:

1. Neurological examination: The righting and equilibrium reflexes, the flexion reflex, the grasping reflex, toe spreading, and the visual placement reaction were investigated according to refs. [6] and [34].
2. Orienting responses: To test the ability of the animals to react to sensory stimuli, tactile, olfactory, auditory, and visual stimuli were applied on both sides according to refs. [20] and [21].

Table 1. ICI: Experimental groups for the behavioral tests. In parentheses are the number of animals considered in the histological investigation

Age	Reperfusion (hours)	Experimental group and number of animals			
		Sham	Sham+GM_1	ICI	ICI+GM_1
1 year	72	10 (3)	10 (2)	10 (4)	10 (2)
	144	–	10 (5)	10 (4)	10 (3)
2 years	72	10 (2)	10 (3)	10 (7)	10 (2)
	144	–	10 (5)	10 (4)	10 (3)

3. Attention test: The methodological details are reported elsewhere [23]. Briefly, the reaction to low electrical stimuli which caused neither pain nor escape behavior was investigated. The animals' reactions were classified into two different responses depending on the intensity of the electrical stimuli. The first reaction related to sensitivity, the second reaction contained two components: an orienting response and a cognitive-controlled type of discriminative behavior.
4. Step-through-type passive avoidance task: The possible disruptive effect of ischemic injury was investigated on acquisition (the rats received a shock on the 2nd day after the operative procedure and 4 hours after the shock the entrance latency was measured) and on retention (the shock was applied before the operation).
5. Pain-induced aggression – only investigated postoperatively.
6. Escape behavior – again, only investigated postoperatively. The rats were placed on a dark platform and received a painful electrical stimulation for 20 s; at this point in time the animal should escape from the electrified compartment.

Immunocytochemical Techniques

After 72 and 144 h recovery periods the animals were anesthetized, tracheotomized, and artificially ventilated. A steady state of arterial normotension (mean arterial blood pressure 100–120 mmHg), normocapnia (pCO_2 35–45 mmHg), normoxemia, and slight hyperoxemia (100–180 mmHg) was maintained over 10 min. Thereafter the animals were perfused via an intraaortic cannula. After washing with 50 ml saline at 37 °C, 150 ml of fixative (4% w/v paraformaldehyde, 0.4% w/v picric acid in 0.1 M phosphate buffer, pH 6.9) was perfused. The brains were removed, postfixed in the same solution overnight, and washed in buffered sucrose solutions. Serial sections 20 μm thick were produced in a cryotome and kept in a cryoprotective solution at −20 °C until the sectioning of all brains was completed. The sections were made at the Bregma level −0.3 mm and −2.8 mm according to the atlas of Paxinos and Watson [26]. Three sections every 120 μm from each level were immunostained after incubation with the following antibodies: (a) rabbit anti-tyrosine hydroxylase (TH, dilution 1/1500) [22]; (b) rabbit anti-mouse kidney ornithine decarboxylase (ODC, 1/2000) [27]; (c) rabbit anti-glial fibrillary acidic protein (GFAP, 1/1000) [8]; (d) rabbit anti-somatostatin (SRIF, 1/1500) [10]; and (e) mouse monoclonal anti-DARPP-32 (dopamine and cAMP regulated phosphoprotein-32, 1/3000) [24].

Another series of sections was stained with cresyl violet. The antibodies were diluted with 0.3 Triton X-100 and 1% whole normal goat serum. After overnight incubation the sections were washed in PBS and incubated with the respective biotinylated anti-antibody (1/2000) (Vector Labs., Burlingame, USA). After washing in PBS the incubation with a solution of avidin plus biotinylated peroxidase (1/100, Vector) was performed for 45 min. All the incubations were carried out at room temperature. The substrate for the peroxidase was 0.03% w/v 3-3′-diaminobenzidine tetrahydrochloride (Sigma, St. Louis, USA) and 0.02% w/v

H_2O_2, in 50 mM Tris pH 7.4. After 6-min incubation the secions were washed in Tris buffer, mounted on coated slides, dried, and mounted in Entellan (Merck, Darmstadt, F.R.G.). The slides were studied using a NIKON microscope (Microphot-FX) and photographed.

Striatal Lesion with Ibotenic Acid

The role of polyamines in the evolution of brain lesions was investigated in a second animal model. Male Sprague-Dawley (150–170 g) rats, under stereotactic surgery, were injected with 0.3 or 0.5 μg ibotenic acid (Sigma) intrastriatally (one or four injections). Subsequently the animals were treated with α-difluoromethylornithine (α-DFMO) (300 mg/kg, daily, i.p.), a nontoxic ODC inhibitor [15], for 3 days. Control groups received saline. Another group of three animals received a single injection of 0.5 μg putrescine in the striatum and minipumps providing continuous intraventricular administration of putrescine (rate of 1 μg/h) over 14 days. After perfusion, fixation, and handling of the brains as described above, serial coronal sections of the entire striatum were stained with cresyl violet or immunostained for ODC. The areas of the brain lesions in the coronal sections were depicted using a camera lucida and summed up to give an estimation of the size of the lesion.

Results

Ischemic Rats

The mortality of the animals after the experimental procedure was around 50% in both ischemic groups. There was no difference in the mortality of the ischemic rats treated with GMI as compared with controls treated with saline. Frequently the animals died in seizures. These animals were not considered in this study.

The brain areas under study are shown in Table 2. The estimation of the lesions was made by the observation of the Nissl and immunostained sections at each brain level and by scoring the damaged area as a percentage of the total area: O: no apparent lesion; +: <25%; ++: 25%–50%; +++: 50%–75%; ++++: 75%–100%. The results are shown in Table 2. The injured area was larger after 144 h than after 72 h of reperfusion. The most frequently affected areas were the cingulate gyrus (Fig. 1a–c) and the hippocampus (CA1 and the hilus of the dentate gyrus). The frontoparietal somatosensory cortex was much more sensitive than the olfactory cortex to the ischemic insult. The lesions in the nucleus caudatus-putamen were localized more frequently in its dorsolateral aspects (Fig. 1d). In the diencephalon the ventrolateral, ventroposterior, and mediodorsal thalamic nuclei were most affected. In relation to the age of the animals, the 2-year-old rats were much more affected by the ischemic procedure than the 1-year-old animals in all areas, most dramatically in the frontoparietal somatosensory cortex.

Table 2. Scoring the damaged area after ICI

Brain areas	Recovery period (hours)	1-year-old		2-year-old	
		ICI	ICI GM_1	ICI	ICI GM_1
Cingulate gyrus	72	++	++	+++	++
	144	+++	+++	+++	+++/++++
Frontoparietal somatosensory cortex	72	+	+/++	+/++	+++
	144	0/+	++	++++	++/+++
Olfactory cortex	72	0	0	0	0
	144	0	0/+	+/++	++
Amygdaloid complex	72	0	0/+	0	0
	144	0	0/+	++	+/++
Septum	72	+	0	0	0
	144	0	+	+/++	0/+
Preoptic area	72	+	++	+	+/++
	144	0	0	++	0/+
Hippocampus (CA_1 C_4)	72	+	+	++	+++
	144	+++	+++	++	+++
Caudate-putamen	72	++	+	0/+	0/+
	144	+	+	+++	++
Thalamic nuclei	72	0	+	++	+
	144	+	+	++	++
Hypothalamic nuclei	72	+	+/++	+/++	+
	144	0/+	+	+	+

Behavioral Tests

As regards the neurological examination in both age groups after ICI, the equilibrium test proved to be the most sensitive one, followed by the visual placement reactions and the grasping reflex (good performance 10%, 20%, and 80% respectively); the latter two might show the function of the cerebral cortex. With respect to all kinds of orienting responses, the preischemic percentage was 100% in all groups. The values after ICI showed a marked decrease as regards the tactile, olfactory, and auditory stimuli (with 25%, 43%, and 50% achievement respectively) but not the visual orientation (84% achievement). With relation to the attention test – which offers the possibility of measuring much more objectively the somatosensory component of orienting- and attention-related alterations – the preoperative comparison clearly shows that there was no significant difference between the first and second reaction amperage values in the adult group but that they considerably diverged in the aged rats. The comparison of the pre- and postoperative first and second reactions' amperage values revealed no significant change in the sham-operated rats. As regards the second reaction. However, the amperage values of both ICI groups were significantly increased as compared with the preischemic values and the cognition-controlled discriminative behavior after ICI disappeared. The results of the passive avoidance test show that inde-

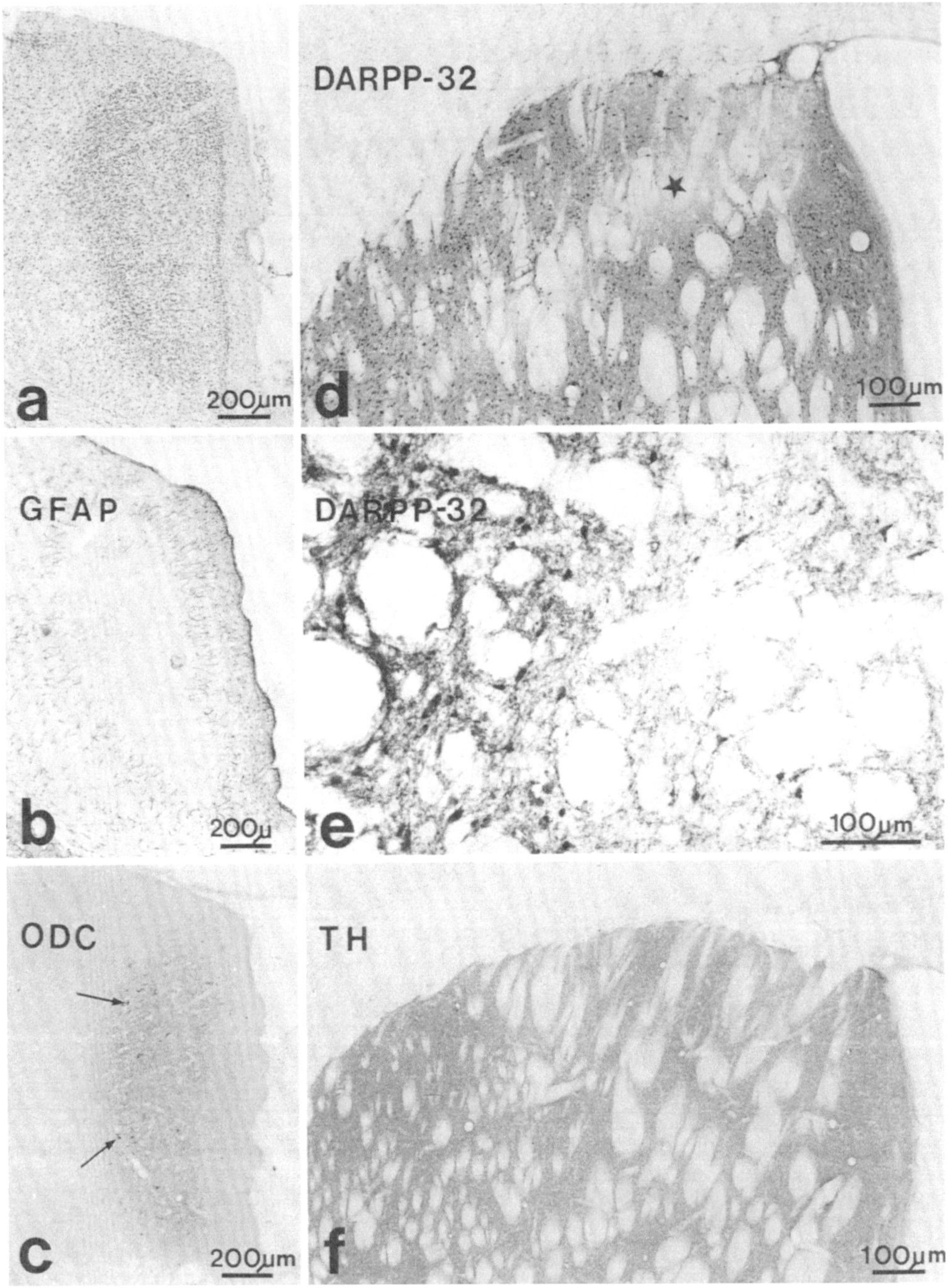

Fig. 1 a–c. Infarction of the cingulate gyrus after ICI (144-h recovery period). **a** Cresyl violet staining: hypercellularity due to microglial proliferation. **b** Fibrillary astrocytes surround the area of infarction. **c** Intense *ODC* IR in the whole area of lesion. *Arrows* indicate degenerated neurons. **d–f** Striatal immunoreactivity after 72-h recovery period. **d** *DARPP-32* IR in dorsal striatum after ICI (72-h recovery period). The lesion is shown as a patch of disappearance of immunoreactivity (*asterisk*). **e** High magnification of **d** shows the lack of medium-sized neurons and processes. **f** *TH* IR in the area of lesion is similar to the rest of the terminal field pendent of whether a pre- or a postoperative footshock was performed, the ICI animals were also able to learn to avoid the dark compartment in the last trial, the retention trial. As regards pain-induced aggression, the ICI rats never produced any attack, and sometimes only a defensive upright posture was observed. Concerning the escape behavior, no ICI rats ran into the current-free illuminated compartment after the high electrical stimulation.

Details of the Ischemic Lesions After 72 Hours of Recovery

After 72 h reperfusion, areas of infarction could already be distinguished, surrounded by proliferating fibrillary astrocytes visualized by glial fibrillary acidic protein immunoreactivity (GFAP IR). There were few astrocytes inside the lesion. In the cerebral cortex common features were shrunken pyramidal neurons with loss of polarity of their axons and dendrites, swollen "ghost cells" (Fig. 2a) [12], and neurons showing ischemic cell changes with incrustations [4]. Other neurons were completely in fragmentation. These damaged neurons and granules were strongly ornithine decarboxylase (ODC) IR. Also strong ODC IR cells, probably representing protoplasmic astrocytes (Fig. 2b), appeared, forming a ring around the lesions. Proliferation of oval and round cells with no ODC IR (probably microglia) was observed inside the lesions. Besides the areas of infarction there were some microinfarcted areas in cingulate cortex, frontoparietal somatosensory cortex, and thalamus. In these areas there appeared some damaged neurons of the types described above and also apparently normal neurons but with strong ODC IR both in their perikarya and processes (Fig. 2c). The striatal lesions looked like patches of disappearance of DARPP-32 IR medium sized neurons and processes (Fig. 1d, e). The staining for ODC revealed the presence of striatal neurons with signs of degeneration. The staining for somatostatin showed the loss of SRIF IR neurons. In the dorsolateral stratum there appeared SRIF IR large cells resembling astrocytes (Fig. 2d). These cells were also observed in the striatum oriens of lesioned CA1 hippocampal field (Fig. 2e) and in the frontoparietal somatosensory cortex. The TH innervation of the neostriatum seemed normal in injured areas (Fig. 1f).

Details of the Ischemic Lesions After 144 Hours of Recovery

The most general feature was hypercellularity due to the intense proliferation of small oval/round cells (microglia, Fig. 1a). There was neither proliferation of fibrillary astrocytes inside the lesions nor an ODC IR cell ring limiting the lesions. No SRIF IR astrocytes were observed in the lesioned areas of the striatum or the hippocampal formation.

The lesions of the striatum visualized by DARPP-32 IR were well defined and the TH IR in these areas was higher than in the normal surrounding terminal field. ODC IR appeared as a homogeneous area in which Nissl staining revealed a reduced number of degenerating neurons and a large number of microglial cells.

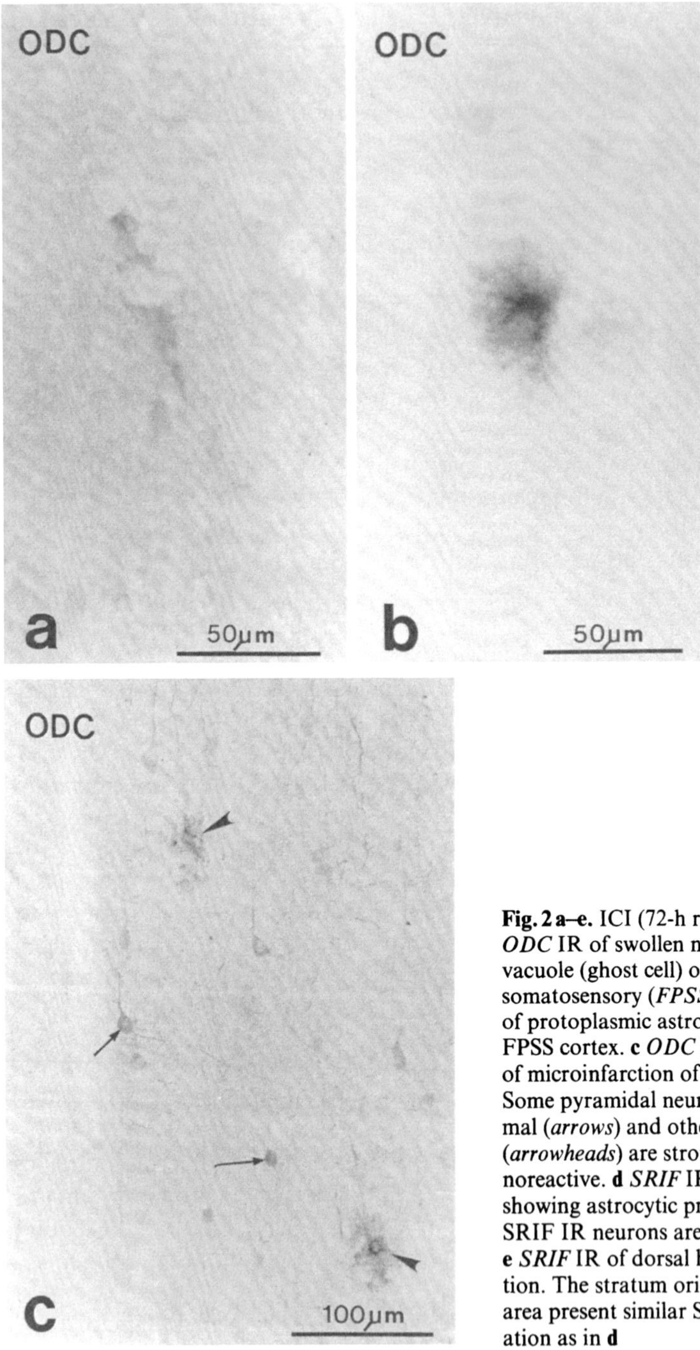

Fig. 2 a–e. ICI (72-h recovery period). **a** *ODC* IR of swollen neuron with large vacuole (ghost cell) of the frontoparietal somatosensory (*FPSS*) cortex. **b** *ODC* IR of protoplasmic astrocyte-like cell of the FPSS cortex. **c** *ODC* IR showing an area of microinfarction of the FPSS cortex. Some pyramidal neurons resembling normal (*arrows*) and others in degeneration (*arrowheads*) are strongly ODC immunoreactive. **d** *SRIF* IR of dorsal striatum showing astrocytic proliferation. Normal SRIF IR neurons are indicated by *arrows*. **e** *SRIF* IR of dorsal hippocampal formation. The stratum oriens (*SO*) and *CA1* area present similar SRIF IR cell proliferation as in **d**

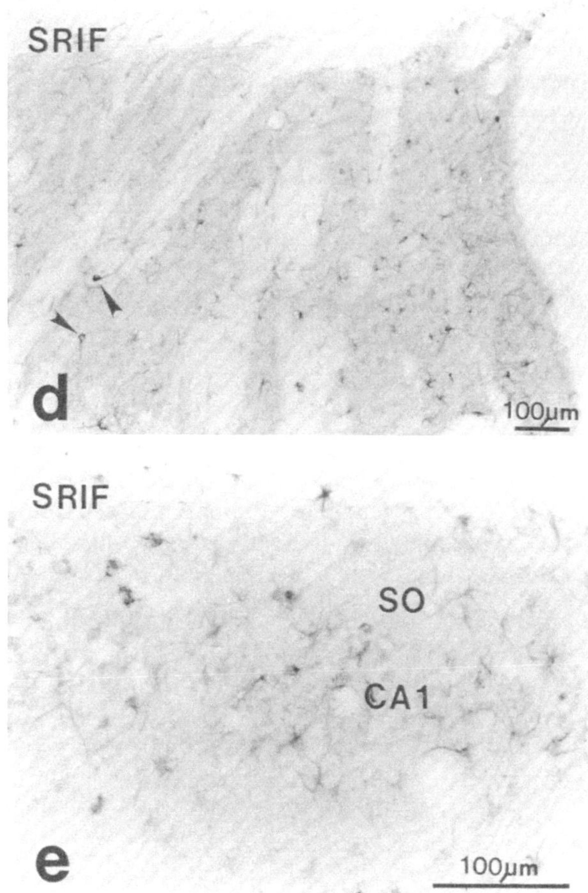

Ibotenic Acid Lesions

Treatment with α-DFMO, which inhibits ODC activity and presumably blocks increases in putrescine, did not change the size of the lesion induced by ibotenic acid (Table 3).

After a 14-day intraventricular perfusion of putrescine (1 µg/h), there was an enlargement of the ventricles (Fig. 3c), vacuolation of neurons (Fig. 3d) like that described in anoxia-ischemia [4], and glial proliferation. There was no apparent difference in the evolution of the ibotenic acid-induced lesions after the putrescine administration.

Discussion

We have studied a model of forebrain ICI in rats after 72- and 144-h recovery periods. We found injuries such as infarction and microinfarction in the areas clas-

Table 3. Effect of α-DFMO administration in the evolution of striatal lesions induced by ibotenic acid

	Ibotenic acid 4 × 0.5 μg Striatum		Ibotenic acid 1 × 0.5 μg Striatum		Ibotenic acid 1 × 0.5 μg Accumbens	
	Saline	α-DFMO	Saline	α-DFMO	Saline	α-DFMO
Number of rats	$n=4$	$n=4$	$n=4$	$n=4$	n = 3	n = 3
Area of the lesion (arbitrary units)	73516	61262	6719	15865	25006	6332
	40132	57275	34134	4245	17679	11719
	53976	46430	26868	19390	12237	14246
	58476	61413	8643	10149		

sically described as being of higher vulnerability, e.g., the hippocampal fields CA1 and CA4, neocortex, and striatum [19, 29]. Our data are similar to the histological features described previously in the rat after 10 min ICI and 7 days of recovery [33]. The pathological observations can be correlated with the findings of the behavior tests:

1. The neurological tests which show the functions of the cerebral cortex, like the visual placement reaction and the grasping reflex, proved to be the most sensitive. This fact may manifest the severity of the lesions of the cingulate gyrus and frontoparietal cortex.
2. After ICI the animals are capable of memorizing and learning simple tasks such as the passive avoidance task. This implies that simple memory mechanisms may still be available despite the large hippocampal and cortical lesions. In our cases the result may be explained by the relatively small ischemic lesion of the striatum. It is known that this structure is capable of assuming some part of the cortical functions in rodents.
3. An impairment of the selective characteristics associated with attention-related mechanisms, and a cognition-controlled type of discriminative behavior as well as a lack of emotional and motivational reaction, could be seen after the ICI. These types of behavior require a very complex interaction between various cortical and subcortical structures – mainly between those which were most severely injured in our model.
4. The fact that the ICI animals were able to respond to visual stimuli at the time when they were not able to orient towards tactile, olfactory, and auditory stimuli, may reflect the variable degree of disruption of different types of sensorimotor integration and cognition.

Also of substantial interest was the finding that the lesion induced by the ICI was substantially larger in the 2-year-old rats than in the 1-year-old rats, indicating an increase of susceptibility to ischemia in the aged rat brain. However, we were not able to detect differences in the behavioral tests between the age groups after ICI.

The increasing size of the lesions in groups with longer recovery periods described in gerbils [17] as the maturation phenomenon and observed in other spe-

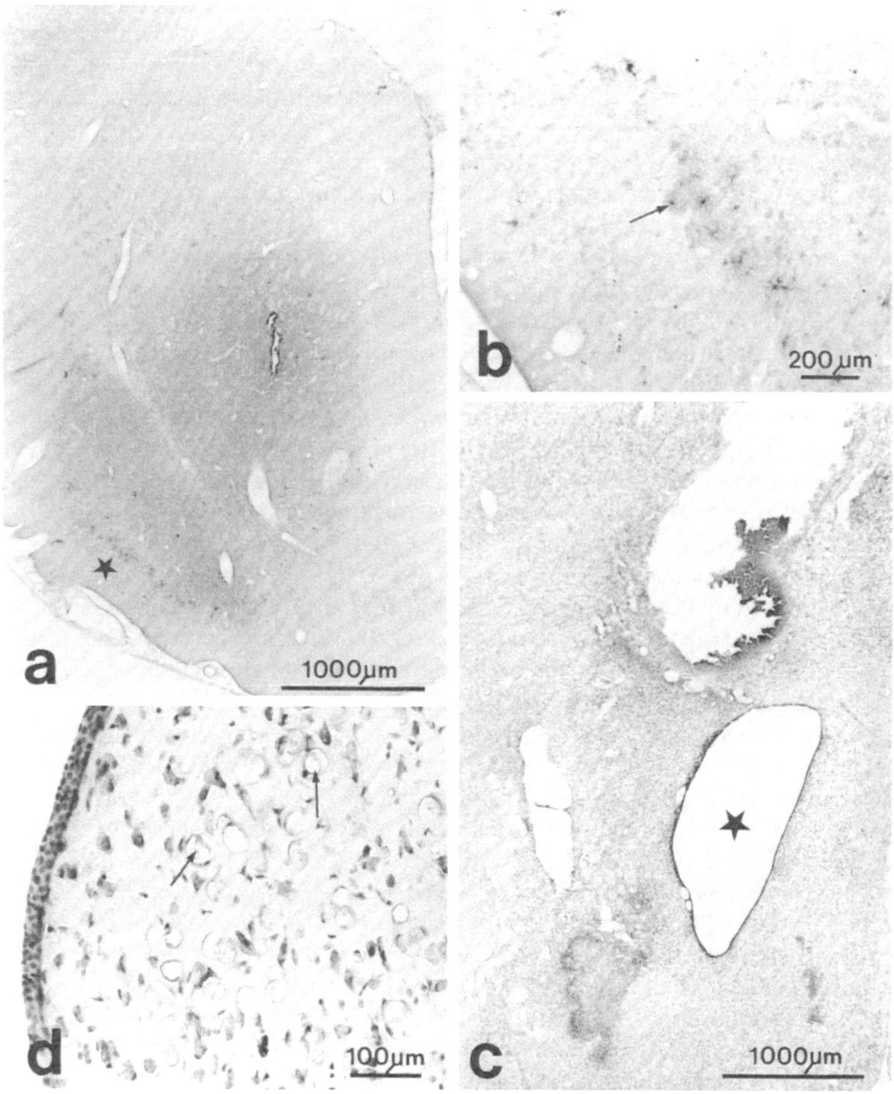

Fig. 3 a, b. Striatal lesion after ibotenic acid injection. **a** Area of strong ODC IR induced in the striatum; the lesion extends to the olfactory cortex. **b** High magnification of the area indicated by *asterisk* in **a**. In the periphery of the lesion there appear cells resembling protoplasmic astrocytes (*arrow*). **c,d** Toxic effect of intraventricular administration of putrescine (1 µg/h) over 14 days. **c** Note the enlarged lateral ventricule (*asterisk*) and the huge cortical lesion (cresyl violet). **d** High magnification of the lateral septus shows neurons with large vacuoles (*arrows*) (cresyl violet)

cies was well noticed in our cases comparing the 72- and 144-h experimental groups. In the areas of microinfarction patterns of ischemic cell change with incrustation, ghost cells, etc. were found which are pathological entities described in the earliest stages of anoxia-ischemia, and which suggest continuous neuronal cell death leading to the coalescence of these areas.

An interesting observation in the present study was the loss of SRIF-IR neurons and the appearance of SRIF-IR glial cells, probably representing astrocytes. These results may reflect the release of somatostatin from neurons to be taken up into glial cells or may alternatively represent the induction of gene expression of a somatostatin precursor within the glial cells as a response to the ischemia. Somatostatin IR astrocytes found in damaged areas of striatum, hippocampal formation, and cortex may be involved in a mechanism to protect GABA neurons from ischemic injury, as reported for the CA1 area [18]. Astrocytic proliferation and swelling appear as soon as 3 h after anoxia-ischemia [5]. GFAP IR astrocyte proliferation is described 24 h after ischemia in gerbils [35].

The lesion in the nucleus caudatus-putamen was demonstrated by performing DARPP-32 immunocytochemistry as the loss of medium-sized neurons and their processes. The striatal lesions were not so extensive as those described in the four-vessel model [28] after 72-h recovery or in the unilateral occlusion of the carotids with short-term reperfusion [13]. TH IR appears normal after 72-h and increased after 144-h recovery. Increase of TH activity and dopamine turnover was described in short recovery periods after ischemic injury [14, 16].

One of the major findings of the present study is the appearance of ODC IR in degenerating neurons and presumably glial cells in the areas surrounding the lesion induced by ICI. It seems possible that the ODC glial cells surrounding the lesion represent protoplasmic astrocytes. After 144 h, ODC also seems to leak out into the neuropil in the infarcted area. These results indicate that after ICI gene expression of ODC may be increased in glial cells. Increased release of polyamines – especially putrescine – may contribute to the development of the ischemic lesion. ODC is ubiquitous in the rat brain. Forebrain areas with a high ODC IR are the frontoparietal somatosensory cortex, the cingulate gyrus, and the horizontal limb of the diagonal band [7], areas which are very affected in the present ischemic model. Increase in ODC activity as soon as 24 h after middle cerebral artery occlusion has been reported in cats [9]. In the four-vessel ischemic model Paschen et al. [25] found that putrescine tissue levels increase 10- to 15-fold in the cerebral cortex, nucleus caudatus-putamen and hippocampus after 24-h recovery, which can be an indication of increased ODC. We have used ibotenic acid lesions of the striatum to mimic localized ischemic lesions. We have found that intraventricular perfusion with putrescine via minipumps in rats with a striatal ibotenic acid induced lesion leads to signs of vacuolation of neurons and glial proliferation. These findings are compatible with the hypothesis that excessive amounts of putrescine may contribute to the lesion development upon ischemia. However, treatment with the polyamine synthesis inhibitor α-DFMO did not significantly change the development of the striatal ibotenic acid induced lesion. Therefore this hypothesis will be tested by treating rats submitted to ICI with α-DFMO.

Our present data showing increasing ODC IR after ischemic injury may support the work to search for a role of polyamines in the pathogenesis of delayed neuronal death by a direct action or by interfering in Ca^{2+} fluxes or release of excitatory amino-acids [3].

Acknowledgments. This work has been supported by a grant from Magnus Bergvalls Stiftelse and by the Hirnliga e. V., Heidelberg. We are grateful for the technical assistance of Roland and Günter Galmbacher, as well as for the excellent secretarial assistance of Mrs. Monia Särd.

References

1. Agnati LF, Fuxe K, Zini I et al. (1985) Effects of lesions and ganglioside GM1 treatment on striatal polyamine levels and nigral DA neurons. A role of putrescine in the neurotropic activity of gangliosides. Acta Physiol Scand 124:277–285
2. Benveniste H, Drejer J, Schousboe A, Diemer NH (1984) Elevation of the extracellular concentrations of glutamate and aspartate in rat hippocampus during transient cerebral ischemia monitored by intracerebral microdialysis. J Neurochem 5:1369–1374
3. Bondy SC, Walker CH (1986) Polyamines contribute to calcium-stimulated release of aspartate from brain particulate fractions. Brain Res 371:96–100
4. Brown AW, Brierley JB (1968) The nature, distribution and earliest stages of anoxic-ischaemic nerve cell damage in the rat brain as defined by the optical microscope. Br J Exp Pathol 1:87–106
5. Brown AW, Brierley JB (1973) The earliest alterations in rat neurons and astrocytes after anoxia-ischaemia. Acta Neuropathol (Berl) 23:9–22
6. Bures J, Buresová O, Huston I (1983) Techniques and basic experiments for the study of brain and behavior. Elsevier, Amsterdam New York, pp 77–133
7. Cintra A, Fuxe K, Agnati LF et al. (1987) Evidence for the existence of ornithine decarboxylase-immunoreactive neurons in the rat brain. Neurosci Lett 76:269–274
8. Dahl D, Bignami A (1976) Immunogenic properties of the glial fibrillary acidic protein. Brain Res 116:150–157
9. Dempsey RJ, Roy MW, Meyer K, Tai HH, Olson JW (1985) Polyamine and prostaglandin markers in focal cerebral ischemia. Neurosurgery 17:635–640
10. Elde R, Hökfelt T (1978) Distribution of hypothalamic hormones and other peptides in the brain. In: Ganong WF, Martini L (eds) Frontiers in neuroendocrinology, vol 5. Raven, New York, pp 1–33
11. Francis A, Pulsinelli W (1982) The response of GABAergic and cholinergic neurons to transient cerebral ischemia. Brain Res 243:271–278
12. Garcia JH, Lossinsky AS, Kauffman FC, Conger KA (1978) Neuronal ischemic injury: light microscopy, ultrastructure and biochemistry. Acta Neuropathol (Berl) 43:85–95
13. Ginsberg MD, Graham DI, Path FRC, Busto R (1985) Regional glucose utilization and blood flow following graded forebrain ischemia in the rat: correlation with neuropathology. Ann Neurol 4:470–481
14. Globus MY-T, Ginsberg MD, Harik SI, Busto R, Dietrich WD (1987) Role of dopamine in ischemic striatal injury: metabolic evidence. Neurology 37:1712–1719
15. Haegele KD, Alken RG, Grove J, Schechter PJ, Koch-Weser J (1981) Kinetics of α-difluoromethylornithine: an irreversible inhibitor of ornithine decarboxylase. Clin Pharmacol Ther 2:210–217
16. Harik SI, Yoshida S, Busto R, Ginsberg MD (1986) Monoamine neurotransmitters in diffuse reversible forebrain ischemia and early recirculation: increased dopaminergic activity. Neurology 36:971–976
17. Ito U, Spatz M, Walker JT, Klatzo I (1975) Experimental cerebral ischemia in mongolian gerbils. I. Light microscopic observations. Acta Neuropathol (Berl) 32:209–223
18. Johansen FF, Zimmer J, Diemer NH (1987) Early loss of somatostatin neurons in dentate hilus after cerebral ischemia in the rat precedes CA-1 pyramidal cell loss. Acta Neuropathol (Berl) 73:110–114
19. Kirino T, Tamura A, Sano K (1985) Selective vulnerability of the hippocampus to ischemia – reversible and irreversible types of ischemic cell damage. Prog Brain Res 63:39–58
20. Ljungberg T, Ungerstedt U (1976) Sensory inattention produced by 6-hydroxydopamine-induced degeneration of ascending dopamine neurons in the brain. Exp Neurol 53:585–600
21. Marshall JF, Gotthelf T (1979) Sensory inattention in rats with 6-hydroxydopamine induced degeneration of ascending dopaminergic neurons: apomorphine-induced reversal of deficits. Exp Neurol 65:398–411
22. Markey K, Kondo A, Schenkmann L, Goldstein M (1980) Purification and characterization of tyrosine hydroxylase from a clonal chromocytoma cell line. Mol Pharmacol 17:79–85
23. Németh Gy, Mayer G, Hoyer S (1989) A new psychometric test of attention-related behavior in rats. Its validity in the aging process. Arch Gerontol Geriatr 8:26–39

24. Ouimet CC, Miller P, Hemmings HC Jr, Walaas SI, Greengard P (1984) DARPP-32, a dopamine- and adenosine 3':5'-monophosphate-regulated phosphoprotein enriched in dopamine-innvervated brain regions. III. Immunocytochemical localization. J Neurosci 4:111–124
25. Paschen W, Schmidt-Kastner R, Djuricic B, Meese C, Linn F, Hossmann K-A (1987) Polyamine changes in reversible cerebral ischemia. J Neurochem 49:35–37
26. Paxinos G, Watson C (1982) The rat brain in stereotaxic coordinates. Academic, Sidney
27. Persson L (1982) Antibodies to ornithine decarboxylase, immunochemical crossreactivity. Acta Chem Scand 36:685–688
28. Pulsinelli WA, Brierley JB, Plum F (1982) Temporal profile of neuronal damage in a model of transient forebrain ischemia. Ann Neurol 5:491–498
29. Pulsinelli WA (1985) Selective neuronal vulnerability: morphological and molecular characteristics. Prog Brain Res 63:29–37
30. Pulsinelli WA (1985) Deafferentation of the hippocampus protects CA1 pyramidal cells against ischemic injury. Stroke 16:144
31. Rothman SM (1983) Synaptic activity mediates death of hypoxic neurons. Science 220:536–537
32. Simon RP, Swan JH, Griffiths T, Meldrum BS (1984) Blockade of *N*-methyl-*p*-aspartate receptors may protect against ischemic damage in the brain. Science 226:850–852
33. Smith M-L, Auer RN, Siesjö BK (1984) The density and distribution of ischemic brain injury in the rat following 2–10 min of forebrain ischemia. Acta Neuropathol (Berl) 64:319–332
34. Tupper DE, Wallace RB (1980) Utility of the neurological examination in rats. Acta Neurobiol Exp 40:999–1003
35. Yoshimine T, Morimoto K, Brengman JM, Homburger HA, Mogami H, Yanagihara T (1985) Immunohistochemical investigation of cerebral ischemia during recirculation. J Neurosurg 63:922–928

Effects of Transient Global Cerebral Ischemia on Peptide Neuronal Populations in Different Brain Regions and on Behavioral Performances in Rats

I. Zini [1], R. Grimaldi [1], E. Merlo Pich [1], V. Solfrini [1], G. Toffano [1], L. F. Agnati [1], and K. Fuxe [2]

Introduction

It is known that neurons are more sensitive to ischemic damage than oligodendroglia, astroglia, and endothelial cells [7]. Furthermore, the neuronal populations in various brain regions show a differential vulnerability to ischemia [2, 13]. The mechanism underlying the regional vulnerability of nerve cell populations to global ischemia is still unknown and does not seem to be simply related to the degree of vascularization, energy requirements, and chemical coding [11]. In previous papers on the aging processes it was shown that various peptide neuronal systems show a marked reduction in the aged rat brain [see, e.g., 1, 15]. In the present study using semiquantitative immunocytochemistry we have therefore studied how neocortical, hippocampal, and hypothalamic neuropeptide-containing neurons respond to global ischemia induced by the Pulsinelli model [12]. The analysis also involves studies on the time course of the recovery of peptide immunoreactivities during reperfusion. This time course was related to the learning deficit induced by the brain ischemia as evaluated by a T-maze test.

Materials and Methods

Male specific pathogen-free Sprague-Dawley rats (250–300 g body wt.) were used. They were kept one per cage under standardized humidity, temperature, and lighting conditions (lights on at 8 a.m. and off at 8 p.m.) and given food pellets and water ad libitum.

Cerebral ischemia was induced by 30 min of four-vessel occlusion [12]. The procedure consists of bilateral temporary clampage of the common carotid arteries in unanesthetized rats with prior permanent occlusion of the vertebral arteries. A detailed description of the characterization of the animal model as far as the EEG pattern, the cardiovascular parameters, the behavioral alterations, and the brain vascular bed are concerned, will be reported elsewhere (Agnati and Fuxe in preparation).

[1] Institute of Human Physiology, University of Modena, Via Campi 287, 41100 Modena, Italy.
[2] Department of Histology and Neurobiology, Karolinska Institute, Stockholm, Sweden.

Cerebral Ischemia and Calcium
Ed. by A. Hartmann and W. Kuschinsky

Immunocytochemical Experiments

Of 70 rats, 35 served as controls. The animals which underwent ischemia were killed at different time intervals: 4 h (sample size: $n=10$), 24 h ($n=10$), 7 days ($n=10$), and 40 days ($n=5$). The rats were sacrificed by intracardiac perfusion with 100 ml of ice-cold 2% paraformaldehyde in 0.1 M acetate buffer (pH 6.5) followed by 300 ml of ice-cold 2% paraformaldehyde + 0.01 glutaraldehyde in 0.1 M borate buffer (pH 8.5). The brains were rapidly dissected out and left in the same fixative overnight. The day after, the brains were cut on a vibratome (Lancer, 50 µm thick sections) at diencephalic (bregma −1.8 mm according to the Paxinos and Watson atlas) and telencephalic (bregma −3.8 mm according to the Paxinos and Watson atlas) levels. Then, the sections were processed according to the indirect immunoperoxidase procedure using the avidine-biotine technique. Some sections were also stained with cresyl violet and graded for damage as reported by Pulsinelli et al. [13]. The immunocytochemical procedure has been described elsewhere [1]. The following antisera were used: rabbit vasoactive intestinal polypeptide (VIP) antiserum (Amersham, U.K.), rabbit neuropeptide Y (NPY) antiserum (Peninsula, U.K.), rabbit somatostatin (SRIF) antiserum [8], rabbit cholecystokinin (CCK) antiserum [14], and rabbit vasopressin (AVP) antiserum, which have been previously characterized. Antisera which had been preadsorbed with their respective peptide (50 µg/ml of diluted antiserum) served as control antisera. These antisera were used in a dilution of 1:1000 for SRIF, CCK, and NPY, 1:150 (after reconstitution with 500 µl) for VIP, and 1:1500 for AVP.

The number of cell bodies positive for VIP, NPY, CCK, and SRIF present in the sampled regions was determined under the light microscope by at least two histologists unaware of treatment groups.

The microdensitometric analysis of the intensity of immunostaining for NPY and AVP in the paraventricular nucleus was performed by means of an automatic image analyzer (IBAS I–II, Zeiss Kontron) according to Zoli et al. [15] (see also Grimaldi et al., this volume).

Statistical analysis was performed by means of the Dunn test [6].

Behavioral Experiments

The water T-maze test was performed according to references 3, 5, and 9.

Each rat received at least 5 days of testing with 12 trials/day, the criterion being ten correct responses in 12 trials. One group of rats ($n=10$) was tested 7–15 days after carotid occlusion. A second group ($n=9$) was tested 30–40 days after reperfusion. Sham-operated rats served as controls.

Statistical analysis was carried out by means of a nonparametric two-way ANOVA.

Results

Immunocytochemical Experiments

VIP-Like Immunoreactivity

As seen in Table 1, a rapid reduction of VIP immunoreactive nerve cell profiles was observed within the neocortex (Fig. 1) and the gyrus hippocampi. In the neocortex the reduction was maintained until the 7th day. By the 40th day after reperfusion a full recovery of the number of VIP immunoreactive profiles had taken place. In contrast, in the gyrus hippocampi the number of VIP nerve cell profiles was still significantly reduced on the 40th day after reperfusion.

SRIF-Like Immunoreactivity

As far as the gyrus hippocampi is concerned (see Table 1) a continuous decline of SRIF immunoreactive nerve cell profiles took place from the first (4 h) until the last (40-day) time interval after reperfusion. By contrast, within the neocortex SRIF immunoreactivity declined until the 7-day time interval but by the 40-day time interval a full recovery in the number of SRIF-containing cell bodies had taken place.

CCK-Like Immunoreactivity

A rapid decline in the number of CCK immunoreactive nerve cells was observed in both the neocortex and the gyrus hippocampi (see Table 1 and Fig. 2). This re-

Table 1. Effects of cerebral ischemia on brain regional peptide immunoreactivity[a]

		Control	4 h	24 h	7 days	40 days
FrPa	VIP	100±4.4	63±0.7*	67±1*	65±1.7*	132±1.6
	SRIF	100±3.8	70±1.8	60±0.9*	51±0.3*	109±1.3
	CCK	100±2.4	66±0.6*	38±0.6*	51±0.6*	111±3
	NPY	100±3.2	48±0.6*	17±0.4*	55±0.6*	85±0.6
HIP	VIP	100±1.8	77±0.8*	61±1*	47±0.5*	59±0.6*
	SRIF	100±1.4	66±0.6*	61±0.01*	43±0.7*	35±0.7*
	CCK	100±2.6	52±0.01*	64±1.4*	58±0.9*	61±0.7*
PA	AVP	100±1.7	30±0.6*	26±0.2*	36±1.2*	–
	NPY	100±0.6	35±0.6*	15±1.7*	49±0.05*	–

[a] Percent values (ischemic rat/control rat) · 100 are given. At cortical and hippocampal levels the immunoreactivity values are expressed as number of cells/9 mm^2 and number of cells/hippocampus, respectively. The control values were as follows: frontoparietal cortex (FrPa): VIP = 134, SRIF = 87, CCK = 43, NPY = 67; hippocampus (HIP): VIP = 30, SRIF = 65, CCK = 62. At the paraventricular nucleus (PA) level the immunoreactivity is expressed as optical density of the staining. Statistical analysis was carried out according to the Dunn test; ischemic vs control, * = $P < 0.05$.

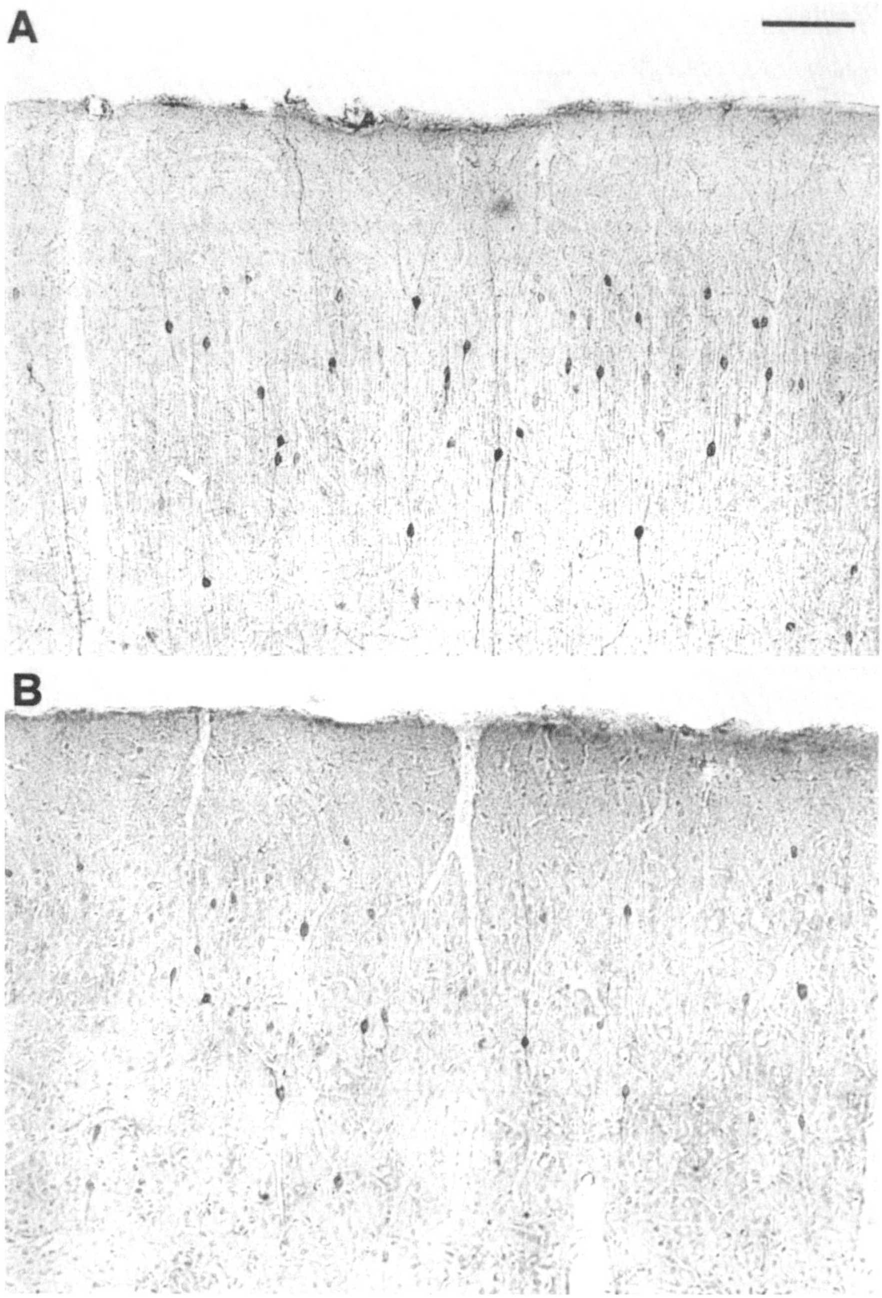

Fig. 1. Microphotographs of VIP immunoreactivity in the frontoparietal cortex of control (**A**) and ischemic (**B**) rats after 7 days from reperfusion (*calibration bar* = 100 μm)

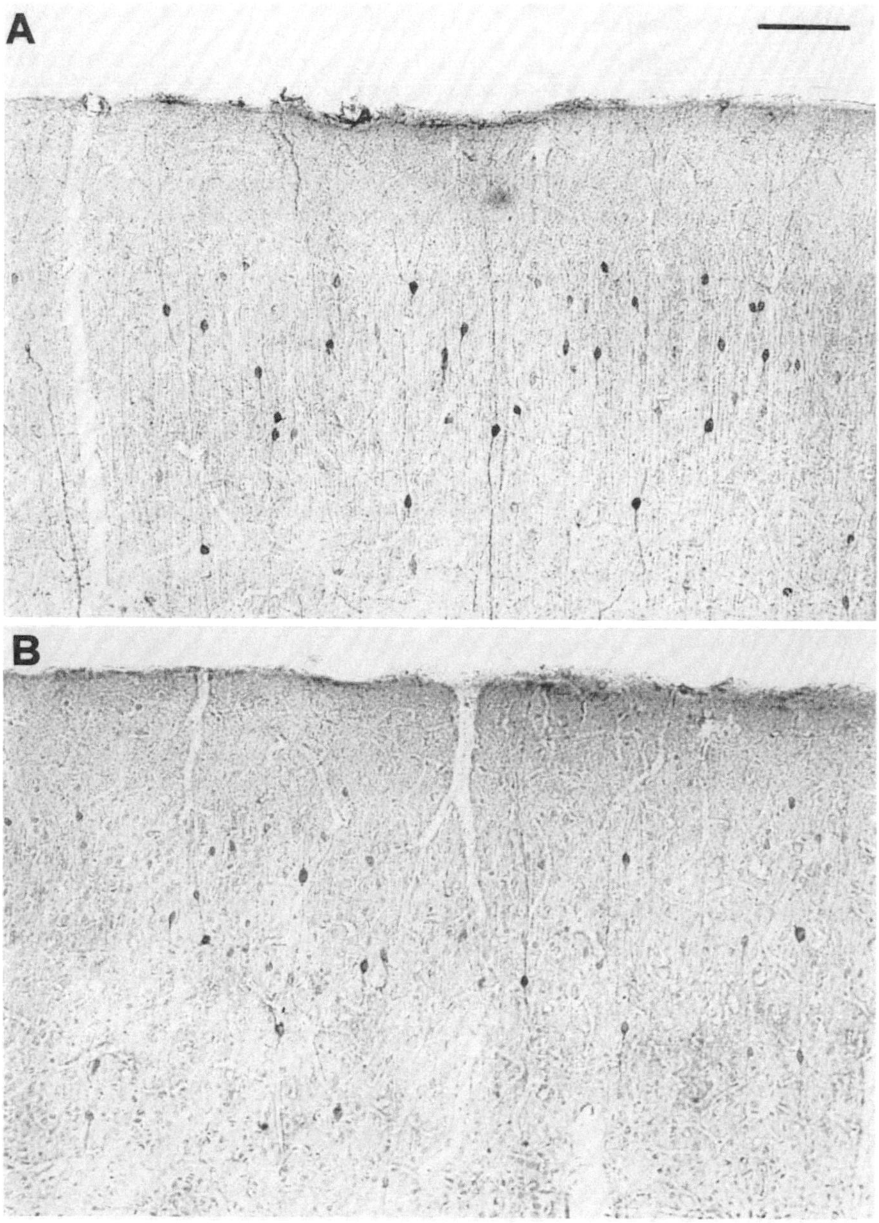

Fig. 2. Microphotographs of CCK immunoreactivity in the CA1 hippocampal field of control (**A**) and ischemic (**B**) rats after 7 days from reperfusion (*calibration bar* = 50 μm)

duction was maintained throughout the period analyzed as far as the gyrus hippocampi is concerned, while in the neocortex a full recovery in the number of CCK immunoreactive nerve cell profiles was observed at the 40-day time interval of reperfusion.

NPY-Like Immunoreactivity

As seen in Table 1, NPY immunoreactive nerve cell profiles were rapidly reduced within the neocortex, peak reduction being observed at the 24-h time interval. Partial recovery of NPY immunoreactive cell profiles was observed at the 7-day time interval, while at the 40-day time interval after reperfusion it was no longer significantly different from the control value. As seen in Table 1, the NPY immunoreactivity nerve terminal profiles in the paraventricular hypothalamic nucleus were also rapidly reduced following brain ischemia, the peak reduction being observed at the 24-h time interval of reperfusion. Partial recovery was again seen at the 7-day time interval.

AVP-Like Immunoreactivity

As seen in Table 1, a rapid reduction in the number of AVP immunoreactive nerve cell profiles was observed after brain ischemia and this reduction was maintained until the 7-day time interval.

Nissl Stainings

Within the hippocampal formation alterations in the nerve cells were also evaluated by means of staining with cresyl violet. The time course of disappearance of nerve cell bodies in the gyrus hippocampi was found to mimic the pattern for the time course of disappearance of peptide immunoreactive nerve cell bodies in this region (data not shown).

Behavioral Experiments

As seen in Fig. 3, the global brain ischemia was found to produce a significant increase in the number of errors in the learning of a water T-maze task as evaluated at the 30- to 40-day time interval. Similar results were also observed at a shorter time interval (7 and 14 days) after brain ischemia.

Discussion

The evaluation of peptide immunoreactivity has to be interpreted with caution as far as the degenerative features of the CNS are studied. In fact, a drop in immunoreactivity may mean either reduced presence of the antigen in the neuron (reduction in synthesis or increase in release) or neuronal death or, finally, both. It is clear that very early changes (4 h and 24 h after reperfusion) should be interpreted as content changes, while late changes (7 days after reperfusion) may represent both content changes and cell degeneration. Very late changes (40 days after reperfusion) very likely represent cell degeneration.

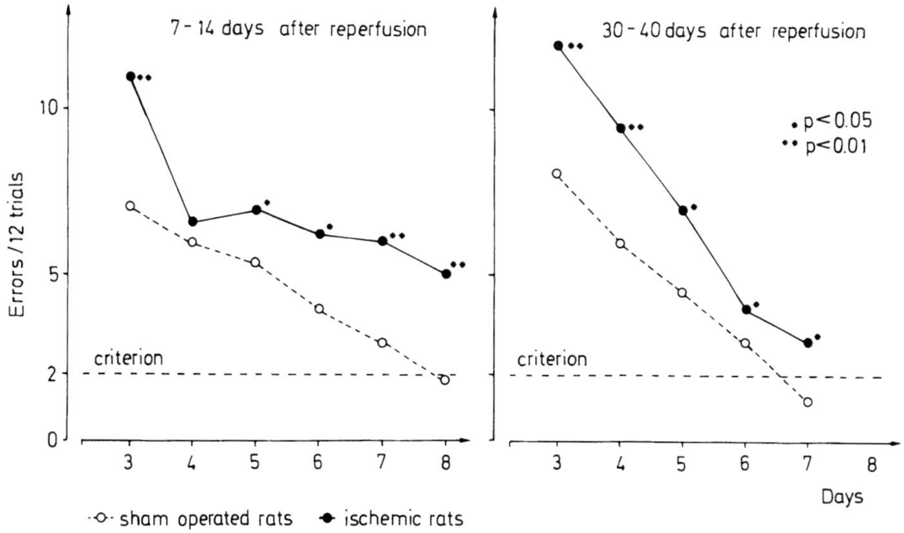

Fig. 3. Effects of cerebral ischemia on water T-maze test at 7–14 days and 30–40 days after reperfusion. Statistical analysis was carried out by means of a nonparametric two-way ANOVA

The regional correlation analysis between Nissl staining and the immunocytochemistry of SRIF, VIP, and CCK immunoreactive neurons in the gyrus hippocampi and the absence of recovery of these immunoreactivities 40 days after ischemia show that at the 7-day time interval the main cause of reduction of immunoreactive profiles, in this region, is the cell death. It seems possible that the putative degeneration of CCK, SRIF, and VIP immunoreactive neurons of the gyrus hippocampi may contribute to the reduced learning of the T-maze task in the rats exposed to cerebral ischemia in view of the role of hippocampal neurons in learning processes. Instead the neocortical peptide neurons showed a full recovery of peptide immunoreactivity at the 40-day time interval, when the learning deficit was still maintained.

Since absence of a clear-cut link between chemical coding and vulnerability to brain ischemia has been reported for classical transmitters [4], the general assumption was made that the chemical coding has no relevance for the survival of neurons. However, this view should be, at least in part, corrected. In fact, when the time course of the disappearance of SRIF, NPY, and CCK immunoreactivities is studied and compared among different telencephalic regions it becomes apparent that some sort of regularity can be observed for a class of peptide neurons, notwithstanding their different neuroanatomical localization. Thus, the SRIF immunoreactivity shows a continuous decline from 4 h until 7 days after reperfusion. The NPY immunoreactivity tends to show a rapid drop followed by a partial recovery. Therefore, it is possible that some aspects of the sensitivity to the ischemic insult can be linked to the chemical coding, at least for peptide neurons. This selective peptide deficit may have profound effects not only on the information handling by neural networks, but also on their trophic capability. Thus, treatments to increase peptide transmission may be needed after ischemic insults.

References

1. Agnati LF, Fuxe K, Zoli M, Zini I, Harfstrand A, Toffano G, Goldstein M (1988) Morphometrical and microdensitometrical studies on phenylethanolamine-*N*-methyltranspherase and neuropeptide Y immunoreactive neurons in the rostral medulla oblongata of the adult and old male rats. Neuroscience 26:461–478
2. Brierly J (1976) Cerebral hypoxia. In: Blackwood W, Corsellis JAN (eds) Grenfield's neuropathology. Edward Arnold, London, pp 43–85
3. Buresova O, Bures J, Oitzl MS, Zahalka A (1985) Radial maze in the water tank: an aversively motivated spatial working memory task. Physiol Behav 34:1003–1005
4. Francis A, Pulsinelli WA (1982) Response of GABAergic and cholinergic neurons to transient cerebral ischemia. Brain Res 243:271–278
5. Hepler DJ, Olton DS, Wenk G, Coyle J (1985) Lesions in nucleus basalis magnocellularis and medial septal area of rats produce qualitatively similar memory impairments. J Neurosci 5:866–873
6. Hollander M, Wolfe DA (1973) Nonparametrical statistical methods. Wiley, New York
7. Jacob H (1963) CNS tissue and cellular pathology in hypoxemic states. In: Schade P, Menemey WP (eds) Selective vulnerability of the brain in hypoxemia. FA Davis, Philadelphia, pp 153–176
8. Johansson O, Hokfelt T, Elde RP (1984) Immunohistochemical distribution of somatostatin-like immunoreactivity in the central nervous system of the adult rat. Neuroscience 13:265–339
9. Olton DS (1979) Mazes, maps, and memory. Am Psychol 34:583–596
10. Paxinos G, Watson C (1982) The rat brain in stereotaxic coordinates. Academic, London
11. Pulsinelli WA (1985) Selective neuronal vulnerability: morphological and molecular characteristics. In: Kogure K, Hossmann K-A, Siesjö BK, Welsh FA (eds) Molecular mechanisms of ischemic brain damage. Prog Brain Res 63:29–37
12. Pulsinelli WA, Brierly JB (1979) A new model of bilateral hemispheric ischemia in the unanesthetized rat. Stroke 10:267–272
13. Pulsinelli WA, Brierly JB, Plum F (1982) Temporal profile of neuronal damage in a model of transient forebrain ischemia. Ann Neurol 11:491–498
14. Vanderhaegen JJ, Lotstra F, De May J, Gilles C (1980) Immunohistochemical localization of cholecystokinin and gastrin-like peptides in the brain and hypophysis of the rat. Proc Natl Acad Sci USA 77:1190–1194
15. Zoli M, Agnati LF, Fuxe K et al. (1988) Morphometrical and microdensitometrical studies on phenylethanolamine-*N*-methyltranspherase and neuropeptide Y immunoreactive nerve terminals and on glucocorticoid receptor immunoreactive nerve cell nuclei in the paraventricular hypothalamic nucleus in adult and old male rats. Neuroscience 26:479–492

Related Responses Between Dopamine Synapses and Astrocytes in the Rat Striatum After Global Cerebral Ischemia: A Morphometric and Microdensitometric Study

R. GRIMALDI[1], I. ZINI[1], M. ZOLI[1], G. BIAGINI[1], M. GOLDSTEIN[3], K. FUXE[2], and L. F. AGNATI[1]

Introduction

In view of the important role of the nigrostriatal neurons in sensory motor integration and behavioral performance, in the present study we analyzed in detail the pre- and postsynaptic features of the nigrostriatal dopamine neurons in the four-vessel model of brain ischemia. We used tyrosine hydroxylase (TH) immunoreactivity as a presynaptic marker and dopamine- and cAMP-regulated phosphoprotein (DARPP-32) immunoreactivity as a postsynaptic marker of the nigrostriatal dopamine neurons. Semiquantitative evaluations of the changes in immunoreactivities were made possible by the use of morphometric and microdensitometric procedures developed in our laboratories [1, 11]. Furthermore, the changes in the dopamine terminals and in dopaminoceptive neurons were related to alterations in astroglial activity within the striatum. To this end alterations in glial fibrillary acidic protein (GFAP) immunoreactivity were measured in the striatum of the ischemic animals.

Materials and Methods

Male specific pathogen-free Sprague-Dawley rats (250–300 g body wt.) were used. They were kept one per cage under standardized humidity, temperature, and lighting conditions (lights on at 8 a.m. and off at 8 p.m.) and given food pellets and water ad libitum.

The cerebral ischemia was performed according to Pulsinelli's four-vessel model [8]. The procedure consists in the clipping (30 min) of the common carotid arteries in rats with prior permanent occlusion of the vertebral arteries. After the ischemic phase various reperfusion time intervals were studied (4 and 24 h and 7 and 40 days; in this paper only the results at the 7-day time interval are shown). The rats (18 control rats, 24 ischemic rats) were sacrificed in the morning by intracardiac perfusion with 50+50 ml of warm saline followed by 300 ml of ice-

[1] Institute of Human Physiology, University of Modena, Via Campi 287, 41100 Modena, Italy.
[2] Department of Neurobiology, Karolinska Institute, Stockholm, Sweden.
[3] Neurochemical Laboratories, Department of Psychiatry, New York University Medical School, New York, N.Y., USA.

Cerebral Ischemia and Calcium
Ed. by A. Hartmann and W. Kuschinsky

cold 4% paraformaldehyde in 0.1 M phosphate buffer (pH 7.2). The brains were rapidly dissected out and postfixed for 1 h in the same fixative. The day after, the brains were cut on a vibratome (Lancer, 50 μm thick sections) and various striatal levels (from bregma 1.7 to −0.3 mm according to the Paxinos and Watson atlas) were selected. Then, the sections were processed according to the indirect immunoperoxidase procedure using the PAP technique [10]. The immunocytochemical procedure has been described elsewhere [2]. The following antisera were used: rabbit polyclonal antiserum against TH [4], mouse monoclonal antiserum against DARPP-32 [6], and rabbit polyclonal antiserum against GFAP (Dako, Denmark), which have been previously characterized. Antisera which had been preadsorbed with their respective peptide (50 μg/ml of diluted antiserum) served as control antisera. TH was purified from rat pheochromocytoma [4]. These antisera were used in a dilution of 1:1000 for TH, 1:2000 for DARPP-32, and 1:300 for GFAP.

Image Analysis Procedure

Morphometric and microdensitometric analyses were performed by means of an automatic image analyzer (IBAS I–II, Zeiss Kontron, Munich, West Germany).

Morphometric Analysis

The quantitative analysis of the areas of DARPP-32 immunoreactivity disappearance was carried out as follows. The analysis was performed at low magnification in order to allow the acquisition of the entire striatum (size of the field 4.8 mm by side). The striatal area was interactively selected by means of a light pen and measured. Subsequently, an interactive objective discrimination procedure was performed in order to separate on a densitometric basis DARPP-32-positive from DARPP-32-negative areas. The percentage of DARPP-32-negative area per striatum as it appears after discrimination was determined. In the analysis of TH immunoreactivity the areas of hyperstaining were discriminated by means of a similar interactive objective procedure.

Analysis of the overlap between the areas of DARPP-32 immunoreactivity disappearance and the areas of hyperstaining for the TH immunoreactivity was carried out as follows. The striatal area of the section stained with DARPP-32 antiserum was marked with an overlay which reported the striatal boundaries and other suitable landmarks, and the area of staining disappearance discriminated (see above). Then, the adjacent section, stained with TH antiserum, was adjusted to fit the previously defined striatal boundaries and landmarks, and the area of TH hyperstaining discriminated (see above). Finally, the area of overlap between the two discriminated images was obtained, using the boolean operator AND [2]. The overlap area was then expressed as percent of the area of TH immunoreactivity hyperstaining. Due to the fact that TH and DARPP-32 were compared in adjacent sections, a correction factor was calculated, comparing couples of adjacent sections both stained with DARPP-32.

Microdensitometric Analysis

The semiquantitative evaluation of the intensity of immunostaining was performed according to Zoli et al. [11].

The analysis was carried out at high magnification, measuring fields of 600 μm by side. The specific profiles were discriminated from the background and the gray tone distribution of the specific profile population was evaluated. Two parameters were considered: the field area, i.e., the area covered by specific profiles, and the mean gray tone value of the specific profiles [2, 11].

Results

DARPP-32 and TH Immunoreactivities

As seen in Fig. 1, global cerebral ischemia induces patches of DARPP-32 immunoreactivity disappearance in the neostriatum as evaluated on the 7th day after reperfusion. The area of disappearance, more evident in the dorsolateral part of the striatum, tends to increase in the rostrocaudal direction (see Fig. 2). An increase in TH immunoreactivity was also observed in certain areas of the neostriatum as a response to the global cerebral ischemia. The area of TH hyperstaining corresponded almost completely to the area of DARPP-32 immunoreactivity disappearance, the corrected overlap being about 99% (see Fig. 3).

GFAP Immunoreactivity

A marked increase in GFAP immunoreactivity within the lesioned striatum was observed at the 7-day time interval after reperfusion (data not shown). Furthermore, the increase was larger in that part of the neostriatum located outside the area of DARPP-32 disappearance (800% vs 500%). The GFAP immunoreactivity increase concerned both the mean gray value and the field area value, being more evident for the last parameter (data not shown).

Discussion

The present study on the pre-and postsynaptic features of the nigrostriatal dopamine neurons, as they appear after 7 days of reperfusion following 30 min of global cerebral ischemia, revealed highly discrete alterations within the neostriatum with regard to TH and DARPP-32 immunoreactivities. These alterations appear to increase in the caudal direction and to reflect the existence of localized striatal lesions that develop upon global cerebral ischemia. Thus, the highly discrete disappearance of DARPP-32 immunoreactivity in nerve cell bodies of the neostriatum indicates an alteration in the dopaminoceptive nerve cells in that area. Fur-

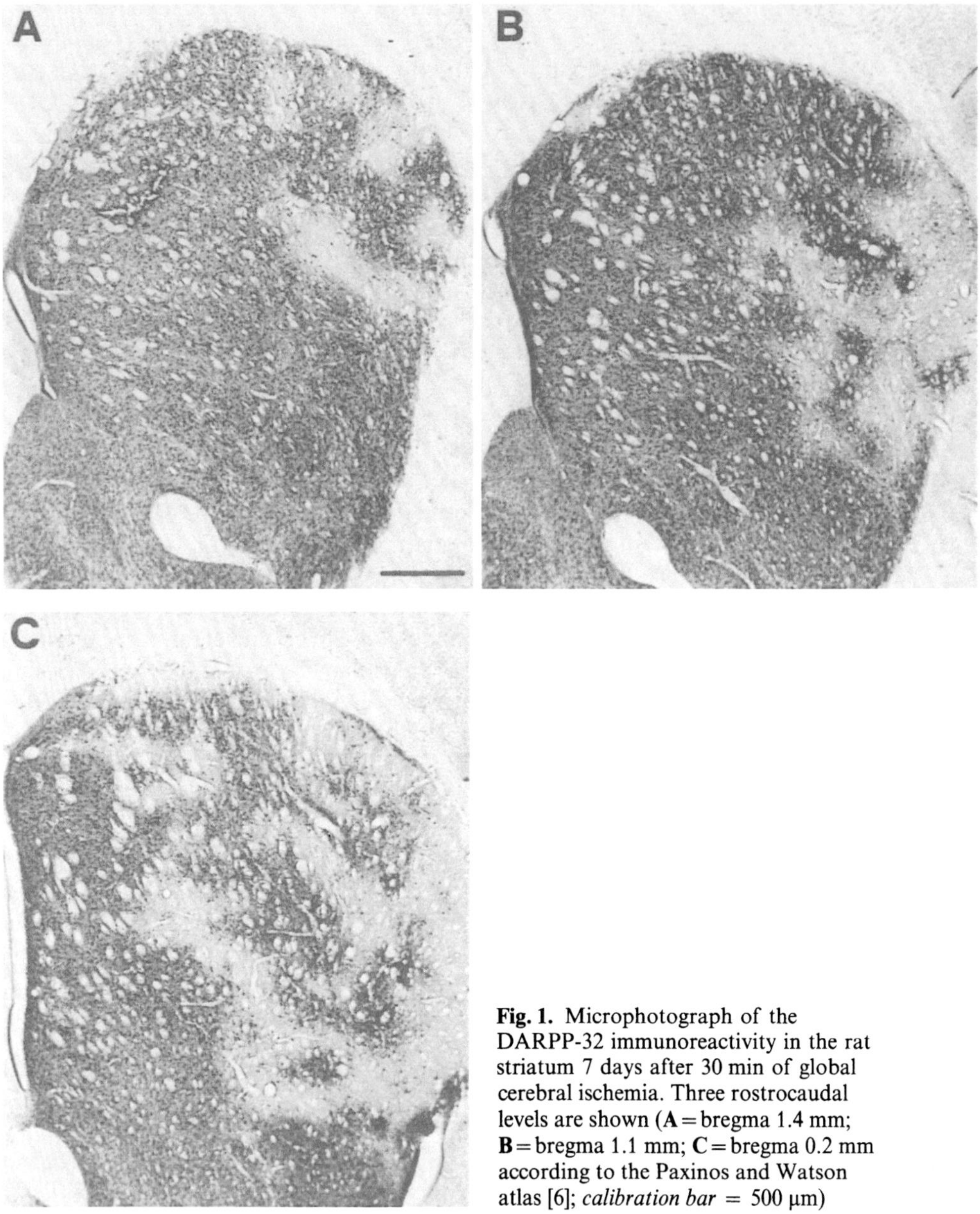

Fig. 1. Microphotograph of the DARPP-32 immunoreactivity in the rat striatum 7 days after 30 min of global cerebral ischemia. Three rostrocaudal levels are shown (**A** = bregma 1.4 mm; **B** = bregma 1.1 mm; **C** = bregma 0.2 mm according to the Paxinos and Watson atlas [6]; *calibration bar* = 500 µm)

thermore, it has been demonstrated that in the same region an increase in TH immunoreactivity takes place within the dopamine nerve terminals, which probably reflects a compensatory activation of the dopamine terminals to the lesioned region. Thus, it seems possible that there exists a mechanism by which the dopamine cell bodies projecting to the lesioned striatal areas can increase the synthesis of TH protein and transport it down into the lesioned areas. The GFAP immu-

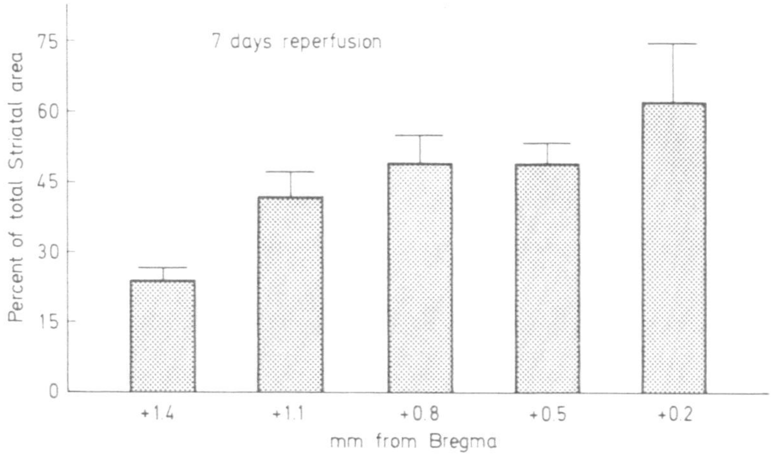

Fig. 2. Rostrocaudal analysis of DARPP-32 immunoreactivity (*IR*) after global cerebral ischemia. The data are expressed as percent of total striatal area. Bregma levels according to Paxinos and Watson [6]

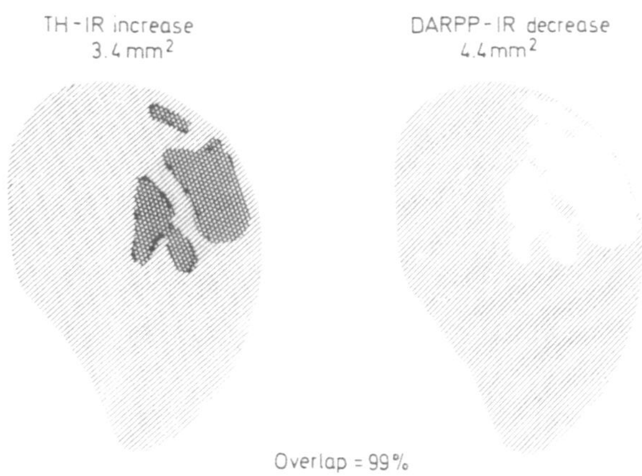

Fig. 3. Overlap area between DARPP-32 and TH immunoreactivities in the rat striatum after 30 min of global cerebral ischemia (7 days reperfusion). For further details see the section "Morphometric Analysis" in the text

noreactivity data demonstrate the massive involvement of astrocytes in the brain responses to the ischemic insult. In fact, a change of more than 500% the basal value is observed in the intralesion area, while an even huger change (more than 800% the basal value) is observed outside the lesion area. This response may be related to the release from lesioned neurons and activated glial cells of trophic factors like polyamines, which via diffusion can affect astrocytes [5] outside the area of lesion. Preliminary data obtained by our group show that the blockade of the lesion-induced putrescine increase in addition prevents the increase of GFAP immunoreactivity in the lesioned regions. These biochemical relationships between neurons and astrocytes have also been studied in other animal models of CNS lesion [3] and shown to be highly relevant for neuronal survival.

References

1. Agnati LF, Fuxe K, Benfenati F, Zini I, Zoli M, Fabbri L, Harfstrand A (1984) Computer assisted morphometry and microdensitometry of transmitter identified neurons with special reference to the mesostriatal dopamine pathway. I. Methodological aspects. Acta Physiol Scand 532:5–36
2. Agnati LF, Fuxe K, Zoli M, Zini I, Harfstrand A, Toffano G, Goldstein M (1988) Morphometrical and microdensitometrical studies on phenylethanolamine-*N*-methyltransferase and neuropeptide immunoreactive neurons in the rostral medulla oblongata of the adult and old rats. Neuroscience 26:461–468
3. Bjorklund H, Dahl D, Olson L (1984) Morphometry of GFA and vimentin positive astrocytes in grafted and lesioned cortex cerebri. Int J Dev Neurosci 2:181–192
4. Markey K, Kondo A, Shenkmann L, Goldstein M (1980) Purification and characterization of tyrosine hydroxylase from a clonal chromocytoma cell line. Mol Pharmacol 17:79–85
5. Morrison RS, de Vellis J, Lee YL, Bradshaw RA, Eng LF (1985) Hormone and growth factors induce the synthesis of glial fibrillary acidic protein in rat brain astrocytes. J Neurosci Res 14:167–176
6. Ouimet CC, Miller PE, Hemmings HJ Jr, Walaas SI, Greengard P (1984) DARPP-32, a dopamine- and adenosine 3′:5′-monophosphate-regulated phosphoprotein enriched in dopamine-innervated brain regions. III. Immunocytochemical localization. J Neurosci 4:111–124
7. Paxinos G, Watson C (1982) The rat brain in stereotaxic coordinates. Academic, London
8. Pulsinelli WA, Brierly JB (1979) A new model of bilateral hemispheric ischemia in the unanesthetized rat. Stroke 10:267–272
9. Pulsinelli WA, Brierly JB, Plum F (1982) Temporal profile of neuronal damage in a model of transient forebrain ischemia. Ann Neurol 11:491–498
10. Sternberger LA (1979) Immunocytochemistry. Wiley, New York
11. Zoli M, Agnati LF, Fuxe K et al. (1988) Effects of chronic uridine treatment on regional neuropeptide and tyrosine hydroxylase immunoreactivities in the brain of the 12 month old male rat. Neurochem Int 413:499–508

The Effects of Transient and Long-Term Ischemia on Tissue PO_2 in the Brain Cortex

A. HAGENDORFF[1], J. GROTE[1], C. HALLER[1], K. ZIMMER[1], C. DETTMERS[2], A. NIERHAUS[2], and A. HARTMANN[2]

Introduction

Ischemic attacks cause drastic vasodilatation in the brain circulation and subsequent hyperemia. Simultaneously the regulation of regional cerebral blood flow is impaired or abolished. Following cerebral ischemia, arterial hypercapnia (Yamaguchi et al. 1972; Seki et al. 1984) and changes in the concentrations of various vasoactive substances in the perivascular space have little or no effect on brain vessel diameter (Berne et al. 1974; Harris and Symon 1984; Strong et al. 1988). During microapplication studies on pial arteries a reduction of the reactivity to local metabolic factors such as H^+ and K^+ was observed (Haller and Kuschinsky 1981; Haller et al. 1986). As a consequence of the postischemic disturbance in cerebrovascular reactivity, an impaired adjustment of cerebral blood flow to the metabolic demand has to be expected.

The present experiments were designed to study the adjustment of brain tissue oxygen supply to oxygen demand following transient and long-term ischemia by measuring the local PO_2 distribution in the brain cortex. In a first series of experiments, performed on cats, ischemia was induced by air embolism. In a second one, performed on baboons, brain infarction was produced by occlusion of the middle cerebral artery.

Methods

In both groups of investigations tissue PO_2 measurements were performed using platinum surface microelectrodes as described by Kessler and Grunewald (1969) and Lübbers et al. (1969). The multiwire electrodes consisted of eight cathodes, each 10–15 μm in diameter. The hemispherical catchment area of a single electrode had a mean diameter of approximately 40–60 μm. The electrodes were calibrated in saline solutions equilibrated with gas mixtures of different PO_2 (approx. 20, 60, and 120 mmHg) at brain surface temperature (Grote et al. 1981). The PCO_2 of the equilibration gases was maintained at 40 mmHg (Söntgerath et al. 1984). Since the electrodes were mounted on a counterbalanced system, no local blood flow restriction occurred in the measuring field.

[1] Institute of Physiology, University of Bonn, Nussallee 11, D-5300 Bonn 1, FRG.
[2] Department of Neurology, University of Bonn, Sigmund-Freud-Str. 25, D-5300 Bonn 1, FRG.

Cerebral Ischemia and Calcium
Ed. by A. Hartmann and W. Kuschinsky

Experiments on Cats

Seven cats weighing 2.2–3.4 kg were anesthetized with glucochloralose (40–50 mg/kg i.v.). After tracheotomy and relaxation with pancuronium (0.133 mg/kg h) the animals were artificially ventilated using a Bird mark 7 respirator. For ventilation control, continuous recordings of end-tidal CO_2 concentration and intermittent measurements of arterial PO_2, PCO_2, and pH were performed. Blood samples were obtained from a femoral artery catheter, which was also used for continuous blood pressure recordings. For infusion of Tyrode's solution one femoral vein was cannulated. Following fixation of the cat's head in a frame, a craniotomy was performed over the right suprasylvian gyrus. The dura was carefully opened and the brain surface was covered with paraffin oil heated to 37° C.

Brain ischemia was induced by blood foam injection via a catheter into the innominate artery according to Fritz and Hossmann (1979). The correct placement of the catheter tip was controlled after each experiment by surgery. Before and after transient brain ischemia simultaneous measurements of tissue PO_2 and pial artery diameter were performed in the measuring field of the suprasylvian gyrus. The vascular diameter was determined using an image-shearing device according to Haller and co-workers (1986).

Experiments on Baboons

Five baboons weighing 12–30 kg were anesthetized with pentobarbital (25 mg/kg i.p. initially and 20 mg/kg i.v. repeated as necessary). After relaxation with pancuronium (0.133 mg/kg h) the animals were artificially ventilated via an endotracheal tube using a Schuler respiratory pump with a 3:1 N_2O/O_2 mixture. Cannulation of one femoral vein allowed the application of medication and fluids and the measurement of central venous pressure. From a catheter placed in one brachial artery, blood samples were obtained for PO_2, PCO_2, and pH analyses. In addition the catheter enabled continuous blood pressure recordings. In all experiments during the PO_2 measurements, the arterial blood gas tensions, the arterial pH, and the arterial blood pressure were found to be in the normal range.

The heads of the baboons were mounted in a stereotactic frame. A bilateral craniotomy was carried out above the frontal and the parietal brain to expose cortical regions supplied by the middle and the anterior cerebral artery. To produce focal brain infarction the middle cerebral artery was embolized with histoacryl-*n*-butyl-cyanoacrylate as described by Brassel et al. (1988). Local cerebral blood flow and tissue PO_2 measurements were performed before and 1 h after ischemia. Tissue PO_2 determinations were repeated 3, 5, and 7 h following vessel occlusion.

Results and Discussion

The results of tissue PO_2 measurements in the brain cortex of cats, determined in seven experiments, are given in Fig. 1. During the preischemic control period mean arterial blood gas tensions and pH remained normal ($P_aO_2 = 109$ mmHg; $P_aCO_2 = 31$ mmHg; $pH_a = 7.36$). The mean blood pressure amounted to 135 mmHg. As can be seen in Fig. 1 A, during control conditions the tissue PO_2 distribution ranged from very low values near 0 mmHg to values near the arterial PO_2. The mean tissue PO_2 was found to be 28.4 mmHg. Comparable tissue PO_2 histograms were found during arterial normoxia and normocapnia in the brain cortex of cats using surface electrodes (Grote et al. 1981) as well as in guinea pigs using needle electrodes (Lübbers 1973).

Air embolism caused a transient interruption of blood flow in the pial arteries and a drastic short-term decrease in tissue PO_2 in the superficial cells of the brain cortex. After complete reperfusion of all visible pial arteries in the measuring field, the obtained mean PO_2 histogram (see Fig. 1 B) was shifted to higher values with a mean tissue PO_2 of 38.1 mmHg. Simultaneous measurements of blood pressure and arterial gas tensions resulted in normal values ($P_aO_2 = 109$ mmHg; $P_aCO_2 = 30$ mmHg; $pH_a = 7.34$; $Pm_a = 125$ mmHg).

The improvement in tissue oxygen supply of the brain cortex (see Fig. 1) can be explained by hyperemia. The pial artery diameter increased by more than 40% from a mean value of 131 μm as measured under control conditions to a mean

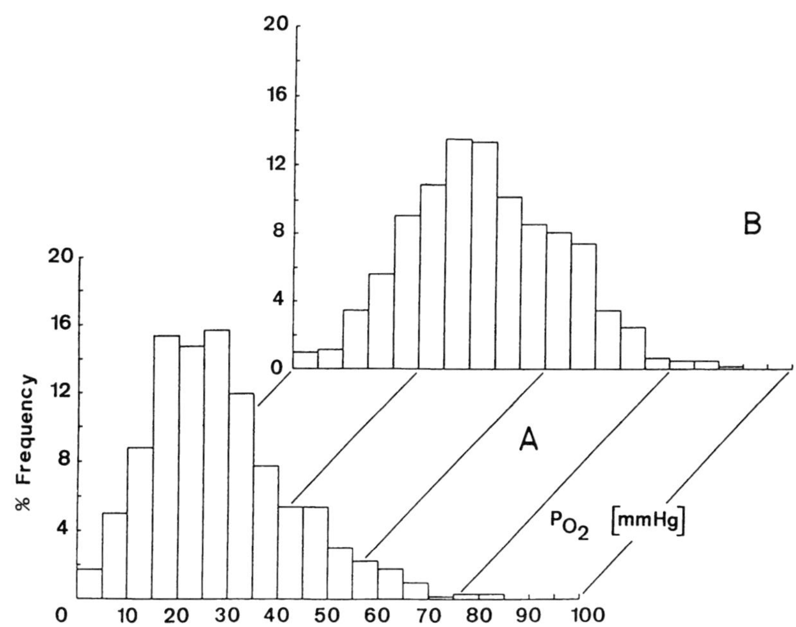

Fig. 1. Tissue PO_2 frequency histograms of the brain cortex (suprasylvian gyrus) of cats during arterial normoxia and normocapnia before (**A:** $n = 639$) and after (**B:** $n = 609$) transient ischemia induced by air embolism

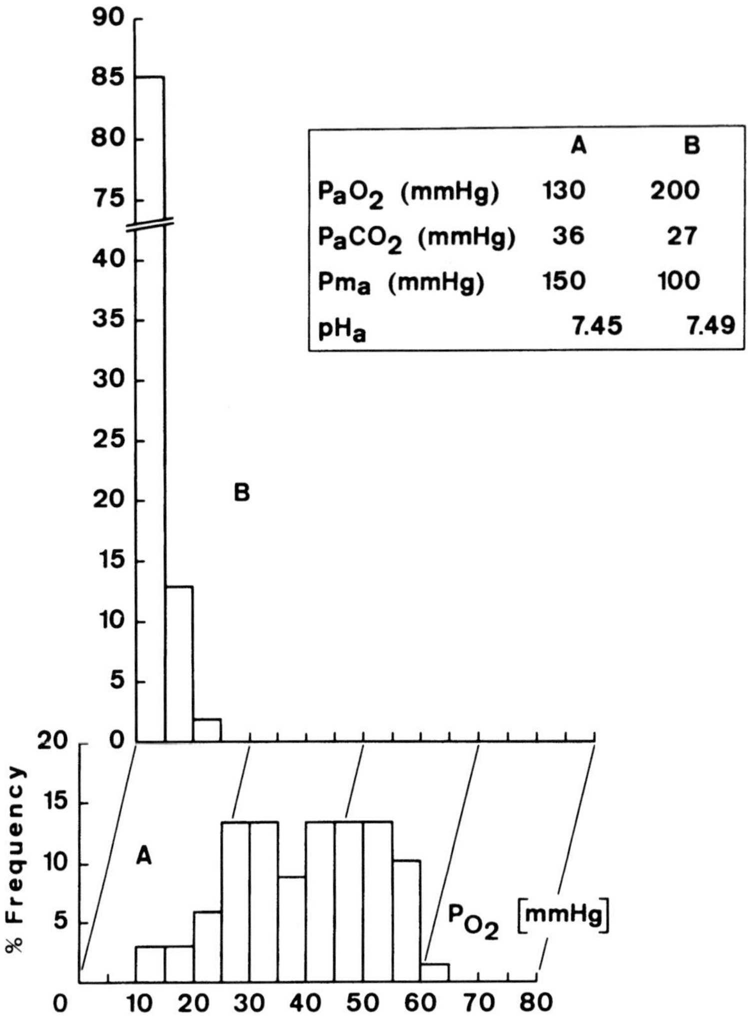

Fig. 2. Tissue PO_2 frequency histograms of the brain cortex (suprasylvian and ectosylvian gyri) of a baboon before (**A:** $n=67$) and after (**B:** $n=54$) complete occlusion of the right middle cerebral artery produced by histoacryl injection

postischemic vessel diameter of 189 μm. The observed vasodilatation may be explained by the release of vasoactive ions, metabolites, and neurohumoral transmitters (Berne et al. 1974; Astrup et al. 1977; Harris and Symon 1984). As an additional stimulus for the vascular relaxation the endothelium-derived relaxing factor has to be discussed (Vanhoutte et al. 1986). In the case of ischemia, induced by air embolism, damage of endothelial cells as described by Johansson (1980) may enhance the postischemic vasodilatation. Besides the pronounced increase in cerebral blood flow the postischemic improvement of brain tissue oxygen supply is due to the decrease in the hemoglobin oxygen affinity, facilitating oxygen diffusion from blood to tissue (Hagendorff et al. 1986). The changes in the blood

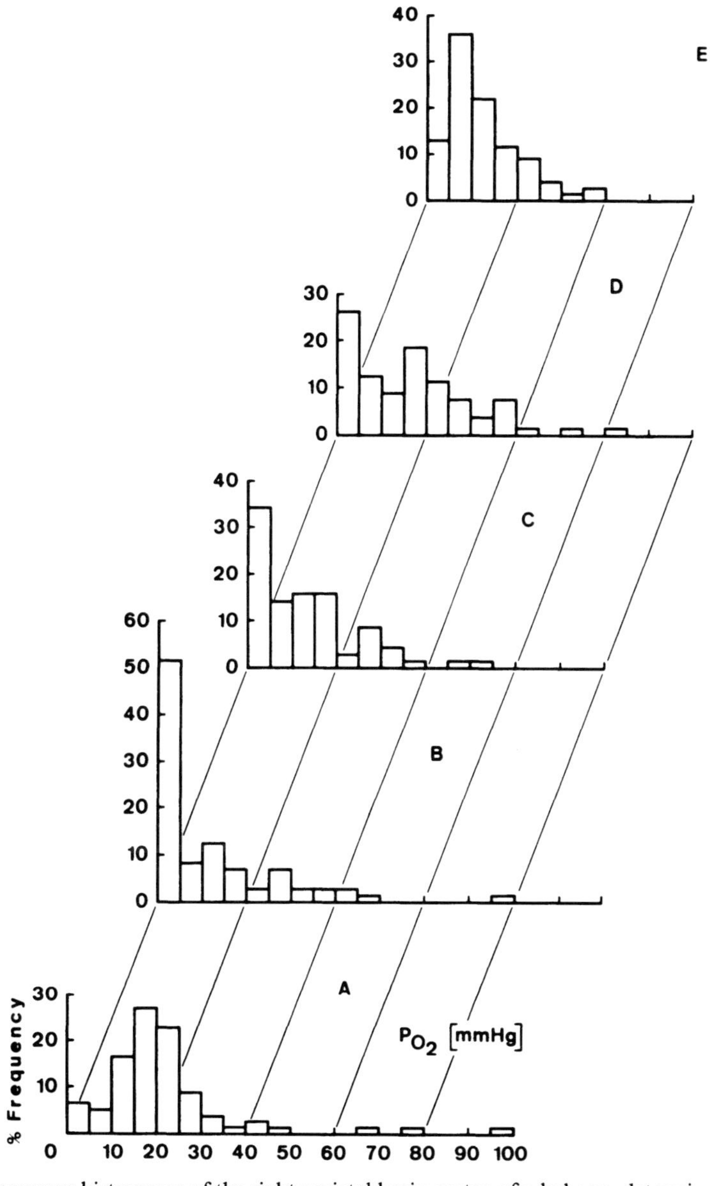

Fig. 3. Tissue PO_2 frequency histograms of the right parietal brain cortex of a baboon, determined before (**A:** $n=78$) and 1 (**B:** $n=72$), 3 (**C:** $n=70$), 5 (**D:** $n=80$), and 7 h (**E:** $n=78$) after distinct proximal occlusion of the right middle cerebral artery

hemoglobin oxygen affinity are to be expected as a result of the severe brain tissue acidosis as observed following ischemia (Fritz and Hossmann 1979; Mabe et al. 1983).

The results of tissue PO_2 measurements in the brain cortex of baboons following focal ischemia are demonstrated by typical examples. Before vessel occlusion in all experiments normal tissue PO_2 distributions were found in the parietal as

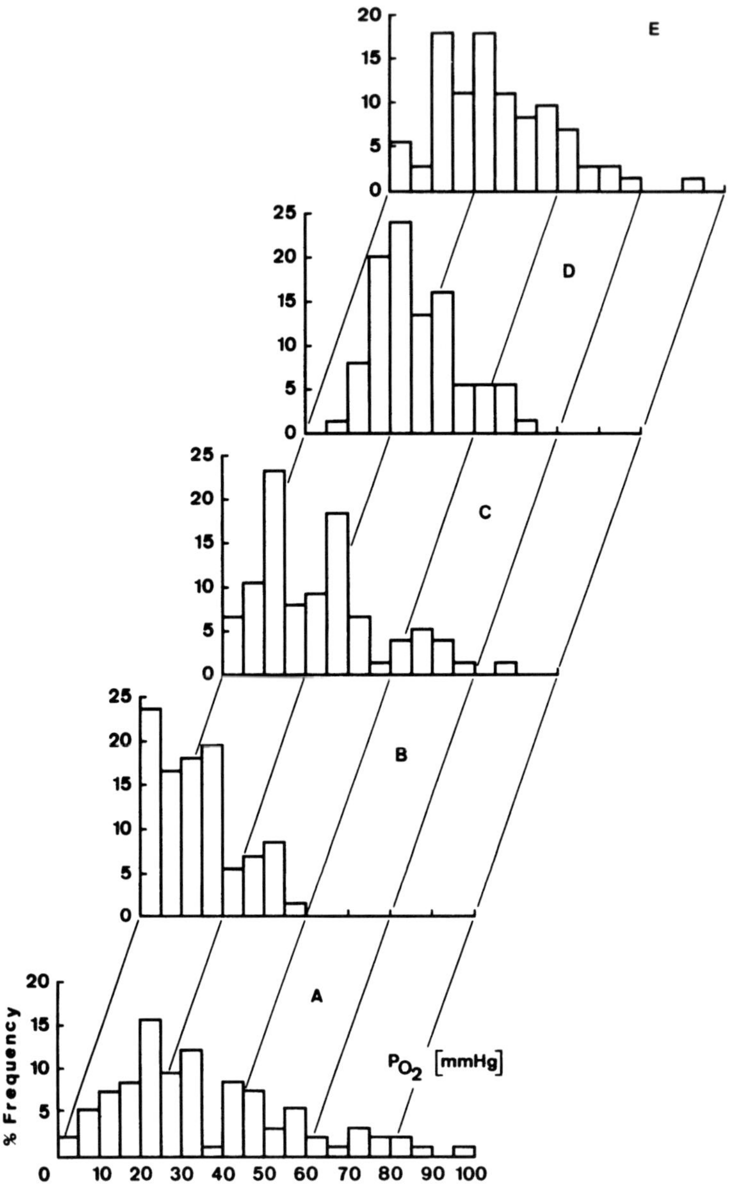

Fig. 4. Tissue PO_2 frequency histograms of the right frontal brain cortex of a baboon, determined before (**A:** $n=96$) and 1 (**B:** $n=72$), 3 (**C:** $n=76$), 5 (**D:** $n=75$), and 7 h (**E:** $n=72$) after distinct proximal occlusion of the right middle cerebral artery

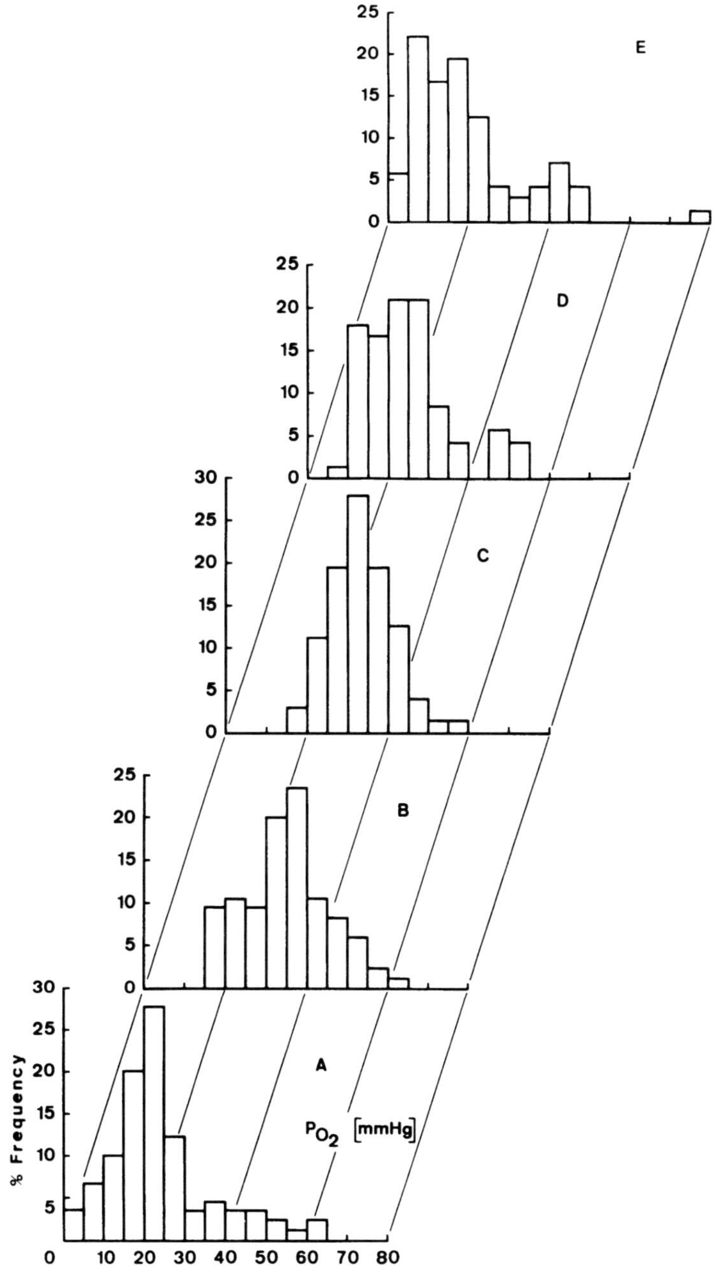

Fig. 5. Tissue PO_2 frequency histograms of the left parietal brain cortex of a baboon, determined before (**A:** $n=90$) and 1 (**B:** $n=86$), 3 (**C:** $n=72$), 5 (**D:** $n=72$), and 7 h (**E:** $n=72$) after distinct proximal occlusion of the right middle cerebral artery

well as in the frontal brain cortex (Figs. 2–5). The corresponding mean tissue PO_2 values of the measuring fields were found to be in the range of 20–35 mmHg. Comparable differences in the regional local PO_2 distribution were observed in the brain cortex as well as in the hippocampus by Nair et al. (1987). The results of the tissue PO_2 measurements are in line with those of blood flow and metabolism studies. Using autoradiography, inert gas clearance, and PET under normal conditions, significant differences were found in local blood flow and local metabolism of the brain cortex (Sokoloff et al. 1977; Sakurada et al. 1978; Phelps et al. 1982). The effect of middle cerebral artery occlusion on tissue PO_2 distribution in the investigated brain areas depended on the extent and site of the vessel embolization. In the majority of the investigations a drastic left shift of the PO_2 histograms was found immediately after histoacryl injection, as can be seen in Fig. 2. In cases of distinct proximal embolization of the right middle cerebral artery a slight improvement of brain tissue oxygen supply could be observed within the first 7 h after infarction in the measuring fields of the affected hemisphere. In Figs. 3 and 4 the results of a typical experiment are given. The observed improvement in tissue oxygen supply of the border zone of the infarct area might have been due to increase in local blood flow through collateral vessels as discussed by Halsey and Clark (1970), Hossmann (1982), and Strong et al. (1988). The decrease in neuronal activity of the penumbra as observed by Strong et al. (1988) is an additional factor influencing the tissue PO_2 distribution. The PO_2 histograms simultaneously determined in the parietal brain cortex of the unaffected hemisphere showed a transient shift to the right (see Fig. 5). It cannot be decided whether the observed increase in tissue PO_2 of the contralateral brain region resulted from an increase in local blood flow or a decrease in local tissue oxygen demand.

References

Astrup J, Symon L, Branston NM, Lassen NA (1977) Cortical evoked potential and extracellular K^+ and H^+ at critical levels of brain ischemia. Stroke 8:51–57

Berne RM, Rubio R, Curnich RR (1974) Release of adenosine from ischemic brain. Effect on cerebral vascular resistance and incorporation into cerebral adenine nucleotides. Circ Res 35:262–271

Brassel F, Dettmers C, Nierhaus A, Hartmann A, Solymosi L (1988) An intravascular technique to occlude the middle cerebral artery in baboons. (in press)

Fritz H, Hossmann KA (1979) Arterial air embolism in the cat brain. Stroke 10:581–589

Grote J, Zimmer K, Schubert R (1981) Effects of severe arterial hypocapnia on regional blood flow regulation, tissue PO_2 and metabolism in the brain cortex of cats. Pflügers Arch 391:195–199

Hagendorff A, Haller C, Grote J (1986) Der Einfluß einer postischämischen Hypokapnie auf die cerebrale O_2-Versorgung. In: Grote J, Witzleb E (eds) Physiologie und Pathophysiologie des Gefäßsystems. Funktionsanalyse biologischer Systeme 17. G. Fischer, Stuttgart, pp 165–173

Haller C, Kuschinsky W (1981) Reactivity of pial arteries to K^+ and H^+ before and after ischemia induced by air embolism. Microcirculation 1:141–159

Haller C, Kuschinsky W, Reimnitz P (1986) Effect of gamma-hydroxybutyrate on the reactivity of pial arteries before and after ischemia. J Cereb Blood Flow Metab 6:658–666

Halsey JH, Clark LC (1970) Some regional circulatory abnormalities following experimental cerebral infarction. Neurology 20:238–246

Harris RJ, Symon L (1984) Extracellular pH, potassium and calcium activities in progressive ischemia of rat cortex. J Cereb Blood Flow Metab 4:178–186
Hossmann KA (1982) Treatment of experimental cerebral ischemia. J Cereb Blood Flow Metab 2:275–297
Johansson BB (1980) Cerebral air embolism and the blood-brain barrier in the rat. Acta Neurol Scand 62:201–209
Kessler M, Grunewald W (1969) Possibilities of measuring oxygen pressure fields in tissue by multiwire platinum electrodes. Prog Resp Res 3:147–152
Lübbers DW, Baumgärtl H, Fabel H et al. (1969) Principles and construction of various platinum electrodes. Prog Resp Res 3:136–146
Lübbers DW (1973) Local tissue PO_2: its measurement and meaning. In: Kessler M, Bruley DF, Lübbers DW, Siver IA, Strauss J (eds) Oxygen supply, theoretical and practical aspects of oxygen supply and microcirculation of tissue. Urban and Schwarzenberg, Munich
Mabe H, Blomqvist P, Siesjö BK (1983) Intracellular pH in the brain following transient ischemia. J Cereb Blood Flow Metab 3:109–114
Nair PK, Buerk DG, Halsey JJ (1987) Comparison of oxygen metabolism and tissue PO_2 in cortex and hippocampus of gerbil brain. Stroke 18:616–622
Phelps ME, Mazziotta JC, Huang SC (1982) Study of cerebral function with positron computed tomography. J Cereb Blood Flow Metab 2:113–162
Sakurada O, Kennedey C, Jehle J, Brown D, Carbin GL, Sokoloff L (1978) Measurement of local cerebral blood flow with iodo(^{14}C)antipyrine. Am J Physiol 234:H59–H66
Seki H, Yoshimoto T, Ogawa A, Suzuki J (1984) The CO_2 response in focal cerebral ischemia – sequential changes following recirculation. Stroke 15:699–704
Sokoloff L, Reivich M, Kennedy C et al. (1977) The (^{14}C)deoxyglucose method for measurement of local cerebral glucose utilization: theory, procedure and normal values in conscious and anaesthetized albino rats. J Neurochem 28:897–916
Söntgerath C, Steuer K, Grote J (1984) Der Einfluß des PCO_2 auf die Messung des O_2-Partialdruckes mit Platin-Mikroelektroden. In: Grote J, Witzleb E (eds) Atemgaswechsel and O_2-Versorgung der Organe. Funktionsanalyse biologischer Systeme 12. Steiner, Stuttgart, pp 172–179
Strong AJ, Gibson G, Miller SA, Venables GS (1988) Changes in vascular and metabolic reactivity as indices of ischaemia in the penumbra. J Cereb Blood Flow Metab 8:79–88
Vanhoutte PM, Rubayi GM, Miller VM, Houson PS (1986) Modulation of vascular smooth muscle contraction by the endothelium. Ann Rev Physiol 48:307–320
Yamaguchi T, Regli F, Waltz AG (1972) Effects of hyperventilation with and without carbon dioxide on experimental cerebral ischemia and infarction. Brain 95:123–132

Vasomotor Response and Blood-Brain Barrier Function of Rat Brain Studied in a Closed Cranial Window Preparation

L. SCHÜRER[1], S. KAWAMURA[2], B. SCHMUCKER[1], A. GOETZ[1], and A. BAETHMANN[1]

Introduction

Studies of the microcirculation of rat brain surface vessels are technically difficult. A high level of microsurgical skill is only one of the prerequisites for the performance of such investigations. Furthermore, after opening of the cranium the rat brain is liable to swell and to herniate out of the craniotomy incision. Brain swelling and herniation lead to an opening of the blood-brain barrier, rendering investigations of blood-brain barrier function difficult. We have developed a technical procedure for the establishment of an improved closed cranial window preparation to investigate the microcirculation and blood-brain barrier function of rat brain surface vessels.

Methods

Male Sprague-Dawley rats were anesthetized intravenously with 120 mg kg^{-1} α-chloralose, tracheotomized, and artificially ventilated under blood gas control and muscle relaxation. Body temperature was kept constant by feedback triggered heating. A tail artery and vein as well as a femoral artery and a jugular vein were cannulated for blood pressure recording, blood sampling, and infusion of drugs and of Na^+-fluorescein – used as a fluorescence dye. The animal's head was then fixed in a stereotactic frame, the skull exposed, and a rectangular window trephined over the left parietal cortex under continuous cooling.

Three polyethylene catheters were fixed in bone ridges at the edge of the window for superfusion and recording of the intracranial pressure and embedded into a circular wall of dental acrylic cement which surrounded the window. A Plexiglas funnel (10 cm high) was then attached above the exposed brain of the rat and filled with paraffin oil. The skull was opened under oil, and the dura mater incised and reflected (Fig. 1).

Finally, a cover glass was placed on top of the window and sealed airtight. After closure of the window, oil and funnel were removed. The window was

[1] Institute for Surgical Research, Ludwig-Maximilians-University of Munich, Klinikum Großhadern, D-8000 München 70, FRG.
[2] Department of Surgical Neurology, Research Institute for Brain and Blood Vessels, Akita, Akita 010, Japan.

Cerebral Ischemia and Calcium
Ed. by A. Hartmann and W. Kuschinsky

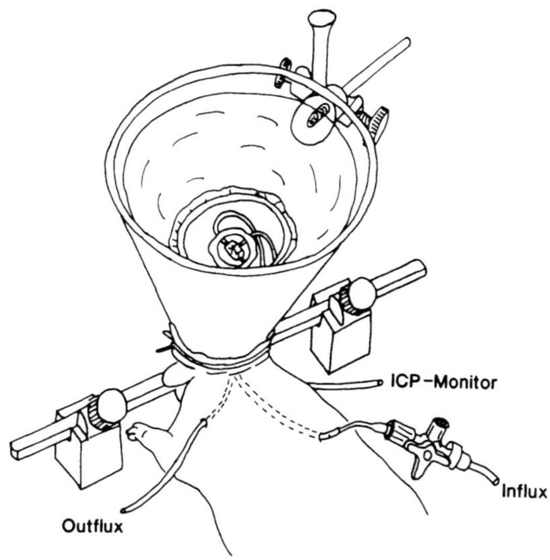

Fig. 1. Schematic drawing showing the experimental animal fixed in a stereotactic frame with a Plexiglas funnel attached to the head of the animal

superfused with artificial CSF at a rate of 5 ml h^{-1}. The level of the outflow catheter was adjusted to obtain an intracranial pressure of 5 mmHg.

Changes in the pial vessel diameters and blood-brain barrier function were studied after intravenous injection of Na^+-fluorescein using a modified Leitz-Orthoplan fluorescence microscope, xenon flash light illumination, a low light TV camera, and a video recorder.

Results

Blood-Brain Barrier Function

Blood-brain barrier function was assessed in 16 rats over a period of 2–6.5 h. Na^+ -fluorescein (MW 376) used as a barrier marker was strictly confined to the intravascular space during the control period, indicating an intact blood-brain barrier. Superfusion of pial vessels with 2000 mosmol l^{-1} D(−)-mannitol or 1 mM histamine dissolved in CSF resulted in diffuse, gross opening of the blood-brain barrier to intravenous Na^+-fluorescein 5 min after the start of the superfusion. Barrier leakage was graded as stage III according to a classification of Wahl et al. [1].

Vascular Reactivity

In four rats, the effect of continuous superfusion of brain surface vessels with buffered artificial cerebrospinal fluid was studied for 180 min. The cerebrospinal fluid had a mean pH of 7.35 and an osmolarity of 309 mosmol l^{-1}. As seen in Fig. 2, continuous superfusion of rat pial arterioles and venules did not affect the

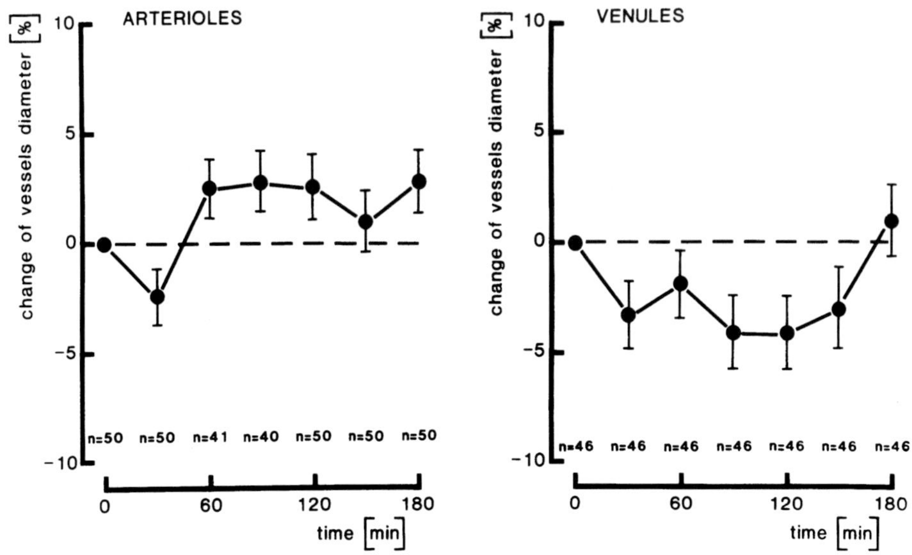

Fig. 2. Effect of continuous superfusion on vessel diameters of rat pial arterioles and venules with artificial cerebrospinal fluid

diameters significantly. At best, slight dilatation of arterioles and a tendency towards temporary venular constriction was seen. The intracranial pressure did not exceed 5 mmHg during the entire experimental period.

Discussion and Conclusion

The establishment of an improved closed cranial window technique in rats is reported. Continuous superfusion of the pial vessels with buffered CSF did not significantly affect the normal vessel diameters, nor was the blood-brain barrier permeable for Na^+-fluorescein used as a low molecular weight barrier indicator. Superfusion of the preparation with a hyperosmolar solution as introduced by Rapoport et al. [2] led to gross and prompt opening of the blood-brain barrier. Barrier disruption occurred as soon as 5 min after the start of superfusion with the hypertonic medium, which is in agreement with the findings of Wahl et al. [1]. Under the currently employed conditions, opening of the barrier seems to become manifest about 1 min later than in corresponding studies in cats [1], indicating a somewhat more resistant blood-brain barrier.

In comparison to existent experimental approaches [3–5], the current model for the first time allows simultaneous studies of the brain surface microhemodynamics and blood-brain barrier function under control of the intracranial pressure. Superfusion of the cerebral surface by use of this model appears suitable for the study of pertinent aspects of the cerebral microcirculation and blood-brain barrier function under various physiological and pathophysiological conditions.

This may provide promising possibilities for a variety of experimental and clinical problems.

Acknowledgments. The excellent technical assistance of Ulrike Goerke and Hildegard Fuderer is gratefully appreciated. This work was supported by the Deutsche Forschungsgemeinschaft.

References

1. Wahl M, Unterberg A, Baethmann A (1985) Intravital fluorescence microscopy for the study of blood-brainbarrier function. Int J Microcirc Clin Exp 4:3–18
2. Rapoport SI, Hori M, Klatzo I (1972) Testing of a hypothesis for osmotic opening of the blood-brain barrier. Am J Physiol 223:323–331
3. Auer LM (1988) Measurement of pial vessel hemodynamics. In: Boulton AA, Baker GB, Boisvert DPJ (eds) Imaging and correlative physicochemical techniques. Neuromethods, vol 8:37–106 (Humana press, Clifton, New Yersey)
4. Haberl RL, Heizer ML, Ellis EF (1987) Effects of the thromboxane A_2 mimetic U46619 on pial arterioles of rabbits and rats. Stroke 18:796–800
5. Morii S, Ngai AC, Winn HR (1986) Reactivity of rat pial arterioles and venules to adenosine and carbon dioxide: with detailed description of the closed cranial window technique in rats. J Cereb Blood Flow Metab 6:34–41

Regional Cerebral Blood Flow After Cryoinjury of the Rabbit Brain

R. MURR[1], L. SCHÜRER[2], S. BERGER[2], O. KEMPSKI[2], and A. BAETHMANN[2]

Introduction

Head injury or ischemia induces focal cerebral lesions with subsequent development of brain edema. The pathological water accumulation may affect CBF in surrounding areas (Frei et al. 1973). Impairments of the local oxygen supply may then result in further damage to the brain. A rabbit model was established in this laboratory to characterize the spatial and time-dependent changes of rCBF and of the brain surface oxygen partial pressure (tissue PO_2) after focal cold injury to the brain.

Materials and Methods

Six New Zealand albino rabbits were anesthetized with 30–40 mg/kg thiopental intravenously. Anesthesia was maintained with α-chloralose (50 mg/kg) intravenously. After tracheotomy and relaxation the animals were artificially ventilated with 30% O_2 and 70% N_2. Arterial and venous lines in femoral vessels were used for monitoring of the mean arterial blood pressure (MAPB), of central venous pressure (CVP), and of arterial blood gases as well as for fluid and drug administration. The left hemisphere was exposed by a rectangular trephination. Four platinum needle electrodes (75 μm diameter, 1 mm free length) were placed close to the lateral bone edge into the cortical gray matter at 4-mm intervals from each other. A fifth electrode was inserted through a burr hole into the contralateral cortex. A layer of paraffin oil was placed over the exposed cerebral surface prior to opening of the dura mater for protection of the brain tissue. rCBF was measured by using the H_2 clearance technique after administration of 3%–7% H_2 to the tidal volume. The clearance kinetics obtained between 60 and 250 s after interruption of H_2 administration were utilized for calculation of the initial slope index (ISI).

The PO_2 of the brain surface was measured by using a multiwire MDO electrode (Mehrdraht-Dortmund-Oberflächenelektrode) according to the method of Kessler and Lübbers (Kessler et al. 1976). PO_2 was measured in four different re-

[1] Institute of Anesthesiology and [2] Institute for Surgical Research, Ludwig-Maximilians-University, Klinikum Großhadern, Marchioninistr. 15, D-8000 München 70, FRG.

Cerebral Ischemia and Calcium
Ed. by A. Hartmann and W. Kuschinsky

gions at the brain surface. Acquisition of the PO_2 histograms was accomplished by stepwise axial rotation of the PO_2 electrode.

Experimental Protocol

After control values of rCBF and brain surface PO_2 were obtained, a focal cold lesion was induced by the method of Klatzo et al. (1958). Measurements of rCBF and PO_2 were subsequently performed every 20 and 60 min, respectively, until 6 h after lesion.

Results

Systemic physiological parameters, e.g., MABP and arterial PCO_2, remained largely constant during the whole experimental course. Focal cold injury was followed by marked hyperemia of the entire ipsilateral hemisphere (Table 1). The hyperemic response of the cerebral cortex appeared to evolve largely independently from the distance of measurement of rCBF from the focal lesion. Cerebral perfusion normalized about 60 min later. rCBF then remained practically unchanged adjacent to the site of injury, whereas it was found to increase continuously at distant areas and in the contralateral hemisphere. The tissue PO_2 was considerably increased initially after trauma in all areas which were studied, and remained high during the subsequent experimental observation period. In some cases, tissue hypoxia developed 2–3 h after trauma in areas close to the lesion, with a tendency towards normalization 3–4 h thereafter, when approaching the end of the experiment (Table 2).

Table 1. rCBF (mean ± SEM) in ml/100 g × min, or percent of control, respectively, at various distances from a focal cryogenic insult and in the contralateral hemisphere 1–6 h after trauma ($n = 6$)

Time (h)	Distance from lesion				Contralateral hemisphere
	4 mm	8 mm	12 mm	16 mm	
Before trauma (ml/100 g × min)	42.6 ± 2.2	44.3 ± 1.9	47.0 ± 3.3	45.3 ± 3.3	41.6 ± 3.1
Maximum increase in					
1st hour	+60.0%	+72.4%	+135.2%	+23.0%	+14.8%
2nd hour	− 0.2%	− 2.8%	− 1.4%	+ 5.5%	+ 9.5%
3rd hour	b 2.3%	ü 3.0%	+ 8.0%	+ 8.9%	+12.5%
4th hour	+ 0.5%	− 3.1%	+ 6.9%	+18.7%	+17.0%
5th hour	+ 0.7%	+ 1.6%	+ 8.6%	+12.5%	+16.9%
6th hour	+ 2.5%	+ 6.2%	+ 11.7%	+16.4%	+27.0%

Table 2. PO_2 (mmHg) at the cerebral surface at various distances from a focal insult 1–6 h after trauma ($n=6$)

Time	Localization of electrode			
	Lesion	Near	Intermediate	Distant
Control period	32.3	34.5	26.5	25.5
% hypox.	6.3	7.7	9.4	6.6
1st hour	53.3	36.6	39.2	30.0
% hypox.	1.9	4.3	4.7	7.1
2nd hour	37.5	39.1	30.5	28.4
% hypox.	5.5	3.1	7.4	6.0
3rd hour	42.1	32.5	32.8	30.6
% hypox.	2.2	10.6	4.6	6.0
4th hour	41.0	32.0	32.0	26.6
% hypox.	1.9	16.9	3.5	10.3
5th hour	45.5	31.5	32.5	28.1
% hypox.	0.6	11.6	1.0	5.1
6th hour	37.0	40.0	32.9	29.9
% hypox.	4.2	4.0	5.2	8.4

The proportion of hypoxic values (<5 mmHg) of the brain tissue PO_2 histograms is given in percent (i.e., % hypox.).

Discussion

Many authors have observed depression of CBF from traumatic lesions of the brain, although in some studies shortlasting early hyperemia is reported in association with pial vasodilatation, e.g., after experimental fluid percussion of the brain (DeWitt et al. 1986). The initial increase in rCBF after trauma in cerebral cortex of the exposed hemisphere observed in the present study may correspond to the above-cited phenomenon. In animals with cryogenic injury, a correlation between the extent of brain edema and reduction of rCBF was shown (Frei et al. 1973). In our experiments we found a relative impairment of blood flow close to the lesion as compared to distant locations or to the contralateral hemisphere. This was concluded from observations that in the vicinity of the insult, rCBF after transient hyperemia was largely maintained at the pretrauma level, whereas in distant areas or the contralateral hemisphere rCBF was gradually increasing.

Early measurements of CBF after a cryogenic injury of the brain have been performed mostly by employing a closed skull preparation (Blasberg et al. 1980; Bruce et al. 1972; Marmarou et al. 1976; Reilly et al. 1977; Yamamoto et al. 1976). An increase in intracranial pressure resulting from the lesion may have prevented an initial posttraumatic hyperemia and, subsequently, contributed to the impairment of the cerebral perfusion. In this study the entire hemisphere was generously exposed, which might have prevented adverse sequelae of posttraumatic brain swelling, such as an increase in intracranial pressure or cerebral herniation.

Simultaneous measurements of rCBF and tissue PO_2 at corresponding distances from the lesion made it possible to assess adequacy of the tissue perfusion after trauma. The presence of tissue hypoxia adjacent to the lesion in some experiments 2–3 h after injury is indicative of a dissociation of perfusion and O_2 consumption of the brain. Since rCBF was not impaired in areas with low PO_2 as compared to the preinjury level, tissue hypoxia might have resulted from a local, transitory increase in the cerebral metabolism without a compensatory increase in perfusion. We have to consider, in addition, whether in experiments with confined tissue hypoxia placement of the H_2 clearance electrodes for assessment of rCBF was, in fact, representative of the locations suffering from hypoxia.

Summary and Conclusion

After a focal cold injury to the brain, rCBF was markedly increased in the cortex of the entire cerebral hemisphere for about 60 min. Subsequently, rCBF close to the lesion fell to the level obtained in the preinjury control phase. In distant cortical areas and in the contralateral hemisphere blood flow was found to increase continuously until termination of the experiment. The development of both hyperemia and tissue hypoxia, as found in the current investigations, may increase intracranial pressure as well as enhance formation and spread of edema, thereby contributing to secondary brain damage. Further studies from this laboratory are concerned with the influence of intravenous and inhalation anesthetics on rCBF and O_2 supply in association with brain edema, since surgery and, consequently, anesthesia is often required in patients with focal cerebral lesions from severe head injury.

Acknowledgments. The technical and secretarial assistance of Hildegard Fuderer and Helga Kleylein is gratefully acknowledged. This work was supported by Deutsche Forschungsgemeinschaft BA 452.

References

Blasberg R, Gazendam J, Patlak C, Fenstermacher J (1980) Freeze lesion of the cerebrum: local capillary permeability, blood flow and bulk flow of edema fluid. In: Shulman K, Miller JD, Becker DP (ed) Intracranial pressure IV. Springer, Berlin Heidelberg New York, pp 314–316

Bruce DA, Vapalahti M, Schutz H, Langfitt TW (1972) rCBF, $CMRO_2$ and intracranial pressures following a local cold injury of the cortex. In: Brock M, Dietz H (ed) Intracranial pressure I. Springer, Berlin Heidelberg New York, pp 85–89

DeWitt DS, Jenkins LW, Wei EB, Lutz H, Becker DP, Kontos HA (1986) Effects of fluid percussion brain injury on regional cerebral blood flow and pial arteriolar diameter. J Neurosurg 64:787–794

Frei HJ, Wallenfang Th, Pöll W, Reulen HJ, Schubert R, Brock M (1973) Regional cerebral blood flow and regional metabolism in cold induced oedema. Acta Neurochirurgica 29:15–18

Kessler M, Höper J, Krumme BA (1976) Monitoring of tissue perfusion and cellular function. Anesthesiology 45:184–197

Klatzo I, Piraux A, Laskowski EJ (1958) The relationship between edema, blood-brain barrier and tissue elements in a local brain injury. J Neuropath Exp Neurol 17:548–564

Marmarou A, Shulman K, Shapiro K, Pöll W (1976) The time course of brain tissue pressure and local CBF in vasogenic edema. In: Pappius HM, Feindel W (eds) Dynamics of cerebral edema. Springer, Berlin Heidelberg New York, pp 113–121

Reilly PL, Farrar JK, Miller JD (1977) Vascular reactivity in the primate brain after acute cryogenic injury. J Neurol Neurosurg Psychiatry 40:1092–1101

Yamamoto L, Soejima T, Meyer E, Feindel W (1976) Early hemodynamic changes at the microcirculatory level following focal cryogenic injury over the cortex. In: Pappius HM, Feindel W (eds) Dynamics of brain edema. Springer, Berlin Heidelberg New York, pp 59–62

Experimental Cerebral Ischemia and Calcium

Calcium Antagonists in Brain Ischemia – Experimental Observations

L. SYMON

There is a considerable literature in relation to the effect of ischaemia on the brain. The occurrence of brain tissue acidosis during total ischaemia or after circulatory rest has been known for many years [9, 24, 30, 32, 38, 40]. This acidosis has been postulated to be one of the factors which influences recovery after an ischaemic insult [37, 39]. Rapid decreases in extracellular pH [1] and increases in extracellular potassium activity have been established, as have the relationships between extracellular pH and potassium and calcium activities in progressive ischaemia in the cortex of the rat or primate [20, 22]. In subarachnoid haemorrhage (SAH) reduced cerebral perfusion secondary to the haemorrhage has been well documented, and widely attributed to spasm of vessels induced by the haemorrhage. Much experimental work has been directed towards the analysis of this phenomenon in the cat [11, 31], the primate [23, 41] and man [13, 27]. It is clear that delayed hypoperfusion in SAH is associated not only with the constriction of vessels on arteriography, but also with raised intracranial pressure and secondary changes in blood vessels [8]. Cerebral vasospasm has become recognised as having its onset around the 1st week, and resolving gradually over some 3 weeks. There has been recent emphasis on earlier operation for aneurysm since it is fairly clear that delayed hypoperfusion occurs predominantly within the 2nd week.

The last decade has seen considerable interest in the use of calcium channel blockers, notably the dihydropyridine derivative, nimodipine (Nimotop, Bayer AG, Wuppertal, FRG), in patients with aneurysmal SAH. A number of studies have suggested that nimodipine may be effective in preventing the common complication of delayed arterial spasm [2, 3, 7] and possibly in protecting the brain from the resultant ischaemia [15, 29]. Our earlier work on the effects of this calcium antagonist on the physiological responses of the cerebral vasculature and its possible influence on focal cerebral ischaemia suggested, however, that while residual blood flow after middle cerebral artery occlusion was significantly higher with nimodipine than in controls, threshold levels of blood flow for the development of cortical oedema and for disturbances of ionic homeostasis were increased. It suggested that nimodipine interfered with cellular energy metabolism and might increase the susceptibility of tissue to damage [19, 44]. There was also at least a theoretical possibility that the evident cerebral arterial dilatation which such dihydropyridine derivatives produce could increase both the frequency and the severity of recurrent haemorrhages from an untreated aneurysm. Work so far

Gough-Cooper Department of Neurological Surgery, Institute of Neurology, The National Hospital, Queen Square, London WC1N 3BG, U.K.

Cerebral Ischemia and Calcium
Ed. by A. Hartmann and W. Kuschinsky

reported, however [15], would suggest no greater incidence of a recurrent haemorrhage in aneurysm patients receiving nimodipine, although there is less information concerning the severity of such recurrent haemorrhage.

This paper reports the results of a study to test the influence of nimodipine in a dose comparable with that in clinical practice on a model of acute SAH caused by vessel avulsion in the anaesthetized primate, and contrasts them with the results obtained by the use of preloading nimodipine before a pure ischaemic episode as previously reported [19].

Materials and Methods

Each portion of the study involved two groups of 12 baboons of comparable weight (between 6 and 8 kg) which were sedated with ketamine and atropine, then anaesthetized with intravenous thiopentone followed by α-chloralose at a dose of 60 mg/kg. They were immobilised with gallamine triethiodide and respiration was maintained via an endotracheal tube with a Starling pump (C.F. Palmer, High Wycombe, Bucks) on 100% oxygen, the stroke volume being adjusted to maintain arterial PCO_2 between 38 and 42 mmHg.

Systemic PCO_2, PO_2, and pH were monitored with an Astrup blood gas analyser through femoral cannulation and the temperature of the animal was maintained with a thermostatically controlled blanket. A large Ag/AgCl electrode was placed in the animal's mouth to serve as a reference for the hydrogen electrode system. The transorbital approach was used in both groups of animals (ischaemic stroke and SAH). In the "ischaemic" group, the animals were prepared for middle cerebral artery occlusion [25] and thereafter divided into two groups. In an open skull preparation a wide craniectomy was performed over the right cerebral hemisphere, and up to 16 platinum-iridium wires 125 μm in diameter inserted into the cortex to measure local CBF by the hydrogen clearance technique [34]. Extracellular potassium and calcium activity were measured using triple-barrelled double ion selective micro-electrodes [16, 17] and the cortex protected by a warm paraffin pool. In the second group of occlusion experiments local CBF was measured in a closed skull preparation [43] to check the effect of nimodipine on cerebral autoregulatory capacity. The middle cerebral artery was not occluded in this group and extracellular activities were not measured.

In the SAH experiments following a transorbital approach a loop of 50 nylon was placed around a suitably sized intracranial artery on the right side (posterior communicating in four animals, posterior communicating plus anterior choroidal in three, anterior cerebral in four and ophthalmic in one). The ends of the loops were then passed through a length of 1 mm external diameter polythene tubing with one end close to the artery and the other protruding from the orbit, the orbit then being sealed thoroughly with silicone rubber and acrylic bone cement. The protruding end of the tube was clamped to prevent continuing cerebrospinal fluid loss as previously described [26].

In the SAH experiments, the scalp and temporal muscles were excised bilaterally and burr holes made anteriorly for an extradural pressure transducer [10] on each side and frontally for the extradural placement of an Ag/AgCl reference for evoked potential measurement.

A group of hydrogen electrodes in areas A, B, and C [42] were placed in each animal's open or closed skull, and in the SAH animals extracellular potassium and hydrogen ions levels were measured adjacent to the SEP focus on the right side using triple-barrelled ion sensitive microelectrodes [16, 17].

The animals in the SAH group were given an intravenous infusion of nimodipine prepared by dissolving crystalline nimodipine in human lutrol solvent (Bayer U.K. Ltd., Newbury, Berks U.K.) in proportions of 5 mg nimodipine per 5 ml solvent to yield a 0.1% stock solution. Ten millilitres of this solution was diluted 200 ml with physiological saline and the dilute preparation infused into a femoral vein by a slow infusion pump at a rate of 1.2 ml/kg/h, giving a systemic dose of nimodipine of 1.0 μg/kg/min, twice the usual therapeutic dose in man [36]. In the ischaemic group nimodipine was infused intra-arterially, into the lingual artery at 0.2 ml/min for 45 min before the ischaemia and continuously thereafter, the nimodipine dosage being 0.6 μg/kg/min. Control animals received similar infusions of lutrol solvent either intravenously or intra-arterially, but without nimodipine.

The usual preparation precautions were taken to exclude light of less than 450 nm wavelength, and sunlight was avoided during administration.

In each series, the animals were allowed to stabilise for 30 min following electrode implantation; basal blood flows were then determined and in the ischaemic group the effect of nimodipine on basal flow was tested for the next 45 min. CO_2 or autoregulation was then tested to aramine infusion or in the autoregulation group of animals to exsanguination, and the test repeated in the ischaemic experiments after middle cerebral artery occlusion, when immediate and delayed post-occlusion flows were recorded.

In the SAH experiments following basal recordings, a further hydrogen saturation was started and after about 1 min of clearance the intracranial artery was avulsed by traction on the nylon thread to create an SAH, the tube being immediately reclamped to prevent loss of blood or CSF. In this group, changes in CBF, systemic arterial pressure, evoked response, extradural pressure and extracellular ions were then followed for about an hour, during which four further hydrogen clearances were carried out where resumption of blood flow had occurred at 15 and 30 min posthaemorrhage. The response to hypercapnia was tested and a final record made at normocapnia.

In the ischaemic experiments following removal of the brain, the water content in various areas was assessed by the gravimetric method.

Results

Nimodipine increased basal blood flow in the open skull experiments and in the group of animals prepared for SAH, although only in the open skull experiments was this increase significant: it amounted to 25% in the open skull group of animals ($P<0.01$) and 12.4% in the closed skull animals for SAH (not significant).

CO_2 reactivity was significantly decreased in the open skull group of animals, but unchanged in the closed skull group. Autoregulation to reduced blood pressure was impaired; autoregulation to increased blood pressure, however, remained.

Ischaemic Stroke

In the group of animals prepared for ischaemic stroke, cerebral blood flow was significantly increased following middle cerebral artery occlusion, consistent with the demonstration by Ott and Lechner [33] that oral nimodipine caused a redistribution of blood flow in an infarcted hemisphere. This suggested an increase in collateral circulation similar to the mechanism of a protective effect found by workers in the field of myocardial ischaemia [4, 14, 35, 45]. Measurement of water content following an ischaemic incident showed that accumulation of water was significantly greater in the nimodipine-treated animals than in untreated controls [6, 18, 43], although the water content values in treated and untreated animals in non-ischaemic cortex were the same.

Blood flow thresholds for the disruption of potassium and calcium haemostasis [1, 5, 17] were also increased by nimodipine treatment. Such an increase in potassium in ischaemia was almost certainly the result of an increase in potassium permeability associated with an impairment of overload of potassium clearance mechanisms, although since there was no change in potassium or calcium extracellular activity when the drug was infused, it rather suggests an increased susceptibility of cerebral energy metabolism to ischaemic damage following nimodipine infusion.

Subarachnoid Haemorrhage

In the SAH cases, there was of course a marked reduction in blood flow, and increases in extradural pressure and systemic blood pressure similar to those previously reported [28].

Cerebral blood flow was severely affected by the haemorrhage, a large decrease being present in all cortical areas ranging from about 70% to 100% of the initial flow. The overall decrease was slightly greater in the nimodipine-treated animals (88%) than in the control group (84%), but not significantly so. In all experiments, the minimal flow was below 10 ml/100 g/min at least in some areas, and the mean duration of this minimum flow in cortical area A on the right side was slightly lower (0.9 ± 0.64 min) in the nimodipine group than in the untreated animals (1.75 ± 1.58 min), although this was not significant statistically.

No attempt was made to study autoregulation in any other way than by studying the fluctuations of systemic pressure which frequently occurred following SAH. These clearly showed that autoregulation of flow to perfusion pressure was disrupted in all animals after the haemorrhage, a linear relationship between flow and pressure being demonstrated over the perfusion pressure ranges 0–100 mmHg; indeed, in the nimodipine-treated animals there was a reasonable linear relationship with a correlation coefficient of 0.92 ($P < 0.001$), this correlation

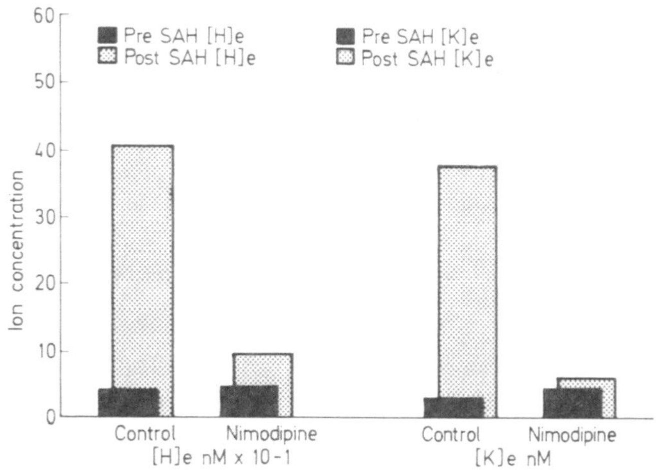

Fig. 1. Effect of SAH on extracellular ions

being much less marked in the control animals without nimodipine infusion ($r = 0.45$, $P < 0.025$). CO_2 reactivity was decreased in all the SAH animals, but slightly more so in the control group, though the difference was not significant.

Ionic activity showed extremely interesting results. Unlike the ischaemic experiments there was clear evidence that nimodipine acted in some protective way following the ischaemia of SAH. Ionic changes after SAH were rapid, with peak levels usually being reached within 3 min, followed by a gradual recovery as flow improved. pHe fell more rapidly than Ke rose, but the peak levels were reached at about the same time. Recovery of pHe also tended to be slower than Ke. Ion changes tended to recover more quickly in the nimodipine-treated animals, though these differences were not significant, but the most notable effect was on the peak ion levels reached. The haemorrhage had a much greater effect in the untreated group, where Ke rose by nearly 20 times as much as in the nimodipine-treated group. Changes in extracellular pH were also significantly larger in the untreated animals than in those receiving nimodipine (Fig. 1).

Changes in extradural pressure were comparable in the two groups.

Discussion

The demonstration in pure ischaemic stroke that nimodipine decreased the autoregulatory capacity to reduced blood pressure is consistent with a general vasodilator effect of the drug. This, together with some protective effect on the collateral circulation, might explain the increased residual blood flow in the ischaemic area. A slight but significant increase in water content in the infarct could, therefore, be explained by an increase in residual blood flow, but a protective effect on pure ischaemic stroke was not brought out by changes in the extracellular ions, where the threshold for ionic haemostasis appeared to be higher, suggesting an increased susceptibility of the brain to ischaemia rather than the reverse. Autoregulation to aramine-induced hypertension was not affected. The positive effect, a doubling of immediate postocclusion flow, would of course tend to maintain flow above threshold levels either of electrical activity loss or of haemostasis in the formation of ischaemic oedema.

In the SAH animals, however, there was more evident appearance of protection. Preloading with nimodipine did not worsen the haemorrhage that occurred in the treated animals, and there was no evidence in these relatively acute experiments of a possibly dangerous rise in the intracranial pressure or fall in perfusion pressure, which has been reported when nimodipine was given to patients with high intracranial pressure [12].

The severity of reduction of CBF in the SAH groups was overwhelming, and nimodipine clearly did not cause cerebral vasodilatation or impede constriction to result in a more severe prolonged haemorrhage from a ruptured vessel. CO_2 reactivity was reduced by about the same proportion whether or not the drug was given. In both groups, the reduction of blood flow was clearly sufficient to cross ionic thresholds, as evidenced by the huge increase in extracellular potassium and hydrogen ion concentration. In untreated animals, profound ion changes occurred after haemorrhage in four out of the five, while in the nimodipine-treated group this happened in only one out of five. There were more minor changes, and more rapid recovery unrelated to differences in depth or duration of ischaemia. The implication of these differences in ion changes is that a protective effect of nimodipine may be due to its acting primarily at a cellular level to minimise the influence of low flow by reducing the adverse effects of ischaemia on ionic haemostasis in SAH.

Although extracellular calcium was not directly measured in the SAH experiments, previous work on ischaemia and hypoglycaemia [17, 21] indicates that inward movement of extracellular calcium occurs only when extracellular potassium is above 13 mm. In four of the five untreated animals potassium reached between 32 and 64 mm, which would almost certainly be accompanied by movements of calcium, but in only one nimodipine-treated animal was there much rise in extracellular potassium, and even then it only reached 12.1 mm. In all others, the peak potassium level was between 2.8 and 5.5 mm, suggesting that the protective action of nimodipine might not be on voltage-dependent calcium channels, but perhaps on energy-dependent channels activated by the short but intense period of ischaemia.

These two groups of experiments, therefore, tend to indicate a difference in the effect of dihydropyridine drugs in pure ischaemia and in the ischaemia associated with SAH, and demonstrate that the detailed neuropharmacology of these compounds is at yet incompletely elucidated.

References

1. Astrup J, Symon L, Branston NM, Lassen NA (1977) Cortical evoked potential and extracellular K^+ and H^+ at critical levels of brain ischaemia. Stroke 8:51–57
2. Auer L (1985) Acute surgical repair of cerebral aneurysms and pharmacological prevention of symptomatic vasospasm improve outcome. In: Meyer JS, Lechner H, Reivich M, Ott EO (eds) Cerebral vascular disease, vol 5. Excerpta Medica, Amsterdam, pp 342–347
3. Auer LM, Brandt L, Ebeling U et al. (1986) Nimodipine and early aneurysm operation in good condition patients. Acta Neurochir (Wien) 82:7–13
4. Bing RJ (1980) Action of anti-anginal drugs on cardiac metabolism. Adv Myocardiol 2:465–476
5. Branston NM, Strong AJ, Symon L (1977) Extracellular potassium activity, evoked potential and tissue blood flow, relationship during progressive ischaemia in baboon cerebral cortex. J Neurol Sci 32:305–321
6. Branston NM, Bell BA, Hunstock A, Symon L (1980) Time and flow as factors in the formation of post-ischaemic oedema in primate cortex. In: Cervos-Navarro T, Ferszt R (eds) Brain edema. Adv Neurol 28:291–298
7. Brawanski A, Gaab MR, Bockhorn J, Haubitz I (1982) A traumatic rCBF measurement: an aid in the timing of surgery and the management of spasm following SAH. Acta Neurochir (Wien) 63:43–51
8. Conway LW, McDonald LW (1972) Structural changes of the intradural arteries following subarachnoid haemorrhage. J Neurosurg 37:715–723
9. Crowell JD, Kaufmann BN (1961) Changes in tissue pH after circulatory arrest. Am J Physiol 200:743–745
10. Dorsch NWC, Stephens RJ, Symon L (1971) An intracranial pressure transducer. Biomed Eng 6:452–457
11. Echlin FA (1942) Vasospasm and focal cerebral ischemia. An experimental study. Arch Neurol Psychiatry 47:77–96
12. Gaab MR, Haubitz I, Brawanski A, Korn A, Czeck TH (1985) Acute effects of nimodipine on the cerebral blood flow and intracranial pressure. Neurochirurgia 28:93–99
13. Gillingham FJ (1958) The management of ruptured intracranial aneurysm. Ann R Coll Surg Engl 23:89–117
14. Gross GJ, Warltier DC, Jolly SR, Hardman HF (1980) Comparative effects of a new calcium antagonist, FR 7534, nitroglycerin, and dipyridamole on regional myocardial blood flow and cardiac contractility during partial coronary artery occlusion in the dog. J Cardiovasc Pharmacol 2:797–813
15. Grotenhuis JA, Bettag W (1986) Prevention of symptomatic vasospasm after SAH by constant venous infusion of nimodipine. Neurol Res 8:243–249
16. Harris RJ, Symon L (1980) A double ion sensitive microelectrode for extracellular cerebral cortical measurements. J Physiol (Lond) 312:3P
17. Harris RJ, Symon L, Branston NM, Bayhan M (1981) Changes in extracellular calcium activity in cerebral ischaemia. J Cereb Blood Flow Metab 1:203–210
18. Harris RJ, Bayhan M, Branston NM, Watson A, Symon L (1982) Modulation of the pathophysiology of primate focal cerebral ischaemia by indomethacin. Stroke 13:17–24
19. Harris RJ, Branston NM, Symon L, Bayhan M, Watson A (1982) The effects of a calcium antagonist, nimodipine, upon physiological responses of the cerebral vasculature and its possible influence upon focal cerebral ischemia. Stroke 13:759–766
20. Harris RJ, Symon L (1984) Extracellular pH, potassium, and calcium activities in progressive ischaemia of rat cortex. J Cereb Blood Flow Metab 4:178–186
21. Harris RJ, Wieloch T, Symon L, Siesjö BK (1984) Cerebral extracellular calcium activity in severe hypoglycaemia: relation to extracellular potassium and energy state. J Cereb Blood Flow Metab 4:187–193
22. Harris RJ, Richards PG, Symon L, Habib AHA, Rosenstein (1987) pH, K^+, and PO_2 of the extracellular space during ischaemia of primate cerebral cortex. J Cereb Blood Flow Metab 7:599–604
23. Harvey J, Rasmussen T (1951) Occlusion of the middle cerebral artery. An experimental study. Arch Neurol Psychiatry 66:20–29

24. Heuser D, Hossmann K-A, Schindler U, Betz E (1975) Changes in extracellular ion activities and brain volume during prolonged cerebral ischaemia and recovery. Pflügers Arch 355 (Suppl 198):R99
25. Hubschmann OR, Kornhauser D (1982) Effect of subarachnoid hemorrhage on the extracellular microenvironment. J Neurosurg 56:216–221
26. Jakubowski J, Bell BA, Symon L, Zawirski MB, Francis DM (1984) A primate model of subarachnoid hemorrhage: change in regional cerebral blood flow, autoregulation, carbon dioxide reactivity, and central conduction time. Stroke 13:601–611
27. Johnson RJ, Potter JM, Reid RG (1958) Arterial spasm in subarachnoid haemorrhage: mechanical considerations. J Neurol Neurosurg Psychiatry 21:68 (abstract)
28. Kamiya K, Kuyama H, Symon L (1983) An experimental study of the acute stage of subarachnoid hemorrhage. J Neurosurg 59:917–924
29. Kazda S, Towart R (1982) Nimodipine: a new calcium antagonist drug with preferential cerebrovascular action. Acta Neurochir (Wien) 63:259–265
30. Kraig RP, Ferreira-Filho CR, Nicholson C (1983) Alkaline and acid transients in cerebellar microenvironment. J Neurophysiol 49:831–850
31. Lende RA (1960) Local spasm in cerebral arteries. J Neurosurg 17:90–103
32. Nemoto EM, Frinak S (1981) Brain tissue pH after global ischemia and barbiturate loading in rats. Stroke 12:77–82
33. Ott E, Lechner H (1980) The influence of nimodipine on cerebral blood flow in patients with subacute cerebral infarction (abstr). In: Betz E, Grote J, Heuser D, Wüllenweber R (eds) Pathophysiology and pharmacotherapy of cerebrovascular disorders. Satellite Symposium of the XXVIII International Congress of Physiological Sciences
34. Pasztor E, Symon L, Dorsch NWC, Branston NM (1973) The hydrogen clearance method in assessment of blood flow in cortex, white matter and deep nuclei of baboons. Stroke 4:556–567
35. Poole-Wilson P (1980) Entry of Ca^{2+} into the ischaemic myocardium. In: Richardson RG (ed) Ca^{++}-antagonism, Abbott Laboratories Ltd, London
36. Raemsch K-D, Ahr G, Tettenborn D, Auer LM (1985) Overview on pharmacokinetics of nimodipine in healthy volunteers and in patients with subarachnoid haemorrhage. Neurochirurgia 28:74–78
37. Rehncrona S, Rosen I, Siesjö BK (1980) Excessive cellular acidosis: an important mechanism of neuronal damage in the brain. Acta Physiol Scand 110:435–437
38. Siemkowicz E, Hansen AJ (1981) Brain extracellular ion composition and EEG activity following 10 minutes ischemia in normo- and hyperglycemic rats. Stroke 12:236–240
39. Siesjö BK (1981) Cell damage in the brain: a speculative synthesis. J Cereb Blood Flow Metab 1:155–185
40. Silver IA (1978) Cellular microenvironment in relation to local blood flow. Ciba Found Symp 56:49–67
41. Symon L (1967) An experimental study of traumatic cerebral vascular spasm. J Neurol Neurosurg Psychiatry 30:497–505
42. Symon L, Pasztor E, Branston NM (1974) The distribution and density of reduced cerebral blood flow following acute middle cerebral artery occlusion: an experimental study by the technique of hydrogen clearance in baboons. Stroke 5:355–364
43. Symon L, Branston NM, Chikovani O (1979) Ischaemic brain oedema following middle cerebral artery occlusion in baboons. Relationship between regional cerebral water content and blood flow at 1–2 h. Stroke 10:184–191
44. Symon L, Harris RJ, Branston NM (1984) Calcium ions and calcium antagonists in ischaemia. Acta Neurochir (Wien) 63:267–275
45. Weishaar R, Ashikawa K, Bing RJ (1979) Effect of diltiazem, a calcium antagonist, on myocardial ischemia. Am J Cardiol 43:1137–1143

In Vivo Fluorometric Measurement of Changes in Cytosolic Free Calcium from the Cat Cortex During Anoxia, Ischemia, and Postischemic Reperfusion

M. REIVICH, D. UEMATSU, and J. H. GREENBERG

Introduction

Cytosolic free calcium concentration ($[Ca^{2+}]_i$) is precisely regulated within specific limits under a variety of conditions in order to maintain proper cellular function. The extracellular free calcium concentration is in the millimolar range and about 10^4 times higher than the intracellular concentration. Recently, calcium has become a focus of attention due to its possible role in mediating anoxic or ischemic cell injury [5, 9, 32]. The increase in $[Ca^{2+}]_i$ may trigger catastrophic enzymatic processes leading to irreversible cell damage. There have been several lines of evidence to indicate that calcium is accumulated in anoxic and ischemic tissue [38] although the causal relationship remains uncertain. In most of the previous studies, the calcium accumulation has been assessed by in vitro histochemistry [16] or estimated from total tissue calcium content by the atomic absorption technique [7]. Extracellular calcium concentration has also been measured by the Ca^{2+} selective electrode [25]. No good correlation exists, however, between the $[Ca^{2+}]_i$ and total cell calcium content, most likely owing to the ability of intracellular organelles such as the endoplasmic reticulum and mitochondria to sequester calcium. In order to clarify the mechanisms of calcium-related cell injury, it is necessary to measure changes in $[Ca^{2+}]_i$ during an anoxic or ischemic insult. Recently, calcium-sensitive fluorescent dyes such as quin-2, fura-2, and indo-1 have been developed [10, 35] and used primarily for measurement of $[Ca^{2+}]_i$ from isolated cells. The present study is designed to assess mean changes in the cytosolic free calcium from the cat brain cortex in situ by using an in vivo fluorometric technique with either quin-2 [36] or indo-1 [37]. We have investigated the changes in $[Ca^{2+}]_i$ in vivo during reversible anoxia and during periods of ischemia and postischemic reperfusion.

Methods

Reversible Anoxia Studies

Adult male cats weighing 2.5–3.6 kg were anesthetized by an intraperitoneal injection of sodium pentobarbital (40 mg/kg). The animals were tracheotomized,

Cerebrovascular Research Center, Department of Neurology, University of Pennsylvania, Philadelphia, Pennsylvania, PA 19104, USA.

Cerebral Ischemia and Calcium
Ed. by A. Hartmann and W. Kuschinsky

and the respiration was controlled within physiological limits using a respirator after immobilization with 5 mg/kg gallamine triethiodide. Polyethylene catheters were inserted into both femoral arteries and one femoral vein for monitoring arterial blood pressure (BP), for analyzing arterial blood gases (pH, pCO_2, and pO_2), and for administration of drugs. The left lingual artery was cannulated so that cerebrovascular hemodilution curves could be obtained by small injections of saline. The rectal temperature was maintained at 37° C by means of a thermostatically controlled heating lamp. A burr hole 13 mm in diameter was drilled over the left middle ectosylvian gyrus and the dura and part of the pia-arachnoid membrane were incised microsurgically to expose the cortical surface. A quartz cranial window equipped with silver EEG electrodes and inlet–outlet tubing was inserted. The space beneath the quartz window was fully filled and superfused continuously during the fluorescence measurement with artificial cerebrospinal fluid (CSF) of the following composition and pH: Na 140 mM, K 2.5 mM, Cl 130 mM, Ca 1.5 mM, HCO_3 14.5 mM, HEPES 10 mM, and pH 7.35 at 37° C.

Quin-2 loading was obtained by superfusing artificial CSF containing 100 μM quin-2 acetoxymethyl ester (quin-2-AM) at a temperature of 37° C continuously over the cortical surface for 60 min. The lipophilic and membrane permeant quin-2-AM dye diffuses into the neuronal cells and is hydrolyzed by the action of cytoplasmic esterase to the membrane impermeant quin-2, which is trapped in the cells as a specific Ca^{2+} indicator [35]. After the loading procedure, the cortical surface was superfused with CSF (without quin-2-AM) at a rate of 5 ml/min for 30 min to wash out the remaining extracellular dye.

Changes in cerebrocortical quin-2-Ca^{2+} fluorescence, NADH fluorescence, and local cortical blood volume (lCBV) were measured by using a microfluororeflectometer [3, 18] as described in detail previously [36]. The illuminated area was approximately 2 mm in diameter and carefully selected so as not to include the large pial vessels.

In order to estimate the contribution of the NADH fluorescence to the total fluorescence signal recorded at 506 nm, the changes in the NADH fluorescence were measured during a period of pure nitrogen inhalation prior to quin-2 loading. The reproducibility of the NADH fluorescence change induced by an identical duration of nitrogen anoxia was examined in six cats, with more than 40 min of recovery between each anoxic interval. After quin-2 loading, the same duration of nitrogen inhalation was repeated in each animal and the net change of the quin-2-Ca^{2+} fluorescence was calculated by subtracting the basal NADH change from the total fluorescence change recorded at 506 nm. The nitrogen inhalation was continued for 120–150 s with a 45 ± 10 s period of isoelectric ECoG in each animal.

Ischemia and Postischemic Reperfusion Studies

Adult male cats weighing 2.4–3.7 kg were anesthetized by inhalation of halothane, immobilized with gallamine triethiodide (3 mg/kg i.v.), and their respiration controlled by the use of a ventilator. A quartz cranial window was inserted over the left middle ectosylvian gyrus. Indo-1 loading was obtained by superfusing artificial CSF containing 7 μM indo-1 acetoxymethyl ester at a temperature of 37° C over the cortical surface for 2 h. During the optical measurements, the cortical surface was superfused with artificial CSF. Fluorescence and reflectance signals from the cortex including changes in indo-1-Ca^{2+} fluorescence, NAD/NADH redox state, and lCBV were measured microfluorometrically as described in detail previously [37]. The left middle cerebral artery (MCA) was exposed microsurgically via a transorbital approach using a modified O'Brien and Waltz technique [26]. Following indo-1 loading, the exposed left MCA trunk was occluded in 14 animals with a miniature Mayfield clip for 1 h. In five animals, a sham insult was produced by lightly touching the MCA with a glass rod. After 1 h of MCA occlusion, the clip was removed and the fluorescence and reflectance signals continued to be monitored over the next 30 min.

Results

Reversible Anoxia

The reproducibility of the NADH fluorescence change at the time of initiation of isoelectric ECoG induced by repeated anoxic insults of the same duration was characterized in six animals. The correlation coefficient for repeated anoxic insults was 0.998, indicating that an almost identical increase in the NADH fluorescence occurs during the repeated anoxic episodes. We used the initiation of isoelectric ECoG point (IP) as a standard point for alignment and subtraction of the basal NADH fluorescence change from the total fluorescence change induced by nitrogen anoxia after the quin-2 loading.

The time course of the net quin-2-Ca^{2+} fluorescence change calculated by subtracting the NADH fluorescence change from the total fluorescence change is shown in Fig. 1 ($n = 6$). The quin-2-Ca^{2+} fluorescence increased significantly by 48.0 ± 13.4 A.U. (arbitrary units) ($P < 0.05$) at 20 s prior to the IP, and remained increased by 54.5 ± 11.0 A.U. ($P < 0.02$) at 10 s prior to the IP and by 59.0 ± 7.4 A.U. ($P < 0.01$) at the IP. It remained significantly increased during nitrogen inhalation, but then decreased slightly following return to room air inhalation although it remained elevated until 7.3 ± 1.7 s prior to the point of reappearance of any continuous ECoG activity (RP), after which it decreased steeply to the control level. The RP was preceded by a sharp decline of quin-2-Ca^{2+} fluorescence in all animals. The physiological data including PaO_2, $PaCO_2$, pH, and plasma glucose values are summarized in Table 1. The decrease in PaO_2 was comparable at the end of each anoxic cycle before and after quin-2 loading.

Ischemia and Postischemic Reperfusion

Physiological data from each group including mean arterial blood pressure, PaO_2, $PaCO_2$, and blood glucose concentration did not differ significantly among the groups (Table 2).

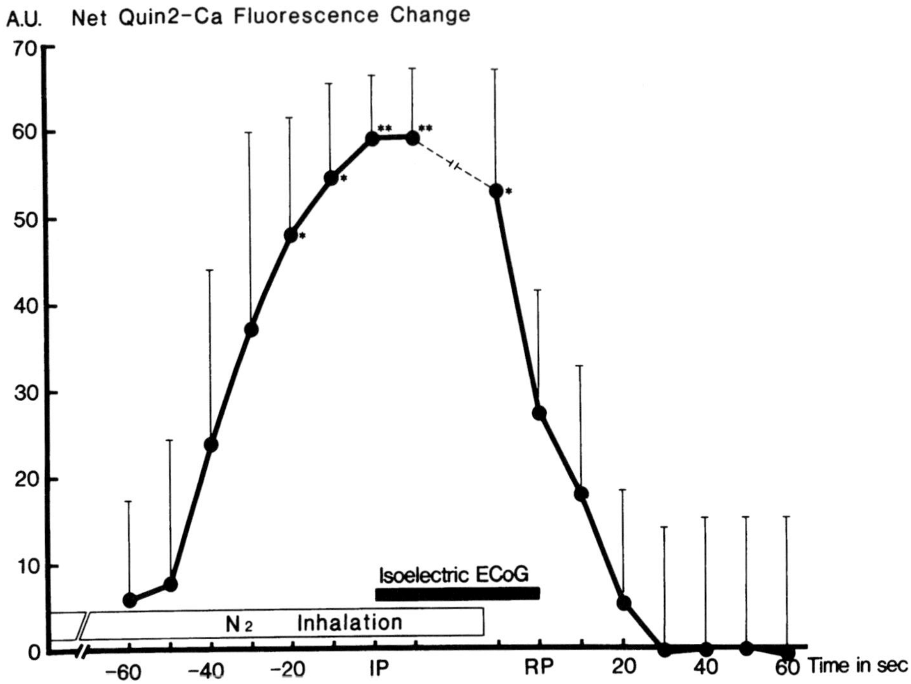

Fig. 1. Time course of the net quin-2-Ca^{2+} fluorescence change calculated by subtracting the NADH fluorescence change from the total fluorescence change. The isoelectric ECoG point (*IP*) was regarded as one standard point for subtraction during anoxia. A significant increase in net quin-2 fluorescence was seen 20 s prior to IP and remained during nitrogen inhalation. The ECoG recovery point (*RP*) was also regarded as a standard point during the recovery period after reoxygenation. Quin-2-Ca^{2+} fluorescence remained at a significant high level at 10 s prior to RP and then decreased steeply preceding the recovery of ECoG activity. *A.U.*, arbitrary unit. Each value is mean ± SEM ($n=6$). *, $P<0.05$; *, $P<0.01$. Uematsu et al. [46]

Table 1. Physiological parameters: anoxia study

	PaO_2 (torr)	$PaCO_2$ (torr)	pH	Glucose (mg/dl)
Control	134 ± 7	31 ± 1	7.31 ± 0.02	112 ± 3
Anoxia				
Before quin-2	13 ± 3	25 ± 3	7.37 ± 0.03	–
After quin-2	14 ± 3	27 ± 3	7.39 ± 0.02	–

Blood samples were taken during the control period and at the end of anoxia before and after quin-2 leading. Values are mean ± SEM.

Table 2. Physiological data: ischemia and reperfusion study

Variables	Sham ($n=5$)	Mild ($n=4$)	Moderate ($n=3$)	Severe ($n=7$)
MABP (mmHg)				
Control	84±10	89± 6	77± 5	79± 5
Occlusion 60 min	89± 9	90± 8	78± 4	78± 5
Reperfusion 30 min	87± 8	78± 5	77± 5	79± 4
PaO_2 (mmHg)				
Control	133±10	121±10	120± 4	134±11
Occlusion 60 min	131± 8	122± 7	120± 4	135±10
$PaCO_2$ (mmHg)				
Control	34± 1	35± 2	35± 3	33± 2
Occlusion 60 min	35± 2	35± 3	34± 2	34± 1
Glucose (mg/dl)				
Control	102± 9	117±11	130±11	116± 9

MABP, mean arterial blood pressure; PaO_2, arterial oxygen tension; $PaCO_2$, arterial carbon dioxide tension. Values are mean±SEM.

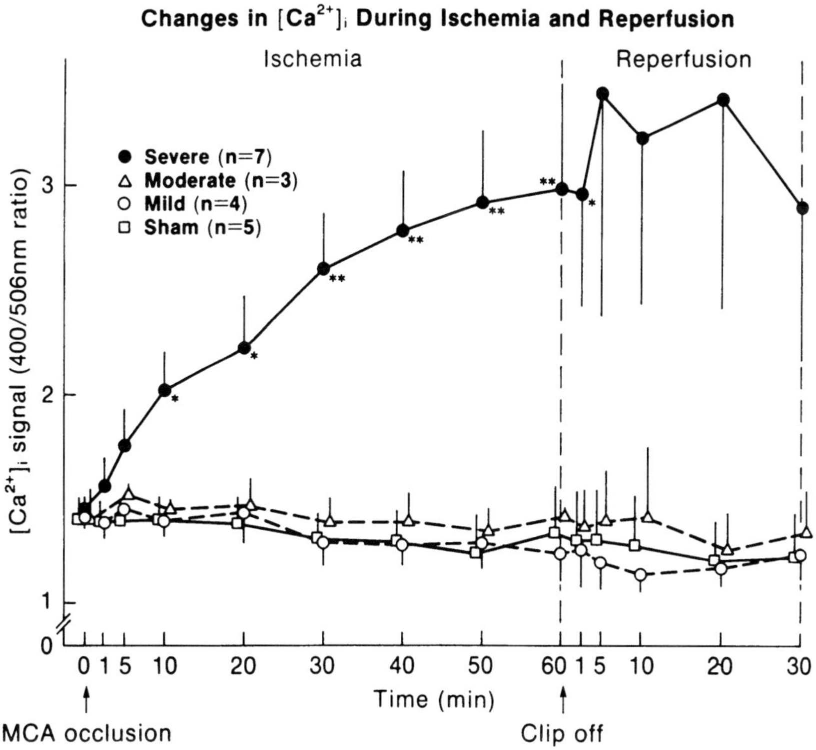

Fig. 2. Time course of changes in $[Ca^{2+}]_i$ in the ischemic and reperfusion periods. The $[Ca^{2+}]_i$ signal increased significantly only in the severe groups, with the profiles of $[Ca^{2+}]_i$ changes in the moderate and mild groups being very similar to that of the sham-operated group. *, $P<0.05$; *, $P<0.01$. (Modified from Uematsu et al. [37]

The animals were divided into three groups according to the reduction of mean EEG amplitude 30 min after the MCA occlusion. Animals in which the EEG amplitude fell to less than 20% of control formed the severe stroke group ($n=7$), animals in which the EEG amplitude was between 20% and 70% of control constituted the moderate stroke group ($n=3$), while animals with an EEG amplitude greater than 70% of control were the mild stroke group ($n=4$). The severe cases were further subdivided into a subgroup with 10% or more recovery of EEG amplitude during 30 min of reperfusion (type 1, $n=4$) and a subgroup with less than 10% recovery (type 2, $n=3$). The mean recovery of EEG amplitude in the type 1 and type 2 groups following the reperfusion period was $16.0 \pm 2.1\%$ and $2.3 \pm 1.5\%$, respectively ($P<0.01$).

There was no significant difference in the mean baseline $[Ca^{2+}]_i$ signal among the groups. Throughout the ischemic and reperfusion periods, the $[Ca^{2+}]_i$ signal increased significantly only in the severe groups, with the profiles of $[Ca^{2+}]_i$ changes in the moderate and mild groups being very similar to that of the sham-operated group (Fig. 2). In the combined severe group (types 1 and 2), the $[Ca^{2+}]_i$ signal started to increase significantly 10 min following MCA occlusion [control: 1.46 ± 0.05; 10 min: 2.02 ± 0.19 ($P<0.05$); 60 min: 2.99 ± 0.37 ($P<0.01$)]. The type 1 and type 2 severe subgroups exhibited distinctly different $[Ca^{2+}]_i$ profiles.

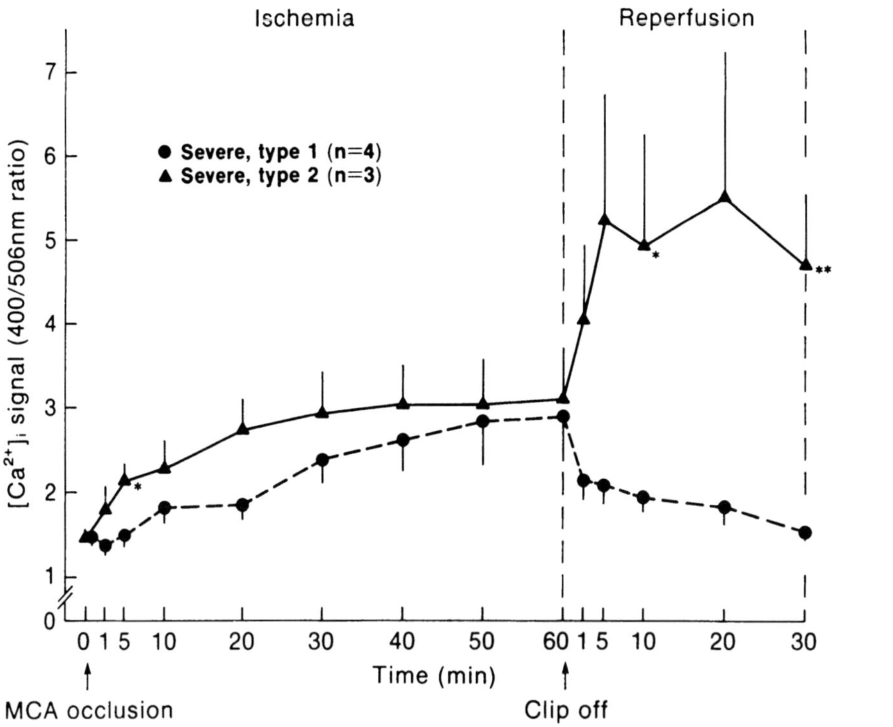

Fig. 3. Time course of changes in $[Ca^{2+}]_i$ in the type 1 and type 2 severe groups. During MCA occlusion, $[Ca^{2+}]_i$ increased earlier and to a greater degree in the type 2 than in the type 1 group. Following reperfusion the type 2 group exhibited a further increase of $[Ca^{2+}]_i$ while the type 1 group showed a decrease. *, $P<0.05$; *, $P<0.01$. (Modified from Uematsu et al. [37])

During MCA occlusion, $[Ca^{2+}]_i$ increased earlier and to a greater degree in type 2 than in type 1 [2.15 ± 0.20 vs 1.48 ± 0.12 at 5 min ($P < 0.05$) and 2.74 ± 0.37 vs 1.86 ± 0.18 at 20 min ($P < 0.05$)]. Following reperfusion, the $[Ca^{2+}]_i$ changes in the type 1 and type 2 subgroups were in opposite directions. Type 2 animals exhibited either a further increase or only slight recovery of $[Ca^{2+}]_i$, while type 1 animals consistently showed $[Ca^{2+}]_i$ recovery. The difference in the $[Ca^{2+}]_i$ signal between those subgroups was statistically significant during the reperfusion period (Fig. 3).

In the severe group, the NADH fluorescence increased significantly for the first 10 min of ischemia, followed by a gradual decrease. Following reperfusion, the NAD/NADH redox state became rapidly hyperoxidized and then by 20 min recovered to the control level. In the moderate group, the NADH fluorescence continued to increase throughout the ischemic and reperfusion periods, while in the mild group, it increased gradually during ischemia and reperfusion, increasing only slightly more than in the sham-operated group. These profiles of the NADH changes are very similar to the previous results from our laboratory [8].

In the severe group, lCBV decreased gradually during MCA occlusion with a tendency toward recovery during reperfusion; it remained, however, at a lower level than in the sham-operated group. The moderate and mild animals showed no significant changes in lCBV during MCA occlusion.

Discussion

It has been reported that during anoxia massive Ca^{2+} influx into cells occurs concomitant with the anoxic depolarization in which the overall ionic permeability of the plasma membrane increases [25]. The depolarization of the plasma membrane is usually preceded by 2–5 min by an isoelectric ECoG [11]. The present results, however, indicate that $[Ca^{2+}]_i$ has already increased significantly 20 s prior to the initiation of an isoelectric ECoG. Possible explanations include:

1. Ca^{2+} channels could be opened prior to the depolarization of the plasma membrane. Receptor-operated Ca^{2+} channels become permeable in response to neurotransmitters such as *N*-methyl-D-aspartate (NMDA) [21]. It has been reported that aspartate and glutamate [14] are released in the brain under anoxic conditions. It has also been reported that extracellular Ca^{2+} activity decreases during application of excitatory animo acids to the cat cortex [15]. It is possible, therefore, that postsynaptic Ca^{2+} channels may be opened by the action of these transmitters during anoxia.

2. Calcium sequestration by the intracellular organelles is impaired during anoxia. The endoplasmic reticulum (ER) can sequester Ca^{2+} by an ATP-dependent mechanism contributing to the fine adjustment of $[Ca^{2+}]_i$ under normal physiological conditions. During anoxia, however, the shortage of ATP might prevent the ER from sequestering Ca^{2+}. The mitochondria can also accumulate Ca^{2+} via an electrophoretic Ca^{2+} uniporter, which requires energy [1]. Cellular acidosis during anoxia may also contribute to the increase in $[Ca^{2+}]_i$ via the Ca^{2+}/H^+ antiporter system in the mitochondria [23].

3. The Ca^{2+} extrusion mechanism of the plasma membrane such as the Ca^{2+}/Na^{2+} antiport system and/or the Ca^{2+}-dependent ATPase (Ca^{2+} pump) is impaired during anoxia [19].

How is the increase in $[Ca^{2+}]_i$ linked to the arrest of ECoG activity? The disappearance of electrical activity from the cortex has been explained by a hyperpolarization accompanied by a reduction in membrane resistance due to a selective increase in K^+ conductance occurring during the early stage of anoxia [11]. The K^+ conductance of the plasma membrane appears to be partially regulated by $[Ca^{2+}]_i$, whereby a rise in $[Ca^{2+}]_i$ opens the Ca^{2+}-sensitive K^+ channels [20, 22]. The early increase in $[Ca^{2+}]_i$ seen in this study is consistent with this hypothesis.

The data from the ischemia studies indicate that with the model used in these studies the intracellular free calcium increases only in the severe stroke cases during ischemia, followed by either a recovery or a further increase in $[Ca^{2+}]_i$ during the reperfusion period. The former cases exhibited a greater recovery in EEG during the reperfusion period than the latter. By contrast, the $[Ca^{2+}]_i$ never increased in the mild or moderate stroke cases throughout the ischemic and reperfusion periods.

Previous studies from our laboratory have indicated that MCA occlusion induces various degrees of EEG deterioration with or without recovery during reperfusion [33] and that the EEG changes observed at 30 min following the occlusion correlate well with the histopathological ischemic changes [34]. The present study may provide a possible explanation for the variable severity and irreversibility in this stroke model. The level of $[Ca^{2+}]_i$ may play an important role not only in the severity of injury during ischemia but also in the recovery during the reperfusion period.

In the severe cases $[Ca^{2+}]_i$ started increasing significantly 10 min following ischemia. This time course is in good agreement with other data in the literature in which the extracellular free calcium concentration measured with an ion-selective electrode began to decrease steeply 10–15 min following initiation of cerebral ischemia in baboons and rats [12, 13]. It is likely that the extracellular calcium ion enters the neurons through voltage-dependent calcium channels during the depolarization of the plasma membrane [17]. Calcium ion can also enter directly through the NMDA-operated channels during ischemia [15, 21]. To counter this Ca^{2+} influx, the neuronal plasma membrane has two different mechanisms for Ca^{2+} extrusion, an ATP-dependent calcium pump and an Na^+-Ca^{2+} exchange system [31]. Severe ischemia will disturb these mechanisms, resulting in an elevation of $[Ca^{2+}]_i$.

Following reperfusion, the $[Ca^{2+}]_i$ signal in the severe animals behaved in two different ways. The type 1 animals showed a noticeable recovery, while the type 2 animals exhibited either a further increase or only a little recovery of $[Ca^{2+}]_i$. The recovery of $[Ca^{2+}]_i$ following reperfusion in the type 1 severe animals may depend on the reversibility of plasma membrane and/or mitochondrial dysfunction produced by the ischemic insult. If these homeostatic mechanisms are irreversibly damaged or their recovery delayed, a further influx of Ca^{2+} into the neurons can be expected during the reperfusion period, since the delivery of Ca^{2+} to the postischemic tissue will be enhanced by the recovery of flow and the disruption of the blood-brain barrier. The massive influx of Ca^{2+} during reperfusion has recently been considered to be injurious to the postischemic cells, triggering irreversible cascades including the generation of oxygen free radicals and the liberation of free fatty acids (reperfusion injury) [2, 6, 9, 27, 30].

The secondary oxidation of NADH observed in the severe stroke group has been explained by an insufficient supply of reducing equivalents into the mitochondria and/or an accumulation of oxidizing substances. The present study suggests another possibility, i.e., the secondary oxidation might be due to increased energy consumption secondary to mitochondrial Ca^{2+} uptake [4, 29, 39]. The excessive Ca^{2+} accumulation in the mitochondria not only causes an inhibition of oxidative phosphorylation [28] but also is injurious to the mitochondria themselves.

In conclusion, we have demonstrated that during reversible anoxia, the arrest and recovery of ECoG are preceded by an increase and recovery of intracellular free calcium, respectively. This observation suggests the possibility that an increase in intracellular free calcium may play a role in the electrical failure during anoxia. Likewise, the present results suggest that the level of intracellular free calcium plays an important role in the progression and recovery of ischemic brain damage.

Acknowledgments. This work was supported by grant from the National Institute of Health NS 10939-15. The authors gratefully acknowledge the expert secretarial assistance of Ms. Emily V. Daniel and Ms. Patricia Galagher.

References

1. Bygrave FL, Reinhart PH, Taylor WM (1985) Mitochondrial calcium fluxes and their role in the regulation of intracellular calcium. In: Marme D (ed) Calcium and cell physiology. Springer, Berlin Heidelberg New York, pp 94–104
2. Chan PH, Fishman RA (1980) Transient formation of superoxide radicals in polyunsaturated fatty acid-induced brain swelling. J Neurochem 35:1004–1007
3. Chance B, Cohen P, Jobsis F, Schoener B (1962) Intracellular oxidation-reduction states in vivo. Science 137:499–508
4. Chance B (1965) The energy-linked reaction of calcium with mitochondria. J Biol Chem 240:2729–2748
5. Cheung JY, Bonventre JV, Malis CD, Leaf A (1986) Calcium and ischemic injury. N Engl J Med 314:1670–1676
6. Demopoulos HV, Falmm ES, Pietronigro DD (1980) The free radical pathology and the microcirculation in the major central nervous system disorders. Acta Physiol Scand [Suppl] 492:91–119
7. Deshpande JK, Siesjo BK, Wieloch T (1987) Calcium accumulation and neuronal damage in the rat hippocampus following cerebral ischemia. J Cereb Blood Flow Metab 7:89–95
8. Dora E, Tanaka K, Greenberg JH, Gonatas NH, Reivich M (1986) Kinetics of microcirculatory, NAD/NADH, and electrocorticographic changes in cat brain cortex during ischemia and recirculation. Ann Neurol 19:546–544
9. Farber JL (1981) The role of calcium in cell death. Life Sci 29:1289–1295
10. Grynkiewicz G, Poenie M, Tsien RY (1985) A new generation of Ca^{2+} indicators with greatly improved fluorescence properties. J Biol Chem 260:3440–3450
11. Hansen AJ (1984) Ion and membrane changes in the brain during anoxia. Behav Brain Res 14:93–98

12. Harris RJ, Symon L, Branston NM, Bayhan M (1981) Changes in extracellular calcium activity in cerebral ischemia. J Cereb Blood Flow Metab 1:203–209
13. Harris RJ, Symon L (1984) Extracellular pH, potassium and calcium activities in progressive ischemia of rat cortex. J Cereb Blood Flow Metab 4:178–186
14. Hauptman M, Nelson D, Wilson DF, Erecinska (1984) Neurotransmitter amino acids in the CNS: II Some changes in amino acid levels in rat brain synaptosomes during and after in vitro anoxia and simulated ischemia. Brain Res 304:23–35
15. Heinemann U, Pumain R (1980) Extracellular calcium activity changes in cat sensorimotor cortex induced by iontophoretic application of aminoacids. Exp Brain Res 40:247–250
16. Hossmann K-A, Ophoff BG, Schmidt-Kastner R, Oschlies U (1985) Mitochondrial calcium sequestration in cortical and hippocampal neurons after prolonged ischemia of the cat brain. Acta Neuropathol (Berlin) 68:230–238
17. Hsen AJ (1985) Effect of anoxia on ion distribution in the brain. Physiol Rev 65:101–148
18. Jobsis FF, O'Connor M, Vitale A, Vreman H (1971) Intracellular redox changes in functioning cerebral cortex. I. Metabolic effects of epileptiform activity. J Neurophysiol 34:735–749
19. Kass IS, Lipton P (1986) Calcium and long-term transmission damage following anoxia in dentate gyrus and CA1 regions of the rat hippocampal slice. J Physiol (Lond) 378:313–334
20. Krnjevic K, Lisiewicz A (1972) Injections of calcium ions into spinal motoneurons. J Physiol (Lond) 225:363–390
21. MacDermott AB, Mayer ML, Westbrook GL, Smith SJ, Barker JL (1986) NMDA-receptor activation increases cytoplasmic calcium concentration in cultured spinal cord neurons. Nature 321:519–522
22. Meech RW (1972) Intracellular calcium injection causes increased potassium conductance in aplasia nerve cells. Comp Biochem Physiol 42A:493–499
23. Meech RW, Thomas RC (1980) Effect of measured calcium chloride injections on the membrane potential and internal pH of snail neurones. J Physiol (Lond) 298:111–129
24. Miller RJ (1987) Multiple calcium channels and neuronal function. Science 235:46–52
25. Nicholson C, Ten Bruggencate G, Steinberg R, Stockle H (1977) Calcium modulation in brain extracellular microenvironment demonstrated with ion-selective micropipette. Proc Natl Acad Sci USA 74:1287–1290
26. O'Brien MD, Waltz AG (1973) Transorbital approach for occluding the middle cerebral artery without craniectomy. Stroke 4:201–206
27. Rehncrona S, Westerberg E, Akesson B, Siesjö BK (1982) Brain cortical fatty acids and phospholipids during and following complete and severe incomplete ischemia. J Neurochem 38:84–93
28. Roman I, Clark A, Swanson PD (1981) The interaction of calcium transport and ADP phosphorylation in the brain mitochondria. Membr Biochem 4:1–9
29. Rossi CS, Lehninger AL (1964) Stoichiometry of respiratory stimulation, accumulation of Ca^{++} and phosphate, and oxidative phosphorylation in rat liver mitochondria. J Biol Chem 239:3871–3980
30. Safar P (1986) Cerebral resuscitation after cardiac arrest: a review. Circulation 74 [Suppl 4]:138–153
31. Schatzmann HJ (1985) Calcium extrusion across the plasma membrane by the calcium-pump and the Ca^{2+}-Na^{2} exchange system. In: Marme D (ed) Calcium and cell physiology. Springer, Berlin Heidelberg New York, pp 18–52
32. Siesjö BK (1981) Cell damage in the brain: a speculative synthesis. J Cereb Blood Flow Metab 1:155–185
33. Tanaka K, Greenberg JH, Gonatas NK, Reivich M (1985) Regional flow-metabolism couple following middle cerebral artery occlusion in cats. J Cereb Blood Flow Metab 5:241–252
34. Tanaka K, Dora E, Urbanics R, Greenberg JH, Toffano G, Reivich M (1986) Effect of the ganglioside GM1, on cerebral metabolism, microcirculation, recovery kinetics of ECoG and histology, during the recovery period following focal ischemia in cats. Stroke 17:1170–1178
35. Tsien RY, Pozzan T, Rink TJ (1982) Calcium homeostasis in intact lymphocytes: cytoplasmic free calcium monitored with a new, intracellularly trapped fluorescent indicator. J Cell Biol 94:325–334

36. Uematsu D, Greenberg JH, Revich M, Kobayashi S, Karp A (1988) In vivo fluormetric measurement of changes in cytosolic free calcium from the cat cortex during anoxia. J Cereb Blood Flow Metab 8:367–374
37. Uematsu D, Greenberg JH, Revich M, Karp A (1988) In vivo measurement of cytosolic free calcium during cerebral ischemia and reperfusion. Ann Neurol
38. Van Reempts J (1984) The hypoxic brain: histological and ultrastructural aspects. Behav Brain Res 14:99–108
39. Vercesi A, Reynafarje B, Lehninger AL (1978) Stoichiometry of H^+ ejection and Ca^+ uptake coupled to electron transport in rat heart mitochondria. J Biol Chem 253:6379–6385

Calcium Antagonists and a Rapid Shift of Water from Extracellular to Intracellular Space After Cerebral Ischemia

Y. Shinohara, M. Yamamoto, M. Haida, and R. Taniguchi

In many respects calcium plays a key role in the central nervous system. Calcium antagonists may serve as protective agents in cerebral hypoxia or ischemia and may be useful in the treatment of cerebrovascular disorders. Although the efficacy of calcium antagonists has already been reported in various models of brain hypoxia or ischemia, the details of the mechanism are still a matter of debate. Among the many possible methods for detecting changes of the brain tissue after exposure to a hypoxic or ischemic insult, nuclear magnetic resonance (NMR) is particularly suitable for detecting ischemic brain edema or a shift of water from extracellular to intracellular space, because of its high sensitivity to the behavior of water molecules. The purpose of the work described here was to obtain information regarding the effect of calcium antagonists on the time course of the shift of water from extracellular to intracellular space after ischemia by means of NMR.

Materials and Methods

The materials used in this study were 6- to 10-week-old male Wistar rats, weighing 200–280 g. Under light anesthesia induced by intraperitoneal urethan (500 mg/kg body weight) and chloralose (50 mg/kg body weight), a small burr hole was drilled in the right frontoparietal area of the rat skull. A brain tissue sample, 2 mm in thickness, which was mainly gray matter, was punched out with a sharp-edged Pyrex glass tube, which was sealed tightly with a waterproof material, Hematoseal, and placed in an NMR spectrometer with a special sample holder. This sample can be regarded as a highly simplified model of brain ischemia. This method is advantageous because the total water content in the sample is constant during measurement (the sampled brain tissue was sealed completely).

NMR measurements were performed by using 100-MHz Fourier-transform NMR equipment (Nihon Denshi JEOL PFT 100) or 90-MHz FT NMR equipment (JEOL FX 90A) at 25° C, and were started within 1 min after samples were obtained. The values of the longitudinal relaxation time, T_1, and the transverse relaxation time, T_2, were measured by using the inversion-recovery method and Hahn's spin-echo method, respectively. However, in this study we mainly made

Department of Neurology, Tokai University School of Medicine, Isehara, Kanagawa 259-11, Japan.

Cerebral Ischemia and Calcium
Ed. by A. Hartmann and W. Kuschinsky

use of T_2, since T_2 is believed to be more sensitive to the environmental changes around water molecules than T_1 [1]. The magnetic field was stabilized by means of a D_2O external lock. The water signal intensity was determined in the frequency spectra obtained by Fourier transformation at intervals of 8 s. As we have already shown [2], the transverse magnetization decay curve of the rat brain obtained from nine different echo times (0, 80, 160, 240, 320, 400, 480, 560, and 640 ms) can be explained in terms of two components, and the fast component of T_2 (T_{2f}) corresponds to the signals from the water of the intracellular space. To estimate one T_{2f} value, four echo times (40, 80, 160, and 240 ms) were used. Then, the average value of four T_{2f} was considered as mean T_{2f} during 2 min. Measurements were continued for approximately 60 min. Details of the method will be described elsewhere [3].

Calcium antagonists used in this study were nimodipine, NC-1100 [1-(3,4-dimethoxyphenyl)-2-(4-diphenylmethylpiperazinyl)ethanol dihydrochloride], and nicardipine. These drugs were injected intravenously for 2 min, and 2 min after administration, the brain was sampled.

Results and Comments

The Basic Experiments

Figure 1 shows an example of the T_{2f} changes in control rat brain after total ischemia. The value of T_{2f} increases with time, suggesting a transfer of water from extracellular space to intracellular space because the total amount of water in the sealed tissue sample should be constant during the measurement. This change was gradual and the maximum increase was 14% within 20 min after total ischemia.

The obtained value of $T_{2f}(t)$ was fitted to the following equation by the least-squares method:

$$T_{2f}(t) = \Delta T_{2f}(1\text{-}\exp(-t/k)) + T_{2f(0)},$$

where ΔT_{2f} is the increment of T_{2f}, k is the time constant, and $T_{2f(0)}$ is the T_{2f} value at the time of sampling. The time constant, k, may be interpreted as a pa-

CALCULATION OF PARAMETERS

$T_{2f}(t) = \Delta T_{2f}(1 - e^{(-t/k)}) + T_{2f(0)}$

$T_{2f(0)}$: Initial T_{2f}(msec)

$T_{2f\,Max}$: Maximum T_{2f}(msec)

k : Time constant (min)

$\%\Delta T_{2f} : \frac{T_{2f\,Max} - T_{2f(0)}}{T_{2f(0)}} \times 100$

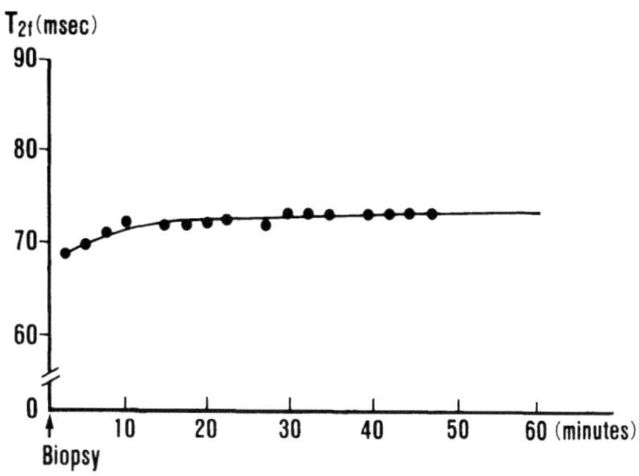

Fig. 1. The time course of T_{2f} in 10-week-old Wistar rat brain at room temperature

rameter of the membrane permeability and/or the driving force which causes cell swelling due to the change of intracellular water content. The time course of changes of the fast component of T_{2f} was also in good agreement with that of the impedance change of the brain tissue after complete cardiac arrest, which has been explained by Hossmann [4] in terms of a shift of the water molecules from extracellular to intracellular spaces.

Bearing these results in mind, the effects of calcium antagonists on the membrane permeability of brain cells were investigated.

Effects of Calcium Entry Blockers

Effect of Nimodipine

Table 1 shows the effect of nimodipine, 0.5 mg/kg, on various NMR parameters. As nimodipine solution for injection contains 15% ethanol, the effect of nimodipine ($n=5$) was compared with that of 15% ethanol (placebo group, $n=6$) to-

Table 1. $T_{2f(0)}$, T_{2f}, $T_{2f\,max}$, and time constant (k) in control ($n=10$), nimodipine placebo ($n=6$), and nimodipine ($n=5$) groups

	$T_{2f(0)}$ (msec)	ΔT_{2f} (msec)	$T_{2f\,max}$ (msec)	k (min)
Control ($n=10$)	63.1±2.6	7.6±2.8	70.0±3.0	9.9±3.5
Placebo ($n=6$)	68.5±2.3	7.4±2.6	75.9±4.3	13.1±4.4
Nimodipine ($n=5$)	66.1±2.9	6.9±4.8	72.4±3.3	19.5±3.3

$T_{2f\,max}=T_{2f(0)}+\Delta T_{2f}$. Mean±SD.

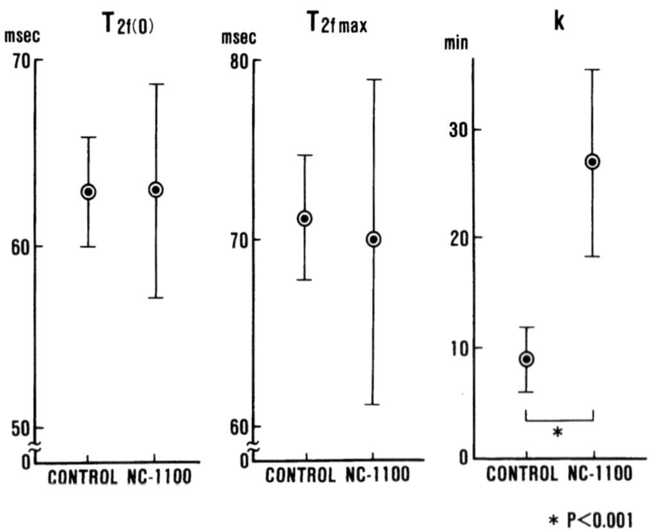

Fig. 2. $T_{2f(0)}$, T_{2max}, and time constant (k) in control ($n=8$) and NC-1100 ($n=6$) groups

gether with controls ($n=10$). As can be seen in Table 1, there were no essential differences in $T_{2f(0)}$, ΔT_{2f}, or T_{2fmax}, where T_{2fmax} is the value of $T_{2f(0)}$ plus ΔT_{2f}, between controls and nimodipine, but the difference of k was statistically significant in these three groups (between control and nimodipine $P<0.01$, between placebo and nimodipine $P<0.05$), suggesting that the preadministration of nimodipine before brain biopsy can delay a shift of water from extracellular to intracellular space after brain biopsy.

Effect of NC-1100

Figure 2 shows the results obtained from control brain ($n=8$) and from brain after NC-1100, 2.5 mg/kg, administration ($n=6$). Again, prolongation of the time constant k was clearly demonstrated (control 8.9 ± 3.0 min and NC-1100 26.7 ± 8.6 min) ($P<0.001$).

Effect of Nicardipine

Administration of nicardipine also seemed to elongate slightly the time constant k [control 15.4 min and nicardipine 17.3 ± 1.8 min ($n=4$)]. But in this study, most of the rats were 6 weeks old (younger than in the other studies); such rats usually have a longer k value, and this made interpretation of the results a little difficult.

Comments and Conclusions

Elongation of T_{2f} can be explained not only by an increase of water contents of the intracellular space but also by changes of the temperature, pH, protein concentration, electrolytes, and so on. However, the temperature was constant at 25° C during measurement and the changes of pH towards acidosis and increased concentration of protein should decrease the T_2 value, if they have any effect, according to our previous experiments. Therefore, it is concluded that the administration of calcium entry blockers can delay a shift of water from extracellular to intracellular space after brain biopsy, which we have employed as a highly simplified model of ischemia. Although we did not measure the intracellular calcium concentration, this result suggests that these drugs may delay the progression of cytotoxic edema due to brain hypoxia. The effects of the drugs investigated were in the order of NC-1100 2.5 mg/kg (most effective), nimodipine 0.5 mg/kg, and nicardipine 1.0 mg/kg (least effective).

References

1. Farrar TC, Becker ED (1971) Pulse and Fourier transform NMR. Academic, New York, pp 46–52
2. Haida M, Yamamoto M, Matsumura H, Shinohara Y, Fukuzaki M (1987) Intracellular and extracellular spaces of normal adult rat brain determined from the proton nuclear magnetic resonance relaxation times. J Cereb Blood Flow Metab 7:552–556
3. Haida M, Shinohara Y, Yamamoto M, Taniguchi R, Ohsuga H, Fukuzaki M (1988) A rapid shift of water from extracellular to intracellular space after total ischemia detected by NMR (submitted)
4. Hossmann KA (1971) Cortical steady potential, impedance and excitability changes during and after total ischemia of cat brain. Exp Neurol 32:163–175

A Fluorometric Technique for the In Vivo Measurement of Intracellular Calcium

J. H. Greenberg, D. Uematsu, and M. Reivich

There is a growing body of evidence indicating that substantial increases in intracellular calcium are harmful to the cell, contributing to cellular damage in a variety of pathological conditions [1, 4]. An increase in intracellular free calcium can activate calcium-dependent phospholipases which can damage plasma membranes and can lead to the production of free fatty acids and lysophospholipids which are also toxic to the cell [4]. In addition, elevation of intracellular free calcium stimulates calcium-dependent ATPase, increasing the need for ATP. If ATP is in short supply, such as occurs in ischemia, this extra demand for ATP can lead to further cellular energy depletion.

Intracellular calcium exists in primarily two forms – a small part is cytosolic free calcium, whereas the majority exists as calcium that is bound to intracellular organelles, primarily the mitochondria and the endoplasmic reticulum, as well as to the plasma membrane and to Ca^{2+} binding proteins. It is the cytosolic free calcium that has been implicated in the chain of deleterious events that leads to cell damage. The evidence for an increase in cytosolic free calcium in such pathological conditions as cerebral ischemia, hypoxia, and hypoglycemia for the most part has been circumstantial since intracellular calcium is quite difficult to measure directly. The observation that extracellular calcium decreases in ischemia suggests that intracellular calcium may increase, but because the amount of the calcium sequestered by the intracellular organelles may vary, the actual change in the cytosolic free calcium cannot be determined. Similarly, it is difficult to infer alterations in free intracellular calcium from measurements of total tissue calcium using atomic absorption.

Recently, a family of calcium-sensitive fluorescent indicators has been developed for the measurement of intracellular free calcium [2, 5, 6]. These indicators, the most widely used being quin-2, fura-2, and indo-1, are administered as acetoxymethyl esters which readily penetrate the cell membrane. Once intracellular, they are hydrolyzed by intracellular esterase to a form that is impermeant to the cell membrane. When excited by ultraviolet light of the appropriate wavelength, they fluoresce with an intensity and wavelength dependent on cytosolic free calcium concentration. The use of these dyes has been primarily restricted to isolated cells [6] due to the relative ease with which such preparations can be loaded and the resulting fluorescence measured. We have extended this technique for use in vivo with the goal of directly examining changes in cytosolic free calcium in a

Cerebrovascular Research Center, University of Pennsylvania, Philadelphia, Pennsylvania 19087, USA.

Cerebral Ischemia and Calcium
Ed. by A. Hartmann and W. Kuschinsky

variety of pathological conditions. Another paper in these proceedings by Reivich and associates discusses in more detail some of the data obtained under conditions of anoxia and focal cerebral ischemia.

The development of the technique occurred in two stages. The initial studies were undertaken using quin-2 as the calcium-sensitive fluorescent dye, while in later studies indo-1 was utilized as the indicator. Quin-2, when excited at a wavelength of 339 nm, fluoresces with an intensity proportional to the cytosolic free calcium concentration [6]. There is negligible shift in the emission spectra, the peak being at approximately 492 nm, independent of calcium concentration.

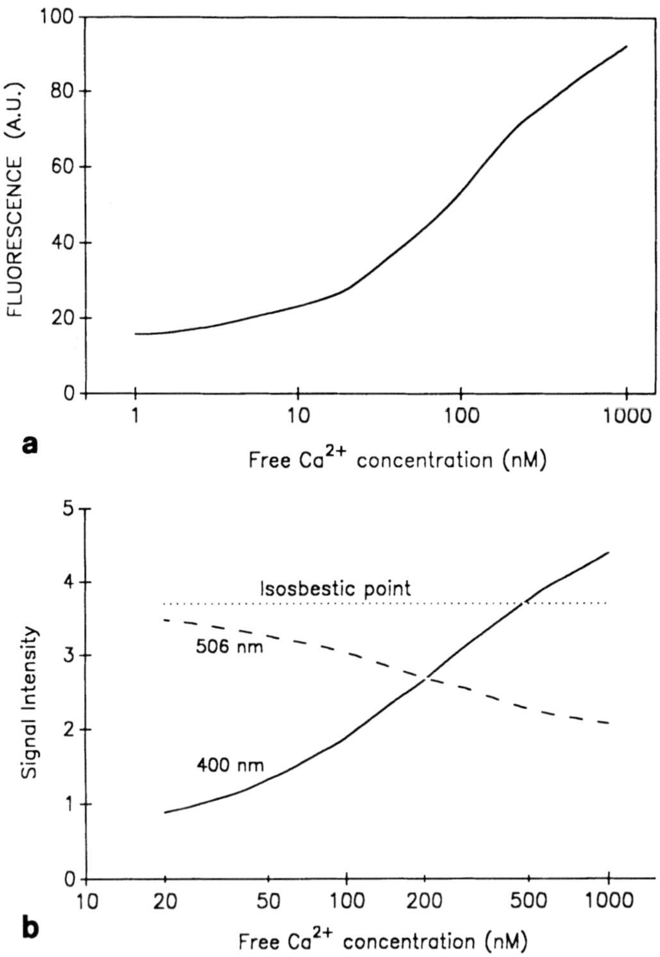

Fig. 1. The emission of **a** quin-2 and **b** indo-1 as a function of free Ca^{2+} concentration. The data for these plots were obtained by Tsien et al. [6] (**a**) and by Grynkiewicz et al. [2] (**b**) using solutions with varying calcium concentrations in a spectrophotometer. For the quin-2 measurements, the concentration was 20 μM and the emission fluorescence was at 492 nm with an excitation at 339 nm. The indo-1 concentration was 6 μM and the excitation wavelength was 355 nm. The emission intensity at two different wavelengths is shown along with the fluorescence intensity at the isosbestic point. The ratio of the signal at 400 nm and 506 nm will be independent of both the excitation and the emission intensities

Figure 1 a shows this relationship in an in vitro preparation. The peak sensitivity of this indicator coincides well with the normal cytosolic free calcium concentration, which is 50–300 nM.

The studies to investigate the use of this tracer in vivo were done in the cat using brief anoxia as an insult [7]. Following anesthesia, the animal was tracheotomized and placed on a ventilator, and femoral arterial and venous cannulas were inserted. A catheter was also placed into the lingual artery for subsequent saline flushes. A burr hole was made over the temporoparietal cortex, and the dura and pia-arachnoid were carefully incised. A quartz cranial window was inserted into the hole and sealed with dental cement. It was through this quartz window that the fluorometric measurements were made. The cranial window had a small inlet and outlet port so that the cortical surface could be superfused with various solutions with ionic concentrations similar to that of cerebrospinal fluid (CSF). The cortical surface of the brain was superfused with the CSF solution containing 100 μM of quin-2 acetoxymethyl ester (quin-2-AM) for 1 h at 37° C. Following this loading procedure, any remaining extracellular dye was washed out by continuing the superfusion for 30 min with the identical solution without the quin-2-AM. The depth of loading was investigated in some animals by examining frozen brain slices under a fluorescence microscope. As can be seen in Fig. 2, the dye penetration was approximately 500 μm into the cortical surface.

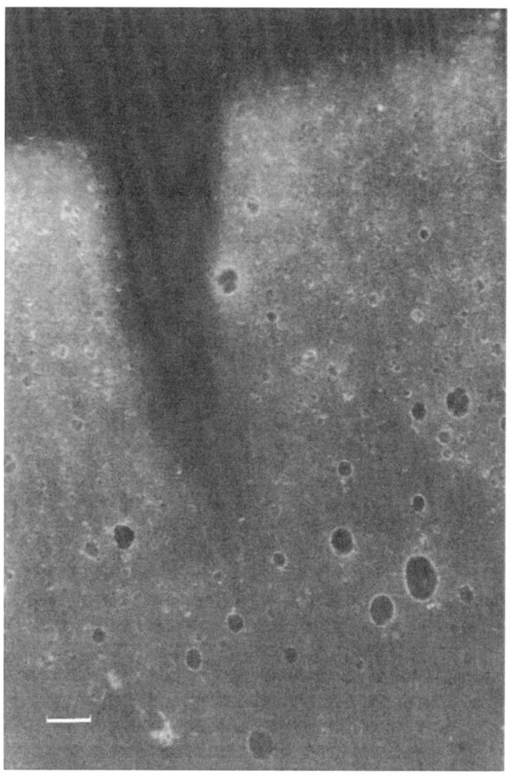

Fig. 2. Fluorescence photograph of a frozen brain slice following loading with quin-2 (100 μM superfusion for 60 min). *Scale bar* is 100 μm. The depth of loading is approximately 0.5 mm. The *circles* on the sections are freezing artifacts

A fluororeflectometer identical to that normally used for NAD/NADH measurements was positioned over the cranial window and the cortical surface was illuminated with a mercury arc lamp at 350 nm (350/30 nm). Using fiber optic light guides, the reflected light (366 nm) and the fluorescence at 506 nm were detected with photomultipliers and appropriate circuitry. Completely analogous to measurements of NAD/NADH fluorometrically [3], changes in local red blood cell concentrations can alter the fluorescence signal independent of actual fluorescence changes. Consequently small boluses of saline were frequently flushed into the lingual arterial catheter so that the effect of local blood volume changes on the fluorescence signal could be determined and electronically subtracted.

It was originally hoped that the NADH fluorescence which is centered at 470 nm would be small in relation to the fluorescence due to the quin-2 and therefore would not contribute significantly to the fluorescence signal at 506 nm. Unfortunately this was not the case, and it was found necessary to produce a short anoxic insult prior to quin-2-AM loading to measure the contribution due to the NADH alone. Following loading, an identical anoxic insult was produced and the fluorescence, along with the reflectance, was measured. By subtracting the two fluorescence signals (after loading and prior to loading) the fluorescence due to quin-2 alone was determined. It is this signal that is a measure of the cytosolic free calcium.

To yield further information concerning the relationship between the fluorescence signal and the intracellular calcium changes, the cortical surface was superfused with two solutions at the end of the experiment. Superfusion with ionomycin at two different concentrations (0.5 and 2 μM) produced a dose-dependent increase in the corrected fluorescence signal. Ionomycin, a Ca^{2+} ionophore, increases the permeability of the cell membrane to calcium, allowing a massive influx of Ca^{2+} into the cells. When the cortex was superfused with EGTA, which chelates the calcium, the fluorescence signal decreased dramatically, supporting the interpretation that this technique is measuring cytosolic free calcium.

The technique as just described for monitoring changes in intracellular free calcium suffers from two limitations. Since the changes in NADH fluorescence can contribute appreciably to the signal at 506 nm, it is necessary to repeat the identical insults before and after the quin-2 loading to assess the NADH change and to subtract it from the calcium signal. Additionally, since the cytosolic free calcium is determined from the intensity of the fluorescence, errors can arise due to other factors which may affect this intensity, including leakage of dye from the cells as well as photobleaching. These limitations can be significantly overcome by the use of another dye with more favorable spectral characteristics. One dye that we have investigated is indo-1 [6], which, when excited at 340 nm, exhibits two fluorescence peaks, near 400 nm and near 500 nm, with an isosbestic point at 464 nm. The amplitudes of these peaks are related to the cytosolic free calcium concentration. Figure 1 b, adapted from the data of Grynkiewicz et al. [2], shows the intensity of fluorescence at 400 nm and at 506 nm for solutions excited with 355 nm light. The ratio (400 nm/506 nm) will be independent of the intensity of both the excitation and the emission signals. Also shown is the emission intensity at the isosbestic point, which is 448 nm when the excitation is 355 nm; at this wavelength the intensity is independent of calcium concentration.

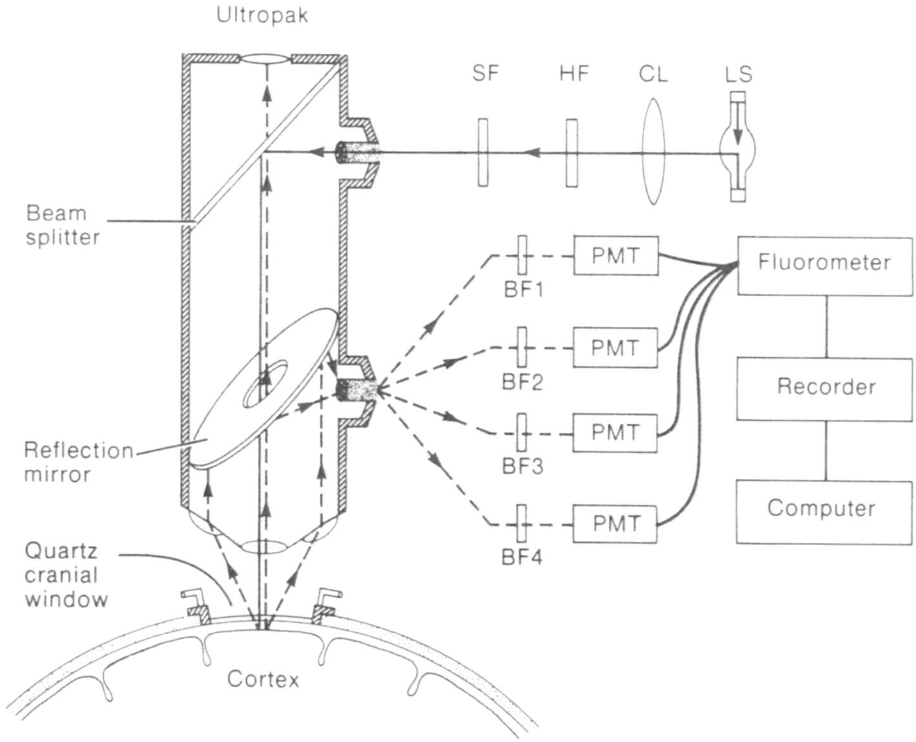

Fig. 3. The experimental setup used in the indo-1 studies. The ultraviolet light travels from the mercury arc lamp through a quartz collecting lens (*CL*), a heat absorption filter (*HF*), and a suppression filter (*SF*) (340/12 nm). It then enters the ultropak of the fluorometer where a beam splitter reflects the rays through an objective lens and a quartz cranial window onto the brain surface. Both the reflected light (340 nm) and the emitted light travel back into the ultropak, where they are reflected by a mirror onto a fiber optic bundle. This bundle splits into four separate fiber optic bundles each going through appropriate barrier filters (*BF*) and then into photomultiplier tubes (*PMT*). The barrier or bandpass filters have peak transmissions at 340 nm (for the reflected light), 464 nm corresponding to the isosbestic point of indo-1 emission when the excitation is 340 nm, and 400 nm and 506 nm for the detection of the two emission peaks of indo-1. (Reproduced with permission from Uematsu et al. [8])

To utilize these spectral characteristics, a four-channel fluororeflectometer was employed [8] (Fig. 3). In this system, the excitation light was 340 nm and one channel of the fluorometer had a bandpass filter centered at 340 nm for measurement of the reflected light intensity, which is proportional to changes in local cerebral blood volume [3]. A second channel had a bandpass filter centered at 464 nm, which corresponds to the isosbestic point of the indo-1 calcium fluorescence. This was used for the measurement of the NADH fluorescence and is independent of calcium concentration. (In these in vivo studies the isosbestic point is slightly higher than that found by Grynkiewicz et al. [2] since the excitation wavelength is different.) In the last two channels, bandpass filters at 400 nm and 506 nm were used. At 400 nm, the intensity of fluorescence increases with an increase in calcium concentration, while at 506 nm the intensity decreases as free intracellular concentration increases. The ratio (400 nm/506 nm) thus becomes a very sensitive indicator of cytosolic free calcium changes. Using the standard bolus saline flush as discussed above and as described in detail by Harbig et al. [3], the NADH channel (464 nm) was corrected for hemodynamic changes to produce a corrected NADH signal. This measurement will be independent of changes in calcium, since the signal at 464 nm is not sensitive to alterations in calcium. Since the NADH fluorescence is seen in the two calcium channels (400 nm and 506 nm) it is necessary to take the corrected NADH signal and subtract it from the signals in those channels. The contribution of the NADH fluorescence to these channels was determined by a short hypoxic insult just prior to indo-1 loading. Any changes in the signal of the calcium channels are due only to NADH changes (after hemodynamic correction) since the dye has not yet been administered. These correction factors permit the NADH contribution to be removed, leaving a signal proportional to cytosolic free calcium only.

Thus, during an insult, changes in cytosolic free calcium, changes in NAD/NADH redox state, and cerebral hemodynamics (including local cerebral blood volume and mean transit time) can be simultaneously monitored. This technique has several advantages over existing methods for assessing changes in calcium. Changes in cytosolic free calcium can be measured in vivo. Except for some very recent work by Silver and co-workers (personal communication), all of the calcium electrode studies have been made with extracellular calcium electrodes, which, as discussed above, are not a good indicator for intracellular calcium. In addition, this fluorometric technique permits the simultancous measurement of NAD/NADH redox state and cerebral hemodynamics, permitting these parameters to be correlated. This technique has excellent time resolution in that the time response to alterations in cytosolic free calcium is of the order of milliseconds, hundreds of times more rapid than electrode measurements. Lastly, this method is less invasive than the use of microelectrodes. Although the duration of measurement is limited due to dye diffusion and photobleaching, measurements can be carried out for up to 2 h in hypoxia, ischemia, and hypoglycemia. The report by Reivich, Uematsu, and Greenberg in these proceedings discusses some of our results in more detail.

References

1. Cheung JY, Bonventre JV, Malis CD, Leaf A (1986) Calcium and ischemic injury. N Engl J Med 314:1670–1676
2. Grynkiewicz G, Peonie M, Tsien RY (1985) A new generation of Ca^{++} indicators with greatly improved fluorescence properties. J Biol Chem 260:3340–3348
3. Harbig K, Chance B, Kovach A, Reivich M (1976) In vivo measurement of pyridine nucleotide fluorescence from cat brain cortex. J Appl Physiol 41:480–488
4. Siesjo BK (1981) Cell damage in the brain: a speculative synthesis. J Cereb Blood Flow Metab 1:155–185
5. Tsien RY (1980) New calcium indicators and buffers with high selectivity against magnesium and protons: design, synthesis, and properties of prototype structures. Biochemistry 19:2396–2404

6. Tsien RY, Pozzan T, Rink TJ (1982) Calcium homeostasis in intact lymphocytes: cytoplasmic free calcium monitored with a new, intracellularly trapped fluorescent indicator. J Cell Biol 94:325–334
7. Uematsu D, Greenberg JH, Reivich M, Karp A (1988) In vivo fluorometric measurement of changes in cytosolic free calcium from the cat cortex during anoxia. J Cereb Blood Flow Metab 8:367–374
8. Uematsu D, Greenberg JH, Reivich M, Karp A (1988) In vivo measurement of cytosolic free calcium during cerebral ischemia and reperfusion. Ann Neurol 24:420–428

Calcium Movements in Brain During Failure of Energy Metabolism

E. Zhang[1], M. Lauritzen[1], T. Wieloch[2], and A. J. Hansen[1]

Introduction

The interstitial ion concentrations in brain are maintained within narrow limits. Changes in plasma concentrations are effectively dealt with by the low ionic permeability of the blood-brain barrier combined with active transport processes located in the brain endothelium and in the choroid plexus [3]. The most important reason for disturbances in brain ion milieu is the fact that the ions are not in electrochemical equilibrium across the cell membrane. Active transport processes are constantly needed to counteract the passive dissipation of the ion gradients.

The level of Ca^{2+} in brain interstitial space (Ca_e) is normally 1–1.5 mM while the intracellular calcium concentration is ca. 10^{-4} mM. Hence the interstitial concentration decreases when either the membrane calcium conductance increases or active extrusion is hampered. During spontaneous activity, changes in Ca_e are hardly noticeable but Ca_e may reach values as low as 0.5 mM when the brain is adequately stimulated [12]. The most significant decrease of Ca_e is seen during spreading depression of Leao (SD) [15] and in conditions with failure of brain energy metabolism. These conditions are accompanied by decreases of Ca_e to values below 0.1 mM, which completely inhibits synaptic transmission [5]. However, this effect is obscured since the calcium concentration changes are part of a stereotyped pattern involving all ions as they move down their electrochemical gradients. The Ca_e decrease is accompanied by an increase of K_e from 3 to 60 mM, a decline of Na_e from 150 to 50 mM, a decline of Cl_e from 130 to 70 mM, and a negative deflection of the local tissue potential [9]. In the case of SD changes start almost abruptly, last ½–3 min, and revert spontaneously. In anoxia-ischemia the large ionic movements are preceded by an initial phase with small but significant ion concentration changes [9]. The ionic events in hypoglycemia start abruptly like SD and may partly and transiently normalize before anoxia-like levels are attained. In both situations the ionic changes are reversible when oxygen/blood/glucose supply to the brain is resumed. The strong similarity between the three situations suggests that the ionic events have the same genesis. In the following we discuss these events, their genesis, and their possible role in the development of brain dysfunction and cellular damage.

[1] Department of General Physiology and Biophysics, University of Copenhagen, Blegdamsvej 3, DK-2200 Copenhagen N, Denmark.
[2] Laboratory of Experimental Brain Research, University of Lund, S-Lund, Sweden.

Cerebral Ischemia and Calcium
Ed. by A. Hartmann and W. Kuschinsky

Interstitial Calcium Measurements

The calcium concentration in the brain was measured by ion-sensitive microelectrodes. It is evident that changes in the interstitial concentration do not express the flux because changes in the size of the interstitial space as a rule accompany ionic fluxes [4, 8, 22]. Even when the interstitial space size is determined, the net flux to the different cellular compartments is unknown and requires direct intracellular measurements.

The calcium content of the interstitial space located in 1 g of gray matter is approximately 0.24 μmol (1.2 μmol/ml $\times$ 0.2 ml) with an interstitial space of 20%. During the stereotyped ionic changes the content decreases to 0.01 μmol (0.1 μmol/ml $\times$ 0.1 ml) since both the concentration and the space size decrease, demonstrating that 95% of the extracellular located calcium moves into other compartments.

Ca_e Changes During SD

Figure 1 (lower part) shows Ca_e during SD elicited in rat brain cortex in response to ligation of the ipsilateral middle cerebral artery. The physiology of SD is described in more detail in the chapter by Lauritzen et al.

Ca_e Changes During Anoxia

The onset of complete cerebral ischemia has little effect on Ca_e (Fig. 1). In fact, there is an initial small but significant increase of 0.1 mM [9]. Although it may be insignificant for neuronal function it reflects an elevated intracellular calcium level leading to extrusion, probably via the Na^+/Ca^{2+} exchanger [14]. It was accordingly shown recently that nerve cells responded to cyanide by increases of intracellular Ca^{2+} level [2].

After a few minutes of anoxia the brain undergoes a so-called anoxic depolarization in which the stereotyped ionic events are elicited. The Ca_e decreases from ca. 1.3 mM to ca. 0.1 mM. During the next minutes the level decreases somewhat. The level remains constant but can be completely normalized within 15 min by resumption of blood flow or oxygen supply [26]. This anoxic depolarization and concomitant cellular calcium influx have been linked to cell damage in anoxia [28].

Ca_e During Hypoglycemia

The variation of Ca_e in hypoglycemia has previously been described by Harris et al. [11]. The decline of Ca_e has been linked to SD because of the abrupt onset (in contrast to anoxia) and because the decrease is often partly and transiently normalized. Figure 2 shows a typical record of Ca_e changes during severe hypoglycemia. It is important to note the glucose level in blood must be near zero in order

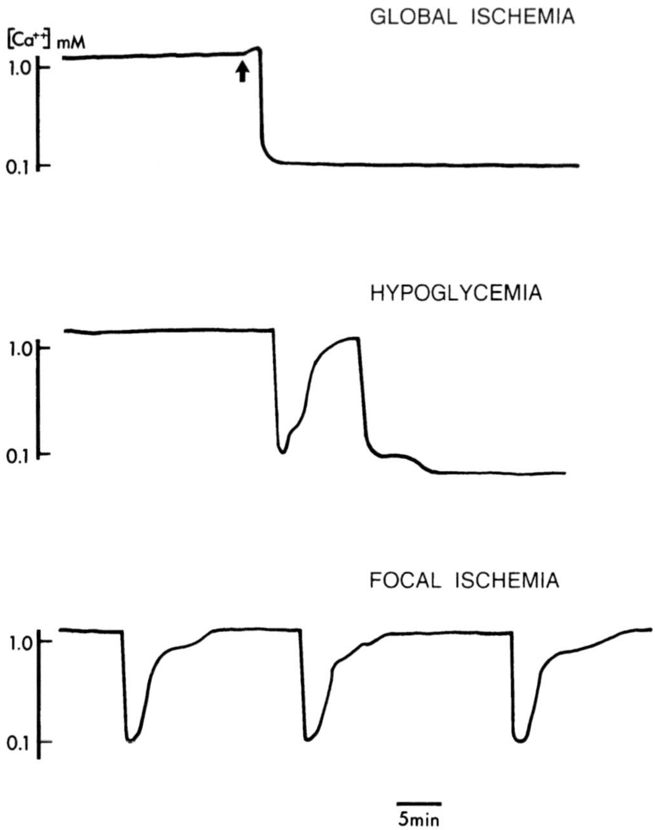

Fig. 1. Interstitial calcium concentration measured in rat parietal cortex with ion-sensitive microelectrodes. The *upper part* shows the changes during acute, global ischemia (*arrow*) induced by cardiac arrest (saturated KCl infused i.v.). The *middle part* shows the concentration change during severe hypoglycemia provoked by insulin (40 IU/kg i.p.) administered 2 h earlier. Note the abrupt decrease and the transient increase. The *lower part* shows the typical, transient decreases elicited by occlusion of the ipsilateral middle cerebral artery

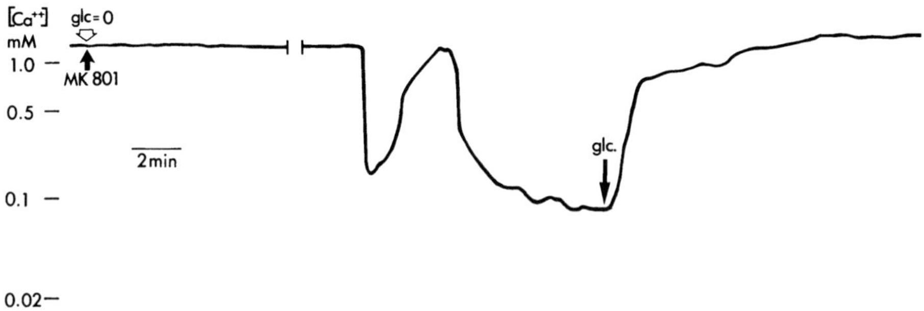

Fig. 2. Interstitial calcium concentration determined by an ion-selective microelectrode positioned in rat parietal cortex. The rat received 40 IU/kg insulin i.p. and when the plasma glucose level was zero the noncompetitive NMDA antagonist MK-801 (a gift from Merck, Sharpe, and Dohme) was given (10 mg/kg). The variation of the calcium concentration was not different from that in untreated rats

to elicit the stereotyped response. When rats are given 40 IU insulin/kg i.p. it takes 2–3 h to reach the sufficient low glucose level. The electroencephalographic activity is arrested near the moment of rapid ion movements. The Ca_e may transiently increase several times before the steady low level is reached. This variation does not signify the presence of SD [11] but most probably fluctuation of the plasma glucose level. We found that plasma glucose levels following the initial Ca_e decline were positively correlated with the Ca_e level. Thus these fluctuations are governed by the small variations in the plasma glucose level allowing the ion pumps to resume their activity for a short while. When larger amounts of glucose are administered the Ca_e is completely normalized (Fig. 2).

Similarities Between the Ionic Movements

It does not seem surprising that large ion fluxes are observed during a situation with failure of brain energy metabolism, since ion transport in brain requires the main part of the energy metabolism [7]. The question is whether the two events (anoxia and hypoglycemia) are elicited by the same mechanisms as SD. When discussing these phenomena we distinguish between the elicitation of the ion fluxes and the events following directly upon their induction.

Energy failure, however, is not the main factor since the ionic fluxes are elicited in situations with normal tissue ATP levels, as in SD and hypoglycemia [17, 23, 32]. On the other hand, the extent of energy depletion following depolarization may determine the type of ion channels activated.

Table 1 shows different parameters in the three events. The important point is that blockade of *N*-methyl-D-aspartate (NMDA) receptors completely inhibits SD, whereas it has no effect on the ionic movements in cerebral cortex during anoxia or hypoglycemia [10]. That is to say that neither the anoxic depolarization

Table 1. Events in brain cortex accompanied by large ionic fluxes

	Interstitial glutamate	pO_2	Glucose content	ATP content		Effect of NMDA antagonists
				I	II	
Spreading depression	Increased [30]	Normal [16]	Normal [13, 17]	Normal [13, 17]	50% [17]	Inhibits [6, 10]
Anoxia/ischemia	Increased [1]	0	0 [27]	10%–20% [27]	0 [27]	No effect [10]
Hypoglycemia	Increased [24]	Normal	0	Normal [11]	50% [11]	No effect

Situations in cerebral cortex with large changes of interstitial ion concentrations, including decreases of calcium concentrations. The table describes parameters within brain cortex near the onset of rapid ion fluxes. The changes in ATP level are given immediately before (I) and immediately following ion fluxes (II). Hypoglycemia means plasma levels near zero. The brain pO_2 in this condition is considered normal since the blood flow is increased without an increase of oxygen consumption [20]. For further details see text.

nor the depolarization in hypoglycemia was inhibited by either competitive or noncompetitive inhibitors of NMDA receptors. In SD, the level of ATP is normal or slightly decreased and the depolarization may only be sufficient to activate the NMDA channel by releasing the voltage dependent magnesium block but not other voltage or second messenger dependent channels. Thus SD can be efficiently blocked by the NMDA receptor antagonists. In contrast, in ischemia the ATP levels are already decreased at the time of depolarization, and in hypoglycemia the ATP levels are decreased by approximately 50% immediately following depolarization. Here it is not sufficient to block the NMDA receptor operated ion channel in order to inhibit the ion fluxes. The energy depletion is so severe that other ion channels (voltage dependent, second messenger regulated or receptor operated) will be activated. Spreading depression thus belongs to a separate category which depends upon NMDA receptor activation whereas the other situations showing the stereotyped response of interstitial ion concentrations, albeit with a different metabolic state, rely on as yet unknown factors. Apparently other channels are opened since an anion conductance also is activated during SD and ischemia [9].

Admittedly this hypothesis does not explain the detailed mechanisms regulating the ion channels during SD, ischemia, and hypoglycemia. The opening of the channels during ischemia and hypoglycemia must, however, directly or indirectly include a highly energy dependent step.

Ionic Movements and Cell Death

The mechanisms outlined above may also determine whether the large influx of calcium during the stereotyped ionic movements is of importance for the development of cell death [25, 28].

However, there is no reliable answer to this problem. A transient calcium influx is not harmful in the otherwise normal brain since continuous elicitation of SD in rat brain cortex for 4 h caused no cellular damage – except for the elicitation site where saturated KCl was applied [19]. However, in the penumbra zone surrounding an ischemic focus, where the neurons are at the threshold of survival, transient depolarization may contribute to cell death [18]. In accordance it was recently suggested that glutamate becomes neurotoxic when the nerve cells are deprived of glucose or oxygen [21].

The fact that NMDA receptor blockers attenuate the cell damage following a hypoglycemic [31] or anoxic challenge [29] strongly suggests the importance of excitotoxins, i.e., glutamate and aspartate, in the process. In ischemia and hypoglycemia calcium entry may also be lethal for the cell, although areas with high glutamate receptor density such as dentate gyrus are resistant to an ischemic insult which is sufficient to destroy the adjacent CA1 region completely. Here differences in intracellular calcium buffering capacities may explain the differences in the vulnerability of the different neuronal populations.

Summary

Calcium movements in brain were investigated by the use of ion-sensitive microelectrodes. The three different pathological situations, spreading depression of Leao, anoxia, and severe hypoglycemia, have in common that almost 95% of the extracellular calcium enters the cellular compartment within seconds. Blockade of the NMDA type of glutamate channels effectively inhibited the ion movements in spreading depression while leaving the two other situations unaffected, showing that the mechanisms behind the ion movements during compromised brain energy metabolism are different from those in spreading depression. We propose that in ischemia and severe hypoglycemia, ion channels, regulated directly or indirectly by an energy dependent step, will be activated. These channels are not blocked by NMDA antagonists.

References

1. Benveniste H, Drejer J, Schousboe A, Diemer NH (1984) Elevation of the extracellular concentrations of glutamate and aspartate in rat hippocampus during transient cerebral ischemia monitored by intracerebral microdialysis. J Neurochem 43:1369–1374
2. Biscoe TJ, Duchen MR, Eisner DA, O'Neill SC, Valdeolmillos M (1988) The effects of glucose removal and cyanide on intracellular Ca in isolated, single mouse dorsal root ganglion cells.
3. Bradbury MWB (1979) The concept of a blood-brain barrier. Wiley, Chichester, UK
4. Dietzel I, Heinemann U, Hofmeier G, Lux HD (1982) Stimulus-induced changes in extracellular Na and Cl concentration in relation to changes in the size of the extracellular space. Exp Brain Res 46:73–84
5. Dingledine R, Somjen G (1981) Calcium dependence of synaptic transmission in the hippocampal slice. Brain Res:218–222
6. Goroleva NA, Koroleva VI, Amemore T, Pavlik V, Bures J (1987) Ketamine blockade of cortical spreading depression in rats. Electroencephalogr Clin Neurophysiol 66:440–447
7. Hansen AJ (1985) Effect of anoxia on ion distribution in the brain. Physiol Rev 65:101–148
8. Hansen AJ, Olsen C-E (1980) Brain extracellular space during spreading depression and ischemia. Acta Physiol Scand 108:355–365
9. Hansen AJ, Zeuthen T (1980) Extracellular ion concentration during spreading depression and ischemia in the rat brain cortex. Acta Physiol Scand 113:437–445
10. Hansen AJ, Lauritzen M, Wieloch T (1988) NMDA antagonists inhibit cortical spreading depression but not anoxic depolarization. In: Frontiers in excitatory amino acid research. Alan R Liss, New York (in press)
11. Harris RJ, Wieloch T, Symon L, Siesjo BK (1984) Cerebral extracellular calcium activity in severe hypoglycemia: relation to extracellular potassium and energy state. J Cereb Blood Flow Metab 4:187–193
12. Heinemann U, Louvel J (1983) Changes in Ca^{++} and K^{+} during repetitive electrical stimulation and during pentetrazol induced seizure activity in the sensorimotor cortex of cats. Pfluegers Arch 398:310–317
13. Krivanek J (1961) Some metabolic changes accompanying Leao's spreading cortical depression in the rat. J Neurochem 6:183–199
14. Krnjević K, Leblond J, Morris ME (1988) Anoxia-evoked increases in extracellular Ca in isolated rat hippocampal slices. J Physiol 400:24P
15. Leão AAP (1944) Spreading depression of activity in cerebral cortex. J Neurophysiol 7:359–390

16. Lehmenkuhler A (1979) Interrelationships between DC-potentials, potassium activity, pO_2 and pCO_2 in the cerebral cortex of the rat. In: Speckmann E-J, Caspers H (eds) Origin of cerebral field potentials. Thieme, Stuttgart, pp 49–58
17. Mies G, Paschen W (1984) Regional changes of blood flow, glucose and ATP content determined on brain sections during a single passage of spreading depression in rat brain cortex. Exp Neurol 84:249–258
18. Nedergaard M, Astrup J (1986) Infarct rim: effect of hyperglycemia on direct current potential and (^{14}C)2-deoxyglucose phosphorylation. J Cereb Blood Flow Metab 6:607–615
19. Nedergaard M, Hansen AJ (1988) Spreading depression is not associated with neuronal injury in the normal brain. Brain Res 449:395–398
20. Norberg K, Siesjo BK (1976) Oxidative metabolism of the cerebral cortex of the rat in insulin-induced hypoglycemia. J Neurochem 26:345–352
21. Novelli A, Reilly JA, Lysko PG, Henneberry RC (1988) Glutamate becomes neurotoxic via the *N*-ethyl-D-aspartate receptor when intracellular energy levels are reduced. Brain Res 451:205–212
22. Pellegrino D, Almquist L-O, Siesjo BK (1981) Effects of insulin-induced hypoglycemia on intracellular pH and impedance in the cerebral cortex of the rat. Brain Res 221:129–147
23. Quistdorff B, Gjedde A, Hansen AJ (1979) Spatial analysis of the freeze-trapped brain provides for temporal resolution of an event. Metabolic, electrical and blood flow changes during spreading depression. Acta Physiol Scand 105:42A
24. Sandberg M, Butcher SP, Hagberg H (1986) Extracellular overflow of neuroactive amino acids during severe insulin-induced hypoglycemia: in vivo dialysis of the rat hippocampus. J Neurochem 47:178–184
25. Schanne FAX, Kane AB, Young EE, Faber JL (1979) Calcium dependence of toxic cell death: a final common pathway. Science 206:700–702
26. Siemkowicz E, Hansen AJ (1981) Brain extracellular ion composition and EEG activity following 10 minutes ischemia in normo- and hyperglycemic rats. Stroke 12:236–240
27. Siesjö BK (1978) Brain energy metabolism. New York, Wiley
28. Siesjo BK (1981) Cell damage in the brain: a speculative synthesis. J Cereb Blood Flow Metab 1:155–185
29. Simon RP, Swan JH, Griffiths T, Meldrum BS (1984) Blockade of *N*-methyl-D-aspartate receptors may protect against ischemic damage in the brain. Science 226:850–852
30. Van Harreveld A, Fifkova E (1970) Glutamate release from the retina during spreading depression. J Neurobiol 2:13–29
31. Wieloch T (1985) Hypoglycemia-induced neuronal damage prevented by an *N*-methyl-D-aspartate antagonist. Science 230:681–683
32. Wieloch T, Harris RJ, Symon L, Siesjo BK (1984) Influence of severe hypoglycemia on brain extracellular calcium and potassium activities, energy, and phospholipid metabolism. J Neurochem 43:160–168

Glutamate Antagonism as a Pharmacological Approach to Prevent Calcium Influx in Focal Cerebral Ischemia

E. T. MACKENZIE[1] and J. MCCULLOCH[2]

Introduction

Cerebral ischaemia can result from a diverse range of pathological states and insults that are all characterised by a decrease in blood flow (and hence metabolite delivery) to levels insufficient to meet the energetic requirements of the brain. When the ischaemia is sufficiently severe and its duration maintained, the energy requirements for the structural integrity of neurones are no longer met, and irreversible neuronal death ensues. The most frequent cause of brain ischaemia in man, namely stroke, is that occasioned by the occlusion of a major cerebral artery, a situation in which neuronal loss precedes a consolidated infarction defined as the loss of all cell elements within the ischaemic territory.

In this review, based essentially on our own – sometimes unpublished – studies, we will address the following points: (a) the development of robust models of focal cerebral ischaemia in which there are clearly defined end-points; (b) the possibilities of tissue-sparing with pharmacological agents that interfere with glutamate-mediated neurotoxicity; and (c) possible directions for future research to better understand the mechanisms of neuronal loss and sparing in focal cerebral ischaemia.

Stereometric Assessment of Cerebral Infarction

Clinical studies on the potential therapy of stroke are laden with methodological difficulties and hence much emphasis has been placed on animal models of human cerebrovascular disease. However, the pharmacology of laboratory stroke models has given rise to conflict and discrepancies which can be related to two principal problems: (a) the relevance of the model to human stroke and (b) the parameters or criteria used to assess brain "protection". It has been argued that experimental *focal* cerebral ischaemia – the consequence of occlusion of a major cerebral artery – is the most pertinent for our understanding of human stroke. Secondly, the most valid criterion of cytoprotection is quantitative histology of

[1] Department of Biology, L.E.R.S.-Synthélabo, 31, Avenue Paul Vaillant-Couturier, F-92220-Bagneux, France.
[2] Wellcome Neuroscience Group, Wellcome Surgical Institute, University of Glasgow, Glasgow G61 1QH, U.K.

Cerebral Ischemia and Calcium
Ed. by A. Hartmann and W. Kuschinsky

either neuronal loss or infarction volume and not secondary parameters or sequelae, such as perfusion, glycolytic metabolism, water content, or electrocortical activity [13].

In this review we have regrouped those studies that concern glutamate antagonism and the quantitative, stereometric evaluation of infarction volume following middle cerebral artery occlusion. However, one should be aware of major differences in these quantitative histological studies. The pathological examination may be either early (in the hours that follow middle cerebral artery occlusion) or late (in the days following). In the former instance one is obliged to delineate the zone of ischaemia cell change (in the belief that this will evolve to the considered infarction); when the pathological examination is made in the days following occlusion, the frontiers of necrosis are clearly defined. Early pathology has the advantage of allowing the monitoring of key physiological parameters (arterial pressure, blood gas status, etc.). However, acute pathology can be criticised on the basis that drug intervention may not reduce the infarction volume but simply retard the evolution of ischaemia-induced neuronal death. An important technical point that relates to both acute and chronic pathology is that infarction volume must be reconstructed from a minimal number of near-equidistant coronal levels; if less than this number, the variability and inaccuracy of the estimate increase considerably [15].

Glutamate, Calcium and Neurotoxicity

The neurotoxicity induced by the excitatory amino acid transmitters, glutamate and aspartate, is mediated by three types of receptor: kainate, quisqualate and *N*-methyl-D-aspartate (or NMDA) receptors. Although the agonists for all three provoke neuronal death, it is the latter (NMDA) receptors that have been the best defined and analysed thanks to the existence of a number of competitive and non-competitive antagonists for this macromolecular complex.

The nature of the neurotoxicity induced by excessive stimulation of the NMDA receptors has been studied in some depth. Receptor stimulation leads to the opening of a cation permeable channel that displays relatively little selectivity. Via this channel there is an increased sodium entry (which entrains a passive chloride and water influx and hence cell swelling and oedema) and increased calcium entry. The subsequent elevation of intracellular calcium concentrations is probably the mechanism responsible for neuronal death as increased cytosolic calcium is associated with a wide range of nefarious processes: proteolysis, mitochondrial uncoupling, the breakdown of phospholipids and the formation of free radicals.

Several lines of evidence have been adduced for the glutamate hypothesis of ischaemic cell change following brain ischaemia. These are, firstly, the close correspondence between area with a pronounced high affinity uptake process for glutamate and those brain regions that display selective neuronal loss following a global ischaemic insult [11]. The second, more telling argument, is that glutamate and its analogues are neurotoxic and that glutamate is released following, again, global ischaemic insults [2, 10, 14]. Thirdly, the administration of NMDA receptor antagonists has been shown to reduce the neuronal loss subsequent to global ischaemia [9, 22], hypoxia [24] or hypoglycaemia [26]. Furthermore, lesions of the glutamatergic corticostriatal projections mitigate against hypoglycaemia-induced striatal necrosis in the rat [27].

If an increased intracellular calcium concentration is the final mechanism of ischaemia-induced and glutamate-induced neurotoxicity then one might argue that calcium entry blockers would be of considerable therapeutic benefit in the protection of neurones against ischaemic injury. However, as pointed out by Rothman and Olney [19], this approach is limited by two major difficulties. Firstly, the NMDA receptor linked channels are little affected by available calcium antagonists which have usually been found on the basis of their selectivity for calcium channels that exist in smooth muscle cells. Secondly, there is no convincing or non-controversial evidence to suggest that available calcium antagonists are of therapeutic benefit in experimental models of cerebral ischaemia [13].

Pharmacological Antagonism of Glutamate (NMDA) Receptors

Competitive antagonists of the NMDA exist and are exemplified by compounds such as 2-amino-7-phosphanoheptanoic acid [20]. As the brain entry of this class of agents is limited, such compounds have been administered intracerebrally and, under such conditions, attenuate the neuronal damage induced by hypoglycaemia and global ischaemia [22, 26]. Kynurenate is a competitive antagonist of the NMDA receptor but, at similar concentrations, also blocks other excitatory receptors [20].

A number of non-competitive antagonists of the NMDA receptor have been identified. The agents include "use-dependent" channel blockers that antagonize the NMDA-induced neuronal depolarisation. This class of compounds is represented by phencyclidine (and its derivative, TCP) and MK-801 [12]. A second group of non-competitive NMDA antagonists is exemplified by ifenprodil (and its derivative, SL 82.0715) and possesses a mechanism of action distinct from those of the reference competitive NMDA antagonists and the non-competitive "use-dependent" NMDA blockers [3].

Glutamate Antagonism and Focal Cerebral Ischaemia

All available studies on the effects of competitive and non-competitive NMDA antagonists have been tabulated (Table 1). These studies are remarkable for their consistency in that: both pretreatment and post-treatment with this class of agents reduce infarction size; the effects are noted in both lissencephalic (mouse, rat) and gyrencephalic (cat) species; comparable tissue-sparing is detected by both acute and chronic pathology; and the results obtained with direct and indirect assessments of infarction volume are entirely concordant.

Table 1. Glutamate antagonism and experimental focal cerebral ischaemia

Compound	Dose, route (mg/kg)	Schedule	Species	Pathology	Effect	Ref.
MK-801	0.5 (i.v.)	Pretreatment (−30 min)	Rat	3 h	−32%**	[18]
	0.5 (i.v.)	Post-treatment (+30 min)	Rat	3 h	−41%**	[18]
	5.0 (i.v.)	Pretreatment (−30 min)	Cat	6 h	−50%**	[16]
	5.0 (i.v.)	Post-treatment (+30 min)	Cat	6 h	−52%**	[17]
	0.1 (i.p.)	Post-treatment[a]	Mouse	96 h	−35%	[7]
	0.3 (i.p.)	Post-treatment	Mouse	96 h	−59%**	[7]
	1.0 (i.p.)	Post-treatment	Mouse	96 h	−93%**	[7]
TCP	1.0 (i.p.)	Pretreatment (−30 min)	Rat	48 h	−27%*	[8]
	1.0 (i.p.)	Post-treatment[a]	Mouse	96 h	−28%	[7]
	3.0 (i.p.)	Post-treatment	Mouse	96 h	−66%**	[7]
	10.0 (i.p.)	Post-treatment	Mouse	96 h	−55%**	[7]
Ifenprodil	0.3 (i.v.)	Post-treatment (+5 min, 3 h)	Cat	96 h	−25%	[8]
	1.0 (i.v.)	Post-treatment (+5 min, 3 h)	Cat	96 h	−24%*	[8]
	3.0 (i.v.)	Post-treatment (+5 min, 3 h)	Cat	96 h	−42%**	[8]
	3.0 (i.p.)	Post-treatment[a]	Mouse	96 h	−32%	[7]
	10.0 (i.p.)	Post-treatment	Mouse	96 h	−58%**	[7]
SL 82.0715	1.0 (i.v.)	Post-treatment (+5 min, 3 h)	Cat	96 h	−17%	[8]
	3.0 (i.v.)	Post-treatment (+5 min, 3 h)	Cat	96 h	−36%**	[8]
	1.0 (p.o.)	Post-treatment (+30 min)	Rat	48 h	−34%*	[8]
	10 (p.o.)	Post-treatment (+30 min)	Rat	48 h	−48%**	[8]
	0.3 (i.p.)	Post-treatment[a]	Mouse	96 h	− 9%	[7]
	1.0 (i.p.)	Post-treatment	Mouse	96 h	−48%**	[7]
	3.0 (i.p.)	Post-treatment	Mouse	96 h	−65%**	[7]
	10 (i.p.)	Post-treatment	Mouse	96 h	−72%**	[7]
Kynurenate	300 (i.p.)	Pretreatment (0 min)	Rat	24 h[b]	−43%**	[6]
	300 (i.p.)	Post-treatment (+60 min)	Rat	24 h[b]	−11%	[6]
	30 (i.p.)	Post-treatment[a]	Mouse	96 h	−25%	[7]
	100 (i.p.)	Post-treatment	Mouse	96 h	−24%	[7]
	300 (i.p.)	Post-treatment	Mouse	96 h	−58%**	[7]

* $P<0.05$, ** $P<0.01$.

[a] First administered 5 min after occlusion, thereafter twice daily. Infarction volume indirectly assessed by proliferation of peripheral type benzodiazepine binding sites (reflecting macrophage invasion and astrocytic response).

[b] Only one coronal slice studied, thus not true change in infarction volume.

A number of strategies may be envisaged for stroke treatment. Currently two have been studied in depth: metabolic depression, as exemplified by barbiturates, and antagonism of glutamate-induced neurotoxicity as discussed in this review. These two strategies have in common the fact that their pharmacodynamic effects are essentially expressed within the central nervous system; in contrast, other pharmacological approaches (calcium entry blockers, drugs acting on osmotic pressure or blood viscosity) are not specifically targeted in that they entail considerable peripheral actions. Nonetheless, there is a major difference between the options of depressing cerebral energetic function and the antagonism of glutamate-mediated neurotoxicity. Inherent in the first approach is the depression of higher cortical, and even vital, functions which render difficult patient evaluation, care and support, at least in the acute phase following a cerebrovascular accident.

From the studies here reviewed it appears that ischaemia-induced glutamate neurotoxicity is a major determinant of the final volume of a focal lesion, at least in the normal brain. However, most cerebrovascular accidents occur in patients with underlying disease processes.

The notion of "risk factors" for stroke arises from those epidemiological studies in which a greater incidence of cerebrovascular accident is associated with a pre-existing or concomitant pathology. While many of those risk factors (atherosclerosis, cardiomyopathies) are undoubtedly at the origin of the thrombi or emboli that initially cause the cerebrovascular accident, we have demonstrated that certain morbid states (arterial hypotension, chronic arterial hypertension, hyperglycaemia) considerably exacerbate the tissue consequences of the defined ischaemic insult [5, 15]. From such observations one could argue that risk factors encompass and determine not just the cause but also the severity of an arterial occlusion within the cerebrovascular bed. It remains to be shown whether excitotoxic mechanisms are of similar importance in focal ischaemic lesions in brains where there is already a diabetes- or hypertension-induced microangiopathy.

Perspectives

The unequivocal demonstration of the feasibility of increasing (with coexisting pathologies) and decreasing (by pharmacological interventions) infarction volume in our models opens new vistas for the future. Three major directions may be envisaged: technological ameliorations, extended pathophysiological and therapeutic investigations and the employment of dynamic measures of cerebrovascular function to better understand the mechanisms of cell sparing and loss in the ischaemic brain. These three areas will be discussed briefly and in turn.

Two technological improvements are desirable. The stereometric reconstruction and calculation of an infarction volume, based on defined coronal sections processed by histology, is long, time-consuming and expensive. What is required is a method that is, of course, reproducible and sensitive but also practical and rapid. One step has been taken in this direction by the introduction of histochemical stains that identify abnormal mitochondrial functions and so delineate precocious ischaemic cell change [1]. Recently, we have adopted an alternative approach (based on a ligand for peripheral benzodiazepine receptors) to quantify the astrocytic and macrophagic reaction to neuronal loss. There is an excellent spatial correlation between the increase densities of peripheral benzodiazepine binding sites (which are selectively located on glia and macrophages) as assessed autoradiographically, and neuronal loss assessed histologically [4]. However, as autoradiography is just as long and as time-consuming as histology, we sought to examine the possibility that the total hemispheric density of peripheral benzodiazepine receptors (measured in membrane preparations) was a faithful reflection of infarction volume after the induction of focal cerebral ischaemia. Our preliminary studies suggest that this is indeed the case and encourage us to continue this research to develop a rapid quantitative and reliable (albeit indirect) marker of brain infarction.

The second technological innovation has yet to be tackled in the rat model of focal cerebral ischaemia. A major criticism of the actual paradigm (with respect to human cerebrovascular disease) is the non-reversibility of the vascular occlusion. In man, it is rare that an intraluminal obstruction persists in the hours that follow the ictus [21, 23]. However, it is not impossible to imagine an experimental model that would allow the re-establishment of blood flow following a temporary clamping of the middle cerebral artery, and indeed such an approach has been described for various species including the mouse [25]. This development is of cardinal importance because it is widely accepted that many of the nefarious factors (oedema, oxygen-derived tree radical formation, endothelial dysfunction and platelet aggregation) that determine final neuronal loss are related to the reperfusion of the ischaemic brain and not to the ischaemia per se.

We intend to pursue our investigations on the importance of risk factors for the size of the necrotic lesion following middle cerebral artery occlusion. Many questions remain unanswered, particularly in relation to chronic arterial hypertension and diabetes. Firstly, does a combination of these two conditions exacerbate even further the increase in lesion volume seen with each singly? In genetically hypertensive and diabetic rats, does the treatment of these states (by antihypertensive drugs and insulin, respectively) reduce the exaggerated lesion size associated with these conditions? Are excitotoxic mechanisms still of major importance when superimposed on an already abnormal cerebral circulation?

The experimental cytoprotection, in focal cerebral ischaemia, obtained with the variety of competitive and non-competitive NMDA antagonists opens promising therapeutic avenues. Already, the fact that considerable tissue-sparing may be achieved with compounds such as SL 82.0715, TCP and MK-801 allows a dissection of factors and conditions that are presently in the realm of the unknown. What delay is permissible between the induction of ischaemia and the first drug administration? How long should one maintain treatment to optimise the cytoprotective benefits? These questions relate to the window of therapeutic opportunity and must be answered first in experimental animals before rational regimens can be applied to man. Another largely unexplored area is whether prophylactic rather than curative therapeutic approaches may result in a similar or greater degree of cytoprotection; some elements of a response have been furnished in this review, which, at least, suggests that prophylactic approaches effect a significant cytoprotection.

Some words of caution are justified. Cerebral ischaemia induces a cascade of haemodynamic, energetic and biochemical disturbances; even more so when ischaemia is followed by reperfusion. It is accordingly unlikely that a single drug or pharmacological approach will provide the universal panacea to completely restitute the threatened brain tissue. In the long run, optimal cytoprotection is likely to be achieved by a combination of therapeutic approaches targeted at the various pathogenic processes that engender neuronal loss in the ischaemic brain.

The final perspective for future research is to better understand the biological mechanisms that subserve cell sparing and loss in the ischaemic brain. In this review we have restricted ourselves to the discussion of pertinent and reproducible models of focal cerebral ischaemia in which, by various interventions, lesion size may be significantly modulated. Yet, although the obligatory initial stage, the measurement of infarction volume alone will not answer all our problems. The ground is now ripe to apply well-established methodologies (e.g. blood flow, glucose use, protein synthesis, receptor distribution, blood-brain barrier permeability) to analyse the dynamic events that explain or mitigate neuronal loss following focal cerebral ischaemia.

Acknowledgements. We are most grateful to our colleagues in L.E.R.S. and the University of Glasgow for allowing us this possibility to review their studies. We thank Mme. M.-L. Pernelet for the typing of this manuscript.

References

1. Bederson JB, Pitts LH, Germano SM, Nishimura MC, Davis RL, Bartkowski HM (1986) Evaluation of 2,3,5-triphenyltetrazolium chloride as a stain for detection and quantification of experimental cerebral infarction in rats. Stroke 17:1304–1308
2. Benveniste H, Prejer J, Schousboe A, Diemer NH (1984) Elevation of extracellular concentrations of glutamate and aspartate in rat hippocampus during transient cerebral ischaemia monitored by intracerebral microdialysis. J Neurochem 43:1369–1374
3. Carter C, Benavides J, Legendre P et al. (1988) Ifenprodil and SL 82.0715 as cerebral anti-ischaemic agents. II. Evidence for NMDA receptor antagonist properties. J Pharmacol Exp Ther 247:1222–1232
4. Dubois A, Benavides J, Peny B et al. (1988) Imaging of primary and remote ischaemic and excitotoxic brain lesions. An autoradiographic study of peripheral type benzodiazepine binding sites in the rat and cat. Brain Res 445:77–90
5. Duverger D, MacKenzie ET (1988) The quantification of cerebral infarction following focal ischemia in the rat: influence of strain, arterial pressure, blood glucose concentration, and age. J Cereb Blood Flow Metabol 8:449–461
6. Germano IM, Pitts LH, Meldrum BS, Bartkowski HM, Simon RP (1987) Kynurenate inhibition of cell excitation decreases stroke size and deficits. Ann Neurol 22:730–734
7. Gotti B, Benavides J, MacKenzie ET, Scatton B (1988) Activité protectrice des antagonistes du récepteur *N*-méthyl-D-aspartate (NMDA) dans un modèle d'ischémie cérébrale focale chez la souris. Circ Métabol Cerveau 5:201
8. Gotti B, Duverger D, Bertin J et al. (1988) Ifenprodil and SL 82.0715 as cerebral anti-ischaemic agents. I. Evidence for efficacy in models of focal cerebral ischaemia. J Pharmacol Exp Ther 247:1211–1221
9. Hagberg H, Andersson P, Butcher S, Sandberg M, Lehmann A, Hamberger A (1986) Blockade of *N*-methyl-D-aspartate-sensitive, acidic amino acid receptors inhibits ischemia-induced accumulation of purine catabolites in the rat striatum. Neurosci Lett 68:311–316

10. Hagberg H, Lehmann A, Sandberg A, Nystrom B, Jacobson I, Hamberger A (1985) Ischaemia-induced shift of inhibitory and excitatory amino acids from intra- to extracellular compartments. J Cereb Blood Flow Metab 5:413–419
11. Jørgensen MB, Diemer NH (1982) Selective neuron loss after cerebral ischemia in the rat: possible role of transmitter glutamate. Acta Neurol Scand 66:536–546
12. Kemp JA, Foster AC, Wong HF (1987) Non-competitive antagonists of excitatory amino acid receptors. TINS 10:294–298
13. Mackenzie ET, Duverger D, Gotti B, Nowicki JP (1986) Cerebral anti-ischaemic agents: problems and approaches. In: J Krieglstein (ed) Pharmacology of cerebral ischaemia. Elsevier, Amsterdam, pp 167–179
14. Olney JW (1978) Neurotoxicity of excitatory amino acids. In: McGeer EG, Olney JW, McGeer PL (eds) Kainic acids as a tool in neurobiology. Raven, New York, pp 95–121
15. Osborne KA, Shigheno T, Balarsky AM, Ford I, McCulloch J, Teasdale GM, Graham DI (1987) Quantitative assessment of early brain damage in a rat model of focal cerebral ischaemia. J Neurol Neurosurg Psychiatry 50:402–410
16. Ozyurt E, Graham DI, Woodruff GN, McCulloch J (1988) Protective effect of the glutamate antagonist, MK-801 in focal cerebral ischemia in the cat. J Cereb Blood Flow Metab 8:138–143
17. Park CK, Nehls DG, Graham DI, McCulloch J (1988) Focal cerebral ischaemia in the cat: treatment with the glutamate antagonist MK-801 after induction of ischaemia. J Cereb Blood Flow Metab 8:757–762
18. Park CK, Nehls DG, Graham DI, Teasdale GM, McCulloch J (1988) The glutamate antagonist MK-801 reduces focal ischaemic brain damage in the rat. Ann Neurol 24:543–551
19. Rothman SM, Olney JW (1986) Glutamate and the pathophysiology of hypoxic-ischaemic brain damage. Ann Neurol 19:105–111
20. Rothman SM, Olney JW (1987) Excitotoxicity and the NMDA receptor. TINS 10:299–302
21. Russel RWR (1983) Transient cerebral ischaemia. In: Ross Russel RW (ed) Vascular disease of the central nervous system. Churchill Livingstone, Edinburgh, pp 204–223
22. Simon RP, Swan JH, Griffiths T, Meldrum BS (1984) Blockade of *N*-methyl-D-aspartate receptors may protect against ischemic damage in the brain. Science 226:850–852
23. Sindermann F, Dichgans J, Bergleiter R (1969) Occlusion of the middle cerebral artery and its branches: angiographic and clinical correlates. Brain 92:607–620
24. Weiss J, Goldberg MP, Choi DW (1986) Ketamine protects cultured neocortical neurons from hypoxic injury. Brain Res 380:186–190
25. Welsh FA, Sakamoto T, McKee AE, Sims RE (1987) Effect of lactacidosis on pyridine nucleotide stability during ischemia in mouse brain. J Neurochem 49:846–851
26. Wieloch T (1985) Hypoglycemia-induced neuronal damage prevented by an *N*-methyl-D-aspartate antagonist. Science 230:681–683
27. Wieloch T, Engelsen B, Westerberg E, Auer R (1985) Lesions of the glutamatergic corticostriatal projections in the rat ameliorate hypoglycemic brain damage in the striatum. Neurosci Lett 58:25–30

Calcium Antagonists and Experimental Focal Cerebral Ischemia

J. McCulloch[1], D. I. Graham[1], A. M. Harper[1], and G. M. Teasdale[2]

Introduction

Calcium channel blockers have the potential to attenuate ischaemic brain damage via a number of distinct mechanisms. Voltage-dependent Ca^{2+} channels are present on cerebrovascular smooth muscle and their blockade results in relaxation. Calcium channel blockers are thus capable of improving cerebral tissue perfusion in ischaemic tissue, to levels above those associated with neuronal damage. Alternatively, or additionally, calcium channel blockers, by limiting the influx of calcium into neurones, might prevent the chain of intracellular events which calcium initiates (protease and lipase activation, mitochondrial dysfunction, etc.) and which leads ultimately to irreversible ischaemic cell damage (for reviews see Siesjö 1981; Greenberg 1987). In this review, we consider the evidence that calcium channel blockers are capable of influencing the amount of ischaemic brain damage in animal models of focal cerebral ischaemia and the extent to which improved perfusion or blockade of neuronal Ca^{2+} entry contributes to any amelioration. The review is based on published reports of the effects of nimodipine in the rat from our laboratory (Mohamed et al. 1984, 1985a, b; Gotoh et al. 1986).

Methods

Surgical Preparation

Adult male Sprague-Dawley rats, weighing approximately 350 g, were anaesthetised with 2% halothane, and a tracheostomy was performed. The animals were mechanically ventilated with a nitrous oxide-oxygen mixture (70% : 30%) together with 0.5% halothane. Polyethylene catheters (15 cm long) were inserted into both femoral arteries and veins to allow the continuous monitoring of blood pressure, the repeated sampling of arterial blood, and the administration of drug and radioactive tracer.

Normocapnia (P_aCO_2 approx. 38 mmHg) and adequate oxygenation (P_aO_2 greater than 100 mmHg) were maintained throughout the experiment. Body tem-

[1] Wellcome Neuroscience Group, Wellcome Surgical Institute, University of Glasgow, Glasgow G61 1QH, U.K.
[2] Institute of Neurological Sciences, Southern General Hospital, Glasgow G51 4TF, U.K.

Cerebral Ischemia and Calcium
Ed. by A. Hartmann and W. Kuschinsky

perature was monitored by a rectal thermometer and the animals were maintained normothermic (37° C) by external heating. Arterial plasma glucose was assessed repeatedly during the experiment.

Induction of Focal Cerebral Ischaemia

The proximal portion of the MCA was permanently occluded by a modification of the microsurgical technique developed in our laboratories (Tamura et al. 1981). To ensure adequate depth of anaesthesia, halothane concentration was increased to 1%–1.5% during the surgery. After partial removal of the temporalis muscle, but without removing the zygomatic arch, the animals underwent a left subtemporal craniectomy. The stem of the MCA was electrocauterised just medial to the olfactory tract, and was transected to ensure the completeness of the vascular occlusion. After the surgery, anaesthesia was maintained with 0.5% halothane. The animals were sacrificed 4 h after MCA occlusion.

Histopathology and Quantification of Ischaemic Damage

The rats were perfusion-fixed with 40% formaldehyde, glacial acetic acid, absolute methanol (1 : 1 : 8, vol/vol/vol) (FAM) for the neuropathological studies. The animals were placed in the supine position and the chest opened. A cannula was passed into the ascending aorta via the left ventricle. The right atrium was incised, and heparinised physiological saline was infused briefly at a pressure equal to MABP in the animal. The saline infusion was followed by perfusion with 150 ml FAM at the same pressure. The animals were decapitated after perfusion, and the head was stored in FAM fixative at 4° C for at least 12 h. The brain was then removed.

Coronal slices of brain were embedded in paraffin wax, sectioned at multiple levels and stained by haematoxylin and eosin or a method combining cresyl violet and Luxol fast blue. The sections were examined by conventional light microscopy by one of us (DIG) who was unaware of the pharmacological treatment received by the animal. Areas of ischaemic damage were drawn on eight, stereotactically determined coronal levels of forebrain drawn to scale using the histological criteria of Brown and Brierley (1968) in FAM-fixed material. The areas of damage were measured on an image analyser and volumes of damage calculated from the known distances between the sections (Osborne et al. 1987).

Measurement of Local CBF

Local CBF (LCBF) was measured using the quantitative autoradiographic technique with [^{14}C]iodoantipyrine as tracer (Sakurada et al. 1978). [^{14}C]Iodoantipyrine (50 μCi in 1.5 ml saline) was infused intravenously over a 30-s period on a "ramp" schedule, i.e. the rate of infusion was increased progressively so that arterial tracer concentration would be highest at the end of the experiments. During administration of the tracer, 15–18 timed samples of arterial blood were collected in preweighed filter paper discs from a free-flowing femoral arterial catheter. At the end of the infusion of the isotope, the animals were killed by decapitation, and the brains were quickly dissected out and frozen in isopentane at $-45°$ C within 2–3 min. The filter discs were tightly capped in scintillation vials to prevent evaporation, and reweighed. Radioisotope levels were measured by liquid scintillation counting. Brain sections (20 µm thick) were cut in a cryostat at $-22°$ C, and three in every 20 sections were mounted on glass coverslips and dried on a hot plate. Autoradiographs were prepared from these sections together with a set of nine precalibrated epoxy resin standards (concentration range 44–1175 nCi/g) by applying them to X-ray film (Kodak SB-5) for 7–10 days in light-tight cassettes.

The radioisotope concentrations in the brain sections were measured by quantitative densitometry either at neuroanatomically defined brain areas or by frequency distribution analysis of defined coronal planes. LCBF was calculated from the concentration of [^{14}C]iodoantipyrine in regions of the brain, from the brain-blood partition coefficient for the tracer (0.8), and from the history of the tracer in arterial blood during the experimental period (Sakurada et al. 1978).

Experimental Protocols

Nimodipine solution was prepared freshly on the morning of each experiment by dissolving the drug in lutrol-glycerine-water solvent (25 µg ml^{-1}). Because the dihydropyridine is photosensitive, the solution was made under sodium light and the infusion catheters and syringes were covered by silver foil. Nimodipine (1 µg kg^{-1} min^{-1}) was administered as a continuous intravenous infusion until sacrifice. Every control animal received an equivalent amount of the lutrol-glycerine-water solvent.

Results

Effects of Nimodipine on Ischaemic Brain Damage After MCA Occlusion

Nimodipine Pre-Occlusion

Ischaemic damage was observed only within the territory of the occluded MCA, i.e. in the dorsolateral cortex and in the neostriatum. These areas showed the morphological characteristics of early infarction-microvacuolation, shrinkage and triangulation of nucleus and cytoplasm, increased basophilia of cytoplasm, and perineuronal swollen astrocytes (Brierley and Graham 1984). The lesion was sharply delineated, with the boundary between the ischaemic area and normal brain being less than 100 µm. The microscopic appearance of ischaemic damage in nimodipine-treated animals was similar to that in vehicle-treated controls.

The intravenous infusion of nimodipine (1 µg kg^{-1} min^{-1}), initiated 30 min prior to MCA occlusion, significantly reduced (by 34%) the volume of ischaemic brain damage present in the cerebral hemisphere. The reduction in ischaemic damage with nimodipine pretreatment occurred principally in the cerebral cortex

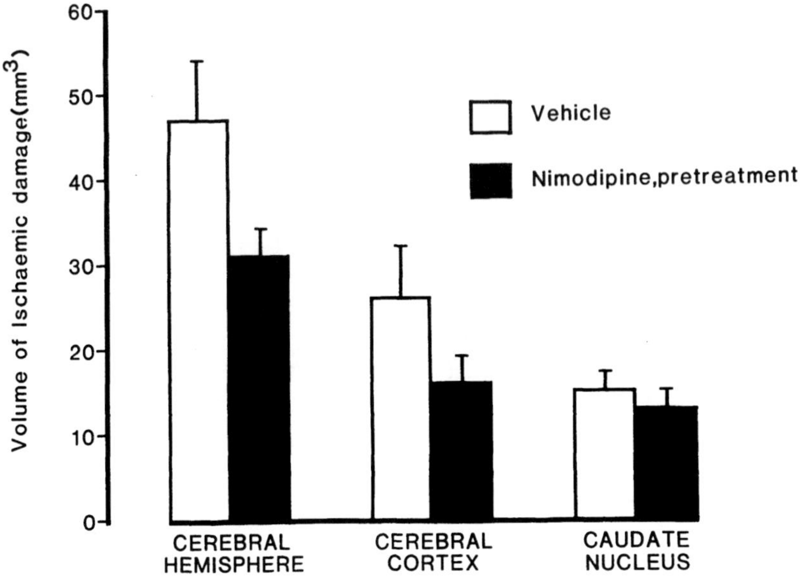

Fig. 1. Nimodipine pretreatment and volumes of ischaemic brain damage after MCA occlusion. Note the reductions with nimodipine in the volumes of damage in the cerebral hemisphere and cerebral cortex. Data are presented as mean ± SEM ($n = 8$ in each group). (Figure redrawn from the data of Mohamed et al. 1985 a)

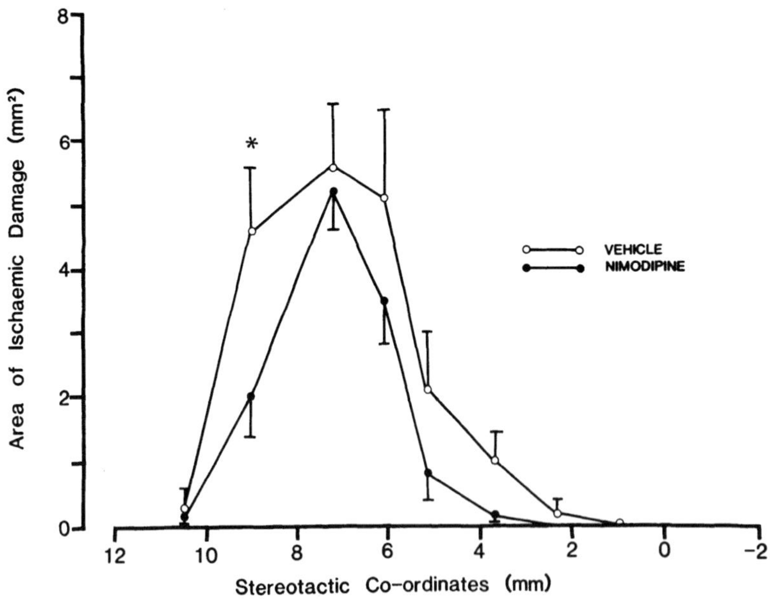

Fig. 2. Nimodipine pretreatment and areas of ischaemic brain damage at various stereotactic planes after MCA occlusion. Note the failure of nimodipine to alter the area in the core of MCA territory (plane 7.2 mm anterior) with the reductions in the areas of damage at levels more peripheral to the core (e.g. planes 8.9 mm and 5.15 mm anterior) with nimodipine pretreatment. Data are presented as mean ± SEM ($n = 8$ in each group). (Modified from Mohamed et al. 1985 a)

(ischaemic volume reduced by 38%), with minimal alteration in the amount of damage present in the caudate nucleus (reduced by 13%) (Fig. 1).

Nimodipine pretreatment had little effect upon the area of ischaemic damage in the core of MCA territory (stereotactic plane 7.2 mm anterior to the interaural line). The neuroprotective effects of nimodipine were most marked in the periphery of MCA territory, e.g. in stereotactic planes 8.9 and 5.15 mm anterior; here the areas of damage were reduced by approximately 60% with nimodipine pretreatment compared to vehicle-treated controls (Fig. 2).

Nimodipine Postocclusion

The intravenous infusion of nimodipine (1 μg kg^{-1} min^{-1}) initiated 5 min after the occlusion of the MCA and maintained to sacrifice had no significant effect on the volume of ischaemic damage in the cerebral hemisphere (ischaemic volume increased by 15% relative to controls), cerebral cortex (increased by 19%) and caudate nucleus (increased by 10%) (Fig. 3). Nimodipine treatment initiated after MCA occlusion was without significant effect on the area of ischaemic damage at each stereotactic plane examined.

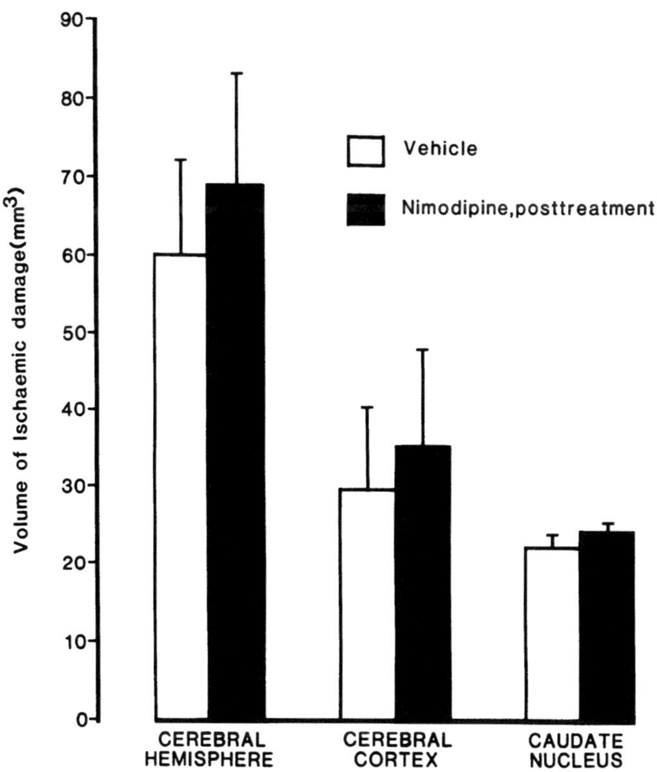

Fig. 3. Nimodipine treatment initiated 5 min after MCA occlusion and volumes of ischaemic brain damage. There are no significant differences in any region. Data are presented as mean ± SEM ($n = 5$ in each group). (Figure redrawn from data of Gotoh et al. 1986)

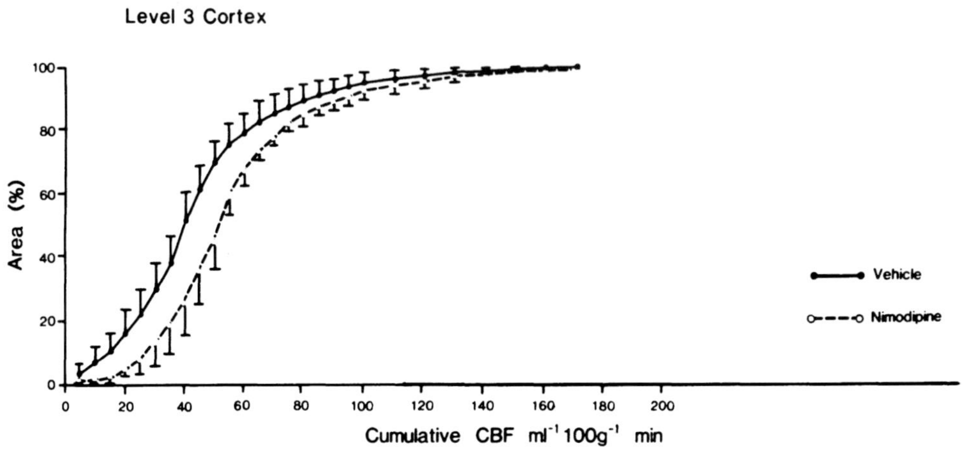

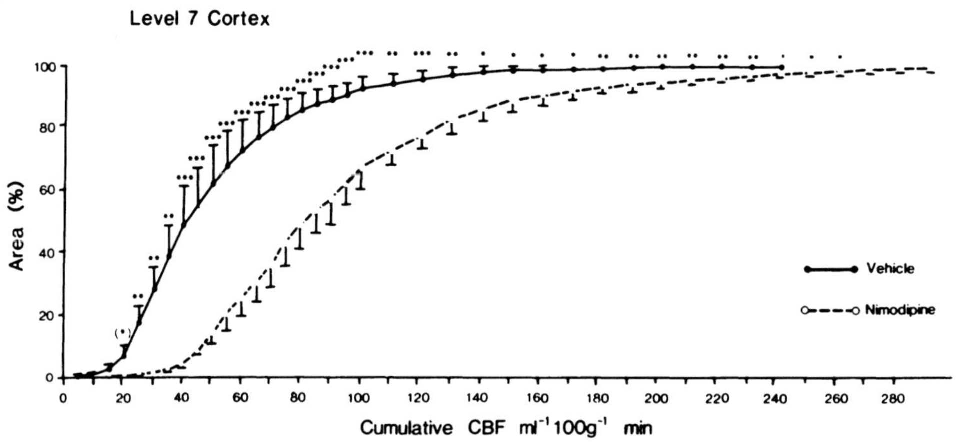

Fig. 4. Effect of nimodipine on CBF (frequency distribution analysis) after middle cerebral artery (MCA) occlusion. Nimodipine administration (1 μg kg^{-1} min^{-1}) was initiated 30 min prior to MCA occlusion and CBF determined 30 min subsequently. Individual sections display the cumulative area of the CNS (*y-axes*) at three defined coronal planes at cumulative CBF intervals (*x-axes*). Data are presented as means ± SEM ($n = 8$). Nimodipine (*open circles*); vehicle (*filled circles*). These data are derived from reanalysis of the CBF investigations of Mohamed et al. (1985a). Only data from the hemisphere ipsilateral to MCA occlusion are shown. Nimodipine did not alter significantly the CBF pattern in cerebral cortex at coronal plane 2.2 mm anterior. By contrast, nimodipine had significant effects upon cortical blood flow at coronal plane 2.2 mm anterior. In both these stereotactic levels, the CBF-area curves are shifted to the right with nimodipine treatment, indicating that low levels of CBF are present in a smaller fraction of the brain with nimodipine pretreatment. The difference in the response to nimodipine at plane 7.2 mm anterior compared with plane 2.2 mm anterior indicates a lack of effect of the drug on CBF in the central core of the ischaemic lesion but with improved cerebral perfusion in more peripheral parts of the lesion in nimodipine-pretreated animals. (Modified from Gotoh et al. 1986)

Effects of Nimodipine on CBF to Ischaemic Cerebral Tissue

The ability of nimodipine to increase CBF in non-ischaemic cerebral tissue has been demonstrated in a number of species and with a variety of measurement techniques. With iodoantipyrine autoradiography in the halothane-anaesthetised rat, nimodipine (1 μg kg^{-1} min^{-1} for 60 min) increases CBF in the frontal cortex by approximately 65% and in the caudate nucleus by approximately 30% (Mohamed et al. 1984). Data such as these are derived from densitometric analysis of the autoradiograph using regions of interest defined by neuroanatomical criteria. Such approaches are poorly suited for the analysis of the neuroanatomically disordered patterns of blood flow which are characteristic of focal cerebral ischaemia. A more objective approach is to delineate an entire brain area and with frequency distribution analysis, define the amount of cerebral tissue within defined blood flow ranges (Gotoh et al. 1986).

In the cortex ipsilateral to MCA occlusion (stereotactic plane 7.2 mm anterior), nimodipine pretreatment had no significant effect upon median CBF (39 ± 5 ml 100 g^{-1} min^{-1} in vehicle-treated animals, 48 ± 4 ml 100 g^{-1} min^{-1} with nimodipine pretreatment). In contrast, in the ipsilateral cortex in the periphery of MCA territory (stereotactic plane 2.2 mm anterior), nimodipine pretreatment significantly increased median CBF (from 43 ± 7 ml 100 g^{-1} min^{-1} in vehicle-treated controls to 83 ± 7 ml 100 g^{-1} min^{-1} with nimodipine pretreatment, $P < 0.01$).

Analysis of hypoperfused tissue revealed a similar pattern. In the core of occluded MCA territory, there was no significant difference in the amount of cortical tissue with very low flow (less than 30 ml 100 g^{-1} min^{-1}), i.e. 30% ± 9% in vehicle-treated controls, 13% ± 7% with nimodipine pretreatment. In contrast, at the periphery of the occluded MCA territory, the amount of cortex with very low flow was markedly and significantly reduced with nimodipine pretreatment (i.e. 28% ± 8% in vehicle-treated controls, 2% ± 1% with nimodipine pretreatment, $P < 0.01$) (Fig. 4).

Nimodipine treatment (1 μg kg^{-1} min^{-1}) initiated 5 min after MCA occlusion had no significant effect at 30 min postocclusion on median CBF at any stereotactic level or on the amount of hypoperfused tissue at any level.

Discussion

A number of conclusions can be drawn from the investigation of effects of nimodipine in experimental models of focal cerebral ischaemia. Firstly, nimodipine is capable of reducing the amount of ischaemic brain damage after permanent MCA occlusion. Secondly, the improved histopathological outcome with nimodipine after MCA occlusion is associated with improved LCBF, i.e. cerebral blood flow is reduced, but not to such low levels as in vehicle-treated animals. Thirdly, these effects of nimodipine are lost if the drug infusion schedule is altered from the most favourable approach (i.e. preloading the tissue with the drug). It should be emphasised that the investigations of the effects of nimodipine in experimental focal ischaemia were initiated only after the cerebrovascular effects of nimodipine had been characterised (Mohamed et al. 1984, 1985 b) and with an appreciation of the pharmacokinetics of nimodipine in the rat (Maruhn et al. 1985). The concentration of nimodipine (1 μg kg^{-1} min^{-1}) was selected on the basis of maximal increases in CBF in the absence of altered cerebral glucose utilisation and without pronounced hypotension or hyperglycaemia, two variables known to have an influence on the extent of ischaemic brain damage. Moreover, the duration of the nimodipine infusion is critical in view of the short plasma half-life (20 min) of the agent in the rat. It is naive to equate studies of the effects of nimodipine in MCA occlusion models in the rat where similar concentrations of the agent are administered for 10 min (Germano et al. 1987) or for 270 min (including 30 min pre-occlusion) (present study; Mohamed et al. 1985). This is particularly important when histopathological outcome after MCA occlusion is assessed at different time points, e.g. 20 h after end of nimodipine infusion (i.e. 24 h after MCA occlusion) (Germano et al. 1987) or after continuous infusion of nimodipine until sacrifice 4 h following MCA occlusion (present study; Mohamed et al. 1985 a).

There is a growing consensus for the efficacy of nimodipine and other calcium antagonists in models of focal cerebral ischaemia; the mechanism appears to be improved perfusion in the periphery of the ischaemic region. Improved outcome after MCA occlusion (assessed histologically or indirectly via the severity of brain acidosis) has been demonstrated with nimodipine treatment in a number of species (Barnett et al. 1986; Meyer et al. 1986; Hakim 1986; Germano et al. 1987; Bartkowski 1988; van Reempts et al. 1987), although in some instances the benefit does not achieve statistical significance (see Barnett et al. 1986). Our data emphasise the importance of assessing the *volume* of ischaemic brain damage; a measurement of the *area* of ischaemic brain damage at a single level in the ischaemic core is poorly indicative not only of the size of the lesion (Osborne et al. 1987) but also of the efficacy of drugs like nimodipine (see Fig. 2).

The NMDA receptor antagonist MK-801 also reduces the volume of ischaemic brain damage after MCA occlusion in the rat (Park et al. 1988) putatively by attenuating glutamate activation and consequently calcium entry into neurones. The anti-ischaemic effects of MK-801 differ from those of nimodipine in two respects, viz. reductions in the areas of ischaemic damage are not confined to the periphery of MCA territory and secondly, the histopathological improvement is *not* associated with any improvement in blood flow to the ischaemic tissue.

There is less convincing evidence of improved morphological outcome with nimodipine when it is administered after MCA occlusion (Gotoh et al. 1986; Sauter and Rudin 1986) and groups reporting benefit with a low concentration of nimodipine administered late (4 h) after MCA occlusion have experienced difficulty replicating their own observations (Germano et al. 1987; Kobayashi et al. 1988). In our studies, nimodipine infusion initiated 5 min after MCA occlusion had no effect on the volume of ischaemic tissue in animals sacrificed at 4 h postocclusion (Gotoh et al. 1986). Whether this is due to our failure to achieve adequate brain concentrations of the drug during the crucial 1st hour after MCA occlusion or whether it is indicative of the obligatory requirement for pretreatment is unclear. The failure of the nimodipine postocclusion treatment schedule to increase CBF in the contralateral hemisphere provides support for the former possibility.

In contrast to focal cerebral ischaemia, there is little reliable evidence in models of global ischaemia where benefit has been demonstrated histologically with nimodipine or other calcium antagonists. However, the experimental data provided here provide a sound basis for the clinical trials of nimodipine in focal cerebral ischaemia, e.g. stroke (see Gelmers et al. 1988) or subarachnoid haemorrhage where prophylaxis is possible (Teasdale et al. this volume).

References

Barnett GH, Bose B, Little JR, Jones SC, Friel HT (1986) Effects of nimodipine on acute focal cerebral ischemia. Stroke 17:884–890

Bartkowski HM (1988) Therapeutic value of a constant intravenous infusion of nimodipine for 24 h after acute cerebral infarction. Stroke 19:147

Brierley JB, Graham DI (1985) Hypoxia and vascular disorders of the central nervous system. In: Adams JH, Corsellis JAN, Duchen LW (eds) Greenfield's neuropathology. Edward Arnold, London, pp 125–207

Brown AW, Brierley JB (1968) The nature, distribution and earliest stages of anoxic ischaemic nerve cell damage in the rat as defined by the optical microscope. Br J Exp Pathol 49:87–106

Gelmers HJ, Gorter K, De Weerdt CJ, Wiezer HJA (1988) A controlled trial of nimodipine in acute ischemic stroke. N Engl J Med 318:203–207

Germano IM, Bartkowski HM, Cassel ME, Pitts LH (1987) The therapeutic value of nimodipine in experimental focal cerebral ischemia. J Neurosurg 67:81–87

Gotoh O, Mohamed AA, McCulloch J, Graham DI, Harper AM, Teasdale GM (1986) Nimodipine and the haemodynamic and histopathological consequence of middle cerebral artery occlusion in the rat. J Cereb Blood Flow Metab 6:321–331

Greenberg DA (1987) Calcium channels and calcium channel antagonists. Ann Neurol 21:317–330

Hakim AM (1986) Cerebral acidosis in focal ischemia: II. Nimodipine and verapamil normalize cerebral pH following middle cerebral artery occlusion in the rat. J Cereb Blood Flow Metab 6:676–683

Kobayashi S, Obana W, Andrews BT, Nishimura MC, Pitts LH (1988) Lack of effect of nimodipine in experimental regional cerebral ischemia. Stroke 19:147

Maruhn D, Siefert HM, Weber H, Ramsch K, Suwelack D (1985) Pharmacokinetics of nimodipine. Arzneimittelforschung 35:1781–1786

Meyer FB, Anderson RE, Yaksh TL, Sundt TM (1986) Effect of nimodipine on intracellular brain pH, cortical blood flow, and EEG in experimental focal cerebral ischemia. J Neurosurg 64:617–626

Mohamed AA, McCulloch J, Mendelow AD, Teasdale GM, Harper AM (1984) Effect of the calcium antagonist nimodipine on local cerebral blood flow: relationship to arterial blood pressure. J Cereb Blood Flow Metab 4:206–211

Mohamed AA, Gotoh O, Graham DI et al. (1985a) Effect of pretreatment with the calcium antagonist nimodipine on local cerebral blood flow and histopathology after middle cerebral artery occlusion. Ann Neurol 18:705–711

Mohamed AA, Mendelow AD, Teasdale GM, Harper AM, McCulloch J (1985b) Effect of the calcium antagonist, nimodipine, on local cerebral blood flow and metabolic coupling. J Cereb Blood Flow Metab 5:26–33

Osborne KA, Shigeno T, Balarsky AM, Ford I, McCulloch J, Teasdale GM, Graham DI (1987) Quantitative assessment of early brain damage in a rat model of focal cerebral ischaemia. J Neurol Neurosurg Psychiatry 50:402–410

Park CK, Nehls DG, Graham DI, Teasdale GM, McCulloch J (1988) The glutamate antagonist, MK-801 reduces focal ischaemic brain damage in the rat. Am Neurol 24:543–551
Sakurada O, Kennedy C, Jehle J, Brown JD, Carbin GL, Sokoloff L (1978) Measurement of local cerebral blood flow with iodo[^{14}C]antipyrine. Am J Physiol 234:H59–H66
Sauter A, Rudin M (1986) Calcium antagonists reduce the extent of infarction in rat middle cerebral artery occlusion model as determined by quantitative magnetic resonance imaging. Stroke 17:1228–1234
Siesjö BK (1981) Cell damage in the brain: a speculative hypothesis. J Cereb Blood Flow Metab 1:155–186
Tamura A, Graham DI, McCulloch J, Teasdale GM (1981) Focal cerebral ischaemia in the rat: 1. Description of technique and early neuropathological consequences following middle cerebral artery occlusion. J Cereb Blood Flow Metab 1:53–60
Van Reempts J, Van Deuren B, Van de Ven M, Cornelissen F, Borgers M (1987) Flunarizine reduces cerebral infarct size after photochemically induced thrombosis in spontaneously hypertensive rats. Stroke 18:1113–1119

Effect of Calcium Antagonists on Cerebral Blood Flow, Protein Synthesis, and Histopathology After Global Ischemia of Gerbil Brain

M. KOCHER, Y. XIE, G. BRINKER, E. DUX, and K.-A. HOSSMANN

Introduction

Under physiological conditions the intracellular concentration of free calcium (10^{-7} mol/l) is about four orders of magnitude lower than the concentration in the extracellular space (10^{-3} mol/l) [16]. If local cerebral flow is reduced to about 10 ml/100 g/min, neurons depolarize and extracellular calcium enters the cell across voltage-dependent channels [12]. The resulting calcium overload of the cytosol has been discussed as triggering a series of pathophysiological processes leading to the manifestation of ischemic neuronal injury [26].

An increase in intracellular calcium activity may also occur in the recirculation period despite reactivation of energy metabolism and ion transport systems, if neurons undergo a phase of increased postischemic excitation. This process is associated with increased calcium fluxes through channels operated by excitatory amino acids and has been suggested to contribute to the phenomenon of the selective ischemic vulnerability of the hippocampal subfield CA1 [24, 31].

However, there are also arguments against the calcium hypothesis of ischemic cell injury. Increase of calcium uptake follows rather than precedes the manifestation of ischemic injury [5, 6, 24, 28], and there are no significant differences between the calcium content of hippocampus and that of more resistant regions during the early phases of postischemic recirculation [13]. There are also divergent opinions about the beneficial effects of calcium antagonists for the prevention of ischemic or postischemic cell injury [34], and about the mechanisms involved [23, 30].

In order to substantiate the postulated therapeutic value of calcium antagonists for the prevention of ischemic cell injury, the following pieces of evidence should be provided: (a) evidence of increased cytosolic calcium activity during and after ischemia, (b) evidence of its interference with vital cellular functions, (c) evidence of the prevention of cytosolic calcium overload by calcium antagonists, (d) evidence of the reversal of the disturbed cellular functions by calcium antagonists, and (e) exclusion of nonmolecular, in particular hemodynamic, effects.

A straightforward investigation of these points is hindered by the difficulty of measuring cytosolic calcium activity and of defining which of the various cellular functions impaired by ischemia is the limiting factor for neuronal survival. In the

Max-Planck-Institute for Neurological Research, Department for Experimental Neurology, Ostmerheimer Str. 200, D-5000 Köln 91, FRG.

Cerebral Ischemia and Calcium
Ed. by A. Hartmann and W. Kuschinsky

present communication, therefore, a more indirect approach has been used to solve some of these problems.

We used the gerbil model of brief global ischemia [14, 15], which results in the selective destruction of CA1 pyramidal cells of the hippocampus after a maturation period of 2–4 days. As a measure of cytosolic calcium activity, mitochondrial calcium sequestration was determined by cytochemical methods. This approach is based on the fact that mitochondrial calcium sequestration is a regulatory process for the maintenance of physiological calcium activity, and that mitochondria will use their entire respiratory capacity for this process when exposed to calcium concentrations of 0.01 mmol/l or more [21].

For the evaluation of vital cellular function, protein synthesis and neuronal injury were studied. It was previously shown in this model that after ischemia protein synthesis is severely inhibited although energy metabolism rapidly recovers, and that this inhibition is irreversible in the selectively vulnerable areas [35]. Prevention of cytosolic calcium overload by calcium antagonists, therefore, should be reflected by a reversal of this disturbance and less severe neuronal damage. Finally, the possibility of a hemodynamic effect of calcium antagonists was evaluated. Since calcium antagonists are potent cerebral vasodilators [11, 19] and were found to diminish postischemic delayed hypoperfusion [18, 27, 29, 30], a potentially beneficial effect might be mediated by an improvement of postischemic hemodynamic disturbances.

The results of the original investigations on which this communication is based have been [8] or will be published elsewhere (Brinker and Hossmann, in preparation; Xie et al., in preparation).

Material and Methods

Animal Preparation

Adult gerbils of both sexes, weighing 55–75 g, were anesthetized with 2%–3% halothane. Both common carotid arteries were exposed and cross-clamped for 5 or 10 min with aneurysm clips. Animals were then processed for one of the following procedures.

Mitochondrial Calcium Uptake

Following 5 min of ischemia and various recirculation times ranging between 5 min and 48 h ($n = 8$ in each group), animals were perfusion fixated with a solution containing 2% paraformaldehyde, 3% glutaraldehyde, and 90 mμ potassium oxalate. Ultrathin sections were cut from the dissected CA1 subfield and from pieces of the parietal cortex and prepared for microscopic visualization of calcium, using the combined oxalate pyroantimonate method [32]. The amount of calcium deposits visible in the mitochondria of CA1 pyramidal cells and in cortical neurons was classified as follows: 1 = few dust-like deposits, 2 = several grains, 3 = numerous massive calcium deposits associated with ultrastructural alteration. In each group, 40 micrographs (magnification 20000) were analyzed.

Protein Synthesis

Following 5 min of ischemia and 2 h or 2 days of recirculation, animals were reanesthetized with 2% halothane. A tracer quantity of ^{3}H-labeled amino acids (mixture of tyrosine, phenylalanine, leucine, isoleucine, and methionine, 1 mCi/kg) was intravenously infused over 30 s and allowed to circulate for 45 min. Animals were then decapitated; brains were rapidly removed from the skull and frozen at $-50°$ C in precooled butanol. Treated animals received a single injection of the calcium antagonist nimodipine (1.5 mg/kg i.p., $n = 6$) or flunarizine (1.5 mg/kg i.p., $n = 6$) at the beginning of recirculation. Untreated animals ($n = 6$) served as controls.

Coronal cryostat sections (20 μm) were prepared at $-20°$ C. In order to remove labeled amino acids not incorporated into protein, sections were wash-incubated in a 10% TCA solution overnight. Uneluted sections and adjacent TCA-eluted sections were exposed to LKB-Ultrofilm.

Autoradiographs were analyzed by an image processing system (ID 2000, De Anza Systems). Regional protein synthesis was expressed as the percentage of total tissue radioactivity that was retained in the TCA-eluted section.

Histopathology

In five gerbils subjected to 10 min of ischemia, nimodipine (1.5 mg/kg i.p.) was given immediately after the restoration of blood flow. Untreated animals ($n = 5$) served as control. After 7 days, animals were decapitated and brains were removed and immersion fixated in 4% formaldehyde solution. Coronal sections (4 μm) passing through the dorsal hippocampus were stained with cresyl violet. To assess the severity of ischemic damage in the hippocampus, the number of intact pyramidal cells per mm CA1 sector were counted under light microscopy.

Cortical Blood Flow

Before production of 10 min of ischemia the dorsal skull was exposed and the probe of a laser Doppler flowmeter (PeriFlux PF 2, PERIMED, Sweden) was positioned 1 mm lateral and 1 mm rostral to bregma. Laser Doppler flow was measured before, during, and up to 30 min after ischemia. In five gerbils, nimodipine (1.5 mg/kg i.p.) was given immediately after the restoration of blood flow. Untreated animals ($n = 5$) served as control.

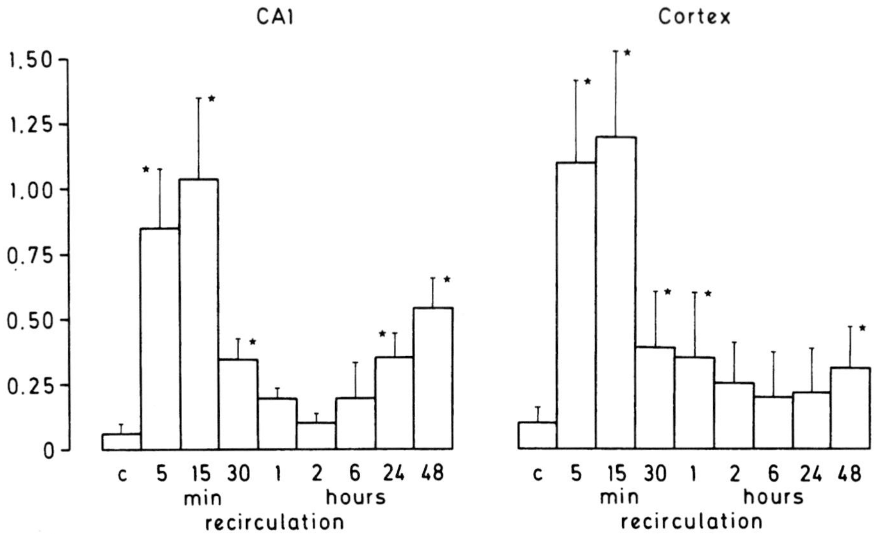

Fig. 1. Mitochondrial calcium uptake index of neurons from CA1 sector and parietal cortex at various recirculation times after 5 min of bilateral common carotid occlusion in gerbils. Uptake index: 0 = no calcium deposits; 1 = few dust-like deposits; 2 = several grains; 3 = numerous massive deposits associated with ultrastructural degeneration. Mean ± SD; *c*, control; $^*P<0.01$

Results

Mitochondrial Calcium Uptake

As illustrated by Fig. 1, calcium deposits were detected in the mitochondria of cortical and of hippocampal CA1 pyramidal neurons at 5–15 min post ischemia. Mean mitochondrial calcium uptake index significantly increased from 0.05 to 1.05 ($P<0.01$) in CA1 neurons and from 0.09 to 1.20 ($P<0.01$) in cortical neurons. At this time, the ultrastructure of mitochondria remained intact. After 2 h of recirculation, the mitochondrial precipitates disappeared. However, after 24–48 h of recirculation, a secondary migration of calcium into mitochondria was observed, but only in neurons which also showed ultrastructural degeneration at this time. This second increase in mitochondrial uptake of calcium was mostly expressed in CA1 pyramidal neurons.

Effect of Calcium Antagonists

Protein Synthesis

The autoradiographic evaluation of protein synthesis in control gerbils revealed that approximately 70%–80% of total tissue radioactivity was incorporated into proteins after circulation of the tracer for 45 min. As shown in Table 1, protein

Table 1. Amino acid incorporation into brain proteins of gerbils. Effect of calcium antagonist treatment on protein synthesis in different brain regions of gerbils subjected to 5 min ischemia and 2 h recirculation

	Control	2 hours recirculation			2 days recirculation		
		Un-treated	Fluna-rizine	Nimo-dipine	Un-treated	Fluna-rizine	Nimo-dipine
Cortex	73±10	40±21	60±14[a]	54±14	64±14	76±23	75±16
CA1	69±14	13±13	53±20[b]	45±20[a]	32±29	44±16	54±17
CA3	71±17	35±26	61±12	64±16[a]	64±20	74±18	77±14
Dentate gyrus	80± 9	69±10	67±14	54±17	77±14	80±16	78±17
Thalamus	56±20	25±11	54±12[b]	48±15[a]	44±20	67±19	62±12

Rate of amino acid incorporation is expressed as percentage of total tissue radioactivity that was retained in sections incubated with trichloroacetic acid. Mean±SD.
[a,b] Significant difference from untreated group after the same recirculation time: (a) $P<0.05$; (b) $P<0.01$.

synthesis was inhibited after 5 min of ischemia and 2 h of recirculation in all brain regions, although to a different degree.

After 2 days of recirculation, protein synthesis recovered in all brain regions except the CA1 sector. Nimodipine significantly ($P<0.05$) improved protein synthesis in hippocampal CA1 and CA3 sectors and in thalamus after 2 h of recirculation. In animals treated with flunarizine, protein synthesis significantly improved in cortex ($P<0.05$), CA1 sector ($P<0.01$), and thalamus ($P<0.01$) after 2 h of recirculation. After 2 days of recirculation, protein synthesis in the CA1 sector was higher in animals treated with nimodipine (54%) or flunarizine (44%) than in untreated animals (32%), but this effect was not significant.

Histopathology

As illustrated in Fig. 2, the number of CA1 neurons that survived after 10 min of global brain ischemia and 7 days of recirculation was significantly ($P<0.05$) higher in the nimodipine-treated than in the untreated group. Flunarizine was not tested in this experiment.

Cerebral Blood Flow

During ischemia, cortical blood flow was less than 5% of preischemic control values. As shown in Fig. 3, the restoration of blood flow resulted in a period of reactive hyperemia that was followed by a period of hypoperfusion starting at 20 min of recirculation. Treatment with nimodipine (flunarizine was not tested) had no effect on blood flow during the initial hyperemic period nor did it prevent or diminish the amount of delayed hypoperfusion.

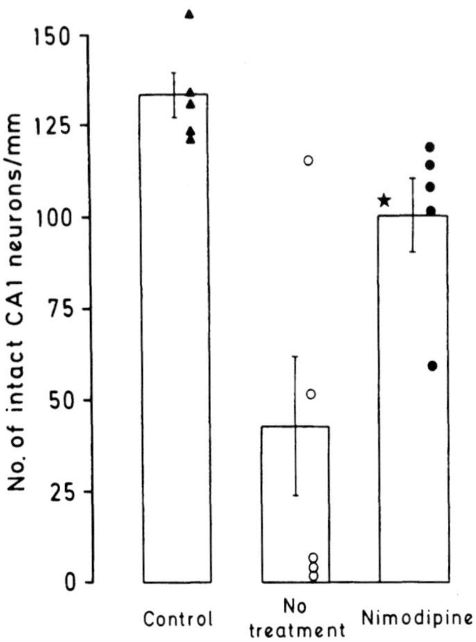

Fig. 2. Effect of nimodipine treatment on neuronal damage in the hippocampal CA1 sector of gerbils subjected to 5 min bilateral carotid occlusion and 7 days recirculation. Mean ± SEM; *$P < 0.05$

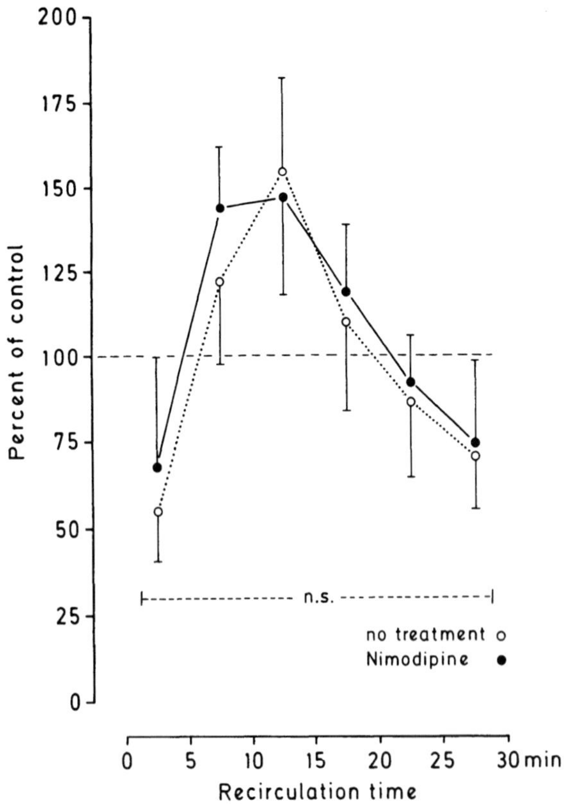

Fig. 3. Effect of nimodipine treatment on cortical blood flow measured by laser Doppler flowmetry at various recirculation times following 10 min of bilateral carotid occlusion in gerbils. Mean ± SEM; *n.s.*, not significantly different from untreated group

Discussion

Our experiments suggest that calcium overload of the cytosol induced by global ischemia occurs in two stages. The initial global mitochondrial calcium uptake probably reflects sequestration of calcium that transiently enters the cells during and shortly after ischemia because it is maximal at 15 min of recirculation and reversible in both vulnerable and resistant brain regions within 2 h. This corresponds to the finding that extracellular calcium returns to preischemic values within 15 min after global ischemia [25]. The second delayed increase in mitochondrial calcium uptake was irreversible; it occurred in parallel with the loss of cellular integrity and was restricted to the vulnerable CA1 sector of the hippocampus.

It is difficult to relate these increases to the metabolic-functional disturbances of the tissue. The early calcium increase apparently does not interfere with energy metabolism because high energy phosphate recovers within 15 min in all brain structures [2, 20]. The second delayed calcium influx, on the other hand, is associated with a decline of ATP in the CA1 sector as soon as morphological damage becomes apparent [17] but it is not clear whether this disturbance is the reason for or the consequence of the ongoing pathological process.

A more likely relationship is an adverse effect of calcium flooding on protein synthesis. This argument is based on the following considerations. A normal protein synthesis requires a normal energy state and a functioning protein synthesizing machinery. During ischemia the machinery remains intact but the energy state breaks down, resulting in severe suppression of the synthesis rate. After ischemia, the energy state recovers but now the machinery is disturbed, as evidenced by the disaggregation of ribosomes, and inhibition of protein synthesis persists [3]. After global ischemia in the gerbil, ribosomes reaggregate within 24 h but remain disaggregated in the CA1 sector [35].

The disaggregation of ribosomes is due to the inhibition of polypeptide chain initiation factors which are in part controlled by calcium-dependent enzymatic processes. It is conceivable that the initial postischemic suppression of protein synthesis is the result of the first calcium flooding, and that recovery of protein synthesis in the resistant parts of the brain reflects the normalization of intracellular calcium homeostasis. However, since the recovery process is slow [7, 22], the secondary increase in calcium activity in the selectively vulnerable CA1 sector may interfere with the reaggregation of ribosomes and, in consequence, prevent the restoration of protein synthesis in this region. Since the neurons are not able to survive if protein synthesis is not resumed, delayed neuronal death could in fact be indirectly linked to the delayed calcium flooding of the CA1 sector. It should be noted, however, that in a previous series, protection of CA1 neurons was achieved by postischemic barbiturate treatment although this drug further enhances the primary suppression of protein synthesis [35].

The therapeutic studies with calcium antagonists do not provide a clear indication for or against the calcium hypothesis, either. Both nimodipine and flunarizine improved protein synthesis and ameliorated selective vulnerability but this effect could not be explained by either an improvement of blood flow or prevention of cytosolic calcium flooding. In fact, the first calcium accumulation started before and the second one long after the drugs were applied, which makes it unlikely that calcium fluxes were reduced. There are, moreover, contradictory reports about the role of calcium antagonists from other laboratories, some authors claiming protection against neuronal damage after global ischemia [1, 4] and others not [10, 34]. In one study, flunarizine did but nimodipine did not ameliorate ischemic injury [33]. As it is also known that calcium antagonists do not affect calcium fluxes across channels operated by excitatory amino acids [9], the pharmacological role of calcium antagonists is far from being solved.

In conclusion, neither the analysis of the early postischemic calcium accumulation nor the interpretation of the therapeutic effects of calcium antagonists provides conclusive evidence for or against the calcium hypothesis of postischemic neuronal injury. The secondary accumulation of calcium in the CA1 sector, on the other hand, may be responsible for the manifestation of selective vulnerability because it correlates with the persistent inhibition of protein synthesis and morphological cell injury. However, it cannot be excluded that in this region calcium accumulation follows rather than precedes the injury, and further studies are needed to clarify this matter.

References

1. Alps BJ, Hass WK (1987) The potential beneficial effect of nicardipine in a rat model of transient forebrain ischemia. Neurology 37:809–814
2. Arai H, Passonneau JV, Lust WD (1986) Energy metabolism in delayed neuronal death of CA1 neurons of the hippocampus following transient ischemia in the gerbil. Metab Brain Dis 1:263–278
3. Cooper HK, Zalewska T, Kawakami S, Hossmann K-A, Kleihues P (1977) The effect of ischaemia and recirculation on protein synthesis in the rat brain. J Neurochem 28:929–934
4. Deshpande JK, Wieloch T (1986) Flunarizine, a calcium entry blocker, ameliorates ischemia brain damage in the rat. Anesthesiology 64:215–224
5. Deshpande JK, Siesjö BK, Wieloch T (1987) Calcium accumulation and neuronal damage in the rat hippocampus following cerebral ischemia. J Cereb Blood Flow Metab 7:89–95
6. Dienel GA (1984) Regional accumulation of calcium in postischemic rat brain. J Neurochem 43:913–925
7. Dienel GA, Pulsinelli WA, Duffy TE (1980) Regional protein synthesis in rat brain following acute hemispheric ischemia. J Neurochem 35:1216–1226
8. Dux E, Mies G, Hossmann K-A, Siklos L (1987) Calcium in the mitochondria following brief ischemia of gerbil brain. Neurosci Lett 89:295–300
9. Greenberg DA (1987) Calcium channels and calcium channel antagonists. Ann Neurol 21:317–330
10. Grotta J, Spydell J, Pettigrew C, Ostrow P, Hunter D (1986) The effect of nicardipine on neuronal function following ischemia. Stroke 17:213–219
11. Harper AM, Craigen L, Kazda S (1981) Effect of the calcium antagonist, nimodipine, on cerebral blood flow and metabolism in the primate. J Cereb Blood Flow Metab 1:349–356
12. Harris RJ, Symon L (1984) Extracellular pH, potassium, and calcium activities in progressive ischaemia of rat cortex. J Cereb Blood Flow Metab 4:178–186
13. Hossmann K-A, Grosse Ophoff B, Schmidt-Kastner R, Oschlies U (1985) Mitochondrial calcium sequestration in cortical and hippocampal neurons after prolonged ischemia of the cat brain. Acta Neuropathol 68:230–238
14. Kirino T (1982) Delayed neuronal death in the gerbil hippocampus following ischemia. Brain Res 239:57–69

15. Kirino T, Sano K (1984) Fine structural nature of delayed neuronal death following ischemia in the gerbil hippocampus. Acta Neuropathol 62:209–218
16. McBurney RN, Neering IR (1987) Neuronal calcium homeostasis. TINS 10:164–169
17. Mies G, Paschen W, Hossmann K-A, Klatzo I (1983) Simultaneous measurement of regional blood flow and metabolism during maturation of hippocampal lesions following short-lasting cerebral ischemia in gerbils. J Cereb Blood Flow Metab 3 (Suppl 1):329–330
18. Milde LN, Milde JH, Michenfelder JD (1986) Delayed treatment with nimodipine improves cerebral blood flow after complete cerebral ischemia in the dog. J Cereb Blood Flow Metab 6:332–337
19. Mohamed AA, McCulloch J, Mendelow AD, Teasdale GM, Harper AM (1984) Effect of the calcium antagonist nimodipine on local cerebral blood flow: relationship to arterial blood pressure. J Cereb Blood Flow Metab 4:206–211
20. Munekata K, Hossmann K-A (1987) Effect of 5-minute ischemia on regional pH and energy state of the gerbil brain: relation to selective vulnerability of the hippocampus. Stroke 18:412–417
21. Nicholls DG, Akerman KEO (1982) Mitochondrial calcium transport. Biochim Biophys Acta 683:57–88
22. Nowak TS Jr, Fried RL, Lust WD, Passonneau JV (1985) Changes in brain energy metabolism and protein synthesis following transient bilateral ischemia in the gerbil. J Neurochem 44:487–494
23. Sakabe T, Nagai I, Ishikawa T, Takeshita H, Masuda T, Matsumoto M, Tateishi A (1986) Nicardipine increases cerebral blood flow but does not improve neurologic recovery in a canine model of complete cerebral ischemia. J Cereb Blood Flow Metab 6:684–690
24. Sakamoto N, Kogure K, Kato H, Ohtomo H (1986) Disturbed Ca^{+} homeostasis in the gerbil hippocampus following brief transient ischemia. Brain Res 364:372–376
25. Siemkowicz E, Hansen AJ (1981) Brain extracellular ion composition and EEG activity following 10 minutes ischemia in normo- and hyperglycemic rats. Stroke 12:236–240
26. Siesjö BK, Wieloch T (1985) Cerebral metabolism in ischaemia: neurochemical basis for therapy. Br J Anaesth 57:47–62
27. Smith M-L, Kagström E, Rosén I, Siesjö BK (1983) Effect of the calcium antagonist nimodipine on the delayed hypoperfusion following incomplete ischemia in the rat. J Cereb Blood Flow Metab 3:543–546
28. Simon RP, Griffiths T, Evans MC, Swan JH, Meldrum BS (1984) Calcium overload in selectively vulnerable neurons of the hippocampus during and after ischemia: an electron microscopy study in the rat. J Cereb Blood Flow Metab 4:350–361
29. Steen PA, Newberg LA, Milde JH, Michenfelder JD (1983) Nimodipine improves cerebral blood flow and neurologic recovery after complete cerebral ischemia in the dog. J Cereb Blood Flow Metab 3:38–43
30. Steen PA, Newberg LA, Milde JH, Michenfelder JD (1984) Cerebral blood flow and neurologic outcome when nimodipine is given after complete cerebral ischemia in the dog. J Cereb Blood Flow Metab 4:82–87
31. Suzuki R, Yamaguchi T, Li C-L, Klatzo I (1983) The effects of 5-minute ischemia in Mongolian gerbils: II. Changes of spontaneous neuronal activity in cerebral cortex and CA1 sector of hippocampus. Acta Neuropathol 60:217–222
32. Van Reempts J, Borgers M, Offner F (1982) Ultrastructural localization of calcium in the cat retina with combined oxalate-pyroantimonate technique. Histochem J 14:517–522
33. Van Reempts J, Haseldonckx M, Deuren van B, Wouters L, Borgers M (1986) Structural damage of the ischemic brain: involvement of calcium and effects of postischemic treatment with calcium entry blockers. Drug Devel Res 8:387–395
34. Vibulresth S, Dietrich WD, Busto R, Ginsberg MD (1987) Failure of nimodipine to prevent ischemic neuronal damage in rats. Stroke 18:210–216
35. Xie Y, Hossmann K-A, Munekata K, Seo K (1987) Prolonged suppression of protein synthesis after brief cerebral ischemia in gerbils. In: Cervós-Navarro J, Ferszt R (eds) Stroke and microcirculation. Raven Press, New York, pp 135–14

The Effects of a Dihydropyridine Calcium Antagonist on Cerebral Blood Flow with and without Middle Cerebral Artery Occlusion

A. M. HARPER[1], J. MCCULLOCH[1], and G. M. TEASDALE[2]

Introduction

The ability of the calcium antagonist nimodipine to dilate cerebrovascular smooth muscle was first recognised by Towart and Kazda [13] and Kazda and his co-workers [4].

This paper summarises some of the work performed in our laboratories over the past few years on the effects of nimodipine on cerebral blood flow (CBF) [1, 3, 5–8] in normal animals and after middle cerebral artery (MCA) occlusion. The initial studies were carried out on anaesthetised baboons using ^{133}Xe clearance. Later studies measured flow with [^{14}C]iodoantipyrine autoradiography in anaesthetised rats.

Studies of the Effects of Nimodipine on CBF in Baboons

Methods

The methods and experimental protocol are fully described in reference 3. In summary, baboons of average weight 8.3 kg were anaesthetised initially with intravenous sodium thiopental and given 70% N_2O and 30% O_2 by positive pressure ventilation in open circuit. A continuous intravenous infusion of phencyclidine (0.01 mg kg^{-1} min^{-1}) was given throughout the experiment.

Catheters were inserted into the aorta via both femoral arteries for the continuous measurement of arterial blood pressure and for the withdrawal of arterial blood samples. Both femoral veins were cannulated, one for the continuous infusion of phencyclidine and the other for the infusion of nimodipine. A catheter was inserted into the right lingual artery so that its tip lay at the carotid bifurcation; the other branches of the external carotid artery were ligated. The scalp and temporalis muscle were removed with cutting diathermy.

A collimated scintillation crystal (dimensions of collimation 5 cm deep, 2.5 cm wide) was placed over the temporoparietal region of the exposed skull and angled in such a way that it viewed only brain and overlying skull.

[1] Wellcome Neuroscience Group, Wellcome Surgical Institute, University of Glasgow, Glasgow G61 1QH, U.K.
[2] Institute of Neurological Sciences, Southern General Hospital, Glasgow G51 4TF, U.K.

Cerebral Ischemia and Calcium
Ed. by A. Hartmann and W. Kuschinsky

Approximately 250 μCi of ^{133}Xe in saline was injected as a bolus into the carotid artery via the catheter in the lingual artery. The rate of clearance was measured from the scintillation detector which was connected to a ratemeter, scaler and chart recorder. CBF was calculated in ml 100 g^{-1} min^{-1} from the height/area equation. In some animals the cerebral metabolic rate for oxygen ($CMRO_2$) was calculated from the flow multiplied by the A-V oxygen difference – cerebral venous oxygen samples being taken from the sagittal sinus.

During the experiments the end-tidal concentration of CO_2 was measured continuously and the ventilating pump was adjusted to maintain an arterial carbon dioxide tension between 37 and 40 mmHg. Body temperature was maintained at 37° C by means of heating lamps and electrically heated blankets.

Ten milligrams of nimodipine was dissolved (under sodium light) in 10 ml of lutrol-glycerine-water solvent. This was diluted a further 24-fold with normal saline for continuous intravenous infusion, or 36-fold where intracarotid infusions were used.

In some experiments the blood-brain barrier was temporarily disrupted by a single intracarotid injection of 10 ml of 2 M hypertonic urea [9].

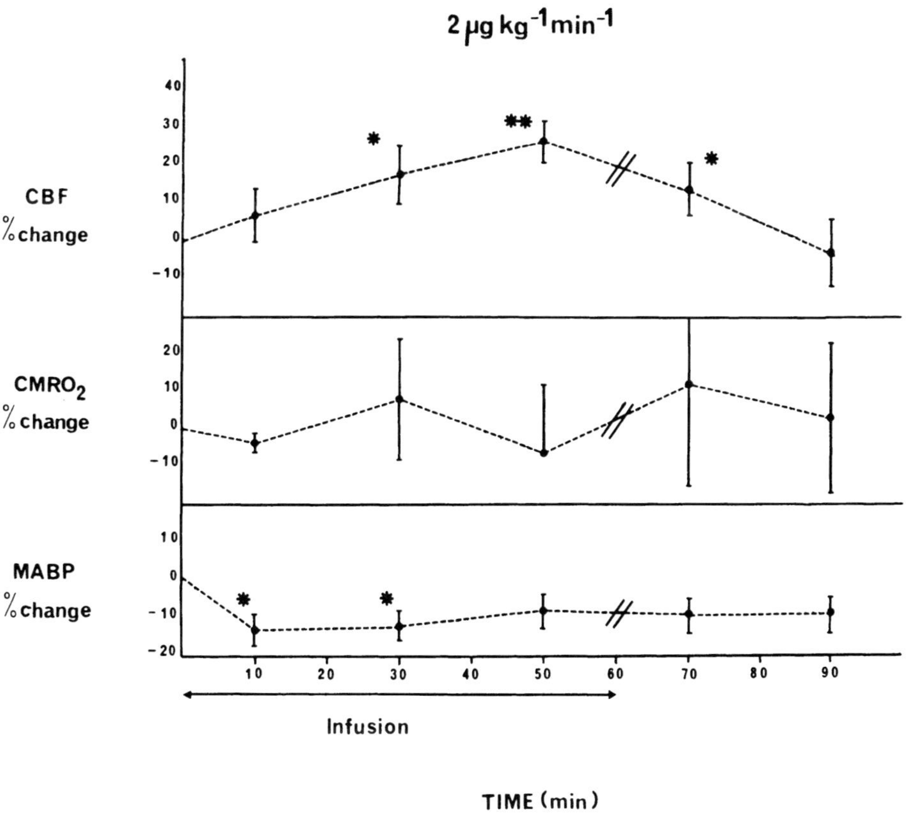

Fig. 1. Effect of intravenous infusion of nimodipine over 60 min on CBF, $CMRO_2$, and MABP. *Bars* indicate 1 SE from the mean. (Harper et al. [3])

Results

Intravenous Infusion of Nimodipine (2 μg kg^{-1} min^{-1})

The continuous intravenous infusion of nimodipine in five animals resulted in a steady, if unspectacular increase in CBF (Fig. 1). After 50 min infusion there was an increase in CBF of 27%. This was accompanied by a drop in MABP of between 12% and 8%. There were no significant changes in $CMRO_2$ or $PaCO_2$.

Intracarotid Infusion of Nimodipine (0.67 μg kg^{-1} min^{-1})

Cerebral blood flow increased by 46% and 57% after 10 and 30 min infusion of nimodipine. This was accompanied by reductions of 8% and 12% in MABP (Fig. 2). $PaCO_2$ was constant.

Ten minutes after the blood-brain barrier had been disrupted with hyperosmolar urea, the CBF had increased to 87% above control. Thirty minutes after urea, when it is expected that the barrier would have closed again, the increase in CBF with continued infusion of nimodipine had dropped to 43% above control. Twenty minutes after stopping the infusion CBF had returned to control values.

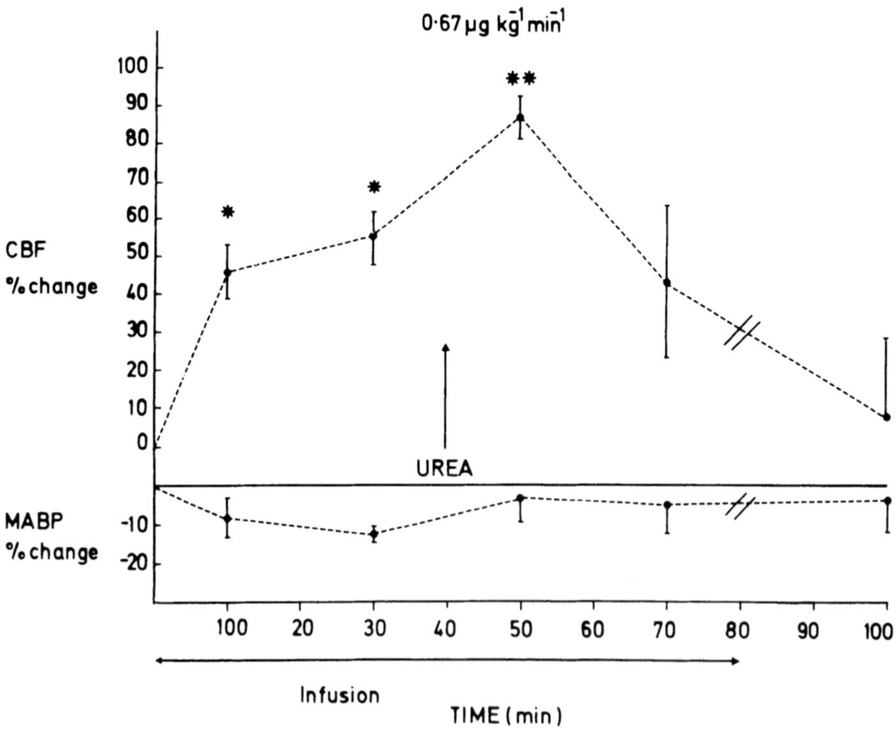

Fig. 2. Effect of intracarotid infusion of nimodipine over 80 min on CBF and MABP. Hyperosmolar urea, 10 ml, was injected into the internal carotid artery indicated. *Bars* indicate 1 SE from the mean. (Harper et al. [3])

The urea infusion was accompanied by a modest rise in arterial blood pressure of 8–10 mmHg. This could have accounted for some of the additional cerebrovasodilatory response with urea but it is unlikely that the rise in blood pressure would have been sufficient to account for a change in CBF, which rose from +57% to +87%. It is likely, therefore, that nimodipine does not cross the blood-brain barrier completely freely.

Studies of the Effects of Nimodipine on CBF in Normal Rats

Methods

The methods and experimental design are fully described in references 5 and 6. In summary, male Sprague-Dawley rats (300–470 g) were anaesthetised initially with 5% halothane. Following tracheostomy, a mixture of 70% N_2O, 30% O_2 and 1% halothane was administered through a respiratory pump. Polythene catheters, 15 cm long and filled with heparinised saline solution (10 units/ml), were introduced retrogradely into both femoral arteries and veins for continuous measurement of arterial blood pressure, for withdrawal of arterial blood (for blood gas measurements, plasma glucose level determination, and blood sampling), and for administration of drugs and radioactive tracers.

Local CBF (LCBF) was measured with the quantitative autoradiographic technique with [^{14}C]iodoantipyrine as tracer [10]. [^{14}C]iodoantipyrine (50 μCi in 1.5 ml saline) was infused intravenously over a 30-s period on a "ramp" schedule; i.e. the rate of infusion was increased progressively so that arterial tracer concentration would be highest at the end of the experiments. During administration of the tracer, arterial blood flow was allowed to drip freely from a partially occluded catheter in the femoral artery, and 15–18 samples of blood were taken separately in preweighed filter discs. The times of sampling were recorded and were used with the radioactive concentration of each blood sample to describe the history of arterial radioactivity during the course of the experiment. At the end of infusion of the isotope, the animals were killed by decapitation and the brains quickly dissected out and frozen in isopentane at −45° C within 2–3 min. The filter discs were placed in tightly capped scintillation vials, to prevent evaporation, and reweighed. Hydrogen peroxide (0.4 ml) and water (1 ml) were added to bleach the blood and extract the radioisotope. Ten millilitres of liquid scintillant (Pico-fluor) was then added. Twenty-four hours were allowed to elapse before liquid scintillation analysis.

Brain sections (20 μm thick) were cut in a cryostat at −22° C, and 3 of every 20 sections were mounted on glass coverslips and dried on a hot plate. Autoradiographs were prepared from these sections, together with a set of plastic standards of known ^{14}C concentration (44–1175 nCi/g), by applying them to X-ray film (Kodak SB-5) for 7 days in light-tight cassettes. Analysis of the resultant images on the X-ray film was performed using a computer-based densitometer (Quantimet 720; Cambridge Instruments) with reference to the precalibrated standards. For each area of the brain, 12 optical density readings were made bilaterally on six consecutive sections in which the structure could be anatomically defined.

In some animals local cerebral glucose utilisation (LCGU) was measured using the [^{14}C]2-DG technique [11]. The isotope (50 μCi per animal), dissolved in saline, was given at a constant rate over 30 s. A total of 14 timed samples of arterial blood were withdrawn from the femoral cannulas and collected into heparinised plastic centrifuge tubes over the subsequent 45 min according to a predetermined time schedule. The volume of the arterial samples was 75 μl of whole blood; this same volume was replaced by saline solution. The arterial blood plasma was separated from the blood sample content by centrifugation within a few minutes of withdrawal. Twenty microlitres of each plasma sample was pipetted from the centrifuge tubes into 1 ml of distilled water in 14 plastic scintillation vials. The glucose level was determined on a 10-μl aliquot of plasma, using a semiautomated glucose analyser (Beckman) with a glucose oxidase method. At the end of 45 min, the rats were killed by decapitation and the brains processed for quantitative autoradiography in the same manner as for the measurement of LCBF.

Following completion of the surgical preparation, at least 30 min was allowed to elapse before the start of infusion of the drug or vehicle. This was done to achieve normoxia, normocapnia, stability of the blood pressure, and normal plasma glucose concentrations. Normocapnia was achieved by adjusting the tidal volume of the respirator. Body temperature was maintained at $37° \pm 0.05°$ C by placing the animals in an automated heating box. In control experiments (seven animals), the lutrol-glycerine-water solvent alone was infused via the femoral vein for 1 h before measuring LCBF. In nimodipine-treated animals the rates of drug administration were 1 μg kg^{-1} min^{-1} (six animals), 2 μg kg^{-1} min^{-1} (seven animals), and 4 μg kg^{-1} min^{-1} (seven animals). CBF was measured 60 min after the start of infusion of the drug. The volume of infusion was 0.0136 ml kg^{-1} min^{-1}.

In these experiments where LCGU was measured, a slightly different time scale was used, LCGU measurements being initiated 15 min after start of infusion, which continued for a further 45 min.

Results

Figure 3 demonstrates the LCBF response of two areas (parietal cortex and medial geniculate) which reacted strongly to nimodipine at doses of 1, 2 and 4 μg kg^{-1} min^{-1}. At an infusion rate of 1 μg kg^{-1} min^{-1}, the parietal cortex increased in flow by 107% and the medial geniculate by 63%. On the other hand some brain areas did not react to nimodipine – the globus pallidus and septal nucleus being illustrated in the figure.

Table 1 shows the effects of the three doses of nimodipine on the parietal and auditory cortex, the hippocampus, cerebellar white matter, and the internal capsule. Arterial blood gases did not change significantly during the experiments but the MABP showed a dose-dependent decrease as the concentrations of nimodipine were increased.

Table 2 shows the effect of 1 μg kg^{-1} min^{-1} nimodipine on LCGU in the same brain areas as shown in Table 1. There were no significant differences between control and nimodipine.

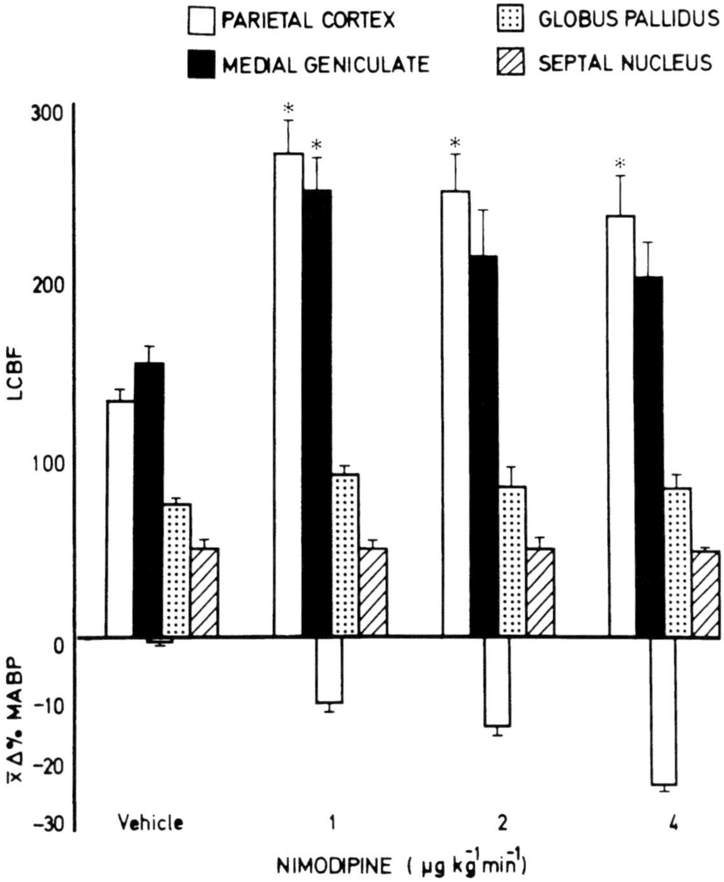

Fig. 3. Effect of vehicle and nimodipine at doses of 1, 2, and 4 μg kg^{-1} min^{-1} on LCBF through parietal cortex, medial geniculate, globus pallidus, and septal nucleus. LCBF is ml 100 g^{-1} min^{-1}. % reductions in MABP are indicated by histograms below the *horizontal line*. *Bars* indicate SEM. *$P<0.05$ using Student's *t*-test for grouped data with Bonferroni correction. (Data for this figure extracted from Mohamed et al. [5] and reproduced from Harper [2])

Table 1. The effect of nimodipine on LCBF (ml 100 g^{-1} min^{-1})

Region	Vehicle ($n=7$)	1 μg kg^{-1} min^{-1} ($n=6$)	2 μg kg^{-1} min^{-1} ($n=7$)	4 μg kg^{-1} min^{-1} ($n=7$)
Parietal cortex	133 ± 7	276 ± 20**	254 ± 21**	239 ± 25**
Auditory cortex	162 ± 17	289 ± 21**	247 ± 21**	256 ± 21**
Hippocampus	122 ± 9	195 ± 18*	188 ± 30	171 ± 18***
Cerebellar white matter	58 ± 7	55 ± 3	58 ± 6	68 ± 83
Internal capsule	50 ± 5	53 ± 5	50 ± 6	49 ± 2
Variable				
$PaCO_2$	34 ± 1	34 ± 1	36 ± 1	36 ± 1
PaO_2	134 ± 4	124 ± 6	134 ± 8	128 ± 7
pH	7.44 ± 0.01	7.44 ± 0.01	7.43 ± 0.01	7.42 ± 0.01
MABP	99 ± 3	93 ± 4	84 ± 4*	75 ± 4*

Mean ± SEM. * $P<0.05$, ** $P<0.01$. Student's *t*-test for grouped data with Bonferroni correction. *** $P<0.05$ (uncorrected *t*-test). (Data extracted from references 5 and 6).

Table 2. The effect of nimodipine (1 μg kg^{-1} min^{-1}) on LCGU

Region	LCGU (μmol 100 g^{-1} min^{-1})	
	Vehicle ($n=6$)	Nimodipine ($n=7$)
Parietal cortex	96 ± 5	89 ± 4
Auditory cortex	127 ± 8	115 ± 4
Hippocampus	77 ± 2	75 ± 2
Cerebellar white	29 ± 1	27 ± 1
Internal capsule	27 ± 1	27 ± 1
Variable		
$PaCO_2$	36 ± 2	35 ± 2
PaO_2	81 ± 1	83 ± 3
pH	7.44 ± 0.04	7.43 ± 0.01
MABP	125 ± 5	114 ± 2

Mean ± SEM. (Data extracted from reference 7).

Studies of the Effects of Nimodipine Given Before and After MCA Occlusion in Rats

Methods

The preparation was the same as described in the previous section with the exception of MCA occlusion using the method of Tamura and co-workers [12]. Briefly, the stem of the MCA was approached through a left subtemporal craniectomy beside the foramen ovale but without removing the zygomatic arch. The stem of the artery and its lenticulostriate branches were occluded by microbipolar coagulation and then divided to ensure the completeness of the vascular occlusion. In the experiments with pretreatment, nimodipine (1 μg kg^{-1} min^{-1}) or the lutrol-glycerine-water solvent used as control was infused for 30 min before MCA occlusion and continued for 30 min when the brain was removed for autoradiography.

In the experiments with post-treatment, the intravenous infusion of nimodipine or control solution was commenced 5 min after MCA occlusion and continued for another 30 min.

Results

Pre-treatment

Figure 4 demonstrates the flow changes induced by nimodipine in contra- and ipsilateral parietal cortex and caudate nucleus. Nimodipine more than doubled the flow in the ipsilateral parietal cortex but had no effect on that in the caudate nucleus. The administration of nimodipine modified the pattern of CBF distur-

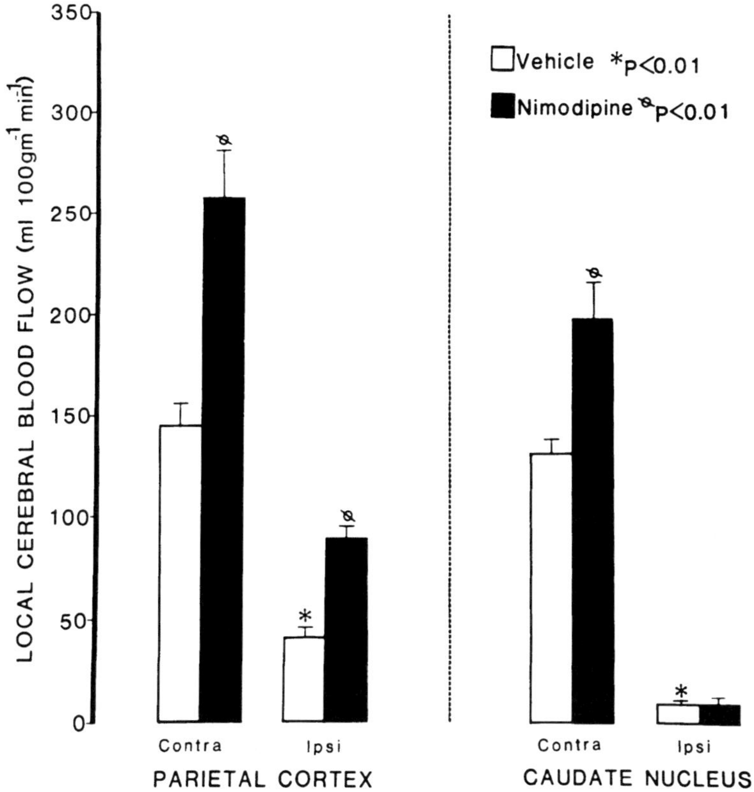

Fig. 4. Effect of pretreatment with nimodipine as compared to vehicle on LCBF following MCA occlusion in ipsilateral and contralateral parietal cortex and caudate nucleus. *Bars* indicate SEM. (Data for this figure extracted from Mohamed et al. [8] and reproduced from Harper [2])

bances in the contralateral and ipsilateral sides in most cortical and subcortical regions examined, compared with the corresponding sides during infusion of the vehicle. However, there was little change in areas in the centre of the ischaemic damage, such as the caudate nucleus.

On the ipsilateral side, the areas showing the greatest increases with nimodipine (as compared to control) were cortex, geniculate and thalamus. In effect, pretreatment and continuing treatment with the drug more than doubled cortical flow in the occluded side, but had no effect in the centre of the ischaemic lesion – namely, the caudate nucleus.

Post-treatment

Figure 5 demonstrates the flow changes in parietal cortex and caudate nucleus when nimodipine infusion was instituted 5 min after MCA occlusion. In contrast to Fig. 4, nimodipine did not affect blood flow.

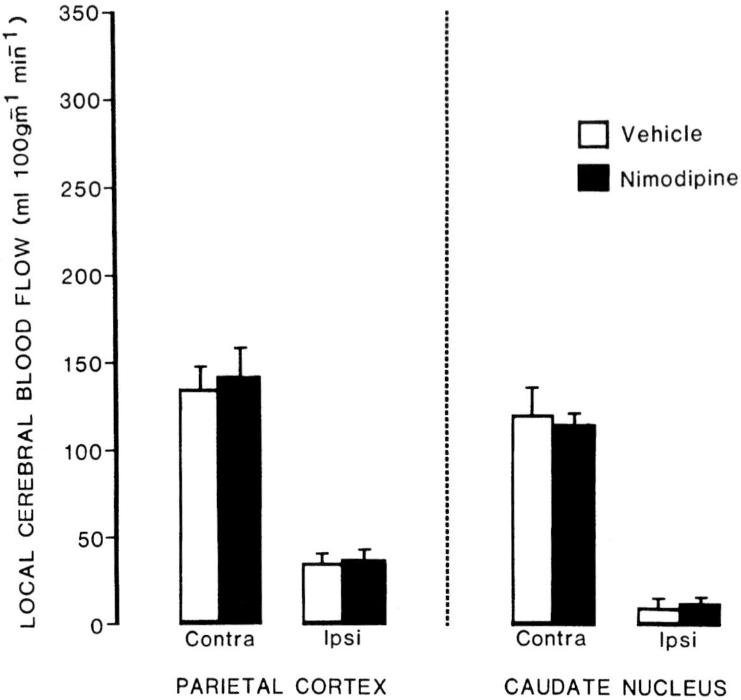

Fig. 5. Effect of post-treatment with nimodipine as compared to vehicle on LCBF following MCA occlusion in ipsilateral and contralateral parietal cortex and caudate nucleus. *Bars* indicate SEM. (Data for this figure extracted from Gotoh et al. [1])

General Discussion

While the ability of nimodipine to increase CBF in the normal animal is without question, its concomitant hypotensive effect must be taken into account. This is best seen in Fig. 3. The maximum effect of nimodipine occurs at a dosage of 1 μg kg^{-1} min^{-1}. Increasing the dosage lowers the blood pressure to a point where the normal autoregulatory process of cerebrovascular dilatation in response to hypotension, counteracts any further propensity of nimodipine to increase CBF. The LCBF increases with nimodipine are most marked in the neocortical areas. The poorest responses are seen in cerebellum and white matter.

Regarding the MCA occlusion experiments in the hemisphere ipsilateral to the MCA occlusion, the control rats showed an area of profound ischaemia that conformed to the pattern previously described in this model. However, nimodipine significantly lessened the fall in LCBF in the ischaemic area of the cortex. Even in sensorimotor cortex, the area of deepest cortical ischaemia, LCBF in nimodipine-treated rats was twice that in vehicle-treated rats (61 ± 9 vs 30ÜWCf 100 g^{-1} min^{-1}). Despite the powerful vasodilatory effects of nimodipine, we did not find evidence to suggest that it adversely influenced LCBF by producing a steal in the neostriatum, the area of maximum ischaemia; here, compared with

LCBF in the contralateral hemisphere of control rats, flow was reduced by 93% in both treated and control groups.

In direct contrast to the pretreatment studies, the administration of nimodipine did not have any beneficial effects on CBF following MCA occlusion. This would appear to suggest that, at least as far as CBF is concerned, any beneficial effects of nimodipine must be prophylactic if given to potential stroke patients before the event.

References

1. Gotoh O, Mohamed AA, McCulloch J, Graham DI, Harper AM, Teasdale GM (1986) Nimodipine and the haemodynamic and histopathological consequences of middle cerebral artery occlusion in the rat. J Cereb Blood Flow Metab 6:321–331
2. Harper AM (1985) Effect of a dihydropyridine-type calcium antagonist on cerebral blood flow and metabolism. In: Fleckenstein A, Van Breemen C, Gross R, Hoffmeister F (eds) Cardiovascular effects of dihydropyridine-type calcium antagonists and agonists. Springer, Berlin Heidelberg New York, pp 353–360
3. Harper AM, Craigen L, Kazda S (1981) Effect of calcium antagonist, nimodipine, on cerebral blood flow and metabolism in the primate. J Cereb Blood Flow Metab 1:349–356
4. Kazda S, Hoffmeister F, Garthoff B, Towart R (1979) Prevention of the post-ischaemic impaired reperfusion of the brain by nimodipine (Bay e 9736). Acta Neurol Scand (Suppl 72) 60:302–303
5. Mohamed AA, McCulloch J, Mendelow AD, Teasdale GM, Harper AM (1984) Effect of the calcium antagonist nimodipine on local cerebral blood flow: relationship to arterial blood pressure. J Cereb Blood Flow Metab 4:206–211
6. Mohamed AA, Mendelow AD, Graham DI et al. (1985) The effect of nimodipine on local cerebral blood flow, glucose use and focal cerebral ischaemia using autoradiographic techniques. In: Betz E, Deck K, Hoffmeister F (eds) Nimodipine: pharmacological and clinical properties. Schattauer, Stuttgart, pp 105–112
7. Mohamed AA, Mendelow AD, Teasdale GM, Harper AM, McCulloch J (1985) Effect of the calcium antagonist nimodipine on local cerebral blood flow and metabolic coupling. J Cereb Blood Flow Metab 5:26–33
8. Mohamed AA, Gotoh O, Graham DI, Osborne KA, Mendelow AD, Teasdale GM, Harper AM (1985) Effect of pretreatment with the calcium antagonist nimodipine on local cerebral blood flow and histopathology after middle cerebral artery occlusion. Ann Neurol 18:705–711
9. Pickard JD, Durity F, Welsh FA, Langfitt TW, Harper AM, MacKenzie ET (1977) Osmotic opening of the blood-brain barrier: value in pharmacological studies on the cerebral circulation. Brain Res 122:170–176
10. Sakurada O, Kennedy C, Jehle J, Brown JD, Carbin GL, Sokoloff L (1978) Measurement of local cerebral blood flow with iodo[^{14}C]antipyrine. Am J Physiol 234:H59–H66
11. Sokoloff L, Reivich M, Kennedy C et al. (1977) The [^{14}C]deoxyglucose method for the measurement of local cerebral glucose utilization: theory, procedure, and normal values in the conscious and anesthetized albino rat. J Neurochem 28:897–916
12. Tamura A, Graham DI, McCulloch J, Teasdale GM (1981) Focal cerebral ischaemia in the rat: I. Description of techniques and early neuropathological consequences following middle cerebral artery occlusion. J Cereb Blood Flow Metab 1:53–60
13. Towart R, Kazda S (1979) The cellular mechanism of action of nimodipine (Bay e 9736), a new calcium antagonist. Br J Pharmacol 67:409–410

Nimodipine Binding in Cerebral Ischemia

A. Hakim, L. Berger, D. Popow, and A. Gjedde

The duration of, as well as the degree of depression in, perfusion influences the eventual clinical and histological outcome of cerebral ischemia. In baboons with middle cerebral artery (MCA) occlusion, Symon et al. [11] showed that CBF above 18 ml/100 g/min sustained normal somatosensory evoked potentials and EEG activity while CBF < 12 caused histologically evident infarction. For CBF values between these limits, interruption of electrical function was reversible. Subsequent work has not modified these values substantially [1, 2, 4, 5, 7]. The time limits within which the functional deficits consequent to ischemia may still be reversible have also been based on CBF measurements. Rosner and Heiss showed that the duration of neuronal viability depended on the magnitude of perfusion within the CBF range that defined the penumbra [9]. Recovery was unlikely after more than 30 min at a CBF of 5 ml/100 g/min, but was possible after more than 2 h with a CBF of 17 ml/100 g/min. It is clear, therefore, that an ischemic brain region may remain viable for several hours in certain circumstances.

Our interest in studying calcium channel activity during ischemia stems from our belief that CBF alone cannot be an adequate criterion for determination of tissue survival. It is known that blood flow can vary quite rapidly after a stroke: one-third of our patients studied within 24 h of the onset of stroke had rates of blood flow to the infarct that exceeded normal [3]. This has been described as a state of luxury perfusion, implying blood flow in excess of metabolic needs. With this in mind we have undertaken a series of experiments to investigate the binding to ischemic tissue of one dihydropyridine calcium channel blocker, nimodipine. There is available literature to support indirectly the hypothesis that nimodipine binding should be altered in ischemia. Wagner et al. have shown in cardiomyopathic hamsters a doubling of maximum nitrendipine binding sites in brain [14]. Magnoni et al. [6] showed a transient change in the binding affinity for a calcium channel blocker in rat hippocampus after 1 h of carotid occlusion. Using cardiac cells Reuter et al. have shown that depolarization increases nimodipine binding [8].

The McConnell Brain Imaging Centre of the Montreal Neurological Institute,
3801 University Str., Montreal, Quebec H3A 2B4, Canada.

Cerebral Ischemia and Calcium
Ed. by A. Hartmann and W. Kuschinsky

Methods

Sprague-Dawley rats weighing ca. 250 g were anesthetized with halothane and a right craniotomy was performed according to the method of Tamura [12]. The right MCA as well as the right common carotid artery (CCA) was isolated. In the first phase of experiments, the effect of duration of ischemia on binding to tritiated nimodipine was determined in separate groups of rats at 5 min, 4 h, and 24 h after occlusion of both MCA and CCA. CBF was determined at these same intervals in comparable animals by the method of Sakurada [10] with infusion of ^{14}C-iodoantipyrine 1 min prior to decapitation of the animal. ^{3}H-nimodipine, 250 μCi (specific activity 135 Ci/mmol, NEN), was injected 30 min prior to decapitation. This duration after injection was based on the work of Van den Kerckoff [13] showing that 30 min was sufficient to establish equilibrium. The validity of this requirement in the setting of ischemia was not verified. Radioactivity in plasma samples obtained after injection of ^{3}H-nimodipine was determined.

In the second phase of experiments, varying mixtures of labeled and unlabeled nimodipine were injected into rats 3.5 h after MCA + CCA occlusion in order to determine saturability, the maximum number of binding sites and their affinity.

Upon decapitation of a rat which had received tritiated nimodipine, the brain was rapidly removed and frozen with liquid nitrogen. Twenty-micron sections were then obtained and exposed to LKB2208-190 ^{3}H-ultrofilm (Bromma, Sweden) for periods reaching 150 days. The activity in various brain regions was then determined using a digitizer (MCID, Imaging Research, St. Catharines, Ontario, Canada). The binding potential (BP) was defined as

$$BP = (V_{BR} - V_{NBR}) / V_{NBR}$$

where V_{BR} and V_{NBR} are the volumes of distribution in binding (BR) and nonbinding (NBR) brain regions. Breakdown products of nimodipine were ignored in this first analysis.

Results

Binding to ^{3}H-nimodipine in the nonischemic hemisphere was fairly uniform and featureless and varied between 300 and 400 nCi/g. In the ischemic hemisphere, the activity often exceeded 1000 nCi/g. When the binding potential in different brain regions was calculated at the three intervals following MCA + CCA occlusion, three patterns became evident (Fig. 1): In one, such as in the cingulate gyrus, the binding potential remained low regardless of the duration of ischemia. In another pattern, such as in the sensorimotor cortex, the medial caudate region, and the watershed zone, the binding potential was much higher at 4 h after occlusion then at 5 min, then either remained high or declined slightly by 24 h after occlusion. In yet a third pattern, such as in the lateral caudate and inferior cortical region, binding potential was highest 5 min after occlusion and subsequently declined sharply, i.e., binding to ^{3}H-nimodipine in these regions at 4 or 24 h after occlusion

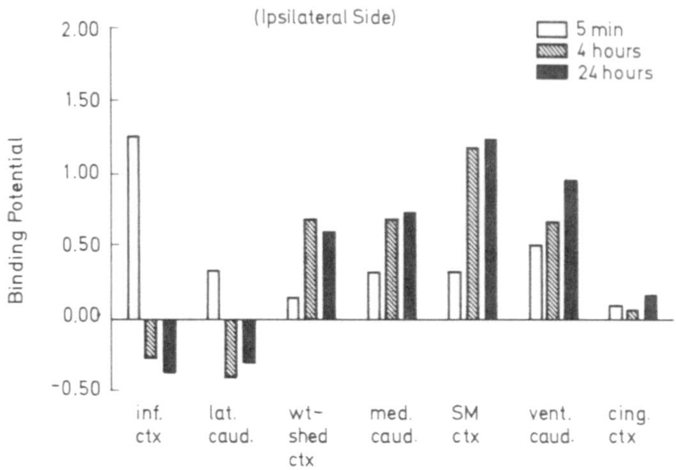

Fig. 1. ^{3}H-nimodipine binding in selected regions of rat brain at 5 min, 4 h, and 24 h after ipsilateral MCA and CCA occlusions. *Inf ctx*, inferior cortex; *lat. caud.*, lateral caudate; *wt-shed ctx*, watershed cortex; *med. caud.*, medial caudate; *SM ctx*, sensorimotor cortex; *vent. caud.*, ventral caudate; *cing. ctx*, cingulate cortex

was actually lower than in the "nonbinding" regions of the nonischemic hemisphere.

When these binding measurements were compared to CBF determined in the same regions, it became evident that at 5 min after occlusion, ^{3}H-nimodipine binding was proportional to the decrease in CBF, but the relationship of binding to CBF at 4 h and 24 h after occlusion was more complex.

Our data are still undergoing analysis, but at least in the sensorimotor cortex, binding to tritiated nimodipine appeared saturable with a B_{max} of 95 pmol/g and a K_D of 68.8 pmol/g.

Discussion

The eventual aim of this work is to relate ^{3}H-nimodipine binding to tissue perfusion and the histological damage consequent to cerebral ischemia. Our preliminary analysis shows that binding to nimodipine is saturable, and when it is observed at various intervals following ischemia, proceeds along patterns that may be used to predict histological outcome. For instance, the cingulate gyrus, which has a low and stable binding pattern, shows no histological damage, while regions in which the binding potential declines following a phase of activation develop histologically evident infarction. The intermediate pattern of slower and more sustained activation up to 24 h may show cellular drop-out and be compatible with the concept of ischemic penumbra. If nimodipine binding in ischemic brain proves to be an accurate and reliable predictor of the histological outcome in the tissue, it may be a useful tool to study the presence of viable brain after ischemia and its responses to therapeutic measures.

References

1. Astrup J, Symon L, Branston NM, Lassen NA (1977) Cortical evoked potential and extracellular K^+ and H^+ at critical levels of brain ischemia. Stroke 8:51–57
2. Branston NM, Symon L, Crockard HA, Pazztor E (1974) Relationship between the cortical evoked potential and local cortical blood flow following acute middle cerebral artery occlusion in the baboon. Exp Neurol 45:195–208
3. Hakim AM, Pokrupa R, Villanueva J et al. (1987) The effect of spontaneous reperfusion on metabolic function in early human cerebral infarcts. Ann Neurol 21:279–289
4. Heiss W-D, Hayakawa T, Walz AG (1976) Cortical neuronal function during ischemia. Arch Neurol 33:813–820
5. Jones TH, Morawetz RB, Crowell RM et al. (1981) Thresholds of focal cerebral ischemia in awake monkeys. J Neurosurg 54:773–782
6. Magnoni MS, Govoni S, Battaini F, Trabucchi M (1988) L-type calcium channels are modified in rat hippocampus by short-term experimental ischemia. J Cereb Blood Flow Metab 8:96–99
7. Morawetz RB, De Girolami V, Ojemann RG, Marcoux FW, Crowell RM (1978) Cerebral blood flow determined by hydrogen clearance during middle cerebral artery occlusion in unanesthetized monkeys. Stroke 9:143–149
8. Reuter H, Porzig H, Prod'hom S (1985) 1,4-Dihydropyridines as tools in the study of Ca^{2+} channels. TINS 8:396–400
9. Rosner G, Heiss W-D (1983) Survival of cortical neurons as a function of residual flow and duration of ischemia. J Cereb Blood Flow Metab 3 (Suppl 1):S393–394
10. Sakurada O, Kennedy C, Jehle J, Brown JD, Carbin GL, Sokoloff L (1978) Measurements of local cerebral blood flow with iodo(^{14}C)antipyrene. Am J Physiol 234:H59–H66
11. Symon L, Lassen NA, Astrup J, Branston NM (1977) Thresholds of ischemia in brain cortex. Adv Exp Med Biol 94:775–782
12. Tamura A, Graham DI, McCulloch J, Teasdale GM (1981) Focal cerebral ischemia in the rat: 1. Description of technique and early neuropathological consequences following middle cerebral artery occlusion. J Cereb Blood Flow Metab 1:53–60
13. Van der Kerckoff W, Drewes LR (1985) Transfer of the Ca-antagonists nifedipine and nimodipine across the blood-brain barrier and their regional distribution in vivo. J Cereb Blood Flow Metab 5 (Suppl 1):S459–460
14. Wagner JA, Reynolds IJ, Weisman HF, Dudeck P, Weisfeldt ML, Snyder SH (1986) Calcium antagonist receptors in cardiomyopaltuc hamster: selective increases in heart, muscle, brain. Science 232:515–518

Calcium Overload and Brain Damage in Stroke-Prone Spontaneously Hypertensive Rats: Prevention by Nimodipine and by Parathyroidectomy

S. Kazda, M. Grunt, C. Hirth, and J.-P. Stasch

Cerebrovascular Lesions and Blood Pressure in Stroke-Prone Spontaneously Hypertensive Rats

Stroke-prone rats were originally derived from the Okamoto strain of spontaneously hypertensive rats (SHRs). Postpuberty they rapidly increase their blood pressure to extremely high values, and at the age of 8–12 months they spontaneously develop cerebrovascular lesions and brain infarctions [8]. Dietary salt load in *young* stroke-prone SHRs (SHRSPs) has been shown to intensify hypertension as well as cerebro- and renovascular lesions [6]. However, if dietary salt loading was introduced in *adult* SHRSPs with already established hypertension, it did not produce any additional increase in blood pressure, though it did drastically increase the mortality and produced severe cerebro- and renovascular lesions at a much earlier age (5–6 months) [4]. Moreover, chronic treatment with the calcium antagonist nimodipine dramatically increased the survival time of salt-loaded SHRSPs, prevented brain damage, and, to a certain extent, reduced renal and heart vascular lesions without affecting the high blood pressure [4] (Fig. 1).

This finding was the first hint that the cerebrovascular lesions in SHRSPs do not necessarily depend on the systemic blood pressure. It was obvious that the high blood pressure alone is not the cause of stroke and high mortality in SHRSPs. It was conceivable that the high salt intake produces fatal tissue damage by inducing a harmful calcium overload of the vascular or neuronal cell. The question was which factors may be responsible for the excessive transmembrane calcium influx resulting in cellular death which is not dependent on the intravascular blood pressure.

Nickerson and Conran [7] found similar dissociation in preventing salt-induced tissue damage without affecting high blood pressure in DOCA-salt hypertensive rats. In DOCA-salt hypertension renal and cardiac lesions were largely reduced by parathyroidectomy (PTX) but the high blood pressure remained uninfluenced. Parathormone (PTH) is known to stimulate calcium uptake into isolated kidney [3] and heart [2] cells and to increase the intracellular calcium concentration, resulting in an early cellular death. The effect of PTH was mimicked by calcium ionophore and prevented by verapamil [2]. In acutely uremic dogs, calcium content in the gray and white matter of brain was increased and this increment was prevented by PTX [1].

Institute of Pharmacology, Bayer AG, P.O. Box 101709, D-5600 Wuppertal 1.

Cerebral Ischemia and Calcium
Ed. by A. Hartmann and W. Kuschinsky

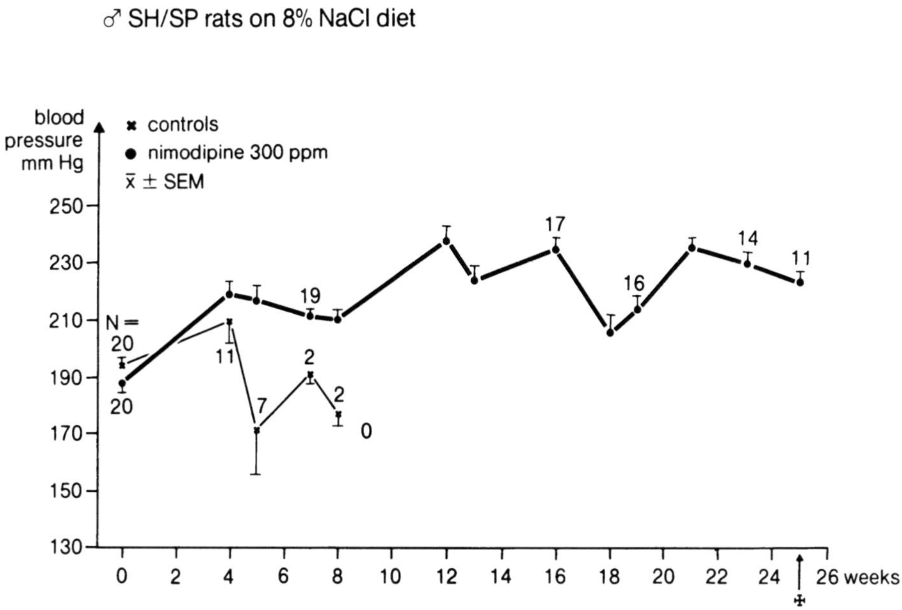

Fig. 1. Nimodipine dramatically increases survival time of salt-loaded (8% NaCl diet) SHRSPs in comparison to that of controls without decreasing high blood pressure. (Kazda et al. [4])

The similarities between the blood pressure independent protection against strokes by nimodipine and the tissue damage caused by PTX in DOCA-salt hypertensive rats prompted us to compare the effects of nimodipine and PTX in salt-loaded SHRSPs.

Parathyroidectomy Mimics the Protective Effect of Nimodipine in Salt-Loaded SHRSPs

Parathyroidectomy was peformed in one of three groups of 12-week-old male SHRSPs; two groups, each of 18 rats, were sham operated. After 17 days all three groups received a diet containing 8% NaCl, and one group of sham-operated animals was treated with nimodipine (300 ppm) added to the high salt diet [5].

The introduction of the high salt diet resulted in an additional increase in blood pressure (more than 250 mmHg during the first 4 weeks) in all groups of these young SHRSPs. This was in contrast to our previous studies with *adult* SHRSPs, in which blood pressure increased only slightly after the same amount of salt was added to the diet.

All control animals died in the course of 4 weeks on a high salt diet. After parathyroidectomy all rats developed tetany after the introduction of the high salt diet, and more than half died in the early weeks. Six animals recovered rather quickly and survived the observation period of 17 weeks, but had extremely high blood pressure (Table 1).

Table 1. Survival and systolic blood pressure in young SHRSPs on a high-salt diet

	Survived at		Systolic blood pressure (mmHg)	
	4 weeks	17 weeks	Initial	Final
Controls	0/18	0/18	211 ± 4	254 ± 9
Parathyroidectomy	14/18	6/18	211 ± 10	276 ± 6
Nimodipine	18/18	18/18	212 ± 5	259 ± 4

In the nimodipine-treated rats, too, blood pressure progressively increased on the high salt diet. However, all rats developed well until the end of the observation (17 weeks).

Morphological lesions typical of hypertensive vascular disease were severe in all control rats but completely absent in the surviving parathyroidectomy group and in all nimodipine-treated rats.

In general, there were some similarities in the effects of parathyroidectomy and nimodipine in this experiment. In contrast to controls, both nimodipine-treated rats and some of the parathyroidectomy group survived for a long period, but had extremely high blood pressure. Obviously, the high blood pressure alone is not the cause of stroke and high mortality in salt-loaded SHRSPs. It is conceivable that a high salt intake produces fatal tissue damage by inducing a harmful calcium overload. The life-saving effect of nimodipine, as mimicked by parathyroidectomy, may be due to inhibition of an excessive increase of cellular calcium concentration.

To answer this question we decided to investigate the influence of parathyroidectomy and of nimodipine on the natural development of stroke and on the tissue calcium content in SHRSPs without the accelerating effect of a dietary salt load.

Cerebral Calcium Overload in Spontaneously Diseased SHRSPs: Effects of Nimodipine and Parathyroidectomy

Twenty-eight-week-old male SHRSPs on normal diet (systolic blood pressure = 252 ± 6.9 mmHg) were divided into three groups: in one group bilateral PTX was performed, in the second group (N) a chronic treatment with nimodipine (300 ppm in the food) was started, and the third group (c) (sham operated) served as controls.

After 14 weeks (age of 42 weeks), when all controls developed neurological symptoms (ataxia, hemiplegia, and/or reversible episodes of clonic convulsions), the experiment was finished by sacrificing all animals. At that time all PTX- and nimodipine-treated animals remained healthy. The final systolic blood pressure was 281 ± 3.0 in C, 278 ± 5.1 in N, and 259 ± 5.2 mmHg in PTX. The serum ion-

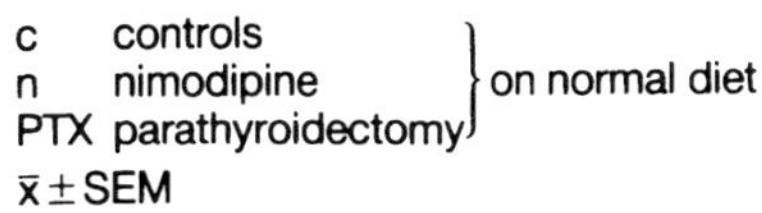

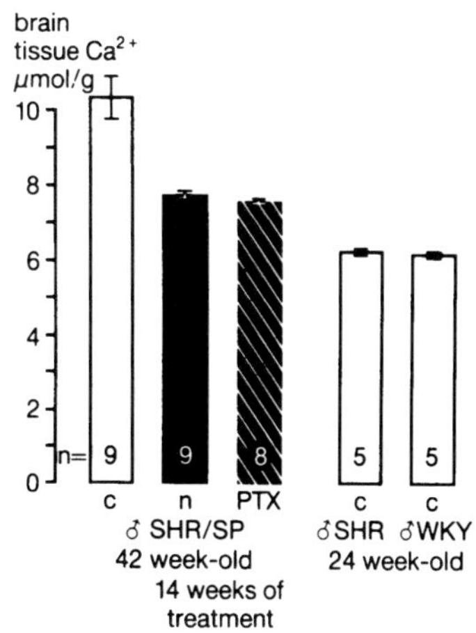

Fig. 2. Calcium overload in old SHRSPs. Effects of nimodipine and PTX

ized calcium concentration was identical in C and N (1.29 ± 0.02; 1.31 ± 0.01) but lower in PTX rats (0.82 ± 0.03 mmol/l). The calcium concentration in whole brain tissue (measured by atomic adsorption) was 10.43 ± 0.55 in C, 7.62 ± 0.11 in PTX, and 7.70 ± 0.11 in N (Fig. 2).

The spontaneous appearance of neurological symptoms in old SHRSPs was accompanied by an elevation of calcium content in the brain. Both PTX and nimodipine prevented cerebral lesions and the increase in brain calcium content without affecting high blood pressure. The prevention of calcium overload in the brain may be the common pathway in the salutary effects of nimodipine and PTX, which is not dependent on the level of systolic blood pressure.

Conclusions

An excessive increase in cellular calcium content is the first step in cellular damage. The breakdown of energy rich phosphates and the activation of lipolytic and proteolytic enzymes by the high calcium concentration result in loss of mitochondrial function, loss of membrane integrity, and cellular death.

In SHRSPs, a dietary salt load resulted in brain and renal lesions. Tissue damage was largely prevented and survival was increased by nimodipine, without a reduction in the high blood pressure. A similar protective effect, independent of blood pressure, was achieved by bilateral PTX in salt-loaded SHRSPs.

The natural appearance of stroke in SHRSPs (even without an additional salt load) was correlated with the increase in calcium content in brain tissue. Nimodipine prevented spontaneous strokes and the increase in brain calcium content without affecting the high blood pressure in SHRSPs. PTX also mimics the protective and life-saving effect of nimodipine.

Presumably, some calcium promoting factors are activated in advanced hypertension, resulting in cellular damage. Excess parathormone may be one of these factors.

Specific calcium antagonistic drug preserves tissue integrity in advanced hypertensive disease, primarily by preventing the harmful calcium overload and independently of their effect on blood pressure.

References

1. Arieff AI, Massry SG (1974) Calcium metabolism of brain in acute renal failure. J Clin Invest 53:387–392
2. Bogin E, Massry SG, Harary J (1981) Effect of parathyroid hormone on rat heart cells. J Clin Invest 67:1215–1227
3. Bork AB (1968) Effects of purified parathyroid hormone on the calcium metabolism of monkey kidney cells. Endocrinology 83:1316–1322
4. Kazda S, Garthoff B, Luckhaus G, Nash G (1982) Prevention of cerebrovascular lesions and mortality in stroke-prone spontaneously hypertensive rats by the calcium antagonist nimodipine. In: Godfraind T, Albertini A, Paoletti R (eds) Calcium modulators. Elsevier Biomedical, New York, pp 156–167
5. Kazda S, Garthoff B, Hirth C, Preis W, Stasch JP (1986) Parathyroidectomy mimics the protective effect of the calcium antagonist nimodipine in salt loaded stroke-prone spontaneously hypertensive rats. J Hypertension 4 (Suppl 3):S482–S485
6. Nagaoka A, Shino A, Inotsuka H (1978) Relationship between hypertension and cerebrovascular lesions in stroke-prone spontaneously hypertensive rats. Jpn Heart J 19:604–605
7. Nickerson PA, Conran RM (1981) Parathyroidectomy ameliorates vascular lesions induced by deoxycorticosterone in rat. Am J Pathol 105:185–190
8. Okamoto K, Yamori Y, Nagaoka A (1974) Establishment of the stroke-prone spontaneously hypertensive rat (SHR). Circ Res 34, 35 (Suppl 1):143–153

Neurotransmitters and Calcium During Hypoxia

G. E. Gibson, L. Toral-Barza, T. Manger, and G. Freeman

Introduction

An interaction of calcium homeostasis and neutrotransmitter metabolism may underlie deficits induced by hypoxia and ischemia. Both in vitro and in vivo results suggest that altered calcium homeostasis has an integral role in hypoxic/ischemic induced cell damage. Cause–effect relations to cell damage in vivo are difficult to determine because so many variables change simultaneously. In addition, the changes in calcium homeostasis differ between presynaptic and postsynaptic processes and these are also difficult to distinguish in vivo. Thus, the studies on calcium that are described here emphasize in vitro approaches. Similarly, numerous studies suggest that altered neurotransmitter metabolism accompanies hypoxia/ischemia. The precise role of neurotransmitters in the production of hypoxic/ischemic induced tissue damage is not well documented. Much of the research on the role of neurotransmitters in ischemic cell damage has emphasized their postsynaptic action. The present studies concentrate on the presynaptic changes that lead to increased extracellular concentrations of these neurotransmitters during hypoxia/ischemia. The interaction of calcium homeostasis with neurotransmitter metabolism appears important to hypoxic-induced alterations.

Differences in the Pre- and Postsynaptic Responses to Hypoxia

Presynaptic and postsynaptic neuronal components respond to hypoxia differently. These responses were distinguished in vitro. Brain slices from various brain regions were used in two treatment paradigms: (a) the response during hypoxia was determined, (b) the response to a hypoxic pretreatment that preceded a second incubation under optimal conditions was determined. Two measures were used to assess the effects of these treatments: $^{14}CO_2$ production from radioactive glucose (i.e., a general measure of oxidative metabolism) and acetylcholine (ACh) production. ACh and CO_2 production are more sensitive to alterations in hypoxia than ATP levels or the adenylate energy charge and may therefore be appropriate markers of hypoxic-induced changes [5–7]. Furthermore, ACh metabolism

Cornell University Medical College, Burke Rehabilitation Center, 785 Mamaroneck Avenue, White Plains, N.Y. 10605, USA.

Cerebral Ischemia and Calcium
Ed. by A. Hartmann and W. Kuschinsky

may reflect early changes in calcium homeostasis, because of calcium's role in ACh release. Three brain regions were selected because of the relative contributions of cholinergic terminals and cell bodies to their metabolism: hippocampus, striatum, and septum. The hippocampus contains only cholinergic nerve terminals; the septum has only nerve cell bodies; the striatum includes both cholinergic nerve terminals and nerve cell bodies.

In the first series of studies, the effects of anoxia on slices from septum (i.e., cholinergic cell bodies) and hippocampus (i.e., cholinergic terminals) were compared [18]. Anoxia reduced hippocampal ACh synthesis by 77% but had no effect on septal ACh formation. The response of oxidative metabolism to anoxia was parallel in the two brain regions. This would be expected since nerve terminals and cell bodies of all neurotransmitter types would be included in the measure.

Differences in pre- and postsynaptic responses were also demonstrated by exposing the tissue to an anoxic preincubation and then examining the effects of this pretreatment in a second incubation under optimal conditions including oxygen. With whole forebrain slices, pretreatment with no oxygen and high potassium reduced ACh synthesis in the second incubation by 63% and CO_2 production by 67% [14]. A variety of preincubation conditions decreased $^{14}CO_2$ and ^{14}C-ACh production in the second incubation in parallel ($r = 0.932$). These effects could be ameliorated if calcium was omitted from the preincubation medium or by the addition of the calcium antagonist verapamil. Curiously, if the preincubation contained the calcium antagonists nimodipine or nifedipine, the anoxic-induced decreases in ACh and CO_2 in the second incubation were even larger [14]. In striatal slices (cholinergic nerve endings and cholinergic nerve cell bodies), an anoxic preincubation reduced ACh production in the subsequent incubation by 46%. However, hippocampal ACh production was only minimally affected by such a pretreatment. In addition, the omission of calcium from the anoxic preincubation increased striatal ACh synthesis by 88% [10]. These results suggest that the nerve cell body rather than the presynaptic process is altered by the anoxic preincubation. This conclusion is further supported by the observation that exposure of synaptosomes to an anoxic preincubation does not alter the CO_2 or ACh production during the subsequent incubation [10]. More severe preincubation conditions (e.g. anoxia and decreased pH) will alter synaptosomal metabolism in the second incubation [24].

The hypoxic-induced changes in calcium uptake in brain slices as compared to synaptosomes also suggest that hypoxia produces different responses in the presynaptic terminal than in the postsynaptic processes. Hypoxia reduces synaptosomal (i.e., presynaptic) calcium uptake; however, exposure of brain slices to anoxia increases calcium uptake [14]. This latter observation is consistent with a postsynaptic increase in calcium uptake. An increase in postsynaptic calcium may explain the large decrease in extracellular calcium that accompanies ischemia [23, 25], since the calcium accumulation that occurs after ischemia is postsynaptic [21, 36, 39]. Furthermore, the damage that occurs in cells after ischemia appears postsynaptically, whereas the presynaptic changes are minimal [36, 39].

Hypoxia and Neurotransmitter Release

Several experimental approaches suggest that hypoxia reduces the release of ACh [5, 17, 25–27, 32] but elevates the release of glutamate [5, 23, 25] and dopamine [2, 4, 5, 19, 25]. Although previous studies left open the possibility that these surprising differences may have been dependent upon the experimental approach, recent results demonstrate these selective alterations in release in a single experimental system (i.e., mouse brain striatal synaptosomes; [13]).

Hypoxic-induced reductions in ACh metabolism have been demonstrated in vivo and in vitro. Histotoxic [6], anemic [6, 16, 32], and hypoxic [8] hypoxia all diminish ACh synthesis in vivo and this decrease appears to be physiologically important [15]. Hypoxia diminishes the release of ACh from synaptosomes [13, 33] and slices [5, 17]. This impairment of release may lead to the hypoxic-induced reduction in synthesis. Hypoxia reduces ACh synthesis in regions that exhibit calcium-dependent ACh release (i.e., hippocampus). However, hypoxia has no effect on regions that do not have calcium-dependent release [18]. Since only calcium-dependent release is diminished, altered calcium homeostasis may play an important role in these deficits. In addition, a reduction of the calcium concentration in the medium from 1.3 to 0.1 mM abolishes the effects of hypoxia and potassium on ACh release [13]. This further supports the suggestion that hypoxia alters only calcium-dependent release. The reduction in ACh release is paralleled by a decline in calcium uptake into the nerve terminal. If calcium uptake is stimulated by the aminopyridines in vitro, the hypoxic-induced deficit in ACh release is overcome [17, 32–34]. Furthermore, administration of the aminopyridines in vivo ameliorates the hypoxic-induced impairment of ACh synthesis and of behavioral measures [32].

Hypoxia increases dopamine release. The effect of hypoxia on dopamine release from rat striatal slices varied with the potassium concentration. Hypoxia increased the extracellular dopamine concentration at the end of a release incubation by twofold under basal conditions and by fivefold in the presence of depolarizing concentrations of potassium [5]. This increase in extracellular dopamine does not appear to be related to impaired dopamine reuptake since dopamine reuptake is stimulated under similar condiitions. Calcium appears to be important to these changes because a reduction of medium calcium from 1.3 mM to 0.1 mM diminishes the effects of KCN on dopamine release by nearly 50% [13]. Elevated extracellular dopamine also occurs in vivo during hypoxia [2] and during ischemia [19, 38]. The increased extracellular dopamine appears important in the pathophysiology of ischemic cell damage, since lesions of the dopaminergic inputs to the striatum protect against ischemic cell damage [20].

Hypoxia and anoxia increase the release of glutamate. In rat brain slices, potassium increased glutamate release by threefold in the presence of 100% oxygen and 29-fold during anoxia. However, anoxia did not alter release under basal conditions [5]. Similar increases in glutamate release are observed in synaptosomes, which suggests that the changes are due to alterations in the presynaptic nerve terminal. The changes in glutamate release with hypoxia are only moderately altered if the medium calcium is reduced from 1.3 to 0.1 mM [13]. Extracellular glutamate increases with ischemia in vivo [22]. That the elevation of extracellular glutamate during ischemia is important in the pathophysiology of hypoxic/ischemic induced cell damage is suggested by the protective effects of calcium antagonists [3, 37] and of lesions of glutaminergic pathways [29, 40].

These selective hypoxic-induced changes in neurotransmitter release can be partially ameliorated by 3,4-diaminopyridine [5, 32–34]. During hypoxia, 100 μM 3,4-diaminopyridine enhanced ACh release but diminished the hypoxic-induced increase in glutamate and dopamine release [5]. In contrast, under nonhypoxic conditions, 3,4-diaminopyridine increased the release of all three neurotransmitters [5]. These results suggest that a common mechanism may be involved in the selective anoxic-induced changes in release. This may be due either to the actions of the aminopyridines on calcium homeostasis as discussed above or to their action on potassium channels [28]. 3,4-Diaminopyridine has been shown to be effective in overcoming lesions produced by the toxin AF64A [30], but its effects on ischemic-induced tissue damage have not been examined.

Hypoxia and Nerve Terminal Calcium Homeostasis

Hypoxia alters nerve ending calcium fluxes. Calcium uptake into synaptosomes declined by 60% and 82% with 2.5% and 0% oxygen, respectively [33]. External calcium binding to the synaptosomal plasma membrane declined by 71% and 88% when the oxygen was reduced to 2.5% or 0%, respectively [33]. Each of these deficits could be ameliorated by the addition of 3,4-diaminopyridine [33]. Calcium uptake by mitochondria within synaptosomes and flux through the cytosolic compartment are reduced by hypoxia [26]. A 10-min hypoxic incubation in normal potassium medium reduced total (−38%), mitochondrial (−28%), and nonmitochondrial (−41%) calcium uptake. In high potassium medium, low oxygen reduced mitochondrial (−41%) and total (−34%) calcium uptake, whereas nonmitochondrial uptake was essentially unaffected. A temporal analysis of the effects of hypoxia on nonmitochondrial calcium uptake during depolarization revealed an initial depression followed by a stimulation [34]. Analogous results were obtained when mitochondrial uncouplers (either FCCP or rotenone plus oligomycin) were added. Both drug and membrane potential studies suggest that hypoxic-induced alterations in the subcellular distribution of calcium may be due to an uncoupling mechanism and the subsequent collapse of the synaptosomal mitochondrial membrane potential [9, 31, 34].

Hypoxia elevates cytosolic free calcium as assessed by the fluorescent probe fura-2 [12, 13]. This method monitors cytosolic free calcium levels continuously [11]. Addition of potassium elevates cytosolic free calcium rapidly (1 s or less), whereas histotoxic hypoxia increases it gradually. Hypoxia increases cytosolic free calcium in both low and normal calcium medium and the elevation occurs under both resting and potassium-stimulated conditions. However, if external calcium is reduced, the potassium- or hypoxic-induced increase in cytosolic free calcium is diminished. Cytosolic free calcium also increases if the glucose in the incubation medium is decreased. Thus, during ischemia low glucose and low oxygen may exaggerate each other's effects.

Interactions of Calcium Homeostasis and Neurotransmitter Release

Changes in either external or internal calcium concentrations selectively alter the release of ACh, glutamate, and dopamine. This suggests that the release of each has a different sensitivity to calcium, which may underlie the selective alteration of neurotransmitter release with hypoxia. These relations can be examined both for their dependence on external calcium as well as cytosolic free calcium.

The effects of external calcium on hypoxic-induced changes varied with the neurotransmitter [13]. With low external calcium concentrations (0.1 mM), the pattern of the hypoxia and potassium effects on glutamate and dopamine release are diminished compared to normal calcium media. However, ACh is unresponsive to either potassium or hypoxia at low external calcium concentrations. The effects of hypoxia on ACh release are abolished in either low calcium or EGTA (i.e., a calcium chelator) containing medium. On the other hand, the effects of hypoxia on glutamate and dopamine release are exaggerated in the presence of EGTA. Thus, of the three neurotransmitters, ACh requires higher medium calcium concentrations to be affected by either potassium or hypoxia.

The relation of cytosolic free calcium to potassium- and hypoxic-induced changes varied with the neurotransmitter (Fig. 1). Cytosolic free calcium was compared to release in medium with high or low calcium, low or depolarizing potassium, with or without KCN. A comparison under all of these conditions revealed that changes in cytosolic free calcium correlated highly with release of glutamate ($r=0.928$) and dopamine ($r=-0.913$). However, ACh release during

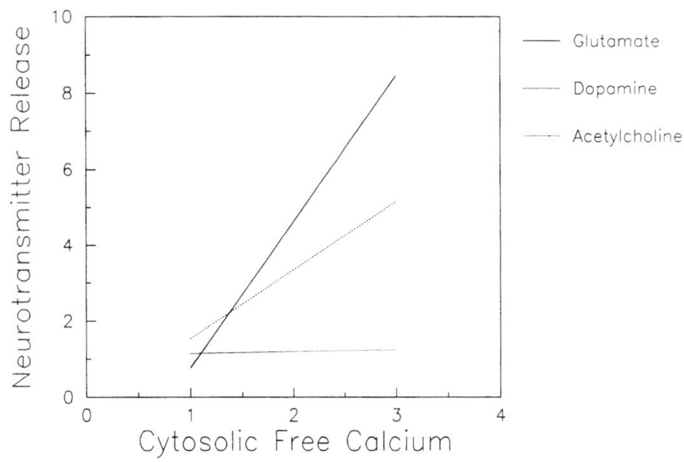

Fig. 1. The relation of cytosolic free calcium to neurotransmitter release during hypoxia or normoxia with normal or reduced calcium. Values from six different conditions were examined for each neurotransmitter: low calcium (basal and K^+-depolarized), normal calcium (basal and K^+-depolarized), KCN + K^+-depolarization (low and normal calcium). The value for each neurotransmitter was divided by its basal value in low calcium and compared to its respective cytosolic free calcium value divided by its basal value in low calcium. The regression was calculated for each comparison: dopamine = 1.807 (Ca_i) − 0.2713 ($r=0.913$); acetylcholine = 0.0466 (Ca_i) + 1.106 ($r=0.081$); glutamate = 3.84 (Ca_i) − 3.068 ($r=0.928$)

hypoxia and variable calcium concentrations showed a very low correlation (r= 0.0814) with cytosolic free calcium. Other investigators [1], as well as the current studies, demonstrate a high correlation between ACh release and cytosolic free calcium with normal oxygenation. Thus, during hypoxia the coupling of ACh release to cytosolic free calcium appears dissociated. Of the three neurotransmitters considered, glutamate seems to be the most sensitive to alterations in cytosolic free calcium: for a given elevation in cytosolic free calcium, the percent increase for glutamate is nearly twice as much as dopamine and nearly 82 times more than ACh (Fig. 1).

Conclusions

Reducing efflux of neurotransmitters from the presynaptic terminals may be an effective approach to limiting ischemic tissue damage. The changes in presynaptic terminals vary from the postsynaptic changes. The neurotransmitter changes in the presynaptic terminal are related to altered calcium homeostasis in a selective manner. Alterations in presynaptic calcium homeostasis may be a useful endpoint to examine potentially useful compounds for the treatment of hypoxic/ischemic damage.

References

1. Ashley RH (1986) External calcium, intrasynaptosomal free calcium and neurotransmitter release. Biochim Biophys Acta 854:207–212
2. Broderick PA, Gibson GE (1986) Extracellular dopamine and serotonin in vivo during repeated episodes of hypoxic-hypoxia. Soc Neurosci Abst 12:565
3. Deshpande JK, Wieloch T (1985) Amelioration of ischemic brain damage by postischemic treatment with flunarizine. Neurol Res 7:27–29
4. Freeman GB, Gibson GE (1986) Effect of decreased oxygen on in vitro release of endogenous 3,4-dihydroxphenylethylamine from mouse striatum. J Neurochem 47:1924–1931
5. Freeman GB, Mykytyn V, Gibson GE (1987) Differential alteration of dopamine, acetylcholine and glutamate release during anoxia and/or 3,4-diaminopyridine treatment. Neurochem Res 12:1019–1027
6. Gibson GE, Blass JP (1976) Impaired synthesis of acetylcholine in brain accompanying mild hypoxia and hypoglycemia. J Neurochem 27:37–42
7. Gibson GE, Blass JP (1976) A relation between [NAD^+]/[NADH] potentials and glucose utilization in rat brain slices. J Biol Chem 251:4127–4130
8. Gibson GE, Duffy TE (1981) Impaired synthesis of acetylcholine by mild hypoxic hypoxia or nitrous oxide. J Neurochem 36:28–33
9. Gibson GE, Freeman GB, Broderick P, Nieslen P (1987) Possible calcium mediated events in selective vulnerability. XIII International Symposium. J Cereb Blood Flow Metab 7S:155
10. Gibson GE, Freeman GB, Mykytyn V (1988) Selective damage in striatum and hippocampus with in vitro anoxia. Neurochem Res 13:329–335
11. Gibson GE, Manger T (1988) Changes in cytosolic free calcium with 1,2,3,4-tetrahydroaminocridine, 4-aminopyridine and 3,4-diaminopyridine. Biochem Pharmacol 37:4191–4196
12. Gibson GE, Manger T (1988) Cytosolic-free calcium and neurotransmitter release during hypoxia. Trans Am Neurochem 19:147

13. Gibson GE, Manger T, Toral-Barza L, Freeman G (1988) Cytosolic-free calcium and neurotransmitter release with decreased availability of glucose or oxygen. Neurochem Res 14:527–543
14. Gibson GE, Mykytyn V (1988) An in vitro model of anoxic induced damage in mouse brain. Neurochem Res 13:9–20
15. Gibson GE, Pelmas CJ, Peterson C (1983) Cholinergic drugs and 4-aminopyridine alter hypoxic-induced behavioral deficits. Pharmacol Biochem Behav 18:909–916
16. Gibson GE, Peterson C, Sansone J (1981) Decreases in amino acid and acetylcholine metabolism during hypoxia. J Neurochem 37:192–201
17. Gibson GE, Peterson C (1982) Decreases in the release of acetylcholine in vitro with low oxygen. Biochem Pharmacol 31:111–115
18. Gibson GE, Peterson C (1983) Acetylcholine metabolism in septum and hippocampus in vitro. J Biol Chem 258:1142–1145
19. Globus MTT, Ginsberg MD, Harik SI, Busto R, Dietrich WD (1987) Role of dopamine in ischemic striatal injury: metabolic evidence. Neurology 37:1712–1719
20. Globus MYT, Ginsberg MD, Dietrich WD, Busto R, Scheinberg P (1987) Substantia nigra lesion protects against ischemic damage in the striatum. Neurosci Lett 80:251–256
21. Griffiths T, Evans MC, Meldrum BS (1982) Intracellular sites of early calcium accumulation in the rat hippocampus during status epilepticus. Neurosci Lett 30:329–334
22. Hagberg H, Lehmann A, Sandberg M, Nystrom B, Jacobson I, Hamberger A (1985) Ischemia-induced shift of inhibitory and excitatory amino acids from intra- to extracellular compartments. J Cereb Blood Flow Metab 5:413–419
23. Harris RJ, Symon L, Branston NM, Bahan M (1981) Changes in extracellular calcium activity in cerebral ischaemia. J Cereb Blood Flow Metab 1:203–209
24. Harvey SAK, Booth RFG, Clark JB (1982) The effects in vitro of hypoglycaemia and recovery from anoxia on synaptosomal metabolism. Biochem J 206:433–439
25. Hirsch JA, Gibson GE (1984) Selective alteration of neurotransmitter release by low oxygen in vitro. Neurochem Res 9:1039–1049
26. Ksiezak HJ, Gibson GE (1981) Acetylcholine synthesis and CO_2 production from variously labelled glucose in rat brain slices and synaptosomes. J Neurochem 37:88–94
27. Ksiezak HJ, Gibson GE (1981) Oxygen dependence of glucose and acetylcholine metabolism in slices and synaptosomes from rat brain. J Neurochem 37:305–314
28. Lechat P, Thesleff S, Bowman WC (1982) Aminopyridines and similarly acting drugs: effects on nerves, muscles and synapses. Pergamon, New York (Advances in the Biosciences 35)
29. Linden T, Kalimo H, Wieloch T (1987) Protective effect of lesion to the glutamatergic cortico-striatal projections on the hypoglycemic nerve cell injury in rat striatum. Acta Neuropathol 74:335–344
30. McArdle JJ, Argentieri TM, Gargano D, Hanin I, Fischer A (1985) Acute effects of the cholinergic neurotoxin AF64A on the secretion of acetylcholine. Pharmacologist 27:534
31. Nielsen PE, Gibson GE (1986) Mitochondrial and plasma membrane potentials during anoxia and normoxia. Soc Neurosci Abstr 12:1403
32. Peterson C, Gibson GE (1982) 3,4-Diaminopyridine alters acetylcholine metabolism and behavior during hypoxia. J Pharmacol Exp Ther 222:576–582
33. Peterson C, Gibson GE (1984) Synaptosomal calcium metabolism during hypoxia and 3,4-diaminopyridine treatment. J Neurochem 42:248–253
34. Peterson C, Nicholls DG, Gibson GE (1985) Subsynaptosomal calcium distribution during hypoxia and 3,4-diaminopyridine treatment. J Neurochem 45:1779–1790
35. Siemkowicz E, Hansen AJ (1981) Brain extracellular ion composition and EEG activity following 10 minutes ischemia in normo- and hyperglycemic rats. Stroke 12:236–240
36. Simon RP, Griffiths T, Evans MC, Swan JH, Meldrum BS (1984) Calcium overload in selectively vulnerable neurons of the hippocampus during and after ischemia: an electron microscopy study in the rat. J Cereb Blood Flow Metab 4:350–361
37. Simon RP, Swan JH, Griffiths T, Meldrum BS (1984) Blockade of *N*-methyl-D-aspartate receptors may protect against ischemic damage in the brain. Science 226:850–852
38. Slivka A, Brannan TS, Weinberger J, Knott PJ, Cohen G (1988) Increase in extracellular dopamine in the striatum during cerebral ischemia: a study utilizing cerebral microdialysis. J Neurochem 50:1714–1718

39. Van Reempts J, Haseldonckx M, Van de Ven M, Borgers M (1984) Morphology and ultrastructural calcium distribution in the rat hippocampus after severe transient ischemia. In: Bes A, Braquet P, Paoletti R, Siesjö B (eds) Cerebral ischemia. Elsevier, New York, pp 113–118
40. Wieloch T, Engelsen B, Westerberg E, Auer R (1985) Lesions of the glutamatergic corticostriatal projections in the rat ameliorate hypoglycemic brain damage in the striatum. Neurosci Lett 58:25

Protective Effects of Calcium Antagonists Against Brain Damage Caused by Ischemia

J. KRIEGLSTEIN [1], D. SAUER [1], J. NUGLISCH [1], C. KARKOUTLY [1], T. BECK [1], G.W. BIELENBERG [1], C. ROSSBERG [2], and H.D. MENNEL [2]

It has been pointed out in numerous papers [4, 6, 7, 18, 19] that the disturbed calcium homeostasis plays a pivotal role in the pathogenesis of neuronal cell damage caused by ischemia. Calcium ions can fulfill their function as second messengers only when the cytosolic concentration of free Ca^{2+} is maintained at a level of approximately 10^{-7} mol/l whereas the extracellular concentration of Ca^{2+} is around 10^{-3} mol/l.

The neurons need a lot of metabolic energy to uphold this transmembrane Ca^{2+} gradient of 10^4. Under ischemic conditions when ATP drops far below the normal level, Ca^{2+} influx occurs along both voltage-dependent and receptor-operated channels. The cytosolic free Ca^{2+} rises to levels above 10^{-4} mol/l that trigger a cascade of reactions. Thus, mitochondria are loaded with Ca^{2+} and energy metabolism is altered. Furthermore, enzyme activities are changed by the enhanced Ca^{2+} level, e.g., phospholipases are activated and disturb various membranes. All these processes may contribute to the neuronal death which is shown to occur several days after ischemia [9, 18, 19]. If the Ca^{2+} entry into the cells can be inhibited, the pathological events caused by the increased cytosolic Ca^{2+} concentration should consequently be reduced.

Calcium antagonists block the voltage-dependent calcium channels. These drugs should be capable of inhibiting Ca^{2+} influx during and after ischemia, at least in part, and they should thereby reduce the damage caused by Ca^{2+} increase in the cytosol. It has been shown in previous work that pretreatment of rats with calcium antagonists accelerates postischemic ATP resynthesis and lactate degradation [2, 11, 12]. The purpose of our present studies was to demonstrate that neuronal necroses caused by ischemia can be reduced by calcium antagonists.

Methods

To test this hypothesis we used the two-vessel occlusion rat model described by Smith et al. [20]. Ischemia was induced in male Wistar rats by clamping both carotid arteries and by lowering blood pressure to 40 mmHg by exsanguination for 10 min. Arterial pO_2, pCO_2, pH, mean arterial blood pressure, and rectal temperature were checked routinely (Table 1).

[1] Institut für Pharmakologie und Toxikologie und [2] Abteilung für Neuropathologie der Universität Marburg, Ketzerbach 63, D-3550 Marburg, FRG.

Cerebral Ischemia and Calcium
Ed. by A. Hartmann and W. Kuschinsky

Table 1. Effects of calcium antagonists on physiological variables

Variables	Flunarizine		Emopamil		Nimodipine	
	Vehicle	2 × 40 mg/kg p.o.	Vehicle	2 mg/kg i.v.	Vehicle	0.3 mg/kg p.o.
PaO_2 (mmHg)	138.1 ± 23.7	135.8 ± 26.5	154.8 ± 22.7	154.6 ± 23.1	125.5 ± 11.9	129.5 ± 34.9
$PaCO_2$ (mmHg)	37.5 ± 1.9	36.9 ± 1.7	36.2 ± 2.2	37.1 ± 2.0	36.5 ± 2.1	36.5 ± 1.5
Arterial pH	7.40 ± 0.03	7.43 ± 0.03	7.43 ± 0.02	7.41 ± 0.02	7.38 ± 0.04	7.35 ± 0.04
Mean arterial blood pressure (mmHg)						
Before injection	–	–	131.4 ± 11.1	130.3 ± 11.1	–	–
Before ischemia	125.5 ± 13.2	121.1 ± 12.5	123.8 ± 12.1	116.0 ± 14.1 ***	126.9 ± 16.5	123.0 ± 15.1
After ischemia	125.2 ± 13.7	120.4 ± 10.8	128.3 ± 11.1	128.6 ± 14.0	130.0 ± 10.0	131.3 ± 13.8
Temperature (°C)	36.9 ± 0.1	37.0 ± 0.1	37.1 ± 0.1	37.0 ± 0.1	37.0 ± 0.1	36.9 ± 0.2

Arterial blood gases and pH were determined 10 min prior to ischemia. Mean arterial blood pressure was measured before emopamil injection as well as 10 min before and 20 min after ischemia. Means ± SD from 8–12 experiments are given. Different from the value before drug administration: *** $P < 0.001$ (paired *t*-test).

Up to 60 min of recirculation the local cerebral flow was measured by the ^{14}C-iodoantipyrine technique [17] and 7 days after ischemia, histological evaluation of brain tissue was performed. Brains were perfusion-fixed, embedded in Paraplast, and sectioned in 5-µm slices. Brain slices were stained with 0.1% cresyl violet or with a mixture of 1% celestine blue and 1% fuchsin [1]. Hippocampal intact and damaged neurons were counted in the entire given structure 6 mm rostral to the tentorial incision.

Flunarizine, emopamil, and nimodipine were selected for the presented experiments because these calcium antagonists are lipophilic enough to permeate through the blood-brain barrier to a sufficient degree.

In preceding experiments it was examined in which brain area and after which postischemic period neuronal damage was most expressed. It was demonstrated that hippocampal CA1 neurons were selectively vulnerable against ischemia, a result in accord with previous findings [10, 15, 16]. Furthermore, it was shown that 2 days after ischemia only very few neurons were damaged in the CA1 subfield and after 7 days a maximum of cell damage was achieved: about 50%–90% in the CA1 subfield, and usually less than 10% in the CA3 and CA4 subfields and in cortical areas. Therefore, further histological evaluation was focused on the hippocampus, especially on its CA1 subfield.

Results

Flunarizine

The calcium antagonist flunarizine has already been studied by numerous investigators. Though its efficacy as a cerebroprotective drug has not been proved unanimously [8, 14], sound evidence has been produced that flunarizine effectively ameliorates ischemic brain damage [5, 13, 21]. However, it is still a matter of debate whether the cerebroprotective effect of flunarizine is mediated by its vascular activity or whether it acts directly on neuronal cells. Therefore, it seemed of interest to estimate to what extent vascular and/or parenchymal effects of this calcium entry blocker contribute to its cytoprotective action.

Flunarizine did not influence the physiological parameters measured, i.e., arterial blood gases, pH, mean arterial blood pressure, and rectal temperature (Table 1). Two minutes after ischemia flunarizine produced a marked increase in local cerebral blood flow of the cortical areas but the blood flow in the hippocampal subfields was not significantly elevated. In addition, no effect of flunarizine on local cerebral blood flow was measured after 10 min (Fig. 1) and 30 min of recirculation. However, neuronal necroses in the CA1 subfield of the hippocampus analyzed 7 days after ischemia were remarkably reduced by flunarizine (Table 2), whereas in the other hippocampal sections and in cortical areas neuronal damage (values not shown) was not significantly influenced by the drug. Thus, in the hippocampus neuronal protection was afforded by flunarizine without an increase in blood flow. This mismatch between cellular protection and flow improvement is partially in line with previous findings. Deshpande and Wieloch [5] also reported a cerebroprotective effect of flunarizine administered to rats pre- and

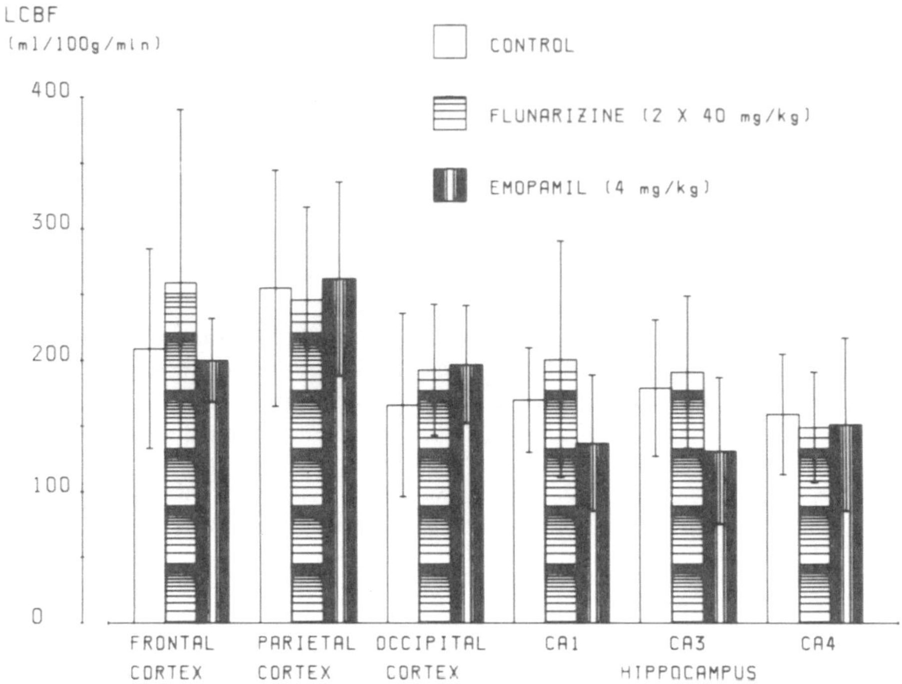

Fig. 1. Postischemic local cerebral blood flow of rats. Measurements were performed after 10 min of recirculation. Values are means ± SD of five or six experiments

Table 2. Inhibition of neuronal damage by calcium antagonists in rat hippocampus

Drug treatment	Percentage of neuronal damage in hippocampal subfields		
	CA1	CA3	CA4
S-Emopamil (2 mg/kg, i.v.)	35.1 ± 8.1***	0.6 ± 0.3**	0.2 ± 0.1
Control	75.9 ± 7.4	2.9 ± 0.8	1.1 ± 0.5
Flunarizine (2 × 40 mg/kg, p.o.)	24.7 ± 4.3*	1.8 ± 0.5	42.7 ± 4.7
Control	59.9 ± 7.7	3.2 ± 0.9	40.2 ± 5.8
Nimodipine (0.3 mg/kg, p.o.)	27.3 ± 6.2*	0.6 ± 0.2	1.2 ± 0.4
Control	45.3 ± 4.6	1.1 ± 0.2	1.2 ± 0,3

After 10 min of forebrain ischemia and 7 days of recovery the rat brains were perfusion fixed and histologically evaluated. Means of 8–12 experiments ± SEM are given.

Different from corresponding control: * $P < 0.05$, ** $P < 0.01$, *** $P < 0.001$ (Mann-Whitney U-test).

postischemically without recording any flow increase. Newberg et al. [14], too, failed to monitor increases in blood flow by flunarizine during the first 10 min postischemia in a canine model of complete cerebral ischemia. However, these authors [14] measured global cerebral blood flow and they could have missed improvements of blood flow confined to distinct brain structures.

Since, in our hands, flunarizine did not increase blood flow during the early postischemic phase but significantly reduced the percentage of necrotic neurons after 7 days of survival in the hippocampal CA1 subfield, we may assume that the drug exerted its cerebroprotective effect by acting directly on the neurons. Vascular effects seem to be secondary if they are involved in brain protection at all.

Emopamil

Emopamil, a calcium antagonist of the verapamil group, also had a protective effect on hippocampal neurons. Neuronal damage in the CA1 and CA3 subfields was markedly reduced by this drug (Fig. 2; Table 2) whereas again in the cortical areas neuronal damage was not diminished following emopamil administration.

Similar to flunarizine, emopamil caused an increase in the blood flow of the parietal cortex at 2 min of postischemic recirculation. After emopamil administration all other cortical structures showed a tendency toward increased blood flow, which nevertheless did not attain significance level probably because of the high standard deviations. However, in the hippocampal subfields the local blood flow remained clearly unchanged. At 10 and 60 min of recirculation emopamil did not produce any alteration in local cerebral blood flow compared to untreated controls (Fig. 1). In addition, emopamil showed only a slight influence on systemic blood pressure. During and shortly after emopamil administration the mean arterial blood pressure slightly decreased but postischemic blood pressure was not influenced by this calcium antagonist (Table 1). Thus, these findings with emopamil, which did not improve postischemic hippocampal blood flow but ameliorated hippocampal neuronal damage, again supported the concept that calcium antagonists act directly on neurons to protect them against ischemic damage. The assumption of such a direct action is further substantiated by previous findings with the isolated rat brain. Lactate was removed from brain cortical tissue independently of the flow rate and the decrease in the postischemic lactate level could be accelerated by emopamil without an increase in the recirculation rate [2]. Similar results were obtained with gallopamil [11] and nimodipine [3]. As postischemic brain lactate clearance is almost entirely due to mitochondrial oxidative degradation, these results strongly favor the concept of a direct effect of the calcium antagonists on brain parenchyma.

Blood flow improvement by calcium antagonists could be important for cell survival mainly during the period of delayed postischemic hypoperfusion, i.e., after around 1 h of recirculation. At this time, however, emopamil and flunarizine did not increase the local cerebral blood flow in our experiments. On the other hand, decrease in systemic blood pressure could reduce cerebral circulation by an extracerebral steal phenomenon. However, the fall in blood pressure caused by

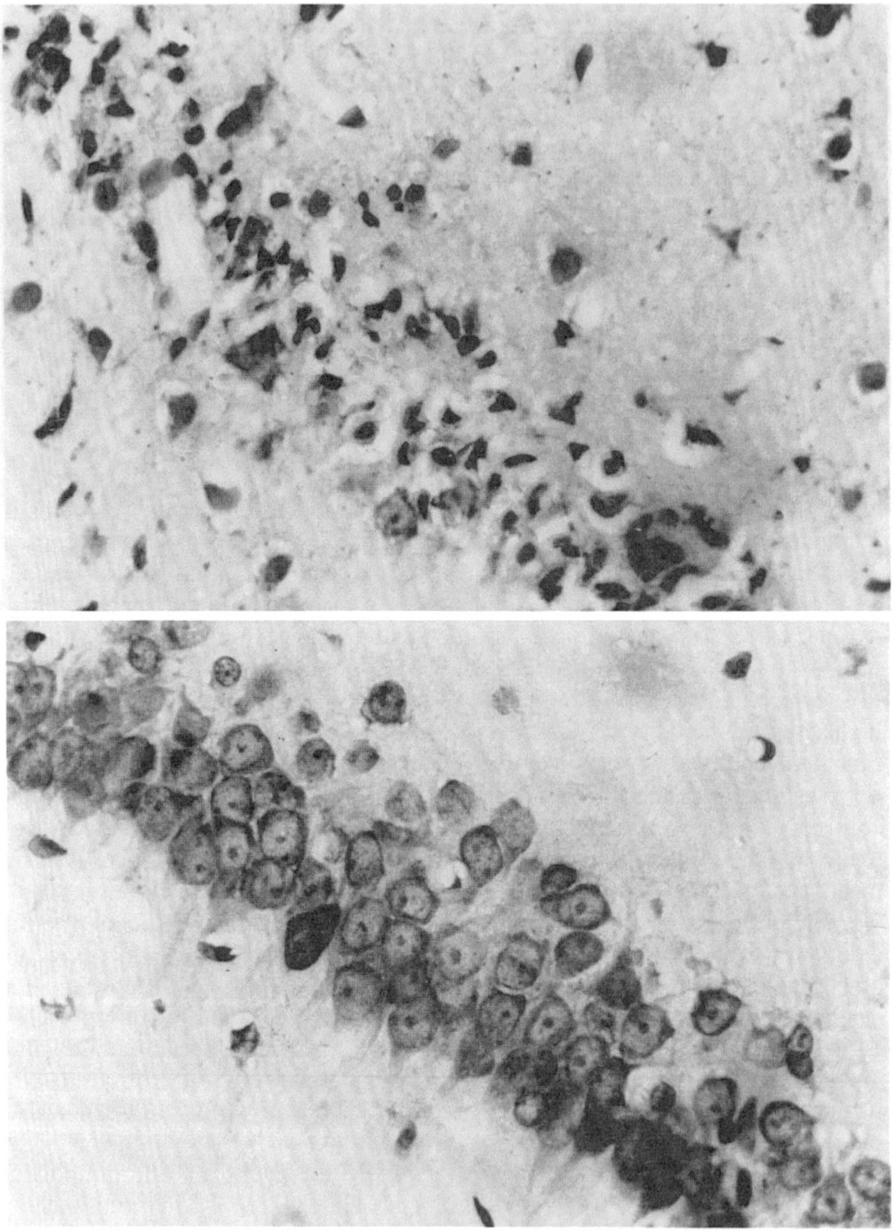

a

b

Fig. 2 a, b. Histological demonstration of neuronal damage in the CA1 subfield of rat hippocampus 7 days after 10 min of ischemia. **a** control; **b** emopamil-treated rat. Cresyl violet, ×400 emopamil (Table 1) was confined to the preischemic period and no change was to be seen postischemia. Therefore, one does not have to expect an unfavorable influence of systemic blood pressure on postischemic cerebral blood flow and, in consequence, on postischemic neuronal outcome in the case of these drugs.

Nimodipine

Interesting results were obtained using nimodipine as a calcium antagonist for protecting rat brain tissue against ischemic damage. In a first series of experiments nimodipine was administered to rats by intravenous infusion over 30 min in a total dose of 10 μg/kg. Infusion was finished 15 min prior to ischemia. This small dose of nimodipine was used to avoid pronounced effects on blood pressure. However, we did not find a protective effect of nimodipine under these conditions. About 70% of pyramidal cells in the CA1 subfield were damaged in con-

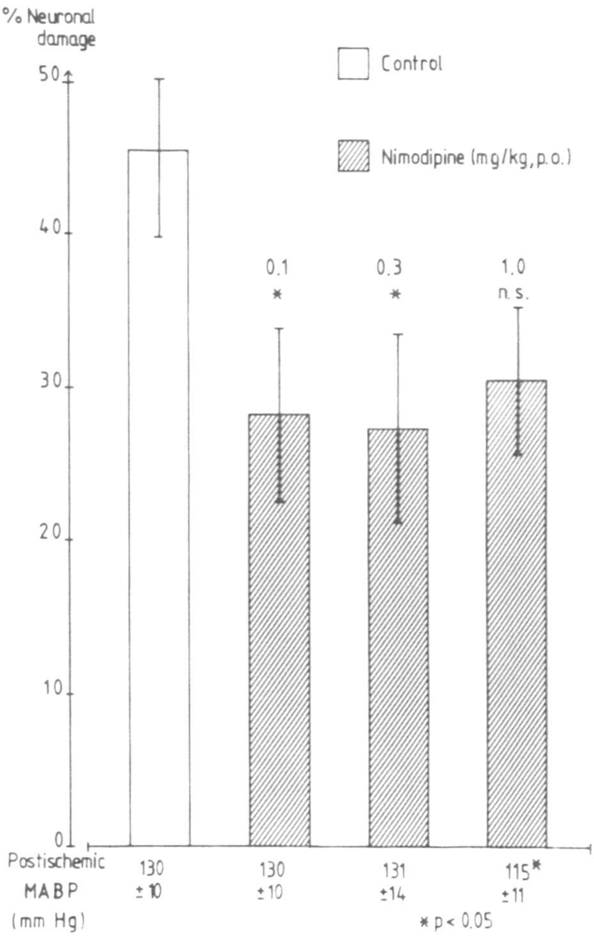

Fig. 3. Reduction of postischemic neuronal damage in the CA1 subfield by nimodipine. Forebrain ischemia was induced in rats for 10 min followed by 7 days of recovery. Nimodipine was administered orally 60 min prior to ischemia. Percentages of neuronal damage are given as means ± SEM of 8–12 experiments. The values of mean arterial blood pressure (*MABP*) are given as means ± SD. *n.s.*, not significant

trols and in drug-treated rats as well. The 10 μg/kg dose of nimodipine was obviously ineffective. Therefore, we decided to repeat these experiments, administering nimodipine to rats orally in different doses (0.1, 0.3, and 1.0 mg/kg).

As is shown in Table 1, postischemic mean arterial blood pressure was really not influenced by the lower oral nimodipine doses of 0.1 and 0.3 mg/kg. Only the higher dose of 1 mg/kg caused a significant decrease in the postischemic mean arterial blood pressure (Fig. 3). When the brains were examined histologically 7 days after ischemia, a significant reduction of neuronal necroses in the CA1 subfield was found after application of 0.1 and 0.3 mg/kg p.o. nimodipine but the reduction of neuronal damage after 1 mg/kg p.o. no longer reached the level of significance (Fig. 3). Thus. these results provided some evidence that a protective effect of a drug against ischemic damage of brain tissue can be neutralized by a postischemic hypotensive effect which could cause an extracerebral steal phenomenon. Therefore, one has to be cautious especially in interpreting a negative effect of a drug on postischemic neuronal damage. An adverse effect of the drug used could counteract its possible cerebroprotective potency.

Conclusion

To sum up, the results presented suggest that the calcium antagonists used are capable of protecting neurons against ischemic damage. This protective effect of the calcium antagonists is probably not mediated by their vasodilating activity. The calcium antagonists seem to act directly on neurons to protect them against ischemic damage. However, the neuronal outcome of cerebral ischemia can be influenced by the experimental procedure as, for instance, by the mode of drug application or by cardiovascular drug effects. A postischemic hypotensive effect of a drug can neutralize its cerebroprotective potency. Therefore, it is advisable to exclude carefully such adverse drug effects and to use more than one dose of a drug to test its cerebroprotective activity.

References

1. Auer RN, Olsson Y, Siesjö BK (1984) Hypoglycemic brain injury in the rat: correlation of density of brain damage with the EEG isoelectric tissue. A quantitative study. Diabetes 33:1090–1098
2. Bielenberg GW, Haubruck H, Krieglstein J (1987) Effects of calcium entry blockers emopamil on postischemic energy metabolism of the isolated perfused rat brain. J Cereb Blood Flow Metab 7:489–496
3. Bielenberg GW, Stierstorfer HJ, Weber J, Krieglstein J (1988) Nimodipine reduces postischemic lactate levels in the isolated perfused rat brain. Biochem Pharmacol (in press)
4. Cheung JY, Bonventre JV, Malis CD, Leaf A (1986) Calcium and ischemic injury. New Engl J Med 314:1670–1676
5. Deshpande JK, Wieloch T (1985) Amelioration of ischaemic brain damage by postischaemic treatment with flunarizine. Neurol Res 7:27–29
6. Dienel GA (1984) Regional accumulation of calcium in postischemic rat brain. J Neurochem 43:913–925

7. Farber JL (1981) The role of calcium in cell death. Life Sci 29:1289–1295
8. Hossmann KA, Grosse Ophoff B, Schmidt-Kastner R, Oschlies U (1985) Mitochondrial calcium sequestration in cortical and hippocampal neurons after prolonged ischemia of the cat. Acta Neuropathol 68:230–238
9. Jennings RB (1984) Calcium ions in ischemia. In: Opie LH (ed) Calcium antagonists and cardiovascular disease. Raven, New York, pp 85–95
10. Kirino T (1982) Delayed neuronal death in the gerbil hippocampus following ischemia. Brain Res 239:57–69
11. Krieglstein J, Weber J (1986) Calcium entry blockers protect brain energy metabolism against ischemic damage. In: Longmuir S (ed) Oxygen transport to tissue VIII. Plenum, New York, pp 243–251
12. Krieglstein J, Bielenberg GW, Stierstorfer HJ (1987) Calcium antagonist and other cerebroprotective drugs. In: Poeck K, Ringelstein EB, Hacke W (eds) New trends in diagnosis and management of stroke. Springer, Berlin Heidelberg New York, pp 129–134
13. Kumar K, Krause G, Koestner A, Hoehner T, White B (1987) Effect of flunarizine on global brain ischemia in the dog: a quantitative morphologic assessment. Exp Neurol 97:115–127
14. Newberg LA, Steen PA, Milde JH, Michenfelder JD (1984) Failure of flunarizine to improve cerebral blood flow or neurologic recovery in canine model of complete cerebral ischemia. Stroke 15:666–671
15. Pulsinelli WA, Brierly JB (1979) A new model of bilaterial hemispheric ischemia in the unaesthetized rat. Stroke 10: 267–272
16. Pulsinelli WA, Brierly JB, Plum F (1982) Temporal profile of neuronal damage in a model of transient forebrain ischemia. Ann Neurol 11:491–498
17. Sakurada O, Kennedy C, Jehle J, Brown JD, Carbin GL, Sokoloff L (1978) Measurement of local cerebral blood flow with iodo [^{14}C]antipyrine. Am J Physiol 234:H 59–H 66
18. Siesjö BK (1981) Cell damage in the brain: a speculative synthesis. J Cereb Blood Flow Metab 1:155–185
19. Siesjö BK (1986) Cell damage in the brain caused by ischemia: an overview. In: Krieglstein J (ed) Pharmacology of cerebral ischemia. Elsevier, Amsterdam, pp 3–11
20. Smith LYL, Bendek G, Dahlgren N, Rosén J, Wieloch T, Siesjö BK (1984) Models for studying long-term recovery following forebrain ischemia in the rat. 2. A 2-vessel occlusion model. Acta Neurol Scand 69:385–401
21. Van Reempts JV, Haseldonckx M, Deuren BV, Wouters L, Borgers M (1986) Structural damage of the ischemic brain: involvement of calcium and effects of postischemic treatment with calcium entry blockers. Drug Dev Res 8:387–395

Ultrastructural Calcium Cytochemistry in the Study of Ischemic Brain Injury

J. VAN REEMPTS and M. BORGERS

Morphological Assessment of Ischemic Brain Injury

Brain structure and function may be compromised in several ways. Hypoxic and ischemic challenges belong to the clinically most important cerebral pathologies. Many experimental models have been described which can be used to evaluate the evolution and the pathophysiological mechanisms which lead to irreversible brain injury. Such injury may be acute but in many cases the damage is a result of slowly progressing mechanisms which only become operative after a certain delay. This phenomenon of delayed neuronal cell death has been recognized by several authors [9, 10, 15]. It is characteristic for transient global ischemia. In other insults such as thrombotic infarction the lesion progressively expands from a peripheral area at risk [7, 24, 26]. It is thus obvious that consequences of cerebral ischemia should be studied in long-term survival models. Although the mechanisms involved in the evolution of brain damage may vary considerably according to the type of insult, the end result is quite predictable. We have studied morphological aspects of the ischemic and hypoxic brain in models of hypoxia–oligemia [18, 20], incomplete ischemia [23], cardiac arrest [25], and thrombotic infarction [24]. The cell changes that occur in these models could be divided into two major groups which we named coagulative and edematous cell change [21]. Coagulative cell change occurs exclusively in neurons while edematous cell change is preferentially found in glial cells (Fig. 1). Glial swelling is a general phenomenon which may be expected very soon after lowering of the oxygen supply. It is also highly dependent upon the degree of tissue acidosis [13, 14, 16]. Coagulative neuronal changes seem to be restricted to selectively vulnerable areas such as cortical lamina III, striatum, and hippocampal CA1 [10, 15, 17]. In focal thrombotic infarction coagulative neuronal necrosis is found in the thrombotic area whereas glial swelling predominates in the penumbral area at risk [24].

The study of morphological aspects of brain injury on its own is not sufficient to explain pathophysiological mechanisms leading to ischemic brain death. During recent years many investigators have stressed the importance of calcium in these destructive processes [8, 14, 16, 27]. We therefore extended our morphological studies to ultrastructural aspects of cellular calcium distribution.

Department of Cardiovascular Pathophysiology, Laboratory of Neuropathology, Janssen Research Foundation, B-2340 Beerse, Belgium.

Cerebral Ischemia and Calcium
Ed. by A. Hartmann and W. Kuschinsky

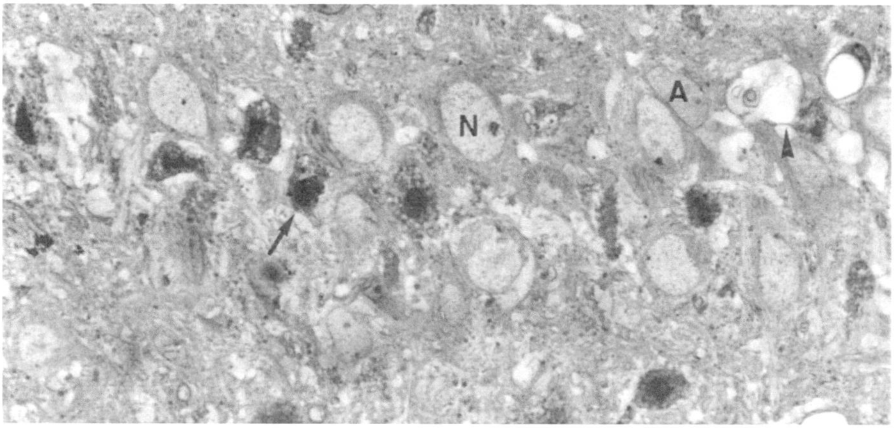

Fig. 1. Light microscopic picture of the ischemic CA1 hippocampus to show two types of irreversible cell change which become prominent when the brain is deprived of oxygen. Coagulative cell change (*arrow*) is exclusively found in neurons while edematous cell change (*arrowhead*) is largely restricted to astrocytes. Both are easily discernible from respectively normal neurons (*N*) and astrocytes (*A*). ×825

Calcium Cytochemistry

Beside the importance of calcium for normal cell functioning, the ion also plays a key role in cerebral injury processes. Compared to the extracellular space the intracellular calcium concentration is about 10000 times lower. Structural integrity of the cell membrane is thus of the utmost importance to preserve this high electrochemical gradient. Lowering of oxygen substrate induces membrane defects and consequently an uncontrolled entry of calcium [2, 16, 21, 27]. However, it remains a matter of debate whether irreversible structural cell injury results from excessive calcium overload or precedes the calcium influx. We have tried to answer this question by using a cytochemical technique which enables us to visualize the subcellular distribution of calcium ions. The method is based upon the precipitation of calcium by oxalate during the initial fixation and the subsequent conversion of this calcium complex into an insoluble antimonate salt during the postfixation step [19].

In the normal brain, calcium is visualized as electron-dense deposits inside synaptic vesicles and multivesicular bodies and occasionally in mitochondria, nuclei, and outer vesicles of the Golgi apparatus (Fig. 2a). Other structures such as the endoplasmic reticulum or the free cytosol do not contain any traces of precipitate. This pattern changes dramatically when the brain is subjected to hypoxic or ischemic insults. Irrespective of the type of initial insult, the pattern of calcium distribution in irreversibly damaged neurons is the same. Dead neurons appear shrunk. Their dense cytoplasm contains a pyknotic nucleus which, after cytochemical staining, was completely free of calcium precipitate. In the cytosol, however, and more particularly in cytoplasmic vacuoles and in certain vesicles of unknown origin, high amounts of calcium deposits were present (Fig. 2b).

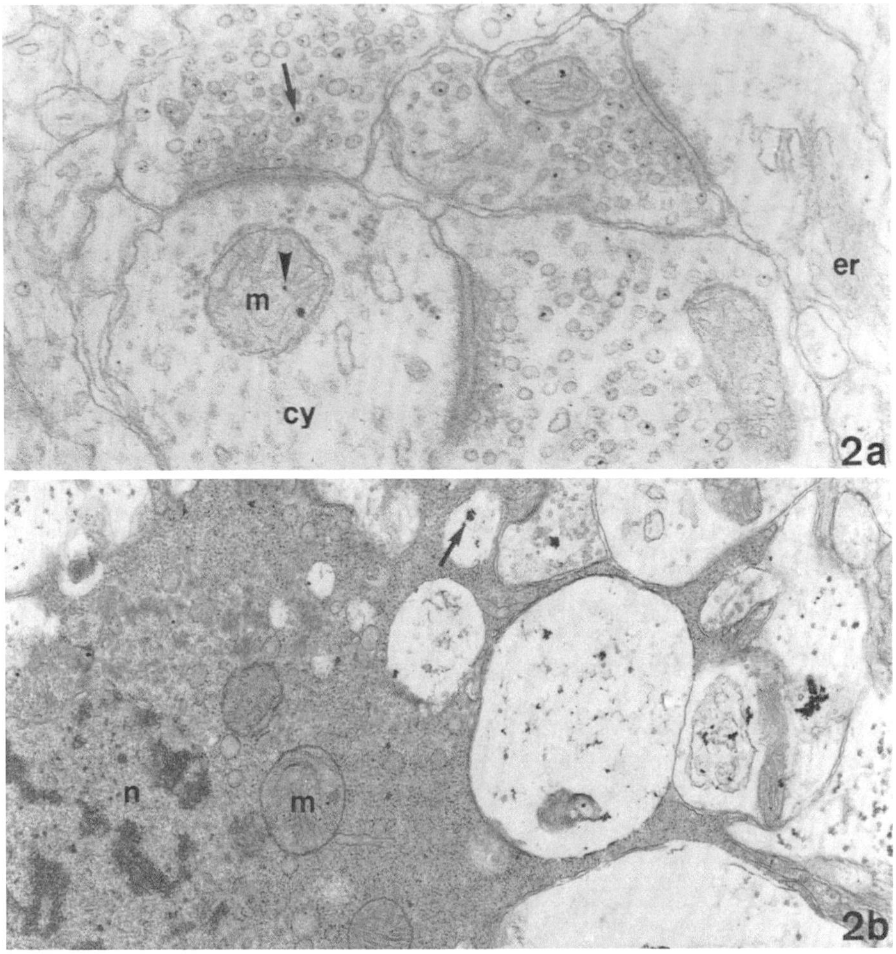
m
cy
er
2a
n
m
2b

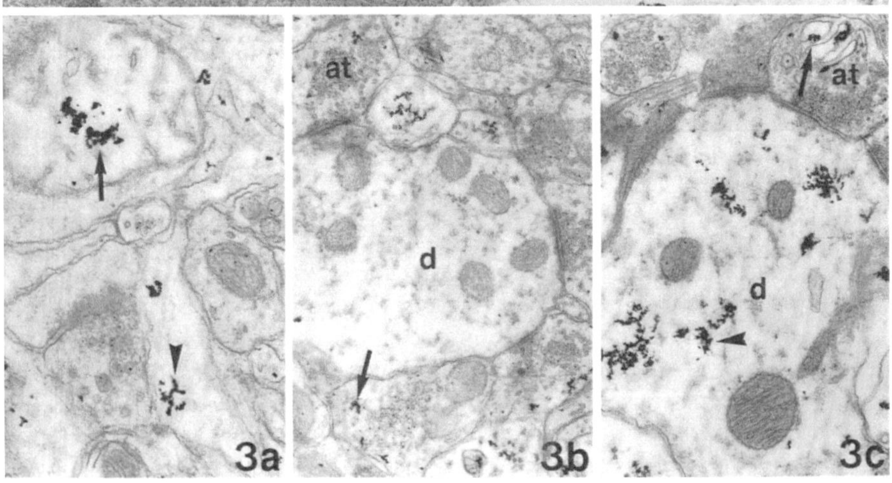
3a
at
d
3b
at
d
3c

In the period which precedes the ultimate cell death the distribution of calcium may be different for each type of insult. Acute hypoxia induced by exposure of rats to pure nitrogen resulted in an early redistribution of calcium [18]. Whereas the calcium content in synaptic vesicles apparently remained normal, increased amounts of precipitate were detected in swollen mitochondria but also in mitochondria which were ultrastructurally normal. Large amounts of calcium precipitate were encountered in many nuclei and in the structureless cytosol of heavily swollen dendritic and astrocytic processes. Within 24 h, a typical coagulative cell change developed, with calcium only present in cytosolic vacuoles (Fig. 2b). Comparable acute changes were observed after photochemical induction of thrombotic stroke [24]. In this model ultrastructural changes and calcium accumulation were studied in the thrombotic core, in the surrounding area at risk, and in the apposing structurally normal tissue. Within 4 h the challenged cortex was severely infarcted. Infarcted neurons again revealed the above-described coagulated aspect. The surrounding parenchyma was characterized by a narrow rim of severe extracellular edema. Calcium influx was restricted to astrocytes, swollen dendritic processes, and cytosolic vesicles. Since it was shown that within 4 h the infarct size gradually increases, thereby including the edematous margins of the thrombotic core [7, 24], observations were also made at shorter time periods after the insult. After 20 min of photochemical agression the edematous area was more extended than after 4 h. A broad rim was seen which contained large numbers of swollen calcium-containing cell processes of both astrocytic and neuronal origin. At some distance from the area at risk the morphological picture was normal, although the calcium content was increased in the neurons as well as in the astrocytes. More particularly, axon terminals revealed a higher cytosolic calcium content. In the above-described conditions of nitrogen exposure and induction of thrombotic stroke, the evolution from early cellular and mitochondrial swelling to irreversible coagulative cell change seems to proceed very fast. Calcium enters the cell during the challenge, where it probably remains during the ensuing degeneration process.

A different pattern of calcium-related cell degeneration was found after incomplete ischemia. When cerebral blood flow was lowered to values close to zero, a typical pattern of postischemic flow became prominent. After a short hyperemic

Fig. 2. **a** Electron micrograph of the subcellular distribution of calcium in the normal brain. Dense calcium precipitates are mainly found as single dots in synaptic vesicles (*arrow*) and in mitochondria (*m, arrowhead*). Cytosol (*cy*) and endoplasmic reticulum (*er*) are devoid of deposits. × 50280. **b** Ultrastructural appearance of an irreversibly damaged ischemic neuron showing typical coagulative cell change and accumulation of calcium in cytoplasmic vacuoles (*arrow*). The pyknotic nucleus (*n*) is free of precipitate. Mitochondria (*m*) appear condensed and contain slightly elevated amounts of deposits. × 20620

Fig. 3 a–c. Subcellular calcium distribution in the ischemic hippocampus. At the end of the ischemic insult (**a**) higher amounts of calcium precipitate were found temporarily in swollen mitochondria (*arrow*) and in astrocytes (*arrowhead*). This calcium disappeared in the early recirculation phase. Between 2 and 6 h after the insult a second wave of calcium influx could be observed in excitatory axon terminals (*at, arrow*) of CA3 (**b**) as well as CA1 (**c**) subfields. Only in the vulnerable CA1 field did this presynaptic influx coincide with postsynaptic calcium overload in dendrites (*d, arrowhead*). **a** × 30070; **b, c** × 19770 phase the flow turned to values far below normal for several hours [15, 16]. Definite damage seemed to develop when the hypoperfusion phase was terminated [15]. However, this hemodynamic pattern alone does not explain the selective damage which develops in hippocampal CA1 after such ischemic insults [10, 17]. Excessive excitatory neurotransmission has been proposed to be responsible for this phenomenon [12]. When the evolution of intracellular calcium distribution was cytochemically followed during the postischemic period, it became clear that two phases of calcium entry were present [22]. Shortly after the induction of ischemia, calcium accumulated in all hippocampal areas. Astrocytes, neuronal mitochondria, and axon terminals were the targets (Fig. 3 a). In the early recirculation period the picture completely normalized. However, after 2–6 h a second wave of calcium influx was observed. This time the highest amounts of precipitate were encountered in axon terminals. No difference in presynaptic accumulation was seen between CA1 and CA3. However, in the postsynaptic parts the picture was completely different since postsynaptic calcium overload was only detected in CA1 (Fig. 3 b, c). The phenomenon preceded the first signs of protein degeneration and the consequent evolution to irreversible coagulative cell change.

In conditions of incomplete ischemia delayed calcium overload thus seems to provoke delayed selective neuronal degeneration. This has been confirmed biochemically by Deshpande et al. [6]. They found that net accumulation of calcium in the dorsal hippocampus 48 h following ischemia preceded marked necrosis of CA1 pyramidal cells. That calcium overload parallels the damage and that extracellular calcium influx is a prerequisite for irreversible necrosis in most instances has been demonstrated in isolated myocardial cells [3]. It is thus clear that cytosolic calcium overload may elicit irreversible brain death. This may be due to the initiation of a series of deleterious processes which compromise cell function and survival, e.g., energy-consuming hyperactivation and membrane lipid and protein degradation.

Therapy

Since calcium accumulation plays a predominant role in the process of structural degeneration of neurons there is certainly room for therapeutic intervention, at least in those conditions where calcium overload occurs with some delay. Experimental studies with the class IV calcium antagonist flunarizine have confirmed this possibility. Posttreatment definitely reduced the amount of necrotic neurons in CA1 hippocampus after incomplete ischemia [1, 5, 23] and cardiac arrest [11, 25] in different animal species. Acute structural damage due to nitrogen exposure or thrombotic infarction could be prevented by pretreatment with flunarizine [20, 24]. With respect to protection against thrombotic cortical infarction it is worthwhile mentioning the recent findings on the effects of posttreatment with flunarizine [4]. Whereas the drug reduced infarct size when given before induction of stroke, no such effect was seen after posttreatment. Nevertheless, the posttreated animals were completely devoid of neurological deficits. These results show that preservation of neuronal functioning does not necessarily depend upon structural preservation at the aggression site itself and that subtle structural, physiological, and metabolic changes may be expected at remote sites as well. In order to clarify this possibility we are at present studying subcellular calcium distribution in areas remote from the cortical infarction.

Conclusions

Using a cytochemical technique for the subcellular localization of calcium we were able to show a definite correlation between intracellular calcium translocation and irreversible neuronal cell degeneration in conditions of cerebral oxygen shortage. In conditions where recirculation occurs, such as in transient incomplete ischemia, the definite toxic calcium overload appeared with a considerable delay. Moreover in vulnerable areas such as hippocampal CA1, postsynaptic calcium overload preceded the first ultrastructural signs of protein degeneration. Biochemical studies on ischemic brain and cytochemical studies on ischemic myocardial cells confirmed these results. Postischemic treatment with flunarizine resulted in a reduction of the number of necrotic neurons after experimental induction of incomplete ischemia and cardiac arrest. Such structural protection was accompanied by the prevention of dramatic increases of cytochemically detectable calcium. Whether calcium blockade is also responsible for the improved neurological outcome in rats treated with flunarizine after experimental stroke has still to be further investigated.

Summary

Calcium is a key factor in the pathophysiology of ischemic brain disease. Pharmacological agents which prevent pathological overload of calcium in cerebral cells are known to reduce damage. An example is flunarizine, which has been shown to protect the brain from hypoxic as well as from global and focal ischemic insults. To demonstrate calcium at the ultrastructural level, the oxalate-pyroantimonate technique has been used. This method has been extensively validated. In the normal brain only synaptic vesicles, multivesicular bodies, and to a lesser extent also nuclei and mitochondria contain calcium precipitate. Striking calcium shifts are seen after different pathological conditions such as epileptic seizures, hypoxia, and ischemia. After an ischemic insult this calcium redistribution occurs in two phases. In an early phase, which is reversible, calcium concentration increases in mitochondria, nuclei, axon terminals, and astrocytic cytoplasm of vulnerable as well as nonvulnerable areas. In a second phase, which occurs with a considerable delay, calcium accumulates in presynaptic endings. Only in vulnerable areas is this presynaptic accumulation associated with postsynaptic calcium overload, which in turn precedes the first signs of irreversible structural decay in the corresponding neurons. Biochemical and electrophysiological studies support the ultrastructural data. The present study shows that cytochemical techniques which enable visualization of calcium at the ultrastructural level may be a valid asset in addition to the morphological examination of the evolution of ischemic brain disease and its therapy.

Acknowledgments. The authors are most grateful to Mr. G. Jacobs for preparing the illustrations and to Mrs. L. Vanherle and K. Donné for respectively reviewing and typing the manuscript.

References

1. Alps BJ, Calder C, Hass WK, Wilson AD (1988) Comparative protective effects of nicardipine, flunarizine, lidoflazine, and nimodipine against ischaemic injury in the hippocampus of the Mongolian gerbil. Br J Pharmacol 93:877–883
2. Borgers M, Thoné F, Van Reempts J, Verheyen F (1983) The role of calcium in cellular dysfunction. Am J Emerg Med 1:154–161
3. Borgers M, Ver Donck L, Vandeplassche G (1988) Pathophysiology of cardiomyocytes. Ann NY Acad Sci 522:433–453
4. De Ryck M, Van Reempts J, Wauquier A, Borgers M, Janssen PAJ (1988) Photochemical neocortical stroke in rats: flunarizine prevents neurologic deficits. Abstr Soc Neurosci 14:1254
5. Deshpande JK, Wieloch T (1986) Flunarizine, a calcium entry blocker, ameliorates ischemic brain damage in the rat. Anesthesiology 64:215–224
6. Deshpande JK, Siesjö BK, Wieloch T (1987) Calcium accumulation and neuronal damage in the rat hippocampus following cerebral ischemia. J Cereb Blood Flow Metab 7:89–95
7. Dietrich D, Ginsberg M, Busto R, Watson BD (1986) Photochemically induced cortical infarction in the rat. 1. Time course of hemodynamic consequences. J Cereb Blood Flow Metab 6:184–194
8. Farber JL (1981) The role of calcium in cell death. Life Sci 29:1289–1295
9. Ito U, Spatz M, Walker JT, Klatzo I (1975) Experimental cerebral ischemia in Mongolian gerbils. I. Light microscopic observations. Acta Neuropathol (Berl) 32:209–223
10. Kirino T (1982) Delayed neuronal death in the gerbil hippocampus. Brain Res 239:57–69
11. Kumar K, Krause G, Koestner A, Hoehner T, White B (1987) Effect of flunarizine on global brain ischemia in the dog: a quantitative morphologic assessment. Exp Neurol 97:115–127
12. Meldrum BS (1981) Metabolic effects of prolonged epileptic seizures and the causation of epileptic brain damage. In: Clifford-Rose F (ed) Metabolic disorders of the nervous system. Pitman, London, pp 175–187
13. Meyers RE (1979) Lactic acid accumulation as a cause of brain edema and cerebral necrosis resulting from oxygen deprivation. In: Korobkin R, Guilleminault C (eds) Advances in perinatal neurology, Spectrum, New York, pp 85–114
14. Plum F (1983) What causes infarction of the ischemic brain? Neurology 33:222–233
15. Pulsinelli WA, Brierley JB, Plum F (1982) Temporal profile of neuronal damage in a model of transient forebrain ischemia. Ann Neurol 11:491–498
16. Siesjö BK (1981) Cell damage in the brain: a speculative synthesis. J Cereb Blood Flow Metab 1:155–185
17. Smith ML, Bendek G, Dahlgren N, Rosén I, Wieloch T, Siesjö BK (1984) Models for studying long-term recovery following forebrain ischemia in the rat. 2. A 2-vessel occlusion model. Acta Neurol Scand 69:385–401
18. Van Reempts J, Borgers M (1982a) Morphological assessment of pharmacological brain protection. In: Wauquier A, Borgers M, Amery W (eds) Protection of tissues against hypoxia. Elsevier Biomedical, Amsterdam, pp 263–274
19. Van Reempts J, Borgers M, Offner F (1982b) Ultrastructural localization of calcium in the rat retina with a combined oxalate-pyroantimonate technique. Histochem J 14:517–522
20. Van Reempts J, Borgers M, Van Dael L, Van Eyndhoven J, Van de Ven M (1983) Protection with flunarizine against hypoxic-ischemic damage of the rat cerebral cortex. A quantitative morphologic assessment. Arch Int Pharmacodyn Ther 262:76–88

21. Van Reempts J (1984) The hypoxic brain: histological and ultrastructural aspects. Behav Brain Res 14:99–108
22. Van Reempts J, Borgers M (1985) Ischemic brain injury and cell calcium: morphologic and therapeutic aspects. Ann Emerg Med 14:736–742
23. Van Reempts J, Haseldonckx M, Van Deuren B, Wouters L, Borgers M (1986) Structural damage of the ischemic brain: involvement of calcium and effects of postischemic treatment with calcium entry blockers. Drug Dev Res 8:387–395
24. Van Reempts J, Van Deuren B, Van de Ven M, Cornelissen F and Borgers M (1987) Flunarizine reduces cerebral infarct size after photochemically induced thrombosis in spontaneously hypertensive rats. Stroke 18:1113–1119
25. Van Reempts J, Lu HR, Borgers M (1989) Morphological study on the cerebroprotective effects of flunarizine in a rat cardiac arrest model. Am J Emerg Med (in press)
26. Welch KMA, Hashi K, Meyer JS (1973) Cerebrovascular response to intracarotid injection of serotonin before and after middle cerebral artery occlusion. J Neurol 36:724–735
27. White BC, Aust SD, Arfors KE, Aronson LD (1984) Brain injury by ischemic anoxia: hypothesis extension. A tale of two ions. Ann Emerg Med 13:862–867

Roles of Calcium-Activated Enzymic Reactions in Ischemia-Induced Neuronal Injury

K. Kogure, H. Onodera, T. Tsuda, T. Araki, and K. Nishioka

Temporal ischemia of the brain injures only the selectively vulnerable brain cells. All of these neurons are constituents of either the limbic system or the basal ganglia, and recipients of the glutamatergic fibers. Recent studies in our laboratories unveiled that postischemic injury of these neurons is caused not by the energy failure but by dysfunction of the intracellular signal transducing system.

The Postischemic Events

Available experimental data indicate that the process of postischemic neuronal injury evolves as follows (Fig. 1): Ischemia-induced release of glutamate from the presynaptic neuron to the synaptic cleft activates phospholipase C in the postsynaptic membrane via glutamate receptors, and also opens the receptor-operated Ca^{2+} gate. Activation of phospholipase C was proven by rapid decomposition

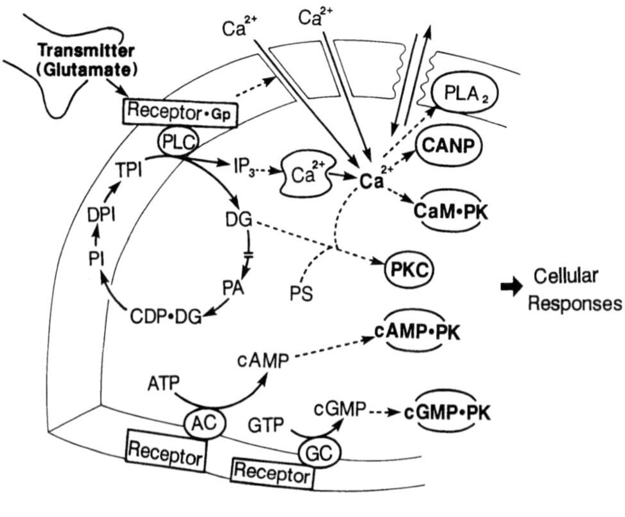

Fig. 1. Schematic view of intracellular responses following ischemia-induced release of the transmitter (glutamate)

Department of Neurology, Institute of Brain Diseases, 1-1 Seiryomachi, Sendai, 980, Japan.

Cerebral Ischemia and Calcium
Ed. by A. Hartmann and W. Kuschinsky

of phosphatidyl inositol and polyphosphoinositides in the rat cerebral cortex as early as 30 s after onset of deprivation of blood supply [1]. The molecular mechanism of the receptor-operated Ca^{2+} gate opening is not yet understood. In in vitro experiments, the intracellular Ca^{2+} level rises by thousands of times within a few seconds after application of glutamate [2], but the exact speed of the Ca^{2+} accumulation in the brain cells in vivo is not known. Our experiment on the rat hippocampus utilizing a particle-induced X-ray emission (PIXE) method indicated that the Ca^{2+} content of the hippocampus increased from 66.3 ± 8.2 to 98.0 ± 17.2 µg/g dry weight 30 min after 20 min global ischemia [5].

These receptor-mediated reactions are followed by activation of protein kinase C by 1,2-diacylglycerol (DG), and on the other hand by mobilization of Ca^{2+} from its endostore(s) by inositol trisphosphate (IP_3). The activation of protein kinase C elicits multifarious cellular responses shortly after ischemia and together with restoration of the energy state. For instance, we recently observed a transient and specific induction of the mRNA for proto-oncogene c-fos in the cerebral cortex of the rat 30–90 min after 20 min global ischemia [3]. Endogenous Ca^{2+} mobilized by IP_3 together with exogenous Ca^{2+} which "flew" into the cell via the voltage-dependent and voltage-nondependent (receptor-operated) Ca^{2+} gate greatly increases the intracellular Ca^{2+} content and causes activation of phospholipase A_2, Ca^{2+}-dependent calmodulin kinase, and protein kinase C.

Phospholipase A_2 cleaves arachidonic acid from structural phospholipids of the cytoplasmic membrane, and perturbs the structural and functional integrity of the membrane. Such membrane perturbation may be accelerated by concomitant decomposition of the lining proteins catalyzed by Ca^{2+}-activated and/or calmodulin-dependent protein kinase-activated proteases. The perturbed, leaky membrane with recovery in the energy state causes a postischemic increase in spike activity in the dying neurons [4], and indeed these neurons fail to maintain the intracellular homeostasis.

Prevention of Postischemic Brain Cell Death by Ca^{2+} Antagonists

Many other enzymic reactions in the cell which are also elicited by these protein kinases may participate in the postischemic loss of homeostasis, and finally cause autolysis of the vulnerable neurons. If the above-mentioned sequence is correct, one may prevent postischemic brain cell injury by antagonizing the Ca^{2+} effect. In order to test this possibility, Tsuda and his collaborators in our laboratory administered Mg^{2+} topically to the hippocampus of the rat subjected to 20 min global ischemia. This divalent is known as an endogenous Ca^{2+} entry blocker. As the result, 50 mmol Mg^{2+} (as $MgCl_2$) exhibited a marked preventive effect against postischemic neuronal death, even when administered 2 h after the ischemic insult (Fig. 2).

Among many pharmacologically registered Ca^{2+} antagonists, the 1,4-dihydropyridine (DHP) derivatives are categorized as nonspecific antagonists: they block Ca^{2+} entry together with Na^+ and other ionic movement, and it is also assumed that these compounds may interfere with mobilization of endostore Ca^{2+}.

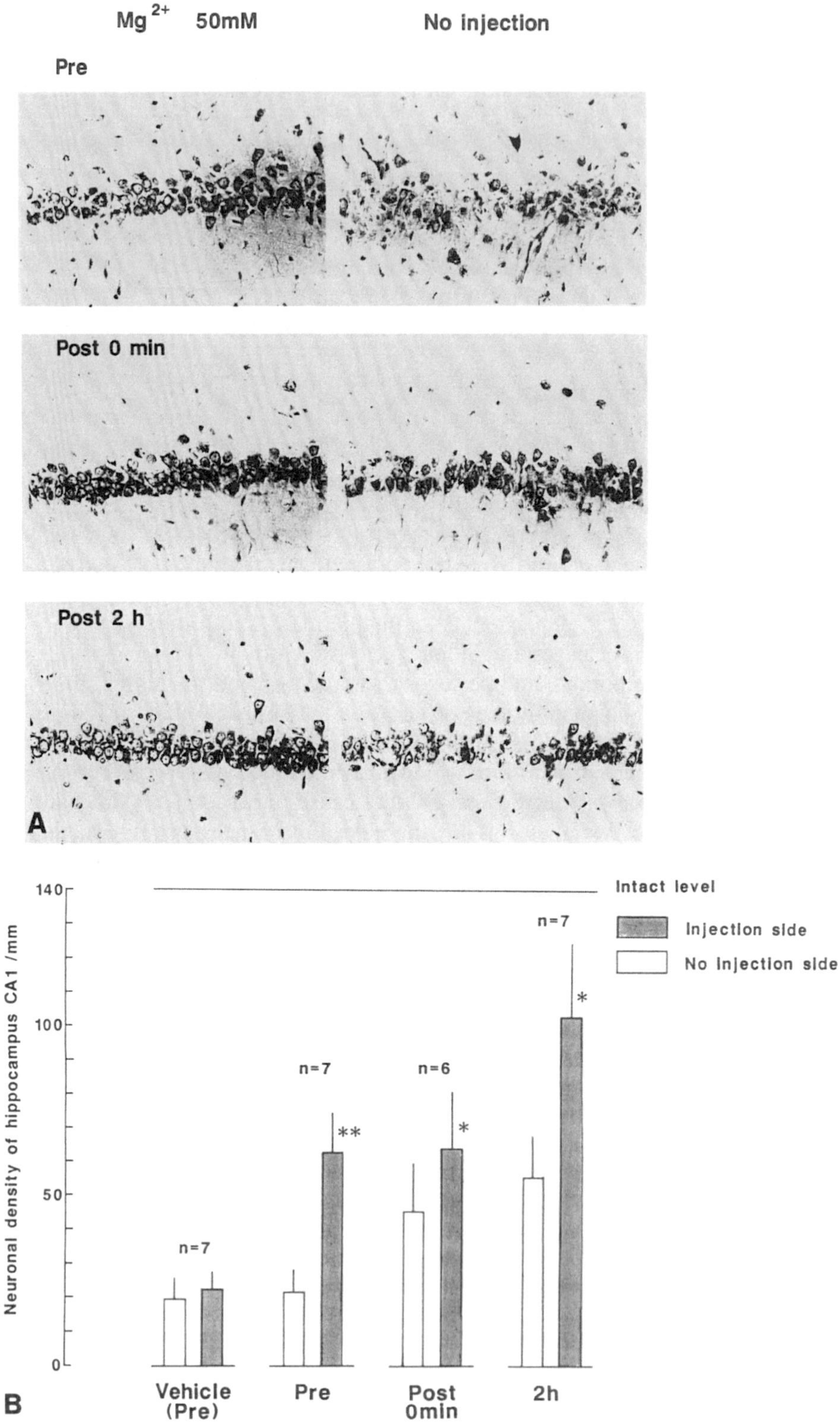

Mg 2+ 50mM
No injection
Pre
Post 0 min
Post 2 h
A
Intact level
Injection side
No injection side
n=7
n=7
n=6
n=7
**
*
*
140
100
50
0
Neuronal density of hippocampus CA1 /mm
Vehicle (Pre)
Pre
Post 0min
2h
B

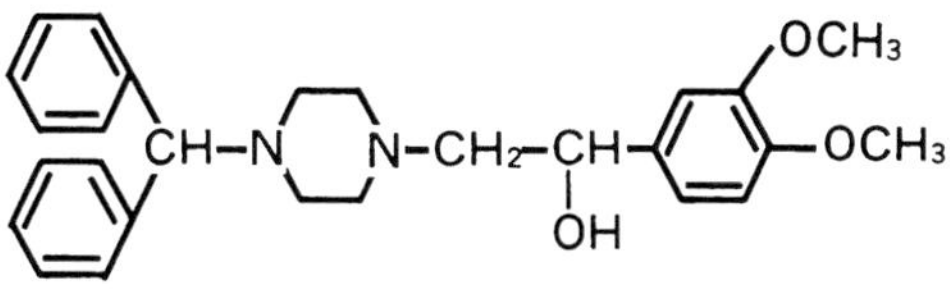

Fig. 3. Chemical structure of NC-1100

The reactive site (binding site studied by radioactive PN200-110) is densely distributed in many specific areas of the brain, e.g., hippocampus CA3, dentate gyrus, and CA1 sector.

For the current communication, we selected a diphenyl piperazine derivative (NC-1100, Fig. 3), which is also categorized as a nonspecific Ca^{2+} antagonist. Even though the reactive site is not necessarily similar to that of the above-mentioned compound, this chemical penetrates the blood-brain barrier rather more easily than the others. Administration of this compound (30 mg/kg i.p.) immediately after 5 min forebrain ischemia in the gerbil preserved pyramidal cells in the CA1 sector of the hippocampus remarkably well (Table 1).

Data obtained in parallel in our laboratory indicated a significant decrease of either the IP_3 binding site (supposed to be the smooth endoplasmic reticulum and other Ca^{2+} endostores) or the DHP binding site (presumably distributed on the cytoplasmic cell membrane and partially on the membrane of the micro-organella) within 3 h after 20 min global ischemia in the rat brain. Such alterations in the important functional unit for calcium metabolism seem to indicate disruption of calcium homeostasis. A deduction from this observations is that administration of Ca^{2+} antagonists as a means of preventing postischemic brain injury may be warranted only in the early postischemic stage

Table 1. Preventive effects of the calcium blocker NC-1100 against neuronal injury in the CA1 sector of the gerbil hippocampus; 5 min ischemia and 7 days reperfusion

	Dose (mg/kg, i.p.)	No.	Neuronal density (per mm)
Control (no carotid ligation)		6	66.03±7.37*
Vehicle		10	7.51±3.26
Calcium blocker	30	10	47.16±4.31*
Calcium blocker	20	10	9.13±3.26

Values are expressed as mean±SEM. * Significantly different from vehicle group by the Mann-Whitney U-test, $P<0.01$.

Fig. 2. Preventive effect of Mg^{2+} against neuronal injury in the CA1 sector of the rat hippocampus; 20 min ischemia and 7 days reperfusion. $MgCl_2$ (50 mM in 1 µl) was injected topically before occlusion and at 0 min and 2 h after reperfusion. **A** Light micrographs of 5-µm coronal sections stained with cresyl violet. **B** Neuronal density of hippocampus CA1. Data are expressed as mean±SEM. Significantly different from the noninjection side by Student's paired *t*-test: $*P<0.05$, $**P<0.01$

Summary

Some chemicals known to be calcium antagonists can prevent ischemia-induced death in the selectively vulnerable brain cells. These neurons are profoundly implicated in the realization of such higher brain functions as memory, emotive expression, spontaneity, and skillfulness. The possibility of preventing ischemic necrosis of these neurons is therefore tantamount to having developed a means of preventing the onset of vascular dementia consequent upon cerebral ischemia.

References

1. Abe K, Kogure K, Yamada H, Imazawa M, Miyamoto K (1987) Mechanism of arachidonic acid liberation during ischemia in gerbil cerebral cortex. J Neurochem 48:503–509
2. Kudo Y, Ogura A (1986) Glutamate-induced increase in intracellular Ca^{2+} concentration in isolated hippocampal neurones. Br J Pharmacol 89:191–198
3. Onodera H, Kogure K, Ono Y, Igarashi K, Kiyota Y, Nagaoka A (1988) Proto-oncogene c-fos is transiently induced in the rat cerebral cortex after forebrain ischemia. Neurosci Lett
4. Suzuki R, Yamaguchi T, Choh-Luh Li, Klatzo I (1983) The effect of 5-minute ischemia in mongolian gerbils: II. Changes of spontaneous neuronal activity in cerebral cortex and CA1 sector of hippocampus. Acta Neuropathol (Berlin) 60:217–222
5. Tsuda T, Kogure K, Ishii K, Orihara H (1988) Post-ischemic changes of calcium and endogenous antagonist in the rat hippocampus studied by proton induced X-ray emission analysis. Brain Res

The Effects of Experimental Anoxia on Glycolysis, Energy Status, Cytosolic Free Calcium, and Glutamate Release from Isolated Nerve Terminals

R. A. KAUPPINEN[1,3], H. T. MCMAHON[2], and D. G. NICHOLLS[2]

Isolated nerve terminals (synaptosomes) from cerebral cortex are used as a model of the presynaptic nerve terminal in order to investigate the effects of experimental anoxia on energy metabolism, membrane potentials, cytosolic free Ca^{2+} concentration ($[Ca^{2+}]_c$), and release of the neurotransmitter glutamate. Synaptosomes function bioenergetically as mini cells, with sealed plasma and mitochondrial membranes. The surface-to-volume ratio of synaptosomes is large, thus ensuring a short diffusion path for oxygen to the mitochondrion. Synaptosomes exhibit an ultimate dependence on oxidative metabolism for the maintenance of energy levels [8, 18, 19]. Apart from oxygen, glucose is required, because overall ATP/ADP ratios in excess of 6 are maintained only in the presence of both these agents [18, 29].

Under aerobic conditions the rates of synaptosomal glycolysis and oxygen consumption are stoichiometric and no lactate is produced [18]. An increase in synaptosomal energy consumption is followed by a proportional activation of both respiration and glycolysis, thus showing coupling between mitochondrial and extra-mitochondrial energy-yielding pathways [18]. From the bioenergetic point of view, synaptosomes exhibit metabolic properties expected in nerve terminals in situ and thus are a valid model for in vitro studies.

A growing body of evidence indicates that an increased activity of postsynaptic excitatory receptors may be involved in development of neuronal damage in anoxia [30]. Glutamate is probably the major excitatory neurotransmitter in mammalian cortex [6] and thus it is plausible that the elevation of brain extracellular glutamate in ischemia [2] might contribute to the brain damage developing upon lack of oxygen.

Glutamate is released from isolated nerve terminals by two different mechanisms [17, 25, 26]. First, there is a Ca^{2+}-independent efflux from the cytoplasmic pool of the transmitter, utilizing the reversal of the plasma membrane uptake pathway [14]. This is activated upon depolarization of plasma membrane potential ($\Delta\Psi_p$) in the presence of ethylene glycolbis(β-aminoethylether)-*N,N'*-tetraacetic acid (EGTA) [25]. Secondly, an energy-dependent and neurotoxin-sensitive exocytotic release is triggered upon depolarization in the presence of Ca^{2+}

[1] Department of Clinical Neurophysiology, University Central Hospital of Kuopio, Kuopio, Finland.
[2] Department of Biochemistry, University of Dundee, Dundee, Scotland, U.K.
[3] Present address: Department of Physics, Royal College of Surgeons of England, 35–43 Lincoln's Inn Fields, London WC2A 3PN, U.K.

Cerebral Ischemia and Calcium
Ed. by A. Hartmann and W. Kuschinsky

[25, 31, 32]. The recent introduction of cell-permeant fluorescent Ca^{2+} indicators has made the estimation of $[Ca^{2+}]_c$ a standard laboratory procedure [9, 10, 20]. Thus methods exist for investigating the interrelationships between synaptosomal energy metabolism and neurotransmitter release and their changes during anoxia in vitro.

Methods

Preparation of Synaptosomes and Incubations

Synaptosomes from the cerebral cortices of Dunkin Hartley strain guinea pigs (aged 3–6 weeks) were prepared in a Ficoll density gradient as described by Nicholls [23]. The incubation medium contained 122 mM NaCl, 3.1 mM KCl, 0.4 mM KH_2PO_4, 5 mM $NaHCO_3$, 1.2 mM $MgSO_4$, 20 mM Na-TES, 16 µM bovine serum albumin, and either 5 mM or 10 mM D-glucose as required, pH 7.4. $CaCl_2$ was added after 5 min preincubation to give a final concentration of 1.3 mM.

Cytosolic Free Calcium and Total Calcium in Synaptosomes

The $[Ca^{2+}]_c$ was determined essentially as described previously, using fura-2 as an indicator [9, 20]. The proportion of the fura-2/Ca^{2+} signal that was quenched by Mn^{2+} did not increase even after 20 min exposure to metabolic inhibitors, indicating that fura-2 did not leak out of the synaptosomes during experiments. The total concentration of Ca^{2+} in synaptosomes was estimated using $^{45}Ca^{2+}$ as an indicator [33].

Glycolysis

The rate of glycolysis was calculated from the formation by aldolase of ^3H_2O from D-3-[^{3}H]glucose [15] as described previously [18]. Amounts of glucose consumed by anoxic synaptosomes are also calculated from the accumulation of lactate, which was measured spectrophotometrically [13].

Isotopic Determination of Membrane Potentials

Synaptosomal $\Delta\Psi_p$ and mitochondrial membrane potential ($\Delta\Psi_m$) were measured using $^{86}Rb^+$ and [^{3}H]tetraphenyl phosphonium respectively as described previously [18, 34].

Adenylates and Creatine Phosphate

Adenosine triphosphate, ADP, and AMP were measured with a luciferin assay [22] as described previously [18]. Creatine phosphate (CrP) was determined with a similar assay [17].

Glutamate Release

Release of endogenous glutamate from synaptosomes was determined fluorometrically with glutamate dehydrogenase in the presence of 1 mM NADP as described previously [26].

Results

Glycolysis in Anoxia

Synaptosomal respiration was inhibited at two different sites of the respiratory chain. Firstly, cytochrome c oxidase was inhibited either by cyanide or by nitrogen purging mimicking bioenergetic consequences of anoxia, and secondly NADH dehydrogenase was inhibited by rotenone. Figure 1 shows that all three methods evoke a Pasteur effect; however, the high activity of anaerobic glycolysis is maintained only in the presence of rotenone. In contrast, glycolysis begins to fall by 6 min both in cyanide-inhibited synaptosomes and in terminals incubated in oxygen-free medium. The failure of glycolysis in the latter experiment is reversible and reoxygenation returns the rate of glucose consumption to preanoxic levels. Subsequent addition of uncoupler caused the same tenfold activation as in control synaptosomes [18]. Thus in nerve terminals where cytochrome oxidase is inhibited a failure to maintain glycolysis develops only some minutes after the onset of anoxia.

Table 1 summarizes the consumption of glucose determined by two independent methods. The glucose consumed in cyanide-inhibited synaptosomes is identical when calculated by either an isotopic method or by lactate accumulation. The formation of lactate in rotenone-treated terminals is 47% greater than in cyanide-treated synaptosomes; however, consumption of D-3-[^{3}H]glucose is 120% greater. Thus in the presence of rotenone the glycolytic activity prior to the aldolase step exceeds the rate of the enzymes distal from aldolase.

Potassium chloride depolarization before the onset of experimental anoxia does not influence anoxic glucose consumption in the presence of cyanide (Table 1). Depolarization, however, significantly reduces the consumption of labeled glucose in rotenone-inhibited synaptosomes, which drops to the same rate as the lactate formation (Table 1).

Potassium chloride depolarization increases $[Ca^{2+}]_c$ [17], suggesting that high $[Ca^{2+}]_c$ inhibits a glycolytic enzyme step which is prior to aldolase. When EGTA was added to chelate external Ca^{2+} before the onset of anoxia, lactate accumulation was identical in cyanide- and rotenone-treated synaptosomes, confirming

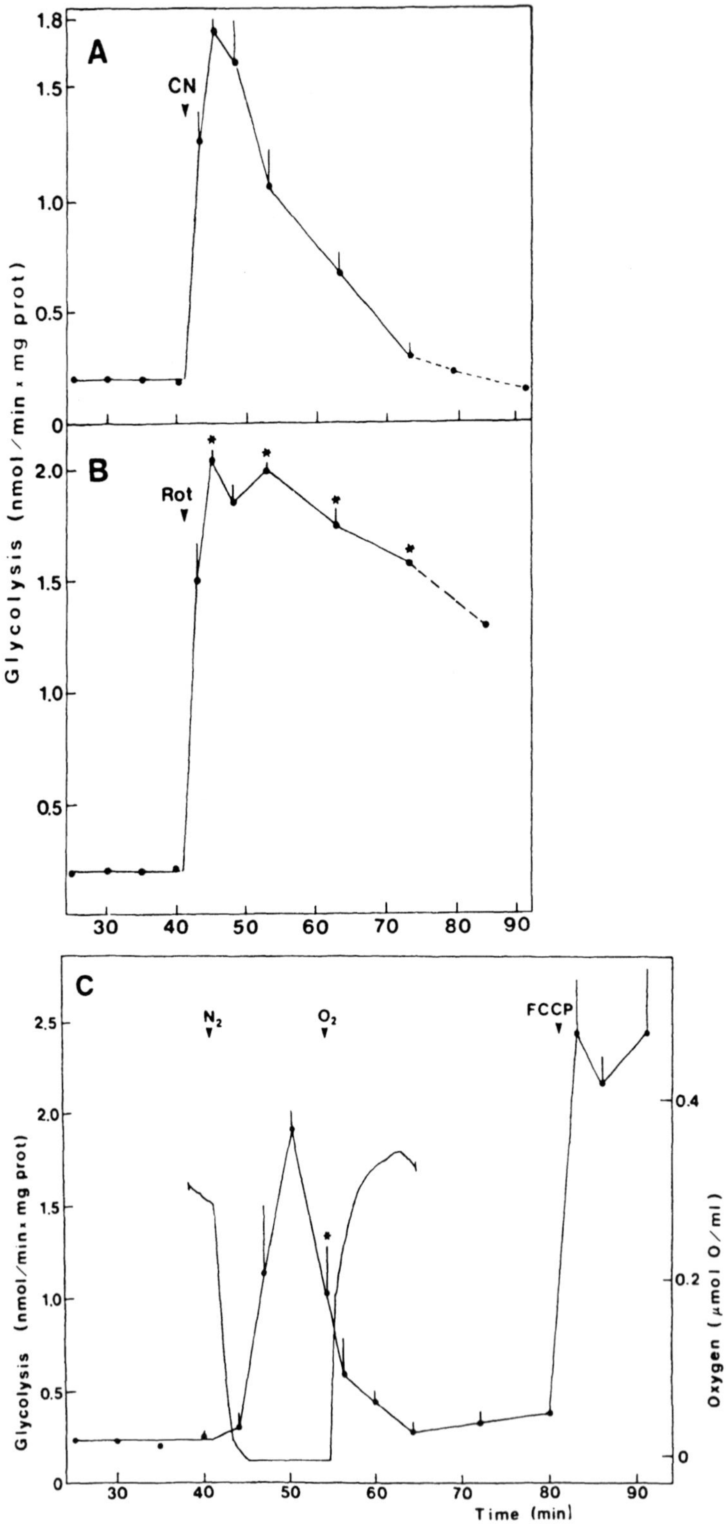
A
CN
B
Rot
Glycolysis (nmol/min x mg prot)
1.8
1.5
1.0
0.5
0
2.0
30
40
50
60
70
80
90
C
N2
O2
FCCP
2.5
Glycolysis (nmol/min x mg prot)
Oxygen (μmol O/ml)
0.4
0.2
Time (min)

Table 1. Glucose consumption in synaptosomes

	Lactate	D-3-[^{3}H]glucose
Control	–	13.4 ± 1.7 (4)
Cyanide	44.8 ± 5.2 (6)	47.0 ± 4.4 (5)
Verapamil + NaCN	49.0 ± 5.2 (4)	–
KCl + NaCN	42.7 ± 7.6 (6)	31.2 ± 3.1 (4)
Veratridine + NaCN	19.9 ± 4.4 (4)	15.9 ± 2.3 (6)
Rotenone	65.7 ± 6.7 (7)	103.9 ± 12.8 (6)
Verapamil + rotenone	66.5 ± 4.7 (3)	–
KCl + rotenone	56.4 ± 5.4 (6)	53.6 ± 4.5 (5)
Veratridine ± rotenone	30.5 ± 5.7 (4)	30.7 ± 6.7 (6)

Synaptosomes (4 mg protein/ml) were incubated for 20 min before addition of 100 μM verapamil, 45 mM KCl, or 75 μM veratridine; 2 mM NaCN or 10 μM rotenone was added 3 min later. Consumption of glucose in 32 min was calculated either from accumulation of lactate or ^3H_2O from D-3-[^{3}H]glucose. Values (nmol) are means ± SE for number of experiments indicated in parentheses.

that Ca^{2+} contributes to the failure to maintain anaerobic glycolysis of synaptosomes. A similar effect of Ca^{2+} chelation has been observed in erythrocytes [4].

Pretreatment of synaptosomes with veratridine, which increases synaptosomal energy consumption and hence lowers ATP levels [34] by allowing continuous Na^+ conductance through the voltage-dependent Na^+ channel, reduces glycolysis by 50% in both cyanide- and rotenone-inhibited nerve terminals (Table 1). Apart from the energetic effects, veratridine has been shown to increase $[Ca^{2+}]_c$ [10]. Glucose is consumed some 40% less in cyanide- than in rotenone-inhibited terminals pretreated with veratridine, indicating that $[Ca^{2+}]_c$ is not the only determinant of the glycolytic failure. The major difference between rotenone and cyanide treatments is that the redox state of the respiratory chain is less negative (reduced) in the former case. Thus a highly reduced respiratory chain in bioenergetic anoxia contributes to the inhibition of glycolysis.

Adenylates and Creatine Phosphate in Anoxic Synaptosomes

Table 2 summarizes the concentration of synaptosomal adenylates and CrP. Synaptosomal ATP/ADP is reduced extensively upon inhibition of respiration and there is a two- to threefold increase in AMP. The ATP/ADP ratio is signifi-

◄

Fig. 1. Influence of cyanide (**A**), rotenone (**B**) and nitrogen anoxia and carbonylcyanide *p*-trifluoromethoxyphenylhydrazone (FCCP) (**C**) on synaptosomal glycolysis. The rate of glycolysis was determined using D-3-[^{3}H]glucose as described in ref. 18. In **A** 2 mM NaCN and in **B** 6 μM rotenone was added as indicated. In **C** nitrogen purging was used to reduce the oxygen tension to zero for a period of 10 min; 1 μM FCCP was added at 80 min. Each *point* is a mean of at least three experiments. SE is indicated by *bars*. Reproduced by permission from the International Society for Neurochemistry

Table 2. Content of metabolites (nmol/mg protein) in synaptosomes

	ATP	ADP	AMP	ATP/ADP	CrP
Control	1.87±0.17 (9)	0.33±0.04 (9)	0.67±0.11 (3)	6.0±0.5 (9)	2.64±0.33 (7)
NaCN	0.36±0.04 (6)	0.31±0.02 (6)	2.30±0.24 (6)	1.1±0.1 (6)	0.27 (2)
Verapamil+NaCN	0.52±0.05 (5)	0.41±0.05 (5)	1.48±0.21 (5)	1.3±0.1 (5)	–
KCl+NaCN	0.39±0.06 (6)	0.32±0.04 (6)	2.47 (2)	1.2±0.1 (6)	–
Veratridine+NaCN	0.12±0.02 (4)	0.24±0.02 (4)	1.14±0.12 (4)	0.5±0.1 (4)	–
Rotenone	0.68±0.01 (5)	0.35±0.04 (5)	1.63±0.21 (5)	2.1±0.2 (5)	0.83±0.17 (4)
Verapamil+rotenone	0.71±0.11 (4)	0.35±0.03 (4)	1.0 (2)	2.1±0.8 (4)	–
KCl+rotenone	0.52±0.06 (6)	0.39±0.04 (6)	1.90±0.19 (5)	1.4±0.1 (5)	–
Veratridine+rotenone	0.20±0.03 (4)	0.29±0.03 (4)	1.63±0.34 (4)	0.7±0.1 (4)	–

Synaptosomes (2 mg protein/ml) were incubated for 20 min before addition of 100 μM verapamil, 45 mM KCl, or 75 μM veratridine. As indicated 3 min later 2 mM NaCN or 10 μM rotenone was added. Samples were quenched with perchloric acid 30 min later and assayed for metabolites. Values (nmol) are means±SE for number of experiments indicated in parentheses.

cantly higher in rotenone- than in cyanide-inhibited synaptosomes, indicating that the failure of glycolysis observed in the latter case is reflected in the energy levels. CrP content, which is a more sensitive measure of cytosolic energy level than the overall ATP/ADP ratio [29], is also lower in cyanide-treated synaptosomes. In synaptosomes which are inhibited with the combination of rotenone and iodoacetate, CrP drops below the limit that can be detected with the luciferase method (Table 2).

As with glycolysis, pretreatment of synaptosomes with KCl or veratridine abolishes the differential effects of cyanide and rotenone upon adenylates (Table 2).

Total and Cytosolic Free Calcium in Anoxic Synaptosomes

Although the failure of glycolysis in cyanide-treated synaptosomes is Ca^{2+} dependent, the content of total $^{45}Ca^{2+}$ does not change significantly. Only after complete inhibition of glycolysis with iodoacetate does a significant increase in total $^{45}Ca^{2+}$ occur (not shown).

The $[Ca^{2+}]_c$ elevates slowly in synaptosomes where respiration is inhibited by rotenone (Fig. 2). After 10 min $[Ca^{2+}]_c$ has increased from 300 nM to 360 nM. After 10 min with cyanide, $[Ca^{2+}]_c$ has risen to 630 nM and a time-dependent increase continues steadily (Fig. 2). Thus although the total Ca^{2+} in synaptosomes is not changed there is an extensive increase in $[Ca^{2+}]_c$ after cyanide. When both respiration and glycolysis are inhibited, $[Ca^{2+}]_c$ rises to 1300 nM after 10 min. The rise in $[Ca^{2+}]_c$ parallels the impairment of energy status of anoxic nerve terminals and thus the failure of glycolysis.

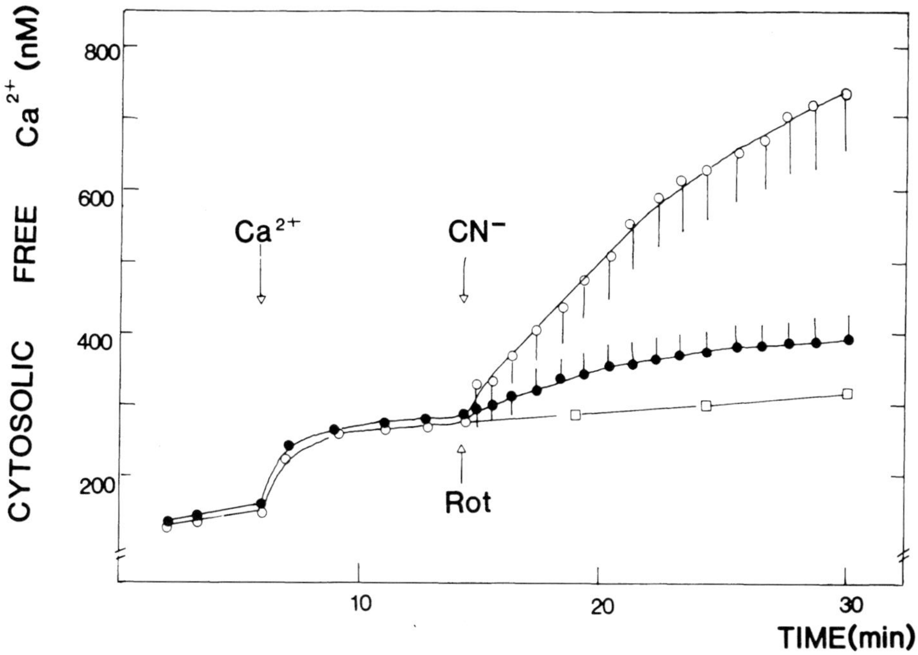

Fig. 2. Influence of metabolic inhibitors on synaptosomal cytosolic free calcium. $[Ca^{2+}]_c$ was determined using intrasynaptosomally trapped fura-2 as described in ref. 20. 2 mM NaCN (*CN^-*) and 10 μM rotenone (*Rot*) were added as indicated

Synaptosomal Membrane Potentials in Anoxia

In the presence of oxygen and glucose, synaptosomes maintain $\Delta\Psi_p$ and $\Delta\Psi_m$ of −73 and −160 mV, respectively. Inhibition of mitochondrial ATP production causes a partial collapse of $\Delta\Psi_m$ within 1 min. In the presence of rotenone, ATP produced by anaerobic glycolysis can support $\Delta\Psi_m$ by the reversal of ATP synthase better than in the presence of cyanide. Inhibition of glycolysis with iodoacetate in rotenone-treated terminals, which produces a bioenergetic condition equivalent to total ischemia, collapses the $\Delta\Psi_m$ by 50 mV, suggesting that anaerobically generated ATP in anoxia might be important in maintaining mitochondrial membrane polarization, and consequently in preventing matrix Ca^{2+} from effluxing to the cytoplasm [24, 33].

$\Delta\Psi_p$ is much better retained in the absence of respiration than is $\Delta\Psi_m$. After 20 min with rotenone and cyanide, $\Delta\Psi_p$ has depolarized 11 and 17 mV, respectively. The combination of rotenone and iodoacetate depolarizes $\Delta\Psi_p$ from −73 mV to −40 mV in 20 min. The voltage-activated Ca^{2+} entry into nerve terminals has been shown to be activated at a $\Delta\Psi_p$ of −40 mV or below [3, 12]. Since in nonstimulated anoxic synaptosomes $\Delta\Psi_p$ does not drop sufficiently (within 20 min) to activate the voltage-gated Ca^{2+} channel or to elevate total $^{45}Ca^{2+}$, the increase in $[Ca^{2+}]_c$ most probably relates to the release of Ca^{2+} from internal stores such as the mitochondrion [33], leading to accumulation of Ca^{2+} in cytosol. Elevated $[Ca^{2+}]_c$ then further inhibits the activity of anaerobic glycolysis, potentiating the drop in ATP.

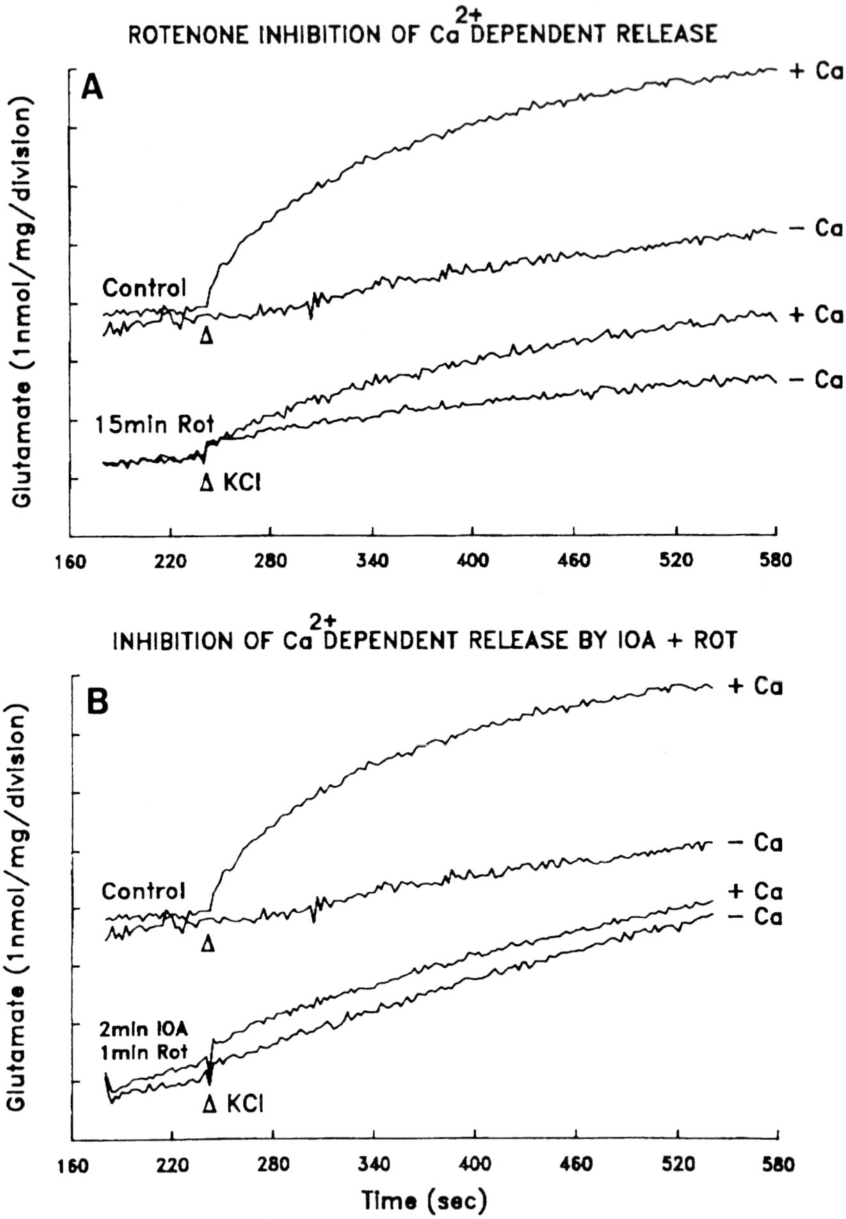

Fig. 3. Influence of rotenone (**A**) and iodoacetate and rotenone (*B*) on glutamate release from synaptosomes. Release of endogenous synaptosomal glutamate was assayed as described in ref. 26. Synaptosomes were incubated in the presence of either 1.3 mM $CaCl_2$ (+*Ca*) or 1.3 mM EGTA (−*Ca*). In two *lower traces* of **A** 10 μM rotenone (*15 min Rot*) was present for 15 min before KCl depolarization (30 mM). In **B** 1 mM iodoacetate and 10 μM rotenone (*2 min IOA, 1 min Rot*) were added 2 and 1 min, respectively, before KCl. Reproduced by permission from IBRO

Release of Endogenous Glutamate from Anoxic Synaptosomes

The excitatory neurotransmitter glutamate is present in at least two compartments within the nerve terminal. Firstly, there is a metabolic pool in the cytoplasm, and secondly, there is a transmitter pool, which is probably located in synaptic vesicles [25, 35). It is possible to distinguish between these two pools with the method used in the present study.

Potassium chloride depolarization in the presence of millimolar Ca^{2+} induces a rapid release of synaptosomal glutamate amounting to 10% of the total content. In the presence of EGTA the efflux of glutamate after KCl depolarization is much slower than in the presence of Ca^{2+} (Fig. 3). The Ca^{2+}-dependent release of the amino acid has been shown to be sensitive to botulinum neurotoxin A [31] and energy dependent [32], whereas the Ca^{2+}-independent release mechanism is not affected.

Rotenone causes an increase in glutamate efflux from polarized synaptosomes of 42% independent of the presence of Ca^{2+} (Fig. 3). After 15 min with rotenone the KCl-evoked release of glutamate was reduced to 30% of the control (Fig. 3). Unfortunately cyanide could not be used in the glutamate assay due to its interference with the reaction; therefore iodoactetate was used to inhibit glycolysis and simulate the profound energetic failure caused by cyanide. The combination of iodoacetate und rotenone increases the rate of Ca^{2+}-independent glutamate release more than rotenone alone. Within 1 min the Ca^{2+}-dependent exocytotic release of the amino acid is completely inhibited (Fig. 3).

Discussion

Isolated nerve terminals compensate for loss of mitochondrial energy production by an instant activation of glycolysis, showing that synaptosomal bioenergetics exhibit properties encountered in intact tissues [36]. When the Pasteur-effect is activated by inhibition of cytochrome c oxidase either by cyanide or by nitrogen, there is an unexpected failure to maintain the high rate of glycolysis. This failure develops concurrently with irreversible brain damage, because the brain is known to recover from anoxia that lasts less than 4–6 min.

The mechanism by which anaerobic glycolysis is inhibited appears to be dependent on two factors: firstly, on the redox state of the respiratory chain, and secondly, on the level of $[Ca^{2+}]_c$. Thus, when the respiratory chain remains oxidized, i.e., after inhibition of electron flow at the substrate terminus by rotenone, the high rate of glycolysis is maintained, preventing a total collapse of the synaptosomal energy level. In the presence of rotenone, the excess accumulation of the end-product of ^3H_2O from the aldolase reaction in comparison to the formation of lactate indicates that there is a rate-limiting step in the Embden-Meyerhof pathway between aldolase and lactate dehydrogenase reactions. In myocardium this step has been identified as glyceraldehyde-3-phosphate dehydrogenase [36].

The reaction that is most effectively activated in anoxic brain is the phosphofructokinase (PFK) [21, 27]. According to our results, synaptosomal glycolysis in anoxia is stimulated in the region of PFK, consistent with the previous reports [21, 27]. The "overactivation" of the hexose region of the pathway is, however, inhibited by high $[Ca^{2+}]_c$ independently of the redox state of the respiratory chain. The precise locus of $[Ca^{2+}]_c$ action remains to be elucidated. The site in glycolysis which is inhibited when the respiratory chain has a high negative redox state could be identified unequivocally.

The loss of cellular energy state in brain during lack of either oxygen [7] or glucose [1] is associated with an inability to maintain the concentrations of neurotransmitter in the extracellular compartment below the levels which activate postsynaptic receptors [2, 5, 37]. Increased extracellular glutamate in ischemic brain has been shown to act as an excitotoxin [30]. Glutamate is present in both neurons and astroglia [11] and there are indications that the vulnerability of neurons and astrocytes to energy depletion differs [16, 28].

A moderate energy depletion in the presence of rotenone significantly increased the rate of Ca^{2+}-independent release of glutamate from the cytoplasmic pool, without completely abolishing the Ca^{2+}-dependent exocytotic release mechanism. Thus in the absence of mitochondrial energy production, a high activity of glycolysis can support energy levels sufficient to allow a partial release of glutamate from the transmitter pool upon depolarization, but it is insufficient to maintain the full plasma membrane Na^+ electrochemical potential, thus allowing the reversal of the Na^+-coupled plasma membrane glutamate carrier with consequent release of glutamate from the cytoplasm. It must be pointed out that $[Ca^{2+}]_c$ did not rise to the level that triggers transmitter release.

In the total absence of energy-yielding reactions, the Ca^{2+}-dependent release of glutamate disappeared within 30 s. The energy failure was completed before $[Ca^{2+}]_c$ rose to a level sufficient to initiate exocytosis. Ca^{2+}-independent leakage of glutamate from the metabolic compartment was greatly enhanced in severely energy-deprived conditions. Thus in severe energy deprivation, such as in genuine anoxia, release of glutamate from nerve terminals might originate from the cytoplasmic (metabolic) pool, while the vesicular transmitter glutamate is largely untouched due to inhibition of exocytosis by the drop in the ATP/ADP ratio.

In conclusion, there are limits to which nerve terminals can tolerate anoxia before suffering a drop in energy level due to a failure to maintain glycolysis. This decay is followed by an inability to restore membrane polarization, by an extensive increase in $[Ca^{2+}]_c$, and by a leakage of glutamate from the cytoplasm of the nerve terminals. These factors might contribute to the development of irreversible brain damage in anoxia.

Acknowledgements. This study was supported by the grants from the Finnish National Academy of Sciences and FEBS (R.A.K.) and from the U.K. Medical Research Council (D.G.N.). Our thanks to Mrs. Eija Rissanen (University of Kuopio) and Mrs. Hellen Murphie (University of Dundee) for expert technical assistance

References

1. Behar KL, den Hollander JA, Petroff OAC, Hetherington HP, Prichard JW, Shulman RG (1985) Effects of hypoglycemic encephalopathy upon amino acids, high-energy phosphates, and pH_i in the rat brain in vivo: detection by sequential ^1H and ^{31}P NMR spectroscopy. J Neurochem 44:1045–1055
2. Benveniste H, Dreyer J, Schousboe A, Diemer NH (1984) Elevation of extracellular concentrations of glutamate and aspartate in rat hippocampus during transient cerebral ischemia monitored by intracerebral microdialysis. J Neurochem 43:1369–1374
3. Blaustein MP (1975) Effects of potassium, veratridine and scorpion venom on calcium accumulation and transmitter release by nerve terminals in vitro. J Physiol (Lond) 247:617–655
4. Brox D, Petermann B, Frunder H (1977) The effects of calcium on glycolysis and ATP concentration in complete and membrane-poor hemolysates of human erythrocytes. Acta Biol Med Germ 36:611–619
5. Butcher SP, Sandberg M, Hagberg H, Hamberger A (1987) Cellular origins of endogenous amino acids released into extracellular fluid of the rat striatum during severe insulin-induced hypoglycaemia. J Neurochem 48:722–728
6. Cotman CW, Foster A, Lanthorn T (1981) An overview of glutamate as a neurotransmitter. In: DiChiara G, Dessa G (eds) Glutamate as a neurotransmitter. Raven, New York, pp 1–27
7. Crockard HA, Gadian DG, Frackowiack RSJ, Proctor E, Allen K, Williams SR, Ross Russel RW (1987) Acute cerebral ischaemia: concurrent changes in cerebral blood flow, energy metabolites, pH, and lactate measured with hydrogen clearance and ^{31}P and ^1H nuclear magnetic resonance spectroscopy. II. Changes during ischaemia. J Cereb Blood Flow Metab 7:394–402
8. Dagani F, Erecinska M (1987) Relationships among ATP synthesis, K^+ gradients, and neurotransmitter amino acid levels in isolated rat brain synaptosomes. J Neurochem 49:1229–1240
9. Grynkiewicz G, Poenie M, Tsien RY (1985) A new generation of Ca^{2+} indicators with greatly improved fluorescence properties. J Biol Chem 260:3440–3450
10. Hansford RG, Castro F (1985) Role of Ca^{2+} in the pyruvate dehydrogenase interconversion in brain mitochondria and synaptosomes. Biochem J 227:129–136
11. Hertz L (1982) Astrocytes. In: Lajtha A (ed) Handbook of neurochemistry, vol 1. Plenum, New York, pp 310–355
12. Heinonen E, Åkerman KEO, Kaila K, Scott IG (1985) Dependence of cytoplasmic calcium transients on membrane potential in isolated nerve endings of guinea pig. Biochim Biophys Acta 815:203–208
13. Hohorst HJ (1963) L-(+)-lactate. In: Bergmeyer H-U (ed) Methods of enzymatic analysis, vol 1. Academic, New York, pp 266–270
14. Kanner B (1983) Bioenergetics of neurotransmitter transport. Biochim Biophys Acta 726:293–316
15. Katz J, Rongstad R (1966) The metabolism of tritiated glucose by rat adipose tissue. J Biol Chem 241:3600–3610
16. Kauppinen RA, Enkvist K, Holopainen I, Åkerman KEO (1988) Glucose deprivation depolarizes plasma membrane of cultured astrocytes and collapses transmembrane potassium and glutamate gradients. Neuroscience 26:283–289
17. Kauppinen RA, McMahon H, Nicholls DG (1988) Ca^{2+}-dependent and Ca^{2+}-independent glutamate release, energy status and cytosolic free Ca^{2+} concentration in isolated nerve terminals following metabolic inhibition: possible relevance to hypoglycaemia and anoxia. Neuroscience 27:175–182
18. Kauppinen RA, Nicholls DG (1986) Synaptosomal bioenergetics. The role of glycolysis, pyruvate oxidation and responses to hypoglycaemia. Eur J Biochem 158:159–165
19. Kauppinen RA, Nicholls DG (1986) Failure to maintain glycolysis in anoxic nerve terminals. J Neurochem 47:1864–1869
20. Komulainen H, Bondy SC (1987) The estimation of free calcium within synaptosomes and mitochondria with fura-2; comparison to quin-2. Neurochem Int 10:55–64

21. Lowry OH, Passonneau JV, Hasselberg FX, Schulz DW (1964) Effects of ischemia on known substrates and cofactors of glycolytic pathway. J Biol Chem 239:18–30
22. Lundin A, Rickhardsson A, Thore A (1976) Continuous monitoring of ATP-converting reactions by purified firefly luciferase. Anal Biochem 75:611–620
23. Nicholls DG (1978) Calcium transport and proton electrochemical potential in mitochondria from guinea-pig cerebral cortex and heart. Biochem J 170:511–522
24. Nicholls DG, Åkerman KEO (1981) Biochemical approaches to the study of cytosolic calcium regulation in nerve endings. Philos Trans R Soc Lond [Biol] 296:115–122
25. Nicholls DG, Sihra TS (1986) Synaptosomes possess an exocytotic pool of glutamate. Nature 321:772–773
26. Nicholls DG, Sihra TS, Sanchez-Prieto J (1987) Calcium-dependent and -independent release of glutamate from synaptosomes monitored by continuous fluorometry. J Neurochem 49:50–57
27. Nordberg K, Siesjö BK (1975) Cerebral metabolism in hypoxic hypoxia. I. Pattern of activation of glycolysis: a re-evaluation. Brain Res 86:31–44
28. Pauwels PJ, Opperdoes FR, Trouet A (1985) Effects of antimycin, glucose deprivation and serum on cultures of neurons, astrocytes and neuroblastoma cells. J Neurochem 44:143–148
29. Rafalowska U, Erecinska M, Wilson DF (1980) The effects of acute hypoxia on synaptosomes from rat brain. J Neurochem 34:1160–1165
30. Rothman SM, Olney JW (1987) Excitotoxicity and NMDA receptor. Trends Neurol Sci 10:299–302
31. Sanchez-Prieto J, Sihra TS, Aston A, Evans D, Dolly JO, Nicholls DG (1987) Botulinum toxin A and the inhibition of glutamate exocytosis from guinea-pig synaptosomes. Eur J Biochem 165:675–681
32. Sanchez-Prieto J, Sihra TS, Nicholls DG (1987) Characterization of the exocytotic release of glutamate from guinea-pig cerebral cortical synaptosomes. J Neurochem 49:58–64
33. Scott ID, Åkerman KEO, Nicholls DG (1980) Calcium-ion transport by intact synaptosomes. Intrasynaptosomal compartmentation and role of the mitochondrial membrane potential. Biochem J 192:873–880
34. Scott ID, Nicholls DG (1980) Energy transduction in intact synaptosomes. Influence of plasma-membrane depolarization on the respiration and membrane potential of internal mitochondria determined in situ. Biochem J 186:21–33
35. Storm-Mathisen J, Leknes AK, Bore AT, Vaaland JL, Edminson P, Haug F-MS, Ottersen OP (1983) First visualization of glutamate and GABA in neurons by immunocytochemistry. Nature 301:517–519
36. Vary TC, Reibel DK, Neely JR (1981) Control of energy metabolism of heart muscle. Ann Rev Physiol 43:419–430
37. Wieloch T (1985) Hypoglycemia-induced neuronal damage prevented by an *N*-methyl-D-aspartate antagonist. Science 230:681–683

Ionized Free Calcium Concentration and Deformability in Platelets

Y. ISOGAI, T. YOKOSE, S. IKEMOTO, T. MAEDA, M. AKIYAMA, J. OGAWA, and K. KUCHIBA

Introduction

Platelets may play an important role in the pathogenesis of cerebral infarction in many ways, for example by forming occlusive thrombi and by releasing vasoconstrictor substances such as serotonin and thromboxane A_2 (TXA_2). It has been found that the platelet function in thrombotic disease is exacerbated, and the significance of this finding is being debated [2, 7].

Thromboxane A_2 is one of the most powerful inducers of platelet function and vasoconstriction, and has been the focus of recent attention [3, 5]. Myocardial infarctions have been reported to occur as a result of injection of TXA_2 into the coronary arteries. However, TXA_2 is unstable in blood and is inactivated very rapidly by being converted to stable TXB_2. For this reason, it was difficult to assess in detail the activity of TXA_2 on platelets. In this study, platelet function and filterability were assessed by the use of STA_2, which is a stabilized derivative of TXA_2. As subjects, 30 healthy normal subjects and 12 patients with cerebral infarction were selected. Observations were made of differences in respect of the above etiological factors.

The shape of platelets has been proved to be changed from discoid to spheroid in the presence of platelet aggregation inducers. This morphological change, which occurs at the initial stage of platelet aggregation, also deserves rheological consideration. The present study was performed to estimate the clinical significance of platelet filterability and its function using STA_2 from the standpoint of the pathogenesis of cerebral infarction.

Patients and Methods

The fluorescent probes fura-2 and fura-2-AM were purchased from Molecular Probes Inc. STA_2 (Ono-11113,9,11-epithio-11,12-methano-TXA_2) was obtained from Ono Pharmaceuticals.

3rd Department of Internal Medicine, The Jikei University School of Medicine, 3-25-8 Nishishinbashi, Minato-ku, Tokyo 105, Japan.

Cerebral Ischemia and Calcium
Ed. by A. Hartmann and W. Kuschinsky

Preparation of Platelet Samples

Whole blood was obtained from the antecubital veins in the 30 normal subjects and 12 patients with cerebral infarction. We selected as the cerebral infarct group only patients in a comparatively early stage (1 week) after crisis; those patients taking drugs which exert an influence on platelet function, such as aspirin and calcium antagonists, were omitted. Coagulation of blood samples was prevented with 1/10 vol. of 3.8% sodium citrate, and the samples were centrifuged for 10 min at 1000 rpm to prepare PRP as specimen.

Platelet Aggregation Studies

Measurement of platelet aggregation was based on the method of O'Brien et al. The degree of aggregation after adding STA_2 under a temperature of 37 °C was recorded as the percentage fall in optical density.

Measurement of Platelet Ca^{2+}

PRP was incubated with 5 μM fura-2-AM for 30 min at 37 °C. Subsequently, PRP with ACD solution was centrifuged for 5 min at 2000 rpm and suspended in PIPES solution (pH 7.1, 1 mM Mg^{2+}, 10 mM PIPES, 1 mM EGTA, ionic strength 0.15 M, PCa 3, 37 °C) after removing supernatant. The suspension was then kept at room temperature in polyethylene tubes prior to use.

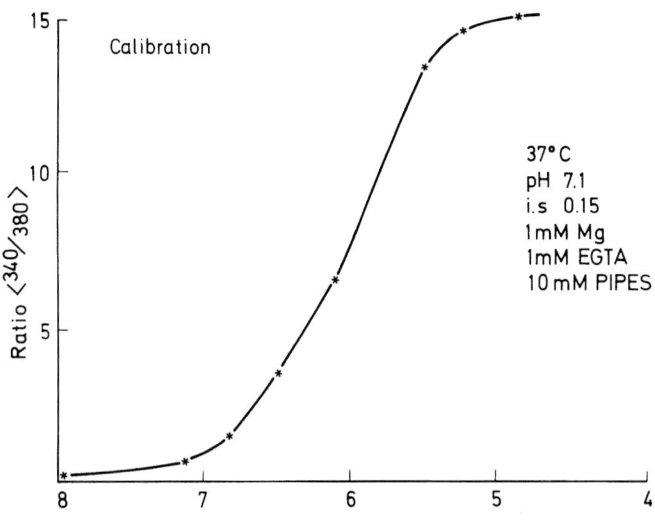

Fig. 1. Calibration curve for fura-2 Ca. The data for the fura-2 fluorescent intensity ratio (340/380 nm), calculated with CAF-100 for each PCa, were plotted. Concentration of fura-2 was 1 μM. Solution conditions: pH 7.1, ionic strength 0.15 M, 37 °C, 1 mM Mg, 1 mM EGTA, 10 mM PIPES

Fluorescent change in fura-2 was determined with a fluorescent spectrophotometer CAF-100 (Nihon Bunko Co. Ltd.). Five hundred nanometers of emission wavelength and two points of 340 and 380 nm of excitation were taken to find the ratio (340/380). Platelet calcium ion level was determined from the fura-2 Ca calibration curve based on the known Ca level (Fig. 1) [4, 9, 10].

Measurement of Platelet Filterability

The measurement was made by the recently devised negative pressure cell filtration system [6]. Platelet was diluted with physiological saline to be adjusted to 3×10^4/ml, to which a regular negative pressure was applied for aspiration at a flow rate of 0.123 ml/min. The diluent was passed through a Nucleopore membrane with a pore diameter of 2 μm. The difference in pressure at the top and bottom of the membrane at this time was recorded to delineate the curve of fluctuation of difference in pressure. High differential pressure indicated a decrease in filterability, and vice versa. Measurement was performed keeping the container for platelet suspension and the Nucleopore membrane at 37 °C in a thermostatic water bath. One milliliter of the platelet suspension was used for the measurement. STA_2 was added to each suspension immediately before measurement. Its final concentration was 0.125 or 1 μM.

Scanning Electron Microscopy

In order to observe the change in shape of the platelets during the filtration process, we used the scanning electron microscope (I). The platelets on the surface of the Nucleopore membrane were fixed with 1.2% glutaraldehyde for 60 min at room temperature. After the fixation, samples were dehydrated through a graded series of alcohol and dried by the critical point method.

They were then evaporated with carbon and gold. Observation was performed with JSM-35 (JEOL).

Results

Platelet Aggregation Studies

Platelet aggregation was significantly enhanced in the patients with cerebral infarction compared with normal subjects when aggregation was induced by low concentrations of STA_2 (0.125 or 0.25 μM). Aggregation reached almost the same levels at a concentration of 0.5 μM, and no difference in the aggregating competence was observed between the two groups at the still higher concentration of 1 μM (Table 1). Table 2 shows the percentage of PRP samples showing more than 50% aggregation following the addition of submaximal and maximal doses of STA_2. There was a significant increase in the frequency of enhanced aggregation at low concentrations of STA_2 when the patient group was compared with normal subjects.

Table 1. Maximum percentage of platelet aggregation following the addition of STA_2 in normal subjects and patients with cerebral infraction

STA_2 (μM)	Normal subjects $n=30$ (max %, mean ± SD)	Cerebral infarction patients $n=12$ (max %, mean ± SD)
0.125	1.73 ± 1.18	12.53 ± 6.50*
0.25	7.28 ± 5.20	60.03 ± 21.66**
0.5	69.50 ± 15.93	73.32 ± 10.91**
1.0	76.44 ± 9.82	78.42 ± 5.79**

Statistical analysis: normal subjects vs cerebral infarction patients: * $P<0.01$; ** not significant.

Table 2. Percentage of PRP samples showing more than 50% aggregation following the addition of STA_2

STA_2 (μm)	Normal subjects $n=30$	Cerebral infarction patients $n=12$
0.125	0%	11.1%*
0.25	0%	60.0%*
0.5	73.1%	80.0%**
1.0	92.0%	100%**

Statistical analysis: normal subjects vs cerebral infarction patients: * $P<0.01$; ** not significant.

Cytoplasmic Ca Response to STA_2

An increase in $[Ca^{2+}]i$ upon adding STA_2 was noted a very short time before the aggregation was started (see Fig. 2). The degree of this elevation was the same in the two groups, and even when the STA_2 level was altered, no change was observed in the elevation of platelet calcium level (Table 3).

Another experiment was performed for the purpose of investigating the nature of STA_2. Once Ca^{2+} was mobilized by STA_2, no subsequent elevation of Ca^{2+} was observed even if STA_2 was added again (Fig. 3). Elevation of $[Ca^{2+}]i$ by STA_2 was observed if the calcium level of the external fluid was lowered to zero, but the level was lower than when calcium was present in the external fluid (Fig. 4).

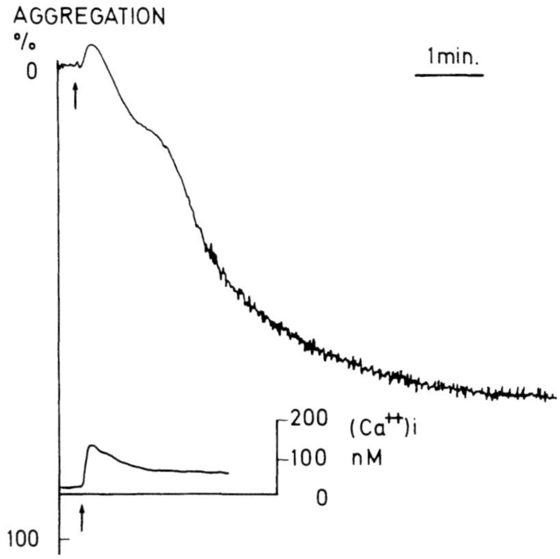

Fig. 2. Effect of STA_2 on $[Ca^{2+}]$i and aggregation. *Upper traces* indicate platelet aggregation as the percentage fall in optical density. *Lower traces* indicate variations in cytoplasmic calcium monitored with fura-2. An aliquot of 50 μM STA_2 was added to 250 μl PRP, and therefore the final concentration of STA_2 was about 0.5 μM. Addition of STA_2 is indicated by *arrows*

Table 3. Increase in levels of $[Ca^{2+}]_i$ in platelets from venous blood of normal subjects and cerebral infarction patients after the addition of STA_2

STA_2	Normal subjects ($n=30$)			Cerebral infarction patients ($n=12$)		
(μM)	Peak $[Ca^{2+}]_i$ (nM)	Time to peak (s)		Peak $[Ca^{2+}]_i$ (nM)	Time to peak (s)	
0.125	140 ± 25 } NS	5.65 ± 1.0 } NS		141 ± 25** } NS	9.62 ± 4.27* } NS	
1.0	141 ± 23	4.61 ± 0.7		143 ± 27**	6.12 ± 3.05**	

Statistical analysis: normal subjects vs cerebral infarction patients: * $P<0.05$; ** not significant. NS also shows not significant.

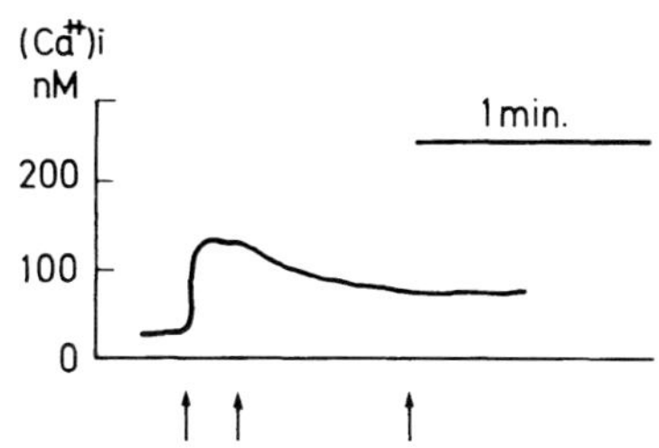

Fig. 3. Mobilization of free calcium ions after the addition of STA_2 in platelet suspension. Each *arrow* indicates the addition of STA_2; the final concentration was 0.5 μM. The second addition of 0.5 μM STA_2 was made soon after the peak $[Ca^{2+}]$i was recorded. The third addition of the same dose of STA_2 was made after the level of $[Ca^{2+}]$i started to decrease

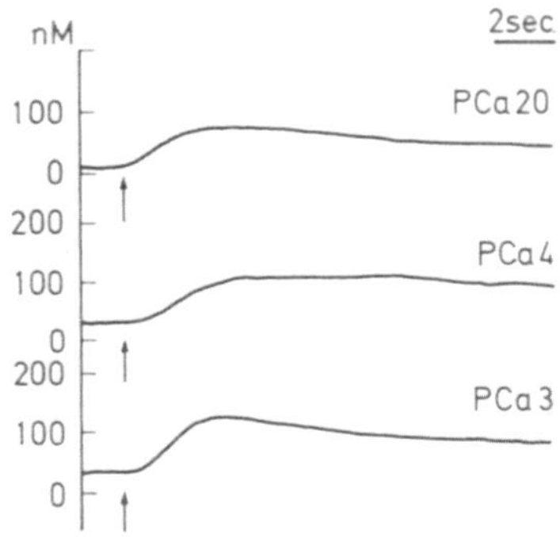

Fig. 4. The effect of calcium concentration in medium on the mobilization of calcium ions monitored with fura-2. Addition of STA_2 is indicated by *arrows*. The final concentration of STA_2 was 0.5 μM for each addition. *Upper traces* indicate variations in cytoplasmic calcium under the condition that the calcium concentration in the medium was free. In the case of the *middle* and *lower traces* the calcium concentration in the medium was 100 μM and 1 mM respectively

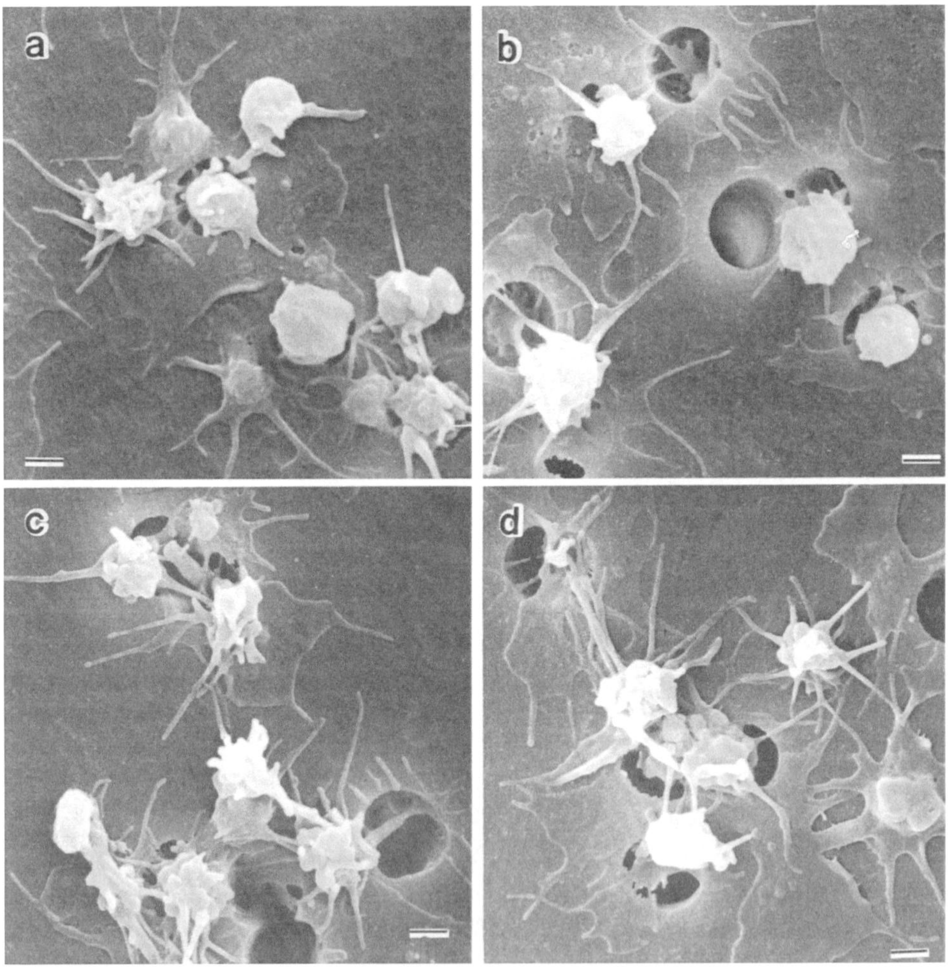

Fig. 5 a–d. Scanning electron micrographs of platelets during the filtration process. Shape change of platelets obtained from normal subjects (**a, b**), and cerebral infarction patients (**c, d**). **a** and **c** show the platelets without STA_2 while **b** and **d** show the platelets with 1 μM STA_2. The *bars* represent 1 μm. × 7,800

Table 4. Comparison of platelet filterability of the patients and the normal subjects. The data shown were collected 1 min and 3 min after the filtration was started

	Control (without STA_2)		With 1 μM STA_2	
	1 min	3 min	1 min	3 min
	(mm H_2O)			
Normal subjects ($n=30$)	35 ±25.1	154.3±45.7	40.5±16.0	149.1±49.9
Cerebral infarction patients ($n=12$)	35.7±14.2	164.0±25.6	44.8±14.7	158.3±53.3

No significant differences existed between these two groups with and without STA_2.

Filterability and Change in the Shape of Platelets

No change was observed in filterability when STA_2 was added at a concentration of 1 μM or when no STA_2 was added (Table 4). Figure 5 shows the observations on the morphology of platelets under these conditions; both groups showed morphological change of the same degree.

Discussion

Platelet aggregation in relation to the blood coagulating system has been studied by many investigators. However, little attention has been paid to the rheological aspects of platelets such as filterability.

We have made assessments of the filterability and morphology of platelets and also of the relationship between platelet aggregation as a platelet function and platelet calcium ions. STA_2, a stabilized derivative of TXA_2, was used as an inducer because TXA_2 is attracting attention at present as one of the causal factors in thrombotic diseases. Aggregation was obviously exacerbated in patients with cerebral thrombosis as compared with the control group but no differences were observed between the two groups in respect of the mobilization of platelet calcium ions measured with fura-2 or filterability and morphological changes.

This suggests that when the activation is by TXA_2, mobilization of calcium ions and the morphological change that follows are not reflected in the difference in aggregation [8]. These two factors may act only as a trigger of aggregation. In other words, the increase in aggregability of platelets from patients may be affected by other factors, for example:

1. Increase of other platelet activators in plasma
2. Decrease of platelet inhibitors released from injured endothelial cells, such as prostacyclins
3. Increase in the number and sensitivity of receptor sites for other inducers, such as ADP
4. Increase in the number of surface glycoprotein complexes
5. Increase in binding proteins

When ADP was used as an inducer of platelet aggregability in diabetes, filterability of platelets was obviously lowered [6]. However, platelet aggregability is exacerbated both in diabetes and in cerebral thrombosis, and the fact that no difference in deformability was observed after STA_2 addition in cerebral thrombosis as compared with normal subjects suggests that the mechanisms of platelet activity may be different in these two patient groups.

The platelet calcium level measured by the use of fura-2 was approximately 140 nM despite the fact that aggregation was induced after STA_2 stimulation. This result also suggests that exacerbation of aggregation is strongly influenced by external factors.

Conclusions

The results of this study were as follows:

1. Platelet aggregability was significantly increased in patients when activated with a low concentration of STA_2. However, no significant differences existed between normal subjects and patients when platelets were activated with a high concentration of STA_2.
2. Platelet calcium level showed no difference between the normal subject group and the patient group.
3. No difference was observed in platelet deformability between the two groups.

These findings suggest that the change in platelet shape and the increase in $[Ca^{2+}]i$ which occurred at the early stage of activation had little effect on platelet aggregation when activated by STA_2. It was further presumed that TXA_2 has some effects on the crisis of cerebral thrombosis since the platelets in cerebral thrombotic patients tend to become more activated and aggregated after STA_2 stimulation.

References

1. Adrian RL (1984) Rapid platelet morphological changes visualized by scanning electron microscopy; kinetics derived from a quanched-flow approach. Br J Haematol 56:387–398
2. Aurelio MV, Maria CG, Roberta L, Alessandva M, Maria PS, Massimo F (1987) Platelet function and thrombin activity in patients with recent cerebral transient ischemic attacks. Stroke 18:892–895
3. Fujioka M, Nagao T, Kuriyama H (1986) Actions of the novel thromboxane A_2 antagonist, ONO-1270 and ONO-3708, on smooth muscle cells of the guinea-pig basilar artery. Naunyn Schmiedebergs Arch Pharmacol 334:468–474
4. Grynkiewirz G, Poenie M, Tsien RY (1985) A new generation of Ca^{++} indicators with greatly improved fluorescence properties. JBC 260:3440–3450
5. Hamano S, Komatsu H, Ujie A, Ikeda S, Hiraku S (1987) Role of TXA_2 in the Ca^{++} mobilization in rabbit platelets. Blood Vessels 18:131–133
6. Isogai Y, Yokose T, Ikemoto S, Maeda T, Akiyama M, Kuchiba K (1987) Filterability and the shape of platelets: the effect of stimulant. In: Hartmann A, Kuschinsky W (eds) Cerebral ischemia and hemorheology. Springer, Berlin Heidelberg New York, pp 90–95

7. Mikhailidis DP, Barradas MA, Mier A, Boag F, Jeremy JY, Havard CWH (1987) Platelet function in patients admitted with a diagnosis of myocardial infarction. Angiology. January: 36–45
8. Nakashima S, Nozawa Y (1986) Platelet activation and Ca^{++} homeostasis. Blood Vessels 17:293–306
9. Pedersen OS, Reighelt KL (1988) Increased calcium response to ADP in blood platelets from women during ovulation compared with menstruation; cytoplasmic calcium measured with the fura-2 technique. Acta Physiol Scand 132:335–339
10. Tsien RY, Pozzan T, Rink TJ (1982) Calcium homeostasis in intact lymphocytes: cytoplasmic free calcium monitored with a new intracellularly trapped fluorescent indicator. J Cell Biol 94:325–334

Calcium 45 Accumulation in the Dentate Hilus: Possible Effect of NMDA Receptor Blockers

H. BENVENISTE [1], P. C. HÜTTEMEIER [2], F. F. JOHANSEN [1], and N. H. DIEMER [1]

Introduction

Among the many complex reactions involved in ischemia-induced neuronal cell injury, intracellular calcium accumulation appears to be particularly cytotoxic [11, 19, 24]. The rise in the intracellular calcium concentration has been postulated to initiate or represent a "final common pathway" for cell death [23]. The mechanisms of intracellular calcium accumulation during neuronal injury are the result of several factors, but one source might be the opening of postsynaptic NMDA receptor-coupled calcium channels by excessive transmitter release of glutamate and aspartate [2].

In the rat the time course of postischemic neuronal death is well described [6, 9, 14, 21]. Following an ischemic insult a characteristic pattern of nerve cell loss is typically seen in the hippocampal region: fast developing damage in the dentate hilus, where somatostatin-positive neurons are extremely vulnerable [15], and delayed neuronal death in the CA-1 region [14, 16, 21]. Several studies have shown that the accumulation of calcium is time dependent and closely related to the progression of ischemic cell change [8, 10, 13]. Recently we detected early ^{45}Ca accumulation in the dentate hilus 1 h following recirculation [4]. Hilar neurons responsible for this specific ^{45}Ca uptake may be somatostatin-positive neurons.

In the following we report preliminary efforts to inhibit postischemic ^{45}Ca uptake in the rat dentate hilus with NMDA antagonists. The possible effect on cell survival was evaluated by morphological studies.

Materials and Methods

General Procedures

Male Wistar rats weighing 300–350 g were used. Before injection of isotope or insertion of microdialysis probes the animals were initially anesthetized with 3% halothane in a N_2O/O_2 mixture. Anesthesia was maintained with 0.75% halo-

[1] PharmaBioTec Research Center, Institute of Neuropathology, University of Copenhagen, DK-2100 Copenhagen, Denmark.
[2] Department of Anesthesiology, Glostrup Hospital, University of Copenhagen, DK-2100 Copenhagen, Denmark.

Cerebral Ischemia and Calcium
Ed. by A. Hartmann and W. Kuschinsky

Table 1. Microdialysis procedures

No.	Perfusion fluid	
	Right probe	Left probe
7 (control)	Mock CSF	Mock CSF
6 (ischemia)	Mock CSF	Mock CSF
5 (ischemia)	Mock CSF	Mg^{2+} (100 mM)
4 (ischemia)	Mock CSF	Ketamine (10 mg/ml)
6 (ischemia)	Mock CSF	2-APH (20 μg/μl)

The left microdialysis probe was continuously perfused with either Mg^{2+}, ketamine, or 2-APH 40 min prior to, during, and 1 h after induction of ischemia.

thane after tracheal intubation. Polyethylene catheters were inserted in the femoral artery and vein for blood pressure measurements, blood sampling, and fluid administration. Blood gases, pH, and plasma glucose were checked at regular intervals. Rectal temperature was maintained at 37 °C with a thermostated heating lamp.

Microdialysis Procedures

Two microdialysis probes (CMA/10, Carnegie Medicine, Stockholm, Sweden) were implanted bilaterally in the dorsal hippocampus. Immediately following implantation, perfusion of the probes started at 2.5 μl/min with 100 μCi/μl ^{45}Ca dissolved in mock CSF and continued for 2 h. This preloading phase was followed by a washout period with perfusion fluids composed as described in Table 1. Seven rats killed 4 h after the start of the washout period served as controls. In the other animals, 20 min of ischemia (four-vessel occlusion model) was induced 2 h after the start of washout. The animals were sacrificed 1 h following recirculation.

Morphological Experiments

Animals were surgically prepared as previously described. Four rats without ischemia served as controls. In four untreated rats ischemia was induced for 20 min. In ketamine-treated rats ($n = 3$) the administration started 40 min prior to the induction of ischemia and continued during and for 24 h after recirculation. Animals were kept alive for 4 days. They were then anesthetized with 3% halothane and sacrificed by intercardial perfusion-fixation with 0.15 M Sorensen phosphate buffer, pH 7.3. After fixation the brains were removed and stored for 4 h in the fixative. The brains were then blocked and sectioned frontally (50 μm) on an Oxford vibratome. Sections were immunohistochemically stained using antibodies against somatostatin (SS) (1:4000, K 18 form, Ferring Arzneimittel GmbH, Kiel, FRG) according to the unlabeled antibody peroxidase-antiperoxidase method of

Sternberger [28]. Finally, sections were counterstained with toluidine blue and mounted. All sections were processed simultaneously to achieve identical conditions during the immunohistochemical reaction.

Results

Calcium 45 Autoradiograms

Control Rats. Typical ^{45}Ca autoradiograms demonstrated that only the cortex, hippocampus, and upper part of the thalamus were exposed to ^{45}Ca. ^{45}Ca was distributed unequally among the brain regions, i.e., gray matter contained more than white. Within the hippocampus the ^{45}Ca distribution pattern in the pyramidal cell layer, stratum oriens, granular cell layer, and dentate hilus were identified from corresponding hematoxylin-eosin stained brain sections. The content of ^{45}Ca was lower in the pyramidal and granular cell layer than in the stratum oriens, which appeared as a dark ^{45}Ca distribution line.

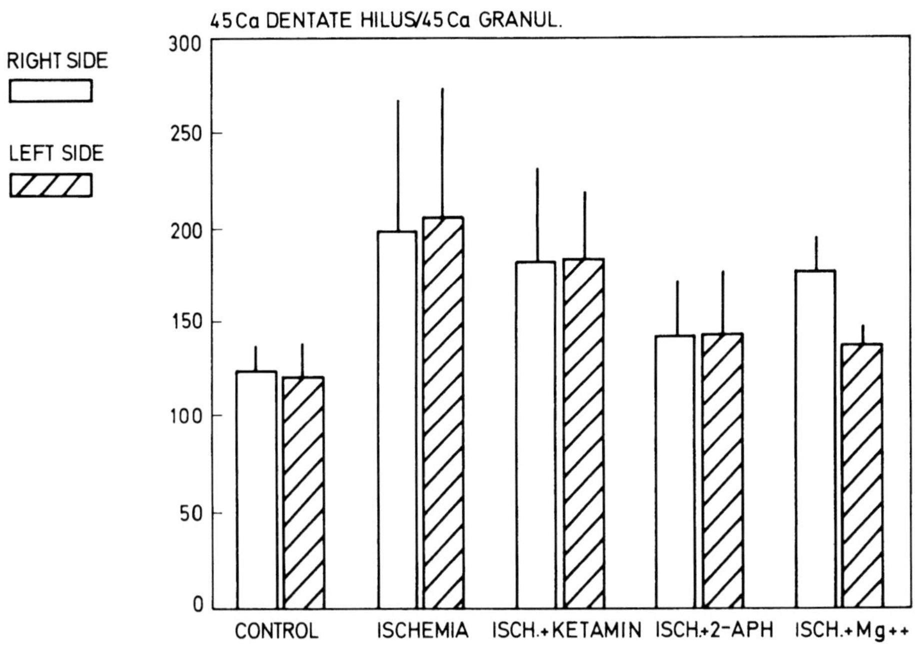

Fig. 1. The ratio between the content (nCi/g) in the dentate hilus and the granular cell layer was measured in the right and left hippocampus. The ^{45}Ca content in the granular cell layer has previously been shown not to change significantly 1 h following ischemia when compared to control ^{45}Ca content [4]. Right- and left-sided ratios are identical in control ($n=7$) and postischemic groups ($n=6$). No significant difference ($P<0.05$) could be demonstrated between left-sided ratios in rats treated with either ketamine ($n=4$) or 2-APH ($n=6$) when these were compared with left-sided ratios in the postischemic group. However, a significant difference ($P<0.03$) was found for rats treated with Mg^{2+} ($n=5$) when these ratios were correlated. Results are average $\pm$SD

The following results are expressed as a ratio between the ^{45}Ca content (nCi/g) in the dentate hilus and the granular cell layer. The ^{45}Ca content in the granular cell layer is used for reference since it remains unchanged [4]. Right- and left-sided ratios are identical during control conditions (Fig. 1, right: 123 ± 8; left: 120 ± 16).

Ischemic Rats. After 1 h of recirculation, a 63% increase in the ratio was demonstrated as a result of a marked ^{45}Ca accumulation in the dentate hilus. Again, no differences could be found between right- and left-sided ratios. Treatment of ani-

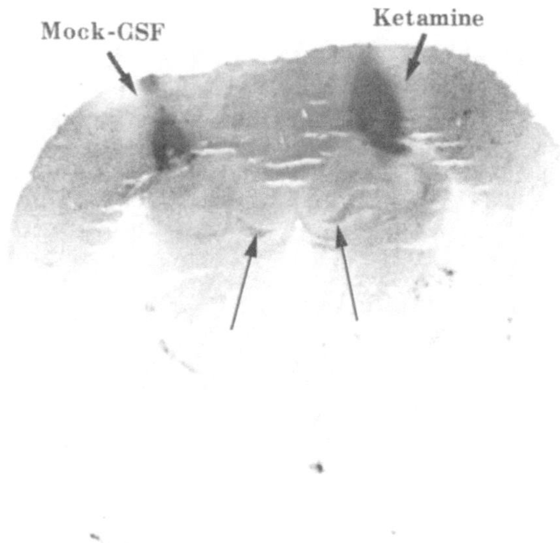

Fig. 2. ^{45}Ca autoradiogram demonstrating the distribution pattern 1 h following ischemia in a rat treated with ketamine via the left microdialysis probe. The ^{45}Ca accumulation in the dentate hilus was still visible on the left side (indicated by *arrows*)

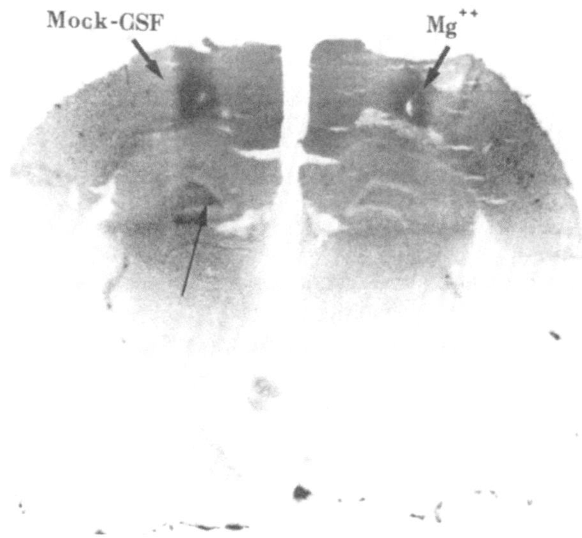

Fig. 3. ^{45}Ca autoradiogram demonstrating the distribution pattern 1 h following ischemia in a rat treated with Mg^{2+}. The ^{45}Ca accumulation in the dentate hilus could only be demonstrated on the right side (indicated by *arrow*)

mals with ketamine via the left microdialysis probe did not significantly change the ^{45}Ca distribution pattern compared to ischemic, untreated controls (Figs. 1, 2). 2-APH treatment produced lower but not significantly different ratios compared to ischemic controls (Fig. 1). However, magnesium treatment significantly reduced left-sided ratios compared to control (Figs. 1, 3).

Morphological Studies

Control brains demonstrated a well-perfused tissue with blood vessels clear of red blood cells. All ischemic rats had CA-1 pyramidal cell loss on the 4th postischemic day. Antibodies to SS clearly marked immunopositive cell somata. The SS neurons were characteristically distributed in stratum oriens of CA-1 and CA-3, within the hilus and the CA-3c subfield close to the hilus. Ischemia induced a significant 80% loss of hilar SS neurons 4 days after the insult, as previously demonstrated [15]. Ketamine treatment did not prevent SS neuronal loss in the dentate hilus.

Discussion

During control conditions the distribution pattern of ^{45}Ca in the hippocampus corresponds to that of ^{3}H-inulin, which is a marker of extracellular space [4]. The distribution pattern could also reflect the fact that pyramidal cells in the hippocampus have strong calcium conductances, with a higher calcium channel density in the cell dendrites than in the soma [26, 27]. A similar control distribution pattern of ^{45}Ca has previously been demonstrated by Dienel [10]. Following ischemia the ratio between ^{45}Ca content in the dentate hilus and the granular cell layer increased by 63% compared to control. This increase is due to a ^{45}Ca accumulation in the dentate hilus because ^{45}Ca content in the granular cell layer is unchanged [4]. Absolute values of ^{45}Ca content (nCi/g) in the dentate hilus have previously been reported and are in agreement with these results [4]. We assume that postischemic ^{45}Ca accumulation is mainly intracellular because we found no change in the distribution pattern of ^{3}H-inulin 1 h after ischemia [4]. From the study by Dienel we know that structures vulnerable to 30 min of ischemia accumulate calcium for days, and that the rate of influx depends on the survival time. However, this study also showed that accumulation of ^{45}Ca in the caudate putamen persisted for at least 2 days after striatal morphological damage had peaked [10]. Therefore, the accumulation may represent detrimental calcium-mediated processes that are activated before irreversible cell damage occurs or may be a "postmortem" event due to ruptured membranes.

Mechanisms responsible for postischemic intracellular calcium accumulation are numerous, but one theory holds that neurons vulnerable to ischemia are those in which Ca^{2+} entry is favored by neuronal depolarization or by bursting behavior during a restitution phase in hypoxia or ischemia [19]. This theory is strongly supported by recordings of enhanced neuronal activity in the CA-1 region 7–10 h after a 5-min episode of ischemia in the gerbil [25] since large influxes of calcium ions accompany burst-firing neurons [22]. Plasma membranes possess special Ca^{2+} channels, the conductances of which are controlled by gates that are opened by critical changes in cellular polarization (voltage-operated channels) or by receptor activation (receptor-operated channels) [7]. One of three well-characterized glutamate receptors – the NMDA receptor – is known to gate calcium ions [17]. This receptor is closely involved with burst firing since iontophoresis of NMDA on rat caudate neurons results in a pattern of neuronal firing (paroxysmal depolarizing shift accompanied by burst firing) [12].

Glutamate release during ischemia [2] may alter NMDA receptor density or affinity so that these postischemically mediate increased calcium conductances. Therefore inhibition of excitatory neurotransmission by 2-APH (competitive) or ketamine (noncompetitive), both of which block NMDA receptor sites, should inhibit burst firing and calcium entry and prevent cell death.

However, we found that dentate hilar ^{45}Ca content in 2-APH treated rats was lower but not significantly different from the content in untreated controls. These results might explain why neuronal protection by administration of 2-APH is incomplete [5]. Likewise, dispite continuous administration, ketamine had no effect on the ^{45}Ca accumulation in the dentate hilus and was not able to prevent loss of SS-positive neurons.

A relevant problem in these experiments is the uncertainty of the actual distributions and concentrations of drugs; for a review, see [3]. Concentration profile studies with the extracellular marker TMA show that the concentration 1 mm from the microdialysis probe is tenfold lower than that of the perfusate (Benveniste and Hansen, unpublished observation). Extrapolating from these data one should expect a ketamine concentration in the range of 1 mg/ml in the dentate hilus, which is sufficient to block NMDA receptor sites [1]. In contrast Mg^{2+} administration significantly reduced the ^{45}Ca content in the dentate hilus. Magnesium may exert this effect through four separate mechanisms: First, by blocking transmitter release, magnesium eliminates activity dependent on synaptic transmission. Secondly, by elevating the threshold for action potential generation, magnesium decreases spontaneous activity (burst firing) independent of synaptic transmission. Third by blocking transmitter release, magnesium may prevent neurotransmitters from exerting direct toxic effects on postsynaptic cells. Finally, magnesium (physiological concentrations) blocks the NMDA receptor coupled calcium channel in a voltage-dependent manner [18, 20]. Unless the physiological concentration of Mg^{2+} decreases during and after ischemia, the latter effect of Mg^{2+} is probably irrelevant. Nevertheless, all these mechanisms may have contributed to a reduction of the dentate hilar ^{45}Ca content.

In conclusion, we have demonstrated an early postischemic calcium accumulation in the rat dentate hilus – an area which is extremely vulnerable to ischemia. We assumed that excessive intracellular calcium accumulation plays a major role in ischemia-induced neuronal injury and death and hypothesized that activated NMDA receptor coupled calcium channels might be involved in this process. However, this hypothesis is not supported by these preliminary results.

References

1. Anis NA, Berry NR, Burton NR, Lodge D (1983) The dissociative anaesthetics, ketamine and phencyclidine, selectively reduce excitation of central mammalian neurones by *N*-methyl-aspartate. Br J Pharmacol 79:565–575
2. Benveniste H, Drejer J, Schousboe A, Diemer NH (1984) Elevation of the extracellular concentrations of glutamate and aspartate in rat hippocampus during transient cerebral ischemia monitored by intracerebral microdialysis. J Neurochem 43:1369–1374
3. Benveniste H (1989) Brain microdialysis, short review. J Neurochem 52
4. Benveniste H, Diemer NH (1988) Early postischemic ^{45}Ca accumulation in rat dentate hilus. J Cereb Blood Flow Metab 8:713–719
5. Block GA, Pulsinelli WA (1987) *N*-Methyl-D-aspartate receptor antagonists: failure to prevent ischemia-induced selective neuronal damage. In: Powers WJ, Raichle ME (eds) Cerebrovascular diseases, 15th Research (Princeton) Conference, pp 37–45
6. Brierley JB (1976) Cerebral hypoxia. In: Blackwood W, Corsellis JAN (eds) Greenfields neuropathology, pp 43–85
7. Cavero I, Spedding M (1983) "Calcium antagonists": a class of drugs with a bright future. Part I. Cellular calcium homeostasis and calcium as a coupling messenger. Life Sci 33:2571–2581
8. Chien KR, Abrams J, Richard G, Farber JL (1977) Prevention by chlorpromazine of ischemic liver cell death. Am J Pathol 88:539–557
9. Diemer NH, Siemkowicz E (1981) Regional neurone damage after cerebral ischemia in the normo- and hypoglycemic rat. Neuropathol Appl Neurobiol 7:217–227
10. Dienel GA (1984) Regional accumulation of calcium in postischemic rat brain. J Neurochem 43:913–925
11. Hass WK (1981) Beyond cerebral blood flow, metabolism and ischemic thresholds: examination of the role of calcium in the initiation of cerebral infarction. In: Meyer JS, Lechner H, Reivich M, Ott EO, Arabinar A (eds) Cerebral vascular disease, vol 3. Proceedings of the 10th Salzburg Conference on cerebral vascular disease. Excerpta Medica, Amsterdam, pp 3–17
12. Herrling PL, Morris R, Salt TE (1983) Effects of excitatory amino acids and their antagonists on membrane and action potentials of rat caudate neurones. J Physiol 339:207–222
13. Hossman KA, Paschen W, Csiba L (1983) Relationship between calcium accumulation and recovery of cat brain after prolonged ischemia. J Cereb Blood Flow Metab 3:346–353
14. Ito U, Spatz JT, Walker Jr, Klatzo I (1975) Experimental cerebral ischemia in mongolian gerbils. I. Light microscopic observations. Acta Neuropathol 68:230–238
15. Johansen FF, Zimmer J, Diemer NH (1987) Early loss of somatostatin neurons in dentate hilus after cerebral ischemia in the rat precedes CA-1 pyramidal loss. Acta Neuropathol (Berl) 73:110–114
16. Kirino T (1982) Delayed neuronal death in the gerbil hippocampus following ischemia. Brain Res 239:57–69
17. MacDermott AB, Mayer ML, Westbrook GL, Smith SJ, Barker LJ (1986) NMDA-receptor activation increases cytoplasmic calcium concentration in cultured spinal cord neurones. Nature 321:519–522
18. Mayer ML, Gary L, Westbrook GL, Guthrie PG (1984) Voltage-dependent block by Mg^{++} of NMDA responses in spinal cord neurones. Nature 309:261–263
19. Meldrum BS (1983) Metabolic factors during prolonged seizures and their relation to nerve cell death. Adv Neurol 34:261–275
20. Nowak L, Bregestovski P, Ascher P, Herbert A, Prochiantz A (1984) Magnesium gates glutamate-activated channels in mouse central neurones. Nature 307:462–465
21. Pulsinelli WA, Brierley JB, Plum F (1982) Temporal profile of neuronal damage in a model of transient forebrain ischemia. Ann Neurol 11:491–498
22. Schwartzkroin PA, Wyler AR (1980) Mechanisms underlying epileptiform burst discharge. Ann Neurol 7:95–107
23. Shanne FAX, Kane AB, Young EE, Farber JL (1979) Calcium dependence of toxic cell death: a final common pathway. Science 206:700–702

24. Siesjö BK (1981) Cell damage in the brain: a speculative synthesis. J Cereb Blood Flow Metabol 1:155–185
25. Suzuki R, Yamaguchi T, Li CL, Klatzo I (1983) The effects of 5-min ischemia in mongolian gerbils. II. Changes of spontaneous neuronal activity in cerebral cortex and CA-1 sector of hippocampus. Acta Neuropathol 60:217–222
26. Wrong RKS, Prince DA (1978) Participation of calcium spikes during intrinsic burst firing in hippocampal neurons. Brain Res 159:385–390
27. Wrong RKS, Prince DA, Basbaum AI (1979) Intradendritic recordings from hippocampal neurons. Proc Natl Acad Sci USA 76:986–990
28. Zimmer J, Sunde N (1984) Neuropeptides and astroglia in intracerebral hippocampal transplants: an immunohistochemical study in the rat. J Comp Neurol 227:331–347

Vascular and Direct Effects on Cerebral Energy Metabolism of Emopamil

G. W. BIELENBERG and J. KRIEGLSTEIN

Introduction

In previous studies the calcium antagonist emopamil was demonstrated to restore more quickly high energy phosphates during the first 5 min of postischemic recirculation following an ischemic episode of 30 min in the isolated rat brain when brains were perfused at constant pressure [3]. When brains were perfused at constant flow rate, however, this effect was no longer present, and hence it was attributed to the vasodilatory effect of the compound. Furthermore, emopamil caused lower cortical lactate levels after 30 min of postischemic reperfusion. This effect was present both at constant pressure and at constant flow rate perfusion and proved to be independent of the flow rate used. As metabolite determinations were performed in cortical tissue and were compared to global flow rates, the purpose of the present investigation was to evaluate whether intracerebral redistribution phenomena might account for the observed diverging results. Furthermore, the question was posed whether the accelerated decrease in cortical lactate levels was due to enhanced oxidative degradation or to an increased loss into the perfusate. In the present study we therefore examined cerebral ATP and lactate levels by techniques allowing for better spatial resolution. On the other hand, flow rate was determined on a local level to investigate whether intracerebral redistribution phenomena of the perfusate might occur with emopamil administration. Furthermore, the time course of lactate accumulation in the perfusate and the effect of emopamil thereon were assessed.

Material and Methods

Materials

(−)-Emopamil was a gift from Knoll (Ludwigshafen, FRG). Enzymes, coenzymes, and substrates were obtained from Boehringer (Mannheim, FRG) or Sigma (St. Louis, MO, USA). All other chemicals were of reagent grade. Male Sprague-Dawley rats (IWV, Geretsried, FRG) weighing 200–230 g were used. The animals were maintained under controlled environmental conditions (12 h dark/light cy-

Institut für Pharmakologie und Toxikologie, Philipps-Universität Marburg, Ketzerbach 63, D-3550 Marburg, FRG.

Cerebral Ischemia and Calcium
Ed. by A. Hartmann and W. Kuschinsky

cle, 23° ± 1 °C, 55% ± 5% rel. humidity) and kept on standard diet (Altromin, Lage, FRG) and tap water ad libitum until the experiment was started.

Perfusion Technique

The rat brain was isolated according to Andjus et al. [1] and perfused with a fluorocarbon (FC 43) emulsion (Green Cross Corp., Osaka, Japan) following a technique described previously [8]. Perfusion rate (5 ml/min) was maintained at a constant level. Bipolar electroencephalograms were recorded from the parietal region of each hemisphere (EEG preamplifier BioAC; Hellige, Freiburg, FRG). Emopamil was added to the perfusion medium at a concentration of 1 μmol/l before perfusion was started. After a perfusion period of 30 min, ischemia was induced for 30 min by switching off medium supply. Following ischemia, perfusion was reinstituted for 5 or 30 min. Metabolism was stopped by immersing brains in liquid nitrogen [14]. Lactate levels in the perfusate were measured at various time points during the experimental procedure.

LCBF Determination

The [^{14}C]iodoantipyrine (IAP) technique [16] was applied to the isolated brain for the determination of local cerebral flow rate (LCBF). First the tissue/blood partition coefficient lambda had to be determined for our perfusion conditions. In a separate series of experiments brains were perfused at a constant flow rate of 5 ml/min. IAP was infused over 15 min at a constant rate of 1 μCi/min per hemisphere into the supplying catheters and mixed with the perfusion medium. Arterial and venous perfusate samples were taken every minute. Radioactivity was measured by liquid scintillation counting (LS 50, Beckman Instruments, Munich, FRG) after addition of Aqualuma (Baker, Gross-Gerau, FRG). At the end of perfusion, medium supply was discontinued. Brains were dissected out and frozen in isopentane chilled to −50 °C. The brains were sectioned in a cryostat (Reichert-Jung, Mod. 2700, Nussloch, FRG) into 20 μm slices and thaw-mounted on cover slips. Together with [^{14}C]methacrylate standards (Amersham-Buchler, Braunschweig, FRG) the slices were exposed to Osray M3 film (Agfa-Gevaert, Leverkusen, FRG) for 10 days. Optical densities of the autoradiograms were measured with a computer-based densitometer (DT 1505 Parry Instruments, Newbury, UK; Commodore 8032/8050, Frankfurt, FRG) and tissue radioactivity was calculated.

For LCBF determinations, brains were perfused for 30 min, then ischemia was induced for 30 min and LCBF was determined after 5 or 30 min of postischemic recirculation. Then 6 μCi IAP per hemisphere was infused at steadily increasing concentrations over 1 min by an infusion pump and mixed with the perfusion medium by an electromagnetic mixer. Ten timed arterial medium samples were taken for radioactivity determination. Medium supply was then discontinued, and brains were dissected out and prepared for quantitative autoradiography as described above.

Bioluminescent and Enzymatic Methods

Brain ATP and lactate levels as well as perfusate lactate levels were measured fluorimetrically [11]. Characterization of regional tissue ATP and lactate content was done by bioluminescence techniques [10, 12]. Because the images obtained revealed little differences between structures in one brain but high variability between different brains, no regional quantification of the bioluminescence images was performed.

Statistics

Values are presented as means ± SEM in the series concerning brain energy metabolism. All other data are given as means ± SD. Comparisons between controls and treated groups were made by Student's *t*-test.

Results

The tissue/plasma coefficient lambda was calculated to equal 1.48 for the isolated rat brain perfused with a fluorocarbon emulsion.

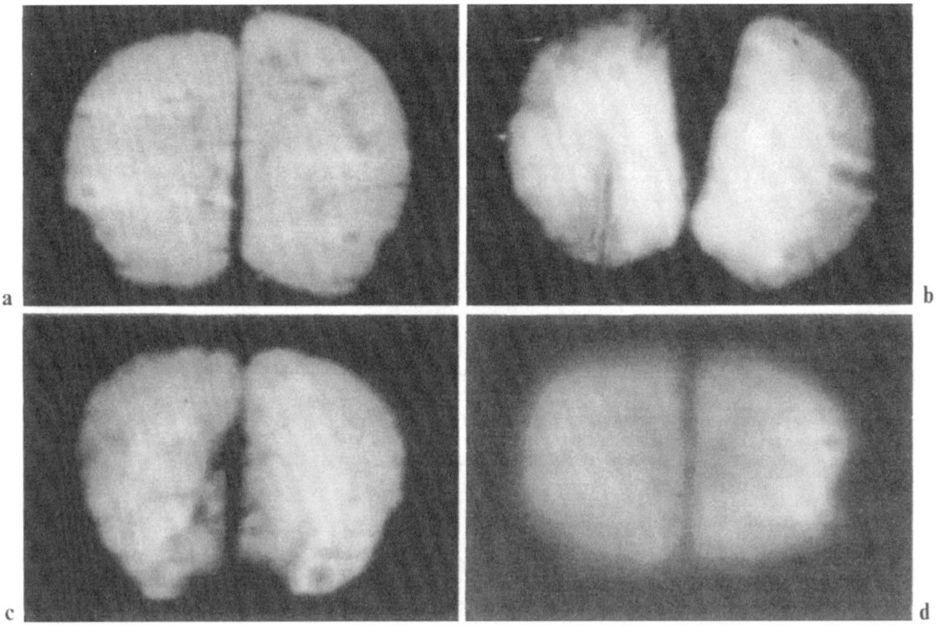

Fig. 1 a–d. Imaging of brain tissue ATP and lactate. The isolated rat brain was perfused at constant flow rate for 30 min, then ischemia was induced for 30 min and brains were reperfused for 5 min. Emopamil was applied to the perfusion medium at a concentration of 1 µmol/l prior to the start of perfusion. **a** ATP image of untreated brain. **b** Lactate image of untreated brain. **c** ATP image of emopamil-treated brain. **d** Lactate image of emopamil-treated brain

The calculated LCBF values varied between 24 and 190 ml/100 g/min for different structures (Table 1). When emopamil was present in the perfusate, LCBF did not differ significantly from controls in any structure or at either time point investigated. Bioluminescence images revealed a rather homogeneous recovery of brain energy metabolism with no pronounced structural differences (Fig. 1). Interanimal differences by far prevailed over interstructural differences within one brain. ATP values in emopamil-treated brains did not differ from those in control brains after 5 min of postischemic reperfusion when brains were per-

Table 1. LCBF (ml/100 g/min) in the isolated perfused rat brain after emopamil administration

Brain structure	5 min reperfusion		30 min reperfusion	
	Control	Emopamil	Control	Emopamil
Gray matter				
Occipital cortex	124±23	148±17	135±64	135±64
Auditory cortex	125±31	158±35	157±55	175±44
Parietal cortex	158±56	164±41	133±53	141±34
Sensorimotor cortex	187±37	190±33	126±20	152±45
Olfactory cortex	144±60	108±24	114±34	124±34
Frontal cortex	131±57	155±33	130±49	141±28
Medial septal nucleus	93±26	76±26	61± 2	71±11
Lateral thalamus	123±52	166±62	91±37	65±21
Ventral thalamus	141±37	155±69	87± 7	80±45
Medial geniculate	124±37	159±47	95±31	76±42
Lateral geniculate	96±30	126±62	71±23	57±21
Hypothalamus	44±10	66±33	60±55	79±27
Substantia nigra	49±15	61±18	55±31	49±35
Caudate nucleus	123±15	109±22	121±66	133±29
Red nucleus	69±14	97±40	76±44	57±18
Nucleus accumbens	114±52	97±33	100±57	144±47
Amygdala	99±23	113±35	60±19	66±21
Hippocampus	88±28	126±43	88±35	77±42
Dentate gyrus	85±29	114±26	60±24	50±25
Superior colliculus	84±38	125±59	78±41	53±23
Inferior colliculus	126±65	166±54	116±41	84±16
Superior olive	73±14	99±43	69±35	76±41
Inferior olive	34±16	33±13	34±13	35±20
Cerebellar cortex	81±30	95±24	102±41	80±27
Vermis	88±26	128±43	85±36	82±11
White matter				
Genu of corpus callosum	41±12	43± 9	33±16	46±11
Corpus callosum	34±11	37± 6	43±27	56±12
Internal capsule	44±19	48±10	47±12	42±17
Cerebellar white matter	24± 8	38±15	27±10	33±10

Values are means ± SD of 4–5 experiments in ml/100 g/min. Brains were perfused at a constant flow rate of 5 ml/min for 30 min; ischemia was induced for 30 min and then brains were reperfused for 5 or 30 min. Emopamil was added to the perfusate prior to the start of perfusion at a concentration of 1 μmol/l.

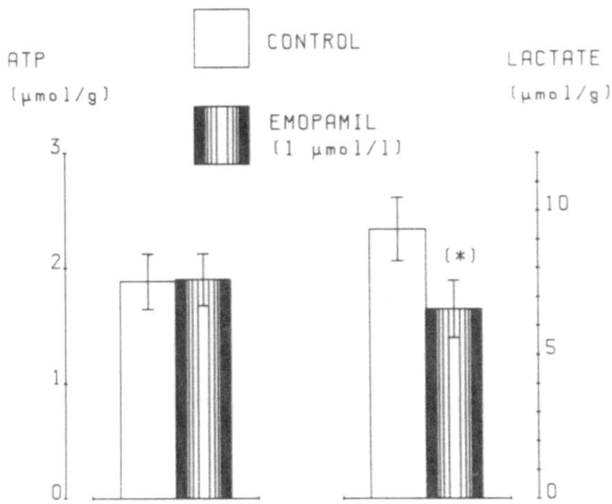

Fig. 2. Brain ATP and lactate after postischemic recirculation. The isolated rat brain was perfused at constant flow rate for 30 min, then ischemia was induced for 30 min and brains were reperfused for 5 min. Emopamil was applied to the perfusion medium at a concentration of 1 µmol/l prior to the start of perfusion. Values are given as means ± SEM of four or five experiments expressed as µmol/g wet weight. * $P<0.1$ (Student's *t*-test)

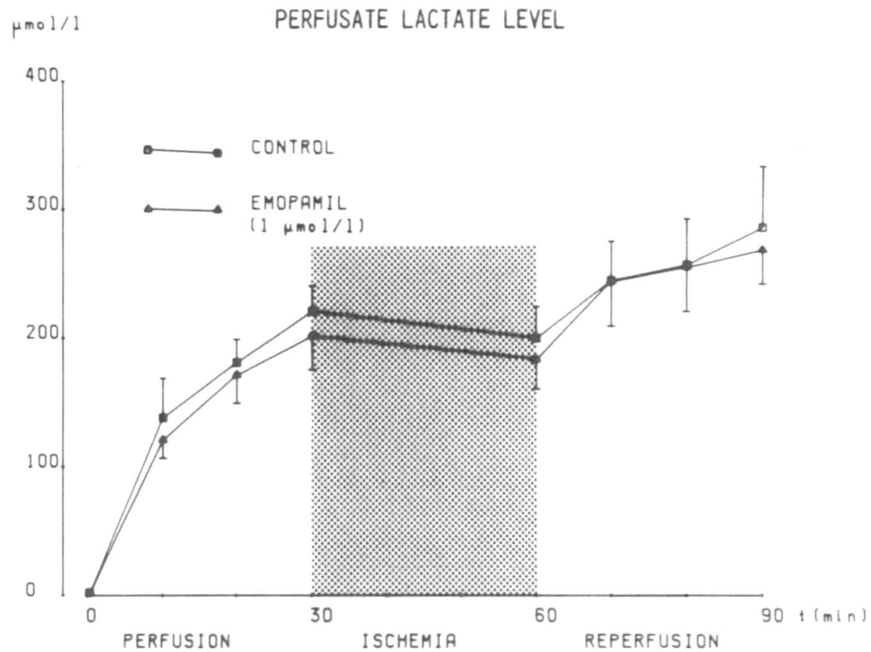

Fig. 3. Brains were perfused at a constant flow rate of 5 ml/min. After 30 min of perfusion and 30 min of complete ischemia, isolated brains were reperfused for 30 min. Values are presented as means ± SD from five experiments

fused at a constant global flow rate (Fig. 2). Lactate levels, however, showed a tendency towards lower values when emopamil was present in the perfusion medium.

Throughout the entire experimental procedure lactate levels in the perfusion medium did not differ from controls in the drug-treated group (Fig. 3).

Discussion

The present work constitutes an extension to and an affirmation of earlier results. Several calcium antagonists have been shown to influence postischemic energy metabolism in different models of cerebral ischemia [2, 3, 5, 7, 9, 15, 18, 19]. In a recent report we demonstrated emopamil to restore more quickly high energy phosphates after global ischemia in the isolated rat brain perfused at constant pressure [3]. This effect was accompanied by an increase in perfusion rate. When brains were perfused at constant flow rate the improvement in high energy phosphate restitution was no longer present. From these results the conclusion was drawn that the faster restoration of energy-rich phosphates was accomplished by the vasodilating properties of emopamil. Since substrate levels were assessed in cortical tissue, the main objection to this conclusion centered on the fact that a global parameter like perfusion rate was compared to locally performed biochemical determinations. The question was posed whether in brains perfused at constant flow rate intracerebral redistribution of the perfusate might have been produced by the calcium entry blocker. Therefore, in the present study we investigated the effect of emopamil on LCBF in the isolated rat brain perfused at constant global flow rate. Our finding of unchanged LCBF in all structures demonstrates that the calcium antagonist does not produce intracerebral redistribution phenomena and thereby affirms our earlier conclusion [3].

In the above cited work emopamil was shown to reduce more quickly cortical lactate levels by mechanisms unrelated to global flow rate. This effect was observed after 30 min of postischemic reperfusion. The fact that LCBF values reported herein are not influenced by emopamil at either 5 or 30 min recirculation lends support to our previous conclusion that the faster degradation of lactate represents a direct action of the calcium entry blocker on cerebral parenchyma. This view is confirmed by the finding that lactate levels in the perfusion medium are unaltered by emopamil. Our results demonstrate a tendency towards lower lactate levels in the emopamil-treated group as early as 5 min after the release of ischemia. This effect is in accordance with earlier results showing accelerated postischemic lactate degradation by calcium antagonists [3, 5, 18]. Since there is no enhanced loss of lactate into the perfusate in the emopamil group we conclude that the faster lactate disappearance caused by emopamil in cerebral tissue is due to accelerated oxidative degradation of this substrate.

The crucial role of lactacidosis as a cause of cell death is well documented [17]. The faster reduction of lactate should improve pH homeostasis during recovery. In a recent study such correction of pH could be demonstrated with nimodipine and verapamil in a model of focal ischemia [6]. The fact that pH changes were not associated with increases in LCBF supports the view of a direct action on brain parenchyma.

Emopamil has been shown to reduce the number of necrotic neurons in the hippocampal CA1 sector after 7 days' recovery in a model of reversible forebrain ischemia in the rat [4]. This effect was not accompanied by blood flow changes during early recirculation of the period of delayed hypoperfusion and hence was ascribed to a direct action of the calcium antagonist on cerebral parenchyma. Our finding of a directly mediated improvement of oxidative metabolism by emopamil, as evidenced by the enhanced lactate degradation, supports this concept.

The bioluminescence study adds some interesting aspects to the discussion. First, the homogeneous restoration of brain energy metabolism following the release of ischemia is in line with similar observations in another model of global ischemia [13]. Furthermore, the results indicate that the improvements in energy metabolism elicited by emopamil obviously are not restricted to cortical tissue but seem to cover total brain tissue in this model of complete global ischemia.

Summarizing, it is concluded that constant volume perfusion of the isolated rat brain seems to be adequate to differentiate between vascular and direct effects of pharmacologically active compounds on cerebral parenchyma. The ameliorative action of emopamil on the postischemic restoration of high energy phosphates is due to its vascular action, while the faster reduction of lactate is accomplished by a direct action of the calcium entry blocker on cerebral parenchyma. This latter effect seems to be due to enhanced oxidative degradation of lactate.

References

1. Andjus R, Suhara K, Sloviter H (1967) An isolated, perfused rat brain preparation, its spontaneous and stimulated activity. J Appl Physiol 22:1033–1039
2. Bielenberg GW, Beck T, Sauer D, Burniol M, Krieglstein J (1987) Effects of cerebro-protective agents on cerebral blood flow and postischemic energy metabolism in the rat brain. J Cereb Blood Flow Metab 7:480–488
3. Bielenberg GW, Haubruck H, Krieglstein J (1987) Effects of calcium entry blocker emopamil on postischemic energy metabolism of the isolated perfused rat brain. J Cereb Blood Flow Metab 7:489–496
4. Bielenberg GW, Sauer D, Nuglisch J, Beck T, Roßberg C, Mennel H-D, Krieglstein J (1989) Effects of emopamil on postischemic blood flow and cell damage in the rat brain. Naunyn-Schmiedeberg's Arch Pharmacol 339:230–235
5. Bielenberg GW, Stierstorfer HJ, Weber J, Krieglstein J (1989) Nimodipine reduces postischemic lactate levels in the isolated perfused rat brain. Biochemical Pharmacology 38:855–858
6. Hakim AM (1986) Cerebral acidosis in focal ischemia: II. Nimodipine and verapamil normalize cerebral pH following middle cerebral artery occlusion in the rat. J Cereb Blood Flow Metab 6:676–683
7. Heffez DS, Passonneau JV (1985) Effect of nimodipine on cerebral metabolism during ischemia and recirculation in the mongolian gerbil. J Cereb Blood Flow Metab 5:523–528
8. Krieglstein G, Krieglstein J, Urban W (1972) Long survival time of an isolated perfused rat brain. J Neurochem 19:885–886
9. Krieglstein J, Weber J (1986) Calcium entry blockers protect brain energy metabolism against ischemic damage. In: Longmuir IS (ed) Oxygen transport to tissue VIII. Plenum, New York, pp 243–251
10. Kogure K, Alonso FO (1978) A pictorial representation of endogenous brain ATP by a bioluminescent method. Brain Res 154:273–284
11. Lowry OH, Passonneau JV (1972) A flexible system of enzymatic analysis. Academic, New York
12. Paschen W (1985) Regional quantitative determination of lactate in brain sections. A bioluminescent approach. J Cereb Blood Flow Metab 5:609–612
13. Paschen W, Hossmann KA, van den Kerckhoff W (1983) Regional assessment of energy-producing metabolism following prolonged complete ischemia of cat brain. J Cereb Blood Flow Metab 3:321–329

14. Ponten U, Ratcheson RA, Salford LG, Siesjö BK (1973) Optimal freezing conditions for cerebral metabolites in rats. J Neurochem 21:1127–1138
15. Proctor HJ, Palladino GW, Fillipo D, Sanders R (1984) Cerebral high energy phosphates after hypoxic-hypotension: effect of verapamil and nifedipine. Circ Shock 14:107–112
16. Sakurada O, Kennedy C, Jehle J, Brown JD, Carbin GL, Sokoloff L (1978) Measurement of local cerebral blood flow with iodo-^{14}C-antipyrine. Am J Physiol 234:H59
17. Siesjö BK (1986) Cell damage in the brain caused by ischemia: an overview. In: Kriegelstein J (ed) Pharmacology of cerebral ischemia. Elsevier, Amsterdam, pp 3–11
18. Weber J, Krieglstein J (1985) Effect of gallopamil on energy metabolism of the isolated perfused rat brain in the postischemic period. Naunyn Schmiedebergs Arch Pharmacol 329:451–454
19. Weber J, Bielenberg GW, Krieglstein J (1988) Effects of phenylalkylamine calcium entry blockers on postischemic energy metabolism in the isolated perfused rat brain: stereoselective action of emopamil. Pharmacology 37:38–49

Cytoprotective Characteristics of Dihydropyridine Calcium Antagonists in a Rat Model of Stroke: Implications for Clinical Trials

A. SAUTER, M. RUDIN, and K.-H. WIEDERHOLD

Introduction

There is good evidence to suggest that calcium plays a key role in the pathophysiology of ischemia [11, 25, 26]. Hence, calcium antagonists have been proposed as candidates for an effective therapy for different types of ischemia, including stroke [7, 12, 17]. Despite their systemic hypotensive effects, calcium antagonists might act in stroke by improving cerebral blood flow (CBF) [15], in particular in regions at risk [14]. An additional, potentially important site of action might be neuronal, since the calcium antagonists (1,4-dihydropyridines), which have been shown to be most efficacious in many animal models of stroke [1, 6, 13, 14, 22, 24, 27] as well as in clinical trials [5], readily cross the blood-brain barrier [28].

Due to the complexity and difficulty of performing clinical trials in stroke, much emphasis is put on the relevance and comprehensiveness of the preclinical data of a drug candidate. Occlusion of an intracranial artery, such as the middle cerebral artery (MCA), in a laboratory animal leads to a pathophysiology similar to that described for embolic stroke in man. Since the cerebral anatomy of man and that of the rat are comparable [30], we have chosen the rat for unilateral permanent MCA occlusion (MCAO), essentially following the methodology described in reference 29. The anatomical location and size of infarcts in these animals have been determined by magnetic resonance imaging (MRI), a valuable diagnostic method used clinically for the study of CNS diseases [2], including stroke [3]. Since MRI is noninvasive it allowed us to follow the evolution of infarcts over days and weeks in the same rat and to correlate these findings with the neurological deficits present [22].

Using this animal model we have shown that dihydropyridine calcium antagonists, such as PN 200-110 [10] and nimodipine, when given before or immediately after MCAO, reduce the infarct size, measured by MRI [19, 22]. These effects were paralleled by improvements in a simple neurological score (posture, movement, paralysis of hindpaws) and several biochemical markers of ischemia, such as brain wet weight and levels of potassium, sodium, dopamine, and norepinephrine. In a subsequent study [24] it was shown that the infarct size, as observed by MRI, corresponded to severe neuronal necrosis revealed postmortem in matching cryostatic sections stained with cresyl violet. From these experiments it can be concluded that dihydropyridine calcium antagonists have cytoprotective effects which are presumed to be permanent.

Preclinical Research, Sandoz Ltd., CH-4002 Basel, Switzerland.

Cerebral Ischemia and Calcium
Ed. by A. Hartmann and W. Kuschinsky

To ensure successful clinical development of a drug in the indication stroke, a number of additional questions need to be addressed, such as duration of cytoprotection, time frame allowed for an effective therapeutic intervention following onset of stroke, comparison with other dihydropyridine calcium antagonists in terms of potency and efficacy, and finally elucidation of the relevant mechanisms involved.

Materials and Methods

Animals

Adult male SHR/KYO rats (250–300 g) were used. The surgical procedure for permanent occlusion of the left MCA is described in detail in references 22 and 29.

Drugs

PN 200-110 [10], PY 108-068 [8], and nimodipine were dissolved in polyethylene glycol 400+ethanol (1+1) and injected subcutaneously, at different doses, in 1 ml/kg solvent. Control animals were injected with solvent alone (1 ml/kg).

Magnetic Resonance Imaging

A Bruker Biospec 47/15 spectrometer with imaging facility was used. Coronal sections 3, 6, 9, and 12 mm anterior to the intraaural line were recorded in the multislice spin-echo acquisition mode (for details see refs. 4 and 22). The infarct area in each section was determined by pixel counting, the sum from the four sections being taken as an estimate of total infarct size.

Histology

Following decapitation the brains were immediately frozen on a block of dry ice. Coronal cryostatic sections of 20 μm thickness were cut at different stereotactic levels [18], mounted on glass slides, and stained with cresyl violet. The areas of infarcts were determined with a calibrated digitizing tablet of a video image analyzer.

Cerebral Blood Flow Determination

Cerebral blood flow was determined 3 h after MCAO, when the animals had recovered from halothane anesthesia used for surgery, by the ^{14}C-iodoantipyrine method [21]. The brain (without brain stem and cerebellum) was cut in eight 2 mm thick coronal slices, which were divided into left and right and subjected to liquid scintillation counting following dissolution and equilibration.

Results and Discussion

Determination of Long-Term Drug Effects by Histology

A key question to be answered for a drug candidate in stroke, such as PN 200-110, is whether the cytoprotective effects of a single subcutaneous injection, observed by MRI up to 3 days following MCAO [22], are transient or whether they will still be present weeks after MCAO. This is an important aspect, because a drug that would simply delay the otherwise inevitable damage would obviously be of minor interest in stroke therapy.

Rats were injected subcutaneously, immediately after MCAO, either with solvent (1 ml/kg) or with PN 200-110 (2.5 mg/kg). Since edema tends to disappear within the first few days, it becomes more difficult with time to delineate exactly the infarcts by MRI. However, histologically the infarcts remain well defined. Therefore, the infarct sizes were determined histologically after 1, 7, and 14 days in coronal sections at three anatomically defined levels [18]. As shown in Table 1, the extent of infarct size reduction elicited by a single injection of PN 200-110 essentially remained constant between days 1 and 14 following MCAO.

We conclude that a single application of PN 200-110 after MCAO has cytoprotective effects which appear to be permanent. Since no further protection can be obtained with subsequent injections of PN 200-110 (unpublished observations), the decisive actions leading to permanent cytoprotection are thought to take place within the first 24 h following MCAO.

Table 1. Evolution of infarct sizes over 2 weeks in controls and PN 200–110 treated rats

Level	Day 1		Day 7		Day 14	
	Control	PN	Control	PN	Control	PN
A4	36±9	3±1*	36±5	16±5*	34±4	7±1*
A6	57±6	26±5*	49±2	34±4*	45±4	13±6*
A8	63±3	41±5*	56±1	42±7	42±2	13±5*
Total infarct size	156	69	141	93	120	33
Drug effect in percent	0%	−66%	0%	−34%	0%	−72%

The infarct size was determined histologically at three different levels (A4, A6, A8) on days 1, 7, and 14 following MCAO in vehicle-injected controls and in rats injected with PN 200–110 (2.5 mg/kg, s.c. injection immediately after MCAO). Values shown are means ± SEM, $n=3$, of infarct area in % of total hemispheric area. * Denotes statistical significance ($P<0.05$, Mann-Whitney U-test).

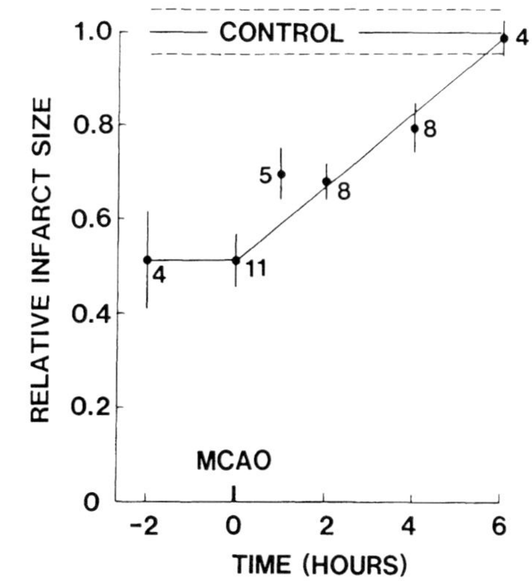

Fig. 1. Effect as function of drug application time. PN 200-110 (2.5 mg/kg) was injected subcutaneously, once, at different times before and after MCAO, as indicated. The infarct size (*I*) was determined from four coronal MR images, taken 24 h after MCAO. Relative infarct size = I(drug)/I(control). Means $\pm$ SEM, number of rats per point as indicated. The absolute infarct size in control rats was 14300 $\pm$ 600 voxels ($n = 12$)

Time Window for Drug Application

The time elapsed between the onset of stroke and drug administration is certainly a critical parameter in stroke trials and is of relevance in the therapeutic success of a drug. We therefore determined the relationship between cytoprotective effect and time point of drug administration with respect to MCAO.

As shown in Fig. 1, PN 200-110 was equally effective in reducing the infarct size when injected either 2 h before or immediately after MCAO. If the drug was administered after MCAO, a linear decrease in efficacy with time was observed. To ensure that the observed effects were statistically significant it was necessary to inject the calcium antagonist not later than 4 h after MCAO. No effects were seen with injections after 6 h.

Dose-Response Curves

Quantitative comparisons of different calcium antagonists require the use of dose-response curves. We compared three structurally similar 1,4-dihydropyridine calcium antagonists in the rat MCAO model.

The infarct size, measured by MRI 24 h after MCAO, was dose-dependently reduced by all three calcium antagonists, as shown in Fig. 2. Differences not only in potency, expected from other in vivo and in vitro data [8–10, 16, 28], but surprisingly also in efficacy were found: PN 200-110 maximally reduced the infarct size at 2.5 mg/kg by approximately 60%, while the maximal effects obtained with higher doses of nimodipine and PY 108-068 were smaller (approximately 40% reduction at 5 mg/kg nimodipine and 25% at 10 mg/kg PY 108-068). A common feature of these calcium antagonists seems to be a reduction of their effects at higher doses, which is probably due to their well known hypotensive properties.

These surprising pharmacological discrepancies might in fact be due to differences in pharmacokinetics (i.e., onset and duration of action), especially since the

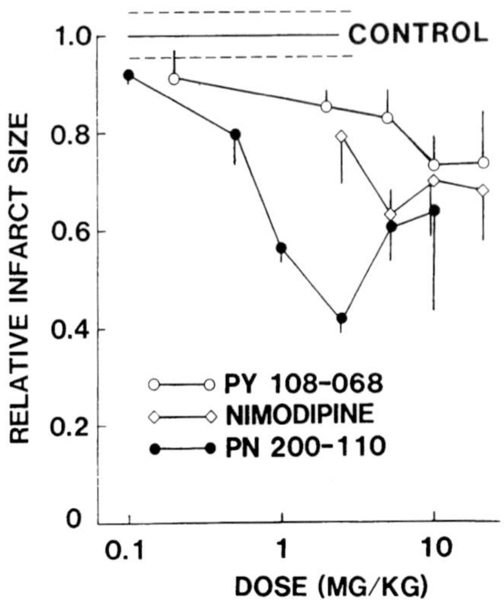

Fig. 2. Dose-response curves of three calcium antagonists. Drugs were injected subcutaneously, once, at different doses immediately after MCAO, as indicated. The infarct size (*I*) was determined from four coronal MR images recorded 24 h after MCAO. Relative infarct size = I(drug)/I(control). Means ± SEM; number of animals per dose, six to ten. The absolute infarct size in control rats was 14300 ± 600 voxels ($n = 12$)

drugs were injected only once, immediately after MCAO, and the effects measured after 24 h.

Cerebral Blood Flow Measurements

Cerebral blood flow was measured in MCA-occluded controls and rats treated with the three calcium antagonists for the following two reasons: first to investigate further a likely mechanism of importance for cytoprotection, and second to determine the duration of action in the CNS of the three calcium antagonists after a single subcutaneous injection, which might explain the observed differences in efficacy.

Cerebral blood flow was reduced 3 h after MCAO (see Fig. 3) in the ipsilateral hemisphere of solvent-injected control rats by at least 20% (in the most rostral slice), and by up to 65% in the most affected slice. Compared to sham-operated animals, the CBF was not changed in the contralateral hemisphere.

As shown in Fig. 3, CBF was increased by the three calcium antagonists tested in both hemispheres with different potency. The relative increase, compared to control values, was larger in the ischemic hemisphere, suggesting that the generally observed effect on CBF did not lead to a decrease in the critically underperfused brain areas (steal effect). As expected, a reversal was noticed at high doses, probably as a result of systemic hypotension.

It is interesting to note that PN 200-110 showed pronounced increases in CBF over a wide dose range (0.1–2.5 mg/kg) while the ranges obtained with nimodipine and PY 108-068 were much narrower (see Fig. 3). Therefore, it might be more difficult with nimodipine and PY 108-068, following one single subcutaneous injection, to reach and maintain for an extended time period, i.e., during the 24 h following MCAO, the drug levels necessary for maximal increases in CBF. If the observed CBF effects are relevant for cytoprotection, the difference in efficacy observed between the three calcium antagonists might be due to inad-

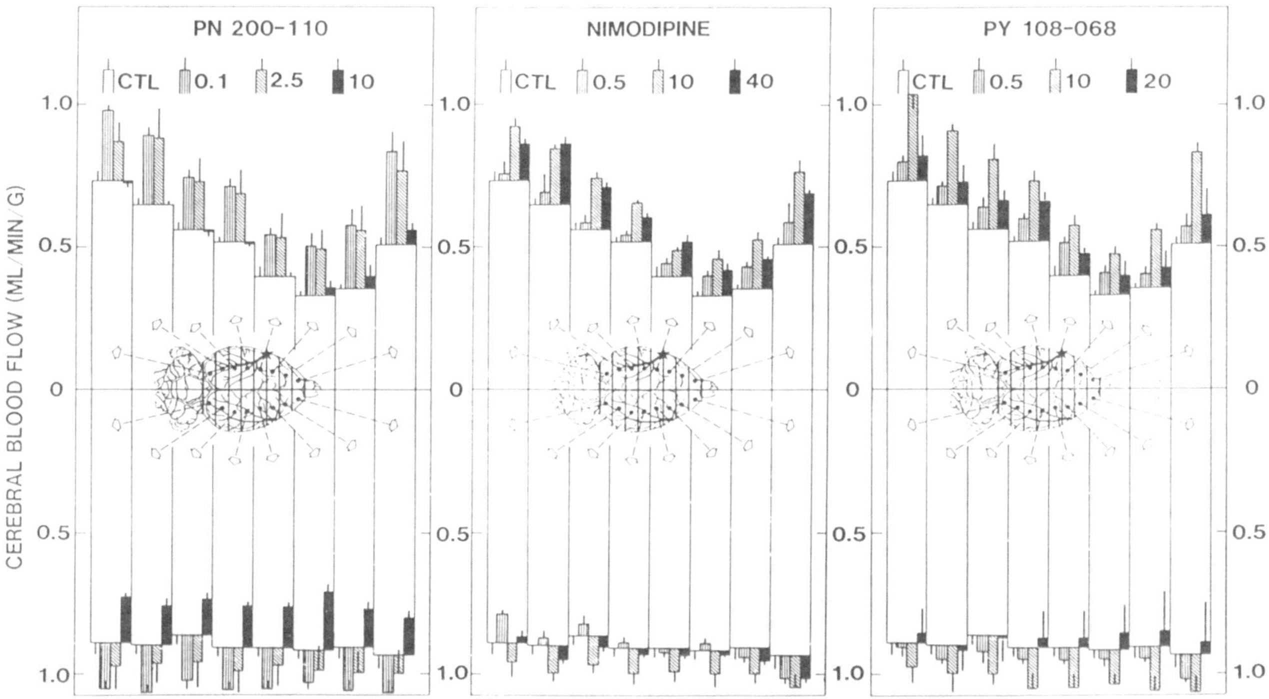

Fig. 3. Effects of three calcium antagonists on CBF. CBF was determined 3 h after left-sided MCAO by the ^{14}C-iodoantipyrine method in eight 2 mm thick slices in the left and right hemispheres, as indicated. Drugs at different doses (in mg/kg, as indicated) or solvent (=controls, *CTL*) were injected subcutaneously, immediately after MCAO. Means ± SEM; number of rats per dose, two to five

equate drug levels being reached and maintained with a single injection. We therefore determined the duration of CBF improvements following single subcutaneous injections.

Duration of Action

Groups of rats were injected subcutaneously with solvent (1 ml/kg, =controls), PN 200-110 (2.5 mg/kg), PY 108-068 (10 mg/kg), or nimodipine (10 mg/kg). CBF was measured 3, 6, and 12 h after injections and 3 h after MCAO, as described. With nimodipine, similar CBF increases in the ischemic hemisphere were obtained at 3, 6, and 12 h, as with PN 200-110 (see Fig. 4). In contrast, PY 108-068 showed a shorter duration: while the 3-h value was similar to that with PN 200-110, the 6-h value was smaller, and after 12 h, values had returned close to those observed for controls.

It is concluded that the duration of CBF increase after a single subcutaneous injection with PY 108-068 is shorter than with either PN 200-110 or nimodipine.

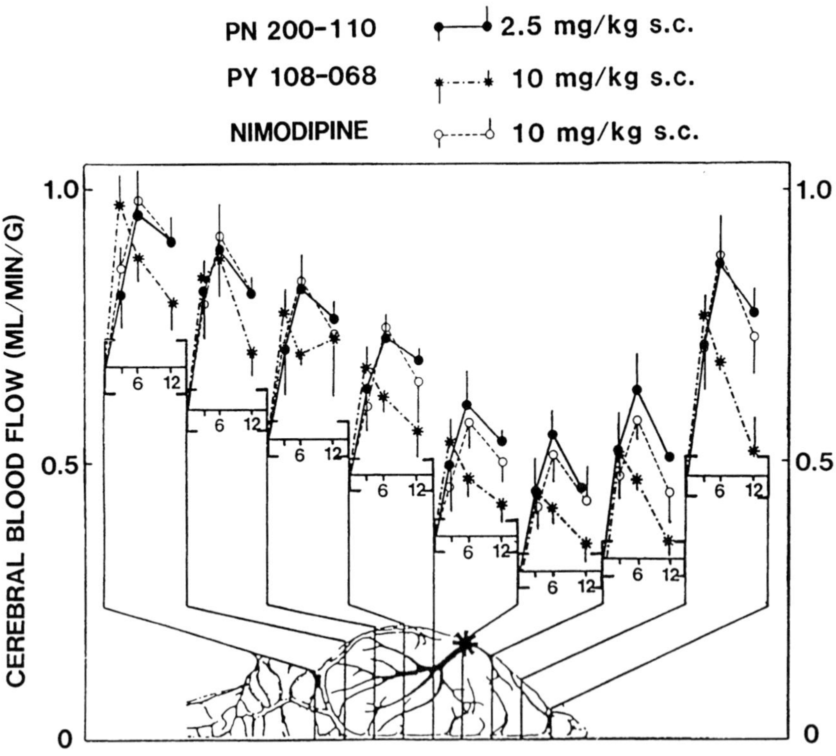

Fig. 4. Duration of action of three calcium antagonists. CBF was determined 3 h after left-sided MCAO by the ^{14}C-iodoantipyrine method in eight 2 mm thick slices in the left hemisphere, as indicated. Drugs at the doses indicated were injected 0, 3, and 9 h before MCAO, i.e., 3, 6, and 12 h before CBF determination. Controls were injected with solvent immediately after MCAO. Means ± SEM; number of rats per time point, three to five

Table 2. Comparison of single and repeated injections of PY 108–068

	No.	24 h	72 h
Controls	6	17200 ± 650	16800 ± 430
Single	6	13000 ± 1400	13200 ± 1300
Repeated	6	13200 ± 1100	13200 ± 1000

Rats were injected subcutaneously immediately after MCAO with solvent (1 ml/kg) (= controls) or PY 108–068 (5 mg/kg) once (= single) or repeatedly (= repeated), i.e., every 4 h 5 mg/kg, total number of injections = 5. The infarct sizes were determined by MRI 24 and 72 h after MCAO. Results are in voxels, means ± SEM.

The lower efficacy found with PY 108-068 in reducing the infarct size, measured by MRI 24 h following a single subcutaneous injection (see Fig. 2), might be due to a shorter duration of action.

We therefore compared the effects of a single subcutaneous injection (5 mg/kg) of PY 108-068 with those of repeated subcutaneous injections (5 × 5 mg/kg every 4 h). As shown in Table 2, the same effect, i.e., a 20% reduction of infarct size, was obtained using either mode of application, suggesting that the differences in efficacy found between PY 108-068 and PN 200-110 are not due to the shorter duration of CBF improvement observed with PY 108-068. This is also consistent with the findings obtained with nimodipine, since it differs from PN 200-110 only with respect to cytoprotective efficacy and not duration of CBF improvement.

Conclusions

PN 200-110, at a single dose of 2.5 mg/kg injected subcutaneously after permanent occlusion of the left MCA, reduces by approximately 60% the acute and chronic pathophysiological consequences of cerebral ischemia in rats. Since the extent of infarction assessed 24 h after MCAO, remains essentially unchanged for 2 weeks, the cytoprotective effects of PN 200-110 are presumed to be permanent.

The extent of cytoprotection afforded by PN 200-110 depends on the time point of application following MCAO. In order to be effective, the drug has to be present at the onset of stroke or administered shortly thereafter. When injected 6 h after MCAO, no protection is obtained. Whether such a time frame would be identical in clinical trials is questionable, since the rapid onset and irreversibility of MCAO in the rat model is different from the clinically observed "stroke in progress." In any case, the time of drug administration will be a decisive parameter in clinical stoke trials.

In the rat, PN 200-110 is more potent and surprisingly more efficacious in reducing the infarct size following MCAO than both nimodipine and PY 108-068, since its maximal effect is 2–3 times greater and its half-maximal effect attained at a 2–4 times lower dose. These apparently pharmacological discrepancies cannot be explained by different pharmacokinetic properties of the compounds, but rather by differing pharmacological profiles. This is an unexpected finding considering the structural similarities of the three 1,4-dihydropyridine calcium antagonists.

This then leads to the question of the putative mode of action relevant for cytoprotection. Calcium antagonists have profound hemodynamic effects not only in the CNS but also in the periphery, which is likely to cause more damage in conditions of cerebral ischemia, e.g., due to hypotension or to the well known "steal effect." Our results suggest that PN 200-110 and, to a lesser extent, PY 108-068 and nimodipine dose dependently increase CBF preferentially in the ischemic parts of the brain. Even at high doses resulting in severe systemic hypotension, these calcium antagonists do not reduce CBF below the values of control animals in the lesioned hemisphere.

We have recently obtained evidence from animal experiments, using magnetic resonance spectroscopy, which suggests that calcium antagonists, in particular PN 200-110, reduce the consumption of high energy phosphates in the CNS at doses similar to those required for cytoprotection [20, 23]. The importance of these possible intraneuronal sites of action for cytoprotection in relation to the cerebrovascular sites has to be further elucidated.

References

1. Barnett GH, Bose B, Little JR et al. (1986) Effects of nimodipine on focal cerebral ischemia. Stroke 17:884–890
2. Brant-Zawadski M (1983) Nuclear magnetic resonance imaging: the abnormal brain and spinal cord. In: Newton TH, Potts DA (eds) Modern neuroradiology, vol 2. Advanced imaging techniques. Clavadel, San Anselmo, pp 159–185
3. Bryan RN, Willcott MR, Schneiders NJ et al. (1983) Nuclear magnetic resonance evaluation of stroke. Radiology 149:189–192
4. Crook LE, Ortendahl DA, Kaufmann L et al. (1983) Clinical efficacy of nuclear magnetic resonance imaging. Radiology 146:123–128
5. Gelmers HJ, Gorter K, Weerdt CJ et al. (1988) A controlled trial of nimodipine in acute ischemic stroke. N Engl J Med 318:203–207
6. Germano IM, Bartkowsky HM, Cassel ME et al. (1987) The therapeutic value of nimodipine in experimental focal ischemia. J Neurosurg 67:81–87
7. Gisvold SE, Steen PA (1985) Drug therapy in brain ischemia. Br J Anaesth 57:82–95
8. Hof RP, Hof A, Neumann P (1982) Effects of PY 108-068, a new calcium antagonist, on general hemodynamics and regional blood flow in anesthetized cats: a comparison with nifedipine. J Cardiovasc Pharmacol 4:352–362
9. Hof RP, Hof A, Scholtysik G et al. (1984) Effects of the new calcium antagonist PN 200-110 on the myocardium and the regional peripheral circulation in anesthetized cats and dogs. J Cardiovasc Pharmacol 6:407–416
10. Hof RP, Scholtysik G, Loutzenhiser R et al. (1984) PN 200-110, a new calcium antagonist: electro-physiological, inotropic and chronotropic effects on guinea pig myocardial tissue and effects on contraction and calcium uptake of rabbit aorta. J Cardiovasc Pharmacol 6:399–406
11. Hossmann K-A (1982) Treatment of experimental ischemia. J Cereb Blood Flow Metab 2:275–297

12. Hossmann V, Goetz J, Schroer K (1985) Kalzium Antagonisten und zerebrale Erkrankungen, eine Standortbestimmung. Springer, Berlin Heidelberg New York Tokyo
13. Milde LN, Milde JH, Michenfelder JD (1986) Delayed treatment with nimodipine improves cerebral blood flow after complete cerebral ischemia in the dog. J Cereb Blood Flow Metab 6:332–337
14. Mohamed AA, Gotoh O, Graham DI et al. (1985) Effect of pretreatment with calcium antagonist nimodipine on local cerebral blood flow and histopathology after middle cerebral artery occlusion. Ann Neurol 18:705–711
15. Mohamed AA, Mendelow GM, Teasdale GM et al. (1985) Effect of the calcium antagonist nimodipine on local cerebral blood flow and metabolic coupling. J Cereb Blood Flow Metab 5:26–33
16. Müller-Schweinitzer E, Neumann P (1983) In vitro effects of calcium antagonists PN 200-110, nifedipine and nimodipine on human and canine cerebral arteries. J Cereb Blood Flow Metab 3:354–361
17. Parrat JR (ed) (1985) Control and manipulation of calcium movement: a biological council symposium. Raven, New York
18. Paxinos G, Watson C (1982) The rat brain in stereotactic coordinates. Academic, Sydney
19. Rudin M, Sauter A, Wiederhold K-H (1987) Application of NMR imaging and spectroscopy to the pharmacology of cerebrovascular disorders. Therapie 42:477–481
20. Rudin M, Sauter A (1988) ^{31}P NMR magnetization transfer in the rat brain: effects of barbiturates and calcium antagonists. Experientia 44:A38
21. Sakurada O, Kennedy C, Jehle et al. (1978) Measurement of local cerebral blood flow with iodo(^{14}C)-antipyrine. Am J Physiol 234:H59–H66
22. Sauter A, Rudin M (1986) Calcium antagonists reduce the extent of infarction in rat middle cerebral artery occlusion model as determined by quantitative magnetic resonance imaging. Stroke 17:1228–1234
23. Sauter A, Rudin M (1987) Effects of calcium antagonists on high energy phosphates measured by ^{31}P NMR spectroscopy. Magn Reson Med 4:1–8
24. Sauter A, Rudin M, Wiederhold K-H (1988) Reduction of neural damage in irreversible cerebral ischemia by calcium antagonists. Neurochem Pathol 9:211–236
25. Siesjö BK (1981) Cell damage in the brain: a speculative synthesis. J Cereb Blood Flow Metab 1:155–185
26. Siesjö BK, Wieloch T (1985) Cerebral metabolism in ischemia: neurochemical basis for therapy. Br J Anaesth 57:47–62
27. Steen PA, Newberg LA, Milde JM et al. (1984) Cerebral blood flow and neurologic outcome when nimodipine is given after complete cerebral ischemia in the dog. J Cereb Blood Flow Metab 4:82–87
28. Supavilai P, Karobath M (1984) The interaction of [H^3]PY 108-068 and of [H^3]PN 200-110 with calcium channel binding sites in rat brain. J Neuron Transm 60:149–167
29. Tamura A, Graham DI, McCulloch J et al. (1981) Focal cerebral ischemia in the rat: 1. Description of technique and early neuropathological consequences following middle cerebral artery occlusion. J Cereb Blood Flow Metab 1:53–60
30. Yamori Y, Horie R, Handa H et al. (1976) Pathogenetic similarity of strokes in stroke-prone spontaneously hypertensive rats and humans. Stroke 7:46–53

Effect of Diltiazem on the Cortical Microcirculation in Acutely Produced Cerebral Ischemia in Cats

M. Tomita, F. Gotoh, N. Tanahashi, M. Kobari, T. Shinohara
Y. Terayama, B. Mihara and K. Ohta

In the literature, the reported changes in cerebral blood flow (CBF) in the ischemic tissue due to calcium (Ca) entry blocker have been diverse and inconclusive. The conflicting results appear to stem from several practical points, as follows. First, the ischemic models produced by investigators have been different: global ischemia due to cardiac arrest or aortic ligation [7, 10], or focal ischemia due to cerebral arterial occlusion [2, 5]. Second, the timing of the administration has been different: preloading [5] or postischemic loading in the acute phase [2, 7, 10] or in the chronic phase of clinical cases [3, 14]. Third, the control flow per se to be compared with the flow affected by the Ca entry blocker has been variable because of difficulties in producing a constant level of cerebral ischemia. Finally, no good methods have been available for measuring CBF repeatedly.

The purpose of the present study was to examine the effect of the Ca entry blocker, diltiazem, on the cortical microcirculation in cerebral ischemia produced acutely by occlusion of the middle cerebral artery in cats. The time-to-time hemodynamic changes were evaluated by the photoelectric method which was developed by us recently [11].

Methods

Our photoelectric method [11] was employed in nine cats to measure the cerebral blood volume (CBV), CBF, and microcirculatory parameters. Each cat was anesthetized with 50 mg/kg body weight of α-chloralose and 500 mg/kg urethane, and ventilated artificially after immobilization with alcuronium chloride. The left lingual artery was catheterized for intracarotid injection of a small amount of biological carbon black. The practical procedure for applying the photoelectric apparatus to the cortex was as described elsewhere [11]. Briefly, a small lamp (0.7 mm in outer diameter) was inserted into the marginal area of the ectosylvian gyrus from a sidewards direction (from the parieto-occipital area), and a photosensor (silicon photodiode with a sensitive area of 3 mm × 5 mm) was fixed on the skull bone surface excavated to the diploë. The light source and photodetector thus constituted a transillumination system of the cortical layer in situ. With a green color filter ($\lambda = 548$ nm), changes in the optical density of the cortical tissue at the isosbestic point of hemoglobin were detected. CBV was evaluated from the

Department of Neurology, Keio University, Shinjuku-ku, Tokyo 160, Japan.

Cerebral Ischemia and Calcium
Ed. by A. Hartmann and W. Kuschinsky

relative changes in the record assuming the control CBV to be 6.3 vol.-% [11]. Following injection of 0.2 ml of 1/40 diluted biological carbon black solution (Pelikan Werke, FRG) into the left carotid artery, the tissue carbon black dilution curve was recorded with the photoelectric apparatus, which also served as a densitometer head in situ. The mean transit time of blood ($\bar{t}$) through the cortical tissue was calculated from the tissue carbon black dilution curve based on the area-over-height principle [12]. The CBF value was calculated from CBV and $\bar{t}$ based on the Steward-Hamilton equation. In four cats, bilateral recordings of these flow parameters were carried out from both hemispheres. In two of the four cats, the flow values obtained by the photoelectric method were compared with those measured simultaneously by the hydrogen clearance method in the same cat. The H_2 electrode, which consisted of a platinum wire 300 μm in diameter with a silver-silver chloride electrode as reference, was implanted into the cortex near the photoelectric apparatus (ca. 3 mm away from the center of the red spot caused by the lamp). Pure H_2 gas was inhaled by the cat for a few seconds through the air inlet of the respirator. Flow values were calculated from the desaturation phase of the H_2 clearance curve. Tissue indicator dilution curves recorded from the cerebral cortex of the left ectosylvian gyrus were analyzed for microcirculatory parameters [12]: i.e., the gamma index (dimensionless dispersion of carbon black indicating the heterogeneity of capillary networks), coefficient of skewness (a measure of the skewness of the curve indicating the relative amount of short circuits), and kurtosis (peakedness of the curve indicating specific channels).

At around 2 h after transorbital occlusion of the left middle cerebral artery (MCA), which corresponded with the phase of low perfusion hyperemia [13], the Ca entry blocker, diltiazem (a 1,5-benzothiazepine derivative), was administered intravenously at a dose of 0.2–0.5 mg/kg. Diltiazem was selected because this drug has been reported to act selectively on cerebral vascular smooth muscle [1, 9]. Before and after the administration of diltiazem, continuous recording of CBV and frequent determinations of CBF were carried out in order to examine the effects of diltiazem on the vessels in the ischemic tissue.

Results

An actual protocol of simultaneous measurements of the control CBF in both hemispheres by the photoelectric method and the hydrogen clearance method is shown in Fig. 1. As illustrated in Fig. 2, intravenous administration of diltiazem (0.3 mg/kg/min) produced a marked decrease in SABP. However, it should be noted that despite the decrease in SABP, two independent sensors in the ischemic area detected increases in CBV by diltiazem. The increases were larger than that in the healthy hemisphere. The average increase in CBF in the ischemic area detected by the two sensors of the photoelectric method was from 13.1 to 13.6 ml/100 g/min while the increase by the hydrogen clearance method was from 11 to 22 ml/100 g/min.

The values for $\bar{t}$, CBV, CBF, MABP, and CVR obtained before and after administration of diltiazem are summarized in the Table 1. MABP decreased from

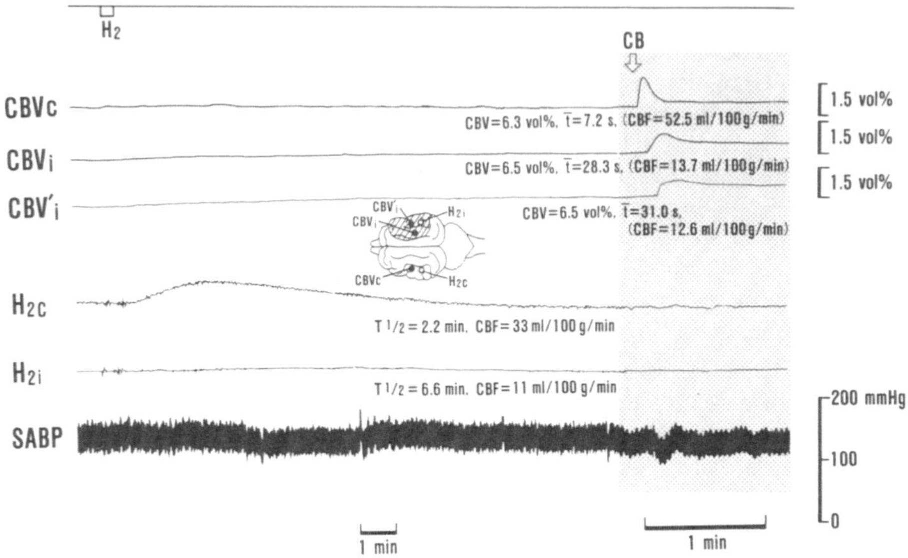

Fig. 1. Actual protocol of simultaneous measurements of control CBF by the photoelectric (PE) method and the hydrogen clearance (H_2) method in both the ischemic (*i*) and healthy (*c*) hemispheres of a cat. The locations of the measurements are indicated. The paper speed was changed from 3 cm/min for the hydrogen clearance curve (short inhalation of 100% hydrogen at the *bar*) to 10 cm/min (*dotted areas*) for the tissue carbon black dilution curve (spike injection at the *arrow*). Values of CBF (PE) were calculated from CBV/t̄ based on the Stewart-Hamilton equation

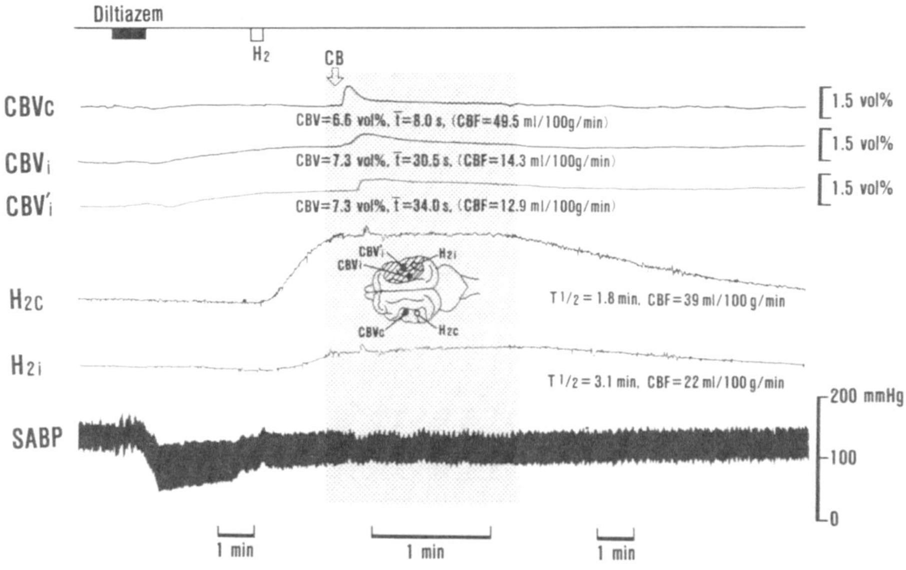

Fig. 2. Intravenous administration of diltiazem (0.3 mg/kg/min) produced a marked decrease in SABP. Despite the decrease in SABP, two independent sensors in the ischemic area detected increases in CBV_i. The average CBF_i (PE) in the ischemic area was increased from 13.1 to 13.6 ml/100 g/min, and the CBF_i (H_2) from 11 to 22/ml/100 g/min, by diltiazem

Table 1. Summary of results

No.	Control					After diltiazem				
	$\bar{t}$	CBV	CBF	MABP	CVR	$\bar{t}$	CBV	CBF	MABP	CVR
Ischemic region (left hemisphere)										
1	12.2	6.2	30.5	125	4.1	14.4	6.3	26.3	125	4.2
2	13.5	5.3	23.5	83	3.5	7.9	5.4	41.0	80	2.0
3	12.7	6.3	29.8	150	5.0	9.8	6.5	39.8	140	3.5
4	45.4	6.3	8.3	125	15.0	31.5	6.3	12,0	100	8.3
5	16.6	5.6	20.2	122	6.0	16.5	5.9	21.2	102	4.8
6	59.7	6.9	6.9	89	12.8	70.0	6.5	5.6	60	10.7
7	93.6	6.4	4.1	180	43.9	43.0	6.6	9.2	150	16.3
8	28.3	6.5	13.7	145	10.6	30.5	7.3	14.3	125	8.7
9	15.8	6.5	24.7	98	4.0	8.8	6.6	44.3	85	1.9
Mean	46.6	6.2	18.0	124	11.7	25.8	6.4	23.7	107*	6.7*
S.E.	13.4	0.2	3.1	10	4.0	6.5	0.2	4.7	9	1.5
Healthy region (right hemisphere)										
3	8.4	6.3	45.0	150	3.3	7.0	6.5	55.7	140	2.5
4	3.1	6.3	121.9	125	1.0	3.4	6.3	111.1	100	0.9
7	10.3	6.3	36.7	180	4.9	8.3	6.3	45.5	150	3.4
8	7.2	6.3	52.5	145	2.8	8.0	6.6	49.5	125	2.5
Mean	7.3	6.3	64.0	150	3.0	6.7	6.4	65.5	129	2.3
S.E.	1.3	–	16.9	10	0.7	1.0	0.6	13.3	9	0.5

* $P<0.05$.
Cases No. 4 and 8: simultaneous measurements of CBF (H_2) showed increases from 23.6 to 35.0, and from 10.5 to 22.4 ml/100 g/min, respectively.

124±10 mmHG to 107±9 mmHg ($P<0.05$). In the ischemic area, the mean transit time ($\bar{t}$) was shortened from 46.6±13.4 s to 25.8±6.5 s. CBV increased in seven cases, remained unchanged in one case, and decreased in one case. The average change in CBV was from 6.2±0.2 vol.-% to 6.4±0.2 vol.-%. CBF increased in seven cases and decreased in two cases. The average change was from 18.0±3.1 ml/100 g/min to 23.7±4.7 ml/100 g/min. CVR decreased from 11.7±4.0 mmHg/ml/100 g/min to 6.7±1.5 mmHg/ml/100 g/min ($P<0.05$).

In the healthy hemisphere, the changes (mean) were 7.3±1.3 s to 6.7±1.0 s for $\bar{t}$, 6.3 vol.-% to 6.4±0.6 vol.-% for CBV, 64.0±16.9 ml/100 g/min to 65.5±13.3 ml/100 g/min for CBF, and 3.0±0.7 mmHg/ml/100 g/min to 2.3±0.5 mmHg/ml/100 g/min for CVR, while the average MABP decreased from 150±10 mmHg to 129±9 mmHg.

The simultaneous measurements of CBF(H_2) in the ischemic area of two cats showed changes from 23.6 ml/100 g/min to 35.0 ml/100 g/min in case No. 4, and from 11 ml/100 g/min to 22 ml/100 g/min in case No. 8. These represented comparable increases in CBF values to those obtained by the photoelectric method.

The microcirculatory parameters obtained from tissue carbon black dilution curves were changed by diltiazem, as follows: 0.54±0.07 to 0.49±0.04 for the gamma index, 1.10±0.19 to 0.80±0.10 for the coefficient of skewness, and

3.95 ± 0.54 to 3.14 ± 0.10 for the kurtosis. None of these changes reached statistical significance. These changes in microcirculatory parameters, although not statistically significant, imply that improvement is accompanied by equalization of capillary perfusion in the ischemic cortex.

Discussion

The decrease in MABP induced by diltiazem indicates that the peripheral arteries were dilated, since cardiac output has been reported to be rather increased by this drug [6]. The increase in CBF in the healthy region by diltiazem suggests that the cerebral arteries were more dilated than the arteries of other organs. These findings agree well with in vitro studies, such as those in which cerebral arteries were found to be more susceptible to Ca entry blockers, including diltiazem, than coronary and mesenteric arteries [9] and diltiazem exerted a selective action on cerebral vascular smooth muscle [1].

One unexpected finding was an increase in CBV even in the ischemic region by diltiazem. Although the difference was not statistically significant, Fig. 2 clearly demonstrates that the CBV increase in the ischemic region was more than that in the healthy region. This is just the opposite to the cases of papaverine. When papaverine hydrochloride (1.5 mg/kg) was administered intravenously to cats, the CBV in the ischemic region was found by us to be decreased while the CBV in the healthy region increased greatly [4]. This papaverine-induced condition was not interpreted by us to be due to the interhemispheric steal phenomenon, but simply to reflect passive shrinkage of the nonresponsive ischemic vasculature by a fall in systemic arterial blood pressure. The present data indicate that diltiazem must have dilated the ischemic vessels. Roy et al. [8] reported that verapamil was associated with worsening of CBF in the ischemic region and inappropriate increases in CBF in the nonischemic region (which were attributed by them to intracerebral steal), while diltiazem increased CBF in the ischemic region. The cerebral vessels in the ischemic region could be responsive to diltiazem, and in some cases be more sensitive than normal vessels. The underlying mechanism could be that the balance between the vasodilating force exerted by vasoactive substances released from the ischemic tissue and the vasoconstrictive force by Ca entry through the depolarized membrane of the smooth muscle of the vessels was eliminated by the Ca entry blocker. However, this idea is highly speculative and must be tested rigorously in the future.

The above findings suggest that the Ca entry blocker, diltiazem, improves the impaired cerebral hemodynamics in the ischemic area produced acutely by occlusion of the MCA despite a decrease in systemic arterial blood pressure. This agent may thus be of value in the treatment of the acute phase of cerebral ischemia.

Summary

Intravenous administration of the calcium (Ca) entry blocker, diltiazem (a 1,5-benzothiazepine derivative, 0.2–0.5 mg/kg), produced hemodynamic changes in the ischemic cerebral tissue of nine cats as detected by our photoelectric method. The mean arterial blood pressure decreased from 124 ± 10 to 107 ± 9 mmHg ($P < 0.05$); the mean transit time of blood was shortened from 46.6 ± 13.4 to 25.8 ± 6.5 s; the CBV increased from 6.2 ± 0.2 to 6.4 ± 0.2 vol.-%; the CBF increased from 18.0 ± 3.1 to 23.7 ± 4.7 ml/100 g/min (increase in seven cases, decrease in two cases); the local cerebral vascular resistance decreased from 11.7 ± 4.0 to 6.7 ± 1.5 mmHg/ml/100 g/min ($P < 0.05$); and the microcirculatory parameters were changed from 0.54 ± 0.07 to 0.49 ± 0.04 (gamma index), from 1.10 ± 0.19 to 0.80 ± 0.10 (coefficient of skewness), and from 3.95 ± 0.54 to 3.14 ± 0.10 (kurtosis). It is concluded that diltiazem improved the impaired microcirculatory conditions in the ischemic tissue of the brain.

Acknowledgement. This work was supported by a research grant for cardiovascular diseases (61-C-1) from the Japanese Ministry of Health and Welfare.

References

1. Bevan JA (1982) Selective action of diltiazem on cerebral vascular smooth muscle in the rabbit: antagonism of extrinsic but not intrinsic maintained tone. Am J Cardiol 49:519–524
2. Date H, Hossmann K-A (1984) Effect of vasodilating drugs on intracortical and extracortical vascular resistence following middle cerebral artery occlusion in cats. Ann Neurol 16:330–336
3. Gelmers HJ (1985) Calcium-channel blockers: effects on cerebral blood flow and potential uses for acute stroke. Am J Cardiol 55:144B–148B
4. Gogolak I, Gotoh F, Tomita M, Tanahashi N, Kobari M, Shinohara T, Mihara B (1983) No intracerebral steal phenomenon in the ischemic brain following papaverine administration. Stroke 16:114–117
5. Harris RJ, Branston NM, Symon L, Bayhan M, Watson A (1982) The effects of a calcium antagonist, nimodipine, upon physiological responses of the cerebral vasculature and its possible influence upon focal cerebral ischemia. Stroke 13:759–766
6. Nagao T, Murata S, Ikezawa K, Ikeo T, Narita H, Sato M (1981) Effects of diltiazem on hemodynamics and His bundle electrogram in the anesthetized dog. Folia Pharmacol Japon 77:195–203
7. Newberg LA, Steen PA, Milde JH, Michenfelder JD (1984) Failure of flunarizine to improve cerebral blood flow or neurologic recovery in a canine model of complete cerebral ischemia. Stroke 15:666–671
8. Roy MW, Dempsey RJ, Meyer KL, Donaldson DL, Tibbs PA, Young AB (1985) Effects of verapamil and diltiazem on acute stroke in cats. J Neurosurg 63:929–936
9. Shimizu K, Ohta T, Toda N (1980) Evidence for greater susceptibility of isolated dog cerebral arteries to Ca antagonists than peripheral arteries. Stroke 11:261–266
10. Steen PA, Newberg LA, Milde JH, Michenfelder JD (1984) Cerebral blood flow and neurologic outcome when nimodipine is given after complete cerebral ischemia in the dog. J Cereb Blood Flow Metab 4:82–87
11. Tomita M, Gotoh F, Sato T, Amano T, Tanahashi N, Tanaka K, Yamamoto M (1978) Photoelectric method for estimating hemodynamic changes in regional cerebral tissue. Am J Physiol 235:H56–H63

12. Tomita M, Gotoh F, Amano T, Tanahashi N, Tanaka K (1980) "Low perfusion hyperemia" following middle cerebral arterial occlusion in cats of different age groups. Stroke 11:629–636
13. Tomita M, Gotoh F, Amano T, Tanahashi N, Kobari M, Shinohara T, Mihara B (1983) Transfer function through regional cerebral cortex evaluated by a photoelectric method. Am J Physiol 245:H385–H398
14. Vorstrup S, Andersen A, Blegvad N, Paulson OB (1986) Calcium antagonist (PY 108–068) treatment may further decrease flow in ischemic areas in acute stroke. J Cereb Blood Flow Metab 6:222–229

Comparison of the Effects of Flunarizine and Nimodipine on Cerebral and Temporal Arteries with Comments on Their Clinical Use in Cerebrovascular Disorders

L. EDVINSSON [1, 2], I. JANSEN [2], J. OLESEN [3], and P. TFELT-HANSEN [3]

Introduction

The diverse chemical structures and pharmacological properties of the calcium antagonists have been extensively reviewed previously (Antman et al. 1980; Braunwald 1982; Ellrodt and Singh 1983; Stone et al. 1980) and a number of pharmacologically distinct subgroups have been identified (Vanhoutte and Paoletti 1987). The dihydropyridines (e.g., nifedipine, nimodipine, nisoldipine, and nicardipine) are vasodilators with some vascular selectivity. The benzothiazepines (diltiazem) and the phenylalkylamines (e.g., verapamil, methoxyverapamil, tiapamil) lack vascular bed selectivity, and demonstrate more negative chronotropic and inotropic effects. Other chemically different calcium antagonists include flunarizine, bepridil, perhexiline, and prenylamine (Vanhoutte and Paoletti 1987).

Specific pharmacological properties may differ even within the subgroups of calcium antagonists. For example, within the dihydropyridine subgroup, differences in selectivity for certain vascular beds are seen; e.g., nifedipine demonstrates selectivity for coronary blood vessels, nimodipine for cerebral blood vessels, and nitrendipine for peripheral blood vessels.

Use of Calcium Entry Blockers in Cerebrovascular Disorders

Migraine

Migraine is classified as "classic" when it begins with a neurological prodrome (aura) and as "common" when such a prodrome is absent (Ad Hoc Committee on Classification of Headache 1988). The theory of classic migraine is that an as yet unidentified substance initiates a series of events which include cerebral vasoconstriction and may involve mediators such as platelet thromboxane A_2, serotonin (5-hydroxytryptamine), prostaglandins, and kinins (Olesen and Edvinsson 1988). This mediator release is thought to result in a calcium influx into vascular smooth muscle with a resultant contraction. The reduction in local cerebral blood flow in conjunction with an attack of migraine with aura is then fol-

[1] Department of Internal Medicine, University Hospital, Lund, [2] Department of Experimental Research, University of Lund, Malmö General Hospital, Malmö, Sweden, and the Department of Neuromedicine, [3] København Amts Sygehus i Gentofte, Hellerup, Denmark.

Cerebral Ischemia and Calcium
Ed. by A. Hartmann and W. Kuschinsky

lowed by late reactive cerebral vasodilatation and sometimes by increased cerebral blood flow (Bousser et al. 1980; Olesen et al. 1982; Skinhøj 1973; Sakai and Meyer 1979).

Prophylactic treatment of migraine has been based on the pharmacological inhibition of specific mediators. However, because there are a variety of mediators, inhibition of a single mediator has not proven satisfactory for prophylaxis. It has been proposed that calcium antagonists may inhibit the onset of migraine by several mechanisms. These include: inhibition of mediator-induced calcium influx, depletion of intracellular calcium stores, and inhibition of platelet aggregation. The main result of a critical review of available controlled clinical trials of calcium entry blockers in migraine prophylaxis is that flunarizine has been demonstrated convincingly to be superior to placebo (Tfelt-Hansen et al. 1988). The two studies published on nimodipine indicate that nimodipine is superior to placebo (Gelmers 1983; Havanka-Kanniainen et al. 1985), but further studies are necessary before a firm conclusion can be drawn. For verapamil and diltiazem available evidence is still insufficient (Tfelt-Hansen et al. 1988).

Stroke and Postresuscitation Ischemia

A few preliminary studies have examined the use of calcium antagonists in the treatment of stroke and postresuscitation ischemia (Gelmers 1982, 1984; Schwartz 1985). In general, some beneficial effects were noted. In a recent prospective, double-blind controlled study Gelmers et al. (1988) reported that in acute ischemic stroke nimodipine reduced the mortality in males with an improvement also in neurological outcome. It was suggested that early treatment with this calcium entry blocker is of benefit in patients with acute ischemic stroke.

Subarachnoid Hemorrhage

Cerebral vasospasm is a phenomenon seen in some patients after subarachnoid hemorrhage (Fisher et al. 1977). Although the exact mechanism behind this proposed vasospasm has not been identified, the role of calcium has been considered to be pivotal. High intracellular calcium concentrations may result in membrane dissolution, the production of vasoconstrictive prostaglandins and thromboxanes, increased neurotransmitter release, and the production of free radicals (Shapiro 1985). All these factors may contribute to the vasoconstriction, the impaired reperfusion, the vessel damage, and the cerebral edema, ultimately leading to neuronal death.

Although a variety of drugs (α-adrenoceptor antagonists, catecholamine and/or serotonin depletors, β-adrenoceptor antagonists, postganglionic antagonists, phosphodiesterase inhibitors, and numerous other agents) have been evaluated in the treatment of cerebral vasospasm, the results have generally proven disappointing (Wilkins 1980). The first multicenter, randomized, double-blind placebo-controlled study of the use of nimopidine reported a significantly decreased occurrence of severe neurological deficits (Allen et al. 1983). This encouraging study led to further tests in patients with subarachnoid hemorrhage (Auer 1984; Ljunggren et al. 1984; Säveland et al. 1986). In general, the studies demonstrated that intravenous administration of nimodipine counteracted delayed ischemic deficits following subarachnoid hemorrhage.

Comparison of the Effects of Calcium Entry Blockers Nimodipine and Flunarizine on Human Cerebral and Temporal Arteries

Methodological Aspects

The calcium entry blockers are not a homogeneous group of drugs. Flunarizine, a piperazine derivate, and nimodipine, a dihydropyridine derivative, have different binding characteristics and pharmacological profiles with regard to tissue selectivity in various test systems (Vanhoutte and Paoletti 1987). The results from animal studies cannot be transferred directly to man since considerable differences exist among species (Spedding 1985), and the doses and/or concentration of the drugs used might induce effects which cannot be obtained with relevant doses in man.

We have used a sensitive in vitro system (Högestätt et al. 1983) for investigating the inhibition by the two calcium entry blockers of depolarization- and agonist-induced contractions (Jansen et al. 1989) of fresh human cerebral arteries obtained in conjunction with neurosurgery. Results regarding potency of various constrictor and dilator agents obtained with this system in vitro on feline cerebral arteries correlate well with in vivo results from microapplication studies on cat pial arterioles (McCulloch and Edvinsson 1984). Similar experimental proof of correlation between in vitro and in vivo results can never be obtained in man, but the results in cats strongly indicate that the in vitro results on viable human cerebral arteries can be used to predict whether a certain concentration of a drug can exert an effect on human arteries also in situ, providing the drug can pass the blood-brain barrier.

The inhibition by the calcium entry blockers of induced contractions of the human temporal artery has also been examined (Jansen et al. 1989) since this artery can serve as "a peripheral control artery" in the evaluation of the postulated selectivity of calcium entry blockers for cerebral arteries, and since this artery traditionally has been implicated in the etiology of migraine pain.

Selectivity of Calcium Entry Blockers for Cerebral Arteries?

Nimodipine was introduced as a selective calcium entry blocker against intracerebral vasoconstriction with a high intrinsic activity, that is a high maximum inhibiting effect (Towart 1981). In our study (Jansen et al. 1989) the maximum inhibition exerted by nimodipine on contractions of pial arteries ranged from about 70% to 90% (Fig. 1). Thus, in man nimodipine has a high intrinsic activity as an inhibitor of induced [potassium (K^+) or norepinephrine (NE)] contractions of

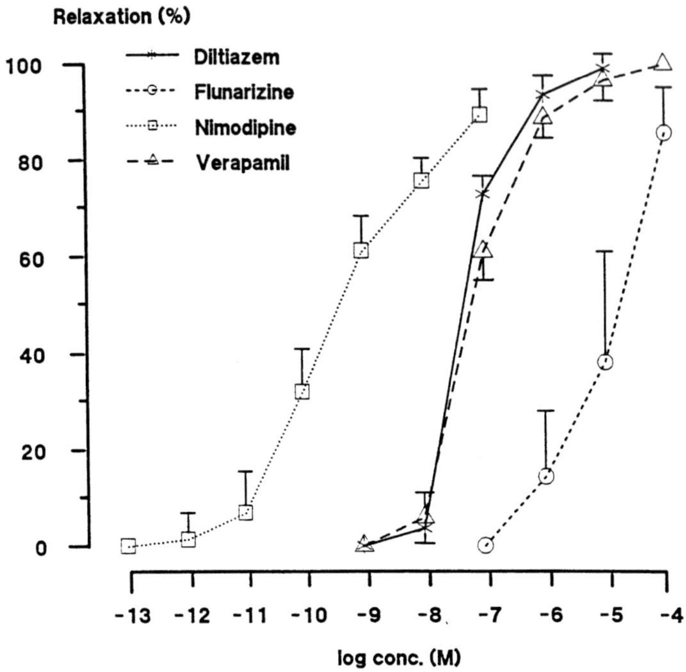

Fig. 1. Relaxation in % of 60 mM K^+-contraction of human cerebral arteries induced by increasing concentrations of diltiazem, flunarizine, nimodipine, or verapamil. Data are given as mean values ± SEM, $n=4$–6

pial arteries. The presumed selectivity of nimodipine for cerebral arteries was, however, based on the very potent inhibition of 5-HT-induced contractions in the rabbit basilar artery in vitro (a vessel with a diameter of about 0.5 mm) versus its relative inefficacy in respect of 5-HT-induced contractions of rabbit saphenous artery (a large peripheral artery with a diameter of about 2–3 mm). In these vessels potassium-induced contractions were inhibited by nimodipine with a difference in potency of about 100 times (Towart 1981). Similar differences in potency for inhibition of 5-HT-induced contractions by nimodipine have been reported for canine basilar and femoral arteries in vitro (Peroutka et al. 1984), again comparing vessels of different size. In a previous study potassium-induced contractions of human cerebral and mesenteric arteries were inhibited by similar concentrations of nimodipine (IC_{50} approximately 10^{-9} M) while a slight selectivity was seen for another dihydropyridine derivative, nifedipine (Brandt et al. 1981). This indicates that nimodipine might not be selective for the voltage-sensitive calcium channels in human pial arteries. In the study by Jansen et al. (1989), the two calcium entry blockers nimodipine and flunarizine were about 10 times more potent inhibitors of potassium-induced contractions in pial than in temporal arteries. Similar differences in potencies of inhibition were found for NE-induced contractions of pial and temporal arteries. Thus, in man, the agonist-operated calcium channels in pial arteries were inhibited with a similar selectivity as the voltage-sensitive calcium channels. Nimodipine still appears to be a selective inhibitor of cerebral vasoconstriction in man, but our results indicate that the selectivity is only in the order of 10 compared with a peripheral artery (Jansen et al. 1989). We must, however, stress the fact that both species differences and the difference in external diameter of the isolated arteries may affect agonist-induced contractions in vascular smooth muscle. Cauvin et al. (1984) have indicated that a decreasing release of intracellular calcium and an increasing dependence on extracellular calcium for NE-induced contractions of rabbit mesenteric artery proceeds from proximal to distal arteries. Thus, to investigate calcium antagonistic actions arteries of similar size should be obtained from cerebral and peripheral tissues.

Prediction from In Vitro to In Vivo Effects

The plasma concentration obtained during therapeutic use of flunarizine is around 100 ng/ml (Holmes et al. 1984), corresponding to 2×10^{-7} M, while for nimodipine it is 20 ng/ml (Vinge et al. 1986), corresponding to 0.5×10^{-7} M. These two hydrophobic drugs have a high protein binding capacity (Allen et al. 1983; Vinge et al. 1986). Following correction for this, the free drug concentration in plasma can be calculated to 2×10^{-9} M for flunarizine and 5×10^{-9} M for nimodipine. These are thus the "therapeutic concentrations" against which the in vitro results should be evaluated. Since both drugs are hydrophobic substances they should readily pass the blood-brain barrier; this has in fact been shown for nimodipine (Allen et al. 1983) and nifedipine (Janicki et al. 1988), and probably holds true for flunarizine as well, particularly since it has some antiepileptic effects (Overweg et al. 1984).

The IC_{50} for flunarizine inhibition of contractions of human cerebral arteries ranges from 4.6×10^{-8} M to 1.1×10^{-6} M (Jansen et al. 1989). With a free plasma concentration of 2×10^{-9} M flunarizine is therefore unlikely to exert a direct vasodilatatory effect on pial arteries in man during therapeutic use. Nimodipine, however, is 100 times more potent than flunarizine as judged from the IC_{50} values. The IC_{50} for nimodipine inhibition of contractions of human pial arteries ranges from 2.8×10^{-10} M to 8.1×10^{-9} M. The calculated free plasma concentration of nimodipine is 5×10^{-9} M, a concentration above or equal to the IC_{50} value. Thus, nimodipine may inhibit cerebral arterial constrictions in man in therapeutically used doses.

Some Clinical and Theoretical Implications

The calcium entry blockers currently examined have been introduced in the treatment of cerebrovascular disorders because of two presumed pharmacological effects: a rather selective action on cerebral vessels (Towart 1981) and a possible cytoprotective effect in ischemia (Siesjö 1981). The results on human cerebral and temporal arteries confirm that nimodipine has an effect on human arteries in therapeutically relevant doses and is a more potent inhibitor of contraction in cerebral than in temporal arteries. The difference in potency is, however, only in the order of 10 times and not in the order of several 100 times as previously noted in animal studies (Towart 1981; Peroutka et al. 1984). The selectivity of nimodipine corresponds to the lack of effect of nimodipine on blood pressure in clinical studies (Allen et al. 1983), but the results indicate that if higher doses of nimodipine are used, problems with a fall in blood pressure will arise and consequently a drop in brain perfusion pressure; this has been observed with high doses of nimodipine in a rat model of cerebral ischemia (Mohamed et al. 1984).

Apart from the effect on pial arteries the calcium entry blockers might theoretically act on neurons and protect them against excessive Ca^{2+} influx/release during cerebral ischemia (Siesjö 1981). Nimodipine did not prevent arterial spasm following subarachnoidal hemorrhage, but was found to be effective in reducing the occurrence of severe neurological deficits in conjunction with the bleeding (Allen et al. 1981). Although nimodipine should be able to inhibit excessive vasoconstriction of pial arteries, the so-called cytoprotective effect of the drug, which has been demonstrated in several animal models of cerebral ischemia (Heffez and Passonneau 1985; Mabe et al. 1986), could theoretically be responsible for its positive effects in subarachnoid hemorrhage of man (Säveland et al. 1986).

The fact that some calcium entry blockers appear to be effective in migraine prophylaxis has seemingly lent support to the traditionally held view of a vascular involvement in the pathogenesis of migraine (Meyer et al. 1985). Flunarizine, which is the calcium entry blocker which has most convincingly been demonstrated to be prophylactic in migraine (Tfelt-Hansen et al. 1988), is, however, based on the present results, probably without any effect on cerebral arterial tone in man. Nimodipine has also been found to have some prophylactic effect in migraine (Tfelt-Hansen et al. 1988), and may in contrast exert an effect on cerebral arteries in man. A possible explanation for the prophylactic effect of flunarizine in migraine is protection against a hypothetical brain hypoxia (Amery 1982). Flunarizine has been found to be effective in protecting against brain ischemia in a rat model (Depshande and Wieloch 1986). Flunarizine is, however, not a pure calcium entry blocker and has, in addition, antihistaminic (Holmes et al. 1984) and antiepileptic effects (Overweg et al. 1984). Similarly, nimodipine binds with high affinity to sites in the brain (Bellemann et al. 1983), has cytoprotective effects in brain ischemia (Heffez and Passonneau 1985; Mabe et al. 1986), and according to some animal studies has considerable behavioral effects (Hoffmeister et al. 1982). Thus, there are several possible mechanisms of action, and the fact that these two calcium entry blockers are effective in migraine and in cerebral ischemic disorders cannot solve the controversy as to whether the disorders are primarily of vascular or neuronal origin.

Acknowledgement. This work was supported by a grant from the Swedish Medical Research Council (05958).

References

Ad Hoc Committee on (1988) Classification and diagnostic criteria for headache disorders, cranial neuralgias and facial pain. Cephalalgia 8 (Suppl 7):1–96

Allen GS, Ahn HS, Presziosi TJ et al. (1983) Cerebral arterial spasm: a controlled trial of nimodipine in patients with subarachnoidal hemorrhage. N Engl J Med 308:619–624

Amery WK (1982) Brain hypoxia: the turning-point in the genesis of the migrain attack? A hypothesis. Cephalalgia 2:83–109

Antman EM, Stone PH, Muller JE, Braunwald E (1980) Calcium channel blocking agents in the treatment of cardiovascular disorders. Part I: Basic and clinical electrophysiologic effects. Ann Intern Med 93:875–885
Auer LM (1984) Acute operation and preventive nimodipine improve outcome in patients with ruptured cerebral aneurysms. Neurosurgery 15:57–66
Bellemann P, Schade A, Towart R (1983) Dihydropyridine receptors in rat brain labeled with ^{3}H nimodipine. Proc Natl Acad Sci 80:2356–2360
Bousser MG, Baron JC, Iba-Zizen MT et al. (1980) Migrainous cerebral infarction: a tomographic study of cerebral blood flow and oxygen extraction with the oxygen^{15} inhalation technique. Stroke 11:145–148
Brandt L, Andersson KE, Edvinsson L, Ljunggren B (1981) Effects of extracellular calcium and of calcium antagonists on the contractile responses of isolated human pial and mesenteric arteries. J Cereb Blood Flow Metab 1:339–347
Braunwald E (1982) Mechanism of action of calcium-channel blocking agents. N Engl J Med 307:1618–1627
Cauvin C, Saida K, van Breemen C (1984) Extracellular Ca^{2+} dependence and diltiazem inhibition of contraction in rabbit conduit arteries and mesenteric resistance vessels. Blood Vessels 21:23–31
Deshpande JK, Wieloch T (1986) Flunarizine, a calcium entry blocker, ameliorates ischemic brain damage in the rat. Anesthesiology 64:215–225
Ellrodt AG, Singh BN (1983) Clinical applications of slow channel blocking compounds. Pharmacol Ther 23:1–43
Fisher CM, Roberson GH, Ojemann RG (1977) Cerebral vasospasm with ruptured saccular aneurysm – the clinical manifestations. Neurosurgery 1:245–248
Gelmers HJ (1982) Effect of nimodipine (Bay e 9736) on postischemic cerebrovascular reactivity, as revealed by measuring regional cerebral blood flow (rCBF). Acta Neurochir 63:283–289
Gelmers HJ (1983) Nimodipine, a new calcium antagonist in the prophylactic treatment of migraine. Headache 23:106–109
Gelmers HJ (1984) The effects of nimodipine on the clinical course of patients with acute ischemic stroke. Acta Neurol Scand 69:232–239
Gelmers HJ, Gorter K, De Weerdt CJ, Wiezer HJA (1988) A controlled trial of nimodipine in acute ischemic stroke. N Engl J Med 318:203–207
Havanka-Kanniainen H, Hokkanen E, Myllylä VV (1985) Efficacy of nimodipine in the prophylaxis of migraine. Cephalalgia 5:39–43
Heffez DS, Passonneau JV (1985) Effect of nimodipine on cerebral metabolism during ischemia and recirculation in the mongolian gerbil. J Cereb Blood Flow Metab 5:523–528
Hoffmeister F, Benz U, Heise A, Krause HP, Neuser V (1982) Behavioral effects of nimodipine in animals. Arzneimittelforschung 32:338–346
Högestätt ED, Andersson K-E, Edvinsson L (1983) Mechanical properties of the rat cerebral arteries as studied by a sensitive device for recording of mechanical activity in isolated small blood vessels. Acta Physiol Scand 117:49
Holmes B, Brodgen RN, Heel RC, Speight TM, Avery GS (1984) Flunarizine. A review of its pharmacodynamic and pharmacokinetic properties and therapeutic use. Drugs 27:6–44
Janicki PK, Siembab D, Paulo EA, Krzascik P (1988) Single-dose kinetics of nifedipine in rat plasma and brain. Pharmacology 36:183–187
Jansen I, Tfelt-Hansen P, Edvinsson L (1989) Comparison of the calcium entry blockers nimodipine and flunarizine on human cerebral and temporal arteries (in preparation)
Ljunggren B, Brandt L, Säveland H, Nilsson PE, Cronqvist S, Andersson KE, Vinge E (1984) Outcome in 60 consecutive patients treated with early aneurysm operation and intravenous nimodipine. J Neurosurg 61:864–873
Mabe H, Nagai H, Takagi T, Umemura S, Ohno M (1986) Effect of nimodipine on cerebral functional and metabolic recovery following ischemia in the rat brain. Stroke 17:501–505
McCulloch J, Edvinsson L (1984) Cerebrovascular smooth muscle reactivity: a critical appraisal of in vitro and in situ techniques. J Cereb Blood Flow Metab 4:129–139
Meyer JS, Nance M, Walker M, Zetusky WJ, Dowell RE Jr (1985) Migraine and cluster headache treatment with calcium antagonists supports a vascular pathogenesis. Headache 25:358–367

Mohamed AA, McCulloch J, Medelow AD, Teasdale GM, Harper AM (1984) Effect of the calcium antagonist nimodipine on local cerebral blood flow: relationship to arterial blood pressure. J Cereb Blood Flow Metab 4:206–211

Olesen J, Edvinsson L (1988) Basic Mechanisms of Headache. Elsevier, Amsterdam

Olesen S, Lauritzen M, Tfelt-Hansen P (1982) Spreading cerebral oligemia in classical and normal cerebral blood flow in common migraine. Headache 22:242–248

Overweg J, Binnie CD, Meijer JWA, Meinardi H, Nuijten STM, Schmaltz S, Wauquier A (1984) Double-blind placebo-controlled trial of flunarizine as add-on therapy in epilepsy. Epilepsia 25:217–222

Peroutka SJ, Banghart SB, Allen GS (1984) Relative potency and selectivity of calcium antagonists used in the treatment of migraine. Headache 24:55–58

Sakai F, Meyers JS (1979) Abnormal cerebrovascular reactivity in patients with migraine and cluster headache. Headache 19:257–266

Säveland H, Ljunggren B, Brandt L, Messeter K (1986) Delayed ischemic deterioration in patients with early aneurysm operation and intravenous nimodipine. Neurosurgery 146–150

Schwartz AC (1985) Neurological recovery after cardiac arrest: clinical sensibility trial of calcium blockers. Am J Emerg Med 3:1–10

Shapiro HM (1985) Post-cardiac arrest therapy: calcium entry blockade and brain resuscitation. Anesthesiology 62:384–387

Siesjö BK (1981) Cell damage in the brain: a speculative hypothesis. J Cereb Blood Flow Metab 1:155–185

Skinhøj E (1973) Hemodynamic studies within the brain during migraine. Arch Neurol 29:95–98

Spedding M (1985) Calcium antagonist subgroups. Trends Pharmacol Sci 6:109–114

Stone PH, Antman EM, Muller JE, Braunwald E (1980) Calcium channel blocking agents in the treatment of cardiovascular disorders. Part II: Hemodynamic effects and clinical applications. Ann Intern Med 93:886–904

Tfelt-Hansen P, Edvinsson L, Olesen J (1989) Treatment of migraine with calcium entry blockers. In: Van Zwieten PA (ed) Clinical Aspects of Calcium Entry Blockers. Prog Basic Clin Pharmacol. Basel,Karger, vol 2, pp 143–154

Towart R (1981) The selective inhibition of serotinin induced contractions of rabbit cerebral vascular smooth muscle by calcium-antagonistic dihydropyridines. Circ Res 48:650–657

Vanhoutte PM, Paoletti R (1987) The WHO classification of calcium antagonists. Trends Pharmacol Sci 8:4–5

Vinge E, Andersson K-E, Brandt L, Ljunggren B, Nilsson L-G, Rosendal-Helgesen S (1986) Pharmacokinetics of nimodipine in patients with aneurysmal subarachnoidal haemorrhage. Eur J Clin Pharmacol 30:421–425

Wilkins RH (1980) Attempted prevention or treatment of intracranial arterial spasm: a survey. Neurosurgery 6:198–210

The Effect of Calcium Antagonists in Experimental Atherosclerosis, the Underlying Cause of Heart and Cerebral Vascular Diseases

J. GRÜNWALD [1], G. N. CHALDAKOV [2], and CH. C. HAUDENSCHILD [3]

Introduction

Migration and proliferation of arterial smooth muscle cells (SMCs) into the intima are considered important events in the development of atherosclerosis, the underlying cause of coronary and cerebral vascular diseases [18, 20]. A variety of different atherosclerotic risk factors stimulate these cellular reactions. In response to experimental stimulations [7, 10, 12], SMCs migrate through the fenestrae of the internal elastic lamina, appear at the luminal surface, and form an intimal thickening that displays many features of an atherosclerotic plaque.

Many stimulators of SMC proliferation and migration, for example PGDF, have in common that they increase temporarily the cytosolic free calcium [4]. This opens the possibility that deleterious effects of untimely growth stimulation, such as are seen in atherogenesis, could be suppressed by the use of calcium antagonists. According to the classification of calcium antagonists [6, 21] the availability of cytosolic calcium can be regulated by the transmembrane influx, by the calmodulin-calcium interaction, or by the blockade of the release of intracellular storage sites.

In contrast to the widely used so-called calcium channel blockers such as verapamil and nifedipine, it has been suggested that flunarizine specifically inhibits the availability of activator calcium due to the exogenous triggering of the vascular SMCs [14]. Furthermore, it was the most effective of several calcium antagonists in inhibiting SMC proliferation both in vivo and in vitro [1, 2]. Compared to other calcium antagonists, flunarizine also had the most sustained effect on the inhibition of KCl-induced vasoconstriction and lacked direct vasodilator effects under basal conditions [16].

The current study was performed to evaluate further the mechanisms of the inhibitory effects of this calcium antagonist. Specifically, we determined whether there is also an inhibition of SMC migration, whether the growth inhibition is due to a shortening of the interdivision times, and whether the target of the inhibiting effect is an actively proliferating SMC or a cell that has been activated to proliferate from a quiescent status by serum stimulation.

[1] Institute for Arteriosclerosis Research at the University of Münster, Domagkstr. 3, D-4400 Münster, FRG. Present address: Lichtwer Pharma, D-1000 Berlin 28.
[2] Department of Anatomy and Histology, Varna Institute of Medicine, BG-Varna, Bulgaria.
[3] Mallory Institute for Pathology, Boston University, 784 Massachusetts Avenue, Boston, MA 02118, USA

Cerebral Ischemia and Calcium
Ed. by A. Hartmann and W. Kuschinsky

We applied the calcium antagonist in vivo to hypertensive rats, using our outgrowth assay [8, 9] to evaluate the migratory behavior of SMCs. In addition we applied the calcium antagonist directly to SMC cultures, analyzing the effect on the cells by the evaluation of growth curves and the quantification of individual interdivision times and migration tracks from time lapse video movies combined with subsequent morphometric analysis [11].

Material and Methods

Groups of 18 male SHR rats (12 weeks old), and age-matched Wistar-Kyoto rats (WKY) were used as a model for genetic hypertension.

As a model for acute renal hypertension, perinephritis was induced by cellophane [17] in 12 male Wistar rats weighing 300–350 g; untreated age-matched Wistar rats served as normotensive controls. Six of the SHR and WKY rats were treated with either 0.5 mg/kg or 1 mg/kg flunarizine intraperitoneally for 8 consecutive days. Half the renal hypertensive rats were treated with 1 mg/kg flunarizine intraperitoneally for 8 consecutive days after stabilization of the high blood pressure. The animals were killed by decapitation and the thoracic aortae were prepared under sterile conditions. Media explants were prepared and the outgrowth assay used for migration analysis.

Subcultured SMCs were used for the analysis of cellular proliferation. The effect of flunarizine was tested on proliferating SMCs as well as quiescent (GO-phase) SMCs with or without the addition of 14 μg/ml flunarizine (14 μg/ml flunarizine showed half-maximal inhibition of SMC proliferation, as shown under "Results"). At that time point the flasks were placed on a Nikon Diaphot microscope which was kept at 37 °C. The cells were continuously videotaped with a Hamamatsu video camera for 72 h and a speed of one frame per min and analyzed as described previously [11].

Cell motility was quantified using a semiautomatic morphometric system (Leitz ASM) to trace a migration track of each individual cell between two cell divisions and afterwards calculating the distance migrated per hour [11].

Results

Migration of SMCs was analyzed in explant cultures under low serum conditions (0.5% FBS). In spontaneously hypertensive rats (SHRs) initial migration from explants began after 24 h, and after 8 days over 50% of the explants showed SMC migration (Fig. 1). Migration of SMCs from explants of normotensive animals began after 4 days, and less than 10% of all explants showed SMC migration. When the SHRs were treated in vivo with 0.5 mg flunarizine/kg body weight only a slight inhibition of SMC migration was observed. In contrast, in explants from SHRs which were treated with 1 mg flunarizine/kg body weight, initial migration occurred after 5 days and only about 10% of all explants showed SMC migration.

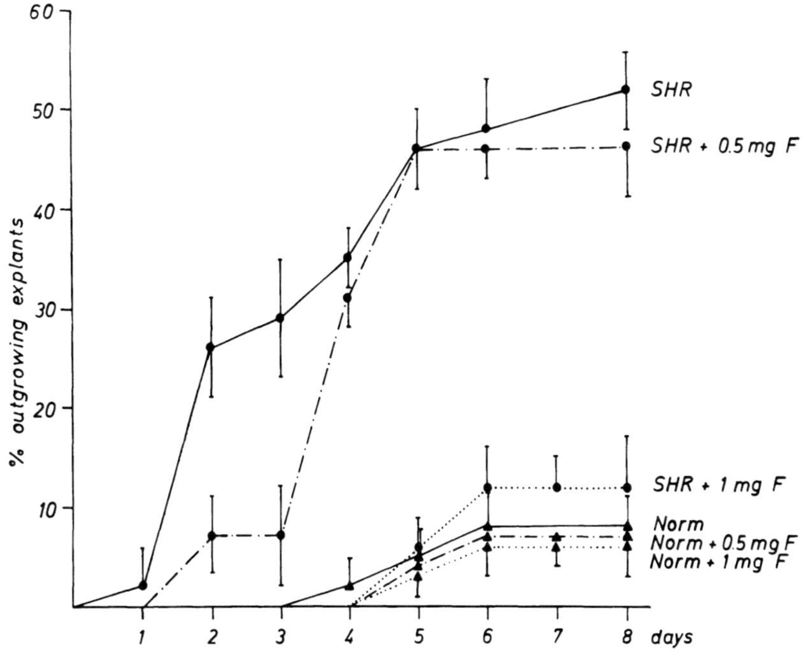

Fig. 1. Cumulative percentages of aortic explants showing migration of SMCs onto the culture dish under low serum (0.5% FBS) concentrations from spontaneously hypertensive (*SHR*) and normotensive (*Norm*) rat aortae. Rats of both groups were treated for 8 days with flunarizine 0.5 mg/kg body wt. and 1 mg/kg body wt. Data represent mean values of six rats per group (18–24 explants analyzed per rat). $P<0.01$, SHR+1 mg flunarizine, days 1–8

Since explants from normotensive animals showed virtually no SMC migration per se, the treatment with flunarizine did not alter this situation either.

The group of rats with renal hypertension was treated with 1 mg flunarizine/kg body weight. Results comparable to those in SHR rats were obtained.

In a second set of experiments, subcultured SMCs were used to study the direct effect of the drug on SMC migration and proliferation. Conventional growth curves were used to analyze the half-maximal growth inhibition of flunarizine (Fig. 2). Flunarizine showed a dose-dependent effect on SMC proliferation with the half-maximal effect at 14 µg/ml. This dosage was used for all further cell culture experiments. Time lapse video recording was used to determine individual interdivision times and migration tracks of SMCs with or without treatment with flunarizine (14 µg/ml) in vitro. Figure 3 shows typical migration tracks of SMCs with or without flunarizine treatment.

Table 1 summarizes the quantification of the effect of flunarizine on migration and proliferation of SMCs analyzed by time lapse video microscopy, differentiating between the effect on quiescent and proliferating SMCs. The table shows that SMC migration of quiescent cells is reduced from 19.2 µm/h to 6.1 µm/h by flunarizine, indicating a reduction of 68%. The migration of proliferating SMCs is reduced from 30.5 µm/h to 10.2 µm/h by flunarizine, indicating a reduction of 67%. In the case of proliferation, flunarizine reacted differently on

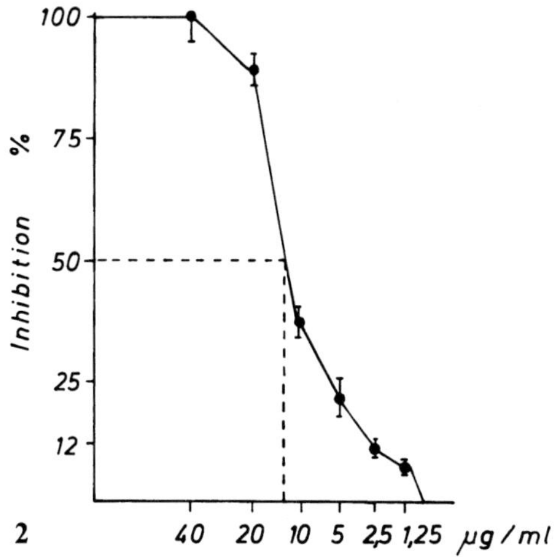

2

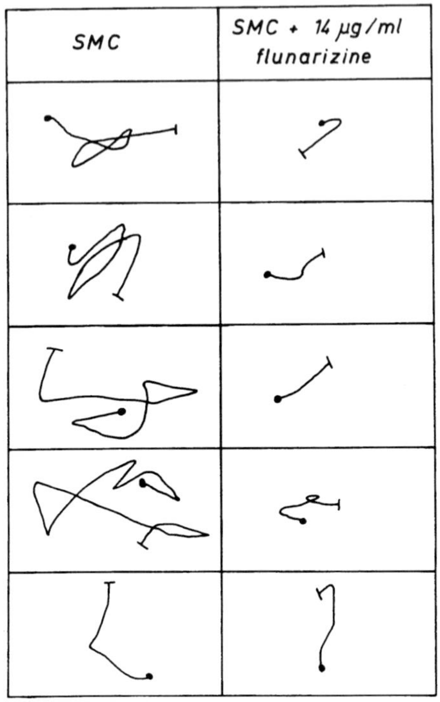

3

Fig. 2. Dose dependence of flunarizine on the proliferation of subcultured rat aortic SMCs. Half-maximal inhibition was obtained at a concentration of 14 µg/ml flunarizine

Fig. 3. SMC migration in vitro. Typical migration tracks between two cell divisions of untreated control SMCs (*left side*) and SMCs treated with 14 µg/ml flunarizine (*right side*)

Table 1. Effect of flunarizine on migration and proliferation of quiescent and proliferating SMCs

	SMC	SMC + flunarizine	[%]
Migration			
1. Quiescent SMCs	19.2 µm/h ± 1.8	6.1 µm/h ± 0.6	−68
2. Proliferating SMCs	30.5 µm/h ± 1.8	10.2 µm/h ± 0.6	−67
Proliferation			
1. Quiescent SMCs IDT	18.6 h ± 0.8	20.9 h ± 1.5	+12
2. Proliferating SMCs IDT	12.7 h ± 0.5	20.5 h ± 0,9	+61

IDT, interdivision time.

quiescent and proliferating SMCs. While in quiescent SMCs flunarizine increased the time between serum stimulation and first division from 18.6 h to 20.9 h, indicating a slight elongation of 12%, in proliferating SMCs the elongation was more pronounced. In this case the interdivision time of untreated SMCs was 12.7 h and in the case of flunarizine treatment, 20.5 h, indicating an elongation of 61%.

Discussion

The importance of SMC migration and proliferation in the development of atherosclerotic lesions is well documented [18, 20]. Today drug treatment of atherosclerosis is directed against single risk factors such as high blood pressure or a high serum cholesterol level. Since all atherosclerotic risk factors lead to comparable cellular reactions in the vessel wall, beginning with the migration of SMCs from the media into the intima and their proliferation, it is suggested that agents capable of interfering with SMC migration and proliferation would be of therapeutic value in the prevention and treatment of atherosclerosis.

Calcium antagonists such as nifedipine [15] and flunarizine [2] have been shown to inhibit SMC proliferation in vitro as well as lesion formation in an experimental model of atherosclerosis [2]. The purpose of this investigation was to analyze specifically the effect of flunarizine on SMC migration and furthermore to analyze the effect of flunarizine on either quiescent or proliferating SMCs with respect to migration and proliferation, using time lapse video microscopy. In this study it is demonstrated that the calcium antagonist flunarizine administered in vivo to hypertensive animals inhibits or reduces the SMC migration stimulated by hypertension. We compared two models of hypertension, firstly the spontaneously hypertensive rat and secondly a form of acute renal hypertension induced by cellophane perinephritis. Flunarizine administered at a dose of 1 mg/kg body weight over a period of 8 consecutive days resulted in a significant reduction of SMC migration in both models of hypertension.

In a further in vitro study we showed that flunarizine added to cultured SMCs at a dose of 14 µg/ml inhibits individual cell motility by approximately 60% independent of whether the cells are in a proliferating or a quiescent state. In contrast, SMC proliferation was inhibited to a higher degree in quiescent SMCs after serum stimulation than in nonarrested proliferating SMCs.

The mode of action of flunarizine in inhibiting SMC migration and proliferation is not known. Since PDGF induces an early elevation of intracellular Ca^{2+} [4] and flunarizine interferes with the calcium availability in the cell, the effect of growth factors might be reduced. Furthermore calmodulin, the intracellular receptor for Ca^{2+}, is believed to interfere in the regulation of cell proliferation [5]. Calmodulin antagonists such as trifluoperazine [13] have also been shown to inhibit SMC proliferation. Flunarizine is therefore of special interest since it interferes not only with calcium, but also with calmodulin metabolism [19].

Our investigation, analyzing predominantly SMC migration and confirming earlier results on SMC proliferation, gives additional support to the theory that flunarizine may inhibit development and progression of atherosclerotic lesions by interfering with migration and proliferation of arterial SMCs.

Acknowledgments. The excellent technical assistance of Roswitha Fischer and Ingrid Meyer is gratefully acknowledged. We also thank Marianne Opalka for the photographic work and Gaby Benning for typing the manuscript. This work was supported by the DFG (Gr 693/3), the BMFT (FIW 0479), and NATO (0039/87-89).

References

1. Betz E, Hämmerle H, Viele D (1984) Ca^{2+}-entry blockers and atherosclerosis. Int Angiol 3:33–44
2. Betz E, Hämmerle H, Strohschneider T (1985) Inhibition of smooth muscle cell proliferation and endothelial permeability with flunarizine in vitro and in experimental atheromas. Res Exp Med 185:325–340
3. Brecher P, Chan CT, Franzblau C, Faria B, Chobanian AV (1978) Effect of hypertension and its reversal on aortic metabolism in the rat. Circ Res 43:561–569
4. Cassel D, Rothenberg P, Zhuang Y, Deuel TF, Glaser L (1983) Platelet-derived growth factor stimulates Na^+/H^+ exchange and induces cytoplasmic alkalinization in NR6 cells. Proc Natl Acad Sci USA 80:6224–6330
5. Chalouleas JG, Bolton WE, Hidaka H, Boyd Ae, Means AR (1982) Calmodulin and the cell cycle – involvement in regulation of cell-cycle progression. Cell 28:41–47
6. Eichelbaum M, Echizen H (1984) Clinical pharmacology of calcium antagonists: a critical review. J Cardiovasc Pharmacol 6:963–967
7. Grünwald J, Robenek H, Mey J, Hauss WH (1982) In vivo and in vitro cellular changes in experimental hypertension. Exp Mol Pathol 36:164–176
8. Grünwald J, Haudenschild CC (1984) Intimal injury in vivo activates vascular smooth muscle cell migration and explant outgrowth in vitro. Arteriosclerosis 4:183–189
9. Grünwald J, Haudenschild CC (1985) The influence of antiplatelet drugs on balloon injury stimulated migration in cultured smooth muscle cells. Artery 12:324–336
10. Grünwald J, Chobanian AV, Haudenschild CC (1987) Smooth muscle cell migration and proliferation: atherogenic mechanism in hypertension. Atherosclerosis 67:215–221
11. Grünwald J (1987) Time-lapse video microscopic analysis of cell proliferation, motility and morphology: applications for cytopathology and pharmacology. Biotechniques 5:680–687
12. Grünwald J, Fingerle J, Hämmerle H, Betz E, Haudenschild CC (1987) Cytocontractile structures and proteins of smooth muscle cells during the formation of experimental lesions. Exp Mol Pathol 46:78–88
13. Kaul D, Kukreja RS (1987) Atherogenesis. Prevention action of trifluoperazine. Atherosclerosis 64:211–214
14. Nakayama K, Kasuja Y (1980) Selective abolition of Ca-dependent responses of smooth and cardiac muscles by flunarizine. Jpn J Pharmacol 30:731–737
15. Nilsson J, Sjölund M, Palmberg L, von Euler AM, Jonzon B, Thyberg J (1985) The calcium antagonist nifedipine inhibits arterial smooth muscle cell proliferation. Atherosclerosis 58:109–122
16. Van Nueten JM, Van Beek J, Janssen PA (1978) Effect of flunarizine on calcium-induced responses of peripheral vascular smooth muscle. Arch Int Pharmacodyn Ther 232:42–52
17. Page JH, Graef J (1939) Hypertension following experimental perinephritis induced by cellophane. Arch Pathol 28:613
18. Ross R, Glomsett JA (1973) Atherosclerosis and the arterial smooth muscle cell. Science 180:1332–1339
19. Stoclet JC, Lugnier C, Follenius A, Scheftel JM, Gerhard D (1985) Calmodulin and calcium regulation: effect of antagonists. In: Te PM, Govoni S, Paoletti R (eds). Raven, New York, pp 31–40
20. Wissler RW (1968) The arterial medial cell, the smooth muscle cell or multifunctional mesenchyme? J Atheroscl Res 8:201–213
21. Van Zwieten PA (1985) Calcium-Antagonisten – Terminologie, Klassifizierung und Vergleich. Arzneimittelforschung/Drug Res 35:298–301

Thrombosis on Stenosis: Mechanisms of Thrombus Growth and Embolization on Carotid Atherosclerotic Plaques

L. BADIMON, J. J. BADIMON, R. LASSILA, V. T. TURITTO, and V. FUSTER

The association between carotid artery disease and stroke has long been recognized. Artery to artery embolism is one of several mechanisms causing either transient ischemic attacks (TIAs) or stroke ipsilateral to atherosclerotic plaques [25]. Thrombus at the site of carotid arterial occlusion [12], atherosclerotic debris, and vertebral artery stenosis or occlusion are well-documented sources of emboli [26]. Fibrin-platelet thrombus forms in carotid artery plaques with severe stenosis. These thrombi, when partially or totally dislodged, can produce distal embolism. Ulceration, a frequent finding in carotid plaques, exposes atherosclerotic plaque matrix to flowing blood and leaves behind cavities (changes in plaque morphology) within the plaques, inducing local flow disturbances; both factors independently or associated can trigger thrombosis.

Pathological studies on carotid plaques removed at carotid endarterectomy have suggested that morphological factors such as intraplaque hemorrhage contribute to embolization from plaque [18, 20]. Changes in plaque morphology and configuration visualized on coronary angiography have also been associated with intraplaque fissuring and thrombosis in coronary arteries and have been implicated in the production of acute myocardial ischemia from coronary artery atherosclerotic disease [1, 8, 9, 13, 16].

Neurological symptoms ipsilateral to atherosclerotic plaque at the carotid artery bifurcation are associated with growth of plaque, particularly when the plaque extends along the wall of the carotid sinus into a mural configuration [29]. Progression of plaque along the wall of the carotid sinus is associated with a higher percentage of ipsilateral symptoms unrelated to hemodynamic factors [29].

Morphological examination of carotid plaques has demonstrated that mural plaques (ultrasonographic examination) have a significantly higher incidence of intraplaque hemorrhage suggestive of plaque fissure than do nodular plaques (ultrasonographic determination) and that only mural plaques are associated with intraluminal thrombus [28]. In brief, plaques which follow the contours of the wall of the carotid sinus and appear to be an integral part of the wall with a length exceeding three times the width or thickness are classified as mural plaques, while plaques which appear to be attached to the wall of the carotid sinus and grow out into the lumen with the width approaching the length are classified as nodular [27]. In coronary artery disease, it has been reported that ischemic symptoms in

---

Division of Cardiology, Mount Sinai Medical Center, One Gustave L. Levy Place, New York, NY 10029, USA.

Cerebral Ischemia and Calcium
Ed. by A. Hartmann and W. Kuschinsky

patients with either unstable angina or myocardial infarction are associated with intraluminal thrombus over a ruptured plaque or fissure [1, 9, 13, 16]. The coronary artery lesions implicated in these forms of myocardial ischemia are usually eccentric and have a crescentic appearance with irregular borders on coronary angiography [1, 8, 11]. These lesions are very similar to lesions at the carotid artery bifurcation designated as mural plaques [27], both in configuration and morphology.

The etiology of cerebrovascular symptoms ipsilateral to plaques at the carotid bifurcation and that of symptoms of myocardial ischemia in the territory of crescentic plaques in the coronary arteries may both be related to intraluminal thrombus. Cerebral ischemic events from a lesion at the carotid artery bifurcation are usually attributed to emboli from the plaque to intracranial arteries [22, 25], while in the coronary circulation, transient narrowing or occlusion of the vessel due to intraluminal thrombosis at the site of stenosis causes ischemia of the myocardium [9, 24]. However, platelet embolization distal to a coronary lesion has been implicated in sudden death following unstable angina [16].

It is, therefore, evident that ruptured atherosclerotic plaque, probably associated with intraplaque hemorrhage or dissection, results in intraluminal thrombus and that this is a common feature in arterial thrombosis. Atherothrombotic strokes, the results of atheroma and subsequent thrombosis in the arteries of the brain, account for a large percentage of the cases of stroke [10].

Symptom-producing occlusions may be primary by atherosclerotic plaque and thrombosis but more often are secondary due to emboli [19]. Primary atherosclerotic occlusion of the carotid or vertebral arteries is not infrequent, but such primary occlusion of branches of the middle cerebral, anterior cerebral, or posterior cerebral arteries is quite rare [17, 19]. The region of the carotid bifurcation is the most frequent site of ulceration and the origin of most emboli producing infarcts [22, 23, 27]. While emboli are probably responsible for most significant strokes, atherosclerotic carotid or vertebral disease may also progressively narrow the arterial lumen [7, 15, 17].

Blood fluid dynamics, exposed vascular substrate, and blood factors significantly influence platelet–vessel wall interaction and the growth of a thrombus [5]. However, specific cell to cell and cell to matrix interactions and their regulation by each of the factors mentioned above in thrombus growth are still a matter of investigation. Thus, the thrombotic complication of an atherosclerotic plaque is determined by the arterial wall, the blood components, and the local fluid dynamics of the area. Our objective is to study the relative contribution of each of these factors to cell–wall, cell–cell, fluid phase–cell, and fluid phase–wall interactions that underlie the thrombotic process. By means of an original perfusion chamber (L. Badimon, 1984) [2–6], we have been able to design studies oriented to discriminate the contributions of each factor to thrombus growth.

Effect of Vascular Wall or Substrate

Following removal of the endothelial lining of a normal blood vessel by a mild or superficial injury, the subendothelium is coated by a layer of adherent platelets (Fig. 1). Adhesion, the process by which platelets attach to the exposed subendothelium, is dependent on platelet receptors, adhesive glycoproteins present in plasma and in platelets, and on exposure of activating sites in the subendothelium [14]. Glycoprotein Ib in the platelet membrane is necessary for normal initial contact of platelet with subendothelial surface. Glycoprotein Ia, by binding directly to collagen, may also be important. Glycoprotein IIb/IIIa in the platelet membrane is the receptor for a variety of circulating proteins, including von Willebrand factor, fibronectin, and vitronectin, and it also mediates platelet adhesion.

Severe injury exposes components of the arterial media, particularly fibrillar collagen (type I being more prevalent in diseased vessels, type III in normal vessels), which in addition to other mediators activates platelets. Activation results in the exposure of glycoprotein IIb/IIIa receptors on the platelet surface that are recognized particularly by fibrinogen and von Willebrand factor, which then

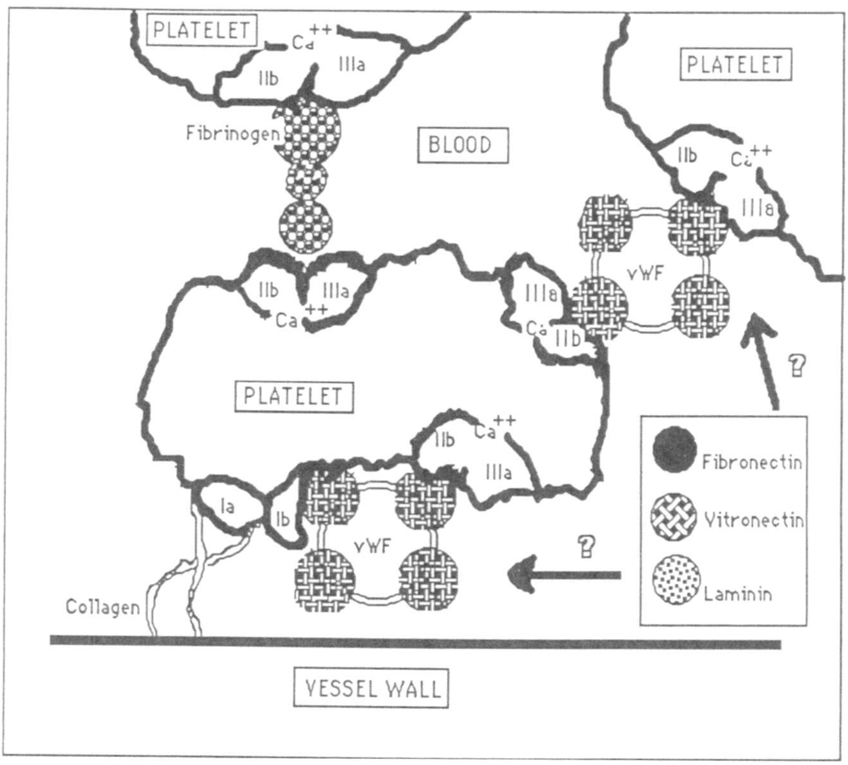

Fig. 1. Diagrammatic representation of platelet–vessel wall and platelet–platelet interaction. Receptors in the platelet membrane and ligands for the receptors are shown

form links between platelets resulting in aggregation (Fig. 1). This activation is promoted by various pathways. There is direct stimulation by collagen and ADP and by thromboxane A_2 (arachidonate pathway). In addition, during platelet adhesion and aggregation the clotting mechanism may be activated by means of exposure of the blood to the negatively charged damaged surface – intrinsic system – and to tissue factor – extrinsic system. Thrombin is then generated which further promotes platelet aggregation (probably through the same pathway as collagen) and, most importantly, leads to the formation and polymerization of fibrin, which stabilizes the platelet mass and allows the arterial thrombus to resist removal by the high intravascular pressure and shear forces.

In the clinical context, thrombosis, related to both platelet aggregation and activation of the coagulation system, is probably stimulated by the deep injury occurring with plaque rupture.

Effect of Blood Flow

The number of platelets which become associated with an injured vessel is determined not only by the degree of injury as described above but also by their transport to the injured area. This is determined by conditions of blood flow, specifically the wall shear rate, which is a measure of the difference in blood velocity between the blood in the center of the vessel and the wall. Shear rate increases with decreasing vessel diameter and with increasing flow. We have explored the relationship between shear rate and platelet deposition using the perfusion chamber [2] in which different vessel wall substrates can be exposed at various shear rates simulating those found throughout the circulation. The exposure of deendothelialized aortic wall, mimicking mild injury to the vessel wall, to blood flow at low shear rates characteristic of large arteries and of veins, results in adhesion of a single layer of platelets. At higher wall shear rates, typical of those locally developed on top of arterial atherosclerotic stenoses, initial platelet deposition rate and maximal extent of deposition are significantly higher than at the lower shear rate, leading to platelet aggregation and thrombosis. The maximum platelet deposition or thrombus formation is observed after 10–20 min but with longer exposure the thrombus is dislodged from the exposed surface by the flowing blood. Thus, the platelet deposition is relatively transient, suggesting labile thrombosis. This transient thrombotic phenomenon on a region with high shear rate mimicking stenosis and with a mildly damaged vessel wall may have clinical implications in the context of episodic embolization.

Platelet deposition on native collagen type I fibrils, a model for the collagen fibrils exposed when the vessel wall is deeply injured, has also been studied in experiments as described above for mildly injured wall in the pig model. At both low and high shear rates platelet deposition is much greater than on subendothelium of mildly injured vessel. Specifically, in contrast with subendothelium, at the higher shear rates locally developed at the peak of stenoses, platelet thrombi are large and remain adherent to the collagen surface. Therefore, in the presence of a powerful thrombogenic stimulus such as collagen type I, the thrombus is more

Table 1. Platelet deposition on areas of stenosis[a]

% Stenosis	0	35	55	80
Total platelet deposition ($\times 10^6/cm^2$)	20±3	23±8	30±9	88±26
% of deposition at apex of stenosis	29	38	60	100

[a] Exposed substrate was aortic tunica media and blood was native (nonanticoagulated). Studies conducted in nonlaminar flow.

Table 2. Platelet deposition on different vessel wall components under different blood flow conditions[a] ($\times 10^6/cm^2$)

Substrate	$212\ s^{-1}/424\ s^{-1}$ [b]	$1690\ s^{-1}/3380\ s^{-1}$
Collagen type I	45 ±12	255±30
Subendothelium	2.5± 0.4	20± 6

[a] Blood was native (nonanticoagulated).
[b] Blood shear rate as defined by $SR = 32Q/60\Pi D^3$; where D is the luminal diameter and Q is blood flow.

permanent. In addition, blood flowing through a vessel is not only suddenly accelerated as it passes by a stenosis, but is then immediately decelerated distal to the stenosis. This deceleration induces flow separations and recirculation zones (vortices) downstream of the stenosis. The combination of a high shear rate area (luminal narrowing at the site of the stenosis) activating platelets, and a low shear rate area (poststenotic recirculation zones) leading to deposition of fibrin, contributes to the "head" of the fixed thrombus composed mainly of platelets and a "tail", which contains a larger amount of fibrin in a meshwork, trapping large numbers of red cells.

As such, the original perfusion chamber has been adapted (L. Badimon, 1987) to study platelet–vessel wall interaction in conditions encountered in the presence of eccentric or mural atherosclerotic lesions. We have studied the role of changes in fluid dynamics on the outcome of thrombus growth on severely damaged arterial wall. Table 1 shows the results obtained with increasing degrees of stenosis using the pig experimental model. Thus, the thrombotic response due to the exposure of similarly thrombogenic material (aortic tunica media) is strongly dependent on the degree of stenosis. The alteration of flow lines at the stenotic lesion seems to lead to increased cell interception and impaction and enhanced local cell activation.

Table 2 shows how the nature of the exposed vascular wall matrix is associated with blood flow factors to stimulate thrombus growth. In spontaneous plaque rupture the residual surface is not homogeneous. The rupture or ulceration leaves areas of medial or atherosclerotic-core exposure and areas of mild injury with exposure of the subendothelial matrix. Studying those variables with the perfusion chamber in laminar flow conditions, we have been able to evidence that platelet deposition is dependent on the reactivity of the vascular matrix ex-

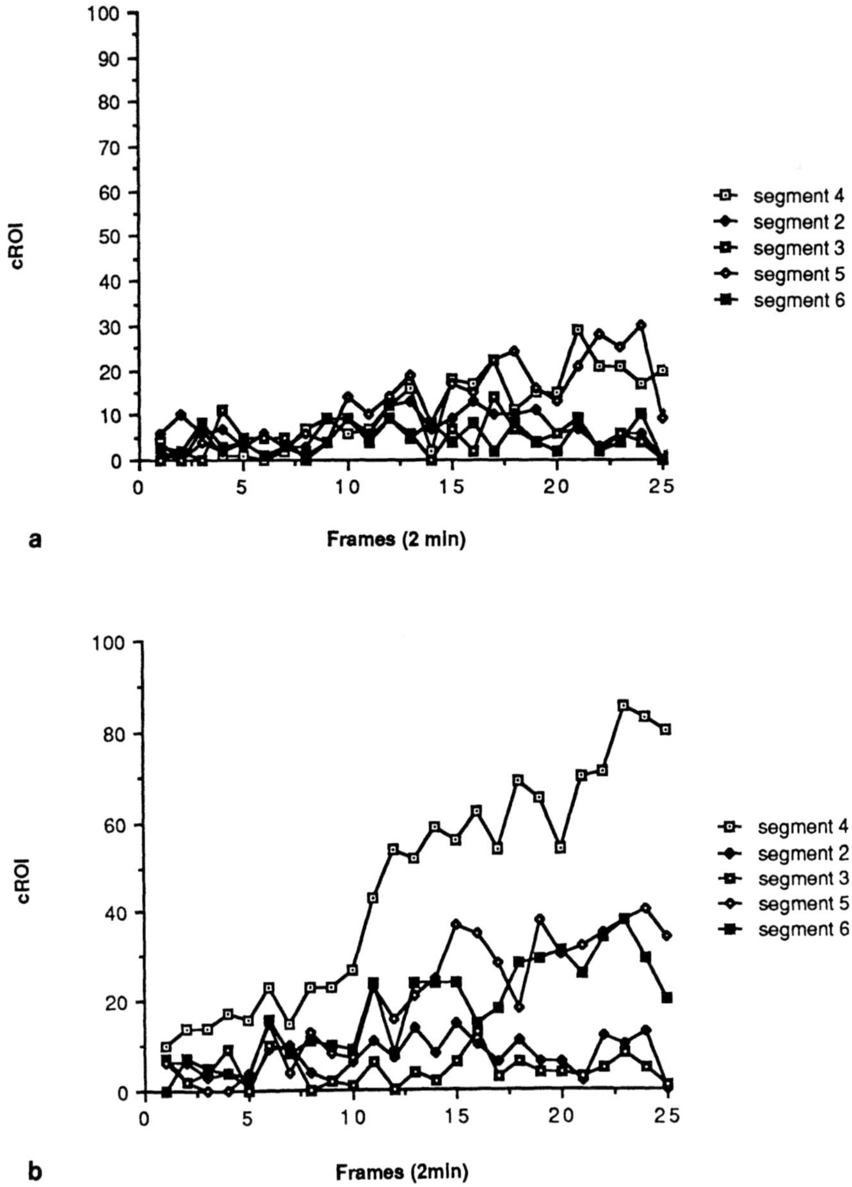

Fig. 2 a, b. Kinetics of platelet deposition by continuous scintigraphic imaging of platelet-vessel wall interaction. Pig heparinized blood perfused pig tunica media in laminar flow conditions (0% stenosis) (**a**) and when there was an 80% stenosis in area 4 (**b**). The stenosis was located in the central part of the perfused surface (area 4); areas 2 and 3 were proximal and 5 and 6 distal to the stenosis. Axial distribution of platelet deposition showed significant platelet accumulation at the peak of the stenosis. *Ordinates* show gamma counts per region of interest (ROI) and *abscissas* frames of acquisition (every 2 min a new frame was acquired). The perfusion time was 50 min. The experiments depicted in **a** and **b** were run in the same conditions (same donor), only changing flow profile posed to blood and on the local flow conditions. The greatest rate of thrombus growth is observed with collagen at high shear rate conditions, mainly when the coagulation pathway is not inhibited.

Studying the dynamics of thrombus growth on vessel wall tunica media by continuous imaging of platelets (scintigraphy), we have been able to evidence the important role of blood flow conditions on thrombus growth. This method of investigation clearly offers an accurate assessment of the kinetics of thrombosis and reveals significant events in thrombus growth and stabilization which would be missed in stop-flow methodology. Figure 2 shows a typical finding when perfusing blood in two flow conditions, laminar (0% stenosis) and 80% stenosis. Figure 2a shows the kinetics of platelet deposition in the absence of stenosis (laminar flow conditions at shear rate typical of medium size unobstructed arteries). There is no significant platelet accumulation in any area. In contrast, Figure 2b shows the kinetics of platelet deposition when there is 80% stenosis (located in area 4). A significant accumulation of platelets takes place on the apex of the stenosis. Areas 5 and 6 distal to the stenosis also show some platelet accumulation. This technique also allows study of embolization phenomena from the growing thrombus (see legend to Fig. 2).

Effect of Blood Factors

Blood components, either cells or fluid-phase components, have an important role in thrombus growth. The effect of blocking coagulation enzymes and zymogens, of enhancing fibrinolytic activity, and of blocking platelet function by platelet inhibitory agents is being exhaustively evaluated by our group [14] and others. In the present article, I will briefly comment on our findings on the role of von Willebrand factor (vWF) a physiological protein present in blood, platelets, and endothelial cells, as an in vivo ligand for thrombus growth.

In an experimental animal with homozygous von Willebrand's disease, we have been able to observe that vWF (absent in this severe form of the disease, similar to type III von Willebrand's disease in humans) is an important in vivo ligand for platelet vessel–wall interaction since thrombus formation on subendothelium (model of mild injury) was significantly inhibited at blood flow shear rates encountered physiologically in the arterial system when heparinized blood was perfused (see Fig. 1) [4]. Collagen type I was also mounted in the tubular (laminar) perfusion chamber placed within an extracorporeal circuit (carotid artery to jugular vein) and the material was exposed to blood from normal pigs and severe homozygous vWF deficient pigs (vWF$<3\%$) [6]. Thrombus formation was measured by autologous ^{111}In-platelet labeling and by ultrastructural morphology (Fig. 3). Heparinized and native blood from these pigs was perfused over the substrate for 3 and 5 min at local shear rates from 212 s^{-1} to 3380 s^{-1}. On collagen type I exposed to nonanticoagulated blood, for all exposure times studied, thrombus formation in the absence of vWF was significantly reduced at a high shear rate typical of stenotic areas but not at a low shear rate typical of unobstructed medium size arteries. A similar inhibition of thrombus formation due

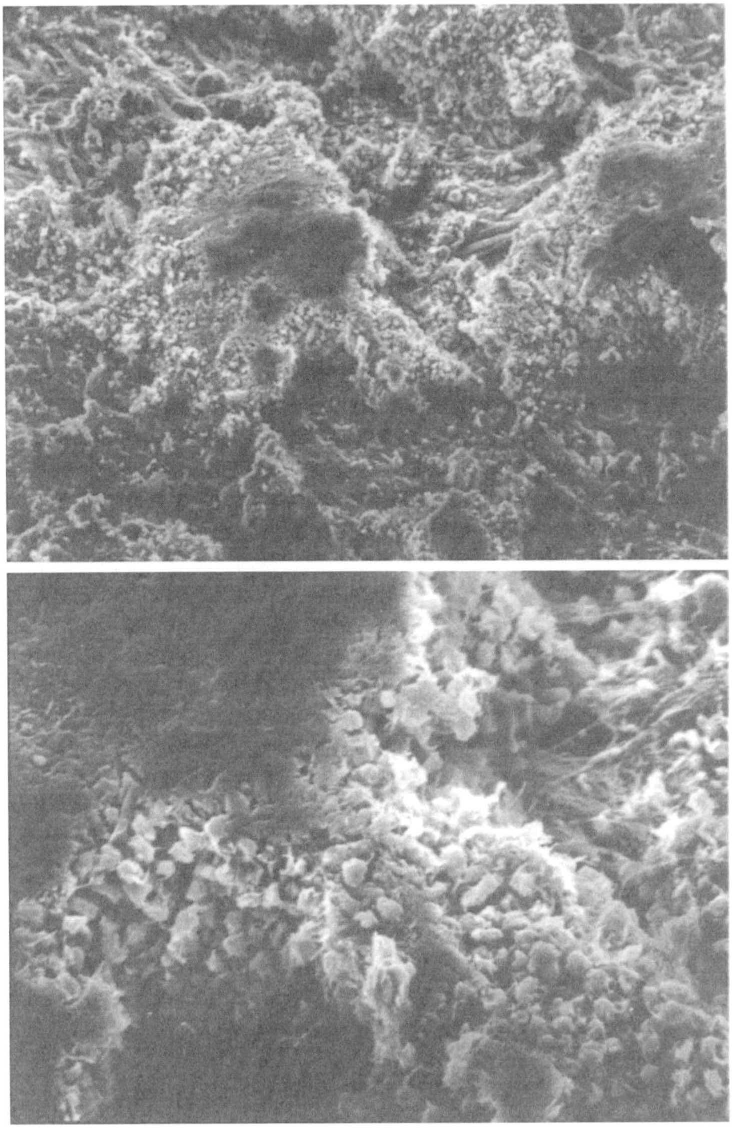

Fig. 3. Scanning electron microscopic micrograph of native collagen type I fibrils perfused by heparinized normal pig blood for 5 min. Observe that platelet aggregates are attached to the substrate in significant amounts (different magnification of the same field; *top*, $\times 450$ and *bottom*, $\times 1800$). Platelet deposition was quantitatively measured by ^{111}In-labeled platelets; this specimen showed an average of 55×10^6 platelets per cm^2 perfused

to vWF deficiency was observed in both heparinized and native blood; however, thrombus formation was significantly more reduced in the presence of heparin, presumably due to the lack of stability of the accumulated platelets in the absence of fibrin formation. Therefore, in severe injury both the activation of blood coagulation proteins and the presence of vWF contribute significantly to the platelet–platelet interactions necessary for thrombus formation.

The thrombotic masses can embolize or be organized into the plaque, accentuating the stenotic narrowing of the vessel and hence the chances of further thrombosis and embolic episodes of greater magnitude (embolic thrombus mass).

In cases of ulcerated "nonstenotic" lesions [21] there will be exposure of atherosclerotic plaque matrix with significant platelet activation and thrombosis [4, 6]. The presence of stenosis will largely increase the chances of thrombosis and embolism due to the additional effect of significant local flow disturbances.

Summary

In this article, we have emphasized that in addition to the exposed substrate, flow and specifically the local shear rate conditions regulate the attachment, growth, and stability of platelet masses on ruptured atherosclerotic plaques. Indeed, rheology mediates significantly the biochemical mechanisms involved in platelet–vessel wall interaction. The high molecular weight glycoprotein von Willebrand factor has an important physiological role on the stabilization of thrombus masses by securing platelet aggregate formation to the underlying substrate.

References

1. Ambrose JA, Winters SL, Arora RR, Ricio A, Gorlin R, Fuster V (1986) Angiographic evolution of coronary artery morphology in unstable angina. J Am Coll Cardiol 7:472–478
2. Badimon L, Badimon JJ, Galvez A, Chesebro JH, Fuster V (1986) Influence of arterial damage and wall shear rate on platelet deposition. Arteriosclerosis 6:312–320
3. Badimon L, Turitto VT, Rosemark JA, Badimon JJ, Fuster V (1987a) Characterization of a tubular flow chamber for studying platelet interaction with biological and prosthetic materials, J Lab Clin Med 110:706
4. Badimon L, Badimon JJ, Turitto VT, Rand J, Fuster V (1987b) Platelet deposition on von Willebrand factor deficient vessels. Extracorporeal perfusion studies in swine with von Willebrand's disease using native and heparinized blood. J Lab Clin Med 110:634
5. Badimon L, Badimon JJ, Turitto VT, Fuster V (1987c) Thrombosis: studies under flow conditions. Ann NY Acad Sci 516:527–540
6. Badimon L, Badimon JJ, Turitto VT, Vallabhajosula S, Fuster V (1988) Platelet thrombus formation on collagen type. I. Circulation 78:1431–1442
7. Busuttil RW, Baker JD, Davidson RK, Machleder HI (1981) Carotid artery stenosis-hemodynamic significance and clinical course. JAMA 245:1438–1441
8. Davies MJ, Thomas AC (1985) Plaque fissuring – the cause of acute myocardial infarction, sudden ischemic death and crescendo angina. Br Heart J 53:363–373

9. DeWood MA, Spores J, Notske R, Mouser IT, Burroughs R, Golden MS, Lang HT (1980) Prevalence of total coronary occlusion during the early hours of transmural myocardial infarction. N Engl J Med 303:896–902
10. Dyken ML, Wolf PA et al. (1985) Special report. Risk factors in stroke. Stroke 6:1105–1111
11. Falk E (1985) Unstable angina with fatal outcome: dynamic coronary thrombosis leading to infarction and/or sudden death. Circulation 71:699–708
12. Finklestein S, Kleinmann GM, Cuneo R, Baringer JR (1980) Delayed stroke following carotid occlusion. Neurology (Minneap) 30:84–88
13. Fuster V, Chesebro JH (1986) Mechanism of unstable angina. N Engl J Med 315:1023–1024
14. Fuster V, Badimon L, Badimon JJ, Adams PC, Turitto VT, Chesebro JH (1987) Drugs interfering with platelet functions: mechanisms and clinical relevance. In: Veerstraete M, Vermylen J, Liujnen R, Arnout J (eds) Thrombosis and haemostasis. Leuven University Press, pp 349–418
15. Gomensoro JB, Maslenikov V, Azambuja N, Fields WS, Lemak N (1973) Joint study of extracranial arterial occlusion. JAMA 224:985–991
16. Gorlin R, Fuster V, Ambrose JA (1986) Anatomic-physiologic links between acute coronary syndromes. Circulation 74:6–9
17. Hass WK, Fields WS, North RR, Kricheff II, Chase NE, Bauer RB (1968) Joint study of extracranial arterial occlusion. JAMA 203:159–166
18. Imparato AM, Riles TS, Mintze KR, Baumann FG (1983) The importance of hemorrhage in the relationship between gross morphologic characteristics and cerebral symptoms in 376 carotid artery plaques. Ann Surg 197:195–203
19. Kricheff II (1987) Arteriosclerotic ischemic cerebrovascular disease. Radiology 162:101–109
20. Lusby RJ, Ferrell LD, Ehrenfeld WK, Stoney RJ, Wylie EJ (1982) Carotid plaque hemorrhage: its role in production of cerebral ischemia. Arch Surg 117:1479–1488
21. Moore WS, Hall AD (1968) Ulcerated atheroma of the carotid artery. Am J Surg 116:237–242
22. Moore WS, Hall AD (1970) Importance of emboli from carotid bifurcation in pathogenesis of cerebral ischemic attacks. Arch Surg 101:708–711
23. Raichle ME (1983) The pathophysiology of brain ischemia. Ann Neurol 13:2–10
24. Rentrop P, Blanke H, Karsch K, Kaiser H, Kostering H, Leitz K (1981) Selective intracoronary thrombosis in acute myocardial infarction and unstable angina pectoris. Circulation 63:307–316
25. Thiele BL, Young JV, Chikos PM, Hirsch JH, Strandness DE (1980) Correlation of arteriographic findings and symptoms in cerebrovascular disease. Neurology 30:1041–1046
26. Wechsler LR (1987) Ulceration and carotid artery disease. Curr Concepts Cerebrovasc Dis 22:19–23
27. Weinberger J, Robbins A (1983) Neurologic symptoms associated with nonobstructive plaque at carotid bifurcation. Analysis by real-time B-mode ultrasonography. Arch Neurol 40:489–492
28. Weinberger J, Marks SJ, Gaul JJ, Goldman B, Schanzer H, Jacobson JJ, Dikman S (1987) Atherosclerotic plaque at the carotid artery bifurcation: correlation of ultrasonographic imaging with morphology. J Ultrasound Med 6:363–366
29. Weinberger J, Ramos L, Ambrose JA, Fuster V (1988) Morphology and dynamic changes of atherosclerotic plaque at the carotid artery bifurcation: sequential imaging by real time B-mode ultrasonography (submitted)

Suppression of Brain and Adrenal Gland Arteriosclerosis by Calcium Antagonism in Hypertensive Rats (Skelton Hypertension)

P. E. SCHWABEDAL

Rats develop hypertension following Skelton's treatment, i.e., (a) unilateral adrenalectomy and contralateral adrenal enucleation, (b) unilateral nephrectomy, and (c) administration of 1% NaCl with the drinking fluid (Skelton 1959). The pathogenesis of Skelton hypertension is still obscure. But in this hypertension model the adrenal operation (a) can be substituted by stress (Deane and McKibbin 1946) due to pantothenic acid deficiency (Schwabedal et al. 1985) and the nephrectomy (b) by means of a slight reduction in the kidney blood perfusion (Schwabedal 1987a, 1988a). Therefore Skelton hypertension can serve as a model for an animal experimental hypertension in which the blood pressure elevation is caused by three factors working simultaneously: (a) stress, (b) a slight reduction in kidney blood perfusion, and (c) high alimentary sodium chloride (cf. Schwa-

Table 1. Mean systolic blood pressure, mean plasma flunarizine concentration, and mean number of arteriosclerotic blood vessels in ten histological sections of the brain and of the adrenal gland (with ten sections/organ and rat and ten rats/group) in experimental groups I–III 7 weeks after Skelton's treatment to induce hypertension

Experimental groups and their treatment		Mean systolic blood pressure (mm Hg)	Mean plasma flunarizine concentration (ng/ml)	Mean number (%) of arteriosclerotic blood vessels in ten histological sections of the:	
				Brain	Adrenal
I	Skelton treatment without flunarizine administration	223±12	0	6.2±7.8 (100%)	31.3±31.5 (100%)
II	Skelton treatment with flunarizine administration	214±16	336±136	0 (0%)	1.5±2.9 (4.8%)
III	No treatment (control)	107±16	0	0 (0%)	0 (0%)

Institute of Anatomy, University of Bonn, Nussallee 10, D-5300 Bonn 1, FRG.

Cerebral Ischemia and Calcium
Ed. by A. Hartmann and W. Kuschinsky

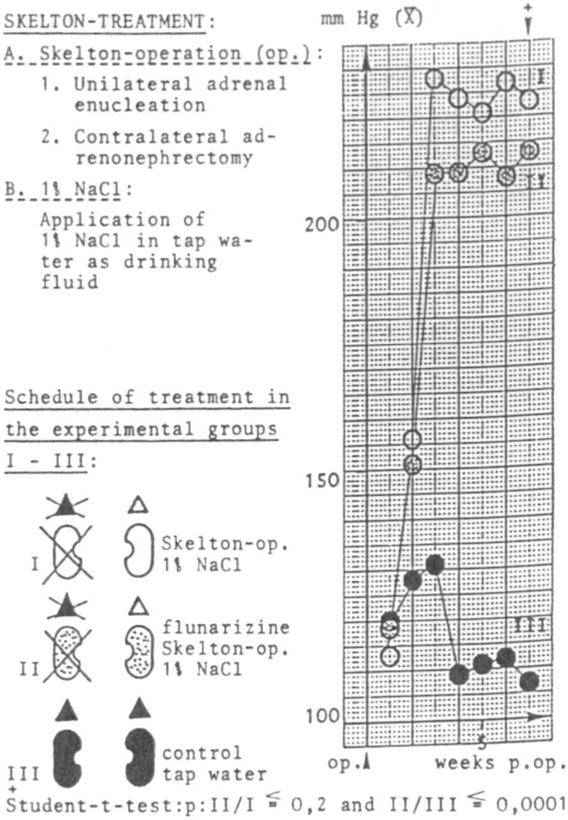

Fig. 1. Mean systolic blood pressure of rats following Skelton treatment with and without application of the calcium-antagonist flunarizine

bedal 1987 b, c). In connection with the findings described here it is interesting to mention that each of these three hypertension-related factors has been discussed repeatedly as being causal in the pathogenesis of essential hypertension (Wollheim and Moeller 1960).

Following Skelton's treatment, experimental group I (ten male rats) developed malignant hypertension within 7 weeks. All further experimental groups had the same number of animals and time course. The mean body weight was 152 g at the start of the experiment (Material and Methods in: Schwabedal 1988 b). In experimental group I mean systolic blood pressure increased from 108 ± 10 to 223 ± 12 mmHg (Fig. 1) and 375 arteriosclerotic vessels could be counted in 200 histological sections (ten from each animal and organ) of the brains and of the adrenal glands (Table 1, Fig. 2). In brain frontal sections of four animals degenerations were evident (Fig. 3 a). In the brain parenchyma of four animals disseminated depositions of extravasated fibrin (Fig. 3 b) and in one animal a submeningeal hematoma could be observed (Fig. 3 c). In the adrenal gland arteries of two animals an aneurysm could be found (Fig. 3 d), in one case with an extravasation of blood.

Following Skelton's treatment and admixture of flunarizine with the food (40 mg/kg rat and day; mean plasma flunarizine value at the end of the experi-

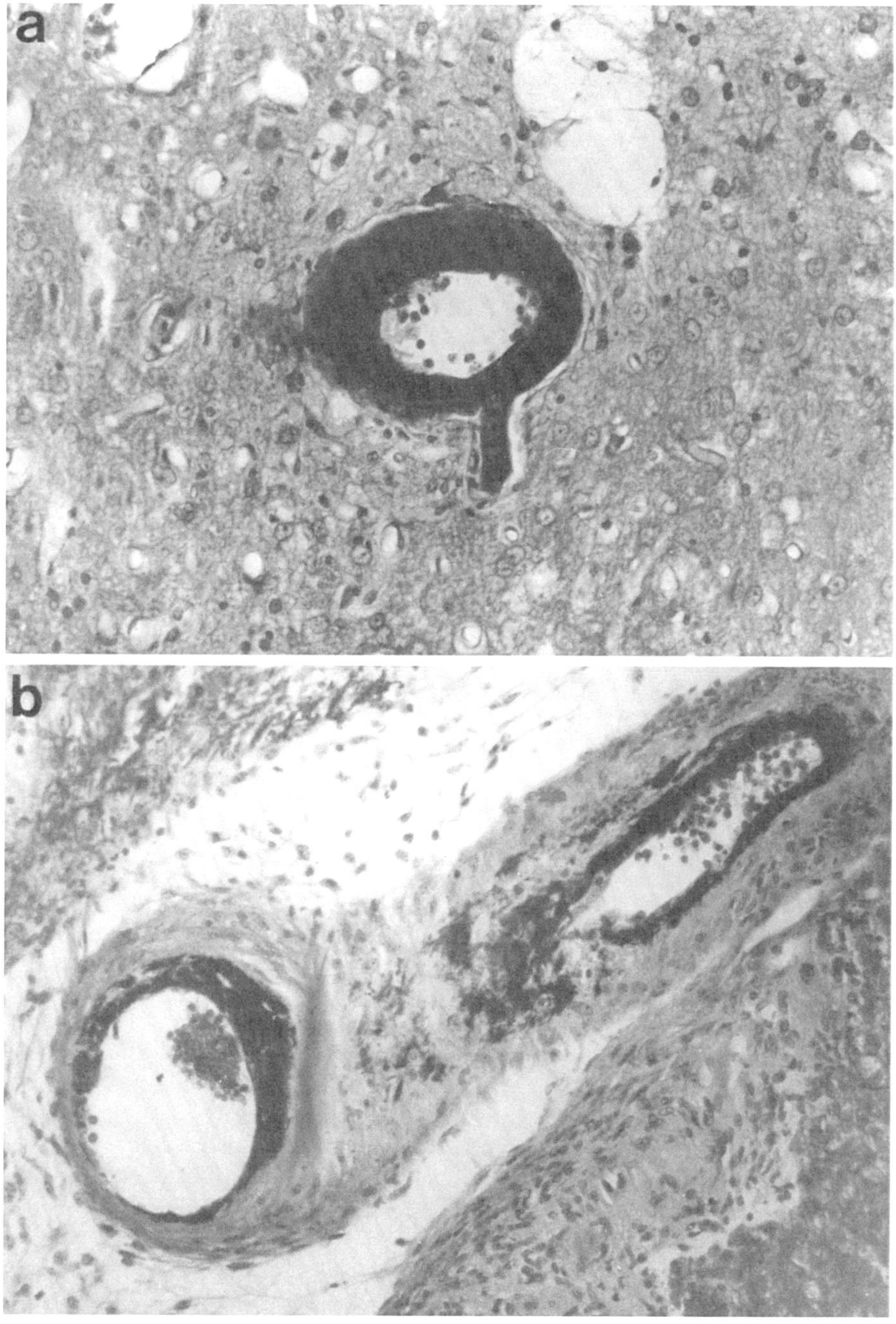

Fig. 2. a Frontal section through the brain of a Skelton hypertensive rat (experimental group I) with a final systolic blood pressure of 239 mmHg and arteriosclerosis. Note the excessive deposition of fibroid in the arterial wall. Azan staining, ×280. **b** Cross-section through the adrenal gland of a Skelton hypertensive rat (experimental group I) with a final systolic blood pressure of 239 mmHg and arteriosclerosis of an adrenal gland artery. Note the massive deposition of fibrin in the tunica intima of an adrenal capsule adjacent artery. Mallory staining, ×280

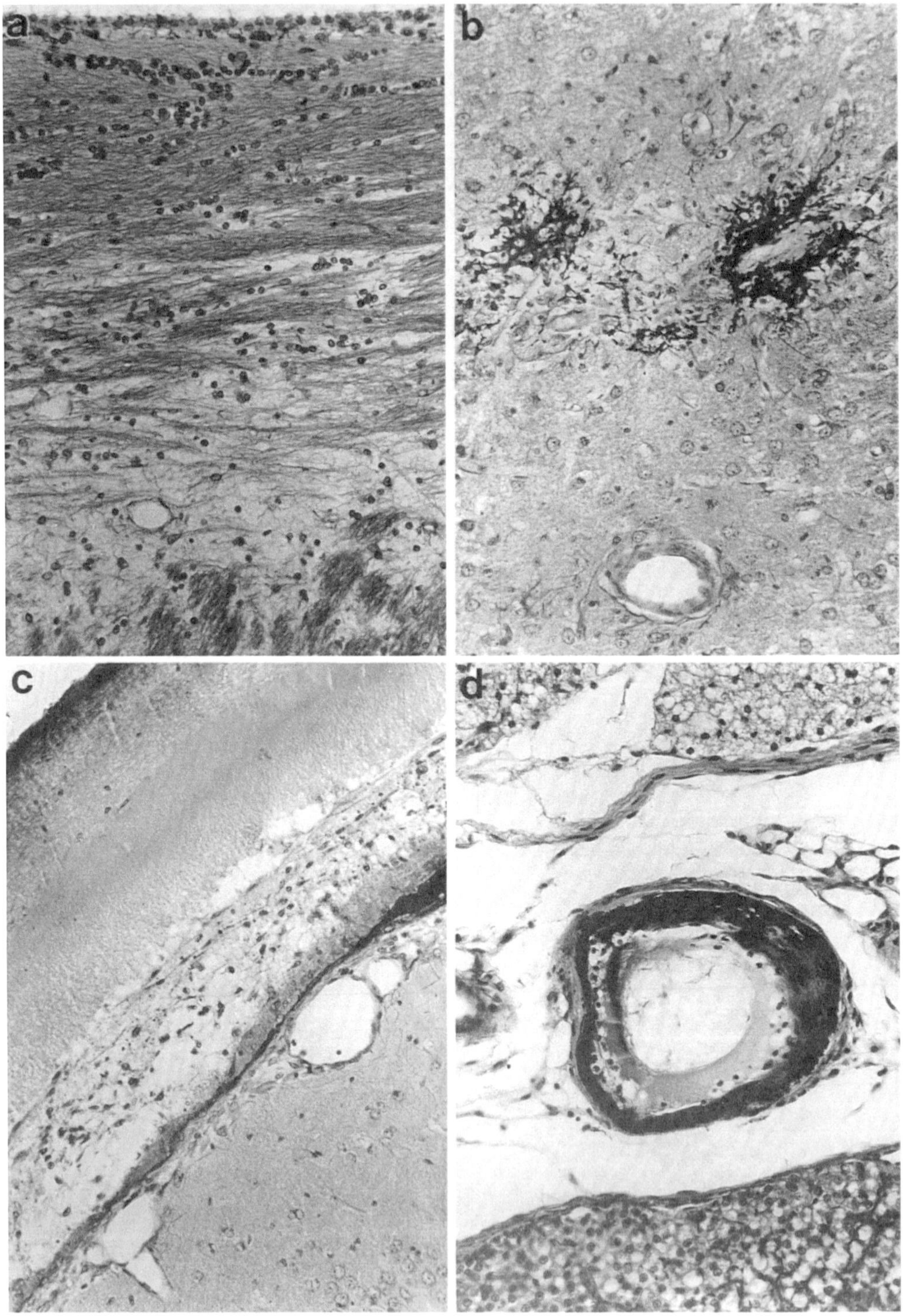
a
b
c
d

ment: 336 ng/ml), in experimental group II all rats developed hypertension, whereby the mean systolic blood pressure increased from 109 ± 10 to 214 ± 16 mmHg (Fig. 1). But in contrast to experimental group I, only 15 arteries with sclerosis could be counted in 200 histological sections from the brains and the adrenal glands (Table 1), and no degenerations, no extravasal depositions of fibrin, no submeningeal hematomas, and no aneurysm could be detected.

The nonoperated, tap water drinking control animals, experimental group III, remained normotensive with mean systolic blood pressure values of 109 ± 10 mmHg at the start and 107 ± 16 mmHg at the end of the experiment (Fig. 1). No pathological findings could be observed in the brains or in the adrenal glands (Table 1).

The findings presented show that the calcium antagonist flunarizine does not abolish hypertension but suppresses arteriosclerosis to a degree of 96%, and thereby is effective in preventing hypertension-induced vascular disease of the brain.

References

Deane HW, McKibbin JM (1946) The chemical cytology of the rat's adrenal cortex in pantothenic acid deficiency. Endocrinology 38:385

Schwabedal PE (1987a) Hypertonie nach extra- und intrarenaler Nebennierenregeneration. Verh Anat Ges 81:649

Schwabedal PE (1987b)Die Skelton-Hypertonie eine „Hypertonia composita". Acta Anat 130/1:84

Schwabedal PE (1987c) Hypothese einer Hypertonie mit definierter multifaktorieller Genese, abgeleitet von der Skleton-Hypertonie. Hochdruck 8:28

Schwabedal PE (1988a) Untersuchungen zur pathogenetischen Bedeutung der unilateralen Nephrektomie für die Entstehung der Skelton-Hypertonie

Schwabedal PE (1988b) Hemmung der hypertoniebedingten Koronarsklerose, Zerebralsklerose und Nephrosklerose durch Calcium-Antagonisten. Münchn Med Wochenschr

Schwabedal PE, Pietrzik K, Wittkowski W (1985) Pantothenic acid deficiency as a factor contributing to the development of hypertension. Cardiology 72 (Suppl 1):187

Skelton RR (1959) Adrenal regeneration and adrenal regeneration hypertension. Physiol Rev 39:162

Wollheim E, Moeller J (1960) Essentielle Hypertonie; Pathogenese. In: Von Bergmann G, Frey W, Schwiegk H (eds) Herz und Kreislauf; Hypertonie. Springer, Berlin Göttingen Heidelberg (Handbuch der Inneren Medizin, vol 9, p 307)

Fig. 3. a Frontal section through the brain of a Skelton hypertensive rat (experimental group I) with a final systolic blood pressure of 213 mmHg and a degeneration of the corpus callosum. Note the rarefaction of the substantia alba. Mallory staining, ×280. **b** Frontal section through the brain of a Skelton hypertensive rat (experimental group I) with a final systolic blood pressure of 212 mmHg and an extravasation of fibrin. Mallory staining, ×280. **c** Frontal section through the brain of a Skelton hypertensive rat (experimental group I) with a final systolic blood pressure of 223 mmHg and a submeningeal hematoma. Mallory staining, ×280. **d** Cross-section through the adrenal gland and the paraadrenal connective tissue of a Skelton hypertensive rat (experimental group I) with a final systolic blood pressure of 239 mmHg and an aneurysm of an adrenal capsule adjacent artery. Note the dilatation of the blood vessel and the deposition of fibroid in the destructed artery wall. Azan staining, ×280

Extracellular Ion Activities During Ischemia and Reperfusion of the Rat Brain: Influences of Drugs

F. Tegtmeier, D. Scheller, H. Dierking, A. Bock, J. Kolb, U. Peters, J. Urenjak, and M. Höller

Membrane failure develops inevitably in all conditions of severe cerebral oxygen shortage. Massive ion fluxes across the cell membrane are associated with a negative shift of the DC potential. A net influx of Ca^{2+} from the extracellular space and liberation from intracellular stores result in a substantial increase in intracellular Ca^{2+}, which may be regarded as a critical factor in the development of ischemic brain damage [4, 5]. It has been suggested that the ischemic membrane failure may trigger processes that cause irreversible structural damage [3]. Hence, delay or prevention of membrane failure may be suggested as one mode of protection of the ischemic brain.

Several observations indicate that brain damage after transient ischemia may also be incurred in the reperfusion period [9, 11], even if flow, ionic gradients, electrical function, and tissue levels of high energy phosphates are restored. Nevertheless, the monitoring of these parameters together with other intermediates of glycolysis and the Krebs' cycle may help to detect transient periods of energy imbalance in the reperfusion period, which may influence the final outcome.

In the present paper, two series of experiments are described. The first one deals with the effects of flunarizine (FLU), nimodipine (NIMO), and diphenylhydantoin (DPH) on the DC potential and extracellular potassium in anoxia. The influence of FLU on extracellular ions, energy metabolism, and organ perfusion in the postischemic reperfusion period of the isolated rat brain was investigated in the second part of the study.

Anoxia

Methods

Cortical extracellular potassium (K_e^+), calcium (Ca_e^{2+}), and the DC signal were measured with ion-selective microelectrodes in anesthetized, ventilated rats. EEG activity, arterial blood pressure, and body temperature were recorded. Anoxia was induced by ventilating the animal with nitrogen. The rats were pretreated with one of the following drugs: FLU (50 mg/kg i.p.), DPH (20 mg/kg i.v.), or NIMO (30 µg/kg i.v.). Animals without treatment served as controls. For statistical analysis the Wilcoxon-Mann-Whitney U-test was used.

Janssen Research Foundation, Raiffeisenstr. 8, D-4040 Neuss 21, FRG.

Cerebral Ischemia and Calcium
Ed. by A. Hartmann and W. Kuschinsky

Results

After treatment, mean blood pressure decreased to 70–80 mmHg without significant differences between the three drugs.

Pretreatment only marginally influenced the time span between onset of anoxia and isoelectricity of EEG, which was near 40 s in treated and control animals.

A typical experiment is given in Fig. 1. The time latency between onset of anoxia and negative DC shift was doubled by FLU (241 ± 30 s vs. 508 ± 104 s;

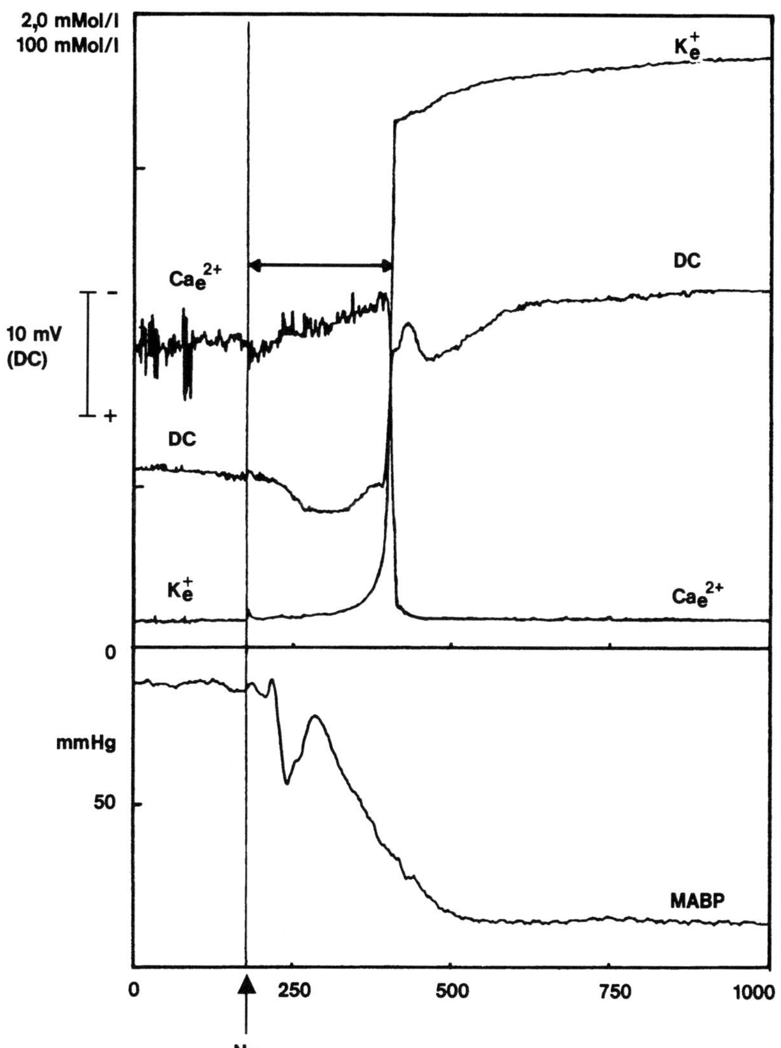

Fig. 1. Typical registration of cortical K_e^+, Ca_e^{2+} and DC signal as well as mean arterial blood pressure (*MABP*) in control rat before and during anoxia. The *arrow* indicates the time span between onset of anoxia (N_2) and negative DC shift/rapid movements of K_e^+ and Ca_e^{2+}

$P < 0.01$). The effect of DPH was less pronounced (354 ± 162 s; $P < 0.1$), while after NIMO the DC latency was not influenced (222 ± 42 s). as already described elsewhere [6], the time course of K_e^+ was biphasic. The threshold for the steep increase in K_e^+ was 12.1 ± 1.5 mmol/l in control brains. It was significantly increased by FLU (30.9 ± 1.9 mmol/l). The effect of DPH was similar (25.3 ± 2.3 mmol/l), while NIMO exhibited a less pronounced effect (16.2 ± 2.3 mmol/l).

Global Cerebral Ischemia and Reperfusion

Methods

Isolated rat brains were perfused at 32 °C with a fluorocarbon emulsion using pulsatory pressure [7]. Cortical extracellular pH (pH_e), calcium (Ca_e^{2+}), and the DC potential were recorded with triple-barreled ion-selective microelectrodes. Additionally, EEG activity and flow rate were registered. Brains treated with FLU (3.14 µmol/l in perfusion medium) and control brains (vehicle in perfusion medium) were investigated. Ischemia was induced by clamping the perfusion cannulas. Reperfusion was started either simultaneously with the DC shift (group I) or 10 min later (group II). Before ischemia, immediately after ischemia, and after 30 min of reperfusion, the brains were deep-frozen and prepared for analysis of ATP, ADP, lactate, and succinate in tissue.

Results

Extracellular Ions

Latency between onset of ischemia and negative DC shift was 165 ± 24 s in control brains. In contrast to experiments under anoxia described above, application of FLU did not prolong the time interval.

Typical registrations of experiments with short-term (group I) and prolonged ischemia (group II) are given in Fig. 2. In both groups, Ca_e^{2+} recovered in a biphasic manner: an initial fast rise followed by a slow increase. There was a latency between onset of reperfusion and increase of Ca_e^{2+}, lasting about 40 s in group I and about 120 s in group II. The rapid increase in Ca_e^{2+} coincided with the definite movement of the DC signal to the positive. The slope of this part of the Ca_e^{2+} curve may characterize the velocity of restoration of ion gradients across the cell membrane. The slope was influenced by the duration of the preceding ischemia,

Fig. 2. Typical registration of cortical pH_e, Ca_e^{2+}, and DC signal in the isolated perfused rat brain (untreated). *Open bars:* ischemic period before DC negativation; *closed bar:* ischemic period after DC negativation. *Top:* experiment with short-term ischemia (group I). Reperfusion was started simultaneously with DC negativation. *Bottom:* experiment with prolonged ischemia (group II). Reperfusion was started 10 min after DC negativation

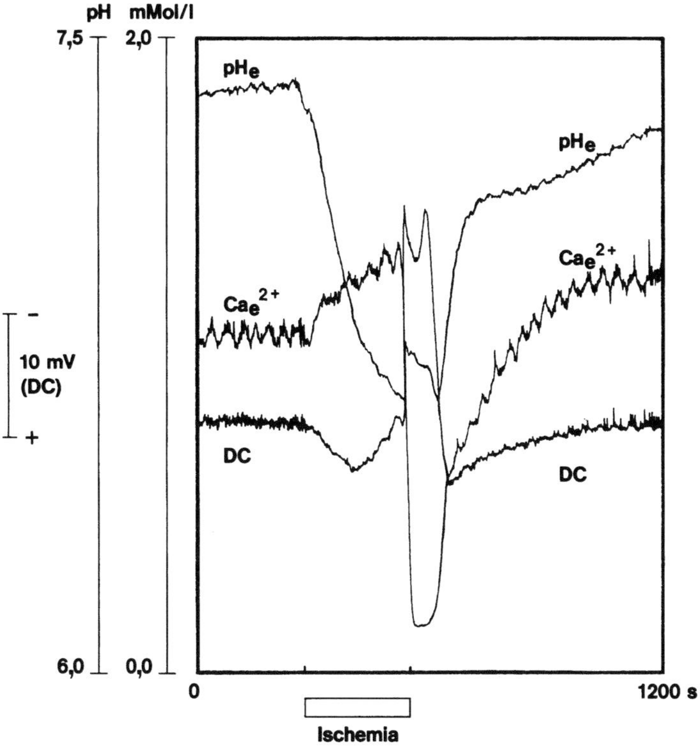
pH
mMol/l
7,5
2,0
6,0
0,0
pHe
pHe
Cae2+
Cae2+
DC
DC
10 mV
(DC)
-
+
0
1200 s
Ischemia

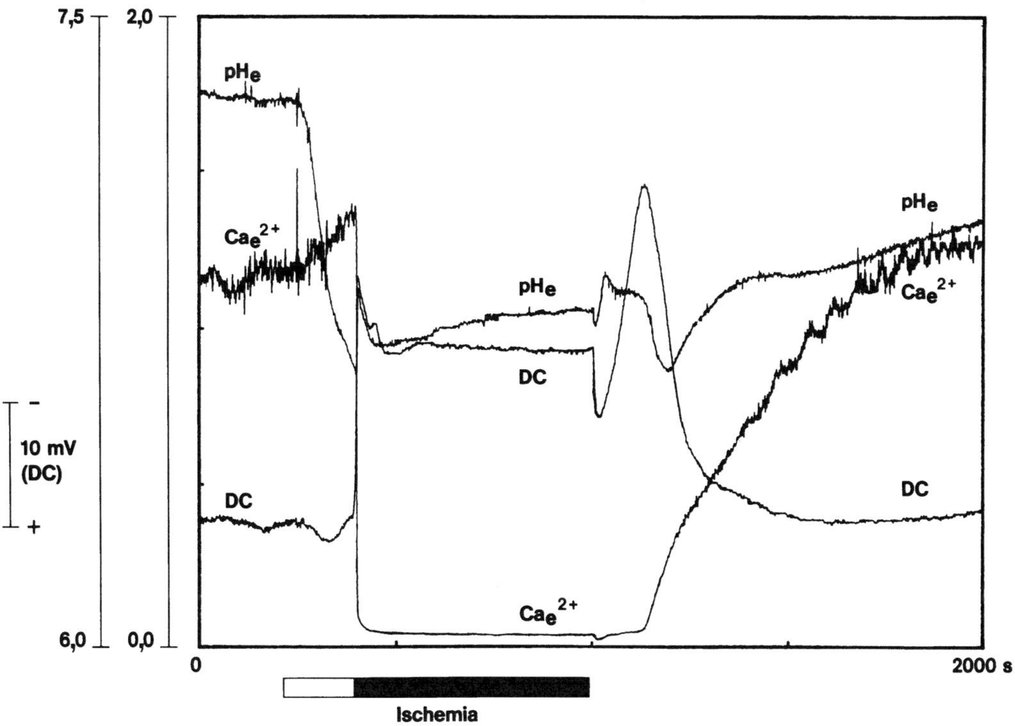
7,5
2,0
6,0
0,0
pHe
pHe
pHe
Cae2+
Cae2+
Cae2+
DC
DC
DC
10 mV
(DC)
-
+
0
2000 s
Ischemia

i.e., the longer the ischemic period, the flatter the slope. In control brains of group I, Ca_e^{2+} recovered at a rate of 0.38±0.15 mmol/l per minute, while in FLU-treated brains the rate was 0.98±0.31 mmol/l per minute ($P<0.05$). The effect of FLU on Ca_e^{2+} recovery in group II was not significant.

Recovery of pH_e was more complex. Simultaneously with the definite movement of the DC signal to the positive, pH_e became more acidic. This acidification was more pronounced after prolonged ischemic periods. Afterwards, pH_e recovered biphasically, with an initial rapid increase followed by a slower one. The slope of the initial increase was taken as a criterion for the velocity of cellular uptake and metabolic turnover of protons accumulated in ischemia; this slope became less steep after prolonged ischemic periods. On the other hand, in group I, the initial fast increase in pH_e was accelerated by FLU (0.25±0.06 pH units/min in controls vs 0.41±0.05 units/min in treated brains; $P<0.05$). The difference between control and treated brains of group II was not significant.

There was no difference in pH_e value at the end of reperfusion between control and treated brains of group I. In contrast, after prolonged ischemia (group II) the pH_e value was less acidic under treatment with FLU (7.15±0.1 vs 7.04±0.1 for controls).

Biochemical Parameters

There was no apparent effect of FLU on the tissue levels of the metabolites in nonischemic brains and in brains which were not subjected to reperfusion (see Table 1). To evaluate the recovery of the energetic state of the brains, ATP, ADP, lactate, and succinate were determined after reperfusion.

Table 1. Concentrations of lactate and succinate (in μmol/g d.w.) as well as the ratio of ATP to ADP before ischemia, immediately after ischemia, and after 30 min of reperfusion. Short-term ischemia: reperfusion simultaneously with DC negativation (group I); prolonged ischemia: reperfusion 10 min after DC negativation (group II). Control and treated brains were investigated. Mean values ±SD are given. $n=5$–8

	Before ischemia	After ischemia		After reperfusion	
		Group I	Group II	Group I	Group II
Lactate					
Control	11.4±3.2	39.2±5.5	39.0±5.9	22.6±4.6	28.4±8.8
Treated	12.8±2.2	36.6±4.1	42.2±5.6	12.4±3.5[c]	12.4±3.8[b]
Succinate					
Control	1.7±1.4	15.9±4.0	60.7±2.5	5.5±2.4	16.9±5.6
Treated	2.2±1.2	17.5±7.5	64.3±3.5[a]	2.7±0.8[b]	12.9±2.2[b]
ATP/ADP					
Control	9.1±1.0	1.1±0.3	0.2±0.1	7.8±0.4	5.8±0.9
Treated	9.7±0.8	0.9±0.3	0.3±0.1	8.9±1.6[a]	6.9±1.6[b]

[a] $P<0.05$
[b] $P<0.01$ } control vs. treated (MWU test).
[c] $P<0.001$

The postischemic recovery of lactate was significantly better in treated brains of both groups; even after prolonged ischemia preischemic levels were reached. Preischemic levels of succinate were only reached in treated brains of group I. However, in group II, restoration of succinate was also better after pretreatment.

Depending on the duration of the preceding ischemic period, the ratio of ATP to ADP was decreased at the end of reperfusion (see Table 1). In both group I and group II, the value was slightly higher in treated brains.

Postischemic Flow

In group I, the flow rate of both control and treated brains recovered within 3–5 min of reperfusion to almost the preischemic level. However, pretreatment with FLU had a remarkable influence on postischemic flow in the prolonged ischemia group (II), which is depicted in Fig. 3. Restoration of flow was considerably faster in FLU-treated brains. Untreated brains of group II exhibited a progressive reduction in global flow (mean decrease at the end of reperfusion: 21%; $P<0.01$ control vs treated) after about 10 min of reperfusion. In treated brains, flow rate remained stable during the whole reperfusion period.

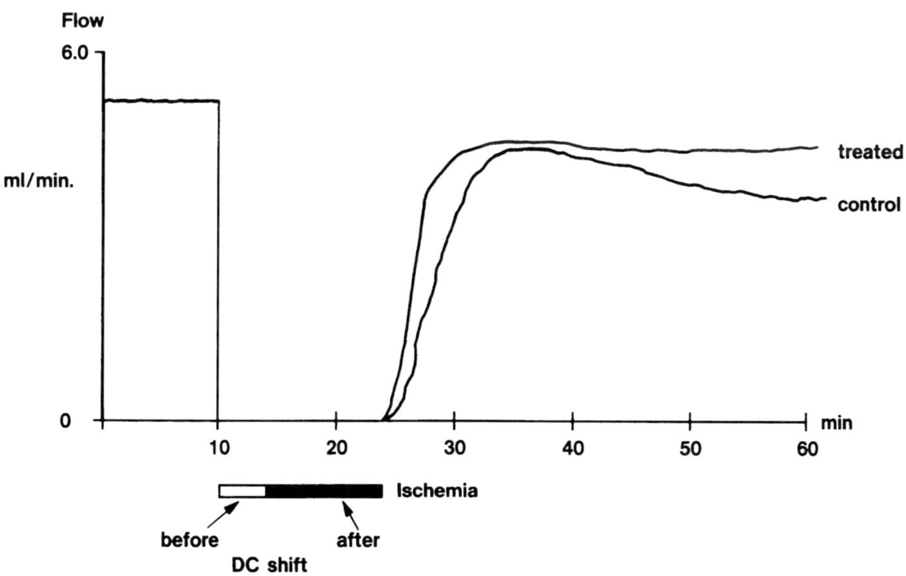

Fig. 3. Typical flow registration in isolated perfused rat brain (control and treated). Flow rate before ischemia was equal in control and treated brains. After ischemia, restoration of flow was retarded in control brains; additionally a delayed hypoperfusion developed. *Open bar:* ischemic period before DC negativation; *closed bar:* ischemic period after DC negativation

Discussion

In the present study, an attempt was made to understand the effects of FLU on extracellular cortical ion activities in ischemia and reperfusion. For this purpose the isolated perfused rat brain was used. This model is strictly controlled and not influenced by extracellular factors. Additionally, the effects of FLU, DPH, and NIMO on cortical K_e^+ and DC potential of the brain of the anoxic rat were compared.

In the experiments with anoxia, pretreatment of the animals with FLU and DPH considerably prolonged the time latency until negative DC shift, while NIMO had no effect on this parameter. Manipulation of preischemic functional activity and metabolic rate of the brain has been shown to influence the time interval between circulatory arrest and terminal ischemic depolarization [2]. Reductions or increases in metabolic rates were associated with inverse changes in the aforementioned time interval. It has also been suggested that the mechanism which reduces metabolic activity is related to a slowing of the initial efflux of K^+ [2]. This corresponds to results of the present study. FLU exhibited the strongest effect both on DC latency and on K_e^+ threshold, while NIMO influenced both parameters only marginally.

Reduction of metabolic rate is not the only mechanism by which negative DC shift can be postponed. It has been shown that the cerebral protection of DPH is linked to a reduction in net outward flux of K^+ from hypoxic cells [1].

The cerebroprotective actions of NIMO have been attributed to its vascular effects [8]. Correspondingly, the compound did not delay negative DC shift in the present experiments. The mechanism by which FLU delays DC shift in anoxia is not elucidated and needs further investigation. Because of the strong effects of FLU on K_e^+ under anoxic conditions, it was of interest to investigate the effects of the compound on ions and energy metabolism in cerebral ischemia and reperfusion . For this purpose, the isolated perfused rat brain was used. No effects of FLU could be shown at the end of the ischemic periods. However, during reperfusion, FLU affected ions as well as energy metabolism. Postischemic restoration of ion gradients needs metabolic energy. A prerequisite is sufficient supply of the brain with O_2 and glucose. Restoration of postischemic flow may be a limiting factor.

In the present experiments, involving the isolated perfused rat brain, postischemic flow rate recovered almost immediately in control and treated brains of group I. Accordingly, effects of FLU in this group cannot be attributed to an improved reflow.

Recovery of Ca_e^{2+} and pH_e was significantly accelerated after application of FLU. This may indicate a faster, energy-dependent restoration of ion gradients across the cell membrane and an unimpaired mitochondrial function. Indeed, at the end of reperfusion, lactate, succinate, and the ratio of ATP to ADP reached preischemic values and were thus significantly better in comparison to control. From these results it may be concluded that FLU improves mitochondrial respiratory activity, resulting in a faster restoration of transmembranal ion gradients.

A distinct effect of FLU on postischemic flow became obvious after prolonged ischemia (group II). In the treated brains of group II postischemic flow switched almost immediately to values which were close to the preischemic levels; in control brains flow restoration was delayed (see Fig. 3). Furthermore, after about 10 min of reperfusion, a progressive reduction in flow developed which could not be seen in the treated brains. It has been shown with intact animals that delayed postischemic hypoperfusion usually occurs 10 min after blood flow has been reestablished [8]. From the present results it may be concluded that the prolongation of ischemia beyond DC shift initiates processes which lead to delayed hypoperfusion, these being independent of the existence of blood constituents in brain vessels.

The substantially increased levels of lactate and the more acidic pH_e in control brains of group II at the end of reperfusion are not necessarily a consequence of the reduced flow rate. As has been shown recently [10], an increase in H^+ production and lactate occurs not only in hypoxia but also when mitochondrial respiration and ATP hydrolysis are not tightly coupled. Indeed, in the present experiments (group II), ATP/ADP was lower in control brains, indicating a stronger impairment of energy metabolism.

To conclude, FLU improves recovery of Ca_e^{2+} and pH_e after short-term ischemia. This is most probably due to improved mitochondrial activity. The delayed hypoperfusion after prolonged periods of ischemia is prevented by FLU, thus indicating that FLU acts also on vessels of the ischemic brain. It remains to be investigated whether the postponement of negative DC shift in anoxia after treatment with FLU is brought about by influences on energy metabolism or whether energy-independent mechanisms play a role.

Acknowlegment. The authors are very grateful to K. Dengler and E. Zacharias for excellent technical assistance and to Dr. Y. Xie for revision of the manuscript.

References

1. Artru AA, Michenfelder JD (1981) Anoxic cerebral potassium accumulation reduced by phenytoin: mechanism of cerebral protection? Anesth Analg 60:41–45
2. Astrup J, Rehncrona S, Siesjö BK (1980) The increase in extracellular potassium concentration in the ischemic brain in relation to the preischemic functional activity and cerebral metabolic rate. Brain Res 199:161–174
3. Astrup J (1982) Energy-requiring cell functions in the ischemic brain. J Neurosurg 56:482–497
4. Chen ST, Hsu CY, Hogan EL, Juan HY, Banik NL, Balentine JD (1987) Brain calcium content in ischemic infarction. Neurology 37:1227–1229
5. Hansen AJ, Zeuthen T (1981) Extracellular ion concentrations during spreading depression and ischemia in the rat brain cortex. Acta Physiol Scand 113:437–445
6. Hansen AJ (1985) Effect of anoxia on ion distribution in the brain. Physiol Rev 65:101–148
7. Höller M, Breuer H, Fleischhauer K (1983) An improved method for perfusion of the isolated brain of the rat – influence of perfusion conditions and application of analeptic and anticonvulsant drugs. J Pharmacol Methods 9:19–32

8. Kazda S, Hoffmeister F, Garthoff B, Towart R (1979) Prevention of the postischemic impaired perfusion of the brain by nimodipine (Bay e 9736). Acta Neurol Scand 60 (Suppl 72):302–303
9. Nordström CH, Rehncrona S, Siesjö BK (1978) Effects of phenobarbital in cerebral ischemia. Part II: Restitution of cerebral energy state as well as of glycolytic metabolites, citric acid cycle intermediates and associated amino acids after pronounced incomplete ischemia. Stroke 9:335–343
10. Paschen W, Djuricic B, Mies G, Schmidt-Kastner R, Linn F (1987) Lactate and pH in the brain: association and dissociation in different pathophysiological states. J Neurochem 48:154–159
11. Simon RP, Griffiths T, EVans MC, Swan JH, Meldrum BS (1984) Calcium overload in selectively vulnerable neurons of the hippocampus during and after ischemia: an electron microscopy study in the rat. J Cereb Blood Flow Metab 4:350–361

Clinical Cerebral Ischemia and Calcium

Density of Functioning Cerebral Capillaries in Stroke Patients, Determined by PET

A. GJEDDE, H. KUWABARA, L. BERGER, C. BEIL, A. C. EVANS, and A. HAKIM

Introduction

The regional density of capillaries varies in the brain. This variation is anatomical and does not necessarily reflect a functional coupling of flow, capillary permeability, or cerebral metabolic rate [6]. The variation merely suggests that the total number of capillaries in any one region is a reflection of the normal average energy requirement of the region, and that individual capillaries have similar properties (radius, length, permeability) in different regions. The regional variation does not address the question of how many capillaries are actually functioning at different times, or how this number varies with the energy requirements of individual regions.

In the present context, we wish to define functioning capillaries as capillaries that permit the transfer of glucose to the tissue. The concept of functioning capillaries does not imply only perfused capillaries. Some capillaries may contain blood that is stationary or moves so slowly that it does not effectively contribute to the nutrition of the tissue. Since the mechanisms of glucose and oxygen transfer are different, it is possible, at least in theory, that capillaries exist which may allow the diffusion of oxygen yet cannot transport glucose. We will ignore this possibility in the present paper.

In order to prove a coupling between the number of functioning capillaries and the energy requirements of the tissue (recruitment), it is necessary to verify parallel adjustments of flow, density of functioning capillaries, and capillary diffusion capacity when metabolism is subject to change in a region of interest.

Positron emission tomography (PET) allows measurements to be made of several variables, including glucose phosphorylation rate (CMR_{glc}), fluorodeoxyglucose transfer coefficients (K_1, k_2, and k_3), net uptake of oxygen ($CMRO_2$), cerebral blood flow (F), and total cerebral blood volume (V_B). However, establishing conditions of activation or deactivation that are sufficiently stable to allow several comparable measurements in the same subject is usually difficult. Alteration of brain function associated with ischemic insults may provide a wide range of flow and metabolism values which are sufficiently stable to allow comparisons in the same subject.

McConnell Brain Imaging Unit, Montreal Neurological Institute, 3801 University, Montreal, Quebec, H3A 2B4 Canada.

Cerebral Ischemia and Calcium
Ed. by A. Hartmann and W. Kuschinsky

Methods

Six patients affected by recent cerebral stroke were studied. Each patient was examined with PET twice. The first study was carried out between 16 and 38 h after the onset of the insult, and the second study 1 week later.

PET Scans and Analysis

The patients were scanned in the Therascan 3128, a three-slice scanner with a resolution of 12 mm (transverse) by 11 mm (axial) [5]. Image reconstruction was performed with correction for random events, detector efficiency variation, attenuation, and scattered events.

In each study, horizontal planes parallel to and 24, 36, and 48 mm above Reid's baseline were scanned so that the same region of interest (ROI) from different studies represented the same brain region.

F was calculated from the brain uptake of radioactive water labeled by inhalation of ^{15}O-labeled carbon dioxide, $CMRO_2$ from the equilibrium distribution of ^{15}O-labeled oxygen, and the V_B from the equilibrium distribution of ^{15}O-labeled carbon monoxide [9, 13]. A dynamic FDG study consisted of a series of 18 scans over a period of 35 min (ten 1-min scans, five 2-min scans, and three 5-min scans) following intravenous administration of FDG. Arterial blood was sampled for analysis of plasma radioactivity.

In the analysis of PET data, ROI templates were used. Each template consisted of 20 square ROIs placed along the cerebral cortex of a PET image (Fig. 1). The ROI templates were applied repeatedly to the functional maps of the same subject as well as to the serial images obtained in dynamic studies. The ability to store ROI templates ensured reproducibility of the regional analysis [8].

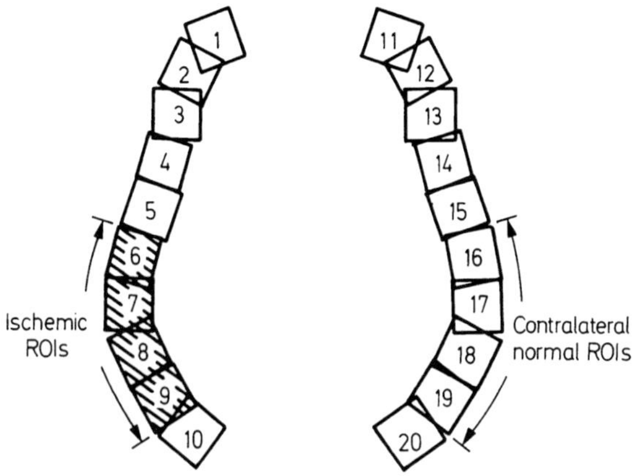

Fig. 1. An example of an ROI template. The same ROI template was applied repeatedly to the functional maps of the same plane of a patient. Each square is 15 by 15 pixels in size

Determination of Rate Constants and Lumped Constant of the Dynamic FDG Method

The lumped constant of the FDG method was defined as the ratio between the net extraction fractions of FDG and glucose in brain (K^*/K) [1].

Assuming that the ratio between the unidirectional transfer rates of FDG and glucose (τ), and the ratio between the phosphorylation rates (φ), are both constant, and also assuming that the ratio of transport rates from blood to the brain and in the reverse direction ($V_e = K_1/k_2$) has the same value for glucose and the glucose analogs [11], a new formula for the lumped constant can be derived in the following manner:

$$\Lambda = \frac{K^*}{K} = K^* \frac{(k_2^*/\tau)+(k_3^*/\varphi)}{(K_1^*/\tau)(k_3^*/\varphi)} = \varphi r_e^* + \tau(1-r_e^*), \tag{1}$$

where

$$r_e^* = \frac{k_2^*}{k_2^*+k_3^*} = 1 - \frac{K}{K_1^*} \tag{2}$$

and k_3^* is the phosphorylation rate of FDG. The transport distribution volume (V_e) is given by:

$$V_e = \frac{V_d}{1+\left(\dfrac{\tau K^* C_a}{K_1^* \Lambda K_t}\right)}, \tag{3}$$

where K_t is the half-saturation concentration of glucose transport, C_a the arterial plasma glucose concentration, and V_d the brain water content.

Using Eq. (2), the sum of k_2^* and k_3^* is a function of K^* and K_1^*:

$$k_2^* + k_3^* = \frac{K_1^*}{V_e r_e^*}. \tag{4}$$

Thus, eliminating k_2^* and k_3^* from the operational equation of the Sokoloff method [15], it is possible to determine the values of K^*, K_1^*, and V_B (blood volume) by a nonlinear least-square method. The values k_2^* and k_3^* were calculated as $k_2^* = K_1^*/V_e$, and $k_3^* = K^*/(r_e^* V_e)$. For the calculation of the rate constants and the lumped constant, we chose the values $\tau = 1.1$, $\varphi = 0.3$, $K_t = 4.8$, and $V_d = 0.77$.

Calculation of Physiological Variables

The capillary permeability-surface area product of fluorodeoxyglucose was calculated from the Bohr equation for capillary exchange [1],

$$\frac{PS}{F} = -\alpha \ln\left(\frac{\Delta C_v}{\Delta C_a}\right), \tag{5}$$

[1] The superscript * refers to FDG. Symbols without an asterisk apply to glucose.

where PS is the permeability-surface area product, F the blood flow, α the solubility of the tracer in whole blood relative to that in brain tissue, and ΔC_v and ΔC_a the concentration difference between tissue and blood at the arterial and venous end of the capillary, respectively. If α is unity, and if the tracer clearance is determined initially when tissue concentrations can be assumed to be negligible, as in the case of tracer fluorodeoxyglucose in human whole blood, Eq. (5) is modified to [7],

$$PS = -F \ln\left(1 - \frac{K_1}{F}\right), \tag{6}$$

where K_1 is the initial clearance of the tracer reflecting unidirectional transport. The number of functioning (i.e., glucose-transporting) capillaries, the capillary fraction of the total blood volume in the brain, and the mean capillary transit time of blood were all calculated from the PS product of fluorodeoxyglucose. The capillary surface area (S) is

$$S = 2\pi r l (d_{CAP}), \tag{7}$$

where r is the radius of a capillary, l the average length of capillaries, and d_{CAP} the average density of functioning capillaries. Since the average length of capillaries is not known with certainty, l was assumed to represent the unit length of 1 mm of capillary bed, and d_{CAP} therefore the average number of capillary cross-sections per square millimeter of tissue. Thus,

$$d_{CAP} = \frac{PS}{2\pi r l P} = 38 PS, \tag{8}$$

where reasonable values of r, l, and P were chosen as 2 μm, 1 mm, and 380 nm/s, respectively. d_{CAP} has unit of mm^{-2} and PS unit of ml/hg/min. According to this calculation, the value of d_{CAP} is, of course, directly proportional to the value of PS.

The capillary fraction of the total volume of the vascular bed (f_{CAP}) was calculated from the ratio between the capillary blood volume, measured with labeled carbon monoxide. The capillary blood volume is:

$$V_{CAP} = \pi r^2 l (d_{CAP}) = \frac{rS}{2}. \tag{9}$$

Hence, the capillary fraction equals,

$$f_{CAP} = \frac{V_{CAP}}{V_B} = \left(\frac{r}{2P}\right)\frac{PS}{V_B} \tag{10}$$

which, with reasonable values for r and P (*vide supra*), equals,

$$f_{CAP} = 4.76 \frac{PS}{V_B}, \tag{11}$$

where f_{CAP} has unit of per cent (%) and V_B unit of ml/hg.

The mean capillary transit time (t_{CAP}) was calculated from the ratio between the capillary volume and blood flow,

$$t_{CAP} = \frac{V_{CAP}}{F} = \left(\frac{r}{2P}\right)\frac{PS}{F} \qquad (12)$$

which, with reasonable values for r and P (*vide supra*), equals,

$$t_{CAP} = 2.86\frac{PS}{F}, \qquad (13)$$

where t_{CAP} has unit of seconds (s).

Results

The clinical characteristics of the patients are summarized in Table 1. All but one patient had associated diseases. The patients were hyperglycemic at the time of PET.

We defined "ischemic regions" as those ROIs which had an average blood flow less than 20 ml/hg/min at the time of the first scan. Each patient had four to ten ischemic ROIs and 32 in all were found. These ischemic ROIs were always adjacent to each other and located in the same area as the cerebral infarct identified by CT scans. Contralaterally matching ROIs were used as normal controls. On PET images of the second study, the same ROIs were analyzed. Thus, the investigation included a total of 128 ROIs.

Table 2 summarizes the findings of the study. The median values of normal elderly controls (mean age 63.0 ± 3.0 years) were also included. Normal ROIs of the first study had higher regional cerebral blood flow and regional oxygen metabolic rate for oxygen than those of the second study, which agreed well with the values of the elderly controls. In the ischemic ROIs, flow and oxygen metabolism

Table 1. Clinical characteristic and plasma glucose levels[a]

	Age	Sex	Neurological features	Associated conditions[b]	Glucose levels[c] (μmol/ml)	
1	65	M	R hemiparesis, aphasia	ASHD	5.5	6.0
2	79	M	Anosognosia	HBP	6.0	6.0
3	43	F	R hemiparesis	–	7.8	8.2
4	51	F	L hemiparesis	DM	6.6	5.1
5	51	M	R hemiparesis, aphasia	ASHD	6.8	6.8
6	57	F	R hemiparesis, aphasia	HBP, DM	9.9	12.4

[a] *Abbreviations:* M, male; F, female; R, right; L, left; ASHD, arteriosclerotic heart disease; HBP, high blood pressure; DM, diabetes mellitus.
[b] Neurological features were those at the time of admission.
[c] Plasma glucose at the time of the first PET study (left-hand column) and at the time of the second study (right-hand column).

Table 2. Summary of variables in ischemic and contralateral matching regions at the first and the second studies[a,b]

Variables	1st study (<38 h)		2nd study (7 days later)		Elderly controls (mean age 63 yrs
	Normal	Ischemic	Normal	Ischemic	
CBF	49.0 ±10.5	12.7 ± 1.3*	37.7 ± 1.5	38.5 ± 5.3	36.8 ±2.8
OEF	0.46 ± 0.04	0.54 ± 0.04*	0.43 ± 0.01	0.16 ±0.03*	0.44 ±0.03
CBV	3.9 ± 0.4	3.9 ± 0.6	3.7 ± 0.3	4.4 ± 0.4	4.4 ±0.5
$CMRO_2$	187 ±28	59 ± 9*	143 ± 5	48 ± 7*	144 ±9
K_1	0.081 ± 0.013	0.041 ± 0.003*	0.065 ± 0.012	0.043 ± 0.007*	0.091 ±0.002
Λ	0.54 ± 0.02	0.59 ± 0.03	0.54 ± 0.02	0.53 ± 0.02	0.58 ±0.01
CMR_{glc}	30.9 ± 4.2	16.1 ± 0.9*	25.4 ± 4.9	16.0 ± 2.0*	29.5±0.57
d_{CAP}	360 ±58	204 ±20	294 ±59	187 ±32	–
f_{CAP}	11.4 ± 0.9	6.8 ± 1.0	9.8 ± 2.0	5.1 ± 0.7	–
t_{CAP}	0.6 ± 0.1	1.3 ± 0.2*	0.6 ± 0.1	0.4 ± 0.1*	–

[a] Mean values ± standard errors of mean are shown.
[b] *Abbreviations and units:* CBF, regional cerebral blood flow (ml/hg/min); OEF, oxygen extraction fraction (ratio) CBV, regional cerebral blood volume (ml/hg); $CMRO_2$, regional cerebral metabolic rate for oxygen (μmol/hg/min) K_1, unidirectional transfer rate of fluorodeoxyglucose (ml/g/min); Λ, lumped constant (ratio); CMR_{glc}, regiona cerebral metabolic rate for glucose (μmol/hg/min); d_{CAP}, number of capillary per cross-section (/mm^2); f_{CAP}, capillar fraction of vascular bed (%), t_{CAP}, mean capillary transit time (s).
* Significant difference from the normal ROIs of the second study ($P<0.05$).

were severely depressed and oxygen extraction fraction (OEF) was slightly elevated at the time of the first study. Seven days later, OEF decreased significantly below normal and oxygen consumption remained low despite normalization of flow.

Normal ROIs of the first study had higher unidirectional uptake of FDG (K_1^*) and regional glucose metabolic rate than those of the second study, although K_1^* was suppressed in both studies in comparison to older controls. In ischemic ROIs, both K_1^* and CMR_{glc} were severely impaired. The regionally calculated lumped constant was elevated in the ischemic ROIs in the early stage.

The relationship between cerebral blood flow, oxygen metabolic rate, and glucose utilization rate is shown in Fig. 2. The lines of regression excluded the ischemic ROIs of the second study except for the relation between CMR_{glc} and $CMRO_2$, where there was no statistical difference between inclusion and exclusion of those ROIs. There was a positive correlation between flow and metabolism over a wide range of F (4.0–123.5 ml/hg/min), $CMRO_2$ (5–351 μmol/hg/min), and CMR_{glc} (7.2–52.2 μmol/hg/min), except for the ischemic regions at 1 week after the first study.

The average number of capillaries actually transporting FDG (d_{CAP}) was 360 mm^{-2} in the normal matching ROIs at the first study. The mean value decreased to 294 mm^{-2} 7 days later. In the ischemic ROIs at the early stage, the capillary density was 204 mm^{-2} and it remained low at the time of the second study.

The capillary density was plotted against blood flow and metabolism in Fig. 3. Lines of regression were calculated from all ROIs except for the relation between

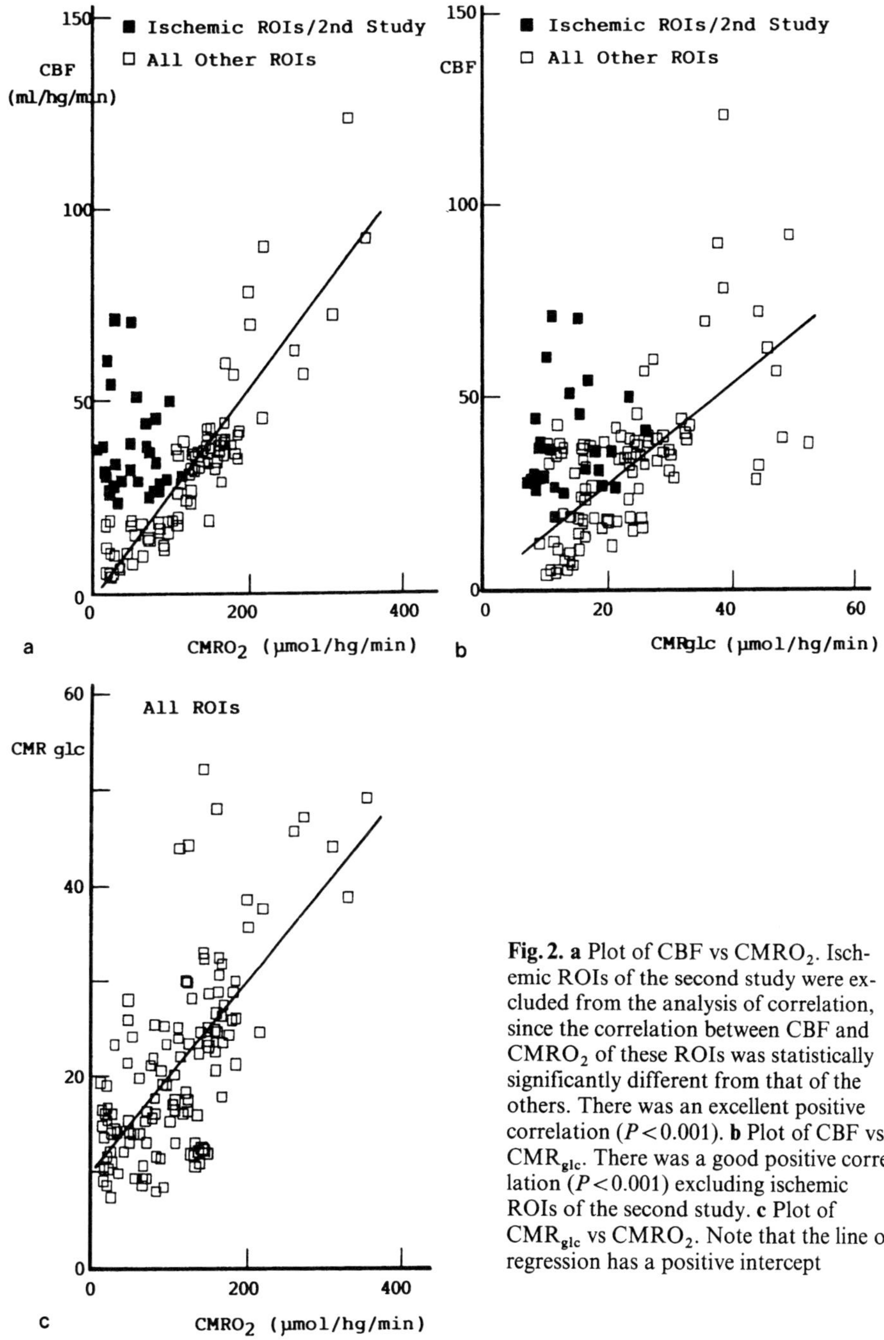

Fig. 2. a Plot of CBF vs $CMRO_2$. Ischemic ROIs of the second study were excluded from the analysis of correlation, since the correlation between CBF and $CMRO_2$ of these ROIs was statistically significantly different from that of the others. There was an excellent positive correlation ($P<0.001$). **b** Plot of CBF vs CMR_{glc}. There was a good positive correlation ($P<0.001$) excluding ischemic ROIs of the second study. **c** Plot of CMR_{glc} vs $CMRO_2$. Note that the line of regression has a positive intercept

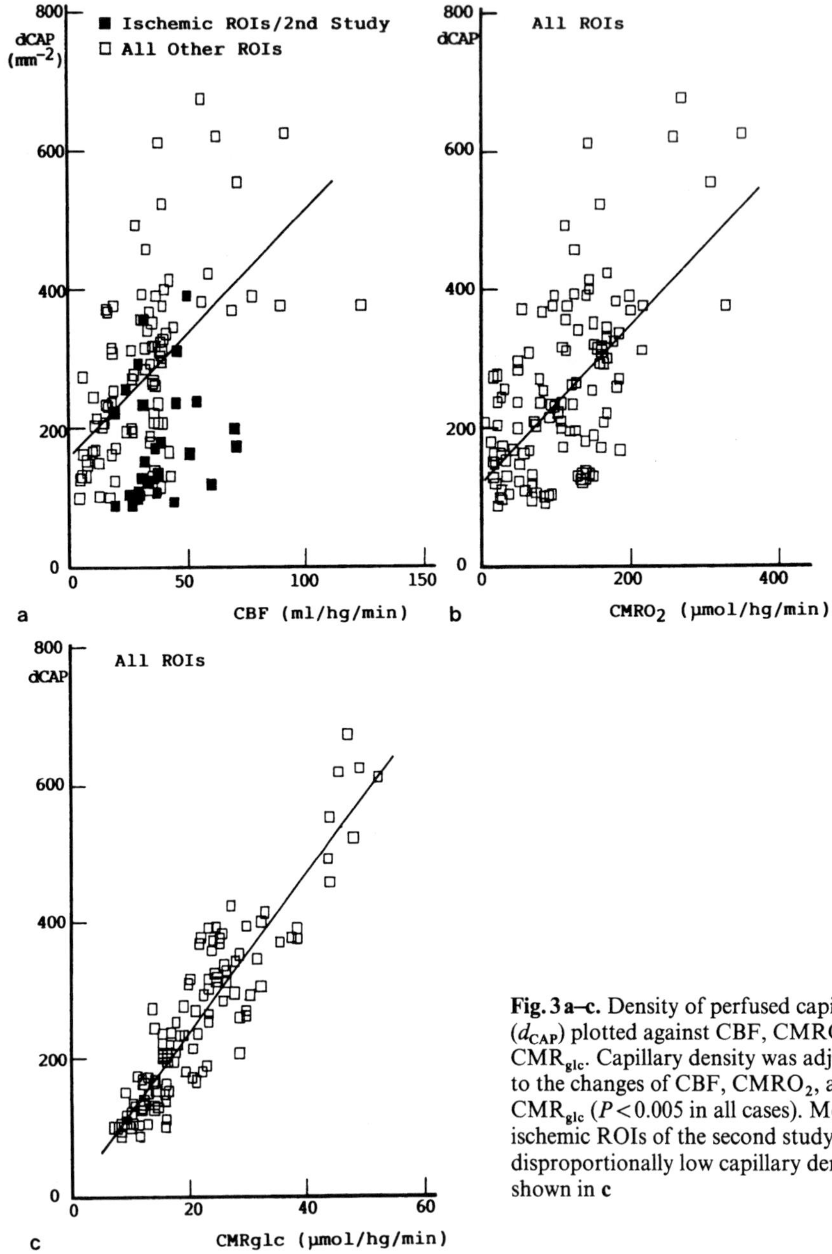

Fig. 3 a–c. Density of perfused capillaries (d_{CAP}) plotted against CBF, $CMRO_2$, and CMR_{glc}. Capillary density was adjusted to the changes of CBF, $CMRO_2$, and CMR_{glc} ($P<0.005$ in all cases). Most ischemic ROIs of the second study had disproportionally low capillary density, as shown in **c**

d_{CAP} and flow, where ischemic ROIs were excluded. We observed parallel changes of the capillary density with flow and metabolism when metabolism was subject to change.

For the contralateral normal ROIs, the average transit time of blood through the capillaries (t_{CAP}) was 0.6 s in both studies. In the ischemic ROIs, t_{CAP} was almost twice as long (1.3 s) in the early stage but, 7 days later, it decreased below normal.

Discussion

Studies on six stroke patients in the early stage of a cerebrovascular accident (16–38 h after the onset) and 1 week after provided a wide spectrum of metabolic states. We confirmed that the density of glucose-transporting capillaries, blood flow, and oxygen and glucose metabolic rates change in parallel (recruitment) over this wide range of metabolic states. We found that the coupling between capillary density, metabolism, and flow was lost in ischemic regions of the second study [10].

In the present study, we measured the density of functioning capillaries on the basis of two simple assumptions, i.e., that capillaries are identical, rigid tubes in all regions of the brain cortex, and that changes of the permeability of the capillary endothelium to hexoses like glucose occur only as a result of changes of the capillary surface area available for facilitated diffusion. While specific changes of the endothelium in ischemia, relevant to glucose transport, cannot be ruled out, we assumed that such changes did not occur.

The average normal density of cortical capillaries actually transporting glucose was 294 ± 114 mm^{-2} of cut surface and ranged from 122 to 675 in the contralateral normal counterparts. Our results suggested that, in resting condition, less than 50% of all capillaries available were actually functioning. Reported values of capillary density of the cerebral cortex vary, ranging from 589 mm^{-2} in rats to 939 mm^{-2} in cats, and 50%–84% of them were shown to be perfused at a given time [3, 4, 14]. Since the situation is dynamic, different capillaries may be perfused at different times, even when the average density stays the same. This implies that methods using dyes which require more than 10 s to circulate may overestimate the density of perfused capillaries [4].

An in vivo microscopic study of blood flow in cortical capillaries of cats by Pawlik et al. reported that flow velocity in capillaries ranged from 0.4–3.9 mm/s with a mean value of 1.5 mm/s [14]. These values correspond to transit time ranging from 0.3–2.0 s with a median of 0.7 s, which agrees very well with the mean capillary transit time observed in the present study.

The hypothesis of recruitment is necessitated by the diffusion-limited transfer of oxygen from blood to brain, which is close to being rate limiting for oxygen utilization in brain and thus responsible for the low oxygen tension in the tissue. Increasing the linear velocity of the blood perfusing the brain does not significantly improve the delivery of oxygen. Hence, decreases of the intercapillary distance are necessary to allow increases of oxygen utilization during activation. We suggested that the density of capillaries was disproportionally higher than expected from the values of flow or oxygen metabolism in these regions, and that decreasing the intercapillary distance was the defence against a low oxygen tension expected when the OEF is elevated.

We have proposed an improved method for the determination of rate constants of the three-compartment model of the FDG method. This method described the data statistically as well as that conventionally used.

In the ischemic regions, the lumped constant was slightly elevated from 0.54 of the normal regions to 0.59 as predicted by Gjedde et al. [12]. This implies that more of the tracer FDG once taken up by brain was phosphorylated in ischemia than in the normal state, while less of the glucose was phosphorylated. Although *F*, OEF, and $CMRO_2$ of the normal regions of the second study were in good agreement with the median values of the older controls, the unidirectional uptake of FDG and the lumped constant were lower than those of the older controls. Brooks et al. studied the transport of hexose across the blood-brain barrier and reported that hyperglycemic subjects had a lower unidirectional brain uptake rate (0.055 ml/g/min) than normoglycemic subjects [2]. Gjedde and Diemer showed that the lumped constant was lower in hyperglycemic rats with the direct enzymatic measurement of brain glucose content [11]. Accordingly, the explanation for the lower unidirectional clearance and lumped constant in these patients is the elevated plasma glucose level (mean 7.3 ± 2.0 vs 5.3 ± 0.7 μmol/ml for older controls).

References

1. Bohr C (1909) Über die spezifische Tätigkeit der Lungen bei der respiratorischen Gasaufnahme und ihr Verhalten zu der durch die Alveolarwand stattfindenden Gasdiffusion. Skand Arch Physiol 22:221–280
2. Brooks DJ, Gibbs JSR, Sharp P et al. (1986) Regional cerebral glucose transport in insulin-dependent diabetic patients studied using [^{11}C]3-*O*-methyl-D-glucose and positron emission tomography. J Cereb Blood Flow Metab 6:240–244
3. Buchweitz E, Weiss H (1986) Alterations in perfused capillary morphometry in awake vs anethetized brain. Brain Res 377:105–111
4. Collins RC, Wagman IL, Lymer L, Matter JM (1987) Distribution and recruitment of capillaries in rat brain. J Cereb Blood Flow Metab 7 (Suppl 1):S336
5. Cook BE, Evans AC, Fanthome EO, Alarie R, Sendyk AM (1984) Performance figures and images from the Therascan 3128 positron emission tomograph. IEEE Trans Nucl Sci NS-31(1):640–644
6. Craigie EH (1945) The architecture of the cerebral capillary bed. Biol Rev 20:133–146
7. Crone C (1963) The permeability of capillaries in various organs as determined by the use of the "indicator diffusion" method. Acta Physiol Scand 58:292–305
8. Evans AC, Beil C, Marrett S, Thompson CJ, Hakim A (1988) Anatomical-functional correlation using an adjustable MRI-based region-of-interest atlas with positron emission tomography. J Cereb Blood Flow Metab (in press)
9. Frackowiak RSJ, Lenzi G-L, Jones T, Heather JD (1980) Quantitative measurement of regional cerebral blood flow and oxygen metabolism in man using ^{15}O and positron emission tomography. Theory, procedure, and normal values. J Comput Assist Tomogr 4:727–736
10. Garcia JH, Conger KA, Briggs L (1981) Brain capillaries in acute and subacute regional ischemia. In: Cervós-Navarro J, Fritschka E (eds) Cerebral microcirculation and metabolism. Raven, New York

11. Gjedde A, Diemer NH (1983) Autoradiographic determination of regional brain glucose content. J Cereb Blood Flow Metab 3:303–310
12. Gjedde A, Wienhard K, Heiss W-D, Kloster G, Diemer NH, Herholz K, Pawlik G (1985) Comparative regional analysis of 2-fluorodeoxyglucose and methylglucose uptake in brain of four stroke patients. With special reference to the regional estimation of the lumped constant. J Cereb Blood Flow Metab 5:163–178
13. Lammertsma AA, Jones T (1983) The correction for the presence of intravascular ^{15}O in the steady-state technique for the measurement of oxygen extraction in the brain. 1: Description of the method. J Cereb Blood Flow Metab 3:416–424
14. Pawlik G, Rackl A, Bing RJ (1981) Quantitative capillary tomography and blood flow in the cerebral cortex of cats; an in vivo microscopic study. Brain Res 208:35–58
15. Sokoloff L, Reivich M, Kennedy C et al. (1977) The [^{14}C]deoxyglucose method for the measurement of local cerebral glucose utilization: theory, procedure and normal values in the conscious and anesthetized albino rat. J Neurochem 28:897–916

Nimodipine Clinical Trial in Acute Ischemic Stroke

J. P. Mohr[1], J. L. Muschett[2], and R. V. Riccio[2]

Introduction

Rationale for Selection of a Calcium Antagonist

At the time of the trial organization, the literature on the subject of calcium antagonists was still limited. Calcium antagonists were thought to be of benefit by virtue of their effects on cellular metabolism and on collateral circulation.

A major rationale for the use of calcium channel blocking agents was the possibility that they may prevent intracellular biochemical events that convert reversible ischemia to infarction. Increases in intracellular calcium during states of ischemia are theorized to interfere with mitochondrial function and initiate irreversible cell damage [9]. Should the main beneficial effects of calcium channel blockers be mediated through this mechanism, their effects in promoting vasodilatation might prove of secondary importance.

The other major potential beneficial effects of calcium-channel blocking agents were thought to reside in their capacity to achieve cerebral arterial dilatation, which may help dislodge embolic material, or to improve collateral circulation, and their suspected capacity to ameliorate intracellular metabolic derangements set in motion by ischemia. In formulating a clinical trial, agents maximizing such effects would be preferred over those having little such effects. Cerebral vessels are sensitive to the concentration of extracellular calcium in the maintenance of contraction controlled by two types of mechanism: first, a potassium mechanism via potential-sensitive channels; second, by certain receptor-sensitive channels controlled by amines, prostaglandin, and some elements in blood [5]. A reciprocal interaction has been shown between calcium and magnesium [14]. Increased calcium or decreased magnesium in subarachnoid hemorrhage caused constriction while increased magnesium caused relaxation [2]; the effect was greater on the larger vessels (i.e., those usually affected in vasospasm). The increase in the calcium-magnesium ratio from the usual 1:1 in spinal fluid to 3:1 after subarachnoid hemorrhage provides a basis for speculating that calcium has a role in vasospasm. Whether the alleged impairment in platelet function attributed to calcium channel blocking agents is due to the direct action of the calcium

[1] Stroke Center, Neurological Institute, Columbia-Presbyterian Medical Center, New York City, NY 10032, USA.
[2] G. H. Besselaar Associates, Princeton, NJ, USA.

Cerebral Ischemia and Calcium
Ed. by A. Hartmann and W. Kuschinsky

antagonists or to increased flow occasioned by the vasodilatation they cause is unclear. It was believed that large concentrations of calcium antagonists are needed in vitro to achieve antiaggregating effects [3], making this role seem less important at doses then known.

A variety of calcium-channel blockers were shown to be effective in producing vasodilatation by both topical application and systemic routes [13]. Topical application, an impractical route of administration (save for intrathecal injection), has more effect in producing vasodilatation, whereas systemic application produces less vasodilatation but achieves the same degree of opposition to arterial constriction.

Many of the calcium channel blocking agents have undesired systemic effects which may limit their utility in treatment of ischemic cerebral vascular disease. Chief among these is hypotension, an effect all agents produce. Others include cardiac chronotropic and negative inotropic effects, which are prominent features of some agents but only slight effects of others.

Then-Current Agents Already Tested in Phase I Studies

A few agents had already been tested partially in the United States and had investigational drug numbers assigned by the Food and Drug Administration. Their use in a trial was therefore preferable to agents not then tested with human subjects or not available in the United States.

Nimodipine was a leading candidate. Among its advantages were less systemic hypotension; fewer cardiac inotropic and chronotropic effects; better penetration into the brain and brain vessels compared with calcium antagonists listed above; evidence that it dilates intracranial vessels in both rats [6] and dogs [11]; and, most important, then-current experience with the agent in a phase II trial indicating considerable safety in doses of 0.35 mg/kg q 4 h. The phase II clinical trial by Allen et al. [1] in a setting of subarachnoid hemorrhage demonstrated what the authors took to be beneficial effects of the drug. At the end of 21 days of treatment, among the 60 taking placebo, 7 deficits occurred that were attributed to spasm, 16 deficits that were attributed to other causes, and 7 patients died. In the 56 cases treated with nimodipine, at doses of 0.35 mg/kg q 4 h, 4 deficits occurred that were attributed to vasospasm, 19 that were attributed to other causes, and 3 patients died. In patients with favorable outcome, the degree of vasospasm was mild and present in both the treated and the placebo group. The degree of vasospasm was more severe in those with a poor outcome. The occurrence of rebleeding was the same in both groups, as was the incidence of other complications, suggesting that the drug did not have major side-effects. The authors concluded that the lower incidence of severe vasospasm in the treated group indicated the effectiveness of the drug in preventing deficits attributable to spasm. This conclusion has been attacked as unwarranted [16]. In defense of the authors, the dose selected for the trial was high enough to be in the range of affinity of the drug for its receptor. Higher doses were avoided because the organizers could not predict whether side-effects would occur. The encouraging results of this trial prompted the organization of another trial to test the issue of dose-response.

A single trial had been reported with nimodipine in ischemic stroke [7]. In this trial, 40 mg t.i.d. oral nimodipine (a dose similar to the 0.35 mg/kg q 4 h= roughly 100 mg/day used in the 1983 Allen et al. trial) was given in a single-blind design to a group of 31 of 60 patients with acute ischemic stroke treated for 28 days. The author selected "only patients with a diagnosed acute ischemic stroke," a disorder not further defined in this protocol. The distribution of cases shown in his tables includes only hemispheral strokes, from which we infer that no brain stem infarcts were included. The Mathew Scale was used to assess disability from the stroke. Compared with the control group, the nimodipine group had a better outcome from the 1st week of treatment both in neurological score ($P=0.0004$) and level of consciousness. This trial showed no side-effects save for two patients who had mild abdominal cramps. No instances of "clinically significant" hypotension were noted. In addition, "low dose" heparin was used (10000 U SQ $\times$ 2) to prevent deep vein thrombosis, with no apparent untoward effects and no report of hemorrhagic infarction. The agent was also reported to have a milligram-per-milligram comparability to amitriptyline as an antidepressant, except that its duration of action was only about 30 min compared with 4 h for amitriptyline [3].

The disadvantage of nimodipine was its failure to influence the functional deficit or the histopathological changes in an experimental spinal stroke model in the rat, despite improvement of local blood flow and dilatation of cerebral vessels [6]. Whether this experimental evidence would be sufficient to predict a similar lack of efficacy in humans was undetermined. Few side-effects had been described, all below a frequency of 5%, including in descending order, muscle pain, nausea, headache, epigastric discomfort, joint pain, and dizziness; in less than 3% there was hypotension, fatigue, chest pain, flushing, constipation, rash, tachycardia, palpitations, depression, or diarrhea.

All told, the agent had enough rationale that it was preferred for a phase II trial in ischemic stroke. The data from a phase II trial for ischemic stroke were also considered useful in expanding the initial observation in the only trial in ischemic stroke and in possibly having a bearing on the interpretation of the phase III trial then just getting underway in a setting of subarachnoid hemorrhage.

Other agents also considered included Sandoz 108-068, nicardipine, and several calcium antagonists used in cardiology. *Sandoz PY 108-068* was structurally similar to nifedipine. Laboratory investigations showed it penetrated well into brain, had beneficial effects on cerebral blood flow, and reduced the severity of the neurological deficit in an animal model embolized by microspheres [17]. In European human volunteers, oral doses of 20 mg produced no hemodynamic effects but 2 h after doses of 40 mg or 80 mg, decreases in total peripheral resistance sufficient to increase cardiac output and heart rate were observed. A half-life of some 11 h had been documented after oral dosing. Although no orthostatic effects were observed in a trial of the agent as an antihypertensive, double-blind doses of 25–200 mg t.i.d. had been shown to reduce supine blood pressure from 173/111 to 155/101 mmHg. The estimated human dose was 75–600 mg/day. Its side-effects closely resembled those of nifedipine: 10% with dizziness, lightheadedness, peripheral edema, nausea, weakness, headache, and flushing; 5% with transient hypotension; 2% with palpitations; 0.5% with syncope. A clinical trial addressing the issue of amelioration of the effects of acute cerebral ischemia was organized in the United States using the agent in oral form, in doses of 25 mg q 6 h for 14 days. Plasma levels were measured on days 3, 7, 10, and 14 on the protocol. Eligible patients had middle cerebral territory ischemic infarction producing incomplete disability, none of the many contraindications, and were started on oral therapy within 24 h of onset of stroke. Due to the severe limitations of the eligibility criteria, too few cases were entered during the 1st year of the trial to gauge even the faintest indication of the effects.

Nicardipine had been tested outside the United States, but was not under FDA approval or IDs. This agent, developed in Japan and available in Europe, produced considerable cerebral vasodilatation in humans at well-tolerated doses of 20 mg p.o., impaired increases in cerebral extracellular K^+ flux following ischemia, suggesting it impeded calcium flux into the cell, and was shown to improve the cortical sensory evoked response in an animal model of stroke from injections of air into the internal carotid artery. For reasons not yet clear to the present writers, a clinical trial of this agent in stroke begun in Europe in 1984 appears to have been canceled.

Some agents with FDA approval for cardiac and systemic diseases were considered but had no advantages over nimodipine and poorer CNS penetration. *Verapamil* had the advantage of wide usage in the medical community and considerable experience with its side-effects. It dilated cerebral vessels in dogs [10]. Its disadvantages were its major cardiac effects, its poor penetration into brain and brain vessels, and its failure to protect brain electric activity, water content, or energy metabolites in rats with pretreatment [4]. Its vasodilator properties were modest compared with nifedipine. *Nifedipine* was attractive because of its relative lack of cardiac chronotropic and negative inotropic effects. Its major disadvantage was its potent peripheral vasodilator properties. Further, it was disappointing in the few limited trials which had been undertaken in its use as a therapy for vasospasm after subarachnoid hemorrhage.

Background for Trial Design

With a plan to determine whether varying doses of nimodipine could prevent worsening or actually improve the outcome in patients with acute cerebral infarction, a decision was made to make ineligible those with hemorrhage, those whose clinical deficit was unstable, and those deemed too late from stroke to show benefit. Accordingly, CT scan showing no hemorrhage was the first prerequisite. The slight delay in obtaining the CT scan was thought worth the effort to exclude those with hemorrhage. An infarct lesion was expected to be identified in roughly 45% of cases within this time frame.

To guard against treatment of cases with hemorrhagic infarction, a repeat CT on day 3 was required. A follow-up CT was needed to identify the infarct if the original CT was negative, to document hemorrhagic infarction if this was a risk of the therapy, and to document the frequency of delayed hemorrhagic infarction for those not receiving the therapy.

Although it would have been ideal to use no ancillary treatment, many centers regularly used intravenous heparin anticoagulation for patients with obvious car-

Table 1. Frequency of worsening by infarct subtype. Preliminary analysis of the NIH Stroke Data Bank done for planning of the nimodipine trial

	Type of infarction				Total
	Unkn	Emb	Athero	Lac	
No worsening	347 68%	254 70%	49 38%	269 82%	919
Worsening, any cause	163 32%	110 30%	81 62%	59 18%	413 31%
Stroke evolution	65 13%	35 10%	28 22%	9 3%	159 12%
Totals	510	364	130	328	1332

Unkn, infarct of undetermined cause despite laboratory studies; Emb, cardiac embolism; Athero, large artery stenosis or occlusion; Lac, deep small infarct with lacunar syndrome.

diac atrial fibrillation, so a standard throughout the study was made on this drug. Heparin was to be given for 3–5 days, during which time coumadin was started for those cardiogenic in origin, the time period depending on the time required for a given case to change his/her coagulation profile in response to the coumadin medication.

Preliminary analysis of the data from both the Harvard Stroke Registry [12] and the population study of the NIH Stroke Data Bank project [8] indicated that worsening occurred in all forms of ischemic stroke and in all vascular territories, but not at a uniform rate (Table 1). Review of these data made it clear that the usual clinical practice of limiting case entry to those with infarction in the middle cerebral territory would exclude many examples of basilar territory infarction, where a high frequency of worsening occurred. These considerations led to the inclusion of all strokes affecting any vascular territory and attributed to any of the mechanisms of ischemic infarction.

Because it was recognized that worsening in such parameters as aphasia, visual field disturbance, ataxia, and the like was poorly quantitated, emphasis was placed on those cases with weakness. By requiring at least some motor deficit, a few patients were eliminated from consideration, especially those with posterior cerebral territory infarction whose deficit was limited to disturbance in visual fields or higher cerebral function.

In selecting the scales to be used for evaluation, the Mathew Scale was deemed necessary because it was used in the Gelmers study and a direct comparison between this and the Gelmers study was desired. Since the Mathew Scale was unfamiliar to many in the United States, the Toronto Scale was also included so the two scales could be compared. Analysis of these two scales left the investigators somewhat unsatisfied with the precision of the correlation between the motor deficit and the CT scan findings, so many elements of the forms created for the NIH Stroke Data Bank project were added to the protocol [15]. The Barthel Index of Disability was also incorporated to assess the functional deficit separate from that found on the neurological examination.

The format of the data collection instrument was derived from the clinical examination forms created for the NIH Stroke Data Bank project, which uses the format familiar to clinical neurologists. The elements, their collectibility, interobserver reliability [15], and utility had been tested first in the Harvard Cooperative Stroke Registry (>800 cases), the pilot phase of the NINCDS Stroke Data Bank (>1200 cases), and the main phase of the Stroke Data Bank project (1807 cases). Instead of trying to teach clinicians how to use a given scale, the advantage of our data sheets is that they simply score the examination using the clinical parameters long familiar to them.

The CT form used for the project was validated in the NIH Stroke Data Bank project on over 2000 cases and proved sensitive to changes in lesion volume.

Methodology

Based on sample size calculations seeking a total of 1272 patients to provide a minimum of 960 patients, 240 in each of the four treatment groups, 53 centers each with a minimum of 24 cases in 2 years were deemed necessary.

Candidates for the study are adults aged 21–89 hospitalized within 48 h of an acute stroke. To be eligible, patients must have at least a quantifiable motor deficit exceeding certain defined values, no evidence of hemorrhage on a CT scan, and not have such a severe stroke as to be in coma.

After initial evaluation and randomization, daily examinations are performed up until day 21 or discharge, repeat CT scan between days 3 and 5, and detailed clinical reevaluation on days 4, 10, and 21. Heparin followed by warfarin has been permitted for all patients whose admission clinical data indicate a cardiac source for embolism.

The test agent, for which the examiner and staff are blinded, is given orally in four tablets in three daily doses according to four separate protocols: For the placebo group, four doses of placebo. For the 60-mg group, three doses of placebo, one dose of 60 mg. For the 120-mg group, two doses of placebo, two doses of 60 mg. For the 240-mg group, no doses of placebo, four doses of 60 mg.

The clinical deficit is estimated at onset and at days 4, 10, and 21 quantitatively by means of the Mathew scale, the Toronto Stroke Scale, a neurological deficit profile and a special Muscle Strength Scale developed for use in the NIH Stroke Data Bank, the Barthel Index of Disability, and an Investigator's Global Assessment of Disability.

Because of the expected delay in the completion of the workup and the wish to include as many cases as possible, a reevaluation on day 5 is undertaken to identify those patients who would have been ineligible for the study had all the information been known at the time of admission. Accordingly, post-randomization exclusion is implemented for those whose investigations have disclosed the presence of brain tumor, intracranial hemorrhage, or other causes of an acute focal deficit inconsistent with acute ischemic infarction and those who have medical exclusions that prevent continued participation.

Results

The trial was organized in July 1986 but actual data collection did not begin until November of that year, given the delays involved in obtaining authorization for the study by the individual institutional Human Subjects Internal Review Boards, as required by the Federal Government. As of June 1988, 12 of the centers have been in the trial over 18 months, and another 26 for over 12 months. Through April 1988, almost 7000 patients had been screened at the 53 institutions, and 600 entered (Fig. 1). Case screening and accrual have remained steady, if anything increasing somewhat. No plateau has yet occurred in the effort by the participants. The original completion target date of September 1988 may require extension.

The four groups in the trial are thus far well balanced in age, sex, and risk factors for stroke, judging from the subset who have thus far experienced fatal complications (Table 2).

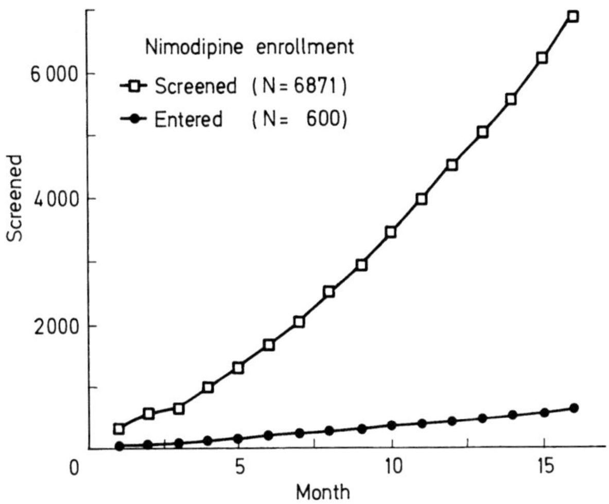

Fig. 1. Rate of patient entry in the nimodipine trial through April 1988

Table 2. Distribution of age, sex, and risk factors in the four treatment groups (placebo, 60 mg, 120 mg, 240 mg daily) who experienced fatility[a]

	Treatment group			
	A	B	C	D
Number, total	119	119	119	119
Age (mean)	66	75	71	65
Sex (male)	6	5	6	4
Prior stroke	2	1	3	0
Hypertension	2	1	5	2
Diabetes	1	1	0	0
Myocardial infarct	1	0	1	2

[a] No infarction is provided as to which letter refers to which subgroup, as the data are still blinded as of June 1988.

Two analyses are planned, an intent-to-treat and another excluding those without acute ischemic stroke. Baseline and demographic characteristics will be compared using an analysis of variance and chi-square tests. The analysis of mortality will focus on the percentage of patients surviving the first 21 days of treatment, with a 3-month and a 6-month posttreatment survival being secondary variables. The Muscle Strength Scale and the Toronto Scale are the two primary indices of neurological function, the Mathew Scale being used to provide comparison with prior studies of nimodipine.

References

1. Allen GS, Ahn HS, Preziosi TJ et al. (1983) Cerebral arterial spasm – a controlled trial of nimodipine in patients with subarachnoid hemorrhage. N Engl J Med 308:619–624
2. Altura BT, Altura BM (1981) Magnesium deficiency induces cerebral vasospasm. Stroke 12:118
3. Battye R (ed) (1984) Calcium antagonists; possible therapeutic uses in neurosurgery. Proceedings of 10–11 June, 1983 workshop held in Dorado Beach, Puerto Rico. Raven Health Care Communications, New York
4. Berger JR, Busto R, Ginsberg MD (1984) Verapamil: failure of metabolic amelioration following global forebrain ischemia in the rat. Stroke 15:1029–1032
5. Brandt L, Ljunggren B, Anderson KE (1983) Effects of topical application of a calcium antagonist (nifedipine) on feline cortical pial microvasculature under normal conditions and in focal ischemia. J Cereb Blood Flow Metab 3:44–50
6. Faden AI, Jacobs TP, Smith MT (1984) Evaluation of the calcium channel antagonist nimodipine in experimental spinal cord ischemia. J Neurosurg 60:796–799
7. Gelmers HJ (1984) The effects of nimodipine on the clinical course of patients with acute ischemic stroke. Acta Neurol Scand 69:232–239
8. Gross CR, Kase CS, Mohr JP, Cunningham SC, Baker WE (1984) Stroke in south Alabama: incidence and diagnostic features. A population-based study. Stroke 15:249–254
9. Hass WK (1981) Beyond cerebral blood flow, metabolism, and ischemic thresholds: an examination of the role of calcium in the initiation of cerebral infarction. In: Lechner H, Reivich M, Oht EO, Arabinar A (eds) Proceedings of the 10th Salzburg conference on cerebral vascular disease. Amsterdam: Excerpta medica, 1981, p 3
10. Leblanc R, Feindel W, Yamamoto LY, Milton JG, Frojmovic MM, Hodge CP (1984) The effects of calcium antagonism on the epicerebral circulation in early vasospasm. Stroke 15:1017–1020
11. McCalden TA, Nath RO, Thiele K (1984) The effects of a calcium antagonist (nimodipine) on basal cerebral blood flow and reactivity to various agonists. Stroke 15:527–530
12. Mohr JP, Caplan LR, Melski JW, Duncan GW, Kistler JP, Pessin MS, Goldstein RJ, Bleich HL (1978) The Harvard cooperative stroke registry: a prospective registry of cases hospitalized with stroke. Neurology 28:754–762
13. Rosenblum WI (1984) Effects of calcium channel blockers on pial vascular responses to receptor mediated constrictors. Stroke 15:284–288
14. Seelig JM, Wei EP, Kontos HA, Becker DP (1983) Effects of changes in magnesium ion concentration on cat cerebral arterioles. Am J Physiol 245:22–26
15. Shinar D, Gross CR, Mohr JP et al. (1985) Inter-observer variability in the assessment of neurologic history and examination in the Stroke Data Bank. Arch Neurol 42:557–565
16. Vermeulen M, van Gijn J, Hijdra A, Wijdicks EFM (1983) Nimodipine for cerebral arterial spasm in subarachnoid hemorrhage. (Letter to the editor). N Engl J Med 309:437
17. Wiernsperger N, Gygax P, Hofmann A (1984) Calcium antagonist PY 108-068: demonstration of its efficacy in various types of experimental brain ischemia. Stroke 15:679–685

Nimodipine in Acute Ischemic Stroke: A Controlled Trial

E. Martínez-Vila, J. Matías-Guiu, F. Guillén, J. A. Villanueva, J. Bigorra, and J. M. Martínez-Lage

Introduction

Ischemic stroke is one of the leading causes of death and chronic disability in the world. Although great progress has been made in recent years in understanding the pathophysiology of cerebral ischemia, the treatment for the patient who has suffered a brain infarction remains controversial. A number of management strategies have been devised to rescue ischemic brain and preserve the nonfunctional penumbra with the goal of reducing the area of infarction as well as minimizing the neurological deficits. An exciting advance in the treatment of brain ischemia has been initiated by the introduction of drugs that block calcium entry into cells.

The 1,4-dihydropyridine calcium antagonist nimodipine has been shown to inhibit calcium influx into muscle cells and neurons and is an effective blocker of the L-type voltage sensitive calcium channel. In animal models of brain ischemia, nimodipine increases hemispherical flow, improves neurological evolution, and reduces the amount of cellular damage [1, 2].

The results of a pilot clinical study [3] and a randomized placebo-controlled trial [4] suggest that nimodipine should provide a novel and effective treatment of neurological deficits due to acute ischemic stroke. The present study was undertaken to evaluate whether early treatment with nimodipine could favorably alter the neurological outcome of patients with acute cerebral infarction.

Patients and Methods

A total of 166 patients suffering from acute cerebral infarction were admitted to the study. There were 86 men and 80 women. Criteria for inclusion were: (a) patients of both sexes aged >45 years; (b) complete acute ischemic stroke diagnosed by clinical examination and cranial CT scan; (c) start of treatment within the first 48 h after the onset of the acute ischemic stroke. Exclusion criteria included transient ischemic attack, progressing stroke, brain-stem infarction, subarachnoid or intracerebral hemorrhage shown by CT scan, severe cardiac, renal, or hepatic dis-

Department of Neurology, Clínica Universitaria, Apartado 192, E-31080 Pamplona, Spain.

Cerebral Ischemia and Calcium
Ed. by A. Hartmann and W. Kuschinsky

ease, and nonstabilized diabetes. Prior to the study, informed consent was obtained from the spouse or legal guardians.

The patients were randomly assigned to treatment with 120 mg nimodipine orally daily (one 30-mg nimodipine tablet every 6 h) for 28 days, or identical placebo, by a central coordinator of the study who was also blinded to the treatment assigned. No calcium channel blockers other than nimodipine were administered.

The clinical deficit was estimated at onset and on days 1, 3, 5, 7, 14, and 21 quantitatively by means of the Mathew Scale. For analysis of the homogeneity of the groups we used a Student's *t*-test for age, a Mann-Whitney test for baseline scores of the Mathew Scale, and a chi-square test for sex. Mortality was compared using a life-table test (Breslow, Kaplan-Meier).

This study was reviewed and approved by the Hospital Committee of Clinical Trials and the Regulatory Board for Drugs and Health Care Products.

Results

One hundred and sixty-four patients were initially admitted to this study between June 1985 and June 1987, and 123 patients were considered valid for the efficacy analysis. The other 41 patients were excluded from the efficacy analysis for several reasons, mainly because of wrong diagnosis or delayed start of treatment. Of the 123 valid patients (69 males, 54 females), 58 received nimodipine and 65 placebo. No significant differences were observed between the two groups with regard to mean age, male–female ratio, and baseline neurological impairment.

Mortality

The 4-week mortality rates were 10.3% (6/58) in the nimodipine group and 15.4% (10/65) in the placebo group (*P*: NS). When only those patients with a moderate to severe neurological impairment were considered (baseline Mathew Scale score 35–65), the mortality rates changed to 13.2% in the nimodipine group and 20% in the control group (*P*: NS).

Neurological Outcome

When considering a reduction of the deficit of 30% with respect to the initial status, in the 123 patients valid for efficacy analysis, a statistically nonsignificant difference in favor of nimodipine was seen ($P<0.15$, Wilcoxon Savage). However, analysis of the subgroups revealed that in patients with a baseline Mathew Scale score below 65 a statistically significant difference in favor of nimodipine was noted ($P<0.05$, Wilcoxon). No statistical differences in systolic and diastolic arterial pressure or heart rate were observed between the treatment groups.

Discussion

Despite intensive effort, there is still no generally recognized therapy for acute ischemic stroke. Synaptic transmission ceases with a reduction in CBF to 15–20 ml/100 g cerebral tissue/min. As soon as the CBF reaches the so-called infarction threshold of about 6–12 ml/100 g/min, membrane depolarization and potassium efflux occur, and sodium, water, and calcium enter the cell [5]. These processes mark the initiation of a complex sequence of metabolic events that, if not corrected, lead to cellular necrosis [6]. A number of different pharmacological approaches to the treatment of cerebral ischemia have been proposed from knowledge gained from animal models of the pathophysiology of acute stroke. Rational approaches to therapy should ideally be directed towards improvement of tissue perfusion and preservation of tissue viability (mainly neurons). Calcium has been implicated in the final common pathway that leads ultimately to neuronal death, not only for contraction of vascular smooth muscle but also for cell death caused by membrane degradation [6, 7].

It would seem highly justified, therefore, to focus attention on calcium entry blockers as potential therapeutic agents in acute focal cerebral ischemia. Among several calcium antagonists, nimodipine has been shown to be one of the most promising with a preferential action on cerebral blood vessels [8]. In our study, nimodipine showed a trend towards a lower 28-day mortality, particularly in the subgroup of patients with moderate to severe baseline neurological deficits (13.2% nimodipine versus 20% placebo). Contrary to the Gelmers et al. study [4], this difference in mortality was not statistically significant, possibly due to an inadequate number of patients and to a relatively low mortality in the placebo group.

For the analysis of neurological evolution three categories of patients were considered according to an initial Mathew Scale score under 35, between 35 and 65, and 65 or above. This is important since previous observations [3, 4], confirmed by our findings, show that nimodipine efficacy is much more marked in patients with a moderate to severe baseline neurological deficit. The maximum interval of 48 h used in our study was chosen because it was clinically feasible and identical to the interval used in other studies with nimodipine [3], and the favorable results achieved with nimodipine suggest it to be adequate. It must be emphasized that nimodipine efficacy is much more marked in patients with a moderate to severe baseline neurological deficit.

The results support the findings of Gelmers et al. [3, 4], and this line of investigation may provide new perspectives in the treatment of acute cerebral infarction.

References

1. Steen PA, Gisvold SE, Milde JH et al. (1985) Nimodipine improves outcome when given after complete cerebral ischemia in primates. Anesthesiology 62:406–414
2. Gotoh O, Mohamed AA, McCulloch J, Graham DI, Harper AM, Teasdale GM (1986) Nimodipine and the haemodynamic and histopathological consequences of middle cerebral artery occlusion in the rat. J Cereb Blood Flow Metab 6:321–331
3. Gelmers HJ (1984) The effects of nimodipine on the clinical course of patients with acute ischemic stroke. Acta Neurol Scand 69:232–239
4. Gelmers HJ, Gorter K, De Weerdt CJ, Wierzer HJA (1988) A controlled trial of nimodipine in acute ischemic stroke. N Engl J Med 318:203–207
5. Astrup J, Symon L, Siesjö BK (1981) Thresholds in cerebral ischemia. The ischemic penumbra. Stroke 12:723–725
6. Siesjö BK (1981) Cell damage in the brain: a speculative synthesis. J Cereb Blood Flow Metab 1:155–185
7. Schanne FAX, Kane AB, Young EE, Farber JL (1979) Calcium dependence of toxic cell death: a final common pathway. Science 206:700–702
8. Kazda S, Towart R (1982) Nimodipine: a new calcium antagonistic drug with a preferential cerebrovascular action. Acta Neurochir 63:259–265

Calcium Blocker Therapy in Acute Ischemic Stroke

H. J. Gelmers

Introduction

Despite major advances in the understanding of the pathophysiology of cerebral circulation, very little progress has been made in the ability to prevent the functional impairment that follows acute ischemic stroke. Recent investigations of the mechanism of ischemic cell injury suggest that increased calcium concentrations are implicated in neuronal death after anoxia-ischemia.

Brain Ischemia

Complete arrest of cerebral blood flow results within seconds in cessation of neuronal electrical activity and within a few minutes in deterioration of the energy state and ion homeostasis [2, 15]. Recent evidence indicates that immediate failure of basic functions is critically dependent on residual blood flow and that these functions fail at a certain critical flow threshold [1]. It appears further that the development of infarction is critically correlated to residual perfusion and that there is a lethal threshold of residual blood flow below which tissue infarction develops after a certain time [21]. The region in the ischemic brain with flow between the two thresholds has been termed the "penumbra" [2]. Brain tissue under the penumbra shows a very complex picture of interacting and cascading events that tend to reinforce the initial ischemic attack and to evoke biochemical reactions that ultimately destroy vital cellular elements. Recently, a disturbance of calcium homeostasis within the ischemic cell has been recognized as essential to the ultimate death of the cell [17].

Role of Calcium in Brain Ischemia

When residual flow drops below the lethal threshold, failure of energy-dependent Na^+-K^+ transport becomes critical and leads to an increase of extracellular K^+ that will cause depolarization of the neuronal membrane. This results in opening

Department of Neurology, Streekziekenhuis Almelo, P.O. Box 7600, NL-7600 SZ Almelo, The Netherlands.

Cerebral Ischemia and Calcium
Ed. by A. Hartmann and W. Kuschinsky

of the voltage-sensitive calcium channels with a rise in intracellular free calcium [20]. This influx is enhanced by failure of the ATP-independent Na^+-Ca^{2+} antiport system. In addition energy-dependent cell systems (endoplasmic reticulum, mitochondrial matrix) for Ca^{2+} sequestration fail, while the mitochondria are overloaded by an electrophoretic accumulation of calcium, the latter further compromising ATP production [14, 21].

The encountered rise in free intracellular calcium is believed to activate both phospholipases A and C, which will attack membrane phospholipids, resulting in production of free fatty acids [14]. This loss of membrane phospholipids will increase the permeability of neuronal and mitochondrial membranes, which will further alter calcium homeostasis with additional detrimental effects, both directly and through their metabolites, on oxidative phosphorylation and on membranes.

Fatty acids, especially arachidonic acid, may be metabolized along the cyclo-oxygenase and the lipo-oxygenase pathways, with accumulation of prostaglandins, thromboxanes, leukotrienes, and free radicals as the end result [5, 18, 23].

Thromboxane A_2 is a potent vasoconstrictor (18); leukotrienes alter membrane permeability and also cause vasoconstriction [23], and free radicals allegedly attack the membrane [24].

Calcium Entry Blockers

Perhaps the most exciting advance in the treatment of brain ischemia has been initiated by the introduction of drugs that, in vitro at least, block calcium entry into cells. Calcium entry blockers comprise a rather heterogeneous group of drugs, containing derivatives of papaverine, dihydropyridine, and piperazine. There is not yet sufficient information on the whole pharmacological profile of most calcium entry blockers, and some have additional potencies not related to the influences on calcium, which also explains the difficulties in classification [8].

Pharmacological and Experimental Data of Nimodipine

Experiments in animals have shown that nimodipine abolishes or reverses the constriction induced in isolated arteries by serotonin, prostaglandin $F_{2\alpha}$, thrombin, or whole blood [22]. The effect of nimodipine is greatest in the cerebral arteries and less marked in the femoral artery [22]. Nimodipine crosses the blood-brain barrier easily and in considerable amounts [13]. Nimodipine administered to laboratory animals after global cerebral ischemia had a more favorable effect on neurological outcome than did placebo [19]: both infarct size (as measured by magnetic resonance imaging) and water content (as measured by brain gravimetry) were markedly reduced [3]. Given with food to spontaneously hypertensive, stroke-prone rats, nimodipine prevented cerebrovascular damage and reduced mortality [12]. In another study involving rats, the local cerebral blood flow, profoundly reduced after middle cerebral artery occlusion, was increased in the animals treated with nimodipine [9]. Neuropathological quantification of the ischemic brain damage showed that pretreatment with nimodipine reduced the amount and extent of cellular damage in the periphery of the lesion [9]. Consequently, studies designed to assess the effect of calcium channel blockers on neurological deficits resulting from acute ischemic stroke appear justified.

Clinical Studies with Nimodipine

Two clinical trials on the effects of nimodipine are available: an open study, conducted as a pilot study [6], and a double-blind, placebo-controlled, multicenter study [7]. In the multicenter study, all the patients were admitted to hospital within 24 h (most of them even within 8 h) and (for ethical reasons) all received a "standard" treatment comprising 500 ml low molecular dextran i.v. 12 h/day for 5 days and intensive supportive care involving intensive nursing, physiotherapy, logopedics, and occupational therapy. In addition, the patients received 120 mg nimodipine or placebo for 28 days according to a random list. Cardiotonics, antihypertensive medications, and antibiotics were administered if required, whereas steroids, vasodilators, and anticoagulants were avoided. Neurological deficits were assessed using the Mathew scale for neurological scoring.

During the 4-week treatment period, mortality from all causes was significantly reduced with nimodipine as compared with placebo (8.6% vs 20.4%, $P<0.05$). However, the improvement in survival was restricted to men (10% vs 27%, $P<0.003$).

A significantly better neurological outcome was also observed in the nimodipine group ($P<0.01$). The improvement in neurological status was greatest in patients with a moderate to severe deficit at baseline. In Table 1 a quantification of the improvement in both treatment groups is given.

There were no important side-effects except in one patient under nimodipine treatment, who developed a mild to moderate renal insufficiency. However, it remains doubtful whether this was the result of nimodipine or low molecular dextran application.

Table 1. Quantification of improvement in neurological deficits in both treatment groups

% Improvement	Nimodipine	Placebo
75% or more	In 18% of the patients	In 15% of the patients
50% or more	In 46% of the patients	In 36% of the patients
25% or more	In 69% of the patients	In 58% of the patients

Comments

The results of the multicenter study suggest first of all that the improvement in survival is restricted to men. It is surely not a reflection of the higher proportion of men in the placebo group. The number of acute medical illnesses as well as the baseline neurological scores were equally distributed between the male treatment and the male placebo group. Although unexpected, sex-related effects in vascular diseases are not unknown. For example, in one study the effect of antiplatelet therapy in the prevention of TIAs was more pronounced in men [4], while in another similar results were obtained after measures to prevent thromboembolic complications after hip arthroplasty [10]. Also experimental data depict sex-specific effects for vascular phenomena [11].

The conclusion that nimodipine is responsible for the improvement of neurological outcome is supported by several findings. First, both treatment groups were similar with respect to all known factors that may have influenced outcome, except for a minor imbalance in the distribution of male and female patients. In addition, nimodipine had no apparent side-effects; therefore, the blinding of patients, physicians, and others involved in the patients' care was maintained. Finally, the findings were consistent with the results of experiments in animals [19, 12].

Therefore, it is suggested that patients with acute ischemic stroke may benefit from treatment with nimodipine. The improvement in survival is restricted to men, and the neurological benefit is greatest in patients with a moderate to severe deficit at admission.

References

1. Astrup J, Symon L, Branston NM et al. (1979) Cortical evoked potentials of brain ischemia. Stroke 8:51–57
2. Astrup J, Siesjö BK, Symon L (1981) Thresholds in cerebral ischemia. The ischemic penumbra. Stroke 12:723–725
3. Bederson JB, Bartkowski H, Obana W, Nishizura M, Tsuji M, Pitts L (1985) Nimodipine in rat MCA occlusion (MCAO). Stroke 16:142
4. Canadian Co-operative Study Group (1978) A randomized trial of aspirine and sulphinpyrazone in threatened stroke. N Engl J Med 299:53–59
5. Demopoulos HB, Flamm ES, Pietronigro DD (1980) The free radical pathology and the microcirculation in the major central nervous system disorders. Acta Physiol Scand (Suppl) 492:91–119
6. Gelmers HJ (1984) The effect of nimodipine on the clinical course of patients with acute ischemic stroke. Acta Neurol Scand 69:232–239
7. Gelmers HJ, Gorter K, de Weerdt CJ, Wiezer HJA (1988) A controlled trial of nimodipine in acute ischemic stroke. N Engl J Med 318:203–207
8. Godfraind T (1984) Basic mechanisms and classification of calcium entry blockers. In: Godfraind T, Herman AG, Wellens D (eds) Calcium entry blockers in cardiovascular and cerebral dysfunctions. Martinus Nijhoff, The Hague, pp 9–18
9. Gotoh O, Mohamed AA, McCulloch J, Graham DI, Harper AM, Teasdale GM (1986) Nimodipine and the haemodynamic and histopathological consequences of middle cerebral artery occlusion in the rat. J Cereb Blood Flow Metab 6:321–331

10. Harris WA, Salzman EW, Athanasoulis CA et al. (1977) Aspirin prophylaxis of venous thromboembolism after total hip replacement. N Engl J Med 297:1246–1249
11. Johnson M, Ramey E, Ramwell PW (1978) Sex and age related differences in human platelet aggregation. Nature 253:355–357
12. Kazda S, Garthoff B, Luckhaus G (1983) Calcium antagonists prevent brain damage in stroke-prone spontaneously hypertensive rats (SHR-SP) independent of their effect in blood pressure: nimodipine versus nitrendipine. J Cereb Blood Flow Metab 3 (Suppl 1):526–527
13. Kerckhoff van den W, Drewes LR (1985) Transfer of the Ca-antagonists nifedipine and nimodipine across the blood brain barrier and their regional distribution in vivo. J Cereb Blood Flow Metab 5 (Suppl 1):495
14. Nowicki JP, Mackenzie ET, Young AR (1982) Brain ischemia, calcium and calcium antagonists. Pathol Biol (Paris) 30:282–288
15. Plum F (1981) Potential directions for therapy in cerebrovascular disease. In: Moose J, Reinmuth OM (eds) Cerebrovascular disease. Raven, New York, pp 195–204
16. Rehncrona S, Mela L, Siesjö BK (1979) Recovery of brain mitochondrial function in the rat after complete and incomplete cerebral ischemia. Stroke 10:437–446
17. Schanne FAK, Kane AB, Young EE et al. (1979) Calcium dependence of toxic cell death: a final common path. Science 206:700–702
18. Siesjö BK (1981) Cell damage: a speculative hypothesis. J Cereb Blood Flow Metab 1:155–186
19. Steen PA, Gisvold SE, Milde JH, Newberg LA, Scheithauer BW, Lanier WL, Michenfelder JD (1985) Nimodipine improves outcome when given after complete cerebral ischemia in primates. Anesthesiology 62:406–414
20. Sundt TM Jr, Michenfelder JD (1972) Focal transient cerebral ischemia in the squirrel monkey. Effect on brain adenosine triphosphate and lactate levels with electrocorticographic and pathologic correlation. Circ Res 30:703–712
21. Symon L (1980) The relationship between CBF, evoked potentials and the clinical features in cerebral ischemia. Proc 23rd Scand Neurol Congress, Acta Neurol Scand (Suppl 78) 62:175–190
22. White RP, Cunningham MP, Robertson JT (1982) Effect of the calcium antagonist nimodipine on contractile responses of isolated canine basilar arteries induced by serotonin, prostaglandin $F_{2\alpha}$, thrombin, and whole blood. Neurosurgery 10:344–348
23. Wolfe LS (1982) Eicosanoids: prostaglandins, thromboxanes, leucotrienes and other derivatives of carbon-20 unsaturated fatty acids. J Neurochem 28:1–14
24. Yoshida S, Inoh S, Asano T et al. (1980) Effect of transient ischemia on free fatty acids and phospholipids in the gerbil brain. Lipid peroxidation as a possible cause of postischemic injury. J Neurosurg 53:323–331

A Pilot Study of Nicardipine for Acute Ischemic Stroke. The Nicardipine Study Group

D. M. ROSENBAUM[1], J. C. GROTTA[1], F. M. YATSU[1], C. M. PICONE[1], L. C. PETTIGREW[1], P. BRATINA[1], J. ZABRAMSKI [2], R. SPETZLER[2], L. LOPEZ[2], J. MARLAR[3], and D. ELLIS[4]

Introduction

Interest in the therapy for acute cerebral ischemia has been stimulated by the demonstration of threatened but potentially viable tissue in experimental models [11, 12] and in stroke patients [1, 3, 6]. One strategy for preserving function in these brain regions is to elevate cerebral blood flow (CBF) because the oxygen extraction fraction is high in some areas of brain for 24–72 h after the onset of a cerebral infarct, which indicates that cerebral perfusion is barely adequate to meet the metabolic needs of ischemic neurons during this period [1, 3, 6]. Theoretically, if this tissue is not perfused, the diminished oxygen supply will cause metabolic failure, which changes ischemia into infarction [2]. Another strategy is to block the damaging and possibly irreversible disruption of cellular hemostasis that results from the failure of energy-dependent membrane pumps. Cerebral ischemia and the consequent reduction of oxygen supply cause conversion to anaerobic glycolysis and depletion of cellular ATP, resulting in an accumulation of lactate and failure of the tissue to maintain the normal extracellular-to-intracellular gradient of calcium. The concentration of extracellular calcium diminishes as this ion enters the cell and intracellular stores of calcium are liberated [13, 15–18]. The consequent increase in intracellular free (liberated) calcium activates neurofilament proteases and membrane phospholipases, resulting in irreversible membrane damage, and metabolism of arachidonic acid to cytotoxic leukotrienes, superoxide and hydroxyl-free radicals, and hydrogen peroxide [13, 15–18].

Calcium antagonists are a group of agents that prevent the intracellular accumulation of calcium. The result might be amelioration of ischemia by either or both mechanisms described above. One effect of their action is relaxation of vascular smooth muscle [5, 9], which results in increased CBF. In addition, preliminary studies suggest that calcium antagonists may prevent intraneuronal accumulation of calcium during reperfusion after cerebral ischemia [8, 18]. By maintaining the intra-extracellular gradient of calcium, membrane changes and their destructive effects may be prevented. Evaluation of calcium antagonists in the

[1] University of Texas, Medical School, 6431 Fannin, Houston, TX 77225, USA.
[2] Department of Neurology, Barrows Neurological Institute, Phoenix, AZ, USA.
[3] National Institute of Neurological and Communicative Disorders and Stroke (NIH), Bethesda, MD, USA.
[4] Syntex Research, Palo Alto, CA, USA.

Cerebral Ischemia and Calcium
Ed. by A. Hartmann and W. Kuschinsky

treatment of acute cerebral ischemia is both timely and appropriate for the following reasons: preliminary data indicate the efficacy of these drugs in animal models [4, 8, 10, 14], and in addition there has been preliminary experience with these agents in patients with acute stroke [7].

The goal of this study was to perform an open labeled, nonrandomized study of nicardipine, a prototype dihydropyridine calcium antagonist, in patients with moderately severe acute cerebral ischemia. Our specific objectives were:

1. To determine the optimal dose and correlate this dose with blood levels.
2. To determine the risks and side-effects, in particular, the effect of nicardipine on blood pressure in patients with cerebrovascular disease.
3. To determine the effect of this agent on patient morbidity, as measured by a standard grading scale, mortality, and infarct size (as measured by C7).

Patients

Patients between the ages of 21 and 90 years of age, admitted to The University of Texas Medical School at Houston or the Barrows Neurological Institute with the sudden onset of cerebral hemispheric symptoms, were studied. Patients with thrombotic/thromboembolic occlusions of large arteries, cardiogenic emboli, small vessel disease, or ischemia of uncertain origin were eligible. Treatment was started within 12 h of onset of ischemic symptoms.

Patients were excluded from the study if they had an entry neurological grading score >80 (i.e., mild deficit), if they were unarousable to voice, or if the stroke was likely due to vasculitis. Patients were also excluded if they had (a) a preexisting neurological deficit that would have interfered with the investigators' ability to determine the outcome of the acute cerebral infarction, (b) comorbid medical or psychiatric conditions that might have limited their survival or follow-up for the next 3 months, or (c) an unstable cardiovascular condition. Also excluded were patients who were already treated with calcium blockers on admission or with hypersensitivity to these drugs and women known to be pregnant or nursing.

Once the patient arrived in the emergency room and qualified for entry, baseline neurological grading was performed and therapy initiated. A CT scan of the brain (contrast optional) was obtained, and an intraarterial line for blood pressure monitoring and blood drawing was placed as medically indicated. Chest roentgenogram, electrocardiogram, and the following laboratory tests were performed: complete blood count with differential and platelet count, PT, PTT, blood glucose, serum electrolytes, blood chemistries, urinalysis, and arterial blood gas (where indicated).

Treatment and Assessment

After baseline measurements of mean arterial blood pressure (MABP) were obtained and averaged, intravenous nicardipine mixed in half normal saline was administered by a continuous infusion rate calculated to deliver 3 mg nicardipine per hour. The rate was increased by 1 mg/h every 30–90 min until a maximum dose of 7 mg/h was reached. MABP and pulse were measured every 15 min during the upward titration of the drug. If two consecutive measurements of MABP were 10% lower than baseline measurements, and/or the pulse rate rose by 20/min above baseline, the rate of administration was stabilized.

If there was a greater than 10% decrease in MABP (or MABP became lower than the investigator considered safe for that particular patient) or a greater than 20 beat/min increase in pulse, the infusion rate was decreased by 1 mg/h. If MABP and/or pulse rate did not return to previous levels in 15 min, then reductions by 1 mg/h were continued every 15 min until previous values were once again reached. If downward titration as far as 1 mg/h was necessary, a nicardipine blood level was obtained, the infusion discontinued for 4 h, and a new baseline MABP established. The infusion was then restarted at 3 mg/h.

Throughout the infusion, the patient remained at bed rest with head elevated 30°. Adequate hydration was maintained and recorded in all patients by supplemental intravenous fluids if necessary. In the last nine treated, in order to correct any hypovolemia which might have contributed to hypotension, plasmanate up to 500 cc was given daily if MABP fell more than 10%. After 72 h, the nicardipine blood level was obtained; following this the drug was tapered by 1 mg/h every 60 min and then discontinued. As soon as the intravenous drug was discontinued, oral nicardipine therapy (30 mg, t.i.d.) was begun. In those patients unable to swallow, the drug was given by emptying the capsule contents and administering through an endogastric tube. Other treatment was considered routine such as antiplatelet/anticoagulants, antibiotics (where indicated), supplemental oxygen (where indicated), and continuation of pretreatment medications. Early intervention by physical therapy, speech therapy, and occupational therapy was standard. Neurological grading was performed daily during the 72 h of intravenous drug administration, and at 1 week, 2 weeks, 4 weeks, and 3 months after the onset of the stroke or at any time neurological change occurred. A Barthel disability index was calculated at 1 and 3 months, and length of hospitalization determined for each patient.

Results

Demographics

During the 18-month period of July 1986 through December 1987, 58 patients were enrolled in the study. Thirty-three patients were treated in Houston, and 25 in Phoenix. The mean age was 64.8 ± 13.9 with a range from 26 to 87. Thirty patients were men and 28 were women. The mean interval from onset of symptoms to commencement of therapy was 7.1 ± 3.2 h.

The stroke was attributed to cardiac emboli in 42% of the patients, atherothrombotic large vessel in 21%, atheroembolic large vessel in 21%; small vessel in 8%; and other in 8%.

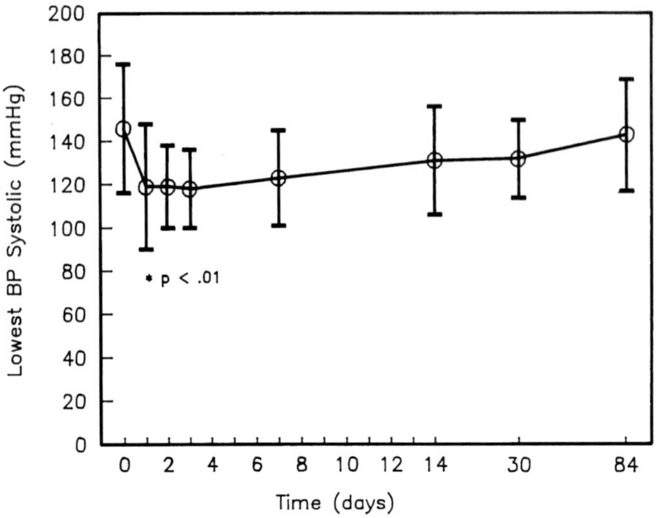

Fig. 1. Change in systolic blood pressure (lowest)

Safety

There were a total of nine deaths, ranging from 4 days to 2 months poststroke. All were due to either the stroke itself or medical complications related to the stroke. None were felt to be secondary to nicardipine.

Adverse reactions consisted of hypotension in five patients requiring decrease or discontinuation of the drug. There was one patient each who developed atrial flutter, nausea and vomiting, and liver function test abnormalities in whom it was felt the condition in question might be related to the drug.

There was a drop in systolic blood pressure from 146 ± 30 at baseline to 119 ± 29 at 24 h, then reaching 118 ± 18 at 72 h and returning to 143 ± 26 at 3 months (Fig. 1).

Dosimetry

In the nicardipine-treated group, the total intravenous dose of nicardipine was 277 mg over the 72-h period versus 194 mg ($P = 0.005$) in the nicardipine plus plasmanate treated group in the same period. Adding plasmanate did not result in hemodilution since there was no drop in hematocrit (baseline 39 ± 7 vs 72 h = 38 ± 5).

Efficacy

Of 38 patients completing 3 months of follow-up to date, the graded neurological examination score on a 100-point scale improved from a baseline of 41 ± 20 to 63 ± 28 ($P = 0.00009$) at 3 months (Fig. 2). At 3-month follow-up, 17% had com-

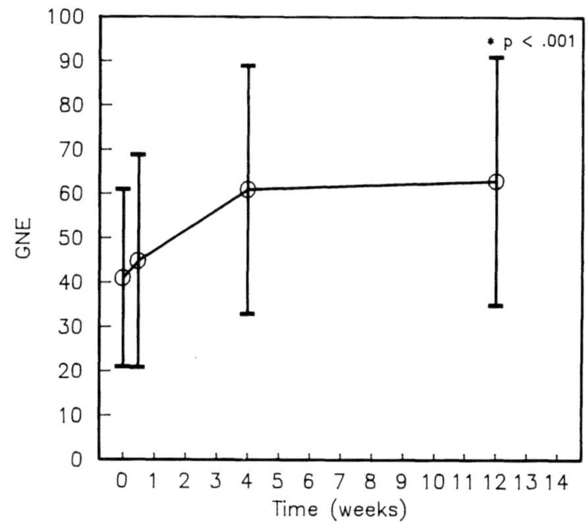

Fig. 2. Absolute change in graded neurological exam (GNE)

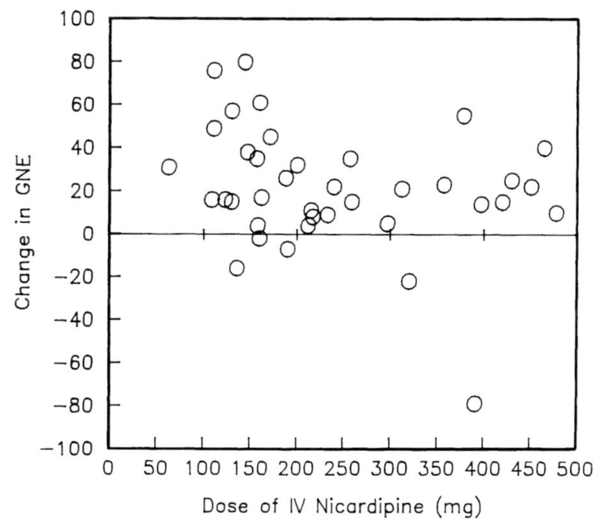

Fig. 3. Correlation of total IV dose of NC and change in graded neurological exam (GNE)

pletely improved, 63% partially improved, 20% stabilized, and 0% worsened. Patient placement class at 3-month follow-up was as follows: 37% no limitation, 26% mild limitation, 29% moderate limitation, and 9% severe limitation. Overall 83% of patients were at home and the mean Barthel's index was 72.

There was no correlation between graded neurological examination at 3 months and total intravenous dose of nicardipine received during the first 72-h period, indicating that there is no advantage to risking hypotension by pushing the intravenous dose above 3–4 mg/h (Fig. 3).

Conclusion

Nicardipine, at a dose of 3–4 mg/h, can be safely and effectively administered to patients with acute cerebral infarction, maintaining euvolemia by volume replacement if necessary. A larger controlled study is warranted to determine the efficacy of this therapy.

References

1. Ackerman RH, Correia J, Alpert N et al. (1982) Positron imaging in ischemic stroke disease using compounds labeled with oxygen 15. Arch Neurol 38:537–743
2. Ackerman RH, Alpert N, Correia J et al. (1984) Positron imaging in ischemic stroke disease. Ann Neurol 15 (suppl):S-126–130
3. Baron JC, Bousser M, Rey A, Guillard A et al. (1981) Reversal of focal "misery-perfusion syndrome" by extra-intracranial arterial bypass in hemodynamic cerebral ischemia. Stroke 12:454–459
4. Brandt L, Ljunggren B, Andersson K-E et al. (1983) Effects of topical application of a calcium antagonist (Nifedipine) on feline cortical pial microvasculature under normal conditions and in focal ischemia. J Cereb Blood Flow Metab 3:44–50
5. Fleckenstein A (1977) Specific pharmacology of calcium in myocardium, cardiac pacemakers, and vascular smooth muscle. Ann Rev Pharmacol Toxicol 17:149–166
6. Frackowiak R, Wise R (1983) Positron tomography in ischemic cerebrovascular disease. Neurology Clin 1:183–200
7. Gelmers HJ, Gorter K, deWeerdt CT, Wiezer HJA (1983) A controlled trial of nimodipine in acute ischemic stroke. N Engl J Med 318:203–207
8. Grotta J, Spydell J, Pettigrew C et al. (1986) The effect of nicardipine on neuronal function following ischemia. Stroke 17:213–219
9. Henry PD (1980) Comparative pharmacology of calcium antagonists: nifedipine, verapamil and diltiazem. Am J Cardiol 46:1047–1058
10. Kazda S, Hoffmeister F, Garthoff B, Towart R (1979) Prevention of the post-ischemic impaired reperfusion of the brain by nimodipine. Acta Neurol Scand (Suppl) 72:302–303
11. Pulsinelli WA, Levy DE, Duffy TE (1982) Regional cerebral blood flow and glucose metabolism following transient forebrain ischemia. Ann Neurol 11:499–509
12. Pulsinelli WA, Brierley JB, Plum F (1982) Temporal profile of neuronal damage in a model of transient forebrain ischemia. Ann Neurol 11:491–498
13. Raichle ME (1983) The pathophysiology of brain ischemia. Ann Neurol 12:2–10
14. Steen PA, Newberg LA, Milde JH, Michenfelder JD (1983) Nimodipine improves CBF and neurologic recovery after complete cerebral ischemia in the dog. J Cereb Blood Flow Metab 3:38–43
15. Vanhoutte PM (1981) Calcium entry blockers and cardiovascular failure. Fed Proc 40:2282–2886
16. Vanhoutte PM (1982) Effects of calcium entry blockers on tissue hypoxia. J Cereb Blood Flow Metab 2:S42–44
17. VanNueten JM, Vanhoutte PM (1980) Improvement of tissue perfusion with inhibitors of calcium ion influx. Biochem Pharmacol 29:479–481
18. Wieloch T, Harris RJ, Siesjö BK (1982) Brain metabolism and ischemia: mechanism of cell damage and principles of protection. J Cereb Blood Flow Metab 2:S5–9

An Open Study with Intravenous Application of Flunarizine in Acute Stroke

P.-J. HÜLSER, H. BERNHART, C. MARBACH, and H. H. KORNHUBER

In acute ischemic stroke increased free intracellular calcium accumulation plays a central role in the pathogenesis of neuronal death [9, 20]. The calcium overload blocker flunarizine has the ability to prevent the pathological influx of calcium ions but has only a very low affinity for the low-voltage-dependent calcium channels. This selectivity explains why flunarizine does not exert significant negative inotropic effects on normal heart muscle and does not influence the myogenic tone of blood vessels [3, 13, 22]. In experimental settings flunarizine reduces post-ischemic hypoperfusion [22], scavenges free radicals [14], and lowers blood viscosity [4, 19]. In various animal models of hypoxia and local or global brain ischemia a favorable effect of flunarizine on functional and morphological parameters was found when the drug was given before [21, 23–26] or after [5, 6] the injury. The formation of atheromas is inhibited [2]. Therefore we decided to study the effect of flunarizine in acute stroke. In preparation of a multicenter double-blind controlled study now on the way, we first did an open pilot study, mainly to investigate possible side-effects of intravenous high dose treatment in patients with cerebrovascular disease.

Patients and Methods

Fifty-five patients (37 males, 18 females) were included in the study, fulfilling the following criteria: age between 50 and 85, time lag between stroke and onset of therapy less than 24 h, no coma at onset of therapy, no earlier stroke in the patient's history (except for transient ischemic attacks), and no severe systemic disease (acute myocardial infarction, acute pulmonary embolism, malignant disease). After a complete neurological examination the patient's mental status, sensorimotor functions, and activities in daily life were quantitatively scored following the Mathew scale [12]. This scoring was done on the day of admission, on the following day, on days 4, 7, 21, and 28 and, if the patient stayed longer in the hospital, every consecutive week. Blood pressure and heart rate were measured before and after each dose of flunarizine. The complete blood count, electrolytic levels, serum creatinine and urea, hepatic enzymes, and blood coagulation were carefully controlled. Additionally, a series of enzymes were investigated in the urine (Dr. Junge, Kiel), a highly sensitive test to detect any adverse effect on kidney function.

Department of Neurology, University Hospital, D-7900 Ulm, FRG.

Cerebral Ischemia and Calcium
Ed. by A. Hartmann and W. Kuschinsky

Treatment

In the first 7 days the patients received 25 mg flunarizine (in 0.9% saline, drug vehicle cyclodextrine, complete volume 110 ml) intravenously in the morning and in the evening, each infusion lasting about 10 min. In addition hemodilution was done with hydroxyethyl starch 2 × 250 ml per day, if necessary combined with phlebotomy to reduce the hematocrit below 40% within 2 days. After 7 days flunarizine was dispensed in oral doses of 10 mg in the morning and 20 mg in the evening for 3 weeks, then it was reduced to 10 mg per day. Acetylsalicylic acid 200 mg and, in bedridden patients, low dose heparin were given. Physiotherapy was started as early as possible. All patients with diabetes, high blood pressure, or overweight received a special diet to reduce cardiovascular risk factors [1].

Results

Eighteen patients reported slight weariness during the infusion of flunarizine and sometimes for a short period thereafter. Its degree usually declined within the days of intravenous treatment. The patients' activities were not impaired in any case. There were no observations of parkinsonism, tardive dyskinesia, akathisia, or depression being induced by flunarizine despite the high dosage. Neither blood pressure nor heart rate showed significant change when measured before and after the infusion of flunarizine. Serum urea and serum creatinine exhibited no change, and the investigation of enzymes in the urine revealed no sign of renal alteration. The complete blood counts, blood coagulation, hepatic enzymes, and serum electrolytes were unchanged except for the effect of hemodilution.

Generally the development of the neurological status ran a favorable course, but there were two patients who died. A 56-year-old man exhibited severe bronchopneumonia on the day of admission and suddenly died 2 days later, perhaps due to pulmonary embolism despite low dose heparin. An 84-year-old man suffered from longstanding severe arterial hypertension with sudden exacerbations. In these periods he showed signs of severe cardiac insufficiency. Despite treatment in an intensive care unit he died after 2 weeks due to irreversible left ventricular failure. A causal relationship of these fatal events to the administration of flunarizine could be excluded.

In a 65-year-old man treatment was stopped after 1 day because of complete remission; another 76-year-old patient developed alcoholic delirium and was transferred to a psychiatric unit. Eight patients were discharged from the hospital within the first 2 weeks because of rapid improvement. The number of examinations were too small in this group for detailed analysis; therefore the extensive statistics relate only to those who stayed in the hospital for at least 14 days.

The course of the individual Mathew score shows that in these more severely impaired patients who stayed longer in the hospital there was a remarkable improvement compared to the initial condition (Fig. 1). Patients with a severe deficit tended to show a higher increase. This is at least partly an effect of the scale itself.

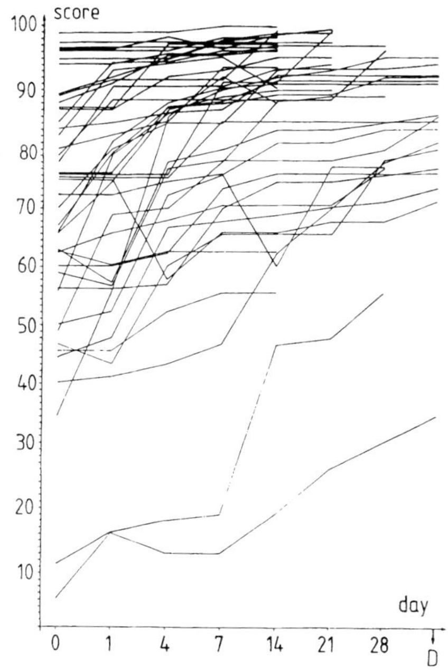

Fig. 1. Course of the individual Mathew scores of 43 patients treated for at least 14 days

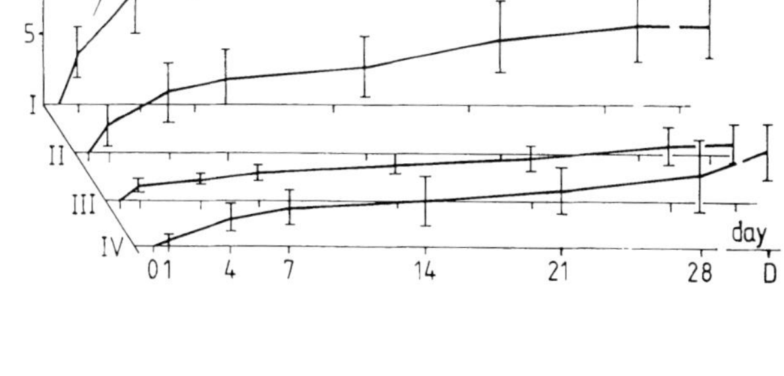

Fig. 2. Increase in the average Mathew score compared with the initial value. The *bar* indicates twice standard deviation. The *curves* depict the average values for 43 patients during the first 2 weeks. Thereafter the number of patients decreases because more and more were discharged from hospital. *I*, total score; *II*, mental status score; *III*, sensorimotor score; *IV*, ability of daily life score

A maximum of 100 points can be reached; therefore patients with an initial high score can make only a small increase. The increase in the total score as well as in the subscores was most pronounced in the 1st week (Fig. 2); the graph resembles an exponential curve. After day 14 more and more patients were discharged, the more severely disabled remaining under control. The slight kinking of the increase in the Mathew scores towards a greater improvement indicates that these patients do profit from the prolonged treatment.

Discussion

Despite strong efforts the drug treatment of acute stroke has been unsatisfactory up to now, as exemplified by the disappointing results of recent studies in hemodilution in acute stroke [7, 10, 15, 16]. An early start of therapy, a better plasma expander, and selection of patients for hemodilution based on sophisticated differentiation of individual pathophysiological changes in rheology [18] might result in better chances for this therapy, but at present there is no reliable basis for it. Possibly, calcium antagonists will provide a new therapeutic approach: the result of a study with nimodipine provides new hope in this field [8].

Possible side-effects of intravenous administration of flunarizine were expected in respect of cardiovascular parameters, the extrapyramidal motor system, and depression. In the examined patients there was no detectable effect on heart rate or arterial blood pressure. This observation is in accordance with the very low affinity of the drug for voltage-dependent calcium channels. This neutral action is favorable as a decrease in arterial blood pressure should be avoided in hypotonia or altered autoregulation. In contrast to some claims in the literature, extrapyramidal side-effects or depression were not noticed despite the high dose treatment. Our own previous experience with this drug in the treatment of migraine and vertigo, as well as the results of a recent study [17], indicates that these side-effects are a very rare complication of flunarizine. Stroke patients, like other patients with severe disease, are frequently depressed about their situation; the emotional situation ameliorates with an increase in their abilities. We did not observe any depression occurring under the therapy. No adverse effect on organ functions could be detected. On the whole the intravenous administration of flunarizine, even in the chosen high dose regimen, seems to be a safe therapy. This is the most important result of our study, for, despite the impression of a favorable clinical outcome in the investigated patients, final conclusions on the effect of intravenous flunarizine in acute stroke will only be possible after completing the multicenter double-blind study, for which we have as yet acquired about two-thirds of the projected number of patients.

Besides the necessity of early initiation of any therapeutic regime in stroke, good general patient care is of outstanding importance, including provision of adequate oxygenation and fluid balance, stable cardiovascular function, prevention of bed sores, urinary retention, deep venous thrombosis, and aspiration pneumonia, provision of physiotherapy, and treatment of cardiovascular risk factors [1, 11].

References

1. Altmann J, Kornhuber AW, Kornhuber HH (1987) Stroke: cardiovascular risk factors and the quantitative effects of dietary treatment on them. Eur Neur 26:90–99
2. Betz E, Hämmerle H, Strohschneider T (1985) Inhibition of smooth muscle cell proliferation and endothelial permeability with flunarizine in vitro and in experimental atheromas. Res Exp Med 185:325–340

3. Borgers M, Ghoss E, Thone F, Van Nueten JM (1980) Effects of flunarizine on the distribution of calcium in vascular smooth muscle. Blood vessels 17:123–130
4. De Clerck F, David JL (1981) Pharmacological control of platelet and red blood cell function in the microcirculation. J Cardiovasc Pharmacol 3:1388–1412
5. Deshpande JK, Wieloch T (1985) Amelioration of ischaemic brain damage following post-ischaemic treatment with flunarizine. Neurol Res 7:27–29
6. Deshpande JK, Wieloch T (1986) Flunarizine, a calcium entry blocker, ameliorates ischemic brain damage in the rat. Anesthesiology 64:215–224
7. Frei A, Cottier C, Wunderlich P, Lüdin E (1987) Glycerol and dextran combined in the therapy of acute stroke. A placebo-controlled, double-blind trial with a planned interim analysis. Stroke 18:373–379
8. Gelmers HJ, Gorter K, De Weerdt CJ, Wiezer HJA (1988) A controlled trial of nimodipine in acute ischemic stroke. N Engl J Med 318:203–207
9. Harris RJ, Symon L, Branston NM, Bayhan M (1981) Changes in extracellular calcium activity in cerebral ischemia. J Cereb Blood Flow Metab 1:203–209
10. Italian Acute Stroke Study Group (1988) Haemodilution in acute stroke: results of the Italian haemodilution trial. Lancet I:318–321
11. Kornhuber HH (1984) Bluthochdruck und Alkoholkonsum. In: Rosenthal J (ed) Arterielle Hypertonie. Springer, Berlin Heidelberg New York, pp 149–162
12. Mathew NT, Meyer JS, Rivera VM, Charney JZ, Hartmann A (1972) Double-blind evaluation of glycerol therapy in acute cerebral infarction. Lancet II:1327–1333
13. Peters T (1984) Wirkung von Kalziumantagonisten auf zelluläre Kalziumbewegungen. Therapiewoche 34:6011–6029
14. Phillis JW, Delong RE, Towner JK (1985) The effect of lidoflazine and flunarizine on cerebral reactive hyperemia. Eur J Pharmacol 112:323–329
15. Scandinavian Stroke Study Group (1987) Multicenter trial of hemodilution in acute ischemic stroke. I. Results in the total patient population. Stroke 18:691–699
16. Scandinavian Stroke Study Group (1988) Multicenter trial of hemodilution in acute ischemic stroke. Results of subgroup analyses. Stroke 19:464–471
17. Schimert GC (1987) Bewertung des zerebralen Calcium-Antagonisten Flunarizin. Wirksamkeit, Verträglichkeit, Erfahrungen aus der Praxis. MMW 129:471–473
18. Schneider R, Wöbker G, Willmes K, Kiesewetter H (1986) Are there hemorheological differences in different ischemic brain lesions. Klin Wochenschr 64:357–361
19. Scott CK, Persico FJ, Carpenter KS, Chasin M (1980) The effects of flunarizine, a new calcium antagonist on human red blood cells in vitro. Angiology 31:320–330
20. Siesjö BK (1981) Cell damage in the brain: a speculative synthesis. J Cereb Blood Flow Metab 1:155–185
21. Silverstein FS, Buchanan K, Hudson C, Johnston MV (1986) Flunarizine limits hypoxia-ischemia induced morphologic injury in immature rat brain. Stroke 17:477–482
22. Van Nueten JM (1982) Selectivity of calcium entry blockers. In: Albertini A, Paoletti R (eds) Calcium modulators. Elsevier Biomedical, Amsterdam, pp 199–208
23. Van Reempts J, Borgers M, Van Dael L, Van Eyndhoven J, Van de Ven M (1983) Protection with flunarizine against hypoxic-ischaemic damage of the rat cerebral cortex. A quantitative morphologic assessment. Arch Int Pharmacodyn 262:76–88
24. Van Reempts J, Van Deuren B, Van de Ven M, Cornelissen F, Borgers M (1987) Flunarizine reduces cerebral infarct size after photochemically induced thrombosis in spontaneously hypertensive rats. Stroke 18:1113–1119
25. Wauquier A, Ashton D, Clincke G, Fransen J (1985) "Calcium entry blockers" as cerebral protecting agents: Comparative activity in tests of hypoxia and hyperexcitability. Jpn J Pharmacol 38:1–7
26. Wauquier A, Ashton D, Edmonds HL (1984) Protection against cerebral hypoxia by antimigraine drugs. In: Amery WK, Van Nueten J, Wauquier A (eds) The pharmacological basis of migraine. Pitman, London, pp 134–148

CBF Decreases in Ischemic Areas After Calcium Antagonist Treatment in Acute Stroke

S. Vorstrup, O. B. Paulson, and A. Andersen

Introduction

Over recent years, much effort has been devoted to the investigation of the pathophysiological mechanisms that lead to irreversible cell damage following ischemia. Several studies have indicated that disturbances in calcium ion homeostasis may be responsible for the initiation of a series of events leading to irreversible cell damage following cerebral ischemia [e.g., 15, 20]. It was therefore anticipated that treatment with a calcium antagonist in the early phase after ischemia might prevent the metabolic disturbances leading to intracellular calcium ion accumulation. In addition, it was hoped that these drugs, by a selective vasodilatory effect on the cerebral vessels, would increase regional CBF in the ischemic areas.

A potent calcium antagonist of the dihydropridine type chemically related to nifedipine, PY 108–068, has been developed (by Sandoz Ltd., Basel, Switzerland). In vitro studies have documented the calcium antagonistic properties of this drug as a selective inhibitor of calcium-induced contraction of the aorta (rabbit) [9] and cerebral arteries (cat) [10]. Furthermore, experimental studies have shown a beneficial effect of PY 108–068 in three different models of cerebral ischemia [21].

Consequently, we decided to study the effect of PY 108–068 on the cerebral circulation in patients with acute stroke, measuring CBF by the atraumatic xenon-133 inhalation technique and single photon emission tomography.

Materials and Methods

Cerebral blood flow was measured using xenon-133 inhalation and a rapidly rotating detector system, the Tomomatic 64, previously described in detail [18]. One study lasts 4.5 min, and the technique permits repetition of the measurement after a minimum interval of 20 min. Three slices of brain tissue are studied simultaneously. These slices are 2 cm thick and have an interslice distance of 2 cm. The resolution is 1.5–1.7 cm full width at half-maximum in the plane. The input function is taken from the lung curve profile obtained from a single narrowly collimated crystal placed above the upper part of the lung. Calculation of CBF is then

Department of Neurology, State University Hospital, Rigshospitalet, 9, Blegdamsvej, DK-2100 Copenhagen, Denmark.

Cerebral Ischemia and Calcium
Ed. by A. Hartmann and W. Kuschinsky

performed, applying the bolus distribution principle on the sum of the first two pictures. Scaling and linearization are performed as previously described [4].

Direct comparison of the tomographic flow pictures with the computed tomography (CT) scanning at the corresponding level was performed. This allowed for the delineation of low flow areas corresponding to the infarcted areas as well as the surrounding low flow (peri-infarct) areas where no hypodense lesions were present. The symmetrically placed region in the contralateral hemisphere was obtained as well, and the degree of side-to-side asymmetry was calculated by taking the CBF difference of the symmetrical regions as a percentage of the higher value. This parameter was used to estimate whether a significant change in regional flow distribution was induced [19].

Patient Material

Two subsequent series of patients with acute ischemic stroke were studied, excluding patients with previous cerebrovascular disease. Only patients showing a focal low flow area on the tomographic CBF measurement in consonance with the clinical symptoms were further investigated using PY 108–068. The patients were studied as soon as possible after admittance to the hospital, but no patient could be included later than day 10. Informed consent was obtained from all patients, and the study was approved by the Ethical Committee in Copenhagen.

The first series comprised three men and three women (mean age 56 years, ranging from 20 to 81 years). CBF was measured from 1 to 9 days after onset of the stroke (average 5 days). A baseline CBF measurement was performed and repeated twice 50 min after doses of 1.5 and 2.5 mg, respectively. The drug was given by intravenous infusion at a rate of 1 mg PY 108–068 over 10 min.

In the second series, three men and two women (mean age 56 years, ranging from 43 to 71 years) were studied. CBF was measured from 2 to 7 days after onset (average 4 days). In these patients, two baseline studies were performed with a minimum time interval of 20 min. Then a continuous intravenous infusion of PY 108–068 was given at a rate of 1 mg/10 min and CBF was measured when total doses of 2.5 and 7.5 mg had been reached.

Results

In the first six cases, only one patient showed a significant change in flow pattern. This patient showed an increase of CBF from 29 to 35 ml/100 mg/min in part of the peri-infarct area, improving the side-to-side asymmetry from 31% to 20%. In the remaining patients, neither mean CBF nor flow distribution changed significantly. End-expiratory CO_2 ($FeCO_2$) values were unchanged. MABP just after the CBF measurement was on average 102, 100, and 102 mmHg, and the heart rate was stable around 100 beats/min.

In the second series, where the higher doses of PY 108–068 were used, mean hemispheric CBF increased slightly by 5% and 3% in the nonaffected and af-

Table 1. CBF values in the peri-infarct regions in the five patients given the higher doses of PY 108-068 (series no. 2). The values indicate the degree of side-to-side asymmetry

	Baseline flows 1 and 2	Following PY 108-068	
		Flow 3	Flow 4
1	11%	11%	33%*
2	21%	33%*	17%
3	21%	32%*	26%
4	33%	39%	28%
5	16%	16%	19%

* Significant change ($P < 0.05$) in the degree of side-to-side asymmetry as compared with the baseline value.

fected sides, respectively. However, the flow distribution changed significantly in three patients: A further decrease in CBF in the ischemic region was induced, with clear-cut aggravation of the side-to-side asymmetry. CBF was reduced in both the infarcted tissue and the peri-infarct regions (Table 1).

The $FeCO_2$ values remained stable. MABP measured just after the first CBF measurements was 109 ± 12 mmHg, and decreased to 95 ± 11 and 98 ± 15 mmHg (1 SD) during infusion. The corresponding heart rates were on average 76, 84, and 82 beats/min.

Discussion

In the stroke patients having larger focal low flow areas, concomitant measurements of CBF and $CMRO_2$ using positron emission tomography have suggested that potentially viable brain tissue may be present for up to 3 days after onset of an acute stroke [22]. This was based on the observation of an increased oxygen extraction ratio with preserved $CMRO_2$ in low flow areas, a finding only observed in one-third of the patients studied within day 3. In such patients, early treatment with a calcium antagonist could theoretically be expected to improve the clinical outcome. Consequently, only patients with focal low flow areas were included in the present series.

The vasodilating properties of calcium antagonists in nonischemic animal models have been demonstrated in several studies by direct measurements of the vessel diameter [2] or by measurement of the resultant flow [3, 6, 8]. However, Harris et al. [7] found only an insignificant increase of the basal blood flow level following nimodipine treatment in a closed-skull primate model. These results correspond to our findings, where only a slight increase of CBF was noted in the nonaffected hemisphere. The study of Harris et al. also demonstrated that responses to arterial PCO_2 changes and autoregulation to reduced blood pressure were severely impaired during treatment.

In experimental ischemia conflicting findings have been reported, but most studies have failed to demonstrate any beneficial effect upon the cerebral circulation. Promising results were obtained in a dog model of complete cerebral ischemia where pretreatment by nimodipine gave an improved neurological outcome in the treated group [17]. The authors suggested that this effect was caused by the increase in CBF noted during the phase of delayed hypoperfusion. Dilatation of constricted cerebral vessels in ischemic territories [3], as well as improved tissue oxygenation, has also been found [21].

In contrast, Hossmann et al. (1983) [11] could not show any positive effect on electrophysiological changes, on metabolic recovery, or on the postischemic accumulation of calcium in the brain tissue during flunarizine treatment in the cat exposed to prolonged complete ischemia. Likewise, Phillis et al. [14] found no effect upon the basal flow rate following anoxia using either nifedipine or felodipine in a rat model. In a study by Smith et al. [16], incomplete ischemia was induced in the rat. Here pretreatment with nimodipine increased mean CBF in the postischemic phase, but this increase resulted from very heterogeneous flow values, i.e., zones of hypoperfusion and evident hyperemia within the same structures. No improvement of the electrophysiological parameters was noted in this series. Harris et al. [7] also found an increase in the residual flow following middle cerebral artery occlusion during nimodipine treatment in an open-skull primate model. However, their results seemed to indicate that the threshold levels of blood flow for the development of cortical edema and for disturbances of ion homeostasis were increased, thereby possibly enhancing the susceptibility of the tissue to ischemic damage.

In the present study, treatment with a calcium antagonist did not improve CBF in the ischemic area or its surroundings, even in the case studied 1 day after onset of clinical symptoms. In fact, further CBF decrease was observed in some of the patients from the group given the higher doses of calcium antagonist. On the presumption that the vessels in the ischemic areas are vasoparalytic, this could be explained by a "steal effect": the decrease in perfusion pressure caused by cerebral vasodilatation elsewhere along with the reduction in systemic blood pressure noted in these patients could further decrease CBF in the ischemic area and its surroundings, where CBF autoregulation and hence vasodilatory effects most likely are abolished. These assumptions are supported by experimental studies on nonischemic rat models, where acute administration of nimodipine nearly suspended cerebral autoregulation, as only a small autoregulatory plateau remained [12]. However, the same authors found that nimodipine preserved CBF during moderate hypotension below the lower limit of CBF autoregulation. Mohamed et al. (13) also showed that increasing doses of nimodipine may lead to a decrease of the MABP.

In conclusion, the present study showed a further CBF decrease in the focal low flow areas after the higher doses of the calcium antagonist PY 108–068 in the early phase after an acute stroke. It could be expected, however, that under different pathophysiological conditions where cerebral ischemia occurs owing to severe vasoconstriction of the cerebral vessels, as seen after a subarachnoid hemorrhage, treatment with a calcium antagonist might prove useful [1]. Still, it cannot be excluded that patients with acute stroke might benefit from treatment with a calcium antagonist owing to a protective effect at the cellular level, in particular if treatment could be given very early or even before onset of the symptoms. A recent randomized study on nimodipine in acute ischemic stroke comprising 186 patients showed a decreased mortality and a better neurological outcome during the initial 4-week period, but the long-term effect was not evaluated [5]. Further carefully planned large-scale studies on the clinical effect in ischemic stroke are needed.

References

1. Allen GS, Ahn HS, Preziosi TJ et al. (1983) Cerebral arterial spasm – a controlled trial of nimodipine in patients with subarachnoid hemorrhage. N Engl J Med 308:619–624
2. Auer LM (1981) Pial arterial vasodilation by intravenous nimodipine in cats. Drug Res 31:1423–1425
3. Brandt L, Ljunggren B, Andersson KE, Edvinsson L, MacKenzie E, Tamura A, Teasdale G (1983) Effects of topical application of a calcium antagonist (nifedipine) on feline cortical pial microvasculature under normal conditions and in focal ischemia. J Cereb Blood Flow Metab 3:44–50
4. Celsis P, Goldman T, Henriksen L, Lassen NA (1981) A method for calculating regional cerebral blood flow from emission computerized tomography of inert gas concentrations. J Comput Assist Tomogr 5:641–645
5. Gelmers HJ, Gorter K, de Weerdt CJ, Wiezer HJA (1988) A controlled trial of nimodipine in acute ischemic stroke. N Engl J Med 318:203–207
6. Harper AM, Craigen L, Kazda S (1981) Effect of calcium antagonist, nimodipine, on cerebral blood flow and metabolism in the primate. J Cereb Blood Flow Metab 1:349–356
7. Harris RJ, Branston NM, Symon L, Bayhan M, Watson A (1982) The effects of a calcium antagonist, nimodipine, upon physiological responses of the cerebral vasculature and its possible influence upon cerebral ischaemia. Stroke 13:759–766
8. Haws CW, Gourley JK, Heistad DD (1983) Effects of nimodipine on cerebral blood flow. J Pharmacol Exp Ther 225:24–28
9. Hof RP, Vuorela HJ, Neumann P (1982) PY 108–068, a new, potent, and selective inhibitor of calcium-induced contraction of rabbit aortic rings. J Cardiovasc Pharmacol 4:344–351
10. Hof RP, Vuorela HJ, Neuman P (1982) PY 108–068, a new calcium antagonist, on general hemodynamics and regional blood flow in anesthetized cats: a comparison with nifedipine. J Cardiovasc Pharmacol 4:352–362
11. Hossmann K-A, Paschen W, Csibal (1983) Relationship between calcium accumulation and recovery of cat brain after prolonged cerebral ischemia. J Cereb Blood Flow Metab 3:346–353
12. Kerchoff van den W, Kazda S (1987) The autoregulation of cerebral blood flow is influenced by calcium antagonists. J Cereb Blood Flow 7 (Suppl 1):S169
13. Mohamed AA, McCulloch J, Mendelow AD, Teasdale G, Harper AM (1984) Effect of the calcium antagonist nimodipine on local cerebral blood flow: relationship to arterial blood pressure. J Cereb Blood Flow Metab 4:206–211
14. Phillis JW, DeLong RE, Towner JK (1986) The effects of nifedipine and felodipine on cerebral blood flow during anoxic episodes. Stroke 17:229–234
15. Siesjö BK (1981) Cell damage in the brain: a speculative synthesis. J Cereb Blood Flow Metab 1:155–185
16. Smith M, Kågstrøm E, Rosén I, Siesjö BK (1983) Effects of the calcium antagonist nimodipine on the delayed hypoperfusion following incomplete ischemia in the rat. J Cereb Blood Flow Metab 3:543–546

17. Steen PA, Newberg LA, Milde JH, Michenfelder JD (1983) Nimodipine improves cerebral blood flow and neurological recovery after complete cerebral ischemia in the dog. J Cereb Blood Flow Metab 3:38–43
18. Stokely EM, Sveinsdottir E, Lassen NA, Rommer P (1980) A single photon dynamic computer assisted tomograph (DCAT) for imaging brain function in multiple cross section. J Comput Assist Tomogr 4:230–240
19. Vorstrup S, Andersen A, Blegvad N, Paulson OB (1986) Calcium antagonist (PY 108–068) treatment may further decrease flow in ischemic areas in acute stroke. J Cereb Blood Flow Metab 6:222–299
20. Wieloch T, Siesjö BK (1982) Ischemic brain injury: the importance of calcium, lipolytic activities, and free fatty acids. Pathol Biol 30:269–277
21. Wiernsperger N, Gygax P, Hofmann A (1984) Calcium antagonist PY 108–068: demonstration of its efficacy in various types of experimental brain ischemia. Stroke 15:679–685
22. Wise RJS, Bernardi S, Frackowiak RSJ, Legg NG, Jones T (1983) Serial observations on the pathophysiology of acute stroke. The transition from ischaemia to infarction as reflected in regional oxygen extraction. Brain 106:197–222

The Effect of Nimodipine on ICP and CBF in the Awake Patient

J. F. Schmidt, G. Waldemar, M. Albeck, O. B. Paulson, and F. Gjerris

In human studies it has been shown that the calcium antagonist PY 108–068 induces a considerable increase in cerebral blood flow (CBF) with a concomitant flow reduction in focal ischemic areas [9]. In experimental studies in rat, nimodipine increased the CBF and impaired the autoregulation [3, 5]. Different calcium antagonists might, however, have different effects on the CBF. Whether this increase in CBF has any adverse effect on intracranial pressure (ICP) has not been studied. In the present study we have investigated these aspects using nimodipine.

Method

1. Four patients with normal pressure hydrocephalus (NPH) and aged 38, 51, 71, and 73 years were studied. The resting mean arterial blood pressure (MABP) was 104, 115, 126, and 140 mmHg and the ICP was 4, 10, 11, and 12 mmHg. ICP and MABP were continuously measured intraventricularly and intraarterially, respectively. Regional CBF (rCBF) was measured by xenon-133 inhalation and single photon emission computerized tomography (SPECT) using the Tomomatic 64 [1, 8]. The rCBF was measured before and 1 h after peroral ingestion of 60 mg nimodipine. MABP was lowered stepwise approximately 30 mmHg using a combination of ganglion blockade and vacuum application on the lower abdomen and the legs, and MABP was increased approximately 30 mmHg by norepinephrine. Hypo- and hypertension were induced in the control period and 1 h after 60 mg nimodipine perorally. CBF was measured for every 5 mmHg by drawing blood samples from the venous jugular bulb and the radial artery (see below). The lower limit of CBF autoregulation was determined according to a computer program [7].

2. Two patients with subarachnoid hemorrhage (SAH) were investigated. Their ages were 26 and 60 years. Three and 5 days after the acute hemorrhage and after an acute X-ray (arteriography) had shown severe spasms of the intracerebral vessels, a catheter for continuous measurements of ICP was placed in the intraventricular system. Then a xenon-133 examination was performed in the young patient (see above). During a transcutaneous puncture on the neck a

Departments of Neurosurgery and Neuromedicine, Rigshospitalet, DK-2100 Copenhagen, Denmark.

Cerebral Ischemia and Calcium
Ed. by A. Hartmann and W. Kuschinsky

catheter was placed with the tip in the venous jugular bulb on the contralateral side of the bleeding. An intraarterial catheter was placed for continuous monitoring of the blood pressure. Intravenous nimodipine was administered via an indwelling catheter with a dose of 15 μg/kg/h in the first 2 h, after which the dose was increased to 30 μg for the next 2 h.

Every 10 min blood samples were drawn from the jugular vein and from the artery for oxygen saturation estimation and measurement of PCO_2.

From the oxygen saturations the relative CBF was determined according to cerebral metabolic rate $(CMR) = (aO_2 - vO_2) \times CBF$. Considering a constant CMR this determination gives a good estimate of CBF [6].

After 4 h treatment in the young patient another CBF measurement was performed using the xenon-133 technique (see above).

Results

Ad 1. Following nimodipine MABP was slightly reduced, by 7 mmHg (0–16 mmHg). The ICP was almost unchanged, i.e., 4, 10, 11, and 12 mmHg before and 7, 9, 11, and 11 mmHg after treatment. CBF was changed by −3%, +5%, +7%, and +10% 1 h after the ingestion of nimodipine. No changes were found in rCBF after nimodipine. The lower limit of CBF autoregulation showed a decrease in two of the patients of 6 and 8 mmHg, respectively. In one patient the lower limit was not reached in spite of a decrease in perfusion pressure of 29 mmHg (from 86 to 57 mmHg). In the final patient a steady increase in CBF [35% (Fig. 1)] was found during the lowering of the perfusion pressure from 126 mmHg to 89 mmHg.

Ad 2. The CBF increased in both patients receiving nimodipine intravenously, by 27% and 49% respectively (Fig. 2). No changes were registered in the ICP. The young patient had normal ICP (10–12 mmHg) and the older patient had increased ICP (33–38 mmHg). The MABP was almost unaltered in the young patient (89 mmHg before and 94 mmHg after 4 h nimodipine administration). In

Table 1. CBF in different regions before and after nimodipine in a patient with vasospasm after SAH[a]

	Before		After	
	Left	Right	Left	Right
Hemispheric flow	52	48[b]	55	54[c]
Frontal area	47	47	53[c]	52[c]
Sylvian area	64	56[b]	59	60
Temporoparietal area	45	40[b]	53[c]	47[b, c]
Occipital area	53	47[b]	57	60[c]
Cerebellum	42	50[b]	53[c]	58[c]

[a] Values are given in ml/100 g/min.
[b] Significant side-to-side difference.
[c] Significantly higher flow after treatment.

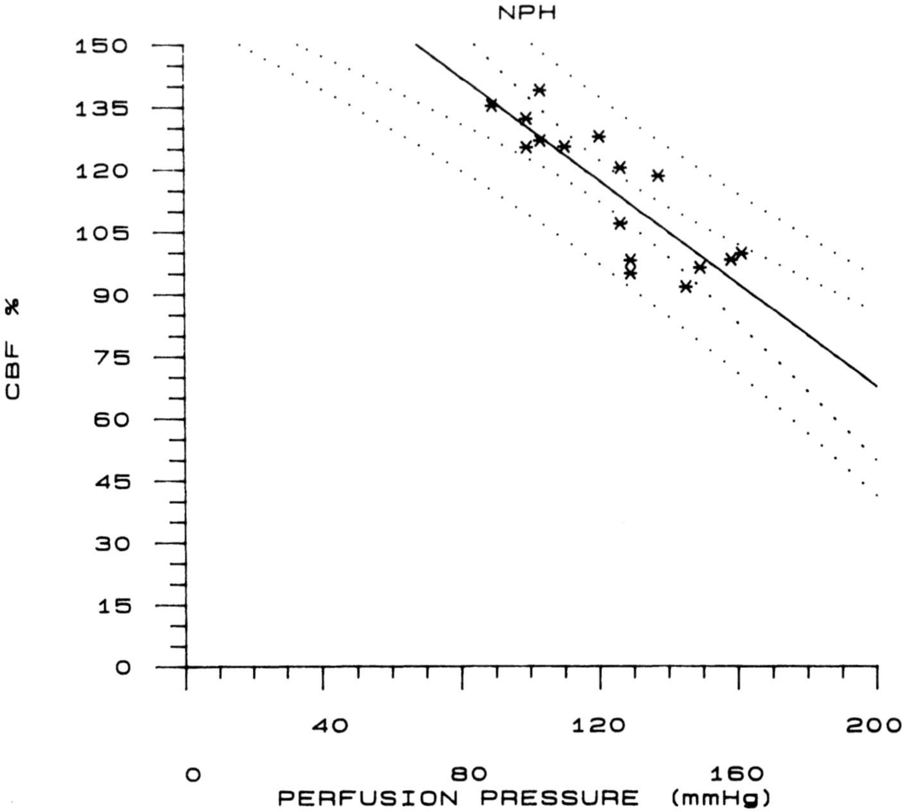

Fig. 1. CBF in relation to perfusion pressure in a patient with NPH after treatment with 60 mg nimodipine orally

the older patient a minor fall was registered, i.e., from 89 mmHg to 75 mmHg. The rCBF in most of the former low perfusion areas was significantly increased after nimodipine administration (Fig. 3; Table 1) in the young patient. This patient was afterwards operated for a lobulated saccular aneurysm of the right middle cerebral artery. Postoperatively the patient was fully awake without any additional neurological deficit. Nimodipine was continued orally at a dose of 60 mg 6 hourly for 2 weeks. Three months later the patient was in good clinical condition without any deficit at all. The older patient had a drop in systolic blood pressure during the night and nimodipine administration was terminated. Finally she was judged to be in too bad a condition for operation and was transferred to a local neurological department without further neurosurgical treatment.

Comments

None of the four investigated patients with NPH had any alterations in ICP after treatment with nimodipine orally. The data from the literature are conflicting.

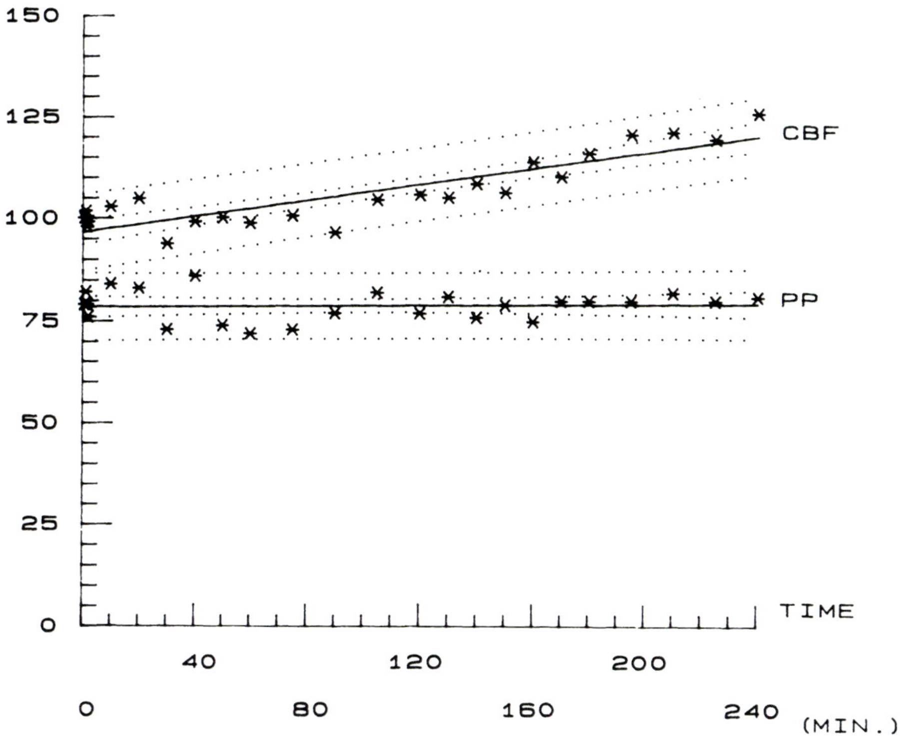

Fig. 2. CBF (% of resting CBF) and perfusion pressure (mmHg) in a patient with angiographically proven vasospasm after SAH during treatment with nimodipine 15 and 30 μg/kg/h

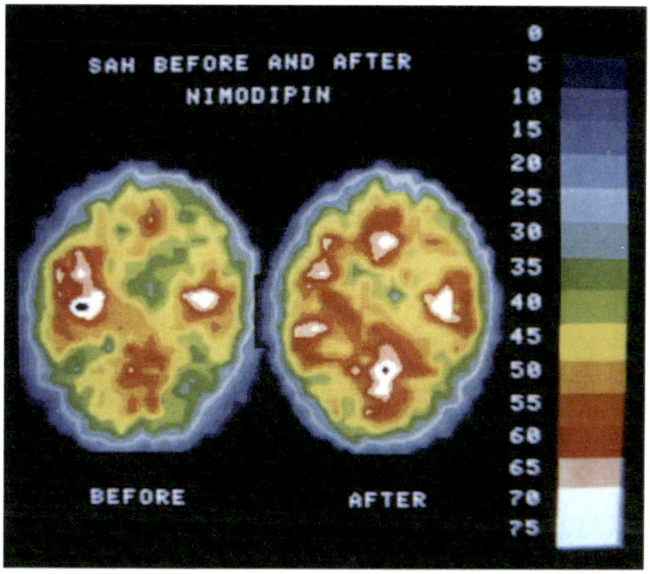

Fig. 3. rCBF before and after treatment with nimodipine in the patient from Fig. 2. The absolute values are given in Table 1

Guggiari et al. [4] found an increase in ICP after nimodipine treatment; in contrast Gaab et al. [2] found no such increase. Both studies concerned patients suffering from an acute cerebral disease. In the present study no increases in ICP during nimodipine infusion were found in the two patients with acute SAH.

In three of the NPH patients only a slight increase in CBF was demonstrated, without regional disturbances. One of the patients showed an increase despite a fall in the cerebral perfusion pressure, suggesting that the autoregulation of CBF had been abolished. The autoregulation was not abolished in the three other patients after treatment with nimodipine; the lower limit was in fact lower after administration of nimodipine. This might be due to methodological circumstances, i.e., repeating the study at too short an interval [7].

In the two patients with SAH and cerebrovascular vasospasm, a significant increase in CBF was found during nimodipine administration. The increase in CBF was predominantly in the low flow area, with no signs of steal in the one patient studied with the rCBF method.

In conclusion, the present study suggests a slight increase in CBF after nimodipine administration. However, in one patient the autoregulation of CBF was abolished. No changes in the ICP were demonstrated. In patients suffering from SAH with arterial vasospasm a significant increase in rCBF in the ischemic areas was found, leaving the ICP unaffected. These results have been obtained in only a few patients, and further prospective studies will be needed to confirm them.

References

1. Celsis P, Goldman T, Henriksen L, Lassen NA (1981) A method for calculating regional cerebral blood flow from emission computed tomography of inert gas concentration. J Comput Assist Tomogr 5:641–645
2. Gaab MR, Haubitz I, Brawanski A, Korn A, Czech T (1985) Acute effects of nimodipine on the cerebral blood flow and intracranial pressure. Neurochirurgia 28:93–99
3. Gaab MR, Høllerhage HG, Zumkeller M, Trost HA (1987) The effect of the Ca-antagonist nimodipine on cerebral blood flow autoregulation. J Cereb Blood Flow Metab 7 (Suppl 1):170
4. Guggiari M, Guillaume A, Dagreou F, Philippon J, Viars P (1983) Intracranial pressure and hemodynamic effects of a new calcium channel blocking agent: nimodipine. Anesthesiology 59:A357
5. Kerckhoff van den W, Kazda S (1987) The autoregulation of cerebral blood flow is influenced by calcium antagonists. J Cereb Blood Flow Metab 7 (Suppl 1):169
6. Lassen NA (1959) Cerebral blood flow and oxygen consumption in man. Physiol Rev 39:183–238
7. Schmidt JF, Waldemar G, Andersen AR, Vorstrup S, Paulson OB, Gjerris F (1988) Autoregulation of cerebral blood flow in man: evaluation of a method applicable for PET and SPECT. J Cereb Blood Flow Metab (submitted)
8. Stokely EM, Sveinsdottir E, Lassen NA, Rommer PA (1980) Single photon dynamic computer assisted tomography (DCAT) for imaging brain function in multiple cross sections. J Comput Assist Tomogr 4:230–240
9. Vorstrup S, Andersen A, Blegvad N, Paulson OB (1986) Calcium antagonist (PY 108–068) treatment may further decrease flow in ischemic areas in acute stroke. J Cereb Blood Flow Metab 6:222–229

Migraine

Calcium Entry Blockers in Migraine: The State of the Art

E. L. H. Spierings

Introduction

The calcium entry blockers or calcium antagonists are a group of medications that share the potential of inhibiting the movement of calcium into cells and cell organelles. They do so through a number of different mechanisms, such as, for instance, the blockade of potential- or receptor-operated calcium channels or the blockade of calcium influx under conditions of overload. Chemically the calcium entry blockers represent medications of different groups with little or no relationship to each other and little or no predictability with regard to activity and/or mechanism [24].

The first calcium entry blocker demonstrated to be effective in the treatment of migraine is *flunarizine* [10], a difluorinated piperazine derivative. In experimental studies, it has been shown to possess antivasoconstrictor and cerebral hypoxia protective effects [1]. To date, its beneficial effect in migraine has been shown in a great number of double-blind placebo-controlled and comparative studies [21]. Other calcium entry blockers which have been studied in the treatment of migraine are the dihydropyridines, *nifedipine* and *nimodipine*, and *verapamil* [21]. However, the evidence for their effectiveness in migraine is by far not as impressive as that for flunarizine. Nifedipine has been investigated in migraine in merely one double-blind placebo-controlled study which included eight patients only [7], and of both nimodipine and verapamil two double-blind placebo-controlled studies are available [21]. However, in particular in the verapamil studies, the number of patients included was relatively small, which raises the issue of power, i.e., the chance with which a study can demonstrate a certain difference as statistically significant. When the power of a study is low, it casts doubt upon whatever findings the study generates, whether statistically significant or not. This issue has recently been addressed with regard to the double-blind studies in which the efficacy of flunarizine in migraine prophylaxis was compared with that of pizotifen [23].

Recently, more information has also become available on the possible efficacy of the calcium entry blockers in the treatment of cluster headache. Cluster headache is a migraine-related condition which probably shares with migraine the mechanism by which the headache is caused, i.e., cranial noncerebral vasodilatation [18]. Its treatment is relatively well defined and differs from the treatment of

The Headache Research Foundation and John R. Graham Headache Centre, The Faulkner Hospital, and Department of Neurology, Tufts University School of Medicine, Boston, Massachusetts 02130, USA.

Cerebral Ischemia and Calcium
Ed. by A. Hartmann and W. Kuschinsky

migraine in several respects, including the use of medications such as oxygen and lithium, which are generally not effective in migraine [19]. Conversely, medications which are effective in migraine prophylaxis, such as propanolol and amitriptyline [20], generally do not offer relief from cluster headache.

Efficacy in Cluster Headache

In a series of open studies involving between five and eight patients each, Meyer and Hardenberg [12] found nifedipine, nimodipine, and verapamil to be effective in the prophylactic treatment of cluster headache. With regard to verapamil's prophylactic efficacy in cluster headache we have recently completed an open study involving 48 patients, 43 males and 5 females [22]. Of these patients 33 suffered from episodic and 15 from chronic cluster headache. Their age averaged 41.2 years and they suffered from cluster headache for an average of 14.5 years. Thirty-three of the 48 patients, i.e., 69%, improved more than 75% on prophylactic treatment with verapamil. Six of the 48, i.e., 12%, benefited between 25% and 75%, and nine, i.e., 19%, including the three patients who discontinued treatment because of side-effects, improved less than 25%. No differences of significance were observed between the patients with episodic and those with chronic cluster headache.

Although the probable mechanism shared by migraine and cluster headache is that of cranial noncerebral vasodilatation, it is unlikely that this is the mechanism addressed by the calcium entry blockers because the calcium entry blockers themselves may cause vasodilatation [24]. One of the studies from which this conclusion was drawn is where verapamil, administered intravenously, was shown to abort nitroglycerin-induced attacks of cluster headache [3]. In the same study vasodilatation was also questioned as the underlying mechanism of cluster headache, because administration of nitroglycerin resulted in vasodilatation within several minutes, while the pain did not occur until an average of almost half an hour later. Verapamil administered at the peak of the pain, which was about 1 h after the nitroglycerin was given, relieved the headache in more than 50% of cases in a little over half an hour. An interaction with the release of nociceptive factors was hypothesized, and possibly one such factor is substance P, a neurotransmitter which also possesses vasodilator properties. The release of substance P from the primary sensory nerve fibers, which is a calcium-dependent process, has been implicated in the pathogenesis of cluster headache [6].

Mechanisms of Action in Migraine

The calcium entry blockers are introduced into the treatment of migraine on the basis of Wolff's vasoconstriction-cerebral hypoxia-vasodilatation concept of the pathogenesis of the migraine attack (Fig. 1). Their relevant modes of action are considered to be antivasoconstriction and protection of the brain against the deleterious effects of hypoxia [1]. The antivasoconstrictor effect is mediated through blockade of potential- and receptor-operated calcium channels in arterial smooth

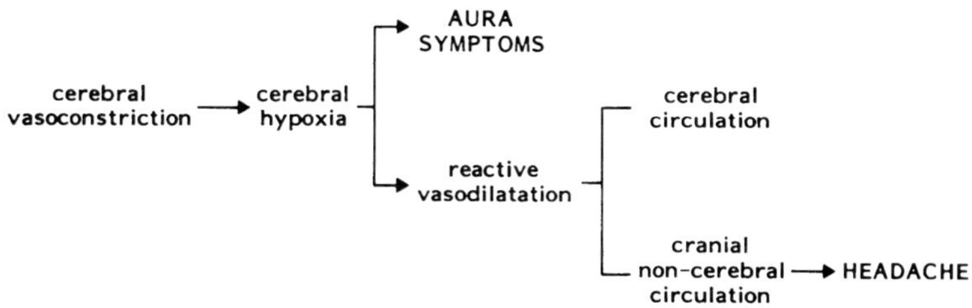

Fig. 1. Schematic representation of Wolff's vasoconstriction–cerebral hypoxia–vasodilatation concept of the pathogenesis of the migraine attack

muscle, and the cerebral hypoxia protective effect through prevention of neuronal damage from cellular calcium overload [24].

Wolff's concept concerning the involvement of vasoconstriction and cerebral hypoxia in the pathogenesis of the migraine attack is based on experiments with the cerebral vasodilators carbon dioxide and amyl nitrite, administered during the migraine aura [20]. The migraine aura is the part of the migraine attack which consists of focal neurological symptoms of transitory nature, such as the scintillating scotoma and digitolingual paresthesias. It is present only in attacks of so-called *classic migraine*, and in these attacks it usually precedes the onset of the headache. In *common migraine* the headache develops without the preceding aura and in this case, according to Wolff's hypothesis, the vasoconstriction-induced cerebral hypoxia occurs in a so-called clinically silent area of the cerebral cortex.

Wolff looked upon the two conditions of migraine, i.e., common and classic migraine, as related conditions sharing the pathogenetic mechanisms of vasoconstriction, cerebral hypoxia and vasodilatation. He hypothesized that the vasodilatation, as a reactive hyperemia, follows the vasoconstriction-induced cerebral hypoxia. The headache, according to Wolff, originates from the involvement of the cranial *non*cerebral circulation in the process of vasodilatation. This supposition he based on experiments with ergotamine. However, for the pathogenesis of the common migraine attack, there is no evidence for the occurrence of vasoconstriction-induced cerebral hypoxia, except for a study by Gelmers [5].

Gelmers, using the intraarterial Xenon injection technique, measured cerebral blood flow in two patients with common migraine and found a localized *decrease* in cerebral blood flow in the occipitoparietal and parietocentral areas, respectively. The localized decrease in blood flow was preceded by an *increase* in blood flow in the parietal and occipital areas, respectively, and was followed by the headache. These particular cerebral blood flow changes, i.e., occipitoparietal hyperemia followed by "oligemia," had been reported shortly before by Olesen et al. [14] in patients with classic migraine during the development of the focal neurological symptoms, i.e., the aura (vide infra). In common migraine, Olesen et al. [15] had found no such changes in cerebral blood flow, which had led them to suggest that classic and common migraine are different conditions with different pathogenetic mechanisms. As there is also no evidence for the involvement of vasoconstriction-induced cerebral hypoxia in cluster headache, the mechanism of action of the calcium entry blockers in common migraine may well be the same as that suggested for cluster headache, i.e., interference with the release of nociceptive substances (vida supra).

Cerebral Hypoxia in Classic Migraine

Wolff hypothesized that vasoconstriction-induced cerebral hypoxia occurs in classic *and* common migraine, but with the exception of Gelmer's study evidence for this is only available for classic migraine. With regard to the cerebral blood flow changes occurring during the migraine aura, the initial studies applying the Xenon clearance technique had shown the decrease in cerebral blood flow to be generalized and bilateral despite the focal and lateralized character of the symptoms [11, 12]. Later studies, using a much more sophisticated application of the technique, revealed this generalized decrease to be but the final step in a series of changes [9, 14]. The changes include an *increase* in blood flow in the occipitoparietal area followed by a *decrease* and a gradual spreading of the decrease toward the frontal pole (Fig. 2).

The decrease in cerebral blood flow of the migraine aura was calculated to be in the order of 20%–30%, which is generally considered not to be enough to cause neuronal dysfunction by hypoxia and, therefore, was referred to as oligemia. The forward spreading of the oligemia was measured to occur at the rate of a little over 2 mm/min [9]. This particular rate of propagation brought back a hypothesis, formulated by Milner in 1958, that the migraine aura is caused by spreading depression rather than by vasoconstriction-induced cerebral hypoxia. Spreading depression is a neurophysiological phenomenon which spreads over the cerebral cortex at a rate of 2–5 mm/min and is associated with a decrease in cerebral blood flow in the order of magnitude of 25% [8]. In the spreading depression hypothesis of migraine the vascular changes are looked upon as secondary to the neuronal dysfunction and, for example, the relieving effect of carbon dioxide on the migraine aura is attributed to an effect of carbon dioxide on the spreading depression rather than on the vascular system [4]. Similarly, the effects of one of the calcium entry blockers, flunarizine, in migraine is related to an interference of the medication with the initiation and propagation of spreading depression [25].

Recently, however, Olsen et al. [16], of the same laboratory as Olesen and Lauritzen, brought back the idea of vasoconstriction-induced cerebral hypoxia as the possible mechanism involved in the migraine aura. Having corrected for scattered radiation in the Xenon clearance recordings, they found the spreading de-

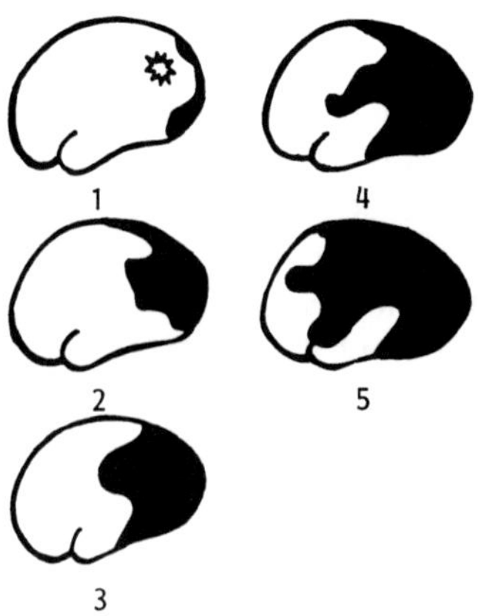

Fig. 2. Schematic representation of the cerebral blood flow changes of the migraine aura as observed by Olesen and Lauritzen [9, 14]. The changes consist of a short-lasting *increase* in cerebral blood flow in the occipitoparietal area (*), followed by a *decrease* and a gradual spreading of the decrease toward the frontal pole

crease in cerebral blood flow as described above to be at least 50% in depth, bringing it into the level of ischemia. As further evidence for the actual occurrence of hypoxia during the migraine attack they considered the impairment of cerebrovascular reactivity observed in the affected areas of the brain [17], as well as the occurrence of delayed hyperemia following the decrease in cerebral blood flow [2].

Wolff's Concept Rehabilitated?

Was Wolff then right in hypothesizing that the focal neurological symptoms of migraine are caused by vasoconstriction-induced cerebral hypoxia? And, in line with this, are antivasoconstriction and protection against cerebral hypoxia indeed the relevant modes of action of the calcium entry blockers, at least in classic migraine?

In their recent paper on cerebral blood flow changes in classic migraine, Andersen et al. [2] state that their "results suggest that a state of arteriolar vasoconstriction is present, and that transient focal cerebral ischemia offers an explanation to aura symptoms." This is a statement clearly in support of Wolff's concept. However, at the same time Andersen et al. consider Wolff's hypothesis "obsolete," because the *increase* in cerebral blood flow following the *decrease*, which they ascribe to reactive hyperemia following ischemia, does *not* coincide with the onset of the headache. Wolff obviously tried to bring his observations with regard to the pathogenesis of the migraine aura and headache together in a way that would inherently also explain the typical time relationship between these two components of the migraine attack, i.e., the headache following the aura. Hence, he hypothesized that the mechanism involved in the aura is at the same time also the cause of the headache. However, he may have been wrong here and that is possibly what the results of Andersen et al. suggest.

As mentioned, the observation describing the decrease in cerebral blood flow as oligemia as well as that of the particular rate of spreading of the oligemia brought back the hypothesis that the migraine aura is caused by spreading depression rather than by vasoconstriction-induced cerebral hypoxia. Spreading depression, as a cause of the migraine aura, stands, in my view, closer to the clinical characteristics of the migraine aura than does vasoconstriction-induced cerebral hypoxia. In neurology, the clinical symptomatology of (transient) cerebral hypoxia is very well known in the phenomenon of "transient ischemic attacks." Transient ischemic attacks are caused by cerebral hypoxia due to thromboembolic vascular occlusion. The difference from migraine, therefore, is only in the process leading to the hypoxia, i.e., vasoconstriction versus thromboembolic occlusion. As it is the cerebral hypoxia that ultimately determines the symptomatology, it is hard to understand that the mentioned difference in the process leading to the hypoxia can explain the dissimilarities in symptomatology between the two conditions. Hence, from a clinical point of view, the involvement of different mechanisms would be more plausible. Also, for other reasons of a clinical nature, a mechanism akin to spreading depression is favored. It is against the background of these statements and also taking Gelmer's earlier mentioned findings into consideration that one wonders whether the cerebral blood flow changes, indicating oligemia or ischemia, are anything but artifacts of the technique applied or haphazard findings that have little or nothing to do with the pathogenesis of migraine.

Epilogue

As Wolff and many others have assumed, classic and common migraine are clinically related conditions and to suppose completely different pathogenetic mechanisms for them seems unrealistic. Also from a clinical perspective I would say, in contrast to Wolff, that it is probably *not* the aura that is the central symptom of migraine but the headache. The aura may be looked upon as a feature which may or may not accompany (or precede) the headache, as is also the case for the other symptoms of migraine such as the nausea, vomiting, and photo- or phonophobia. Looking at it from that perspective, it is much more likely that the calcium entry blockers exert their prophylactic effect in migraine through an interference with the pathogenesis of the headache rather than with the pathogenesis of any of the other migraine symptoms, including the aura. In addition, one has to realize that, for example, the effect of the calcium entry blockers in protecting the brain against the deleterious effects of hypoxia has to date only been demonstrated for one of these medications effective in migraine, i.e., flunarizine (Table 1). As I have stated before [21], the actual relevant modes of action of the calcium entry blockers in migraine may well still have to emerge from the wealth of pharmacological properties these medications possess. Also the medications may well address different aspects of the pathogenesis of the migraine attack and the aspects they address may vary for the different medications which so far have been grouped together as calcium entry blockers or calcium antagonists.

Table 1. Qualitative pharmacological profiles of the calcium entry blockers (Vanhoutte [24])

	Verapamil-like	Nifedipine-like	Diltiazem	Flunarizine-like
Inhibition of potential-operated calcium channels of arterial smooth muscle	+	+	+	+
Inhibition of receptor-operated calcium channels of arterial smooth muscle	+	+	+	+
Protection against structural damage resulting from cellular calcium overload in brain	o	o	o	+

+, experimentally proven effect; o, experimental data not available.

References

1. Amery WK, Wauquier A, Van Neuten JM, DeClerck F, Van Treempts JV, Janssen PAJ (1981) The anti-migrainous pharmacology of flunarizine (R14950), a calcium antagonist. Drugs Exp Clin Res 7:1–10
2. Andersen AR, Friberg L, Olsen TS, Olesen J (1988) Delayed hyperemia following hypoperfusion in classic migraine. Arch Neurol 45:154–159
3. Boiardi A, Gemma M, Porta E, Peccarisi C, Bussone G (1986) Calcium entry blocker: treatment in acute pain in cluster headache patients. Ital J Neurol Sci 7:531–534
4. Gardner-Medwin AR (1981) The effect of carbon dioxide and oxygen on Leão's spreading depression: evidence supporting a relationship to migraine. J Physiol 316:23P–24P
5. Gelmers HJ (1982) Common migraine attacks preceded by focal hyperemia and parietal oligemia in rCBF pattern. Cephalalgia 2:29–32
6. Hardebo JE (1984) The involvement of trigeminal substance P neurons in cluster headache. A hypothesis. Headache 24:294–304
7. Kahan A, Weber S, Amor B, Guerin F, DeGeorges M (1983) Nifedipine in the treatment of migraine in patients with Raynaud's phenomenon. N Engl J Med 308:1102–1103
8. Lauritzen M, Jorgenesen MB, Diemer NH, Gjedde A, Hansen AJ (1982) Persistent oligemia of rat cerebral cortex in the wake of spreading depression. Ann Neurol 12:469–474
9. Lauritzen M, Olsen TS, Lassen NA, Paulson OB (1983) Changes in regional cerebral blood flow during the course of classic migraine attacks. Ann Neurol 633–641
10. Louis P (1981) A double-blind placebo-controlled prophylactic study of flunarizine (Sibelium) in migraine. Headache 21:235–239
11. Mathew NT, Hrastnik F, Meyer JS (1976) Regional cerebral blood flow in the diagnosis of vascular headache. Headache 15:252–260
12. Meyer JS, Hardenberg J (1983) Clinical effectiveness of calcium entry blockers in prophylactic treatment of migraine and cluster headache. Headache 23:266–277
13. O'Brien MD (1971) Cerebral blood flow changes in migraine. Headache 10:139–143
14. Olesen J, Larsen B, Lauritzen M (1981) Focal hyperemia followed by spreading oligemia and impaired activation of rCBF in classic migraine. Ann Neurol 9:344–352
15. Olesen J, Tfelt-Hansen P, Henriksen L, Larsen B (1981) The common migraine attack may not be initiated by cerebral ischemia. Lancet II:438–440
16. Olsen TS, Friberg L, Lassen NA (1987) Ischemia may be the primary cause of the neurologic deficits in classic migraine. Arch Neurol 44:156–161
17. Olsen TS, Friberg L, Ronager J (1987) Cerebrovascular reactivity is completely abolished during attacks of classic migraine. Cephalalgia 7 (Suppl 6):103–105
18. Spierings ELH (1987) Cluster headache: distinguishing features. Drug Therapy 17(5):56–67
19. Spierings ELH (1987) Cluster headache: relieving and preventing attacks. Drug Therapy 17(5):68–72
20. Spierings ELH (1987) Migraine headaches: mechanisms and management. New Dev Med 2:59–76
21. Spierings ELH (1988) Clinical and experimental evidence for a role of calcium entry blockers in the treatment of migraine. Ann NY Acad Sci 522:676–689
22. Spierings ELH, Gabai IJ (1989) Prophylactic treatment of cluster headache with verapamil. Headache 29:167–168
23. Spierings ELH, Messinger HB (1988) Flunarizine versus pizotifen in migraine prophylaxis: a review of comparative studies. Cephalalgia (in press)
24. Vanhoutte PM (1987) The expert committee of the World Health Organization on classification of calcium antagonists: the viewpoint of the raporteur. Am J Cardiol 58:3A–8A
25. Wauquier A, Ashton D, Marranes R (1985) The effects of flunarizine in experimental models related to the pathogenesis of migraine. Cephalalgia 5 (Suppl 2):119–120

Nimodipine, a Calcium Entry Blocker, Platelets, and Migraine

P. T. G. Davies, T. J. Steiner, and F. Clifford Rose

Introduction

Any theory of migraine pathogenesis must account for, or predict, the mechanism of action of drugs used effectively in prophylactic treatment. The platelet theory of migraine pathogenesis proposes that such drugs act by inhibiting platelet aggregation and/or the release reaction, or by interfering with the pharmacological activity of the release products [1], notably 5-HT. The importance of the platelet as a prime mover in the migrainous process is now doubted [2, 3]. To some extent the efficacy of the major prophylactic drugs, pizotifen and methysergide, can be explained within the platelet theory of migraine, but our work has shown that this theory cannot accommodate the actions of beta-blockers on platelet activity measured ex vivo [2].

There is increasing current interest in calcium channel blockers in migraine prophylactic treatment. The suggestion that platelets may be a model for the study of calcium-linked neuronal vesicular secretion [4] prompted us to study the action of these drugs on platelet behaviour. Gelmers [5] first found nimodipine to be significantly superior to placebo in reducing the frequency and severity of migraine attacks, and further evidence of this followed [6]. Studies on platelets suggested that this drug either had no effect on platelet activity in normals or else inhibited it, while other drugs of the dihydropyridine group (e.g. nifedipine) inhibited platelet aggregation [7]. In migraineurs, in whom platelet activity was measurably different [8, 9], effects might not be the same, and it was of interest to correlate platelet activity in migraine sufferers with symptomatic change during prophylaxis with nimodipine.

Methods

Patients

Ten patients (four males, six females) presenting to the clinic for treatment with migraine according to Vahlquist's criteria [10] and requiring prophylaxis, were selected on the following basis:

The Princess Margaret Migraine Clinic, Charing Cross Hospital, Fulham Palace Road, London W6 8RF, U.K.

Cerebral Ischemia and Calcium
Ed. by A. Hartmann and W. Kuschinsky

1. Recognised attacks for at least 2 years with, currently, two to five attacks each month
2. Aged 18–55
3. Non-smokers
4. On no regular medication and able, during the study, to limit drug treatment of acute attacks to paracetamol, codeine, and metoclopramide (specifically avoiding other medications likely to affect platelet function).

Informed consent was given by all patients in terms approved by the Charing Cross Hospital Ethical Sub-committee.

Treatment

After 1 month's untreated run-in to ensure withdrawal from all other medication, treatment with nimodipine was opened: 40 mg t.d.s. for the 2nd month of the study and continuing, or increasing in clinical non-responders to 60 mg t.d.s., for the 3rd month.

Assessment of Clinical Response

Clinical assessments after entry were monthly at subsequent visits to the clinic. Diary cards issued monthly asked patients to record attack frequency, duration and severity; the last scored 1–3 according to curtailment of daily activities (not at all, partial, complete). The "sum of severities score" was the product of the mean of the daily severity scores in the month and the number of days affected, corrected for 28 days; this was used comparatively.

Platelet Activity Tests

Platelets were studied monthly after entry. Venous blood was taken into sodium citrate (final concentration 0.31%) and spun at 150 *g* for 10 min to obtain platelet-rich plasma (PRP) and again at 400 *g* for 15 min for platelet-poor plasma (PPP). The latter was used as diluent to adjust the manual platelet count in the PRP sample to $250 \pm 10 \times 10^9$/l. Aggregation of platelets in PRP at 37 °C to each of two inducers, ADP and epinephrine (EPI), was measured optically in a Payton dual channel lumi-aggregometer, with one channel simultaneously recording ATP release by the luciferin-luciferase reaction. Aggregation was automatically calibrated against a baseline set as the amount of light transmission through the disaggregated PRP, and 100% aggregation, measured as the amount of light transmission through the same patient's PPP. ATP release was calibrated directly by reference to a curve derived from ATP standard concentrations.

Response indices were, of primary aggregation, initial slope (PS) and height of the aggregation curve at 30 s (H30); of secondary aggregation, height at 60 s minus height at 30 s (60–30) and final height; and of ATP release, peak height of the release curve. Dose-response curves for each of these were constructed for

ADP and EPI. Readings were derived in arbitary units for each patient for an inducer concentration initially selected to be submaximal for platelet aggregation and then fixed for the duration of the study. These figures expressed platelet behaviour and allowed within-patient comparisons before and during treatment.

At the time of each platelet study, 10 ml of blood was taken into lithium heparin containing indomethacin and a further 10 ml into lithium heparin alone. Plasma from the former was stored at −70 °C for thromboxane B_2 (TXB_2) estimation using a radioimmunoassay kit (Amersham). Plasma from the latter was stored at −70 °C in darkness for later analysis of nimodipine and nimodipine pyridine metabolite levels by Bayer (UK).

Statistical Analysis

Statistical analyses used Student's paired *t*-test.

Results

There was between-patient variability in measured indices of platelet activity at each stage of the study (Table 1). During treatment with nimodipine these indices essentially remained within the pretreatment range. While fluctuations were noted within individuals, these had no consistent direction within the group as a whole. In individual patients increases or decreases in some indices did not correlate with clinical response to treatment nor with plasma levels of nimodipine or its pyridine metabolite (Table 2). Similarly, TXB_2 levels varied both between patients and within patients during treatment but without any trend (Table 3).

Table 1. Indices of platelet activity in the nimodipine-treated patients. Baseline results (obtained after the 1-month run-in) are compared with those obtained after 1 and 2 months' treatment. Mean values only are shown (see text). NS, not significant at the 5% level

Index	Baseline		After 1 month			After 2 months		
	Mean	SD	Mean	SD	*P*	Mean	SD	*P*
ADP-induced aggregation (2.5 μM)								
PS	9.5	2.06	10.7	2.05	NS	9.9	1.98	NS
H30	7.0	1.60	7.2	1.63	NS	7.2	1.3	NS
60–30	2.5	1.21	2.9	0.79	NS	2.4	1.21	NS
Final Ht	16.4	1.86	16.4	3.56	NS	14.9	2.90	NS
ATP release	27.9	13.17	22.9	12.58	NS	29.7	13.54	NS
EPI-induced aggregation (1 μM)								
PS	2.8	0.99	2.8	0.69	NS	2.4	0.75	NS
H30	2.5	0.95	2.6	0.73	NS	2.2	0.68	NS
60–30	2.0	0.99	1.7	0.39	NS	1.4	0.63	NS
Final Ht	17.6	2.69	16.9	4.29	NS	16.9	1.90	NS
ATP release	24.4	7.70	21.4	10.09	NS	20.3	6.32	NS

Table 2. Plasma levels (ng/ml) of nimodipine (N) and nimodipine pyridine metabolite (NPM) at 1 and 2 months after start of treatment, and reported clinical response after 2 months. Limit of detection of N = 0.5 ng/ml, NPM = 0.1 ng/ml; ms, missing sample

Patient	At 1 month		At 2 months		Clinical response
	N	NPM	N	NPM	
1	1.5	0.8	3.1	1.7	Moderate
2	2.2	0.6	7.4	1.3	Good
3	<0.5	0.2	0.9	0.5	Good
4	9.8	5.0	15.1	7.6	Moderate
5	0.6	0.5	0.5	0.6	Moderate
6	0.9	0.7	1.6	1.1	Good
7	0.5	0.3	0.5	0.4	Moderate
8	3.2	1.9	ms	ms	Good
9	<0.5	0.2	ms	ms	Moderate
10	0.8	0.4	1.0	0.7	Poor

Table 3. Plasma TXB_2 levels (pg/ml) after 1-month drug-free run-in (baseline) and after 1 and 2 months' treatment with nimodipine (% change from baseline in parentheses). ms, missing sample; NS, not significant at the 5% level (Student's paired *t*-test)

Patient	Baseline		After 1 month		After 2 months
1	86		58 (− 33)		48 (− 44)
2	37		32 (− 14)		38 (+ 3)
3	52		48 (− 8)		26 (− 50)
4	12		25 (+108)		27 (+125)
5	19		25 (+ 32)		33 (+ 74)
6	61		53 (− 13)		67 (+ 10)
7	100		144 (+ 44)		166 (+ 66)
8	59		116 (+ 97)		191 (+224)
9	140		26 (− 81)		ms
10	26		59 (+127)		38 (+ 46)
Mean	59		59 (0)		70 (+ 19)
		NS		NS	

Thus, there were no consistent changes in any of the measured indices of platelet activity after 1 or 2 months' treatment with nimodipine at either 40 or 60 mg t.d.s. Meanwhile, benefits in terms of symptomatic relief for migraine were described by nine of the ten patients (good in four and moderate in five); only one had a poor response.

Discussion and Conclusion

These experiments need to be considered in the context of others in our laboratory using similarly selected patients and the same methods. These have shown consistent effects of drugs used for migraine on platelet activity measured ex vivo [2] and on TXB_2 [11]. In this platelet study, measured changes in platelet activity with treatment did not differ from those spontaneously occurring in patients we have followed up regularly whilst on placebo therapy. Furthermore, changes were random, with none of the statistical tests supporting rejection of the null hypothesis with a probability greater than 50%. In other words these changes were wholly compatible with physiological variation, and nimodipine has no measurable effect on platelet activity in migraineurs in doses used for prophylaxis.

At the same time, there was evidence that this group of ten patients were clinically responsive: the 95% confidence interval for nine responders out of ten is 55%–100%, not embracing 40%, the maximum expected placebo response. While this is not claimed as evidence of a generally beneficial effect of nimodipine in the treatment of migraine, it does indicate rather clearly a dissociation between platelet and clinical effects of treatment. These findings are in accord with other recent studies [2] which have failed to couple any consistent trend in effect on platelets with the prophylactic efficacy of other useful agents [11].

Any remaining credence for the platelet theory of migraine pathogenesis rests on possible drug interference with the pharmacological actions of the platelet release products [1]. The likelihood of this is remote in view of the range of pharmacological activities apparently associated with successful prophylaxis of migraine. The suitability of platelets as a model for the study of neuronal calcium channels is now also questionable, particularly as the diversity of calcium channels in different tissues becomes increasingly appreciated [12].

References

1. Hanington E (1986) The platelet and migraine. Headache 26:411–415
2. Steiner TJ, Joseph R, Clifford Rose F (1985) Migraine is not a platelet disorder. Headache 25:434–440
3. Steiner TJ, Joseph R, Clifford Rose F (1987) Migraine is not a platelet disorder (Letter to the Editor). Headache 27:400
4. Holmsen H (1975) Biochemistry of the platelet release reaction. CIBA Foundation Symp 35:175–201
5. Gelmers HJ (1983) Nimodipine, a new calcium antagonist in the prophylactic treatment of migraine. Headache 23:106–109
6. Havanka-Kanniainen H, Hokkanen E, Myllyla VV (1985) Efficacy of nimodipine in the prophylaxis of migraine. Cephalalgia 5:39–43
7. Dale J, Landmark KH, Myhre E (1983) The effects of nifedipine, a calcium antagonist, on platelet function. Am Heart J 105:103–105
8. Couch JR, Hassenein RS (1977) Platelet aggregability in migraine. Neurology 27:843–848
9. Kalendovsky Z, Austin JH (1975) "Complicated migraine" – its association with increased platelet aggregability and abnormal plasma coagulation factors. Headache 15:18–35
10. Vahlquist B (1955) Migraine in children. Int Arch Allergy Appl Immunol 7:348–355
11. Joseph R, Steiner TJ, Poole CJM, Littlewood J, Clifford Rose F (1985) Thromboxane synthetase inhibition: potential therapy in migraine. Headache 25:204–207
12. Spedding M (1985) Calcium antagonist subgroups. Trends Pharmacol Sci 3:109–114

Treatment of Acute Migraine Attacks with Intravenous Flunarizine

D. SOYKA

Until now ergotamine has been considered the most effective drug in the treatment of acute migraine attacks. Nevertheless, about 20% of patients thus treated are nonresponders. The consumption of more than 1 mg ergotamine daily for 3 months or longer may be the cause of ergotamine-induced headache and high daily doses over a long time may induce ergotism. So alternatives are desirable.

In this multicenter randomized double-blind study, efficacy and tolerance of intravenous flunarizine administrations were compared with placebo in migraine attacks of the "common" or "classical" type. Patients with other disorders causing migrainous headache had to be excluded by differential diagnosis.

For the diagnosis of migraine the following criteria had to be met:

1. Exclusion of an organic cause of the headache
2. Paroxysmal uni- or bilateral attacks of headache lasting as a rule between 4 (minimum) and 48 (maximum) h
3. Definite migraine history with attacks starting at least 1 year previously
4. In the case of common migraine, additionally at least one of the following criteria:
   a) photophobia
   b) phonophobia
   c) nausea
   d) vomiting
5. In the case of classical migraine, additionally at least one of the following symptoms:
   a) initial scintillating scotoma
   b) transient neurological symptoms, such as
      - unilateral paresthesias
      - hemiparesis
      - speech disorders
      - hemianopsia.

According to the average duration of attacks during the preceding 6 months and the duration of the present attack so far, only patients in the first half of the migraine attack qualified for inclusion in the study. In this manner short-term spontaneous remissions were excluded.

Neurological Clinic, University of Kiel, Niemannsweg 147, D-2300 Kiel, FRG.

Cerebral Ischemia and Calcium
Ed. by A. Hartmann and W. Kuschinsky

Table 1. Patient and disease characteristics

	Flunarizine ($n=31$)	Placebo ($n=29$)
Men	6 (19.4%)	8 (27.6%)
Women	25 (80.6%)	21 (72.4%)
Age* (years)	41 ± 10	38 ± 10
Weight* (kg)	68 ± 13	70 ± 16
Common migraine	23 (74.2%)	19 (65.5%)
Classical migraine	7 (22.6%)	10 (34.5%)
Duration of the disease* (months)	154 ± 115	154 ± 120

* Mean values ± standard deviations.

Patients and Methods

A total of 66 patients of both sexes were included in the study; all 66 were evaluable with regard to safety aspects, whereas 60 were statistically evaluable. Thirty-one attacks were treated with flunarizine and 29 with placebo. The distribution by sex, age, and weight was comparable in both groups (Table 1). In the flunarizine group as well as in the placebo group patients with common migraine were predominant. In the patients' history the average duration of the migraine disease was 154 months (median 10 years). Before entering the study, eight flunarizine patients and seven of the placebo group had already tried to stop the attack by self-medication.

After patient selection and detailed patient history the acute migraine attack characteristics were recorded, with special emphasis on the exact documentation of the accompanying symptoms. The patients then received 4 ml of a solution with or without 20 mg flunarizine by slow intravenous injection.

The efficacy of the treatment was evaluated by means of a visual analogue scale, on which the patient was asked to record the intensity of his acute pain at 5-min intervals during an observation period of 60 min after the injection. Only those patients were considered responders whose baseline values during the 60-min observation period had been reduced on the pain scale by at least 50%. Furthermore the therapeutic success with regard to accompanying symptoms as well as efficacy and tolerance of the test substances was evaluated by doctor and patients – also in comparison with former treatments of attacks. Fisher's exact test with Bonferroni's correction was used for statistical evaluation [14]. Given *P*-values are one-tailed.

Results

At the end of the 60-min observation period all patients (100%) of the flunarizine group recorded a reduction in pain, whereas under placebo only 20/29 patients (69%) mentioned an improvement of their complaints towards the end of the ob-

Table 2. Reduction of pain at the end of a 60-min observation period after drug injection, as evaluated by means of a visual analogue scale

	Flunarizine ($n=31$)		Placebo ($n=29$)	
Complete pain relief	11	74.2%	1	27.6%
Marked pain reduction (>50%)	12		7	
Limited pain reduction (<50%)	8		12	
No pain reduction	0		9	
		$P<0.017$		

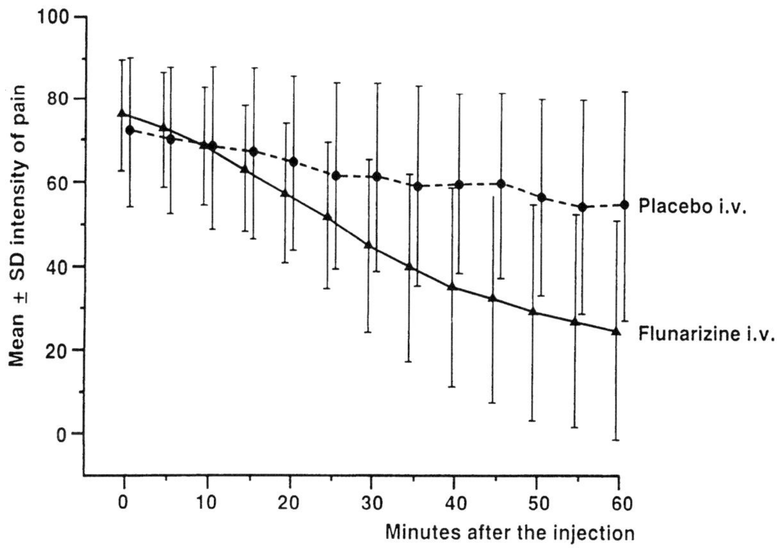

Fig. 1. Intensity of headache assessed with a visual analogue scale during the 60-min observation period

servation period. This difference between the treatment groups is statistically significant ($P<0.017$).

A complete remission of pain within 60 min was reported by 11 patients (35.5%) under flunarizine, but only by one patient of the placebo group (3.5%). Another 12 flunarizine patients (38.7%) but only seven patients receiving placebo (24.1%) displayed a reduction in pain symptoms by at least 50%. If the patients considered responders are combined, the responder rate in the flunarizine group amounts to 74.2%, but only to 27.6% in the placebo group (Table 2). These differences between the two treatment groups are statistically significant ($P<0.017$). The evolution of pain during the observation period in both groups is shown in Fig. 1.

The development of intensity of pain, subdivided in patients with common and classical migraine, was similar to the results in the total study, taking the fact

Table 3. Remission rates of accompanying symptoms (≥10 patients affected at the start)

	Flunarizine (n=31)			Placebo (n=29)		
		Complete remission			Complete remission	
	n	n	[%]	n	n	[%]
Nausea	16	10	62.5*	22	4	18.2
Photophobia	20	9	45.0*	22	2	9.1
Vertigo	17	10	58.8*	20	4	20.0

* $P<0.05$ (Fisher's exact test).

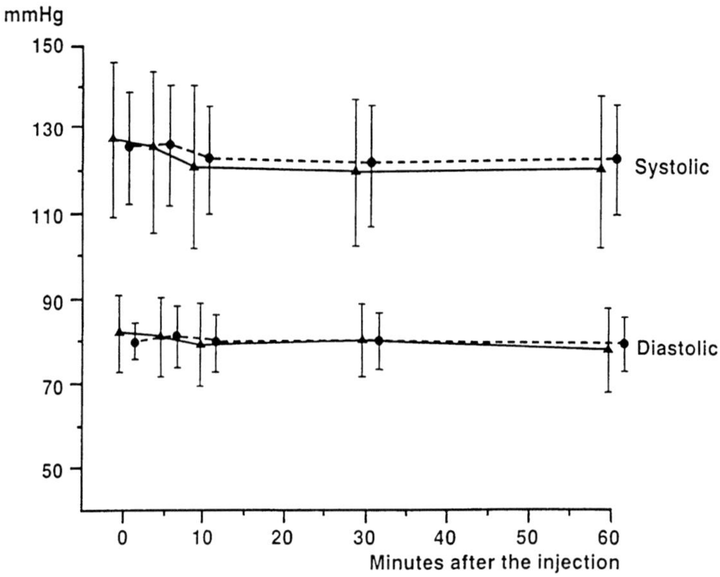

Fig. 2. Mean (±SD) blood pressure in flunarizine- and placebo-treated patients

into consideration that the sample in classical migraine is too small for selective statistical statements.

In the patients of the flunarizine group the accompanying symptoms of migraine, especially drowsiness, paling, fatigue, and phonophobia, disappeared much more than in those receiving placebo. Statistically significant remission rates versus placebo were observed with regard to nausea, photophobia, and vertigo under flunarizine ($P<0.05$, Table 3).

The investigators evaluated the efficacy of the treatment with flunarizine as excellent in 29.0% of the cases, good in 48.4%, moderate in 9.7%, and unsatisfactory in 12.9%. Placebo figures were 13.8%, 13.8%, 13.8% and 58.6% respectively. Flunarizine was rated significantly better than placebo ($P<0.05$).

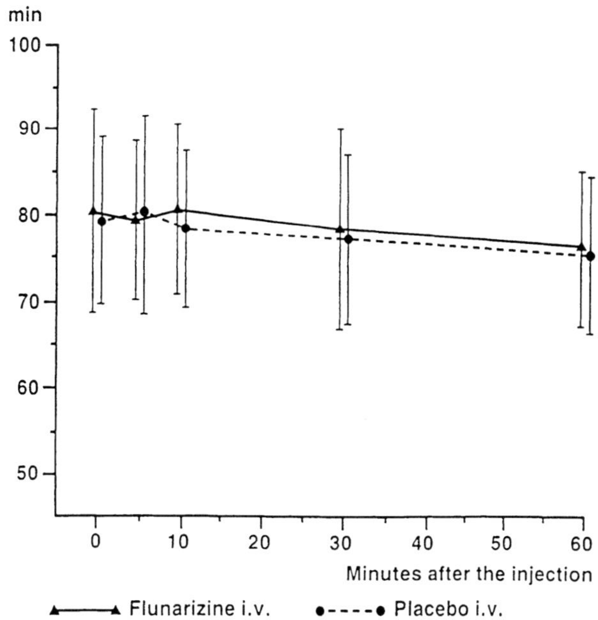

Fig. 3. Mean (±SD) heart rate in flunarizine- and placebo-treated patients

Table 4. Adverse reactions observed during 1 h after injection (patients evaluable for safety aspects; $n = 66$)

Adverse reactions	Flunarizine ($n = 36$)	Placebo ($n = 30$)
Somnolence	9 (25.0%)	1 (3.3%)
Nausea	2 (5.6%)	3 (10.0%)
No. of patients with adverse reactions	11 (30.6%)	4 (13.3%)

The separate evaluation of the target parameters in 15 patients who had already taken a drug to stop the attack before injection of the test substance showed that these patients derived less benefit from the treatment. These patients, who came to one of the study centers after unsuccessful self-medication, suffered from particularly heavy and refractory pain.

In the 60 min following the administration of flunarizine or placebo, blood pressure (Fig. 2) and heart rate (Fig. 3) remained largely unchanged in both groups. Of the 36 flunarizine patients, 11 (30.6%) reported undesirable side-effects, mainly fatigue (Table 4).

Discussion

The treatment of the acute migraine attack aims to shorten the duration and reduce the severity of a migraine attack after it has started. Nonnarcotic analgesics, possibly preceded by a gastro-prokinetic drug if taken orally, and ergotamine have both proven to be effective [1]. Not all migraine attacks can be controlled with these drugs. Especially patients looking for emergency help have exhausted accepted therapies and are therefore in need of alternative acute treatments [2].

The efficacy of certain oral calcium entry blockers in the prophylactic long-term treatment of migraine has been demonstrated [3]. Especially under prophylactic treatment with flunarizine a decrease in migraine frequency was observed [4, 5], and in several studies there was an additional reduction in duration and intensity of the remaining attacks [6–8, 17]. On the other hand, no calcium entry blocker was effective in the treatment of the acute migraine attack. Neither nimodipine, 40 mg sublingually [9], nor verapamil, given intravenously [10], showed very beneficial effects versus placebo.

This study indicates that one single intravenous dose of 20 mg flunarizine can successfully arrest an acute migraine attack. These positive results emphasize, once again, that calcium entry blockers are a heterogeneous group of substances. This has recently been confirmed by a WHO expert committee which divided calcium entry blockers into six different substance classes [11]. Flunarizine is different from verapamil and nimodipine in that it does not interfere with myocardial slow calcium channels or with myogenic activity in vascular smooth muscles. All calcium entry blockers, including flunarizine, inhibit the potential-operated calcium channels of arterial smooth muscles. Flunarizine, consequently, behaves as an antivasoconstrictor [12]. The efficacy of flunarizine is not necessarily due to its vascular effects, though. The drug protects brain cells against hypoxia at a cellular level [13]. Brain hypoxia has been suggested to play a pivotal role in migraine [14]. Ischemic cerebral blood flow has been observed during classical but not common migraine attacks [15]. The antivasospastic or brain protective effect of flunarizine might therefore explain the efficacy of the drug in migraine attacks. However, a similar effect in our study in common and classical migraine attacks prevents us from drawing definitive conclusions about the mechanism of action as long as the pathophysiology in particular of the common migraine attack remains unsolved.

The use of an intravenous formulation guarantees absolute bioavailability but greatly limits the usefulness in clinical practice. Its value for the study of the pathophysiology of migraine attacks, on the one hand, and its therapeutic potential as an alternative to the parenteral administration of ergotamine, on the other, justify further examinations of intravenous flunarizine in migraine.

Summary

Flunarizine, 20 mg by slow intravenous injection, was studied in the acute treatment of migraine attacks in a multicenter double-blind placebo-controlled study.

At the end of the 60-min observation period, 23 (74.2%) of the 31 flunarizine patients reported complete relief or a pain reduction of more than 50%, versus 8 (27.6%) of 29 placebo patients ($P < 0.017$). Accompanying symptoms also improved significantly better in the flunarizine group. The investigators evaluated the therapy as good or excellent in 77.4% of the flunarizine and in 27.6% of the placebo patients respectively. Tolerance of the therapy was good and comparable in the two groups. Somnolence was the only flunarizine-related adverse reaction. Blood pressure and heart rate were not affected. At the moment it looks as if flunarizine, given intravenously, might by a very good alternative to the parenteral administration of ergotamine in the treatment of the acute migraine attack.

References

1. Peatfield RC, Fozard JR, Clifford Rose F (1986) Drug treatment of migraine. In: Clifford Rose F (ed) Handbook of clinical neurology, vol 13. Headache. Elsevier Science, Amsterdam, pp 173–216
2. Lane PL, Ross R (1985) Intravenous chlorpromazine. Preliminary results in acute migraine. Headache 25:302–304
3. Greenberg DA (1986) Calcium channel antagonists and the treatment of migraine. Clin Neuropharmacol 9:311–328
4. Louis P (1981) A double blind placebo-controlled prophylactic study of flunarizine (Sibelium) in migraine. Headache 21:235–239
5. Soelberg Sørensen PS, Hansen K, Olesen J (1986) A placebo-controlled, double-blind, crossover trial of flunarizine in common migraine. Cephalalgia 6:7–14
6. Wörz R, Drillisch C (1983) Migräne-Prophylaxe durch einen Kalziumeintrittsblocker. Ergebnisse einer Doppelblindstudie Flunarizin vs Pizotifen. MMW 125:711–714
7. Louis P, Spierings ELH (1982) A comparison of flunarizine (Sibelium) and pizotifen (Sandomigran) in the treatment of migraine. A double-blind study. Cephalalgia 2:197–203
8. Rascol A, Montastruc JL, Rascol O (1986) Flunarizine versus pizotifen. A double blind study in the prophylaxis of migraine. Headache 26:83–85
9. Jensen K, Tfelt-Hansen P, Lauritzen M, Olesen J (1985) Clinical trial of nimodipine for single attacks of classic migraine. Cephalalgia 5:125–131
10. Molaie M, Olson CM, Koch J (1987) The effects of intravenous verapamil on acute migraine headache. Headache 27:51–53
11. Vanhoutte PM (1987) The expert committee of the World Health Organization on classification of calcium antagonists: the viewpoint of the reporteur. Am J Cardiol 59:3A–8A
12. Van Nueten JM, Wellens D (1979) Mechanisms of vasodilation and antivasoconstriction. Angiology 30:440–446
13. Wauquier A, Ashton D, Edmonds HL (1984) Protections against cerebral hypoxia by antimigraine drugs. In: Amery WK, Van Nueten JM, Wauquier A (eds) The pharmacological basis of migraine therapy. Pitman, London, pp 134–148
14. Amery WK (1982) Brain hypoxia: the turning point in the genesis of the migraine attack? Cephalalgia 2:83–109
15. Olsen TS, Friberg L, Lassen NA (1987) Ischemia may be the primary cause of the neurologic deficits in classic migraine. Arch Neurol 44:156–161
16. Tfelt-Hansen P, Olesen J (1985) Methodological aspects of drug trials in migraine. Neuroepidemiology 4:204–226
17. Soyka D, Oestreich W (1987) Flunarizin versus Propranolol in der Intervallprophylaxe der Migräne – eine multizentrische Doppelblindstudie in 12 Kliniken. Nervenheilkunde 6:177–183

Treatment of Migraine and Cluster Headache with Calcium Antagonists

J. S. MEYER, J. LOTFI, and M. KOBARI

A number of pharmacological agents have recently become available that are capable of limiting or blocking calcium passage through cellular membrane channels. These drugs are termed calcium antagonists or calcium channel or calcium entry blockers. Cell membranes of smooth muscle and nerve cells have several different calcium channels which may be selectively altered by calcium channel blockers [28, 29]. Since the cellular entry of calcium initiates contraction of vascular smooth muscle, calcium channel blockers exert both antispasmodic and slight vasodilator actions [6]. Calcium channels are susceptible to ischemia resulting in calcium entry with intracellular "overload", which may be responsible for cell damage and eventually cell death [7, 26]. Calcium antagonists also exert protective actions against ischemic and anoxic damage to both nerve cells and smooth muscle.

The history of the discovery of calcium channel blockers is one of the most interesting pharmacological advances of the last decade. Fleckenstein [10–12] was the first to describe calcium channel blockers and to demonstrate how they modify the activity of contractile tissues of the myocardium and smooth muscle and exert effects on the cardiac pacemaker and conduction systems, leading to their widespread therapeutic application in numerous cardiovascular disorders. Discussion here will be confined to their usefulness in the prophylaxis of migraine and vascular headaches.

Flunarizine (Sibelium, 14950), was one of the first and most extensively investigated calcium channel blockers (CCBs). It appears to be less potent and less selective for blood vessels than the more recently developed CCBs [29]. Flunarizine has a membrane-stabilizing effect and protects laboratory rats from brain hypoxia [3]. It does not alter resting vascular smooth muscle and myogenic activity, but renders smooth muscle cells less responsive to vasoactive stimuli.

Louis, in 1981 [19], was the first to evaluate flunarizine in a double-blind placebo controlled parallel group study of its usefulness in migraine prophylaxis. Twenty-nine patients received flunarizine 5 mg twice daily for 3 months and results were compared with 29 migraineurs who received placebo. The frequency of migraine attacks was reduced by 57% in the flunarizine group, compared to a 14% reduction in the control group. Therapeutic responsiveness was delayed by 2 months. By the 3rd month, decreases in the migraine attacks were seen in 86% of flunarizine-treated patients. Younger patients appeared to respond better

Cerebral Blood Flow Laboratory, Veterans Administration Medical Center, 2002 Holcombe Boulevard, Houston, Texas, TX 77211, USA.

Cerebral Ischemia and Calcium
Ed. by A. Hartmann and W. Kuschinsky

than older patients to flunarizine treatment. Louis and Spierings [20] in 1982 reported another study comparing the effects of flunarizine and pizotifen. Ten milligrams of flunarizine given orally daily were compared to 1 mg pizotifen per day, administeredover a 4-month interval. Of 38 patients, 92% improved with flunarizine. By the 4th month, migraine frequency had decreased by 80% with an average decrease of 54% over the 4-month treatment interval.

Amery et al. [2] pooled data obtained from seven different studies testing flunarizine in migraine, including the two studies mentioned above. All studies were comparable according to dose and duration of treatment. All studies were conducted in four European countries and included 326 patients, 202 of whom were treated with flunarizine. There were 58 male and 144 female patients aged between 13 and 65 years, and 54 had classic and 148 common migraine, as defined by the criteria of the Ad Hoc Committee for the Classification of Headaches (1962) [1]. At a dose of 10 mg daily, flunarizine reduced the frequency of attacks by 83%. Improvement was more marked when the disease was of recent onset and was more rapid when the pretreatment frequency of attacks was high. Somnolence and increased weight gain (usually occurring in patients with low pretreatment weight) were the main adverse effects. One hundred twenty-five of these patients were followed for over a year and the effect of flunarizine was optimally achieved after 4 months (51%–60% average reduction in attacks).

Analysis of the data in the seven studies showed that at the end of the 3-month treatment period, the frequency of attacks was reduced by at least 50% in 68% of patients, which compares favorably with reported average responses for other commonly used prophylactic agents (methysergide 58.3%, pizotifen 50.9%, and propranolol 51.6%). Severity of migraine headaches was also reduced. Long-term therapy showed that efficacy was maintained and there were no rebound headaches after withdrawal. Therapeutic benefits were equally good for both classic and common migraine.

In a study in the United States by Diamond and Schenbaum in 1983 [7], flunarizine was evaluated as a prophylactic agent in the treatment of migraine in 20 patients utilizing a dose of 10 mg per day. After a 1- to 3-month placebo stabilization period, a single-blind cross-over study was begun in which 17 patients experienced statistically significant reductions in the incidence of headache and/or their severity (average 53%).

Nattero et al. [27] studied 176 patients with headache and found that flunarizine was useful in "tension vascular headache" (83.9%), as well as migraine of both the common and classic types (81.4%) tested over a 3-month period.

Verapamil was one of the first calcium channel blockers to be investigated for possible benefit in the prophylaxis of migraine in the United States. It also appears to inhibit arterial vasospasm and to inhibit platelet aggregation and block platelet serotonin release [21].

Solomon et al. [32] gave verapamil in doses of 80 mg four times daily to 12 patients who showed an overall reduction in migraine frequency of 49%, while 8 out of the 12 (67%) noted a 50% or greater reduction in the frequency of migraine.

Markley et al. [22] carried out a double-blind cross-over study of the prophylactic efficacy of verapamil in a group of patients with migraine. The number of patients was small (28) and only 14 out of 28 completed the study. The dose of verapamil was 80 mg three times daily and the duration of the investigation was for 4 months. It was concluded that verapamil was significantly superior to placebo in prophylaxis of chronic migraine headaches. Mean migraine frequency decreased from 6.7 migraine attacks per patient per month during placebo treatment to 3.8 migraines per patient per month in the treated group. Constipation was the only important side-effect of this drug and the headaches returned some time after cessation of treatment.

Meyer et al. [23] carried out an open trial of long-term prophylactic therapy with verapamil (240 mg/day) in 44 patients with different types of headache (classic migraine = 15, common migraine = 11, cluster headache = 14, and muscle contraction headache = 4). Results were compared to a similar trial using the calcium channel blockers nifedipine and nimodipine. Cerebral blood flow was measured before and after treatment. All three drugs were effective in controlling migraine and cluster, but not in tension headache. Within 2–3 weeks, verapamil abolished or attenuated migraine prodromes, decreased headache frequency in cluster and classic migraine, but appeared to be less effective in common migraine. Postural hypotension and constipation were the only important side-effects.

Jonsdottir et al. [17] reported combined long-term (5 years) results of prophylactic trials with verapamil, nimodipine, and nifedipine. One hundred eighteen patients were treated with verapamil (160–480 mg/day) for a mean interval of 5.6 months (range = 1–28 months). The diagnostic categories in these patients were classic migraine 25, common migraine 26, chronic cluster 34, and combined headaches 33. Verapamil was efficacious in treatment of 81% of patients with common and 72% of those with classic migraine.

Nifedipine (Procardia, Adalat) and nimodipine have also been evaluated for their prophylaxis of migraine. Nimodipine (Nimotop), another dihydropyridine drug, was found to share this property. Cerebral hemodynamic studies [23] among patients before and during treatment with nimodipine, verapamil, and nifedipine confirmed that cerebrovascular resistance or tone was reduced by all three drugs. A double-blind cross-over study of nifedipine was carried out by Kahan et al. [18] in eight migraine patients who also had Raynaud's phenomenon. Ten milligrams three times a day were prescribed for 1 month and results were compared to placebo. Nifedipine reduced the frequency of migraine headache by 74% and the severity was reduce by 71% in these patients.

Meyer and Hardenberg, in 1983, compared the effects of two other calcium channel blockers (nifedipine and verapamil) in 35 headache patients (classic, common, and cluster types) utilizing a double-blind cross-over method [24]. The study was primarily concerned with evaluating the efficacy of nimodipine therapy, but nifedipine and verapamil were found to have comparable results in treating cluster headaches but were less effective in common and classic migraine and produced more side-effects. The side-effects, although not major, occurred in a large percentage of patients in the nifedipine group (60%) and consisted mainly of flushing, dizziness, irritability, fatigue, mild nausea, and abdominal cramps.

In a later study, nifedipine was studied by Jonsdottir et al. [17] in 66 patients (26 males and 40 females) with different types of vascular headache for a mean interval of 4.5 months (range = 1–26 months). The group treated included 13 classic, 13 common, 17 cluster, and 17 combined headaches. a 65% decrease in frequency and severity occurred with a total daily dosage of 30–120 mg nifedipine divided into three doses. Best results were obtained in classic migraine (77%) and least response in combined headache (41%). There was a 65% decrease in frequency of common migraine and a 60% decrease in chronic cluster headaches. In the same study, nifedipine was compared to verapamil and nimodipine. The authors concluded that nifedipine appeared to be more effective than verapamil and nimodipine for classic migraine, whereas verapamil was most effective in controlling mixed headaches.

Cerebral hemodynamic measurements in patients before and during treatment with nimodipine, verapamil, and nifedipine [23] confirmed that cerebrovascular resistance or tone is reduced by all three drugs. However, only nimodipine reduced cerebral vasoconstriction responses induced by 100% oxygen inhalation, confirming its marked pharmacological selectivity for cerebral vessels. As mentioned earlier, nimodipine and nifedipine were the most selective for the cerebral vasculature when compared to other calcium channel blockers [28]. Nimodipine is the most potent inhibitor of voltage-sensitive calcium channels [13] and has the highest selectivity for cerebral vasculature [9].

Nimodipine therapy was investigated by Gelmers [14] in a double-blind placebo-controlled parallel design study in 60 patients, of whom 50 completed the trial. Patients had classic or common migraine and at least two to three attacks per month. The dose of nimodipine was 40 mg three times daily. The percentage improvement (69%) showed that after the 8th week through the 12th week of treatment, therapeutic benefits were significantly higher during nimodipine therapy than during placebo treatment ($P<0.01$). Side-effects were negligible and consisted mainly of abdominal complaints. During nimodipine treatment, headaches were less frequent and severe and their duration was shortened.

Nimodipine was found to be the most effective calcium channel blocker for patients with both migraine and cluster headache in a study by Meyer and Hardenberg [24], who compared its effects with those of verapamil and nifedipine. Thirty-five patients, 13 with classic, 14 with common, and 8 with cluster headaches, were evaluated using two dosage schedules of 60 and 120 mg/day in a double-blind randomized cross-over study versus placebo. The duration of study was 18 weeks, comprising two 1-week washout intervals and two 8-week treatment periods. The first reported change was a statistically significant reduction ($P<0.01$) in the frequency of prodromes in patients with classical migraine. The percentage reduction in the frequency of prodromes was more marked with the higher dosage (70.6% vs 55.1%). Common and classic migraine were also reduced significantly 2–4 weeks after the start of therapy, leveled off during the intratrial washout period, and continued to be significantly more effective throughout the second 8-week treatment interval ($P<0.005$). Typical dose-response curves during treatment of migraine (Fig. 1) and cluster headaches (Fig. 2) are illustrated. Serial CBF measurements with testing of cerebral vasoconstrictive responses during oxygen inhalation showed significant reduction of cerebral vasoconstrictive responses after high dose nimodipine.

Verapamil and nifedipine also produced significant relief from headache, but were less effective than nimodipine and produced more side-effects.

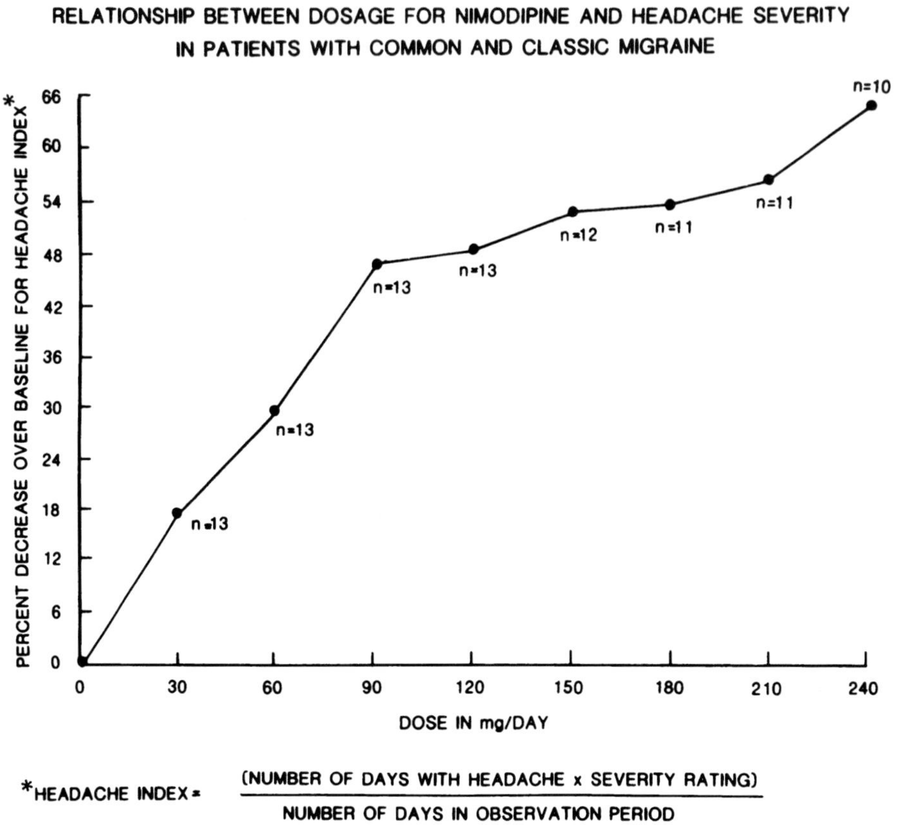

Fig. 1. Typical dose-response curve showing percentage decrease in headache index (number of days with headache × severity/number of days observed) in patients suffering from common and classic migraine during treatment with the calcium channel blocker, nimodipine. The dose-response curve illustrated shows that the therapeutic effectiveness of nimodipine is greatest between 90 and 240 mg daily

Nimodipine has been tested against pizotifen (Sandomigran, a serotonin antagonist) in two double-blind studies [16, 25]. Both studies found nimodipine to be at least as effective as pizotifen with a shorter latency and producing less side-effects. Bussone et al. [5], in a comparison made between nimodipine and flunarizine treatment among 30 patients with common migraine (25 completed the trial), confirmed the shorter latency of nimodipine, with the efficacy of flunarizine being delayed until after the 1st month. Nimodipine was equally effective and did not produce the relatively frequent side-effect of flunarizine (e.g., weight gain, drowsiness, and depression).

Diltiazem (cardizem) is a calcium channel blocker with a marked antiplatelet aggregation effect. However, diltiazem is a relatively weak calcium channel blocking agent with poor selectivity for vascular smooth muscle. There are few reports of its use in the prophylaxis of migraine. In a pilot study, Riopelle and McCans [30], using diltiazem 90 mg/day in 15 patients with frequent migraine attacks (>2/

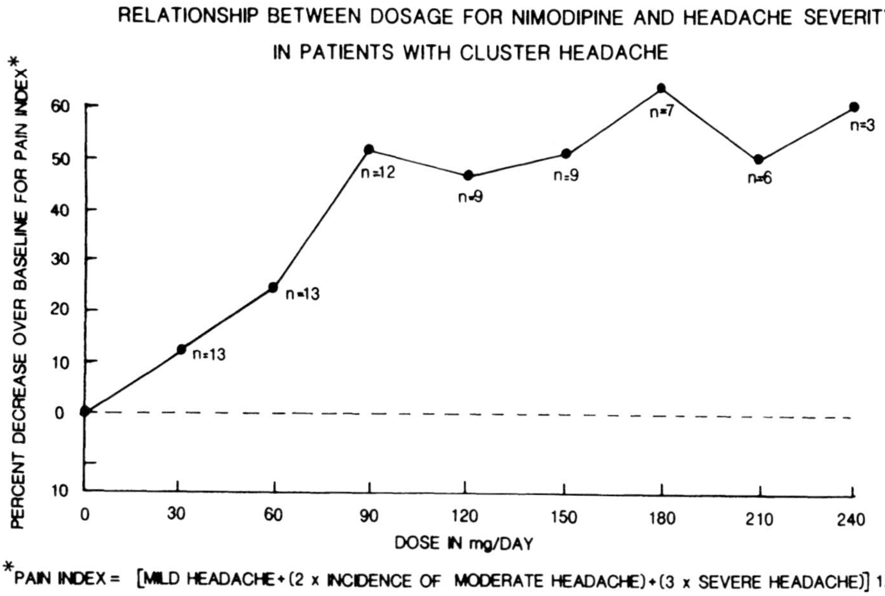

Fig. 2. Typical dose-response curve showing percentage decrease in headache index (number of days with headache × severity/number of days observed) in patients suffering from chronic cluster headache during treatment with calcium channel blockers. The dose-response curve illustrated shows that the therapeutic effectiveness of nimodipine is greatest between 90 and 240 mg daily

week), recorded an 86% reduction in frequency and a 55% reduction in severity. There is also a short report by Smith and Schwartz [31] of its use in refractory migraine in nine patients (eight women and one man) – seven with classic and two with common migraine. All had been nonresponsive to 8 weeks of treatment with a β-adrenergic blocker, naldolol 40–160 mg/day utilizing a mean dose of 75 mg four times daily. A statistically significant improvement in headache scores was obtained after 8 weeks of treatment.

The side-effects of CCBs are dose dependent and are mainly due to undesirable extensions of their calcium blocking effects on the central nervous system, the smooth muscle of the heart, peripheral blood vessels, and bowel. The incidence of side-effects in various studies has been between 20% and 60%, but all these side-effects are usually transient, clinically unimportant, and seldom require discontinuation of therapy.

The most important side-effects reported with flunarizine were daytime sedation and weight gain. Daytime sedation and somnolence was seen in 7.1% (decreasing to 6.1% by the 3rd month). The weight gain was significantly more common in females ($P<0.05$) and in patients with a lower pretreatment weight ($P<0.05$), resulting in some cases acquiring normal weight. Reviewing the studies not included in Amery's pooled results, the results are similar, with weight gain, sedation, and fatigue being the most important side-effects. Mild vertigo, dry mouth, anxiety and some nonspecific gastrointestinal symptoms were uncommon.

The most commonly reported side-effects of verapamil are constipation and cardiac abnormalities including bradycardia, hypotension, and atrioventricular conduction defects. In a study by Meyer et al. [23] of 44 patients with various types of headache, 33% developed side-effects (20% after long-term treatment). Vascular changes included edema of the hands and feet, flushing, and lightheadedness among 23%, constipation in 17%, headaches in 10%, and skin rash in 10%. In another study of verapamil reported from the same laboratory, Jonsdottir et al. [17] noted side-effects in 49 out of 118 patients with various types of headache (41%). In this group, 53% had gastrointestinal problems, especially constipation, 26% vascular abnormalities, especially postural hypotension and transient cardiac dysrhythmias. Twenty-two percent gave peripheral vascular complaints, mainly flushing. The side-effects were usually mild and transient but severe enough in 3 out of 49 patients to discontinue treatment.

The incidence of side-effects with nifedipine has varied among different reports. Most authorities agree that this drug has little or no effect on the atrioventricular conduction systems. Flushing, headache, tachycardia, and diarrhea are often mentioned. In a study with 127 patients treated by nifedipine for coronary artery spasm [4], 39% had side-effects, mainly dizziness, flushing, hypotension, and edema. The incidence of adverse effects in the study of Meyer et al. [23] in an open trial of nifedipine (60 mg/day) in 20 migraine patients was also 39% (gastrointestinal complaints 13%, fatigue 13%, tachycardia 10%, and headache 6%). Jonsdottir et al. [17] found the incidence of side-effects to be higher (71%) in 118 patients (23 with migraine). Four patients had to discontinue treatment because of these effects. The most common complaint (53%) was of hypotension and cardiovascular complaints, with peripheral vascular problems ranking second with 45%. In a report from Sweden, the side-effects of nifedipine, which is the most commonly used calcium channel blocker in that country, were edema, tachycardia, headache, and skin rash [26].

Nimodipine appears to be the calcium channel blocker having the least number of reported adverse effects. These are usually central nervous system and/or gastrointestinal complaints. Meyer and Hardenberg [24] found nimodipine had fewer adverse effects when compared to verapamil and nifedipine. Muscular complaints and menstrual discomfort, behavioral changes (e.g., irritability, fatigue), and lightheadedness were the most frequently reported side-effects. Havanka-Kanniaiken et al. [15] reported no side-effects in 33 patients treated for migraine prophylaxis and Jonsdottir et al. [17] found the incidence of side-effects to be only 21% among 18 patients, and these were transient and mild (compared to 71% for nifedipine and 41% for verapamil).

References

1. Ad Hoc Committee on Classification of Headache (1962) Classification of headache. JAMA 179:717–718
2. Amery WK, Caers LI, Aerts TJL (1985) Flunarizine, a calcium entry blocker in migraine prophylaxis. Headache 25:249–254
3. Amery WK, Wauquier A, Va Nueten JM, DeClerck F, Van Reempts JV, Janssen PAJ (1981) The anti-migrainous pharmacology of flunarizine (R 14950), a calcium antagonist. Drugs Exp Clin Res VII (1):1–10
4. Antman E, Muller J, Goldberg S et al. (1980) Nifedipine therapy for coronary-artery spasm. N Engl J Med 302:1269–1273
5. Bussone G, Baldini S, D'Andrea G et al. (1987) Nimodipine vs flunarizine in common migraine. A controlled pilot trial. Headache 27:76–79
6. Cauvin C, Loutzenhiser R, Van Breemen C (1983) Mechanism of calcium antagonist induced vasodilation. Ann Rev Pharmacol Toxicol 23:373–396
7. Diamond S, Schenbaum H (1983) Flunarizine, a calcium channel blocker, in the prophylactic treatment of migraine. Headache 23:39–42
8. Farber JL (1981) The role of calcium in cell death. Life Sci 29:1289–1295
9. Flaim SF (1982) Comparative pharmacology of calcium blockers, based on studies of vascular smooth muscle. In: Flaim SF, Zolis R (eds) Calcium blockers: mechanism of action and clinical application. Urban & Schwarzenberg, Baltimore, pp 155–178
10. Fleckenstein A (1964) Die Bedeutung der energiereichen Phosphate für Kontraktilitat und Tonus des Myokards. Verh Dtsch Ges Inn Med 70:81–89
11. Fleckenstein A (1977) Specific pharmacology of calcium in myocardia, cardiac pacemakers and vascular smooth muscle. Annu Rev Pharmacol Toxicol 17:149–166
12. Fleckenstein A, Tritthart H, Fleckenstein B, Herbst A, Grun B (1969) A new group of competitive calcium antagonists (Iproveratril, D600, prenylaine) with highly potent inhibitory effect on excitation/conduction coupling in mammalian myocardium. Pflügers Arch 307:R25
13. Freedman SB, Dawson G, Villareal ML, Miller RJ (1984) Identification and characterization of voltage-sensitive calcium channels in neuronal clonal cell lines. J Neurosci 4:1453–1467
14. Gelmers HJ (1983) Nimodipine, a new calcium antagonist in the prophylactic treatment of migraine. Headache 23:106–109
15. Havanka-Kanniaiken H, Hokkanen E, Myllyla VV (1985) Efficacy of nimodipine in the prophylaxis of migraine. Cephalalgia 5(1):39–43
16. Havanka-Kanniaiken H, Myllyla VV, Hokkanen E (1982) Nimodipine in the prophylaxis of migraine, a double-blind study. Acta Neurol Scand 65 (Suppl 90):77–78
17. Jonsdottir M, Meyer JS, Rogers RL (1987) Efficacy, side effects and tolerance compared during headache treatment with three different calcium blockers. Headache 27:364–369
18. Kahan A, Weber S, Amor B, Guerin F, Degeorge M (1983) Nifedipine in the treatment of migraine in patients with Raynauds' phenomenon (letter). N Eng J Med 308:1102–1103
19. Louis P (1981) A double-blind placebo-controlled prophylactic study of flunarizine (Sibelium) in migraine. Headache 21:235–239
20. Louis P, Spierings ELH (1982) Comparison of flunarizine (Sibelium) and pizotifen (Sandomigran) in migraine treatment. Cephalalgia 2:197–203
21. McGoon M, Vlietstra R, Holmes D, Osborn JE (1982) The clinical use of verapamil. Mayo Clin Proc 57:495–510
22. Markley HG, Cheronis JCD, Piepho RW (1984) Verapamil in prophylactic therapy of migraine. Neurology 34:963–976
23. Meyer JS, Dowell R, Mathew N, Hardenberg J (1984) Clinical and hemodynamic treatment of vascular headaches with verapamil. Headache 24:313–321
24. Meyer JS, Hardenberg J (1983) Clinical effectiveness of calcium entry blockers in prophylactic treatment of migraine and cluster headaches. Headache 23:266–277
25. Micieli G, Trucco M, Agostinis C, Mancuso A, Papalia F, Sinforiani E (1985) Nimodipine vs. pizotifen in common migraine. Results of double-blind cross-over trial. Cephalalgia 5 (Suppl 3):532–533

26. Myrhed M, Wilholm B-E (1986) Nifedipine – a survey of side effects: four years' reporting in Sweden. Acta Pharmacol Toxicol 133–136
27. Nattero G, Savi L, Delorenzo C (1985) Flunarizine in the treatment of headache with or without neurological symptoms. Cephalalgia 5 (Suppl 2):141–143
28. Peroutka CJ (1983) The pharmacology of calcium channel antagonists: a novel class of anti-migraine agents? Headache 23:278–283
29. Peroutka CJ, Banghart SB, Allen GS (1984) Relative potency and selectivity of calcium antagonists used in the treatment of migraine. Headache 24:55–58
30. Riopelle RJ, McCans JL (1982) A pilot study of the calcium antagonist diltiazem in migraine syndrome prophylaxis. Can J Neurol Sci 9:269
31. Smith R, Schwartz A (1984) Diltiazem prophylaxis in refractory migraine (letter). N Engl J Med 310:1327–1328
32. Solomon GD, Griffith Steel MC, Spaccavento LJ (1983) Verapamil prophylaxis of migraine. JAMA 250:250–252

Calcium Entry Blockers in Migraine Prophylaxis – A Critical Review of Therapeutic Trials

P. SOELBERG SØRENSEN [1], P. TFELT-HANSEN [1], L. EDVINSSON [2], and J. OLESEN [1]

Introduction

Calcium entry blockers are powerful antivasoconstrictive drugs with specific binding sites in vascular smooth muscle cells, in brain, and in cardiac and striated muscle membranes. Although the role of vasoconstriction and cerebral hypoxia as the instigators of a migraine attack is still under debate (Olesen et al. 1981 b, c; Amery 1982), calcium blockers have been introduced for migraine prophylaxis on the theoretical basis of being protective against cerebral vasoconstriction and brain hypoxia.

The efficacy of several calcium entry blockers, belonging to different chemical families, as prophylactic antimigraine drugs has been evaluated in therapeutic trials. In the present paper the evidence for a prophylactic effect in migraine will be critically reviewed for the calcium entry blockers flunarizine, nimodipine, nifedipine, verapamil, and diltiazem, and the possible mode of action of these drugs in migraine will be discussed.

Clinical Trials

Flunarizine

Flunarizine belongs to the group of diphenylpiperazines and has a strong resemblance to the antihistamine drug cinnarizine. Among the calcium entry blockers, flunarizine has been the most extensively studied in migraine prophylaxis.

Placebo-Controlled Trials

To date four double-blind studies have been published in which the prophylactic effect of flunarizine has been compared to that of placebo. Three studies used parallel group comparison (Louis 1981; Frenken and Nuijten 1984; Mendenopoulous et al. 1985) and one study employed the crossover design (Soelberg Sørensen et al. 1986). An overview of the designs and results of these studies is

[1] University Departments of Neurology, Rigshospitalet and Gentofte Hospital, DK-2100 Copenhagen, Denmark.
[2] Department of Internal Medicine, University of Lund, S-22185 Lund, Sweden.

Cerebral Ischemia and Calcium
Ed. by A. Hartmann and W. Kuschinsky

Table 1. Placebo-controlled trials of flunarizine (10 mg/day) in migraine prophylaxis

Study	No. of patients	Type of migraine		Design	Duration (weeks)	Attack frequency reduction
		Common	Classic			
Louis	58	+	+	Group	13	57%, $P<0.01$
Frenken and Nuijten	35	+	+	Group	12	53%, $P<0.05$
Mendenopoulos et al.	20		+	Group	13	?[a]
Soelberg Sørensen et al.	27	+		Cross-over	16	38%, $P<0.01$

[a] Significant difference for headache index.

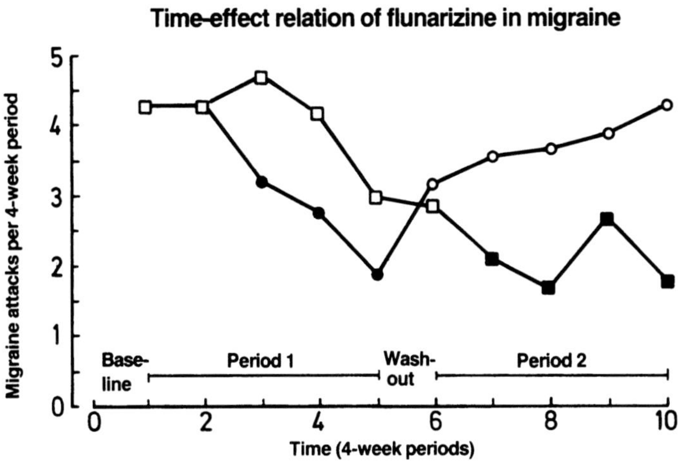

Fig. 1. Evolution in number of migraine attacks per 4 weeks in a cross-over study of flunarizine 10 mg daily vs placebo. Mean values are shown. (Soelberg Sørensen et al.)

given in Table 1. In all four studies flunarizine was convincingly superior to placebo. The effect was predominantly on the number of migraine attacks, whereas the duration and severity of attacks were almost unaltered. The onset of effect was slow and increased gradually over 3 months. In two studies using group comparison (Louis 1981; Frenken and Nuijten 1984) the attack frequency was reduced to half compared to placebo after 2 months of treatment. In the crossover study of patients with common migraine (Soelberg Sørensen et al. 1986), flunarizine reduced the migraine attack frequency by 40% compared with placebo, and during the 4th month of treatment the attacks were reduced by 50% (Fig. 1). Interestingly, the reduction of migraine frequency was correlated to the pretreatment frequency, being most prominent in patients with four or more attacks per month. While convincing evidence of an effect has been shown for flunarizine in common migraine, none of the present major studies have included a sufficient number of patients with classic migraine to show statistically significant efficacy

Table 2. Comparative trials of flunarizine (10 mg/day) in migraine prophylaxis

Study	No. of patients	Type of migraine		Design	Comparative drug	Attack frequency reduction
		Common	Classic			
Louis and Spierings	75	+	+	Group 18 weeks	Pizotifen 2–3 mg	NS
Cerbo et al.[a]	18	+	+	Crossover 12 weeks	Pizotifen 1.5 mg	NS
Rascol et al.	35	+	+	Group 18 weeks	Pizotifen 2.2 mg	NS
Soyka and Oestreich	69	+	+	Group 16 weeks	Propranolol 120 mg	NS

[a] Dose of flunarizine 15 mg/day.

in that subclass of migraine. It therefore remains uncertain whether flunarizine acts only in common migraine or in classic migraine as well.

Comparative Trials

Flunarizine has been compared to the standard drugs pizotifen and propranolol (Table 2). In two studies using parallel groups and 4-month treatment periods, flunarizine 10 mg daily was compared with pizotifen 2–3 mg daily (Louis and Spierings 1982; Rascol et al. 1986). No significant difference in efficacy was found between the two drugs. Neither of the studies included placebo or gave confidence limits or power estimates, and both studies were probably too small to attain reasonable statistical power with parallel groups. One study using the crossover design compared a rather high dose of flunarizine (15 mg) with a low dose of pizotifen (1.5 mg) (Cerbo et al. 1986). Both drugs decreased the frequency of attacks compared to a run-in period, but the study was weakened by a high drop-out rate. In a double-blind study with parallel group design flunarizine 10 mg was compared with propranolol 120 mg in 69 patients with common or classic migraine (Soyka and Oestreich 1986) (Table 2). No significant difference was found between the two drugs, both of which decreased the frequency of migraine attacks during a 4-month treatment period. The study was probably too small to have sufficient statistical power to demonstrate comparability between the two drugs. Moreover, propranolol was used in a smaller dose (120 mg) than is normally used in prophylactic trials. Further studies comparing flunarizine with beta-blocking agents are needed.

Other Studies

The long-term effect and side-effects of flunarizine have been evaluated in two large open trials (Bono et al. 1985; Leandri et al. 1985). It appeared that the beneficial effect of the drug obtained after 3 months of treatment was unweakened after 1 and 2 years. The drug was generally well tolerated. The major side-effects were weight gain, daytime sedation, and depression.

Table 3. Trials of nimodipine (120 mg/day) in migraine prophylaxis

Study	No. of patients	Type of migraine		Design	Comparative drug	Attack frequency reduction
		Common	Classic			
Gelmers	50	+	+	Group 13 weeks	Placebo	[a] $P<0.05$
Havanka-Kanniainen et al. 1985	29	+	+	Crossover 8 weeks	Placebo	28%, $P<0.01$
Havanka-Kanniainen et al. 1987	43	+	+	Crossover 12 weeks	Pizotifen 1.5 mg	NS
Micieli et al.	20	+		Crossover 13 weeks	Pizotifen 1.5 mg	NS

[a] Significant difference for headache index.

Nimodipine and Nifedipine

Both nimodipine and nifedipine are dihydropyridines, a group of calcium entry blockers which, like verapamil and diltiazem, are used in cardiovascular disease and hypertension.

Placebo-Controlled Trials

Nimodipine 120 mg daily has been compared with placebo in two studies comprising patients with common migraine as well as classic migraine. The study by Gelmers (1983) was a group comparison in which nimodipine significantly reduced the headache index. The other study (Havanka-Kanniainen et al. 1985) was a crossover study showing a significant 28% attack frequency reduction by nimodipine compared with placebo (Table 3). Interestingly, the aura phase in patients with classic migraine was significantly shorter during treatment with nimodipine than during placebo treatment. In a very small placebo-controlled study of nifedipine as a prophylactic antimigraine drug, Kahan et al. (1983) found that nifedipine significantly reduced the number of migraine attacks per month when compared to placebo.

Comparative Trials

Two studies have compared 120 mg nimodipine with 1.5 mg pizotifen in crossover studies (Havanka-Kanniainen et al. 1987; Micieli et al. 1985). Both studies found no difference in efficacy of the two drugs, and both drugs decreased the number of attacks compared with a placebo run-in period. However, both trials included only a limited number of patients and no confidence limits were given. Hence, the claimed effect might be an expression of the "time effect", that is, a decrease in frequency with time irrespective of treatment (Olesen et al. 1981a). Statements regarding the value of nimodipine compared to other prophylactic drugs in migraine must await further studies.

Verapamil and Diltiazem

In two small studies using the crossover design verapamil was compared to placebo in migraine prophylaxis (Solomon et al. 1983; Markley et al. 1984). Both studies suggested a prophylactic effect of verapamil but were invalidated by drop-out rates of 48% and 57%, respectively. Diltiazem has not been examined in controlled clinical trials, but prophylactic effect in migraine has been claimed in two open trials (Riopelle and McCans 1982; Smith and Schwartz 1984).

Discussion

Of the different calcium entry blockers, flunarizine has been most extensively investigated in the prophylactic treatment of migraine. There is good evidence that flunarizine is superior to placebo at least in patients with common migraine (Soelberg Sørensen et al. 1986). Based on the experience from two studies (Gelmers 1983; Havanka-Kanniainen et al. 1985), nimodipine seems superior to placebo, but more studies are needed before a final conclusion can be drawn. For other calcium entry blockers evidence of efficacy in migraine prophylaxis is still insufficient. Regarding the question of comparability of flunarizine and nimodipine with currently used standard treatment in migraine prophylaxis (beta-blockers and 5-HT antagonists), a final statement must await further controlled trials. All hitherto published studies using group comparison have been too small to have sufficient power to demonstrate equipotency (Olesen et al. 1981a; Tfelt-Hansen and Olesen 1985). Only three comparative crossover studies have been published so far, one comparing flunarizine with pizotifen (Cerbo et al. 1986) and two comparing nimodipine with pizotifen (Micieli et al. 1985; Havanka-Kanniainen et al. 1987). Only one study (Soyka and Oestreich 1987) has evaluated a calcium entry blocker against propranolol, which currently is probably the most extensively used drug for prophylactic treatment of migraine. Hence, controlled trials comparing calcium entry blockers with beta-blockers are needed to evaluate the place of calcium entry blockers in our armamentarium for the treatment of migraine. Ideally, comparative trials with two presumably active drugs should include a placebo control in order to eliminate the well known "time effect" which is often observed in trials of migraine prophylaxis (Tfelt-Hansen and Olesen 1985). It is possible that one advantage of calcium entry blockers is that they have fewer side-effects than the currently used standard treatments for migraine prophylaxis. Weight gain, daytime sedation, and depression have been reported in open long-term trials of flunarizine (Bono et al. 1985; Leandri et al. 1985). Flunarizine may provoke symptoms of parkinsonism in patients with imminent basal ganglia dysfunction. The frequency of these side-effects has not yet been determined but seems to be low. The frequency of side-effects for other calcium blockers used in migraine prophylaxis remains unknown.

The mechanism by which calcium entry blockers exert their prophylactic effect in migraine is not yet known. Calcium entry blockers are potent vasodilators in many vascular beds. Precontracted human brain vessels are thus effectively relaxed by calcium entry blockers (Edvinsson et al. 1979; Brandt et al. 1981). An effect on the vasculature of the brain might be responsible for the effectiveness of some calcium blockers in migraine. Flunarizine was introduced for migraine prophylaxis on the theoretical basis of being protective against brain hypoxia (Amery 1982). However, changes in cerebral blood flow and hypoxia have not yet been demonstrated in common migraine, where flunarizine has been most convincingly demonstrated to be effective (Olesen et al. 1981 b). Another possible mode of action could be an effect on blood flow changes during attacks of classic migraine resembling the experimental phenomenon spreading depression of Leao (Lauritzen 1987). Studies on the effect of flunarizine on spreading depression have, however, been inconsistent (Hansen et al. 1984; Marannes and Wauquier 1985). There is also experimental evidence that some of the calcium entry blockers may have an effect on neurotransmitter systems (Holmes et al. 1984; Itil and Itil 1985).

In conclusion, calcium entry blockers are interesting newcomers to the firmament of antimigrainous drugs. Placebo-controlled trials have shown that flunarizine, and possibly nimodipine, has a prophylactic effect in migraine patients. However, the question of comparability of the efficacy of calcium entry blockers to the currently standard treatments remains open.

References

Amery WK (1982) Brain hypoxia: the turning-point in the genesis of the migraine attack? Cephalalgia 2:83–109

Bono G, Manzoni GC, Martucci N et al. (1985) Flunarizine in common migraine: Italian cooperative trial II. Long-term follow-up. Cephalalgia 5 (Suppl 2):155–158

Brandt L, Andersson K-E, Edvinsson L, Ljunggren B (1981) Effects of extracellular calcium and of calcium antagonists on the contractile responses of isolated human pial and mesenteric arteries. J Cereb Blood Flow Metab 1:339–347

Cerbo R, Casacchia M, Formisano R, Feliciani M, Cusimano G, Buzzi MG, Agnoli A (1986) Flunarizine-pizotifen single-dose double-blind cross-over trial in migraine prophylaxis. Cephalalgia 6:15–18

Edvinsson L, Brandt L, Andersson K-E, Bengtsson B (1979) Effect of a calcium antagonist on experimental constriction of human brain vessels. Possible efficacy in cerebrovascular spasm. Surg Neurol 11:327–330

Frenken CWGM, Nuijten STM (1984) Flunarizine, a new preventive approach to migraine. Clin Neurol Neurosurg 86:17–20

Gelmers HJ (1983) Nimodipine, a new calcium antagonist, in the prophylactic treatment of migraine. Headache 23:106–109

Hansen AJ, Lauritzen M, Tfelt-Hansen P (1984) Spreading cortical depression and antimigraine drugs. In: Amery WK, van Nueten JM, Wauquier A (eds) The pharmacological basis of migraine therapy. Pitman, London, pp 161–170

Havanka-Kanniainen H, Hokkanen E, Myllyla VV (1985) Efficacy of nimodipine in the prophylaxis of migraine. Cephalalgia 5:39–43

Havanka-Kanniainen H, Hokkanen E, Myllyla VV (1987) Efficacy of nimodipine in comparison with pizotifen in the prophylaxis of migraine. Cephalalgia 7:7–13
Holmes B, Brogden RN, Heel RC, Speight TM, Avery GS (1984) Flunarizine. A review of its pharmacodynamic and pharmacokinetic properties and therapeutic use. Drugs 27:6–44
Itil TM, Itil KZ (1985) The comparative CNS pharmacology of nimodipine in humans. In: Betz E, Deck K, Hoffmeister F (eds) Nimodipine. Pharmacological and clinical properties. Schattauer, Stuttgart, pp 185–202
Kahan A, Weber S, Amor B, Guerik F, Degeorges M (1983) Diltiazem prophylaxis in refractory migraine. N Engl J Med 308:1102–1103
Lauritzen M (1987) Cortical spreading depression as a putative migraine mechanism. TINS 10:8–13
Leandri M, Parodi C, Bacigalupo F, Farinini D (1985) Long-term study on effectiveness of flunarizine in headache prophylaxis. Cephalalgia 5 (Suppl 3):556–557
Louis P (1981) A double-blind placebo-controlled prophylactic study of flunarizine in migraine. Headache 21:235–239
Louis P, Spierings ELH (1982) Comparison of flunarizine and pizotifen in migraine treatment: a double-blind study. Cephalalgia 2:197–203
Markley HG, Cheronis CD, Piepho RW (1984) Verapamil in prophylactic therapy of migraine. Neurology 34:973–976
Marrannes R, Wauquier A (1985) Influence of flunarizine on spreading depression. Cephalalgia 5 (Suppl 3):208
Mendenopoulos G, Manafi T, Logothesis I, Bostantjopoulou S (1985) Flunarizine in the prevention of classical migraine: a placebo-controlled evaluation. Cephalalgia 5:31–37
Micieli G, Trucco M, Agostinis C, Mancuso A, Papalia F, Sinforiani E (1985) Nimodipine vs pizotifen in common migraine: results of a double-blind cross-over trial. Cephalalgia 5 (Suppl 3):532–533
Olesen J, Krabbe AE, Tfelt-Hansen P (1981 a) Methodological aspects of prophylactic drug trials in migraine. Cephalalgia 1:127–141
Olesen J, Larsen B, Lauritzen M (1981 b) Focal hyperemia followed by spreading oligemia and impaired activation of rCBF classic migraine. Ann Neurol 9:344–352
Olesen J, Tfelt-Hansen P, Henriksen L, Larsen B (1981 c) The common migraine attack may not be initiated by cerebral ischemia. Lancet II:438–440
Rascol A, Montastruc J-L, Rascol O (1986) Flunarizine versus pizotifen: a double-blind study in the prophylaxis of migraine. Headache 26:83–85
Riopelle RJ, McCans JL (1982) A pilot study of the calcium antagonist diltiazem in migraine syndrome prophylaxis. J Scan Sci Neurol 9:269
Smith R, Schwartz A (1984) Diltiazem prophylaxis in refractory migraine. N Engl J Med 310:1327–1328
Soelberg Sørensen P, Hansen K, Olesen J (1986) A placebo-controlled, double-blind, cross-over trial of flunarizine in common migraine. Cephalalgia 6:7–14
Solomon GD, Steel JG, Spaccavento LJ (1983) Verapamil prophylaxis of migraine. JAMA 250:2500–2502
Soyka D, Oestreich W (1987) Flunarizine versus Propranolol in der Intervall Prophylaxe der Migräne – eine multizentrische Doppelblindstudie in 12 Kliniken. Nervenheilkunde 6:177–183
Tfelt-Hansen P, Olesen J (1985) Methodological aspects of drug trials in migraine. Neuroepidemiology 4:204–226

Epilepsy

Calcium, Neuronal Hyperexcitability, and Ischemic Neuronal Injury

F. B. MEYER, R. E. ANDERSON, and T. M. SUNDT, JR.

Introduction

In the last decade a wealth of information has emerged regarding the pivotal role of Ca^{2+} in neuronal function. It is also apparent that abnormal calcium metabolism will initiate pathophysiological states leading to neuronal injury. Considering that extracellular calcium concentration is approximately 10^{-3} versus an intracellular concentration of 10^{-7}, subtle alterations in membrane permeability to calcium can have profound effects on neuronal function. Although there are expert analyses on the detrimental role of calcium in various pathophysiological states [40, 63, 66], the goal of this article is to offer a new perspective on one potential role of abnormal calcium influx – the precipitation of neuronal hyperexcitability. After briefly reviewing evidence which indicates that calcium is critical for seizure initiation and discharge, two hypotheses are offered for consideration: first, an integrative mechanism which integrates calcium metabolism with epileptic phenomena, and second, that calcium influx during ischemia precipitates neuronal hyperexcitability which escalates ischemic neuronal injury.

Evidence for the Role of Calcium in Neuronal Hyperexcitability

Briefly, there are multiple avenues of research which indicate that Ca^{2+} influx is essential for seizure discharge.

First, intrinsic bursting of CA3 and CA1 pyramidal neurons as studied in hippocampal brain slice preparations is thought to be analogous in part to the sequence of events leading to repetitive neuronal discharge [35, 55]. There are at least two calcium currents: slow Ca^{2+} depolarizations and transient fast Ca^{2+} spikes [10, 30, 36, 62, 77]. There are also calcium-activated ionic currents: K^+ and Cl^- currents [9, 53]. A speculative mechanism to explain intrinsic bursting might be that inward Ca^{2+} currents generate slow depolarizations which trigger Na^+ spikes and higher threshold transient Ca^{2+} spikes. The Ca^{2+} activated K^+ and Cl^- currents along with the Na^+/K^+ pump then cause rapid repolarization.

Second, epileptic discharge requires synchronization of a population of discharging neurons. This synchronization may be synaptically mediated. From the

Department of Neurosurgery, Mayo Clinic, Rochester, Minnesota 55905, USA.

Cerebral Ischemia and Calcium
Ed. by A. Hartmann and W. Kuschinsky

classic studies of Katz and Miledi, it is apparent the calcium entry into the presynaptic terminal is essential for neurotransmitter release [37, 38].

Third, extracellular Ca^{2+} decrease occurs during epileptic activity in vivo induced through electroconvulsive shock, penicillin, and metrazole [31]. This decrease in Ca_e^{2+} occurs just prior to the onset of burst firing and reaches a stable decline within the first few seconds of seizure discharge. During status epilepticus, there is accumulation of Ca^{2+} in the mitochondria of bursting neurons, predominantly in hippocampal CA3 and CA1 neurons when seizures are induced through bicuculline, allylglycine, or kainic acid [28]. Furthermore, during progressive seizures there is accumulation of fatty acids thought to be due to activation of phospholipase A_2 by calcium influx [11].

Fourth, current anticonvulsants directly affect neuronal calcium metabolism. Phenytoin interferes with synaptosomal calcium uptake and also inhibits binding of some calcium channel blockers like dihydropyridines to neuronal membranes [26, 54]. Barbiturates have been demonstrated to block both neuronal L and N channels in dorsal root ganglion neurons [29]. Furthermore, phenytoin, carbamazepine, and benzodiazepines inhibit Ca^{2+}-calmodulin mediated protein phosphorylation in the presynaptic terminal [18, 19].

Fifth, some excitatory amino acids like glutamic and aspartic acid, which gate a receptor-operated calcium channel, are also epileptogenic [3, 51]. Antagonists to these excitatory amino acids are anticonvulsant in seizure-prone DBA/2 mice

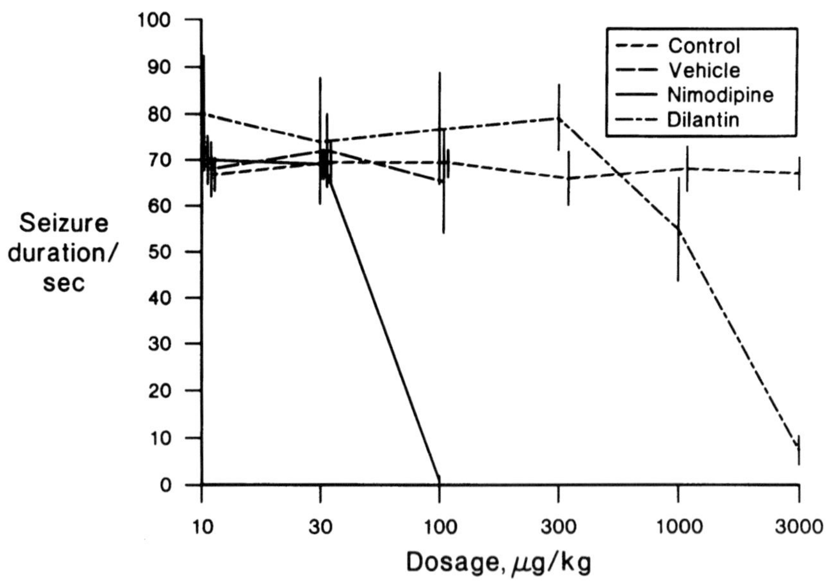

Fig. 1. Graph of seizure duration versus cumulative log dosage in white New Zealand rabbits subjected to direct electrical cortical shock. After a control seizure, increasing doses of the test drug were administered 2 min prior to shocks separated by 15-min intervals until an anticonvulsant effect was obtained in ten animals each. Nimodipine at 140 μg/kg prevented induction of seizure discharge. The vehicle polyethylene glycol was ineffective. To obtain a similar anticonvulsant effect, phenytoin 4440 μg/kg was required. In nontreated control animals there was no cortical fatigue during repetitive seizures with 15-min recovery. (Meyer et al. [44])

[16]. When used to elicit epileptic activity there is an influx of calcium into CA3 and CA1 pyramidal neurons [24, 27]. In vitro hippocampal slices have confirmed these in vivo observations by demonstrating that some excitatory amino acids lead to a depletion of extracellular calcium and an increase in calcium uptake in brain slice preparations [14, 78].

Sixth, there is now strong evidence that some calcium channel blockers have potent anticonvulsant properties [1, 20, 42–46, 48]. For example, dihydropyridine calcium antagonists like nimodipine which penetrate the blood-brain barrier have anticonvulsant properties in experimental seizures induced through ischemia, bicuculline, pentylenetetrazole, electrical cortical shock, kainic acid, DBA/2 mice, nitrous oxide withdrawal, and cefusolin-induced seizures (Fig. 1). In fact, the dihydropyridine agonist BAY K 8644 will induce seizures in mice and dogs. Since the voltage-dependent calcium L channel is modulated by dihydropyridine agonists and antagonists [47, 49, 50], it can be proposed that the observed anticonvulsant effect of dihydropyridines is due to inhibition of Ca^{2+} influx through neuronal L channels [45]. The density of dihydropyridine receptor sites in the rat brain is highest in the molecular layer of the dentate gyrus, followed by the anterior amygdala, hippocampal CA3 and CA1, and pyriform cortex. There are also abundant receptor sites in the olfactory bulb, cerebral cortex lamina I–IV, and thalamus [2, 15]. It is also apparent that there is widespread distribution of functional dihydropyridine-sensitive Ca^{2+} channels in cultured neurons from various brain regions, with hippocampal voltage-dependent Ca^{2+} channels demonstrating the greatest sensitivity to dihydropyridines [73]. In addition to the apparent anticonvulsant effects of dihydropyridines, the diphenylalkamine verapamil [17, 75, 76] and diphenylpiperazine flunarizine [4] have been shown to be effective as anticonvulsants. However, both of these drugs also antagonize Na^{+} channels, which may explain in part the observed antiepileptic effect.

Hypothesis: Integrative Approach for Ca^{2+} in Seizure Discharge

A hypothetical schema for the role of Ca^{2+} in seizure initiation and propagation is depicted in Fig. 2. This mechanism attempts to integrate current information on neuronal calcium channels, calcium regulation, data obtained from hippocampal brain slice preparations, and in vivo work with Ca^{2+} antagonists as anticonvulsants. First, the initiating event is Ca^{2+} influx through either voltage-dependent L channels or receptor-operated channels. This is based on experimental evidence in which seizure initiation can be prevented by preadministration of dihydropyridine Ca^{2+} antagonists which are known to block only voltage-dependent L channels [42–46]. However, preadministration of these same calcium antagonists has variable effects on seizure initiation induced through excitatory amino acids (Meyer, in press). With this initial influx of calcium there is then membrane depolarization which leads to opening of higher threshold transient Ca^{2+} currents and Na^{+} currents. It is proposed that this transient Ca^{2+} current corresponds to the voltage-dependent T channel. This would conform to hippocampal brain slice preparations in which a longlasting Ca^{2+} current (L channel) leads to

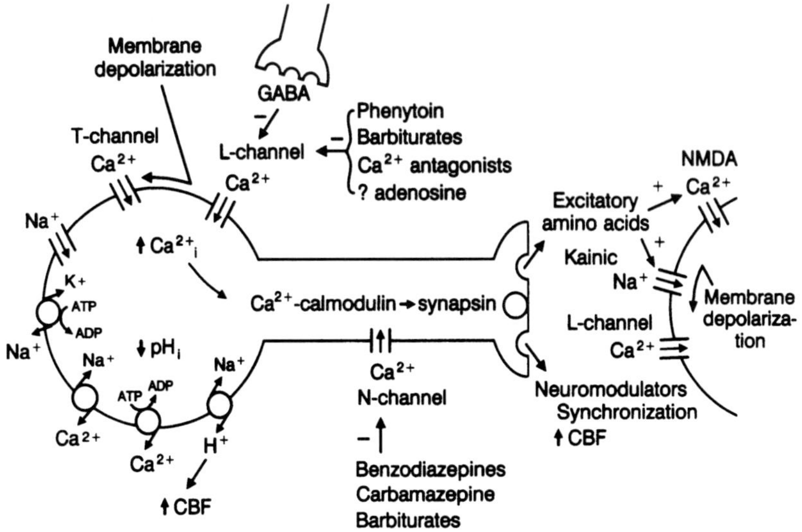

Fig. 2. Hypothetical model of the role calcium in seizure discharge. See text for explanation

rapid higher threshold Ca^{2+} currents (T channel) and Na^+ currents. Thereafter, rapid repolarization occurs through the outward Ca^{2+} driven K^+ current and inward Cl^- current and through the Na^+/K^+ pump.

Synaptic mediation of a discharge would follow by opening of voltage-dependent presynaptic N channels due to a propagating action potential. Ca^{2+} influx would then bind with calmodulin, leading to exocytosis of neurotransmitters with recruitment of additional neurons. Exocytosis of excitatory amino acids would therefore be dependent on Ca^{2+} influx.

In this postulate, excitatory amino acids could initiate and propagate epileptic discharge through calcium influx mediated through receptor-operated Ca^{2+} channels. However, binding to the kainic acid receptor gates a Na^+ channel. It can be envisioned that an influx of Na^+ would lead to membrane depolarization, leading to opening of voltage-dependent L channels. Therefore, excitatory amino acids would lead to Ca^{2+} influx both indirectly through the kainic acid receptor and directly through the NMDA receptor.

To continue a seizure discharge, in addition to rapid membrane repolarization, neurons would be forced to regulate intracellular Ca^{2+} through rapid Na^+/Ca^{2+} exchange, the Ca^{2+} ATPase, and endoplasmic reticulum sequestration. It might be anticipated that since intracellular Na^+ would be elevated, the Na^+/Ca^{2+} pump might be of secondary importance as compared to the latter two mechanisms. During prolonged status epilepticus, mitochondria would also electrophoretically accumulate Ca^{2+}.

Although this model is speculative it serves the useful purpose of suggesting sites at which current anticonvulsants could beneficially modulate seizure discharge. First, the initiating mechanism would explain why GABA neurons are inhibitory since GABA will block L channels [22]. Second phenytoin, some organic Ca^{2+} channel blockers like dihydropyridines, and the endogenous Ca^{2+} blocker adenosine would primarily block Ca^{2+} influx through L channels [23, 26, 54]. Third, carbamazepine and benzodiazepines would block Ca^{2+} influx through presynaptic N channels [18, 19]. Fourth, barbiturates would block both L and N channels [29].

This schema also indicates two mechanisms by which cerebral blood flow would increase during epileptic discharge: extracellular acidosis and release of vasodilators at the synapse [33, 74].

Hypothesis: Ca^{2+}, Neuronal Hyperexcitability, and Ischemic Injury

The well documented influx of Ca^{2+} during ischemia is thought to have several deleterious effects as authoritatively discussed elsewhere in this book. However, given the fact that Ca^{2+} influx is essential for seizure initiation, it is reasonable to propose that Ca^{2+} influx during ischemia may precipitate a state of neuronal hyperexcitability. The occurrence of overt seizures during experimental ischemia is well described. For example, in Wistar rats subjected to four-vessel occlusion, the incidence of seizures during a 72-h follow-up correlated well with the duration of ischemia: 10, 20, and 30 min of ischemia caused an incidence of seizures of 0%, 8%, and 40% respectively [57]. If neuronal hyperexcitability does occur during ischemia, then several detrimental effects may result.

First, intracellular acidosis would be enhanced. In nonhypoxic experiments, lactic acid concentrations increase from approximately 1.3 nmol/kg to 4.98 and 6.91 nmol/kg during the first 15 and 30 s of fluorothyl-induced seizures and are then relatively stable even during status epilepticus [12, 21, 34]. This is reflected in a decline in brain pH_i by approximately 0.15 pH units [65, 72]. Under hypoxic conditions, lactic acid production increases from a control of 1.42 nmol/kg to 23.22 nmol/kg during 2 h of status epilepticus [5]. The primary means by which neurons normalize pH_i during seizure discharge is through the Na^+/H^+ antiport exchange [64]. The driving force for this pump is the Na^+ gradient maintained by the Na^+/K^+ dependent ATPase. However, during ischemia Na^+ gradients are quickly lost due to ATP depletion. Therefore, neuronal hyperexcitability during ischemia would not only lead to greater lactic acid accumulation, but neurons would be less capable of extruding H^+ in an attempt to normalize pH_i.

Second, neuronal hyperexcitability would be predicted to accelerate depletion of high energy phosphates during ischemia. During seizure discharge, cortical metabolic rate increases by a factor of 1.5–2.5 [41]. This increased demand for substrate delivery is matched by marked increases in cerebral blood flow. Accordingly, energy charge is maintained during both short seizure discharges and status epilepticus. However, hypoglycemic animals will demonstrate deterioration in energy charge during seizures due to a lack of substrate for increased glycolysis [25]. Therefore, during ischemic seizures, with decreased O_2 and glucose delivery, there should be rapid deterioration in energy charge along with increases in lactic acid concentrations. The accelerated depletion of ATP would be reflected in a more rapid loss of ionic homeostasis due to failure of ATP-dependent mechanisms, including the Na^+/K^+ pump and Ca^{2+}-ATPase.

Selective Vulnerability

Neuronal hyperexcitability during transient ischemia may explain in part the phenomenon of selective vulnerability as first suggested by Brown and colleagues [8]. Rat brain regions which demonstrate irreversible neuronal injury after 2–5 min of ischemia followed by 1 week of recovery include the CA1 and subicular region of the hippocampus, the dentate gyrus, olfactory cortex, amygdala, entorhinal cortex, and reticular nucleus of the thalamus [6, 69]. Since these limbic structures are elements of an integrated neuronal circuit, seizure activity may play an essential role for their vulnerability [13]. In fact similar pathological changes can be found in kindling models of epilepsy or prolonged stimulation of the perforant pathway [39, 52, 68]. During the early recovery period after transient ischemia there is an increase in both the firing rate and 2-deoxyglucose uptake of CA1 neurons indicative of hyperexcitability [70, 71].

There are several current theories to explain selective vulnerability, including the potential deleterious effects of excitatory amino acids [59–61, 66]. The excitotoxic theory of neuronal injury postulates that an influx of Na^+ through the kainic acid receptor promotes early injury through neuronal swelling [60]. Subsequently an influx of calcium through the NMDA receptor precipitates a cascade of detrimental metabolic events [32, 58]. In this schema, Ca^{2+} influx could occur through two mechanisms: voltage-dependent channels opened by membrane depolarization from the inward Na^+ current and through the NMDA receptor-operated channel. In separate experiments NMDA antagonists prevent both ischemic injury and also exhibit potent anticonvulsant properties [16, 67]. Perhaps the anti-ischemic effects are secondary in part to inhibition of neuronal hyperexcitability.

Cultured rat hippocampal neurons are more susceptible to anoxia once synapses are formed [59]. This increased vulnerability is blocked by addition of $MgCl_2$ to the culture medium, which can be explained by the fact that Mg^{2+} blocks the NMDA receptor. Therefore, this is in vitro support for the excitotoxic theory. However, it also suggests that synaptic integration of neuronal circuits, a prerequisite for seizure discharge, increases vulnerability. Pulsinelli examined the effects of deafferentation of the CA1 when subjected to 30 min of forebrain ischemia [56]. They noted that elimination of excitatory input to CA1 offered a protective effect. Although this is evidence for the importance of neurotransmitter release in selective vulnerability, an additional explanation is the severing of neuronal circuits important for seizure discharge emanating from this region.

Calcium and Ischemic Seizures

If neuronal hyperexcitability does indeed complicate ischemic pathophysiology, is calcium the cause? Experimental support for the importance of Ca^{2+} in ischemic seizures is depicted in Fig. 3, in which a reperfusion seizure occurring after 30 min of transient focal ischemia was arrested by administration of a Ca^{2+} blocker. It has also been noted that the incidence of seizures occurring during pro-

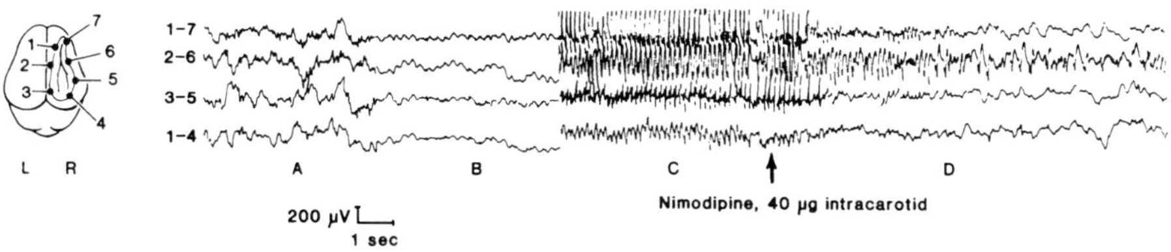

Fig. 3 A–D. Electrocorticogram of a reperfusion seizure in the white New Zealand rabbit. **A** Preocclusion recording was a 6- to 10-Hz rhythm under light halothane anesthesia. **B** Thirty seconds after middle cerebral artery occlusion there is attenuation in the EEG with a decrease in both amplitude and frequency. **C** Fifteen seconds after restoration of flow there is seizure activity primarily in the frontotemporal cortex. **D** One second after intracarotid injection of 40 µg/kg nimodipine there is suppression of seizure activity except for one small focus which persists for approximately 12 s. The remainder of the EEG demonstrated postictal attenuation. (Meyer et al. [42])

longed focal ischemia is significantly less in animals treated with a dihydropyridine calcium antagonist [42].

Meldrum demonstrated that prolonged status epilepticus will produce a pathological picture of hippocampal injury similar to that observed in ischemia [40]. Additionally, both entities exhibit a similar deposition of calcium. Therefore, they proposed that burst firing precipitated by Ca^{2+} influx might play a role in ischemic injury.

Suzuki examined single nerve recordings of CA1 and cortical neurons in gerbils subjected to forebrain ischemia [70, 71]. They noted that immediately after the onset of ischemia there was a brief period of neuronal hyperactivity followed by electrical silence. Upon flow restoration many CA1 neurons demonstrated significant spike discharges when assessed at 7–10 or 24 h postocclusion. Thereafter, at 2 days postocclusion CA1 neurons were electrically silent, which corresponded with histological evidence of neuronal demise. In a separate set of experiments there was a decline in Ca_e^{2+} which chronologically corresponded to the hyperactivity noted after vessel occlusion.

One difficulty with this concept of calcium flux precipitating neuronal hyperexcitability is that many brain regions will demonstrate calcium accumulation without apparent seizure discharge. Only specific regions like CA1 and CA3 will demonstrate calcium fluxes, seizure-like activity, and selective vulnerability. This may be due to the fact that spontaneous burst firing is a unique characteristic of hippocampal neurons – perhaps secondary to differences in the number and conductance of dendritic calcium channels. The fact that in terms of selective vulnerability CA1, CA4 > CA3 > cortical layers 3, 4, 6 suggests that seizure-like phenomena are indeed critical for differences in ischemic vulnerability. However, it should be noted that experiments in which depth cortical neuronal recordings during ischemia were performed are limited. Therefore, neuronal hyperactivity emanating from cortical neurons may be present but undocumented. Surface electrocorticograms may be insensitive to neuronal excitability without overt seizure discharge.

Conclusion

It is apparent that calcium influx plays a pivotal role in neuronal hyperexcitability as evidenced by the anticonvulsant actions of calcium channel blockers in a variety of experimental models. A prerequisite for the potential clinical use of organic calcium channel blockers in the treatment of epilepsy is that they must penetrate the blood-brain barrier and modulate neuronal calcium channels. Our own research would suggest that the dihydropyridine receptor plays a facilitatory role in seizure initiation. A speculative mechanism for epileptic discharge is constructed which places emphasis on calcium fluxes and calcium channels. This model provides sites of action for current and potential anticonvulsants. Given the fact that calcium is critical for seizure discharge, it is hypothesized that calcium influx during ischemia leads to neuronal hyperexcitability which escalates lactic acid production and the depletion of high energy phosphates. This hyperexcitability may explain in part the phenomenon of selective vulnerability. Therefore, calcium channel blockers may have a modulatory role in the treatment of cerebral ischemia by the inhibition of neuronal hyperexcitability irrespective of their effects on cerebral blood flow.

Acknowledgments. This work was supported in part by the New York Academy of Medicine and NIH RO1 NS25374-01.

References

1. Ascioti C, Sno GB, Meldrum BS et al. (1986) Calcium entry blockers as anticonvulsant drugs in DBA/2 mice. Br J Pharmacol 80:379–394
2. Bellemann P, Schade A, Towart R (1983) Dihydropyridine receptor in rat brain labelled with H^3-nimodipine. Proc Natl Acad Sci USA 80:2356–2360
3. Ben-Ari Y, Lagowski J, Tremblay E et al. (1979) A new model of focal status epilepticus: intraamygdaloid application of kainic acid elicits repetitive secondarily generalized convulsive seizures. Brain Res 163:176–179
4. Binnie CD, de Beukelaar F, Meijer JWA et al. (1985) Open dose ranging trial of flunarizine as add-on therapy in epilepsy. Epilepsia 26:424–428
5. Blennow G, Brierly JB, Meldrum BS et al. (1978) Epileptic brain damage. The role of systemic factors that modify cerebral energy metabolism. Brain 101:687–700
6. Blomquist P, Wieloch T (1985) Ischemic brain damage in the rat following cardiac arrest using a longterm recovery model. J Cereb Blood Flow Metab 51:420–432
7. Bolger GT, Weissman BA, Skolnick P (1984) The behavioral effects of the calcium agonist BAY k8644 in the mouse: antagonism by the calcium antagonist nifedipine. Naunyn Schmeidenbergs Arch Pharmacol 328:373–377
8. Brown AW, Levy DE, Kublik M et al. (1979) Selective chromatolysis of neurons in the gerbil brain: a possible consequence of epileptic activity produced by common carotid artery occlusion. Ann Neurol 5:127–138
9. Brown DA, Griffith WH (1983) Calcium activated outward current in voltage clamped hippocampal neurons of the guinea pig. J Physiol (Lond) 337:287–301
10. Brown DA, Griffith WH (1983) Persistent slow inward current in voltage clamped hippocampal neurons of the guinea pig. J Physiol (Lond) 337:303–320
11. Chapman AG (1981) Free fatty acid release and metabolism of adenosine and cyclic nucleotides during prolonged seizures. In: Morselli PL, Lloyd KG, Loscher W, Meldrum BS (eds) Neurotransmitters, seizures, and epilepsy. Raven, New York, pp 165–173

12. Chapman AG, Meldrum BS, Siesjo BK (1976) Cerebral blood flow and cerebral metabolic rate during prolonged epileptic seizures in rats. J Physiol 254:61–72
13. Collins RL, Lothman EW, Olney JW (1983) Status epilepticus in the limbic system: biochemical and pathological changes. Delgado-Escueta AV, Wasterlein CG, Treiman DM, Porter RJ (eds) Advances in neurology, vol 44. Raven, New York, pp 277–288
14. Cooke WJ, Robinson JD (1971) Factors influencing calcium movement in rat brain slices. Am J Physiol 221:218–225
15. Cortes R, Supavilai P, Karobath M et al. (1984) Calcium antagonist binding sites in the rat brain: quantitative autoradiographic mapping using the 1,4-dihydropyridine H^3-PN200-110 and H^3-PY 108-068. J Neural Transm 60:169–197
16. Croucher MJ, Collins JF, Meldrum BS (1982) Anticonvulsant action of excitatory amino acid antagonists. Science 216:899–901
17. Deisz RA, Prince DA (1987) Effect of D890 on membrane properties of neocortical neurons. Brain Res 422:43–55
18. DeLorenzo RJ (1982) Calmodulin in neurotransmitter release and synaptic function. Fed Proc 41:2265–2272
19. DeLorenzo RJ (1984) Calmodulin systems in neuronal excitability: a molecular approach to epilepsy. Ann Neurol 16 (Suppl):S104–114
20. Dolin SJ, Little HJ (1986) The dihydropyridine nitredipine prevents nitrous oxide withdrawal seizures in mice. Br J Addict 81:708
21. Duffy TE, Howse DC, Plum F (1975) Cerebral energy metabolism during experimental status epilepticus. J Neurochem 24:925–934
22. Dunlap K, Fischbach GD (1981) Neurotransmitters decrease the calcium conductance activated by depolarizations of embryonic chick sensory neurons. J Physiol (Lond) 317:519–535
23. Dunwiddie TV, Werth J (1982) Sedative and anticonvulsant effects of adenosine analogs in mouse and rat. J Pharmacol Exp Ther 220:70–76
24. Evans MC, Griffiths T, Meldrum BS (1984) Kainic acid seizures and the reversibility of calcium loading in vulnerable neurons in the hippocampus. Neuropath Appl Neurobiol 10:285–302
25. Folbergrova J, Ingvar M, Siesjo BK (1981) Metabolic changes in cerebral cortex, hippocampus, and cerebellum during sustained bicuculline induced seizures. J Neurochem 37:1228–1238
26. Greenberg DA, Cooper EC, Carpenter CC (1984) Phenytoin interacts with calcium channels in brain membrane. Ann Neurol 16:616–617
27. Griffiths T, Evans MC, Meldrum BS (1982) Intracellular sites of early calcium accumulation in the rat hippocampus during status epilepticus. Neurosci Lett 30:329–334
28. Griffiths T, Evans MC, Meldrum BS (1983) Intracellular calcium accumulation in rat hippocampus during seizures induced by bicuculline of L-allyglycine. Neuroscience 10:385–395
29. Gross RA, MacDonald RL (1988) Barbiturates and nifedipine have different and selective efects on calcium currents of mouse DRG neurons in culture. Neurology 38:443–451
30. Halliwell JV (1983) Caesium loading reveals two distinct Ca currents in voltage clamped guinea pig hippocampal neurons in vitro. J Physiol (Lond) 341:10–11
31. Heinemann U, Louvel J (1983) Changes in Ca_0^{2+} and K_0^+ during repetitive electrical stimulation and during pentylenetetrazole induced seizure activity in the sensorimotor cortex of cats. Pflugers Arch 398:310–317
32. Heinemann V, Pumain R (1980) Extracellular calcium activity changes in cat sensorimotor cortex induced by iontophoretic application of amino acids. Exp Brain Res 40:247–250
33. Heuser D (1978) The significance of cortical extracellular H^+, K^+, and Ca^{2+} activities for regulation of local cerebral blood flow under conditions of enhanced neuronal activity. Ciba Found Symp 56:339–353
34. Howse DC, Caronna JJ, Duffy TE et al. (1974) Cerebral energy metabolism, pH, and blood flow during seizures in the cat. Am J Physiol 227:1444–1451
35. Johnston D, Brown TH (1984) Mechanism of neuronal burst generation. In: Schwartzkroin PA, Wheal HV (eds) Electrophysiology of epilepsy. Academic, New York, pp 207–301
36. Johnston D, Hablitz JJ, Wilson WA (1980) Voltage clamp discloses slow inward current in hippocampal burst firing neurons. Nature 286:391–393

37. Katz B, Miledi R (1969) Spontaneous and evoked activity of motor nerve endings in calcium ringer. J Physiol (Lond) 203:689–706
38. Katz B, Miledi R (1970) Further study of the role of calcium in synaptic transmission. J Physiol (Lond) 207:789–801
39. McIntyre DL, Nathanson D, Edson N (1982) A new model of partial status epilepticus based on kindling. Brain Res 250:53–63
40. Meldrum BS (1986) Cell damage in epilepsy and the role of calcium in cytotoxicity. In: Delgado-Escueta AV, Ward AA (eds) Advances in neurology. Raven, New York, pp 813–847
41. Meldrum BS, Nilsson B (1976) Cerebral blood flow and metabolic rate early and late in prolonged epileptic seizures induced in rats by bicuculline. Brain 99:523–542
42. Meyer FB, Anderson RE, Sundt TM et al. (1986) Selective central nervous system calcium channel blockers – a new class of anticonvulsants. Mayo Clin Proc 61:239–247
43. Meyer FB, Anderson RE, Sundt TM et al. (1987)Suppression of pentylenetetrazole seizures by oral administration of a dihydropyridine calcium antagonist. Epilepsia 28:409–414
44. Meyer FB, Tally P, Anderson RE et al. (1986) Inhibition of electrically induced seizures by a dihydropyridine calcium channel blocker. Brain Res 384:180–183
45. Meyer FB, Anderson RE, Sundt TM (1988) Anticonvulsant effects of dihydropyridine calcium antagonists on electrically induced seizures (Epilepsia, in press)
46. Meyer FB, Anderson RE, Sundt TM (1988) Anticonvulsant properties of dihydropyridine calcium antagonists. Drugs of Today 24:175–186
47. Miller RJ (1987) Multiple calcium channels and neuronal function. Science 235:46–53
48. Morocutti C, Pierelli F, Sanarelli L et al. (1986) Antiepileptic effects of a calcium antagonist (nimodipine) on cefusolin induced epileptogenic foci in rabbits. Epilepsia 27:498–503
49. Nowycky MC, Fox AP, Tsien RW (1985) Three types of neuronal calcium channels with different calcium agonist sensitivity. Nature 316:440–443
50. Nowycky MC, Fox AP, Tsien RW (1985) Long-opening mode of gating of neuronal calcium channels and its promotion by the dihydropyridine calcium agonist BAY k 8644. Proc Natl Acad Sci USA 82:2178–2182
51. Olney JW (1985) Excitatory transmitters and epilepsy-related brain damage. In: Smythies JR, Bradley RJ (eds) International review of neurobiology. Academic, London, pp 337–362
52. Olney JW, Gubareff de T, Labruyere J (1982) Seizure related brain damage induced by cholinergic agents. Nature 301:520–522
53. Owen DG, Segal M, Barker JL (1984) A Ca^{2+} dependent Cl^- conductance is present in cultured mouse spinal neurons. Nature 311:567–570
54. Pincus JH, Hsiao K (1981) Calcium uptake mechanisms affected by some convulsant and anticonvulsant drugs. Brain Res 217:119–127
55. Prince DA (1985) Physiological mechanisms of focal epileptogenesis. Epilepsia 26:S3–S14
56. Pulsinelli WA (1985) Selective vulnerability: morphological and molecular characteristics. In: Kogure K, Hossmann K-A, Siesjo BK, Welsh FA (eds) Progress in brain research, vol 63. Elsevier Scientific, Amsterdam, pp 29–38
57. Pulsinelli WA, Brierly JB, Plum F (1982) Temporal profile of neuronal damage in a model of transient forebrain ischemia. Ann Neurol 11:491–498
58. Pumain R, Heinemann V (1985) Stimulus and amino acid induced calcium and potassium changes in rat neocortex. J Neurophysiol 53:1–16
59. Rothman SM (1983) Synaptic activity mediates death of hypoxic neurons. Science 220:536–537
60. Rothman SM (1985) The neurotoxicity of excitatory amino acids is produced by passive chloride influx. J Neurosci 5:1483–1489
61. Rothman SM, Olney JW (1986) Glutamate and the pathophysiology of hypoxic ischemic brain damage. Ann Neurol 19:105–111
62. Schwartzkroin PA, Slawsky M (1977) Probable calcium spikes in hippocampal neurons. Brain Res 135:157–161
63. Siesjo BK (1984) Cerebral circulation and metabolism. J Neurosurg 60:883–908
64. Siesjo BK (1985) Acid-base homeostasis in the brain: physiology chemistry, and neurochemical pathology. In: Kogure K, Hossmann K-A, Siesjo BK, Welsh FA (eds) Progress in brain research, vol 63. Elsevier Scientific, Amsterdam, pp 121–154

65. Siesjo BK, Hanwehr R, Nergelius G et al. (1985) Extra- and intracellular pH in the brain during seizures and in the recovery period following the arrest of seizures activity. J Cereb Blood Flow Metab 5:47–57
66. Siesjo BK, Wieloch T (1986) Epileptic brain damage pathophysiology and neurochemical pathology. In: Delgado-Escueta AV, Ward AA (eds) Advances in neurology. Raven, New York, pp 813–847
67. Simon RP, Swan JH, Griffith T et al. (1984) Blockade of *N*-methyl-D-aspartate receptors may protect against ischemic damage in the brain. Science 226:850–852
68. Slovite R (1983) "Epileptic" brain damage in rats induced by sustained electrical stimulation of the perforant pathway. I. Acute electrophysiological and light microscopic studies. Brain Res Bull 10:665–697
69. Smith M-L, Auer RN, Siesjo BK (1984) The density and distribution of ischemic brain injury in the rat following 2–10 minutes of forebrain ischemia. Acta Neuropathol (Berl) 64:319–332
70. Suzuki R, Yamaguchi T, Kirino T et al. (1983) The effects of 5-minute ischemia in mongolian gerbils: I. Blood-brain barrier, cerebral blood flow, and local cerebral glucose utilization changes. Acta Neuropathol (Berl) 60:207–216
71. Suzuki R, Yamaguchi T, Li L-L et al. (1983) The effect of 5-minute ischemia in mongolian gerbils: II. Change of spontaneous neuronal activity in cerebral cortex and CA1 sector of hippocampus. Acta Neuropathol (Berl) 60:217–222
72. Tenny RT, Sharbrough FW, Anderson RE et al. (1980) Correlation of intracellular redox states and pH with blood flow in primary and secondary seizure foci. Ann Neurol 8:564–573
73. Thayer SA, Murphy SN, Miller RJ (1986) Widespread distribution of dihydropyridine-sensitive calcium channels in the central nervous system. Mol Pharmacol 30:505–509
74. Wahl M (1985) Local chemical, neural, and humoral regulation of cerebrovascular resistance vessels. J Cardiovasc Pharmacol 7:S36–S46
75. Walden J, Speckmann E-J, Witte OW (1985) Suppression of focal epileptiform discharges by intraventricular perfusion of a calcium antagonist. Electroencephalogr Clin Neurophysiol 61:299–309
76. Witte OW, Speckmann E-J, Walden J (1987) Motor cortical epilectic foci in vivo: actions of a calcium channel blocker on paroxysmal neuronal depolarizations. Electroencephalogr Clin Neurophys 66:43–55
77. Wong RKS, Prince DA (1978) Participation of calcium spikes during intrinsic burst firing in hippocampal neurons. Brain Res 159:385–390
78. Zanotto L, Heinemann U (1983) Aspartate and glutamate induced reductions in extracellular free calcium and sodium concentration in area CA1 of "in vitro" hippocampal slices of rats. Neuroscience Lett 35:79–84

Calcium Accumulation in Ischemia and Seizures: Cause or Result of Cell Death?

K. Inamura [1], E. Martins [2], K. Themner [3], K. G. Malmqvist [3], A. Terashi [1], and B. K. Siesjö [2]

Introduction

The mechanisms leading to neuronal cell damage in the brain in various pathological conditions are not yet fully understood. Nevertheless, it has been hypothesized that elevation of free cytosolic calcium concentration may have an important role in mediating cell damage; this is because calcium ions, which play a vital role in cellular activities serving as a second messenger, can also trigger deleterious reactions (Schanne et al. 1979; Farber et al. 1981; Siesjo 1981; Trump et al. 1981; Raichle 1983). The resting intracellular free calcium level (Ca_i^{2+}) is maintained at about 10^{-7} M. Physiological responses probably occur within the range 100–500 nM, while pathological responses are elicited at Ca_i^{2+} values exceeding 1 μM. It requires energy to maintain intracellular calcium homeostasis; hence anoxia/ischemia will always lead to raised cytosolic concentrations.

Accumulation of calcium in mitochondria occurs in pathological conditions provided that Ca_i^{2+} is raised and energy is available (see Rasmussen and Waisman 1983; Carafoli and Sottocasa 1984). Large calcium loading results in decreased energy production from the mitochonchria. Finally, the damaged mitochondria release their calcium into the cytosol, thus further increasing the Ca_i^{2+}. Since the mitochondria have a high storage capacity, calcium loading will result in an increase in the total calcium concentration.

The extracellular free calcium level (Ca_e^{2+}) is kept to around 10^{-4} M, which decreases in pathological conditions, i.e., ischemia, hypoglycemia, etc. (Nicholson et al. 1977; Harris et al. 1984; Hansen 1985). The mechanism of this decrease is unknown, but calcium entry into neurons can be suspected. The blood-brain barrier is relatively inpermeable to calcium ions, thus composing a major barrier for calcium entry into the tissue (see Katzman and Pappius 1973; Bradbury 1979; Murphy et al. 1986). This is probably why hypoglycemic coma of 60 min duration does not cause the total tissue content to increase in spite of the low extracellular level (Siesjö and Deshpande 1987).

The total tissue level of calcium in the hippocampus following transient ischemia has been previously studied with atomic absorption methods (Dienel 1984;

[1] Second Department of Internal Medicine, Nippon Medical School, Daiichi Hospital, 3-5-5 Iidabashi, Chiyodaku, Tokyo 102, Japan.
[2] Laboratory for Experimental Brain Research, Lund Hospital, Lund University, S-221 85 Lund, Sweden.
[3] Department of Nuclear Physics, Lund Institute of Science and Technology, Sölvegatan 14, S-223 62 Lund, Sweden.

Cerebral Ischemia and Calcium
Ed. by A. Hartmann and W. Kuschinsky

Deshpande et al. 1986). The results reported by Deshpande et al. (1986) revealed that calcium accumulation may precede cell death. We have tried to extend the study by detecting calcium in a more precise manner with a proton microprobe (PIXE), asking the additional question of whether calcium accumulation preferentially occurs in dendritic or soma regions.

In epilepsy, seizures of a long duration result in damage of substantia nigra pars reticulata (Nevander et al. 1985). The proton microprobe method has also been applied to study calcium accumulation in sustained seizures.

The present study thus reports the accumulation of calcium in ischemia and epilepsy in relation to the tissue damage incurred in these pathological conditions. The main purpose of the study is to discuss whether this accumulation of calcium is the result of cell damage or perhaps the cause of it.

Part of this study has been presented elsewhere (Martins et al. 1987; Inamura et al. 1988).

Materials and Methods

All experiments were performed on male Wistar rats, weighing 300–350 g (Mollegaard's Breeding Center, Copenhagen, Denmark). For studying ischemia, the rats were divided into the following groups: sham-operated controls and ischemic animals with recovery of 1, 2, 3, and 4 days. For studying seizures, the rats were divided into the following groups: sham-operated controls, 20-min seizure group, 60-min seizure group, 20-min seizure with recovery, and 60-min seizure with recovery (15 min and 120 min).

The ischemia model used in this study has been previously reported by Smith et al. (1984). In summary, ischemia was induced by bilateral occlusion of the common carotid arteries, combined with hypotension (50 mmHg) induced by withdrawal of blood from the vena cava. Trimethaphan (Arfonad, Hoffman-La Roche, Basel, Switzerland) was administered at the beginning of ischemia. Ischemia was continued for 10 min and recovery was subsequently induced by release of the vascular clamps and by reinfusion of blood.

Seizures were induced by the convulsant gas flurothyl as described previously (Nevander et al. 1985). Following operation, seizure was induced by injection of flurothyl (65 μl) into the rebreathing system. Phentolamine (Regitine) was administered to decrease the initial elevation of blood pressure. Seizure was terminated by infusion of thiopental and changing back to the normal breathing system.

The brains were frozen in situ for PIXE analysis. The freezing was performed by the in situ freezing method of Pontén et al. (1973). The frozen brains were sectioned in a cryostat (−20 °C) to 6 μm and subsequently freeze-dried. Adjacent sections were prepared for histology.

PIXE Analysis

Particle induced X-ray emission (PIXE) methodology was employed to analyze calcium contents in the brain tissue in this study not only because of its high sensitivity but because of its high spatial resolution where limitation depends on the beam diameter (approximately $20 \times 20\ \mu m^2$). The method allows precise measurement of the spectrum of compounds, including calcium and potassium, in the dehydrated tissue (Johansson and Johansson 1976; Malmqvist 1986).

Results

The physiological variables before inducing ischemia or seizures were within normal limits.

Analysis of calcium content was performed in the dorsal hippocampal CA1, where delayed neuronal damage appeared 2–4 days after the ischemic insult. Therefore, the analysis was performed on rats recovered for 1, 2, 3, and 4 days, with sham-operated controls serving as the reference.

To investigate whether calcium accumulation occurs in dendritic fields or in the cell soma, the calcium analysis was performed separately in str. oriens layer, pyramidal cell layer, str. radialis layer, and str. lacunae molecularis. The regions investigated are illustrated in Fig. 1. Delayed neuronal necrosis became manifest on the 2nd–4th day after instituting recovery in our model, and also in the similar series of experiments performed in the same period in which histopathology was studied following perfusion fixation. It should be emphasized, though, that many animals showed damage to the CA4 and subiculum sectors only, with no or only very sparse damage to the CA1 sector to which the PIXE analysis was applied. The results of the measurements are shown in Fig. 2. The calcium levels had a tendency to increase after the 2nd day but the increase was statistically insignificant.

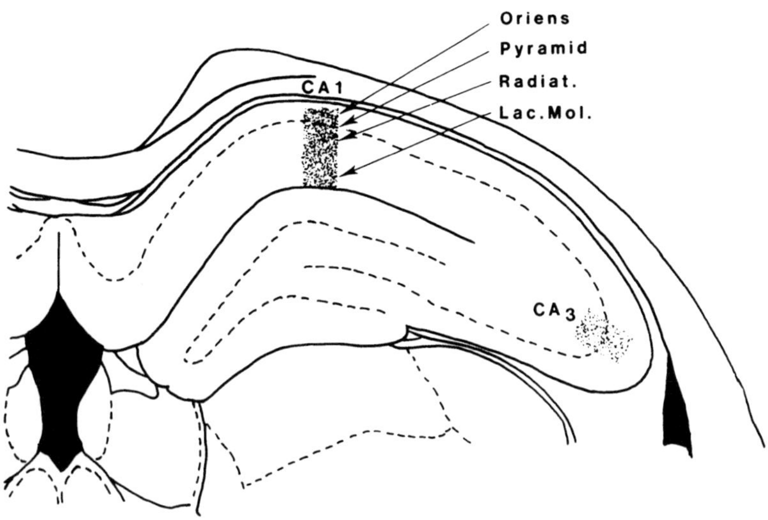

Fig. 1. Diagram illustrating areas of calcium analyses by PIXE (*shaded areas*) in the hippocampus. (Reproduced with permission from Martins et al., J Cereb Blood Flow Metab, Raven Press, New York, 1988)

The calcium level was significantly elevated on the 3rd day, when the damage started to appear. When calcium levels of the postischemic hippocampal CA1 region were plotted in individual samples with the damage, some samples showed elevated calcium contents without apparent damage (Fig. 3).

In the seizure groups, the regions investigated were the caudate nucleus and substantia nigra pars reticulata (SNPR). In our model of reversible epilepsy induced by flurothyl inhalation, dense damage was noted in the SNPR with no damage in the caudate (Nevander et al. 1985). The PIXE measurements were performed after 20 and 60 min of seizure, and after 20-min and 60-min seizures with recovery (induced by infusion of thiopental). The results of the PIXE measurements are shown in Fig. 4. Calcium content in the 20-min seizure group showed no increase in either the caudate of the SNPR. In the 60-min seizure group, there was no increase in the caudate but an apparent increase in the SNPR. This change

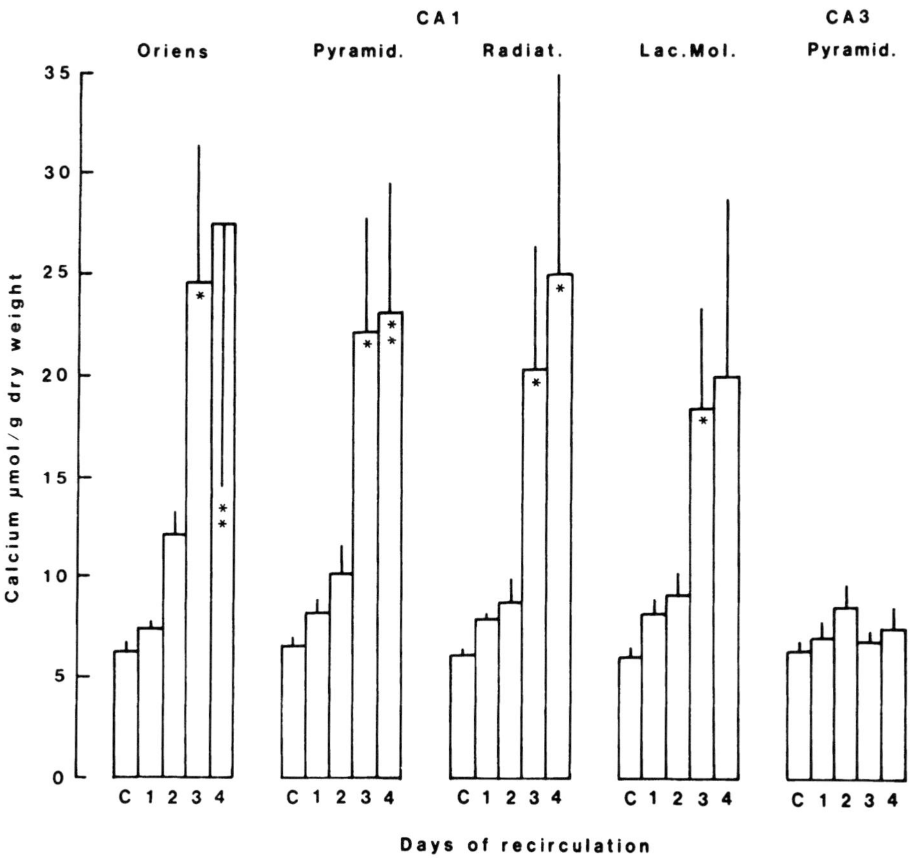

Fig. 2. Mean calcium contents in the hippocampal CA1 subfields and in the CA3 soma regions in control animals, as well as in animals allowed 1, 2, 3, and 4 days of recirculation, following 10 min of forebrain ischemia. The values are means ± SEM $*P<0.05$, $*P<0.01$ in comparison to control. (Reproduced with permission from Martins et al., J Cereb Blood Flow Metab, Raven Press, New York, 1988)

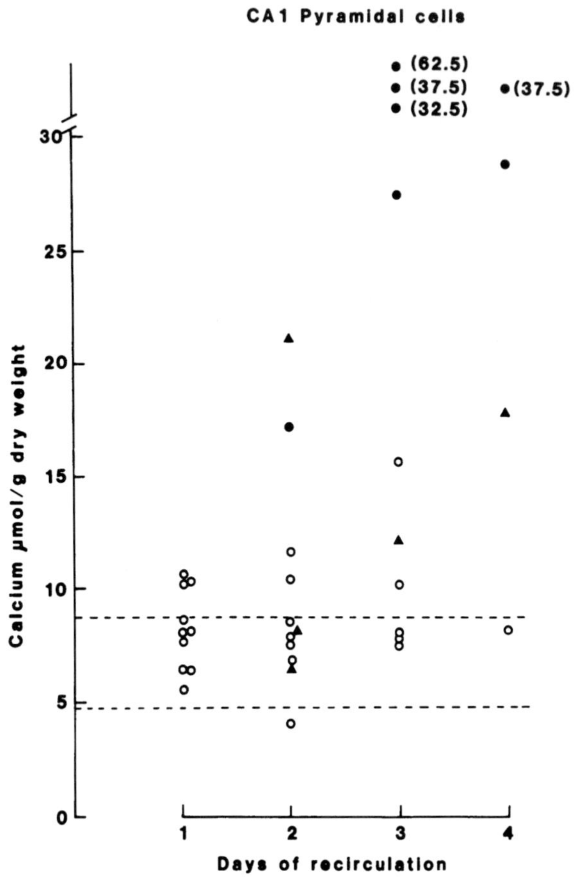

Fig. 3. Calcium content in pyramidal cell layer of CA1 as a function of time of recirculation, following 10 min of ischemia. Symbols are individual values in animals without detectable CA1 damage (○), with grade 1 damage (△), or with more extensive damage (●). The *stippled horizontal lines* demonstrate the range of control values. The damage scores were as follows; 0 for no damage, 1 for <5%, 2 for 5%–10%, 3 for 10%–50%, 4 for 50%–90%, 5 for >90%. (Reproduced with permission from Martins et al., J Cereb Blood Flow Metab, Raven Press, New York, 1988)

was statistically significant. The calcium level continued to rise after recovery in the SNPR.

The calcium accumulation was correlated with the morphological development of damage in the SNPR. The beginning of net calcium accumulation appeared to coincide with the damage incurred to this structure (see Auer et al. 1986; Inamura et al. 1987). The tendency toward calcium accumulation is notable since it occurred before gross damage was evident. The histological investigation of the SNPR during seizures showed accentuation of damage after recovery and this agrees with the finding that calcium accumulation largely occurs after instituting recovery.

A simple comparison of present results between the hippocampus and substantia nigra is, however, difficult. One reason is the structural difference. The hippocampus has a demarcated lining of cells and dendritic layers; this organization is lacking in substantia nigra. Substantia nigra is known for its paucity of neurons since the nucleus is mainly composed of neuropil. It is difficult to pinpoint the neuronal cell soma and measure its calcium content by the present technique. It is more likely that the present results reflect calcium levels in the neuropil, primarily dendritic structures.

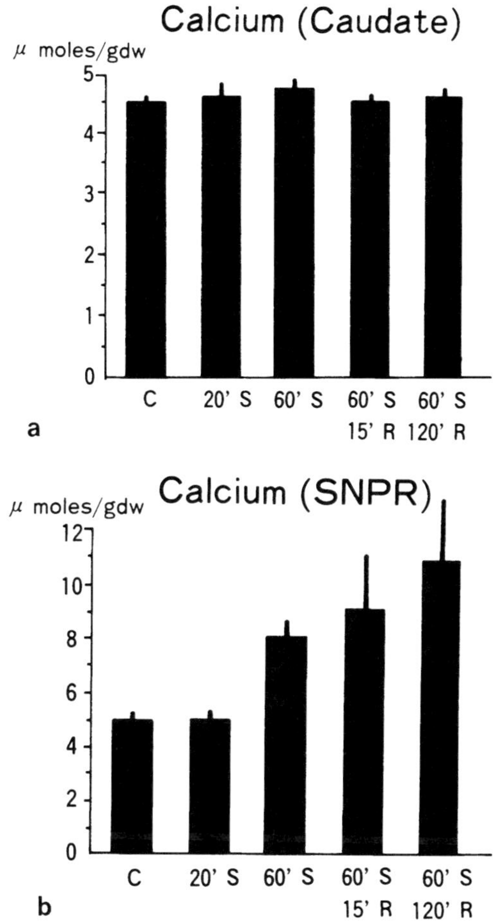

Fig. 4. Calcium content in **a** the caudate and **b** the SNPR as a function of time of seizure and recovery. The values are means ± SEM

Discussion

Before discussing the present result, it seems justified to recall some facts which facilitate interpretation of the results. The Ca_e^{2+} is about 10000 times higher than the Ca_i^{2+}. Energy failure therefore causes calcium influx. However, the calcium load represented by the extracellular pool is relatively small (about 300 μM). This is not an amount which, by itself, could account for irreversible damage to neurons. The intracellular calcium precipitates shown by Dux et al. (1987) may represent redistribution of extracellular calcium into the cell, which accumulates in mitochondria. These authors found a biphasic accumulation of calcium in the mitochondria following ischemia in the CA1 region of the hippocampus by the oxalate-pyroantimonate technique. The initial elevation occurred just following recirculation, whereas the secondary elevation occurred after 1–2 days. In the present study, the early recovery period was not studied; however, the time course of the secondary calcium accumulation was followed with a quantitative technique which yielded total tissue calcium (mainly mitochondrial) contents.

The blood-brain barrier is thought to be only sparingly permeable to calcium (see Katzman and Pappius 1973; Bradbury 1979; Murphy et al. 1986). However, the integrity of this barrier in pathological conditions is still unclear. At least, the influx to the tissue seems to be elevated in the postischemic period (Dienel et al. 1984). The low extracellular level in the interstitial fluid during the insult may thus facilitate the influx of calcium into the tissue through the blood-brain barrier, raising its total calcium content. If such redistribution should occur, the total amounts of calcium detected by the present PIXE method will increase.

The present results clearly demonstrated that calcium accumulates in the damaged tissue. The present result on the delayed neuronal damage in hippocampus showed that calcium accumulation may even precede the damage, thus indicating the possibility that calcium accumulation may be a cause rather than the result of damage.

Another important feature from this study is that there was no "preferential" site of calcium accumulation, indicating that the dendrites and soma accumulate calcium at the same rate. The site of calcium entry is expected to be through the postsynaptic sites where calcium channels reside. It cannot be predicted at present that calcium, after entering the channels, will be distributed by diffusion and convection, which is a relatively slow process (see Abercrombie and Hart 1986; Baker and Umbach 1987).

In seizure damage to substantia nigra, calcium accumulation was not seen in the 20-min seizure group, where no damage could be observed (Nevander et al. 1985; Auer et al. 1986). However, in the 60-min seizure group, calcium values were significantly elevated. The calcium seemed to accumulate continuously in substantia nigra following recovery, since the level was clearly higher in rats 2 h after recovery.

The precise route for calcium entry remains elusive. Recent results indicate that the agonist-operated channels may be responsible for calcium entry into postsynaptic structures, especially the glutamate receptor, NMDA (Rothman and Olney 1986; Choi 1987).

We investigated regional calcium values in regions subjected to damage in ischemia and seizures. Calcium levels increased temporally in these tissues as damage proceeded. In some individual values, calcium accumulation preceded the damage. We also observed calcium accumulation during prolonged seizures but due to the rapid cell death in this condition it cannot be stated whether calcium accumulation precedes or follows irreversible cell damage. We think that the increase in tissue calcium content observed in this study in hippocampus and substantia nigra involves precipitated calcium in mitochondria.

Summary

Calcium accumulation was studied in the hippocampal CA1 and substantia nigra regions which suffer distinct damage in ischemia and seizures. Calcium contents of these regions were investigated with a proton microprobe (PIXE) which allows precise regional evaluation of tissue elements. In delayed neuronal damage of hippocampal CA1, calcium accumulated with progression of damage, and in some samples calcium accumulation was evident before any histologically detectable damage was seen. In seizures, calcium accumulation was not seen in the 20-min seizure group, in which no damage was expected. Calcium accumulation occurred during seizures of 60-min duration and was statistically significant, as well as when recovery was instituted. The results show that calcium accumulates with damage. There is some indication that calcium accumulation may precede the damage in ischemia. It will be necessary to study the subcellular distribution to find out whether this is the case in the rapid cell death during seizures.

Acknowledgments. The authors are grateful to the skillful assistance of Lille-Mor Lindeström and Birgit Olsson. This study was supported by the Swedish Medical Research Council, project 14X-263, and by the U.S. Public Health Service via the N.I.H. (Grant No. 5 ROI NS-07838).

References

Abercrombie RI, Hart CE (1986) Calcium and proton buffering and diffusion in isolated cytoplasm from *Myxicola* axons. Am J Physiol 250:C391–405

Auer RN, Ingvar M, Nevander G, Olsson Y, Siesjö BK (1986) Early axonal lesion and preserved microvasculature in epilepsy-induced hypermetabolic necrosis of the substantia nigra. Acta Neuropathol (Berl) 71:207–215

Baker PF, Umbach JA (1987) Calcium buffering in axons and axoplasm of loligo. J Physiol 383:369–394

Bradbury M (1979) The concept of a blood-brain barrier. Wiley, New York

Carafoli E, Sottocasa G (1984) The uptake and the release of calcium by mitochondria. In: Ernster (ed) Bioenergetics. Elsevier, Amsterdam, pp 269–289

Choi DW (1987) Ionic dependence of glutamate neurotoxicity. J Neurosci 7:369–379

Deshpande JK, Siesjö BK, Wieloch T (1987) Calcium accumulation and neuronal damage in the rat hippocampus following cerebral ischemia. J Cereb Blood Flow Metab 7:89–95

Dienel GA (1984) Regional accumulation of calcium in post-ischemic rat brain. J Neurochem 13:913–925

Dux E, Mies G, Hossmann K-A, Siklos L (1987) Calcium in the mitochondria following brief ischemia of gerbil brain. Neurosci Lett 78:295–300

Farber JL (1981) The role of calcium in cell death. Life Sci 29:1289–1295

Hansen AJ (1985) Effects of anoxia on ion distribution in the brain. Physiol Rev 65:101–148

Harris RJ, Wieloch T, Symon L, Siesjö BK (1984) Cerebral extracellular calcium activity in severe hypoglycemia: relation to extracellular potassium and energy charge. J Cereb Blood Flow Metab 4:187–193

Inamura K, Martins E, Siesjö BK (1987) Tissue pH in substantia nigra during and after recovery from status epilepticus: relationship of acidosis to developing hypermetabolic necrosis. J Cereb Blood Flow Metab 7 (Suppl 1):S426

Inamura K, Martins E, Themner K, Tapper S, Malmqvist KG, Siesjö BK (1988) Accumulation of calcium in substantia nigra lesions induced by status epilepticus. (submitted)

Johansson SAE, Johansson TB (1976) Analytical application of particle induced X-ray emission. Nucl Intr Meth 137:473–516

Katzman R, Pappius HM (1973) Brain electrolytes and fluid metabolism. Williams & Wilkins, Baltimore

Malmqvist KG (1986) Proton microprobe analysis in biology. Scan Electron Microsc 3:821–845

Martins E, Inamura K, Themner K, Malmqvist KG, Siesjö BK (1987) Accumulation of calcium and loss of potassium in the hippocampus following transient cerebral ischemia: a study with proton microprobe. J Cereb Blood Flow Metab 7:S116

Murphy VA, Smith QR, Rapoport SI (1986) Homeostasis of brain and cerebrospinal fluid calcium concentrations during chronic hypo- and hypercalcemia. J Neurochem 47:1735–1741
Nevander G, Ingvar M, Auer R, Siesjö BK (1985) Status epilepticus in well-oxygenated rats causes neuronal necrosis. Ann Neurol 188:281–290
Nicholson C, Bruggencate GT, Steinberg R, Stockle H (1977) Calcium modulation in brain extracellular microenvironment demonstrated with ion-selective micropipette. Proc Natl Acad Sci USA 74:1287–1290
Pontén V, Ratcheson R, Salford LG, Siesjö BK (1973) Optimal freezing conditions for cerebral methods in the rat. J Neurochem 21:1127–1128
Raichle M (1983) The pathophysiology of brain ischemia. Ann Neurol 13:2–10
Rasmussen H, Waisman DM (1983) Modulation of cell function in the calcium messenger system. Rev Physiol Biochem Pharmacol 95:111–148
Rothman SM, Olney JW (1986) Glutamate and pathophysiology of hypoxic-ischemic brain damage. Ann Neurol 19:105–111
Schanne FAX, Kane AB, Young EE, Farber JL (1979) Calcium dependence of toxic cell death: a final common pathway. Science 206:700–702
Siesjö BK (1981) Cell damage in the brain: a speculative synthesis. J Cereb Blood Flow Metab 1:155–185
Siesjö BK, Deshpande J (1987) Electrolyte shifts between brain and plasma in hypoglycemic coma. J Cereb Blood Flow Metab 7:789–793
Smith M-L, Bendek G, Dahlgren N, Rosen I, Wieloch T, Siesjö BK (1984) Models for studying long-term recovery following forebrain ischemia in the rat. 2. A 2-vessel occlusion model. Acta Neurol Scand 69:385–401
Themner K, Malmqvist KG (1986) Determination of local areal densities of heterogeneous biological samples by elastically backscattered protons in PIXE microanalysis. Nucl Instr Meth B15:404–406
Trump BF, Beregesky IK, Osornio-Vargas AR (1981) Cell death and disease process. The role of calcium. In: Bower ID, Lockshin RA (eds) Cell death in biology and pathology. Chapman and Hall, London, pp 209–242

Aspects of Calcium Ions in Cortical Spreading Depression

M. LAURITZEN [1], M. SHEARDOWN [2], and A. J. HANSEN [1]

Introduction

Cortical spreading depression (CSD) is a transient perturbation of nerve cell activity which propagates across the cortex at a rate of 2–5 mm/min [15]. CSD is a highly stereotyped stimulus-independent event in which different stimuli are believed to trigger complex chain reactions converging on a common link which is at the same time the first step of the autocatalytic reaction [3]. The initiation and spread of CSD are probably intimately related and can be examined by procedures that either block or facilitate the phenomenon.

The understanding of CSD has advanced in recent years due to the application of ion-selective microelectrodes. It has become clear that the extracellular microenvironment undergoes a unique sequence of changes with respect to the ionic composition and water concentration: $[Na^+]_o$ declines from 150 mM to 50–60 mM, $[Cl^-]_o$ decreases from 120 mM to about 70 mM, while $[K^+]_o$ rises from 3 to about 30 mM, $[Ca^{2+}]_o$ declines from about 0.8–1.2 mM to about 0.2 mM, and the volume fraction of the extracellular space decreases by about 50% [8, 9, 19]. Simultaneously, nerve and glial cells depolarize as revealed from intracellular measurements [22]. The perturbation is *transient*, the return to normal taking place within a few minutes, and the disturbance *propagates*. If these two criteria for CSD are not fulfilled, one is usually dealing with hypoglycemia or cerebral ischemia, not CSD [19].

Our knowledge of the nature of the membrane conductance changes of nerve cells and glial cells during CSD is minimal: Nicholson [18] showed that the time base, form, and amplitude of the membrane potential change of the glial cells of the turtle cerebellum followed very closely $[K^+]_o$ measured in the immediate vicinity of the cell body. This suggests activation of K^+ channels of glial cells during CSD, but is the only fact on the issue. We are ignorant of the intracellular changes of ions during CSD, and it is unknown whether the large extracellular $[Ca^{2+}]$ changes are matched by fluctuations of intracellular $[Ca^{2+}]$. We do know, however, that the decline of $[Ca^{2+}]_o$ commences *after* $[K^+]_o$ rises, suggesting that the influx of Ca^{2+} from the extracellular space is probably not involved in eliciting the K^+ movements [9, 19].

[1] Department of General Physiology and Biophysics, University of Copenhagen, Blegdamsvej 3c, DK-2200 Copenhagen, and [2] A/S Ferrosan, Research Division, DK-Søborg, Denmark.

Cerebral Ischemia and Calcium
Ed. by A. Hartmann and W. Kuschinsky

The role of extracellular Ca^{2+} in CSD can tentatively be related to four mechanisms: (a) presynaptic influx of Ca^{2+} in response to depolarization and release of transmitter substance; (b) dendritic Ca^{2+} spikes; (c) post- or extrasynaptic Ca^{2+} influx; and (d) charge screening. Previous studies [3, 24, 25] have shown that divalent cations inhibit CSD, but that the capability of the tissue to produce CSD returns with addition of glutamate to the superfusion medium. This suggests that the blocking effect of divalent cations is due to presynaptic inhibition, but other Ca^{2+} conductances might also play a role. The new tools in studies of the role of Ca^{2+} in CSD are drugs that block post- or extrasynaptic Ca^{2+} channels. The focus has been on:

1. Blockers of the glutamate-operated Ca^{2+} channel, which preferentially is operated by the NMDA receptor [4, 12].
2. Classic Ca^{2+} channel antagonists (verapamil, nifedipine, flunarizine), which probably affect voltage-dependent more than transmitter-operated Ca^{2+} channels [recently reviewed in 7].

Glutamate is believed to be a main factor in the triggering and propagation of CSD in the following way: Release of glutamate at one locus depolarizes neighboring neurons, which in turn release glutamate, and so on away from the site of elicitation [23]. Agents which blocked the glutamate receptor also blocked the occurrence of CSD in the chicken retina [26]. Mody et al. [17] showed that in the hippocampal slice lowering the $[Mg_2{}^{+}]_o$ triggered CSD which was blocked by D-APV, a specific blocker of the NMDA (*N*-methyl-D-aspartate) subtype of glutamate receptors [27]. This suggested that NMDA receptors might play a special role in the generation of CSD, since the voltage-dependent block by Mg^{2+} of the Ca^{2+} conductance is removed under these conditions [2].

Studies of CSD in the Turtle Cerebellum

In the isolated turtle cerebellum CSD may be electrically triggered provided the brain has been conditioned for about 30 min with anion-substituted medium containing reduced amounts of chloride. The purpose of the study was to examine whether activation of all subtypes of the glutamate receptor, kainate (KA), quisqualate (QA), and NMDA-operated, produced CSD. Firstly it was confirmed that reducing $[Mg^{2+}]_o$ concentration to 0 mM increased the susceptibility to CSD, which appeared spontaneously, and the CSD produced in this fashion was blocked by D-APV. Similarly, kynurenic acid, a nondiscriminating glutamate antagonist [21], and D-APV blocked electrically induced CSD. Glutamate, aspartate, NMDA, QA, and KA all triggered CSD. Moreover, the time course, amplitude, and waveform of change of the DC potential associated with CSD in response to the excitatory amino acids and electrical stimulation were virtually identical. Thus, activation of glutamate receptors triggered cerebellar CSD and was inhibited by glutamate blockers [14].

Studies of CSD in the Chick Retina

The results from the retina have previously been reported in abstract form [20]. The purpose of the study was to examine the effects of blockers of NMDA and non-NMDA (KA and QA) receptors and the classic Ca^{2+} antagonists on CSD produced by KCl and the three agonists of glutamate receptors: NMDA, KA, and QA. Chicks were decapitated, and the eyes removed and sectioned along the equatorial plane. After removal of the anterior chamber and the vitreous body, the posterior chamber of the eye was placed in a small Petri dish containing mock-CSF. The eyes were initially incubated in mock-CSF for 30 min and then transferred to mock-CSF containing either KCl (20 mM), NMDA (100 μM), QA (5 μM), or KA (5 μM). These "stimulating solutions" triggered waves of CSD which were readily observed by eye. Prior to drug test the latency for CSD start in each eye was measured.

After a further 15 min incubation in normal mock-CSF the eyes were transferred to normal mock-CSF containing the blocker for another 15 min. Thereafter, the eyes were transferred to a stimulating solution containing the same concentration of the potential blocker. The latency for a CSD to start was again measured. The eyes were then taken back in the drug-free mock-CSF, and after 15 min the time taken for a CSD to start again in response to a stimulating solution was measured in order to assess the degree of recovery from any drug effects. An increase in the time taken for CSD to start of 30 s or more was taken to represent 100% inhibition of CSD. (The retinas rapidly deteriorated when kept for longer periods in the stimulating solutions.) Drug effects are expressed as the percentage maximum inhibition obtained for a given dose. Thus the test value can be quoted as the concentration of test substance which produces 50% maximal inhibition (IC_{50}).

The results showed that KCl, NMDA, QA, and KA consistently produced CSD in the chick retina, presenting itself as a "milky" area spreading slowly over the retina at 2–3 mm/min [16]. These changes eventually spread over the entire retina and were easily visible to the naked eye. The latency of the CSD was very consistent in response to each of the stimulating solutions (KCl 6.4 ± 0.96 s; NMDA 10.4 ± 1.76 s; QA 11.26 ± 1.93 s; KA 11.4 ± 1.26 s – means ± 1 standard deviation) and the preparations were stable over a period of 5–8 h. The short latency from application of the stimulus to onset of CSD suggests that the CSD is triggered close to the inner layer of the retina.

D-APV suppressed CSD evoked by NMDA, QA, and KA equipotentially (Fig. 1). Also the specific noncompetitive NMDA antagonists ketamine and phencyclidine [1] were approximately equipotent in inhibiting CSD triggered by NMDA, KA, QA, and KCl. The new selective non-NMDA antagonists FG 9041 and FG 9065 [5, 11] were more potent against QA- and KA- than against NMDA- and KCl-evoked CSD, FG 9065 being approximately 17 times more potent against QA- and KA- than NMDA- and KCl-evoked CSD (Fig. 2). The IC_{50} values are shown in Table 1. The conclusion was that blockade of NMDA receptors inhibit CSD more efficiently than blockade of non-NMDA receptors.

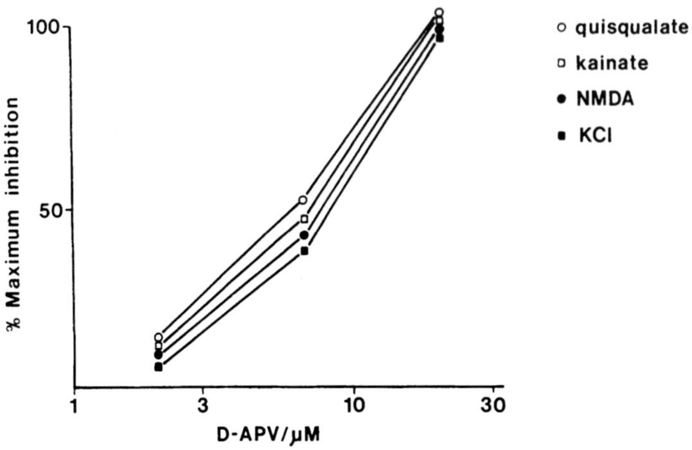

Fig. 1. Effect of D-APV on spreading depression (*S.D.*) in the chick retina triggered by short exposure (up to 30 s) to one of the glutamate agonists kainate (*KA*), quisqualate (*QA*), and *N*-methyl-D-aspartate (*NMDA*) and KCl. Subsequently the retinas were transferred to mock-CSF containing a known concentration of D-APV and the stimulation was repeated. D-APV inhibits S.D. triggered by the three glutamate agonists and KCI equipotentially

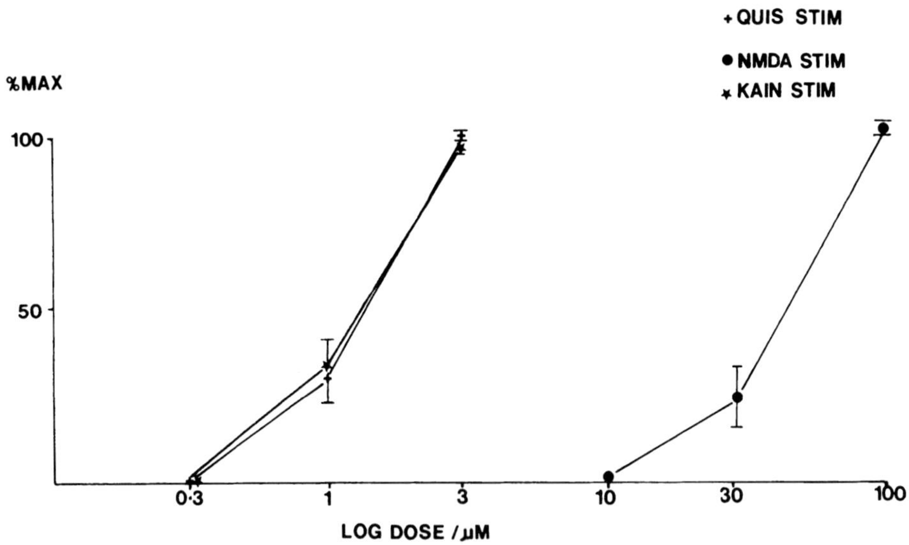

Fig. 2. Effect of the non-NMDA blocker FG 9065 on spreading depression (*S.D.*) in the chick retina evoked by short exposure (up to 30 s) to one of the three glutamate agonists quisqualate (*QA*), kainate (*KA*), and *N*-methyl-D-aspartate (*NMDA*) or KCl. Subsequently the retinas were transferred to mock-CSF containing a known concentration of FG 9065 and the stimulation was repeated. FG 9065 inhibited S.D. triggered by QA and KA much more efficiently than NMDA- and KCl-mediated S.D.

Table 1. Inhibition of spreading depression

	Quisqualate IC_{50} (μM)	Kainate IC_{50} (μM)	NMDA IC_{50} (μM)	KCl IC_{50} (μM)
D-APV	6.8	7.5	8.0	9.6
Ketamine	75	92	42	52
Phencyclidine	20	13.6	6.4	10.2
FG 9065	0.9	1.3	17.2	18
FG 9041	1.64	1.44	4.2	4.5

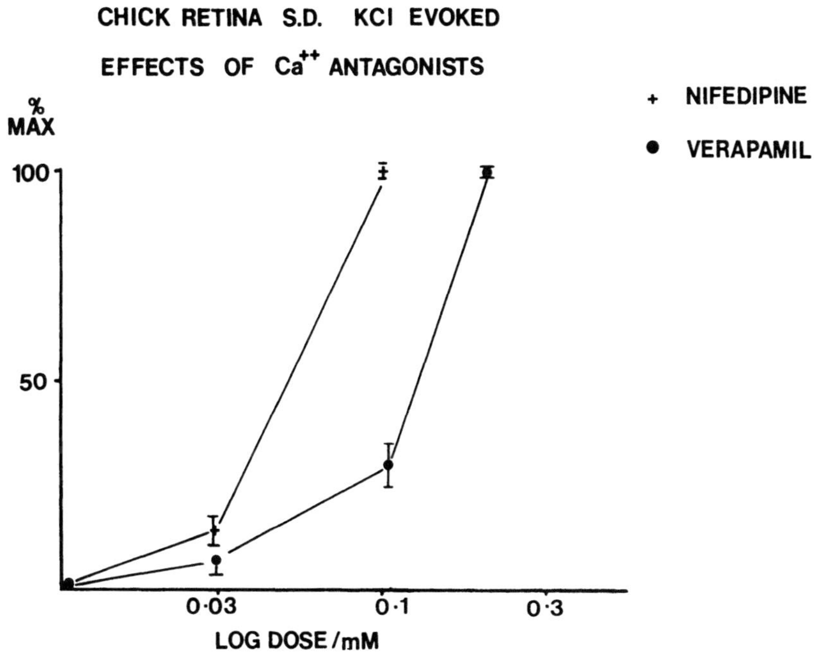

Fig. 3. Effect of nifedipine and verapamil on spreading depression (*S.D.*) in the chick retina evoked by short exposure (up to 30 s) to KCl. Subsequently the retinas were transferred to mock-CSF containing a known concentration of either of the blockers, and the stimulation was repeated. Nifedipine and verapamil were both effective in suppressing KCl-evoked S.D., but at high concentrations

In addition the calcium antagonists nifedipine and verapamil were tested against CSD (Figs. 3, 4). Both were effective blockers at high concentrations, the IC_{50} value against KCl-evoked CSD being 50 μM and 125 μM respectively. Interestingly, when tested against NMDA-, QA-, and KA-evoked CSD, nifedipine showed a greater effect against QA (IC_{50} 1.4 μM) and KA (IC_{50} 1.1 μM) than against NMDA (IC_{50} 46 μM). Verapamil was also more potent in suppressing QA- (IC_{50} 50 μM) than NMDA-evoked CSD (IC_{50} 130 μM). Flunarizine was inactive against CSD evoked by excitatory amino acids.

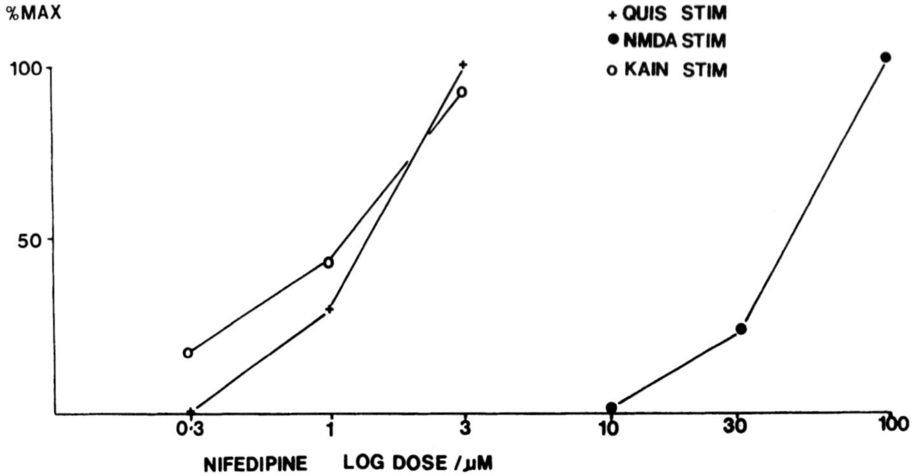

Fig. 4. Effect of nifedipine on spreading depression (*S.D.*) in the chick retina evoked by short exposure (up to 30 s) to one of the two glutamate agonists quisqualate (*QA*) and *N*-methyl-D-aspartate (*NMDA*). Subsequently the retinas were transferred to mock-CSF containing a known concentration of nifedipine and the stimulation was repeated. Nifedipine showed a significantly greater potency against QA-evoked S.D. than against NMDA-evoked S.D.

Studies of CSD in the Rat Cortex

This subject is dealt with in detail in the chapter by Hansen et al., and similar experiments have been carried out by Roger Marannes from the Janssen Company (Beerse, Belgium). In brief, the purpose was to elucidate whether CSD was blocked by antagonists of the glutamate receptor given systemically. CSD was triggered electrically in the rat parietal cortex and the local $[K^+]_o$ and $[Ca^{2+}]_o$ measured with ion-selective microelectrodes at a depth of 300–400 μm below the surface. Spread to the frontal cortex was monitored by a single-barreled microelectrode measuring the local DC potential. Prior to intravenous injection of drugs, CSD was elicited 3 times and a threshold stimulus was defined. D-APH and MK-801, competitive and noncompetitive NMDA blocking agents respectively, were injected intravenously at 10 mg/kg. Subsequently (5–10 min later) the electrical stimulus was applied at threshold and while increasing stimulus parameters 10 times (strength, duration, frequency), but CSD was not triggered at all. The conclusion was that NMDA blockers given *systemically* are effective blockers of CSD. The result confirms the observation of Goroleva et al. [6] that ketamine inhibits CSD in the rat brain. We have previously found flunarizine to be ineffective in blocking CSD in rat cortex [10], while other putative Ca^{2+} antagonists remain to be tested.

Discussion

Both competitive and noncompetitive antagonism of glutamate-mediated synaptic transmission appear to be effective blockers of CSD. Thus, the CSD-glutamate hypothesis of Van Harreveld [23] has gained renewed credibility.

In the isolated turtle cerebellum it was shown that Mg^{2+}-free medium increased the susceptibility to CSD markedly, as indicated by spontaneous CSD waves, and nonspecific glutamate antagonists as well as D-APV blocked CSD effectively. It was further shown that activation of all three subtypes of glutamate agonist induced CSD at very low concentrations [14].

In the isolated chick retina it was found that KCl and all three glutamate agonists triggered CSD. Nifedipine and verapamil were both effective in suppressing the triggering of CSD by these substances, but the concentrations of the drugs required were rather high. Interestingly, the concentration of nifedipine in particular was much more potent (35–40 times) in suppressing the response to QA and KA than to KCl and NMDA. Verapamil, though less potent, showed an increased potency against QA- than NMDA-evoked CSD. Nifedipine has no significant AMPA binding activity (Honoré, unpublished data). Therefore, it is possible that the increased potency of nifedipine in suppressing QA- and KA-evoked CSD compared to that evoked by NMDA and KCl reflects that QA and KA exert their CSD-triggering effect in part by activating voltage-dependent Ca^{2+} channels.

D-APV and the specific NMDA channel blockers ketamine and phencyclidine inhibited the triggering of CSD by NMDA, QA, KA, and KCl in the retina. Similarly, D-APH and MK-801 were effective blockers of CSD in the rat cortex. It might be that the final common pathway in triggering CSD in all three preparations primarily involves activation of the NMDA receptor–ionophore complex causing a Ca^{2+} influx, irrespective of the source of depolarization.

In the retina FG 9065 and FG 9041 (non-NMDA antagonists) were more potent (FG 9065 by a factor 17) in suppressing QA- and KA-evoked CSD than NMDA- or KCl-evoked CSD. It appears that non-NMDA blockade is of limited utility in blocking KCl-evoked CSD, while NMDA blockade produces inhibition.

For the clinician it is of potential interest that drugs that effectively block the NMDA receptor and CSD can be given systemically. It has recently been suggested that CSD might be a trigger mechanism of classic migraine attacks [13]. Therefore, with the further development of this category of drugs, glutamate antagonism could be a new strategy for migraine therapy. Nevertheless, both CSD and migraine should be viewed as cascade reactions rather than single events with one dominant cause-effect relation. Therefore, despite the present optimism further studies are needed before the two phenomena can be firmly linked, and progress of CSD pharmacology and mechanisms can be transferred to the clinical condition.

Summary

The mechanism of triggering and spreading of cortical spreading depression (CSD) was examined in (a) the isolated turtle cerebellum, (b) the chick retina, and (c) the in vivo rat cortex with respect to the effect of calcium and glutamate blockade. Glutamate blockade, especially of the NMDA receptor and less so of non-NMDA receptors, effectively abolished CSD. In addition, verapamil and nifedipine blocked CSD in the chick retina, suggesting that voltage-dependent Ca^{2+} channels probably contribute to the development of CSD. It is concluded that CSD represents a cascade reaction rather than a single event with one dominant cause-effect relation.

Acknowledgments. The research of ML and AJH was supported by the University of Copenhagen, Medical Research Council, The Carlsberg Foundation, Foundation for Experimental Neurological Research, Fonden af 1870, and the Danish Migraine Society.

References

1. Anid NA, Berry SC, Burton NR, Lodge O (1983) The dissociative anesthetics ketamine and phencyclidine selectively reduce excitation of central mammalian neurones by *N*-methyl-D-aspartate. Br J Pharmacol 79:565–575
2. Ault B, Evans RH, Francis SS, Oakes DJ, Watkins JC (1980) Selective depression of excitatory amino acid induced depolarization by magnesium ions in isolated spinal cord preparations. J Physiol 307:413–428
3. Bures J, Buresova O, Krivanek J (1974) The mechanism and applications of Leao's spreading depression of electroencephalographic activity. Academic, New York
4. Cull-Candy SG, Usowicz MM (1987) Multiple conductance channels activated by excitatory amino acids in cerebellar neurons. Nature 325:525–528
5. Drejer J, Honoré T (1988) New quinoxalinediones show potent antagonism of quisqualate responses in cultured mouse cortical neurones. Neurosci Lett 87:104–108
6. Goroleva NA, Koroleva VI, Amemori T, Pavlik V, Bures J (1987) Ketamine blockade of cortical spreading depression in rats. Electroencephalogr Clin Neurophysiol 66:440–447
7. Greenberg DA (1987) Calcium channels and calcium channel antagonists. Ann Neurol 21:317–330
8. Hansen AJ, Olsen CE (1980) Brain extracellular space during spreading depression and ischemia. Acta Physiol Scand 108:355–365
9. Hansen AJ, Zeuthen T (1981) Changes of brain extracellular ions during spreading depression and ischemia in rats. Acta Physiol Scand 113:437–445
10. Hansen AJ, Lauritzen M, Tfelt-Hansen P (1984) Spreading cortical depression and anti-migraine drugs. In: Amery W, Wauquier A, Van Nueten JH (eds) The pharmacological basis of migraine therapy. Pitman Medical, London, pp 161–170
11. Honoré T, Davies SN, Drejer J, Fletcher EJ, Jacobsen P, Lodge O, Nielsen FE (1987) Potent and competitive antagonism of non-NMDA receptors by FG 9041 and FG 9065. Soc Neurosci (abstr) 13 109.7
12. Jahr CE, Stevens CF (1987) Glutamate activates multiple channel conductances in hippocampal neurones. Nature 325:522–525
13. Lauritzen M (1987) Cortical spreading depression as a putative migraine mechanism. Trends Neurosci 10:8–13
14. Lauritzen M, Rice M, Okada Y, Nicholson C (1988) Spreading depression by activation of glutamate receptors in turtle cerebellum in vitro. Brain Res 475:317–327

15. Leao AAP (1944) Spreading depression of activity in the cerebral cortex. J Neurophysiol 7:359–390
16. Martins-Ferreira H, Oliveira-Castro G (1966) Light scattering changes accompanying spreading depression in isolated retina. J Neurophysiol 37:773–783
17. Mody I, Lambert JDC, Heinemann U (1987) Low extracellular magnesium induces epileptiform activity and spreading depression in rat hippocampal slices. J Neurophysiol 57:869–888
18. Nicholson C (1984) Comparative neurophysiology of spreading depression in the cerebellum. An Acad Bras Cienc 56:481–494
19. Nicholson C, Kraig RP (1981) The behaviour of extracellular ions during spreading depression. In: Zeuthen T (ed) The application of ion-selective microelectrodes. Elsevier, Amsterdam, pp 217–238
20. Sheardown MJ (1987) Effects of excitatory amino acid antagonists and calcium entry blockers on spreading depression. Proceedings of the 2nd World Congress of Neuroscience, Budapest, Hungary, 1987, Neuroscience 22 (Suppl)
21. Stone TW, Connick JH (1985) Quinolinic acid and other kynurenines in the central nervous system. Neuroscience 15:597–617
22. Sugaya E, Takato M, Noda Y (1975) Neuronal and glial activity during spreading depression in cerebral cortex. J Neurophysiol 38:822–841
23. Van Harreveld A (1959) Compounds in brain extracts causing spreading depression of cerebral cortical activity and contraction of crustacean muscle. J Neurochem 3:300–315
24. Van Harreveld A, Fifkova E (1970) Glutamate release from the retina during spreading depression. J Neurobiol 2:13–29
25. Van Harreveld A, Fifkova E (1971) Effects of glutamate and other amino acids on the retina. J Neurochem 18:2145–2154
26. Van Harreveld A (1984) The nature of the chick's magnesium-sensitive retinal spreading depression. J Neurobiol 15:333–344
27. Watkins JC, Evans RH (1981) Excitatory amino acid transmitter. Ann Rev Pharmacol Toxicol 21:165–204

Depression of Focal and Generalized Tonic-Clonic Epileptic Activity by the Organic Calcium Antagonist Verapamil

E.-J. SPECKMANN [1], J. WALDEN [1], H. POCKBERGER [2], A. LEHMENKÜHLER [1], and D. BINGMANN [3]

Transmembranous calcium currents have been described to be involved in the generation of epileptic activity of single nerve cells in different vertebrate and invertebrate preparations (cf. Heinemann et al. 1986; Speckmann et al. 1986a; Walden et al. 1988). In accordance with that, calcium channel blockers have been found to depress neuronal epileptic discharges (Bingmann and Speckmann 1986a, b; Bingmann et al. 1988; Speckmann et al. 1986b; Witte et al. 1987). In the present investigations the action of the calcium antagonist verapamil on focal and on generalized tonic-clonic seizures was investigated. Since verapamil is able to cross the blood-brain barrier only to a small extent, the drug was perfused intracerebroventricularly. With this technique relatively high local concentrations were reached in brain tissue and effects on the systemic circulatory system were avoided (cf. Walden et al. 1985).

The experiments were carried out in the anesthetized and artificially ventilated rat. Focal epileptic activity and tonic-clonic seizures were induced by local epicortical application of penicillin and repeated intraperitoneal injections of pentylenetetrazol (PTZ), respectively. The conventional EEG and the DC potential were led from the surface of the motor and somatosensory cortex. The calcium antagonist verapamil was applied into a lateral cerebral ventricle by means of the push-pull technique.

Focal epileptic activity was reduced in amplitude and in frequency of occurrence during intracerebroventricular perfusion of verapamil (Fig. 1 A). In some experiments the depression of seizure activity was preceded by a transient enhancement of epileptic discharges. After termination of the drug perfusion focal epileptic activity was often reestablished (Fig. 1 A; Walden et al. 1985, 1986).

Generalized tonic-clonic seizures were reduced in intensity to a great extent and often abolished during intracerebroventricular perfusion of verapamil (Fig. 1 B). With the suppression seizures decreased in their repetition rate. The negative displacement of the epicortical DC potential which had been evoked by the PTZ injections before verapamil application turned over to a positive shift during verapamil perfusion (Fig. 1 B; Walden and Speckmann 1988).

Control experiments revealed (a) that intracerebroventricular perfusion with drug-free cerebrospinal fluid did not change focal and generalized tonic-clonic

[1] Institut für Physiologie, Universität Münster, Robert-Koch-Str. 27a, D-4400 Münster, FRG.
[2] Institut für Neurophysiologie, Universität Wien, Währinger Gürtel 17, A-1090 Wien, Austria.
[3] Institut für Physiologie, Universität Essen, Hufelandstr. 55, D-4300 Essen, FRG.

Cerebral Ischemia and Calcium
Ed. by A. Hartmann and W. Kuschinsky

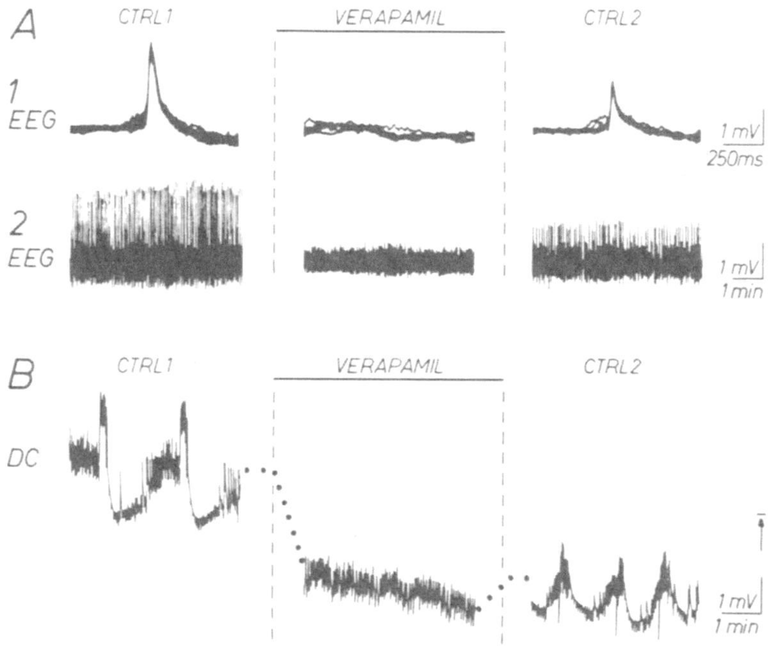

Fig. 1. Depressive actions of the calcium antagonist verapamil on focal interictal (**A**) and generalized tonic-clonic epileptic discharges (**B**) in the epicortical EEG and DC potential. *A1*: single epileptic potentials, six sweeps superimposed. *A2, B*: inkwriter tracings. Anesthetized and artificially ventilated rat, motor cortex. Verapamil (1 mmol/l artificial cerebrospinal fluid) was applied into a lateral cerebral ventricle (push-pull perfusion; duration 30 min). Recordings 27 min (**A**) and 21 min (**B**) after start of perfusion. *CTRL1, CTRL2*: Recordings before start [3 min (**A**), 10 min (**B**)] and after termination [50 min (**A**), 25 min (**B**)] of verapamil perfusion. (Experiments in **A** in collaboration with O. W. Witte)

seizure activity (Walden et al. 1985; Walden and Speckmann 1988), and (b) that, in experiments without epileptic activity, verapamil perfusion exerted no depressive effect on the power of the spontaneous EEG (Fig. 2 A) or on somatosensory evoked potentials (Fig. 2 B) and failed to shift the DC potential to the positive side (Fig. 2 C; Walden et al. 1985; Walden and Speckmann 1988).

In animals treated and not treated with an epileptogenic drug an intracerebroventricular perfusion of verapamil did not change the action of the heart in frequency and rhythm.

As a whole, systemic administration of the calcium antagonist verapamil suppressed focal and generalized tonic-clonic seizure activity and did not depress nonepileptic cerebral activity. Along with other results already published, this justifies the hope that calcium antagonists might be useful in the treatment of human epilepsies (cf. Speckmann et al. 1986 a; Overweg et al. 1984).

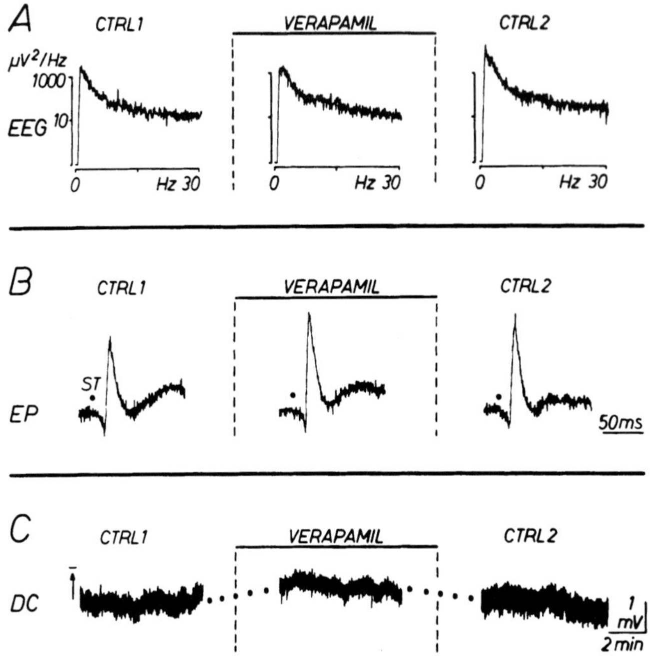

Fig. 2. Power spectra of spontaneous waves of the EEG (**A**), somatosensory evoked potentials (*EP;* **B**), and DC potential (**C**) at the cortical surface before, during, and after intracerebroventricular perfusion of the calcium antagonist verapamil. Anesthetized and artificially ventilated rat, motor (**A, C**) and somatosensory (**B**) cortex. Animals not treated with epileptogenic agents. Verapamil (1 mmol/l artificial cerebrospinal fluid) was applied into a lateral cerebral ventricle [push-pull perfusion; duration: 90 min (**A, B**), 30 min (**C**)]. Recordings 85 min (**A**), 60 min (**B**), and 22 min (**C**) after start of perfusion. *CTRL1, CTRL2*: Recordings before start [11 min (**A**), 55 min (**B**), 5 min (**C**)] and after termination [48 min (**A**), 50 min (**B**), 55 min (**C**)] of verapamil perfusion. **A** Power spectral analysis of the spontaneous EEG. Epochs of analysis: 2 min; frequency range: 1–30 Hz. **B** EP elicited by stimulation (*ST*) of the sciatic nerve. Average potentials. $n = 100$. **C** Inkwriter tracing of the DC potential. (Experiments in **B** in collaboration with O. W. Witte)

References

Bingmann D, Speckmann E-J (1986a) Actions of pentylenetetrazol (PTZ) on CA3 neurons in hippocampal slices of guinea pigs. Exp Brain Res 64:94–104

Bingmann D, Speckmann E-J (1986b) Depression of pentylenetetrazol induced epileptiform discharges in CA3 neurons of hippocampal slices by flunarizine and verapamil. In: Speckmann E-J, Schulze H, Walden J (eds) Epilepsy and Calcium. Urban & Schwarzenberg, Munich, pp 301–317

Bingmann D, Speckmann E-J, Baker RE, Ruijter J, de Jong BM (1988) Differential antiepileptic effects of the organic calcium antagonists verapamil and flunarizine in neurons of organotypic neocortical explants from newborn rats. Exp Brain Res 72:439–442

Heinemann U, Klee M, Neher E, Singer W (eds) (1986) Calcium electrogenesis and neuronal functioning. Exp Brain Res 14

Overweg J, Binnie CD, Meijer JWA, Meinardi H, Ningten STM, Schmalz S, Wauquier A (1984) Double-blind placebo-controlled trial of flunarizine as add-on therapy in epilepsy. Epilepsia 25:217–222

Speckmann E-J, Schulze H, Walden J (eds) (1986a) Epilepsy and Calcium. Urban & Schwarzenberg, Munich

Speckmann E-J, Witte OW, Walden J (1986b) Involvement of calcium ions in focal epileptic activity of the neocortex. In: Heinemann U, Klee M, Neher E, Singer W (eds) Calcium electrogenesis and neuronal functioning. Exp Brain Res 14:386–395

Walden J, Speckmann E-J (1988) Suppression of recurrent generalized tonic-clonic seizure discharges by intraventricular perfusion of a calcium antagonist. Electroencephalogr Clin Neurophysiol 69:353–362

Walden J, Speckmann E-J, Witte OW (1985) Suppression of focal epileptiform discharges by intraventricular perfusion of a calcium antagonist. Electroencephalogr Clin Neurophysiol 61:299–309

Walden J, Speckmann E-J, Witte OW (1986) Depression of focal interictal epileptiform discharges by intracerebroventricular perfusion of the calcium channel blocker verapamil. In: Speckmann E-J, Schulze H, Walden J (eds) Epilepsy and Calcium. Urban & Schwarzenberg, Munich, pp 335–355

Walden J, Speckmann E-J, Witte OW (1988) Membrane currents induced by pentylenetetrazol in identified neurons of *Helix pomatia*. Brain Res 473:294–305

Witte OW, Speckmann E-J, Walden J (1987) Motorcortical epileptic foci in vivo: actions of a calcium channel blocker on paroxysmal neuronal depolarizations. Electroencephalogr Clin Neurophysiol 66:43–55

Flunarizine, a Useful Adjuvant for Therapy-Resistant Epilepsy Patients

J. Overweg [1] and C. D. Binnie [1, 2]

Introduction

Remarkably, the natural history of untreated epileptic seizures is almost completely unknown. Drug treatment has been available since 1857 with the introduction of bromide therapy. Phenobarbitone, worldwide still the most commonly prescribed antiepileptic drug (AED), was introduced into clinical practice in 1912, and phenytoin (PHT) was introduced in 1938, both before well designed prognostic investigations were possible. There are therefore no major studies of the course of untreated epilepsy and no such studies are now likely to be carried out. Studies of the epidemiology of epilepsy can provide the usual insight into the course and prognosis of this seizure disorder. The traditional view is that, in the majority of cases, epilepsy is a chronic condition with a continuing tendency to seizure relapse and in which treatment is usually lifelong. This is incorrect and indeed, the majority of patients in whom epilepsy is diagnosed do not develop chronic seizures. The incidence of epilepsy is usually found to be in the order of 50 new cases per 100000 persons per year in a western population [13]. Including febrile convulsions, the cumulative incidence of all types of epileptic seizure is 2.5% [16]. Excluding febrile convulsions this will fall to 2%, whereas for patients with the diagnosis of epilepsy (more than two unprovoked epileptic seizures within a short period making antiepileptic treatment necessary), the cumulative incidence rate is 1.3%. If epileptic seizures never recovered, prevalence rates should approximate to these cumulative incidence rates, but in fact the prevalence rates of epilepsy are much lower, in the order of one person per 150.

From these epidemiological figures it is clear that in most cases the seizure disorder is not permanent. Approximately 75% of the patients with newly diagnosed epilepsy can be well controlled with currently available AEDs. For the remaining 25% the existing antiepileptic armamentarium is inadequate. The first 2 years of treatment are crucial for the longer term prognosis of epilepsy. Patients who continue to have seizures during the first 2 years of treatment generally develop chronic epilepsy. About 3% of those patients will require institutional care, not only for their current seizures but also because of associated mental and neurological handicaps.

---

[1] Instituut voor Epilepsiebestrijding Meer en Bosch/De Cruquiushoeve, Achterweg 5, NL-2103 SW Heemstede, The Netherlands.
[2] Present address: The Maudsley Hospital, Denmark Hill, London SE5 8AZ, U.K.

Cerebral Ischemia and Calcium
Ed. by A. Hartmann and W. Kuschinsky

The response to treatment is influenced by a number of factors. Seizure type and organic brain damage are the most important ones. Typical absences (petit mal) and idiopathic tonic-clonic (grand mal) seizures have the best treatment prognosis. Partial seizures, with and without generalization, are more difficult to control. Seizures associated with organic brain damage and/or mental retardation to some extent are the most resistant to antiepileptic therapy. For obvious ethical reasons efficacy evaluations of potential AEDs will be add-on trials in therapy-resistant patients. By definition these patients do not respond to the most effective drugs available. Such trials are therefore biased against showing efficacy.

Numerous drugs are currently under investigation for the potential management of epilepsy (the screening project of the NIH has to date screened and evaluated approximately 10000 compounds for anticonvulsant activity). In view of the known role of calcium in the paroxysmal depolarization shifts of epileptic neurons, the possibility arises that certain calcium entry blockers (calcium antagonists) possess antiepileptic activity. Calcium entry blockers may be divided into (a) dihydropyridines (nifedipine, nitrendipine, nimodipine), (b) diphenylalkylamines (flunarizine, cinnarizine), (c) verapamil-like compounds (verapamil, gallopamil, or D-600), (d) diltiazem, and (e) bepridil. The diphenylalkylamines are known to enter the brain and in anticonvulsant tests in animals, flunarizine (FNZ) inhibits tonic seizures like PHT and carbamazepine (CBZ).

Flunarizine was introduced into clinical practice in several countries for its beneficial effect in peripheral and cerebral circulatory disorders and more specifically vertigo. It was also successfully used as a prophylactic against migraine. It is suggested that FNZ therapeutic effects are due to its ability to reduce transmembranous calcium flux into tissue cell [15]. A growing number of observations indicate that transmembranous calcium flux contributes to the generation of epileptic activity [24, 25].

One of the most characteristic electrophysiological phenomena in experimental epilepsy is the paroxysmal depolarization shift underlying bursting. As this depends largely on neuronal calcium influx, it is conceivable that calcium antagonists might have antiepileptic properties [5, 25]. Due to FNZ's ability to cross the blood-brain barrier and its ability to suppress calcium influx, FNZ may prevent neuronal hyperexcitability and exert its apparent antiepileptic activities.

Anticonvulsant Effects in Animals

Flunarizine's antiepileptic activity was first described against maximal seizures induced by metrazole in rats and maximal electroshock seizures in mice [11]. FNZ was active only against tonic seizures. This profile of activity was also shown by PHT and CBZ. In the bicuculline seizure threshold test, a striking similarity between FNZ, PHT, and CBZ was also found [1].

Flunarizine has been shown to antagonize clonic seizures in amygdaloid kindled rats as PHT and CBZ do. In DBA/2 mice, genetically prone to audiogenic seizures, FNZ antagonized tonic seizures and mortality but had no effect on agitation and clonic seizures [27]. In photosensitive baboons FNZ (2 mg/kg i.v.)

completely suppressed myoclonic responses to stroboscopic stimulation for 3 h [10]. In the in vitro hippocampal slice FNZ at antiepileptic brain concentrations did not increase GABA-ergic inhibition or pulse-paired facilitation [2]. However, FNZ did suppress spontaneous epileptic bursts induced by incubation of hippocampal slices in low Ca^{2+} and high Mg^{2+} medium [3]. PHT and CBZ were also active in this model at low doses; however, other calcium entry blockers were inactive. These results suggest that FNZ reduces the spread of epileptic seizures by reducing the synchronicity of burst discharges.

Pharmacokinetics

Steady-state for daily 10-mg doses of FNZ is reached within 5–6 weeks at plasma levels fluctuating between 40 and 100 mg/ml. Serum levels of FNZ in patients with epilepsy, treated with comedication, are considerably lower than those reported after similar doses in nonepileptic patients. This may reflect an increased metabolism due to the induction of liver enzymes by comedication. For a daily dose of 10 mg FNZ, serum concentrations ranged from 3 to 32.5 mg/ml [20]. After discontinuation of treatment, the plasma levels decay biphasically, with a terminal half-life of approximately 19 days, similar to that found after single oral dosing [14].

Clinical Data

First clinical evidence for the antiepileptic activity of FNZ was given by Declerck and Wauquier [9]. Encouraged by these findings, Overweg and colleagues [20] performed a double-blind placebo-controlled crossover trial. FNZ was given as "add-on" therapy in 30 therapy-resistant adult patients in order to evaluate its effect on seizure incidence and on plasma levels of FNZ and associated antiepileptic drugs. The results of this study have been published in detail elsewhere [20]. The influence of FNZ on seizure incidence is shown in Table 1. There was no systematic difference in seizure incidence between phase I and phase II of the trial and thus no serial effect could be demonstrated. There was, however, a lower seizure incidence during the FNZ period as compared with placebo in 20 patients, no difference in one, and a higher incidence in nine. This result is significantly in favor of FNZ (Wilcoxon matched-pairs signed-rank test, $P = 0.005$; one-tailed). Eleven patients improved by 25% or more with FNZ and two with placebo. During the FNZ phase, seven patients had a seizure reduction of over 50%; none showed a 50% improvement on placebo. All patients had epileptiform discharges in their EEGs. Paroxysmal activity was not significantly different in the three phases of the study.

Prolonged administration of FNZ did not increase the plasma levels of the other antiepileptic drugs. The clinical efficacy demonstrated in this study was therefore considered not to be due to a metabolic interference with the comedi-

Table 1. Lower seizure incidence in patients[a]

	Phase I	Phase II	Placebo	Flunarizine
All subjects	13	16	9	20
With $\geq 25\%$ difference	5	8	2	11
With $\geq 50\%$ difference	4	3	0	7

By the Wilcoxon matched-pairs signed-ranks test, the significance of the difference between flunarizine and placebo was $z = 2.573$, $P \simeq 0.005$ (one-tailed).
[a] No difference in 1 of the 30 patients.

Table 2. Seizure frequency relative to baseline

	Decreased	Unchanged	Increased	P^a
All subjects	33	5	9	<0.001
25% change	24		7	<0.01
50% change	16		6	<0.05

[a] Sign test, one-tailed.

cation. Plasma levels of FNZ were lower than expected after a daily dose of 10 mg taken for 3 months. Furthermore, there seemed to be a correlation between FNZ plasma levels and clinical response since highest plasma levels were found in those patients who gave the best response. Because of this correlation, it was decided to do an open dose-ranging trial to investigate whether higher FNZ dosages would result in higher plasma levels and in a further decrease in seizure incidence.

Forty-seven adult patients with therapy-resistant epilepsy who had at least three seizures per month participated in the study [6]. After a baseline period of 3 months, FNZ was added to the existing antiepileptic medication in an increasing dosage over successive 3-month periods up to a maximum of 25 mg/day or up to the limit of tolerance or therapeutic effect. Comedication remained unchanged throughout. As shown in Table 2, 16 patients showed a 50% or more reduction of seizure incidence and three became seizure-free; six, however, had a 50% increase. Twenty-four patients showed a more than 25% increase. Overall, 33 patients had less seizures on FNZ, nine had more, and five were unchanged. With four exceptions, patients who responded showed the lowest seizure incidence at the highest FNZ dose. Further differentiation among the patients showed that all subjects with the diagnosis of secondary generalized epilepsy (symptomatic epilepsy) responded, as did all patients with neurological deficits and those with some degree of mental subnormality.

Interactions

Flunarizine had no apparent effect on blood levels of the comedication. FNZ blood levels were lower in patients receiving enzyme-inducing antiepileptic drugs. Consequently, neither the apparent efficacy nor the side-effects of FNZ therapy could be attributed to changes in the concentration of the other drugs. However, other calcium entry blockers screened for antiepileptic efficacy have shown such interactions. Verapamil increases the area under the CBZ concentration/time curve and within a few days symptoms of CBZ neurotoxicity will develop [18].

Side-Effects

The add-on therapy with FNZ was well tolerated on the whole. The most frequently observed side-effect was drowsiness. Of the patients, 16% complained about this side-effect [23]. The second most frequently observed side-effect was weight gain in 4% of the cases. These adverse reactions were mild and dose dependent. Recently the occurrence of extrapyramidal symptoms and depression during chronic FNZ treatment has been reported in several publications [4, 7, 8, 19]. These adverse reactions did not occur during the trials with FNZ in epilepsy [23].

The EEG

The EEG is included as an outcome variable in most antiepileptic drug trial protocols. The problem of evaluating changes in EEG recordings may be due to the limited knowledge about spontaneous fluctuations in EEGs, not related to treatments [12]. It is widely assumed that the amount of epileptiform activity in the interictal EEG reflects the severity of epilepsy, and should be predictive for the seizure frequency, for the results of drug treatment [26], and in cases of seizure freedom for the results of drug withdrawal. These assumptions generally appear to break down [22].

Analysis of findings in FNZ trials shows that changes in background activity were difficult to assess on visual inspection of the recordings, but could be demonstrated with power spectral analysis [26]. As EEG changes on hyperventilation are attributed to cerebral vasoconstriction and FNZ protects against both vascular spasm and cerebral anoxia [28], the failure of this drug to reduce the EEG responses to overbreathing was unexpected. However, we have found that FNZ does not affect either the normal physiological EEG slowing on overbreathing in healthy volunteers or the reduction of cerebral blood flow induced in anesthetized dogs by hypocapnia. Apparently FNZ prevents cerebrovascular spasms without effecting autoregulation. FNZ did not produce any consistent changes in epileptiform discharges [26].

Incidental Beneficial Effects Due to FNZ

Two findings set the investigators speculating on the apparent analogies between migraine and epilepsy. The initial events in both conditions are termed an "aura." Perhaps the most common epileptic aura is the rising epigastric sensation seen in seizures of temporal lobe origin, presumed to represent viscerosensory hallucinosis. Epileptic auras do not in general resemble those of migraine, with the single striking exception of the scintillating scotomata in the occipital form of benign childhood epilepsy. Of the 77 patients in our trials of FNZ in epilepsy, 28 reported a seizure pattern commencing with an aura. Twenty-two of these patients found that the aura was abolished on FNZ irrespective of whether the seizures were reduced in frequency [21]. The aura in partial epilepsy corresponds to a localized epileptiform discharge recordable with depth electrodes at the site of the focus and often undetectable in the scalp EEG. This initial discharge may remain localized or may spread, giving rise to more florid symptoms, together with more widespread discharges recordable at both depth and surface electrodes. Antiepileptic drugs often alter the seizure pattern by preventing spread from the primary focus, but abolition of the aura appears to be unique to FNZ. Possibly initial focal discharges are prevented from producing subjective symptoms. Alternatively the ictal discharge, once initiated, could spread so rapidly that more obvious symptoms develop before the patient has experienced the aura. A third possibility is that the interpretation of the aura as a sensory ictal event is incorrect; conceivably the rising epigastric sensation is due to a visceromotor effect. Given the known action of FNZ on smooth muscle, a peripheral action might block this feedback.

Fourteen patients had postictal headaches [21]. This symptom is probably unrelated to the rare ictal epileptic manifestation of headache [29], and is seen following most tonic-clonic convulsions and often after partial seizures. Of the 14 patients, 13 reported that their postictal headache was relieved during FNZ treatment. Given the efficacy of FNZ as an antimigrainous agent, it is tempting to postulate a common mechanism for migrainous and postictal headache, probably related to spreading depression [17].

Conclusion

The benefits of an adjuvant therapy with FNZ are clear-cut. It produces a clinically significant and longlasting improvement in seizure incidence in one-third of the patients otherwise resistant to the existing antiepileptic drugs. No interactions with the existing antiepileptic medications are to be expected. Besides of its effect on seizure incidence, FNZ produces relief of postictal headache in quite a number of patients.

The risks seem to be quite acceptable. Side-effects are mild and dose dependent, and long-term safety seems to be warranted.

References

1. Ashton D (1983) Diazepam, pentobarbital and D-etomidate produced increases in bicuculline seizure threshold; selective antagonism by RO-15-1788, picrotoxin and DMBB. Eur J Pharmacol 94:319–325
2. Ashton D, Marrannes R, Pauwels PJ, van Belle H, Reid K, Wauquier A (1986) Possible mechanisms of flunarizine's anticonvulsant activity. In: Van den Bussche G, Wauquier A (eds) Proc. Int. Workshop on Flunarizine in Epilepsy, Beerse 1985. Health Sci Rev, p 96
3. Ashton D, Wauquier A (1986) Antagonism of hyperexcitability phenomena by Ca^{2+}-entry blockers in vivo and in in vitro experiments. In: Speckmann ED, Walden J (eds) Calcium and epilepsy. Urban & Schwarzenberg, Munich, pp 319–334
4. Benvenutti F, Baroni A, Bandinelli S, Sommazzi P, Corradetti R, Pantaleo T (1986) Letter to The Lancet. (Perhaps it is reduction of neurotransmission by Ca^{2+} channel blockade that produces extrapyramidal side-effects.) Lancet 3:463–464
5. Bingmann D, Speckmann EJ (1986) Depression of pentylenetetrazol-induced epileptiform discharges in CA3 neurons of hippocampal slices by flunarizine and verapamil. In: Speckmann EJ, Schulze H, Walden J (eds) Epilepsy and calcium. Urban & Schwarzenberg, Munich, pp 301–318
6. Binnie CD, de Beukelaar F, Meyer JWA, Meinardi H, Overweg J, Wauquier A, van Wieringen A (1985) Open dose-ranging trial of flunarizine as add-on therapy in epilepsy. Epilepsia 26:424–428
7. Chouza Z, Scaramelli A, Caamaño JL, de Medina O, Aljanati R, Romero S (1986) Parkinsonism, tardive dyskinesia, akathisia, and depression induced by flunarizine. Lancet 3:1303
8. D'Allessandro R, Benassi G, Morganti G (1986) Side-effects of flunarizine. Lancet 3:463–464
9. Declerck AC, Wauquier A (1978) Double-blind study on the effectiveness on flunarizine in therapy-resistant epilepsy in mentally retarded children. In: Wada JA, Penry JK (eds) Advances in epileptology: The Xth Epilepsy International Symposium. Raven, New York, p 356
10. De Sarro GB, Nisticó G, Melrum BS (1986) Anticonvulsant properties of flunarizine on reflex and generalised models of epilepsy. Neuropharmacol (submitted)
11. Desmedt LKC, Niemegeers CJE, Janssen PAJ (1975) Anticonvulsive properties of cinnarizine and flunarizine in rats and mice. Arzneimittelforschung 25:1408–1413
12. Gram L, Drachman Bentsen K, Parnas J, Flachs H (1982) Review of controlled trials of antiepileptic drugs. In: Betts TA, Hoeppener RJ, Pedersen B (eds) Epilepsy: British-Danish-Dutch Meeting. Br J Clin Pract (Suppl):133–135
13. Hauser WA, Kurland LT (1975) The epidemiology of epilepsy in Rochester, Minnesota, 1935 through 1967. Epilepsia 16:1–66
14. Heykants J, de Cree J, Hörig Ch (1979) Steady-state plasma levels of flunarizine in chronically treated patients. Arzneimittelforschung 29:1168–1171
15. Holmes B, Brogden RN, Heel RC, Speight TM, Avery GS (1984) Flunarizine. A review of its pharmacodynamic and pharmacokinetic properties and therapeutic use. Drugs 27:6–44
16. Juul-Jensen P, Foldsprang A (1983) Natural history of epileptic seizures. Epilepsia 24:297–312
17. Leao AAP (1944) Spreading depression of activity in the cerebral cortex. J Neurophysiol 7:359–390
18. Macphee GJA, McInnes GT, Thompson GG, Brodie MJ (1985) Verapamil potentiates carbamazepine neurotoxicity: a clinically important inhibitory interaction. Lancet 3:700
19. Meyboom RHB, Ferrari MD, Dieleman BP (1986) Parkinsonism, tardive dyskinesia, akathisa, and depression induced by flunarizine. Lancet 2:292
20. Overweg J, Binnie CD, Meijer JWA, Meinardi H, Nuijten STM, Schmaltz S, Wauquier A (1984) Double-blind placebo-controlled trial of flunarizine as add-on therapy in epilepsy. Epilepsia 25:217–222
21. Overweg J, Ashton D, de Beukelaar F, Binnie CD, Wauquier A (1986) Add-on therapy in epilepsy with calcium entry blockers. Eur Neurol 25 (Suppl 1):93–101
22. Overweg J, Binnie CD, Oosting J, Rowan AJ (1987) Clinical and EEG prediction of seizure recurrence following antiepileptic drug withdrawal. Epilepsy Res 1:272–283

23. Overweg J (1988) Flunarizine as an antiepileptic drug. Advances in Epileptology. Proceedings of the 17th Epilepsy International Congress, Jerusalem
24. Prince DA, Deisz RA (1986) Calcium-meditated events during epileptogenesis in cortical neurons. In: Speckmann EJ, Schulze H, Walden J (eds) Epilepsy and calcium. Urban & Schwarzenberg, Munich, pp 165–183
25. Speckmann EJ, Walden J, Witte OW (1986) Epileptic neuronal activity in the neocortex of the rat: involvement of calcium ions. In: Van den Bussche G, Wauquier A (eds) Proc. Int. Workshop on Flunarizine in Epilepsy, Beerse 1985. Health Sci Rev, pp 24–29
26. Van Wieringen A, Binnie CD, de Boer PTE, van Emde Boas W, Overweg J, de Vries J (1987) Electroencephalographic findings in antiepileptic drug trials: a review and report of 6 studies. Epilepsy Res 1:3–15
27. Wauquier A, Ashton D, Clincke G, Fransen J, Gillardin JM, Janssen PAJ (1986) Anticonvulsant profile of flunarizine. Drug Dev Res 7:49–60
28. Wauquier A, Clincke G, Van den Broeck WAE, Hermans C, Melis W, Van Loon J (1984) Global incomplete ischaemia in dogs assessed by quantitative EEG analysis. In: Pfurtscheller G, Jonkman EJ, Lopes da Silva FH (eds) Brain ischemia: quantitative EEG and imaging techniques. Prog Brain Res 62:217–235
29. Wieser HG, Isler H (1983) Headache as an epileptic manifestation. Schweiz Rundschau Med 24:844–848

Review of Flunarizine in Epilepsy

C. D. Binnie [1], F. de Beukelaar [2], and J. Overweg [3]

Introduction

A feature of epileptogenesis is burst firing of individual neurons. The burst comprises a slow paroxysmal depolarizing ion shift (PDS) upon which is superimposed a train of action potentials [6, 36]. Bursting depends largely on influx of Ca^{2+}, although other ionic currents are also involved [37, 56].

In experimental foci the extracellular activity of calcium ions decreases during epileptic discharges [11, 24]. Both intracellular injection of the calcium antagonist D890 and intraventricular application of verapamil reduce depolarization shifts and epileptiform discharges [50, 54]. These drugs suppress neither excitatory postsynaptic potentials [55] nor somatosensory evoked potentials [49]. In pentylenetetrazole-treated hippocampal slices, flunarizine suppresses spontaneous rhythmic burst activity without changing postsynaptic potentials, excitability, or action potentials of CA3 neurons [9]. By contrast, Bay K8644, which stimulates calcium entry, increased liability to tonic-clonic convulsions in DBA/2 mice [20].

These findings suggest that calcium currents may play an important role in epileptogenesis and that calcium entry blockers deserve consideration as potential antiepileptic drugs.

Anticonvulsant Action of Flunarizine in Animal Models

Flunarizine suppresses selectively tonic seizures in several rodent models – maximal metrazole and electroshock [21], D,L-allylglycine [2], and intravenous bicuculline [51] –, a profile of activity shared by carbamazepine and phenytoin. A similarity with flunarizine is also shown in the bicuculline seizure threshold test [1]. These drugs antagonize tonic forepaw seizures, not by producing a progressive threshold increase but rather by total suppression so that increased doses of bicuculline fail to produce tonic seizures. Like phenytoin and carbamazepine, flunarizine antagonizes clonic seizures in amygdaloid kindled rats and dogs [3, 51]. Flunarizine is also effective in genetic models of epilepsy, suppressing audiogenic tonic seizures in DBA/2 mice [52] and photomyoclonus in photosensitive baboons [38].

[1] The Maudsley Hospital, Denmark Hill, London, SE5 8AZ, U.K.
[2] Janssen Pharmaceutica B.V., B-Beerse, Belgium.
[3] Instituut voor Epilepsiebestrijding, Achterweg 5, NL-2103 SW Heemstede, The Netherlands.

Cerebral Ischemia and Calcium
Ed. by A. Hartmann and W. Kuschinsky

GABA-ergic compounds exhibit a dose-related elevation of seizure threshold [29, 30, 32, 40], unlike the antagonism of tonic seizures by flunarizine, carbamazepine, and phenytoin, which shows an all-or-none-effect. On stimulation of hippocampal slices with paired pulses at separations of 12–30 ms, the population spike response to the second pulse normally shows facilitation which is converted into inhibition by GABA-ergic agents [4] but not by flunarizine [5]. The anticonvulsant profile of flunarizine resembles those of phenytoin and carbamazepine, suggesting a possibly similar mechanism of action, on voltage-sensitive sodium channels [53]. Ashton et al. [5] showed that flunarizine and carbamazepine both inhibit the binding of batrachotoxinin A 20-α-benzoate, which implicates binding site II of the sodium channel [8]. Although flunarizine is a potent calcium entry blocker, it may owe its anticonvulsant action to an effect on sodium channels.

The ED_{50} of flunarizine in those models where it is effective is typically in the range 6–40 mg/kg [47]. The duration of action ranges from 3 to 24 h, being greatest in the amygdaloid kindling model.

Studies in Humans

Pharmacokinetics in Patients with Epilepsy

When flunarizine was added to existing antiepileptic medication [10, 22, 34, 45] it reached steady-state concentrations 4–5 times lower than in patients receiving the same doses as monotherapy [25]. Mean flunarizine plasma concentrations at 10 and 20 mg daily were 18.1 and 59.3 ng/ml respectively [10]. These findings indicate that antiepileptic drugs such as carbamazepine and phenytoin induce flunarizine liver metabolism.

With increasing oral dosages, from 5 to 105 mg daily, Treiman et al. [45] found almost linearly increasing steady-state plasma concentrations within patients. Thus flunarizine administered as add-on therapy exhibits linear kinetics over the therapeutic range.

Drug Interactions

There is no evidence that flunarizine causes either enzyme induction or inhibition or has any other pharmacokinetic influence on other antiepileptic drugs [10, 15, 22, 28, 34]. Although strongly bound to plasma proteins (99.1%), flunarizine does not affect binding of comedication [22].

Efficacy

Twelve open trials have been reported (Table 1), all of add-on flunarizine therapy in resistant patients. In 11 of the trials 74 of 197 patients (38%) showed a 50% seizure reduction and 11 (6%) a similar increase on flunarizine. Eight patients became seizure free. In the 12th trial 22 of 30 patients achieved 50% seizure reduc-

Table 1. Open trials

	Patients	No. evaluated	Age		Seizure types					Flunarizine dosage (mg/day)	Dosing period (mths)	50% change in seizure frequency		Withdrawals	
			Mean	Range	Partial	Generalized			Other or not specified			Decrease	Increase	Number	Reason
						Abs	T–C	Other							
Binnie et al. (1985)	A	47	35	17–64	47		6		4	10–25	6–12	16	6	5	2 wanted pregnancy 2 noncompliant 1 lost aura
Debandt et al. (1985)	A	8	33	25–38	3	5	6			10–25	4	2	2	2	1 too few Sz in baseline 1 intercurrent illness
Sorel (1987)	A	20	31	20–51	18	1	2	1		10–20	4	10	2	–	–
De Domenico et al. (1987)	A	11	28	18–71	6		3	1	1	>70 kg : 15 ≤70 kg : 10	3	1	0	–	–
Curatolo et al. (1986)	C + Ado	21	5	0.5–17	12				9	>10 kg : 5 ≤10 kg : 2.5	3	8	1	5	5 drowsiness
Giroud et al. (1987)	C	10	12	18–15	6		4			5–10	3	3	0	–	–
Pothmann et al. (1986)	C + Ado + A	21	12	1–20					21	5–20	< 3	10	0	3	1 noncompliant 2 moved away
Declerck et al. (1986)	C + A	14	?	5–36					14	10		1	0	–	–
Tata et al. (1987)	A	30		18–50	13		2	15		10–20	< 6	22	?	–	–
Takahashi (1988)	A	13	36	17–53	12		1			10	7	4	0	–	–

Umezawa (1987)	A+C	25	?	10–54					25	10	<22	15	0	1	1 depressed
Konishi et al. (1987)	C+ Ado	7	?	6–16					7	4–10	1	4	0	–	–

A, adult 17 years of age or more; C, child not more than 13 years; Ado, adolescent 13–17 years old; Abs, absence; T–C, tonic–clonic; Sz, seizures.

Table 2. Controlled trials

	Patients	No. evaluated	Age		Seizure types					Flunarizine dosage (mg/day)	Dosing period (mths)	50% change in seizure frequency		Withdrawals	
			Mean	Range	Partial	Generalized			Other or not specified					Number	Reason
						Abs	T–C	Other				Decrease	Increase		
Overweg et al. (1984)	Ado +A	30	33	15–56	29		8	3		10	3	7	0	3	1 drowsy 1 intercurrent illness 1 too few Sz in baseline
Touquet and Van Cappel (1985)	Ado +A	5	21	15–27	4		1			10 after loading	2	2	0	–	–
Fröscher et al. (1988)	Ado +A	22	32	14–51	22		2			15 after loading	4	2	3	8	All noncompliance or protocol violation
Mancia (1987)	Ado +A	29	34	16–57	22		5	1	1	>70 kg 15 ≤70 kg 10	3	20% mean Sz reduction on flunarizine		–	–

Table 2 continues on p. 474

Table 2 (continued)

	Patients	No. evaluated	Age		Seizure types					Flunarizine dosage (mg/day)	Dosing period (mths)	50% change in seizure frequency		Withdrawals	
			Mean	Range	Partial	Generalized			Other or not specified			Decrease	Increase	Number	Reason
						Abs	T–C	Other							
Moglia et al. (1986)	Ado +A	20	36	15–73	17				3	>70 kg 15 ≤70 kg 10	3	33% mean Sz reduction on flunarizine		–	–
Bergamasco et al. (1987)	A	27	33	17–64	21		1		5	>70 kg 15 ≤70 kg 10	3	34% mean Sz reduction on flunarizine		3	?
Cavazzuti et al. (1986)	Ado +C	14	13	3–17	0		2	2	10	>12 yr 10 ≤12 yr 5	3	5	0	–	–
Curatolo et al. (1986)	C	16	5	0.5–13	6				10	>10 kg 5 ≤10 kg 2.5	3	1	0	–	–
Starreveld et al.	A	25	32	17–47	25	1	5	5		15	4	7	2	9	4 hospitalized 1 died (before flunarizine) 3 noncompliant 1 too few Sz
Keene et al. (1987)	C+ Ado +A	29	11	0.2–24					29	≤20 kg 5 20–40 kg 10 >40 kg 15	4	6	1	5	3 did not start phase I 1 side-effects (on placebo) 1 progressive degenerative disorder
Declerck and Wauquier (1978/1983)	Ado +C	PS F7	12	9–17	3				9	10	3	P0/5 F3/7		8	8 protocol violations

A, adult 17 years of age or more; C, child not more than 13 years; Ado, adolescent 13–17 years old, Abs, absence; T–C, tonic–clonic; Sz, seizures.

Table 3. Adverse experiences during flunarizine treatment and probably related to flunarizine

Authors	No.[a]	Drowsiness	Weight gain	Others
Binnie et al.	47	15	8	2 memory impairment 1 aggression
Debandt et al.	8		2	
Sorel	20	2	3	1 memory impairment
De Domenico	11	–	–	
Curatolo et al.	26	5		
Giroud et al.	10	3		
Pothmann et al.	21	5	1	
Declerck et al.	14	–	–	
Tata et al.	30	–	–	
Overweg et al.	31	1	–	
Touquet and Van Cappel	5	–	–	
Fröscher et al.	22	1	–	1 more alert
Mancia	29	–	–	6 asthenia
Moglia et al.	20	9	–	1 dizziness
Bergamasco et al.	27	6	–	
Cavazzuti et al.	14	1	–	2 torpor and apathy
Curatolo et al.	16	–	–	
Starreveld et al.	25	–	–	
Keene et al.	29	–	–	
Declerck and Wauquier	12	–	–	
Takahashi	13	4	7	
Umezawa et al.	26	–	–	1 depression
Konishi et al.	7	3	3	

[a] Patients completing trial or withdrawn because of adverse experiences.

tion but how many showed increases was not stated. Three trials included both adults and children and three were confined to patients under 17 years of age ($n = 38$). The efficacy in children was similar to that in adults. The optimal adult add-on dose for efficacy and avoidance of adverse reactions was generally 15 mg daily.

Eleven controlled trials are reported (Table 2), two single- and eight double-blind trials with a two-period crossover design and one parallel trial. Three used mean seizure incidence as the outcome variable and found that on flunarizine to be significantly lower than that on placebo (differences in the range 20%–34%). In the other seven crossover studies, 30 patients (21%) were 50% improved on flunarizine vs. six (9%) having 50% less seizures on placebo. This effect is highly significant across trials but achieved statistical significance in only two individual studies. In the parallel trial three of seven patients showed a 50% reduction of seizures with respect to baseline on flunarizine and none of five did so on placebo. Two studies concerned children and adolescents ($n = 30$): six patients (20%) showed a 50% reduction of seizures on flunarizine and none showed a similar improvement on placebo. Doses employed were 10–15 mg daily in adults, 2.5–5 mg in children up to puberty, and 10 mg in adolescents.

Most patients studied had partial seizures; flunarizine showed efficacy in both partial and generalized seizures and in both partial and secondary generalized epilepsy. A more detailed analysis of the effects of seizure type and epilepsy syndrome is not possible due to inconsistencies in the classifications employed.

Tolerance did not appear in long-term therapy from 6 to over 22 months [23, 46, 48].

Human Safety Data

Flunarizine has been taken for up to 14 years for indications other than epilepsy. No drug-related changes in hematological, biochemical, renal, or hepatic variables have been found. The add-on trials in epilepsy have demonstrated no flunarizine-related changes in laboratory values over 1 year [33]. Elevation of liver enzymes due to induction by the other medication was not increased by addition of flunarizine.

Adverse Experiences in Clinical Trials

Drowsiness was reported in 26% of patients in open trials of flunarizine in epilepsy, but in only 8% under placebo-controlled conditions. This was dependent on blood levels and dosage, occurring chiefly above 50 ng/ml or 20 mg/day [10]. In 16 patients who showed a 50% seizure reduction, this could be maintained without concomitant drowsiness in all but two after dose adjustment [10]. Ten of the 12 patients in a tolerability study by Treiman et al. [44, 45] developed symptoms of sedation but eight did so only at plasma levels above 100 ng/ml. The second most common adverse experience was weight gain (5.2%). Occasional extrapyramidal symptoms and depression have occurred during flunarizine treatment for conditions other than epilepsy. In the epilepsy trials only one case of depression and no instances of extrapyramidal symptoms have been reported.

Conclusions

Flunarizine as add-on therapy produces a clinically significant reduction in seizure frequency in a third of therapy-resistant patients with epilepsy. It is effective in both adults and children and no particular type of epilepsy or seizure has yet been shown to be especially sensitive or resistant. Safety is good; side-effects are generally mild, dose related, and avoidable at therapeutic blood levels. In adults the optimal blood level for seizure control without side-effects is about 50 ng/ml. With enzyme-inducing comedication this is generally achieved at a daily dose of 15–20 mg. Without enzyme-inducing comedication 10 mg daily will probably suffice. In children weighing less than 20 kg, 5 mg daily is a suitable initial dose.

The eventual place of flunarizine in the treatment of epilepsy can be assessed only when monotherapy trials are available, including studies of patients who are not therapy resistant.

References

1. Ashton D (1983) Bicuculline seizure thresholds, actions of diazepam, pentobarbital and etomidate isomers alone and in combination with Ro 15-1788, picrotoxin and (±) DMBB. Eur J Pharmacol 94:319–325
2. Ashton D, Wauquier A (1979a) Effects of some anti-epileptic, neuroleptic and gabaminergic drugs on convulsions induced by D,L-allylglycine. Pharmacol Biochem Behav 11:221–226
3. Ashton D, Wauquier A (1979b) Behavioural analysis of the effects of 15 anticonvulsants in amygdaloid kindled rats. Psychopharmacology 65:7–13
4. Ashton D, Wauquier A (1985) Modulations of a gaba-ergic inhibitory circuit in the in vitro hippocampus by etomidate isomers. Anaesth Analg 64:975–980
5. Ashton D, Marannes R, Pauwels PJ, Van Belle H, Reid K, Wauquier A (1986) Possible mechanisms of flunarizine's anticonvulsant activity. Health Science Review: International Workshop on Flunarizine in Epilepsy, Beerse, 24 May 1985, p 45–59
6. Ayala GF, Dichter M, Gumnit RZ, Matsumoto H, Spencer WA (1973) Genesis of epileptic interictal spikes: new knowledge of cortical feedback systems suggest a neurophysiological explanation of brief paroxysms. Brain Res 211:227–234
7. Bergamasco B, Cantello R, Riccio A (1987) Utilita della flunarizina come coordiuvante nel trattamento dell'epilessia (confronto in doppio cieco con un placebo in disegno cross over) (unpublished report)
8. Bidard JN, Vijverberg HPM, Frelin C, Chungue E, Legrand A-M, Bagnis R, Ladunski M (1984) Cignatoxin is a novel type of Na^+ channel toxin. J Biol Chem 259:8253–8357
9. Bingmann D, Speckmann EJ (1984) Paroxysmal depolarization shifts in neurons of hippocampal slices: depression by calcium antagonists. Pflugers Arch S402, R29
10. Binnie CD, de Beukelaar F, Meijer JWA, Meinardi H, Overweg J, Wauquier A, van Wieringen A (1985) Open dose-ranging trial of flunarizine as add-on therapy in epilepsy. Epilepsia 26:424–428
11. Caspers H, Speckmann EJ, Lehmkühler A (1980) Electrogenesis of cortical DC potentials. In: Kornhuber HH, Deecke L (eds) Motivation motor and sensory processes of the brain: electrical potentials, behaviour and clinical use. Prog Brain Res 54:3–15
12. Cavazzuti GB, Galli V, Benatti A (1986) The use of flunarizine in pediatric epilepsy. Funct Neurol 1:551–554
13. Curatolo P, Bruni O, Brindesi I, Pruna D, Cusmai R (1986) La flunarizina nelle epilessie formaco-resistenti dell'infanzia e dell'adolescenza. Riv Neurol 56:25–38
14. Debandt R, De Smedt R, Jorens H, Leysen M, Rodrigus E, Van Eycken L (1985) Flunarizine as "add-on" therapy in institutionalized patients with refractory epilepsy. Effect on seizure incidence and behaviour. Janssen Research Product Information Service: CRR R 14950/65
15. Declerck AC, Wauquier A (1978) Double-blind study on the effectiveness on flunarizine in therapy-resistant epilepsy in mentally retarded children (abstr). 10th Epilepsy International Symposium, p 169
16. Declerck AC, Wauquier A (1983) Flunarizine, a double-blind study in therapy resistant mentally retarded epileptic children. Suppl Boll Chim Farm 122:9CL–16CL
17. Declerck AC, Oei LT, Wauquier A (1986) Flunarizine in mentally retarded children with therapy-resistant epilepsy. Health Science Review: International Workshop of Flunarizine in Epilepsy, Beerse, 24 May 1985, p 68
18. De Domenico P, Gallitto G, Musolino R, Di Perri R (1987) Saggio dell „Utilita" condiuvante della flunarizina in epilettia gravi (unpublished report)
19. De Sarro GB, Nistico G, Meldrum B (1986) Anticonvulsant properties of flunarizine on reflex and generalised models of epilepsy. Neuropharmacology 25:695–701
20. De Sarro GB, Meldrum BS, Nisticó G (1988) Anticonvulsant effects of some calcium entry blockers in DBA/2 mice. Br J Pharmacol 93:247–256
21. Desmedt LKC, Niemegeers CJE, Janssen PAJ (1975) Anticonvulsant properties of cinnarizine and flunarizine in rats and mice. Arzneimittelforschung 25:1408–1413
22. Fröscher W, Bülau P, Burr W et al. (1988) Double-blind, placebo-controlled trial in flunarizine in therapy-resistant persons with epilepsy. Clin Neuropharmacol (in press)

23. Giroud M, Nivelon JL, Dumas R (1987) Antiepileptic activity of flunarizine in 10 therapy-resistant children with epilepsy. Presse Méd 16:913
24. Heinemann U, Lux HA, Gutnick MJ (1977) Extracellular free calcium and potassium during paroxysmal activity in the cerebral cortex of the rat. Exp Brain Res 27:237–243
25. Heykants J, De Cree J, Hörig C (1979) Steady-state plasma levels of flunarizine in chronically treated patients. Arzneimittelforschung 29:1168–1171
26. Keene D, Whiting S, Humphreys P, Findlay HP, Cumming B (1987) Flunarizine, an add-on drug in the treatment of refractory childhood epilepsy. XXIInd Canadian congress of neurological sciences. Vancouver, June 1987
27. Konishi T, Murakami M, Yamatani M, Koda O, Okada T, Seto H, Nitani T (1987) Efficiency of flunarizine (a calcium entry blocker in therapy resistant epileptic children with partial epilepsy). The 20th Congress of the Japanese Society of Epilepsy: Abstract Book, p 131
28. Mancia D (1987) La flunarizina come coordiuvante della terapia di base nel trattamento dell'epilessia (Studio clinico controllato) (unpublished report)
29. Meldrum BS (1979) Convulsant drugs, anticonvulsants and GABA-mediated neuronal inhibition. In: Krogsgaard-Larsen P, Scheel-Kruger J, Kofod H (eds) Gaba-neurotransmitters. Munksgaard, Copenhagen, pp 390–405
30. Meldrum BS (1983) Convulsants and anticonvulsant drug mechanisms. In: Baldy-Moulinier M, Ingvar DH, Meldrum BS (eds) Current problems in epilepsy. Libbey, London, pp 324–336
31. Moglia A, Bergamasco B, Di Perri R, Mancia D (1986) Flunarizine as add-on therapy in epilepsy, cross over study versus placebo. Furet Neurol 1:547–550
32. Olsen RW (1982) Drug interactions at the GABA-receptor ionophore complex. Annu Rev Pharmacol Toxicol 22:245–277
33. Overweg J (1987) Safety evaluation of flunarizine after long-term "add-on" administration in epileptic patients. Janssen Research Product Information Service, Clinical Research Report, R14 950/71
34. Overweg J, Binnie CD, Meijer JWA, Meinardi H, Nuijten STM, Schmaltz S, Wauquier A (1984) Double-blind placebo-controlled trial of flunarizine as add-on therapy in epilepsy. Epilepsia 25:217–222
35. Pothmann R, Braumann G, Kröger T (1986) Calciumeintrittsblockade als Zusatz-Behandlung bei therapieresistenten kindlichen Epilepsien. Deutsche Sektion der Internationalen Liga gegen Epilepsie: 26. Jahrestagung, Münster, Abstract, p 56
36. Prince DA (1978) Neurophysiology of epilepsy. Annu Rev Neurosci 1:395–415
37. Prince DA (1982) Epileptogenesis in hippocampal and neocortical neurones. In: Klee MR, Lux HD, Speckman EJ (eds) Physiology and pharmacology of epileptogenic phenomena. Raven, New York, pp 151–161
38. Sorel L (1987) La flunarizine dans l'épilepsie: evaluation d'un traitement „additif" de quatre nous chez des patients resistant au traitement antiépileptique existant. Acta Neurol Belg 87:140–147
39. Starreveld E et al. Flunarizine in epilepsy (in preparation)
40. Study RE, Barker JL (1981) Diazepam and (−) pentobarbital: fluctuation analysis reveals different mechanisms for the potentiation of GABA responses in cultured central neurones. Proc Natl Acad Sci USA 78:7180–7184
41. Takahashi T (1988) An open trial of flunarizine hydrochloride as add-on treatment in therapy-resistant epileptics. Clin Psychiatry (in press)
42. Tata MR, Guizzaro A, Giudici S, Parisi A, Bravaccio F (1987) Flunarizine in pharmacoresistant epilepsies: preliminary experiences in 30 cases. International Symposium on Calcium Antagonists, Pharmacology and Clinical Research, New York City, 10–13 February 1987, p 113
43. Touquet G, Van Cappel F (1985) Flunarizine as adjuvants in therapy resistant epilepsy, a double-blind placebo-controlled trial. Janssen Research Product Information Service, Clinical Research Report No. 14950/64
44. Treiman DM, De Giorgio CM, Cornish JW, Cereghino JJ (1986) Efficacy and toxicity of flunarizine in partial onset seizures at concentrations of 30, 60, 120 ng/ml. Epilepsia 27:649–650

45. Treiman DM, Tsay J-Y, Whitley LP, Seaman CA, Kupferberg HJ, Pledger GW, Cereghino JJ (1987) Final report of flunarizine open tolerability study. Epilepsy Res NINCDS, Bethesda
46. Umezawa Y, Amano K, Kawamurah H et al. (1987) Add-on therapy of flunarizine in therapy-resistant epilepsy. The 20th Congress of the Japanese Society of Epilepsy. Abstract Book p 92
47. Van Bever W (1974) Acute oral toxicity (LD50) of flunarizine dihydrochloride in mice, rats and guinea pigs. Janssen Pharmaceutica, Preclinical Research Report R14950/1 December 1974
48. Van Wieringen A, Ashton D, De Beukelaar F, Binnie CD, Overweg J, Wauquier A (1986) Flunarizine in the treatment of epilepsy. In: Schalse H, Walden J (eds) Epilepsy and calcium. Urban & Schwartzenberg, Munich, pp 387–399
49. Walden J, Speckmann EJ, Witte OW (1984) Effects of the calcium antagonist verapamil on spontaneous and evoked bioelectric activity on the cerebral cortex. Pflugers Arch 402:R36
50. Walden J, Speckmann EJ, Witte OW (1985) Suppression of focal epileptiform discharges by intracerebroventricular perfusion of a calcium antagonist. Electroencephalogr Clin Neurophysiol 61:299–309
51. Wauquier A, Ashton D, Melis W (1979) Behavioural analysis of amygdaloid kindling in Beagle dogs and the effects of clonazepam, diazepam, phenobarbital, diphenylhydantoin and flunarizine on seizure manifestation. Exp Neurol 64:579–586
52. Wauquier A, Ashton D, Clincke G, Fransen J, Gillardin JM, Janssen PAJ (1986) Anticonvulsant profile of flunarizine. Drug Dev Res 7:49–60
53. Willow M, Catterall WA (1982) Inhibition of binding of [^{3}H] batrachotoxin in A-20-α-benzoate to sodium channels by the anticonvulsant drugs diphenylhydantoin and carbamazepine. Mol Pharmacol 22:627–635
54. Witte OW, Walden J, Speckmann EJ, Elger CE (1983) Reduction of penicillin-induced paroxysmal depolarization shifts in neurons of rat cerebral cortex by intracellular injection of calcium channel blocker. Neurosci Lett Suppl 14:S405
55. Witte OW, Speckmann EJ, Walden J (1984) Contribution of calcium and calcium-dependent membrane currents to epileptic discharges in neocortical neurons of the rat. Cell Calcium 5:311
56. Wong RKS, Prince DA, Basbaum AI (1979) Dendritic mechanisms underlying penicillin induced epileptiform activity. Science 204:1228–1231

Brain Protection

The Effects of Flunarizine on Survival Following Prolonged Cerebral Ischemia in the Gerbil

S. L. COHAN [1], D. VON LUBITZ [1], D. REDMOND [1], S. SZATHMARY [2], and J. A. WAKSMAN [3]

Introduction

There is presently no known effective treatment for the global cerebral ischemia which results from cardiac arrest. Successful cardiac resuscitation results in significant brain damage in the majority of surviving patients (Mackintosh et al. 1978; Earnest et al. 1979; Earnest et al. 1980; Levy et al. 1981; Eisenberg et al. 1982; Longstreth et al. 1983). Furthermore, many patients who are successfully resuscitated subsequently die from cerebral damage, particularly patients surviving more than 48 h (Earnest et al. 1979; Longstreth et al. 1983). Flunarizine, a diphenylpiperazine, has been shown to have cerebral protective effects following ischemia in animals, as shown by improved histological survival of neurons (Van Reempts et al. 1983, 1986; Deshpande and Wieloch 1985, 1986; Kumar et al. 1987; Alps et al. 1988), and improved return of neurophysiological activity following cerebral ischemia (Wauquier et al. 1984, 1985). In order to be relevant to the clinical situation encountered in patients being resuscitated from cardiac arrest, a drug suitable for cerebral protection would have to be effective when given after resuscitation had taken place. The demonstration that there is delayed onset of cell death in the postischemic period (Kirino 1982; Pulsinelli et al. 1982; Van Reempts and Borgers 1985; Deshpande et al. 1987) increases the possibility that pharmacological agents administered in the postresuscitation period might be of benefit in enhancing brain survival. Demonstration of avid brain uptake of flunarizine (Michiels et al. 1983) and experiments in which flunarizine enhanced cell survival, even when administered after cessation of ischemia (Deshpande and Wieloch 1985, 1986; Van Reempts et al. 1986; Kumar et al. 1987; Alps et al. 1988), led us to choose flunarizine as a treatment modality in studies of survival following prolonged global cerebral ischemia in the Mongolian gerbil.

Methods

Male gerbils weighing 65–75 g were anesthetized with halothane 2.5% and $N_2O:O_2$ (2:1). Both carotid arteries were exposed under sterile technique

[1] The Department of Neurology, Georgetown University School of Medicine, 3800 Reservoir Road, N.W., Washington, D.C. 20007-2197, USA.
[2] Janssen Pharmaceutica, Raiffeisenstr. 8, D-4040 Neuss 21, FRG.
[3] Janssen Pharmaceutical, 40 Kingsbridge Road, Piscataway, NJ 08054, USA.

Cerebral Ischemia and Calcium
Ed. by A. Hartmann and W. Kuschinsky

through an anterior cervical incision, and a 6-0 polypropylene loop was placed around both carotid arteries. Halothane was decreased to 0.5% and the animals were placed in a prone position. Body temperature was maintained at 36.5°–37.5° C by a heating blanket attached to a rectal probe. Blood pressure and pulse were continuously monitored by a tail cuff manometer, and O_2 saturation continuously monitored transcutaneously. Biparietal monopolar electrodes were inserted subcutaneously for continuous EEG monitoring. Ischemia was produced for 20 min by suspending a 24-g weight from the carotid polypropylene loop. Adequacy of the occlusion was verified by direct inspection of the carotids and by developing an isoelectric EEG. Following release of the weight, at the termination of ischemia, restoration of carotid blood flow was verified by direct inspection. Histological inspection of the carotids revealed no delayed thrombotic occlusion of carotids at the site of the polypropylene loop.

Flunarizine 5 mg/ml was dissolved in hydroxypropyl β-cyclodextrin 100 mg/ml (Janssen Pharmaceutica). Animals were divided into five groups and received either flunarizine (2 mg/kg, 5 mg/kg, or 10 mg/kg), the cyclodextrin vehicle alone, or 0.9% NaCl intraperitoneally 5 min after cessation of ischemia and daily thereafter for 7 days, survival permitting.

A second group of gerbils received diazepam 25 µg/kg (Roche) in combination with flunarizine 10 mg/kg or cyclodextrin intraperitoneally 5 min after cessation of 20 min cerebral ischemia. Each animal subsequently received daily injections of flunarizine or cyclodextrin vehicle intraperitoneally for 7 days.

A third group of gerbils received flunarizine 2, 5, or 10 mg/kg intraperitoneally 5 min after cessation of 20 min cerebral ischemia and were subsequently sacrificed by decapitation at intervals up to 24 h after the drug injections. Brains were rapidly removed from the skulls and frozen until brain flunarizine levels could be determined by gas chromatography (Woestenborghs et al. 1982).

Statistical Methods

Kaplan Meier distributions were fitted separately to the groups in all survival studies. Differences among and between the groups in each survival study were tested with log-rank and generalized Wilcoxon statistics.

Results

All but three gerbils developed an isoelectric EEG within 30 s of initiation of ischemia (Table 1). These three animals were discarded and are not included in the final data. There were no differences between treatment groups in blood pressure, pulse, O_2 saturation, body temperature, or time to develop an isoelectric EEG (Table 1). All animals had generalized convulsions between 30 and 60 min after cessation of ischemia.

Figure 1 demonstrates mean whole brain concentrations of flunarizine following a single intraperitoneal injection, with peak levels being reached within 45–60 min of injection.

Table 1. Physiological variables in gerbils subjected to 20 min bilateral carotid occlusion[a]

Treatment	No.	Pulse (±SD)	MAP (mmHg)	O_2 sat. (%)	Temp. (°C)	Time to isoelectric EEG following ischemia (s)
Saline	20	342±27	62±5.2	99.1±1.4	37.1±0.28	12.6±2.37
Cyclodextrin vehicle	54	351±23	59±4.6	99.3±0.9	37.2±0.22	13.3±2.16
Flunarizine 2 mg/kg	12	338±15	63±6.7	99.3±1.1	36.9±0.21	11.9±1.85
Flunarizine 5 mg/kg	24	364±26	62±5.4	99.2±0.8	37.2±0.27	12.6±1.89
Flunarizine 10 mg/kg	20	357±19	57±6.1	99.4±0.6	37.3±0.24	13.1±1.92
Diazepam + cyclodextrin	18	346±21	63±5.7	99.2±1.1	37.0±0.26	13.4±2.01
Diazepam + flunarizine 10 mg/kg	18	358±24	62±5.2	99.5±0.4	37.2±0.19	12.9±1.74

[a] Three animals are excluded because time to isoelectric EEG exceeded 30 s.

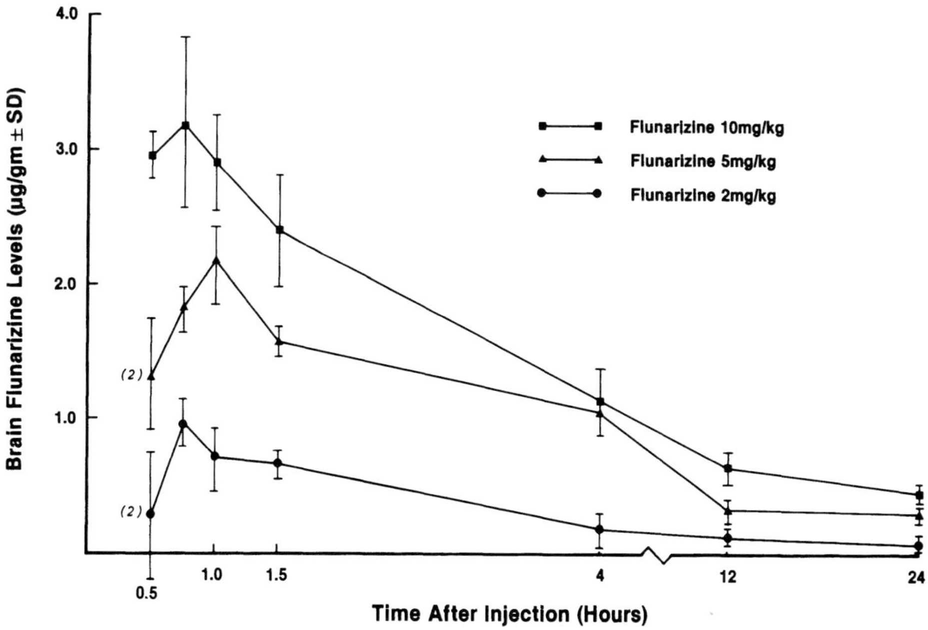

Fig. 1. Brain flunarizine levels in the gerbil at various time intervals following intraperitoneal injection. Doses of flunarizine used were 2 mg/kg (●—●), 5 mg/kg (▲—▲), and 10 mg/kg (■—■). Each *data point* is the mean (±SD) value for three animals except where otherwise specified in parentheses. Flunarizine was injected 5 min after termination of 20 min bilateral carotid ligation

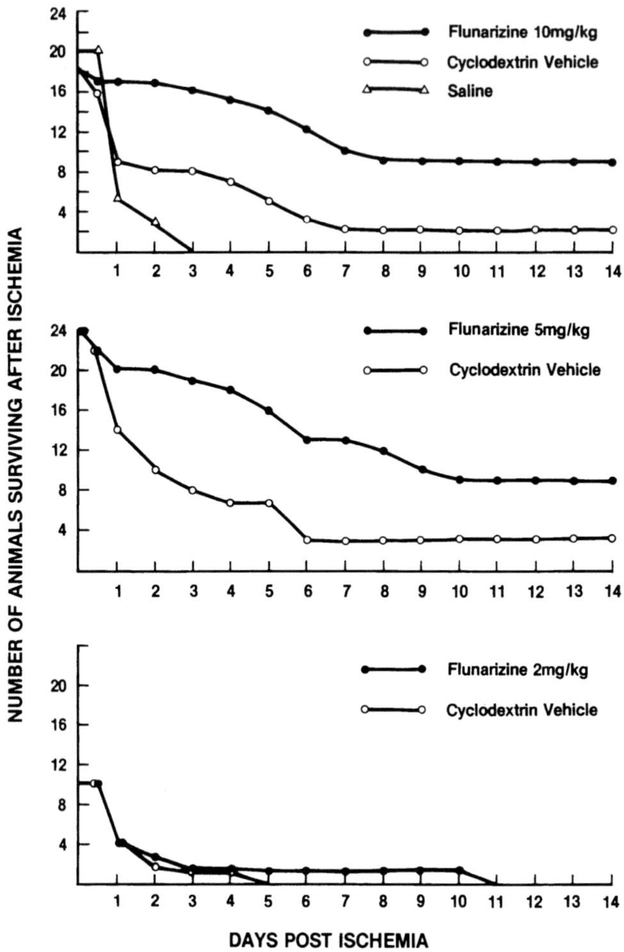

Fig. 2. Duration of survival following 20 min bilateral carotid occlusion in the gerbil. Flunarizine at intraperitoneal doses of 5 mg and 10 mg significantly improved survival as compared to saline or cyclodextrin vehicle controls. There was no improvement in survival with use of flunarizine 2 mg/kg as compared to controls

Table 2. Comparison of survival distributions following 20 min global cerebral ischemia in gerbils

Compared groups	Log-rank test		Wilcoxon test	
	χ-square	*P*-value	χ-square	*P*-value
F 10 mg vs. V, S	30.43	0.0001	24.96	0.0001
F 10 mg vs. V	10.24	0.001	10.91	0.001
F 10 mg vs. S	32.32	0.0001	26.37	0.0001
F 5 mg vs. V	7.61	0.006	8.28	0.004
F 2 mg vs. V	0.27	0.606	0.03	0.865
V vs. S	6.29	0.012	2.55	0.110

F, flunarizine; V, cyclodextrin vehicle; S, saline.

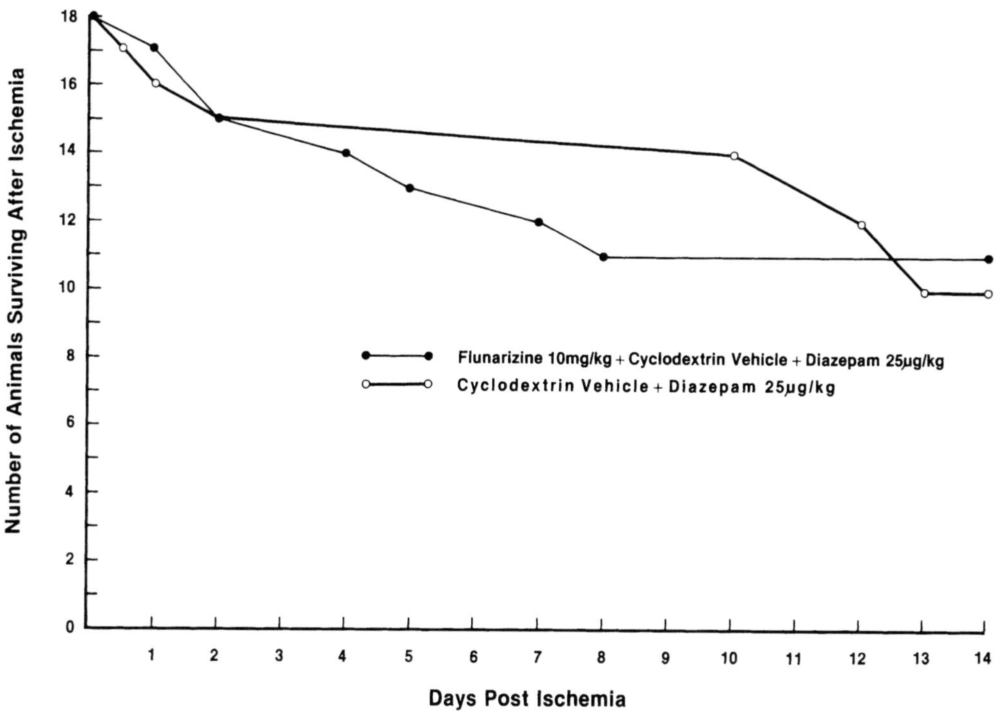

Fig. 3. Duration of survival following 20 min bilateral carotid occlusion in the gerbil. There was no significant difference between flunarizine 10 mg/kg and cyclodextrin vehicle with the addition of diazepam 25 µg/kg to each group

Table 3. Further comparison of survival distributions following 20 min global cerebral ischemia in gerbils

Compared groups	Log-rank test		Wilcoxon test	
	χ-square	*P* value	χ-square	*P* value
F+D vs. V+D	0.01	0.906	0.00	0.986
F vs. F+D	0.01	0.922	0.00	0.974

F, flunarizine; V, cyclodextrin vehicle; D, diazepam 25 µg/kg.

Flunarizine significantly prolonged survival at 5–10 mg/kg intraperitoneally, administered 5 min after cessation of 20 min cerebral ischemia, when compared to cyclodextrin vehicle or saline (Fig. 2, Table 2), but flunarizine did not prolong survival at 2 mg/kg. Cyclodextrin appeared to slightly but significantly prolong survival as compared to saline by the log rank but not the Wilcoxon test (Table 2).

The addition of diazepam to cyclodextrin vehicle prolonged survival obtained with cyclodextrin alone (Fig. 3, Table 3), but addition of diazepam to flunarizine

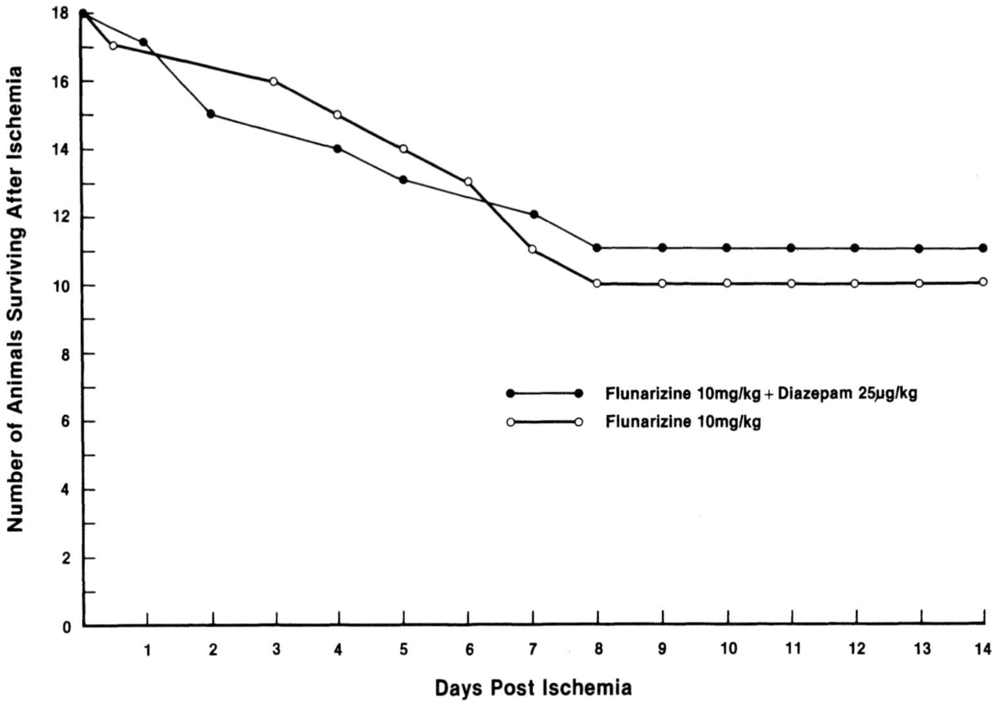

Fig. 4. Duration of survival following 20 min bilateral carotid occlusion in the gerbil. There was no significant difference between flunarizine 10 mg/kg and flunarizine 10 mg/kg + diazepam 25 µg/kg

did not further increase the survival obtained with flunarizine alone (Fig. 4, Table 3). There was no significant difference in survival between treatment with diazepam plus flunarizine and diazepam plus vehicle alone (Fig. 3, Table 3).

Discussion

Global cerebral ischemia, usually arising from cardiac arrest, is a major source of morbidity and mortality. Although heart resuscitation has resulted in many survivors of cardiac arrest (Mackintosh et al. 1978; Earnest et al. 1979, 1980; Levy et al. 1981; Eisenberg et al. 1982; Longstreth et al. 1983), severe brain damage ensues in many patients (Mackintosh et al. 1978; Earnest et al. 1979, 1980; Levy et al. 1981; Eisenberg et al. 1982; Longstreth et al. 1983), with some neurological damage being found in most, if not all, survivors (Earnest et al. 1979, 1980; Levy et al. 1981). Growing evidence of delayed neuronal death following anoxic or ischemic insult (Kirino 1982; Pulsinelli et al. 1982; Van Reempts and Borgers 1985; Deshpande et al. 1987) has suggested the possibility that intervention, even following the ischemic or anoxic event, may improve survival. The possibility that increased intracellular calcium concentration may play a role in delayed neuronal death (Borgers 1981; Farber 1981; Hass 1981; Siesjo 1981; Hossmann et al. 1983;

Dienel 1984; Choi 1985; Van Reempts and Borgers 1985; Deshpande et al. 1987) has led to studies utilizing calcium overload blockers. In many studies these compounds have been administered prior to an ischemic insult. There is a decrease in the size of focal lesions histologically when flunarizine is administered prior to ischemia (Van Reempts et al. 1987) as well as maintenance of cerebral blood flow in the postischemic period with the use of nimodipine (Steen et al. 1983), lidoflazine (Dean et al. 1984) and flunarizine (White et al. 1982), presumably by blocking calcium-stimulated vasospasm. There is also improved clinical survival in dogs when nimodipine is given before (Steen et al. 1983), but not after, the ischemic event (Steen et al. 1984).

For drugs to be useful in brain protection following resuscitation from cardiac arrest, they have to be effective even if given during the resuscitation period or later. In our study brain flunarizine levels did not reach a peak until 45–60 min after cessation of ischemia. [It was necessary to measure brain flunarizine levels following ischemia to be certain that the progressive loss in cerebral blood flow known to occur during the first 1–2 h after termination of ischemia (Levy et al. 1979; White et al. 1982; Steen et al. 1983) did not impair flunarizine uptake by the brain.] At flunarizine doses of 5 or 10 mg/kg significant improvement in survival was observed. It was reported by Newberg et al. (1984) that flunarizine given in the post-ischemic period did not improve survival in the dog, but at the dose employed by these investigators, 6 μg/kg i.v., it is highly unlikely that brain levels of flunarizine achieved were high enough to be protective, as our study demonstrated that prolongation of survival was not observed at peak brain levels of less than 1.0 μg/g. It is extremely unlikely that following a 6 μg/kg dose, brain concentrations approaching a therapeutic level could be achieved. In studies of histological changes in the CA_1 region of the hippocampus, 100 μg/kg i.v. flunarizine was employed and a protective effect was observed (Deshpande 1985; Van Reempts et al. 1986). In one of our tests of statistical significance, the cyclodextrin vehicle alone had some protective effects. We cannot at present propose a mechanism to explain this finding.

The mechanisms by which flunarizine may be protective are incompletely understood. Flunarizine does inhibit calcium-stimulated vasoconstriction (Van Neuten and Wellens 1979) and may increase cerebral blood flow (Beck and Krieglstein 1987), and it has been suggested that amelioration of postischemic fall in cerebral blood flow by flunarizine is a potential protective mechanism (Beck and Krieglstein 1987). Data on this point are conflicting (Newberg et al. 1984). Although prevention of postischemic loss of cerebral blood flow may be an important mechanism for cerebral protection when drugs are administered before an ischemic event (Dean et al. 1984), the fact that the anticytotoxic mechanism of drugs administered in the postischemic period is not a result of maintaining cerebral blood flow is clearly demonstrated by the failure of postischemic administration of nimodipine to improve quality of survival in dogs despite improvement in cerebral blood flow in the postischemic period (Steen et al. 1984). Furthermore, lidoflazine (Winegar et al. 1983; Vaugenes et al. 1984) and flunarizine (Deshpande and Wieloch 1985; Van Reempts et al. 1986) have been demonstrated to have cytoprotective effects in the absence of improved cerebral blood flow in the postischemic period (Dean et al. 1984).

Calcium overload has been proposed as an important mechanism in delayed onset postischemic cell death (Borgers 1981; Farber 1981; Hass 1981; Siesjo 1981; Hossmann et al. 1983; Dienel 1984; Choi 1985; Van Reempts and Borgers 1985; Deshpande et al. 1987), and increased calcium concentration has been observed in dying neurons biochemically, autoradiographically, and histologically (Hossmann et al. 1983; Dienel 1984; Van Reempts et al. 1985; Deshpande et al. 1987), although it is not yet certain whether this calcium accumulation is a cause or the result of cell death. Flunarizine inhibits cellular calcium uptake (Borgers et al. 1984) and prolongs the period necessary for depolarization following anoxia, delays the attendant increase in ionic conductance, and thus retards the influx of calcium and Na^+ (Holler et al. 1986). Recent experiments utilizing guinea pig arterial cells demonstrate that cinnarizine, a compound closely related to flunarizine, may impair calcium influx by blocking "L" type and "T" type calcium channels (Van Skiver et al. 1988).

All of our animals had generalized seizures in the postischemic period. This is a higher frequency than previously reported (Levy et al. 1979). Increased calcium accumulation in neurons has been seen following prolonged seizures (Griffiths et al. 1982). Flunarizine is known to be an anticonvulsant (Binnie et al. 1985; Treiman et al. 1986) and despite the occurrence of seizures in all groups, it is possible that some of the protective effects of flunarizine were due to its anticonvulsant properties. It is for this reason that a separate set of experiments were conducted with the addition of the anticonvulsant diazepam. When diazepam was added there was no difference in survival between flunarizine plus cyclodextrin treated animals and animals receiving cyclodextrin alone. That part of flunarizine's effects may be the result of its anticonvulsant properties cannot be excluded. Protective effects of diazepam in cerebral ischemia have been reported previously (Moursi et al. 1983). If flunarizine's protective effects were by a different mechanism than those of diazepam, one might have seen an additive effect of diazepam plus flunarizine, which was not the case. There is increased release of the facilitative neurotransmitters glutamate and aspartate in the brain following ischemia (Benveniste et al. 1985; Bosely et al. 1983; Rothman and Olney 1986) and the release of those excitatory neurotransmitters is calcium dependent (Nadler et al. 1977). The facilitative agonist *N*-methyl-D-aspartate may also be cytotoxic in the hippocampus (Simon et al. 1984). Thus postsynaptic facilitation may contribute to cell vulnerability to anoxia and ischemia. Drugs such as benzodiazepines, which lead to hyperpolarization by means of their GABA-ergic properties (Haefely et al. 1981; Olsen 1982), and flunarizine, which retards depolarization following anoxia (Holler et al. 1986) and may impair the presynaptic calcium uptake necessary for release of excitatory amino acids, may in part be protective by impairment of postanoxia or ischemia-induced synaptic excitation effects.

Acknowledgments. The authors gratefully acknowledge the devoted assistance of Ms. Scharlean Bland in the preparation of this manuscript. This work was supported in part by a grant from the Janssen Research Foundation and the Bradley Memorial Fund.

References

Alps BJ, Calder C, Hass WK, Wilson AD (1988) Comparative protective effects of nicardipine, flunarizine, lidoflazine and nimodipine against ischemic injury in the hippocampus of the Mongolian gerbil. Br J Pharmacol 93:877–883

Beck T, Krieglstein J (1987) Local cerebral glucose utilization and local cerebral blood flow in conscious rats after administration of flunarizine. Naunyn Schmiedebergs Arch Pharmacol 335:680–685

Benveniste H, Drejer J, Schousboe A, Diemer NH (1985) Elevation of the extracellular concentrations of glutamate and aspartate in rat hippocampus during transient cerebral ischemia monitored by intracerebral microdialysis. J Neurochem 40:189–201

Binnie CD, de Beukelaar F, Meijer JWA, Meinardi H, Overweg J, Wauquier A, Van Wieringen A (1985) Open dose-ranging trial of flunarizine as add-on therapy in epilepsy. Epilepsia 26:424–428

Borgers M (1981) The role of calcium in the toxicity of the myocardium. Histochem J 13:839–848

Borgers M, De Clerck F, Van Reempts J, Xhonneau R, Van Nueten J (1984) Selective blockade of cellular Ca^{2+}-overload by flunarizine. Int Angiol 3:25–31

Bosely TM, Woodhams PL, Gordon RD, Belazs R (1983) Effects of anoxia on the stimulated release of amino acid neurotransmitters in the cerebellum in vitro. J Neurochem 40:189–201

Choi DW (1985) Glutamate neurotoxicity in cortical cell culture is calcium-dependent. Neurosci Lett 58:293–297

Dean JM, Hoehner PJ, Rogers MC, Traystman RJ (1984) Effect of lidoflazine on cerebral blood flow following twelve minutes of total cerebral ischemia. Stroke 15:531–535

Deshpande JK, Siesjo BK, Wieloch T (1987) Calcium accumulations and neuronal damage in the rat hippocampus following cerebral ischemia. J Cereb Blood Flow Metab 7:89–95

Deshpande JK, Wieloch T (1985) Amelioration of ischemic brain damage by post ischemic treatment with flunarizine. Neurol Res 7:27–29

Deshpande JK, Wieloch T (1986) Flunarizine, a calcium entry blocker, ameliorates ischemic brain damage in the rat. Anesthesiology 64:215–224

Dienel GA (1984) Regional accumulation of calcium in post-ischemic rat brain. J Neurochem 43:913–925

Earnest MP, Breckinridge JC, Yarnell PR, Oliva PB (1979) Quality of survival after out-of-hospital cardiac arrest: predictive value of early neurologic evaluation. Neurology 29:56–60

Earnest MP, Yarnell PR, Merrill SL, Knapp GL (1980) Long-term survival and neurologic status after resuscitation from cardiac arrest. Neurology 30:1298–1302

Eisenberg MS, Hallstrom A, Bergner L (1982) Long-term survival after out-of-hospital cardiac arrest. N Engl J Med 306:1340–1343

Farber JL (1981) The role of calcium in cell death. Life Sci 29:1289–1295

Griffiths T, Evans MC, Meldrum BS (1982) Intracellular sites of early calcium accumulation in the rat hippocampus during status epilepticus. Neurosci Lett 30:329–334

Haefely W, Pieri L, Polo P, Schaffner R (1981) General pharmacology and neuropharmacology of benzodiazepine derivatives. In: Hoffmeister F, Stille G (eds) Psychotropic agents. Springer, Berlin Heidelberg New York (Handbook of Experimental Pharmacology, vol 55, pp 13–262)

Hass WK (1981) Beyond cerebral blood flow, metabolism and ischemic thresholds: an examination of the role of calcium in the initiation of cerebral infarction. In: Meyer JS, Lochner H, Reivich M, Ott EO, Aranibar A (eds) Cerebral vascular disease 3. Excerpta Medica, Amsterdam, pp 3–17

Holler M, Dierking H, Dengler K, Tegtmeyer F, Peters T (1986) Effect of flunarizine on extracellular ion concentration in the rat brain under hypoxia and ischemia. In: Battistine N, Courbier R, Fiorani P, Plum F, Fieschi C (eds) Acute brain ischemia. Raven, New York, p 229

Hossmann K-A, Paschen W, Csiba L (1983) Relationship between calcium accumulation and recovery of Ca^{+} brain after prolonged cerebral ischemia. J Cereb Blood Flow Metab 3:346–353

Kirino T (1982) Delayed neuronal death in the gerbil hippocampus following ischemia. Brain Res 239:57–69

Kumar K, Krause G, Koestner A, Hoehner T, White B (1987) Effect of flunarizine on global brain ischemia in the dog: a quantitative morphologic assessment. Exp Neurol 97:115–127

Levy DE, Van Uitert RL, Pik CL (1979) Delayed post-ischemic hypoperfusion. A potentially damaging consequence of stroke. Neurology 29:1245–1252

Levy DE, Bates D, Caronna JJ et al. (1981) Prognosis in non-traumatic coma. Ann Intern Med 94:293–301

Longstreth WT, Inui TS, Cobb LA, Copass MK (1983) Neurologic recovery after out-of-hospital cardiac arrest. Ann Intern Med 98:588–592

Mackintosh AF, Crabb ME, Grainger R, Williams JH, Chamberlain DA (1978) The Brighton resuscitation ambulances: review of 40 consecutive survivors of out-of-hospital cardiac arrest. Br Med J I:1115–1118

Michiels M, Hendriks R, Knaeps F, Woestenborghs R, Heykants J (1983) Absorption and tissue distribution of flunarizine in rats, pigs and dogs. Drug Res 33:1135–1142

Moursi MM, Luyckx BA, D'Alecy LG (1983) The role of ethanol in diluents of drugs that protect mice from hypoxia. Stroke 14:791–796

Nadler JV, White WF, Vaca KW (1977) Characterization of putative amino acid transmitter release from slices of rat dentate gyrup. J Neurochem 29:279–290

Newberg LA, Steen PA, Milde JH, Michenfelder JD (1984) Failure of flunarizine to improve cerebral blood flow or neurologic recovery in a canine model of complete cerebral ischemia. Stroke 15:666–671

Olsen RW (1982) Drug interactions at the GABA receptor-ionophore complex. Annu Rev Pharmacol Toxicol 22:245–247

Pulsinelli WA, Brierly JB, Plum F (1982) Temporal profile of neuronal damage in a model of transient forebrain ischemia. Ann Neurol 11:491–498

Rothman S, Olney JW (1986) Glutamate and the pathophysiology of hypoxic ischemic brain damage. Ann Neurol 19:105–111

Siesjö BK (1981) Cell damage in the brain: a speculative synthesis. J Cereb Blood Flow Metab 1:155–185

Simon RP, Swan JH, Griffiths T, Meldrum BS (1984) Blockade of *N*-methyl-δ-aspartate receptors may protect against ischemic damage in the brain. Science 226:850–852

Steen PA, Gisvold SE, Milde JH, Newberg LA, Scheithauer BW, Lanier WL, Michenfelder JD (1985) Nimodipine improves outcome when given after complete cerebral ischemia in primates. Anesthesiol 62:406–414

Steen PA, Newberg LA, Milde JH, Michenfelder JD (1983) Nimodipine improves cerebral blood flow and neurologic recovery after complete cerebral ischemia in the dog. J Cereb Blood Flow Metab 3:38–43

Steen PA, Newberg LA, Milder JH, Michenfelder JD (1984) Cerebral blood flow and neurologic outcome when nimodipine is given after complete cerebral ischemia in the dog. J Cereb Blood Flow Metab 4:82–87

Treiman DM, DeGiorgio CM, Cornish JW, Cereghino JJ (1986) Efficacy and toxicity of flunarizine in partial onset seizures at serum concentrations of 30, 60, and 120 ng/ml. Epilepsia 27:649–650

Van Neuten JM, Wellens D (1979) Mechanisms of vasodilatation and antivasoconstriction. Angiology 30:440–446

Van Reempts J, Borgers M (1985) Ischemic brain injury and cell calcium: morphologic and therapeutic aspects. Ann Emerg Med 14:736–742

Van Reempts J, Borgers M, Van Dael L, Van Eyndhoven L, Van de Ven M (1983) Protection with flunarizine against hypoxic-ischaemic damage of the rat cerebral cortex. A quantitative morphologic assessment. Arch Int Pharmacodyn Ther 262:76–88

Van Reempts J, Haseldonckx M, Van Deuren B, Wouters L, Borgers M (1986) Structural damage of the ischemic brain: involvement of calcium and effects of postischemic treatment with calcium entry blockers. Drug Dev Res 8:387–395

Van Reempts J, Van Deuren B, Van de Ven M, Cornelissen F, Borgers M (1987) Flunarizine reduces cerebral infarct-size after photo-chemically induced thrombosis in spontaneously hypertensive rats. Stroke 18:1113–1119

Van Skiver DM, Spires S, Cohen CJ (1988) Block of T-type Ca channels in guinea pig atrial cells by cinnarizine. Biophys J 53:223 a

Vaugenes P, Contadore R, Safar P, Moossy J, Rao G, Diven W, Alexander H, Stezoski W (1984) Amelioration of brain damage by lidoflazine after prolonged ventricular fibrillation cardiac arrest in dogs. Crit Care Med 12:846–855

Wauquier A, Clincke G, Van den Broeck WAE, Hermans C, Melis W, Van Loon J (1984) Global incomplete ischemia in dogs assessed by quantitative EEG analysis. Effects of hypnotics and flunarizine. In: Pfurtscheller V, Jonkman EJ, Lopes da Silva FH (eds) Quantitative EEG and imaging techniques in brain ischemia. Elsevier Biomedical, Amsterdam, pp 217–233

Wauquier A, Edmonds HL, Van den Broeck WAE, Melis W, Van Loon J (1985) Effects of flunarizine on recovery of cerebral function following cardiac arrest. Neuropsychobiol 13:147–151

White BC, Gadzinski DS, Hoehner PJ (1982) Correction of canine cerebral cortical blood flow and vascular resistance after cardiac arrest using flunarizine, a calcium antagonist. Ann Emerg Med 11:118–127

Winegar CP, Henderson O, White BC et al. (1983) Early amerioration of neurologic deficit by lidoflazine after 15 minutes of cardiopulmonary arrest in dogs. Ann Emerg Med 12:471–477

Woestenborghs R, Michielsen L, Lorreyne W, Heykants J (1982) Sensitive gas chromatographic method for the determination of cinnarizine and flunarizine in biologic samples. J Chromatogr 232:85–91

Cerebroprotective Effects of a Parenteral Flunarizine Formulation

H. L. Edmonds, Jr. [1], G. Raque [2], P. Y. Zhang [1], S. A. Jenkins [1], and C. B. Shields [2]

Since the synthesis of flunarizine over 25 years ago, there have been over 230 experimental and 150 clinical peer-reviewed descriptions of its pharmacological activity. According to the proposed WHO classification of calcium antagonists, flunarizine is regarded as a type IV compound [9]. However, it has a wide range of inhibitory actions (Table 1) and potential clinical applications (Table 2). Prominent among these is cerebroprotection. Most of the experimental investigations on flunarizine's protectant and resuscitant properties have obtained positive results [13]. Nevertheless, neither the mechanism(s) nor the site(s) of action for this cerebroprotection are known with certainty. The primary site for this effect could reside within neurons, glia, or the cerebral capillary endothelium. Until recently, the lack of an appropriate intravenous formulation limited both mechanistic studies and the clinical evaluation of flunarizine's protectant action.

We examined the potential of this new preparation and its solvent (hydroxypropyl-β-cyclodextrin) for neurovascular protection. First, an 8-min complete global ischemic insult was produced in ketamine-anesthetized adult Wistar rats. Atraumatic hydraulic chest compression completely stopped cerebral perfusion by a readily resuscitable cardiac electromechanical dissociation. Abolition of cerebral cortical electrical activity was assessed by EEG, while the upper facial EMG made parallel determination of brain stem function (Fig. 1). Postischemic sound-induced convulsions (SICs) occurred in 83% of the untreated rats. Both 2-h flunarizine (100 μg/kg i.v.) and solvent (100 mg/kg i.v.) pretreatment afforded significant protection against postischemic convulsions (Table 3).

Flunarizine's anticonvulsant properties are well-established [12]. Therefore, the drug solution and solvent were also tested on rats with fully developed and stable postischemic convulsions. One-minute exposure to a 20-kHz tone (100 dB at 1 m) consistently elicited generalized clonic convulsions in untreated rats. The significant, but partial, anticonvulsant effect (Table 3) may have been due in part to action of the solvent.

Since histological evidence of ischemic damage is lacking in this chest compression model, we sought signs of functional damage in the selectively vulnerable hippocampus. A comparison of in vitro hippocampal slice excitability was made between tissue from normal rats and from those demonstrating postischemic SICs. Fewer (χ^2 11.5, $P<0.001$) postischemic (17/70) than normal (65/133) slices exhibited spontaneous epileptiform bursting when bathed in artificial cerebrospi-

[1] Departments of Anesthesiology and [2] Surgery (Division of Neurosurgery), University of Louisville, Louisville, KY 40292, USA.

Cerebral Ischemia and Calcium
Ed. by A. Hartmann and W. Kuschinsky

Table 1. Some inhibitory actions of flunarizine

Action	Reference
Serotonin	Gen Pharmacol 17:133 (1986)
Histamine	Drug Dev Res 1:107 (1981)
Norepinephrine	J Pharmacol Exp Ther 217:510 (1981)
$PGF_{2\alpha}$	Br J Pharmacol 73:252 (1981)
Neurotensin	Reg Peptides 6(Suppl 4):188 (1985)
Adenosine uptake	Eur J Pharmacol 112:323 (1985)
Dopamine release	Neuropharmacol 25:527 (1986)
Thromboxane A_2 formation	Arch Int Pharmacodyn 274:4 (1985)
Free radical formation	Arch Int Pharmacodyn 272:283 (1984)
Calmodulin-dependent phosphodiesterase	Biochem Biophys Res Comm 124:315 (1984)
Atherogenesis	Int Angiol 3(Suppl):33 (1984)
Drug-induced seizures	Drug Dev Res 8:397 (1986)
Hypoxic/ischemic damage	Biochem Pharmacol 35:2693 (1986)
Vasospasm	Gen Pharmacol 17:133 (1986)
Morphine abstinence syndrome	Eur J Pharmacol 137:9 (1987)
Neoplastic growth	Oncology 44:17 (1987)
Capillary endothelial permeability increase	Res Exp Med 185:325 (1985)
Cardiac arrhythmias	Br J Pharmacol 83:299 (1984)
Postischemic blood viscosity increase	Arch Int Pharmacodyn 230:321 (1977)

Table 2. Disorders in which flunarizine has been evaluated

Disorder	Reference
Vascular	
Raynaud's disease	Acta Therapeut 10:153 (1984)
Varicose veins	Invest Med Internacion 10:337 (1983)
Essential hypertension	Clin Trials J 22:30 (1985)
Neurological	
Dementia	Drugs 33(Suppl 2):90 (1987)
Epilepsy	Epilepsia 26:424 (1985)
Alternating hemiplegia	Neuropediat 18:191 (1987)
Duchenne muscular dystrophy	Muscle Nerve 9:349 (1986)
Neurovascular	
Migraine	Headache 23:70 (1983)
Vestibular syndrome	Cephalgia 5:205 (1985)
Chronic cerebrovascular insufficiency	Clinica E Terapia 110:541 (1987)
Postictal headache	Funct Neurol 1:543 (1986)

nal fluid containing 8 mM potassium [4]. Intravenous treatment before chest compression with solvent (100 mg/kg) prevented the ischemia-associated depression of hippocampal excitability (Table 3).

Oxidative reactive intermediates, like free radicals and products of arachidonic acid biotransformation, are often implicated in the genesis of ischemic brain damage [3]. Thus, we compared the concentration of the important antioxi-

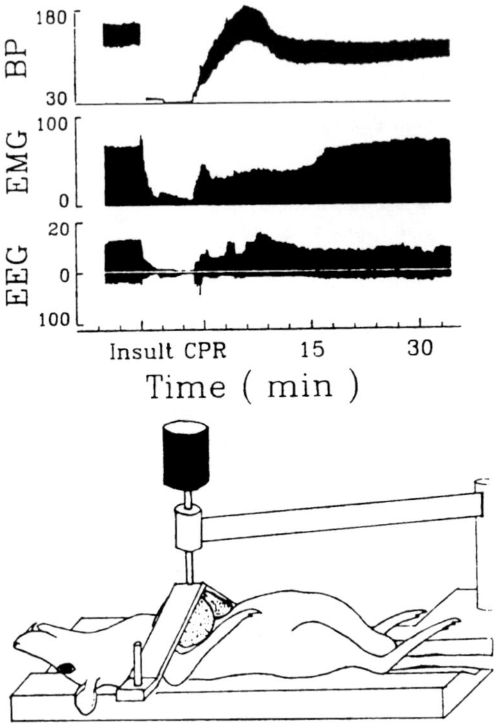

Fig. 1. The chest compression device consists of a curved metal frame that positions two partially fluid-filled balloons at the midthoracic level of a supine, anesthetized rat. The frame is held in place with a 1-kg weight. Fluid expansion of the balloons transfers the force through to the myocardium, resulting in electromechanical dissociation. Absence of flow was initially assessed by absence of carotid outflow and flat EEG and EMG. The later experiments described here eliminated vascular cannulation. Intravenous drug administration was achieved through a hindlimb venipuncture

Table 3. Effects of parenteral flunarizine formulation and its hydroxypropyl β-cyclodextrin solvent

	Untreated	Solvent (100 mg/kg)	Flunarizine (0.1 mg/kg)
48-h survival from chest compression ischemia	67/139	6/13	5/13
Incidence of SICs			
Preinsult dose	56/67	1/6**	1/5**
Postinsult dose	8/8	5/8 (8 mg/kg)	4/8*
Spontaneous epileptiform bursting in slices	17/70	58/80*	
% patent cremaster microvessels ($X \pm 95\%$ CI)	86 ± 12 ($n=7$)	64 ± 17 ($n=5$)	31 ± 16** ($n=6$)
Seconds to asphyxia-induced			
EEG flattening	40 ± 5	47 ± 18	
EMG flattening	93 ± 9	94 ± 7	

* $P < 0.05$ compared to untreated.
** $P < 0.01$.

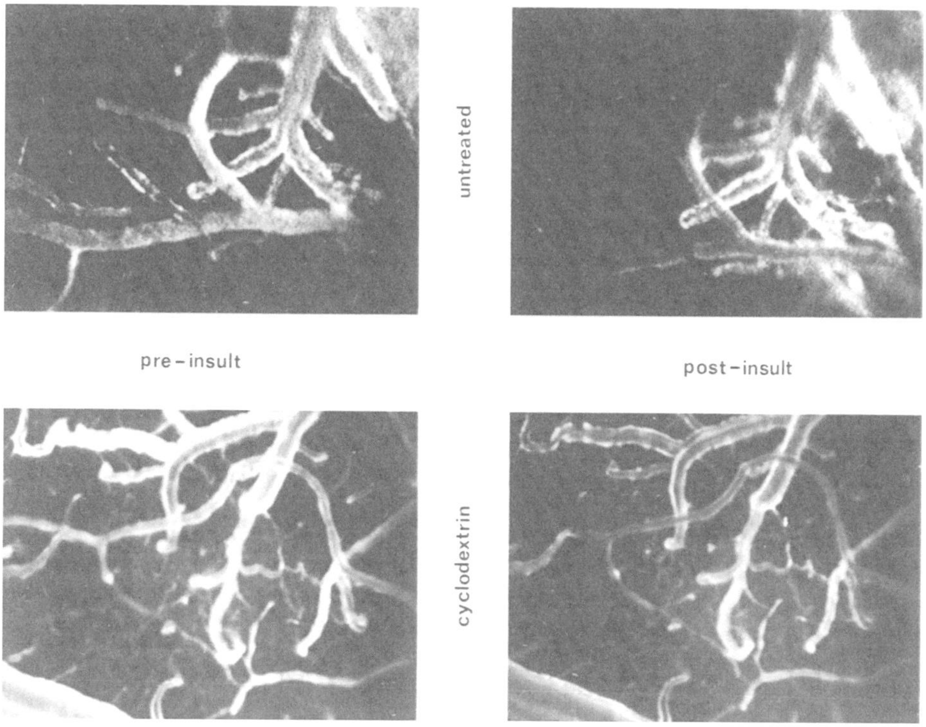

Fig. 2. The *upper left photomicrograph* depicts the normal integrity of cerebral microvessel capillary endothelium. The sharp image indicates that fluorescein-labeled dextran (mol. wt. 2 million) is restricted to the vascular compartment. In contrast, the cloudy appearance of the microvessels in the *upper right picture* results from the leakage of the fluorescent probe into the extravascular space following reactive intermediate attack in untreated rats. Also note the thrombotic occlusion of the small venules in the *left* of the photograph. The *lower left photograph* represents preinsult microcirculatory integrity in a solvent-treated rat. Unchanged appearance of the postinsult *lower right photograph* demonstrates the almost complete protection offered by the cyclodextrin solvent

dant, reduced glutathione, in various brain regions of normal rats with those derived from rats exhibiting postischemic SICs. Glutathione was decreased by 19% in hippocampal tissue from the epileptic rats. No such change was noted in the cortex, midbrain, or cerebellum. Thus, this sign of biochemical injury restricted to the selectivity vulnerable hippocampus is consistent with the electrophysiological evidence of diminished excitability. Although this observation indirectly implicates reactive intermediate-generated oxidative mechanisms in the genesis of ischemic brain damage, it does not define the source – neuronal or vascular.

Ischemia-damaged cerebrovascular endothelial cells may initiate neuronal injury. Earlier histological evidence from our brain slice studies noted extensive pericapillary neuronal damage associated with a different model of global ischemia [7]. De Clerck et al. [1] reported that flunarizine blocked the production of certain reactive intermediates from injured cerebral capillary endothelium. The drug also limits the extent of tissue injury from this source [10]. Therefore, we pro-

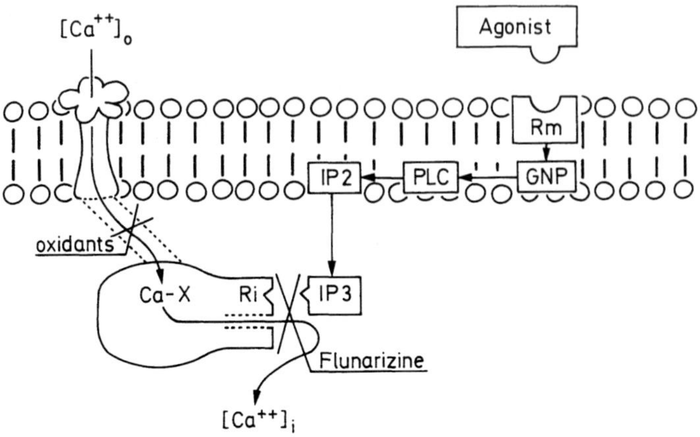

Fig. 3. Calcium may act as a second messenger for many receptor-mediated events. Activation of membrane-bound receptors (*Rm*) by the appropriate agonist (*A* = histamine, eicosanoids, norepinephrine, etc.) induces a guanine nucleotide-dependent regulatory protein (*GNP*) to activate phospholipase C (*PLC*). This enzyme extracts phosphatidyl inositol diphosphate (*IP2*) from the membrane lipid matrix into the cytosol, where it is phosphorylated to the 1,4,5-triphosphate (*IP3*). Its activation of receptors (*RI*) located on the intracellular organelles frees bound calcium (*Ca-X*) which increases cytosolic ionic calcium ($[Ca^{++}]_i$). Calcium then activates effector proteins like calmodulin and tropomyosin that ultimately result in physiological or pharmacological actions. IP3, or related substances, may also influence transport of ionized extracellular calcium ($[Ca^{++}]_o$) to intracellular storage sites. (Scheme and diagram modified from [5])

duced reactive intermediate injury by photoactivation (20-min exposure of high intensity 570 nm light) of the dye, rose bengal (4.5 mg/kg i.v.). The superoxide radical is generated from this reaction and initiates a host of destructive changes. Superoxide liberation caused platelet aggregation and later thrombotic occlusion in 86% of the capillaries and third-order venules in the isolated rat cremaster muscle. Ten-minute pretreatment with the flunarizine solution (100 μg/kg i.v.) maintained the patency in 63% of the microvessels. The cyclodextrin solvent effect was not significantly different from untreated (Table 3). Two hours after insult, at peak drug effect, these ketamine-anesthetized rats were sacrificed by asphyxia. The time to flattening of EEG and upper facial EMG traces was the same in both flunarizine-treated and untreated rats (Table 3). Although flunarizine protected peripheral capillaries from injury, it had no effect on the rate of anoxic depression of cerebral cortical (EEG) or brain stem (EMG) electrical activity.

Superoxide attack of the vascular endothelium may also result in disruption of the tight junctions. The reactive intermediates lead to elevations in cytosolic free calcium which stimulate endothelial cell contractile proteins [2]. We observed integrity of rat pial microcirculation through a temperature-regulated cranial window (Fig. 2, upper left panel).

Fluorescein-labeled high-molecular-weight dextran was restricted to the vascular compartment. Under superoxide-generating conditions similar to those used in the cremaster preparation, little thrombosis was seen. Instead, the fluorescent label leaked profusely (Fig. 2, upper right panel). In a preliminary evaluation, 10-min pretreatment with the flunarizine solution (100 μg/kg) or cyclodextrin (8 mg/kg) prevented thrombosis and extravascular fluorescence (Fig. 2, bottom right panel).

Flunarizine has antihypoxic effects when added to the solution bathing hippocampal slices in vitro [11]. However, this does not prove that the site of cerebroprotective action is neuronal. Indeed, Reid et al. [6] concluded that flunarizine has minimal neuronal effects in undamaged brain tissue. Therapeutic action was demonstrable only in damaged tissue with, we presume, a disruption of the blood-brain barrier. Thus, our results suggest that flunarizine protects the brain against initial damage by maintaining microvascular integrity and perfusion. In addition, we illustrate the efficacy of the new intravenous formulation and note that the cyclodextrin solvent also seems to have some cerebroprotectant properties of its own.

Irreversible cell injury is associated with the well-known accumulation of free cytosolic calcium. Entry of calcium through non-voltage-dependent channels into intracellular storage sites and release of the free ion from these sites into the cytosol (Fig. 3) appear to be controlled by inositol 1,4,5-triphosphate (IP3) [5]. In contrast to other calcium antagonists (classes I, II, III, and VI) and neuroprotectants (trifluoperazine, lidocaine), flunarizine is a potent inhibitor of IP3-induced calcium release [8]. Many of flunarizine's inhibitory actions (Table 1), including blockade of calcium "overload," may be explained by IP3 inhibition.

Acknowledgments. The authors are grateful to Drs. Calvin Lang and Theresa Chen for glutathione determinations and Dr. Dan Jiang for evaluating flunarizine's effect in rat cremaster muscle microcirculation. In vitro investigation was done by Drs. Farouk El-Sabban and Kenneth H. Reid. Development of the chest compression model was greatly aided by Dr. Markku Paloheimo. The work was supported by the Johnson & Johnson Focused Giving program, the Janssen Research Foundation, and the Veterans Administration. Samples of the flunarizine formulation and solvent were generously provided by Drs. T. Peters and A. Wauquier. DATEX Instrumentarium & Puritan-Bennett supplied the ABM and Cardiocap monitors used to monitor EEG, EMG, blood pressure, ECG, and end-tidal carbon dioxide.

References

1. De Clerck F, Loots W, Somers Y, Van Gorp L, Verheyen A, Wouters L (1985) Thromboxane A_2-induced vascular endothelial cell damage and respiratory smooth muscle cell contraction: inhibition by flunarizine, a calcium-overload blocker. Arch Int Pharmacodyn 274:4–23
2. De Clerck F, Van Gorp L, Verheyen F (1986) Microvascular permeability increase in rat skin: inhibition by flunarizine, a calcium entry blocker. Blood Vessels 23:50–52
3. Edmonds HL Jr, Schurr A, West CA, Shields CB (1985) Free radicals and iron toxicity in the genesis of postischemic brain damage. Mol Physiol 8:631–638
4. El-Sabban F, Reid KH, Edmonds HL Jr, Shields CB (1988) Effect of global ischemic insult on electrophysiological responses of rat hippocampal slices (abstr). Proc Soc Neurosci (in press)
5. Putney JW Jr (1987) Calcium-mobilizing receptors. Trends Pharmacol Sci 8:481–486
6. Reid KH, Marrannes R, Wauquier A (1987) Effects of flunarizine on rat cerebral cortex. Drug Dev Res 10:107–116
7. Tseng MT, Shields CB, Reid KH, Schurr A, Edmonds HL Jr (1985) Perivascular modification in postischemic hippocampus, an ultrastructural analysis (abstr). Stroke 16:5

8. Seiler SM, Arnold AJ, Stanton HC (1987) Inhibitors of inositol triphosphate-induced Ca^{2+} release from isolated platelet membrane vesicles. Biochem Pharmacol 36:3331–3337
9. Vanhoutte PM (1987) The expert committee of the World Health Organization on classification of calcium antagonists: the viewpoint of the raporteur. Am J Cardiol 59:3A–8A
10. Van Reempts J, Van Deuren B, Van de Ven M, Cornelissen F, Borgers M (1987) Flunarizine reduces cerebral infarct size after photochemically induced thrombosis in spontaneously hypertensive rats. Stroke 18:1113–1119
11. Wauquier A, Ashton D, Edmonds HL Jr (1984) Protection against cerebral hypoxia by antimigraine drugs. In: Amery WK, Van Nueten J, Wauquier A (eds) The pharmacological basis of migraine. Pitman, London, pp 134–148
12. Wauquier A, Ashton D, Clincke G, Fransen J, Gillardin J-M, Janssen PAJ (1986) Anticonvulsant profile of flunarizine. Drug Dev Res 7:49–60
13. Wauquier A, Edmonds HL Jr, Clincke G (1987) Cerebral resuscitation: pathophysiology and therapy. Neurosci Biobehav Rev 11:287–306

Studies on the Efficacy and Mechanism of Action of a Calcium Blocker After Global Ischemia

D. M. Rosenbaum, J. C. Grotta, F. M. Yatsu, and L. C. Pettigrew

Introduction

After numerous studies of calcium channel blockers, the efficacy and mechanism of action are still debated [1, 2, 7, 10, 11, 16, 17, 21, 22, 24, 28, 29]. Two possible mechanisms have been proposed for their beneficial effect: (a) relaxation of vascular smooth muscle and increase in cerebral blood flow, and (b) prevention of calcium flux into neurons, thereby limiting activation of phospholipases, proteases, and consequent membrane and protein degradation and production of damaging metabolic by-products such as free radicals [3–5, 15, 20, 23, 25, 26, 30].

Despite the uncertainty about the mechanisms of action, the results of animal experiments to date have been sufficiently encouraging that therapeutic trials in human stroke patients are underway [6]. This paper will review our experiments to date using nicardipine (NC), a dihydropyridine calcium channel blocker, and will discuss some of our ongoing studies in this area.

Preliminary Studies

We began our investigations on the effects of NC given before and after ischemia on neuronal function as measured by evoked potentials (EPs) as well as histological outcome [10]. Pulsinelli's four-vessel ischemia model was chosen to produce 20 min of forebrain ischemia [19]. Animals received NC by continuous subcutaneous infusion via an osmotic pump implanted in the dorsum of the neck to provide drug at a rate of 0.05 mg/kg/h for 96 h. In the pretreated group (PR), the pump was inserted 24 h before ischemia and continued for an additional 72 h. In the postischemic animals (PO), the pump was inserted at the end of the 20-min ischemic period.

Plasma NC levels were obtained in both groups at intervals of 4, 24, 48, 72, and 96 h. Somatosensory evoked potentials (SEPs) and brain stem evoked potentials (BEPs) were performed 72 h after ischemia in control, PR, and PO groups. Cortical SEPs were recorded from a platinum subdermal electrode just anterior to the interaural line ("vertex" position). Preliminary studies in our laboratory

Department of Neurology, The University of Texas Medical School at Houston, Houston, Texas 77030, USA.

Cerebral Ischemia and Calcium
Ed. by A. Hartmann and W. Kuschinsky

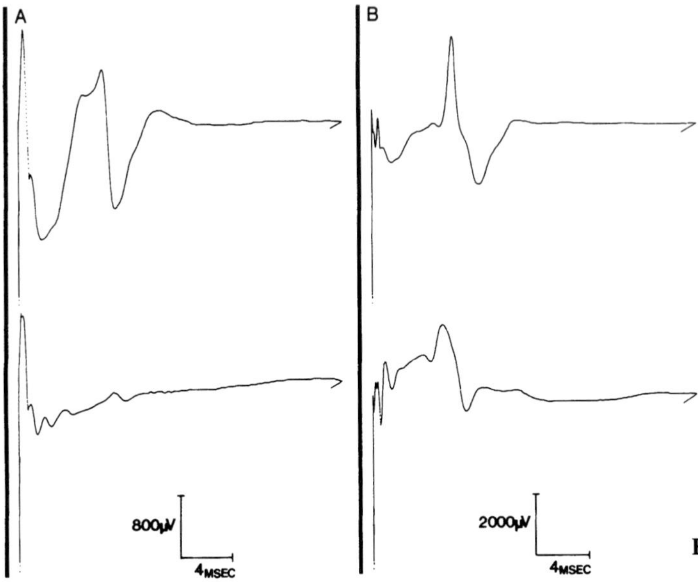

Fig. 1

have shown that a recording electrode at this position overlies the parietal cortex. Stimulation electrodes were inserted bilaterally in the perinasal area. BEPs were recorded between the vertex electrode and a subdermal electrode placed just anterior to the pinna of the stimulated ear. Ischemic cell changes were quantitated histologically (H&E) in the hippocampi, and histological changes were ranked by overall severity of ischemic damage [9].

The results of the study showed no difference in physiological parameters except for a modest but statistically significant reduction of mean arterial blood pressure (MABP) in PR animals. When hippocampi were ranked for density and linear extent of cellular changes, no significant difference was found between the three groups.

Nicardipine was found in the blood within 4 h of placement of the subcutaneous infusion pump; however, a predictable plasma level of 6–10 ng/ml was not reached until 20 h, at a time when MABP was lowest.

Somatosensory evoked potential amplitudes 72 h after ischemia were reduced in all three groups compared to normal; however, the mean amplitude of the N1 was significantly higher in both PR ($P<0.025$) and PO ($P=0.05$) animals compared to ischemic controls (Fig. 1 A, B). In all animals tested, BEPs remained normal at 72 h. The difference between the outcome of SEPs and BEPs was attributed to the sparing of brain stem structures (using only 20 min of ischemia in this model).

Despite the uncertainties in the anatomical localization of SEP recordings from the scalp, it can be concluded that there was a lack of correlation between the amount of histological change and cellular function since similar neuropathological damage occurred in thalamus and in hippocampal and parietal cortices in all groups despite significant differences in SEPs.

The next study done was designed to determine whether biologically active amounts of NC are available to dihydropyridine calcium receptors on channels in neurons as well as those of the vascular receptors. Seven normal rats and seven animals studied 24 h after production of 30 min of forebrain ischemia (using the Pulsinelli model) received a 0.2-ml bolus of 1 μg Ci of ^{14}C-NC HCl and ^{3}H-water via a needle stick into a common carotid artery in the neck. Another five normal rats were treated identically except for the substitution of 1 μg Ci of ^{14}C-inulin (IN) for the radiolabeled NC. Each animal was decapitated 5 s after injection, and its brain was removed. Assuming that ^{3}H-water is completely extracted into brain on its first-pass, the percent first-pass extraction of ^{14}C-NC or ^{14} C-IN by non-ischemic brain was calculated by the formula:

$$\frac{^{14}\text{C (brain)}/^{3}\text{H (brain)}}{^{14}\text{C (infusate)}/^{3}\text{H (infusate)}} .$$

When intravascular NC was corrected by subtracting ^{14}C from ^{14}C-NC and incomplete extraction of water was corrected by multiplying the result by 0.84, the mean first-pass extraction of NC was calculated to be 30.7% into nonischemic cerebral hemisphere and 42.3% into hippocampi. Increased extraction of NC occurred in ischemic hemispheres but not in hippocampus, where ischemic damage is most severe in this model [19].

These data show that NC uptake does occur in normal and ischemic brain and lend support to the concept that NC is available in sufficient quantities to affect neuronal calcium channels.

Recent Work

Our next study was designed to evaluate the effect of NC on learning and to clarify the effect of the drug on cerebral blood flow (CBF) and measurement of mitochondrial function and phospholipid metabolism.

Methods

Induction of Ischemia

Male Wistar rats weighing 300 g were rendered ischemic by a modification of the Pulsinelli method [19]. All rats were monitored for blood pressure, arterial blood gases, temperature, plasma glucose, and EEG throughout the entire ischemic period. Only rats demonstrating loss of all cortical electrical activity on EEG and loss of the righting reflex throughout the entire ischemic period were regarded as having suffered adequate ischemia and were considered for further analyses. For all studies, rats were rendered ischemic by occluding the carotid arteries for 30 min.

In the NC-treated rats, an Alzet osmotic pump (model 2ML1, Alza Corporation, Palo Alto, California) was implanted subcutaneously in the suboccipital region immediately following vertebral artery surgery, 24 h before carotid ligation. Enough NC was dissolved in 3 ml sterile water to deliver a constant infusion of 0.05 mg/kg/h. As in prior studies [10], this dosing paradigm achieved steady state levels of 6–10 ng/ml after 24 h and was associated with a 10% reduction in MABP.

Assessment of Learning

Nicardipine-treated ischemic rats were tested on an eight-arm radial maze (each arm S6V10 cm) and compared to untreated controls, NC-treated sham-operated controls, and untreated, unoperated controls. All rats were given trials for 65 days, after maze conditioning for 15 days. For each trial, the rat was placed on the center platform facing a random direction and allowed to make choices until each of five baited arms had been chosen, until 10 min had elapsed, or until 16 total choices had been made. The same five arms were baited for all 65 trials for any single rat. A reference error occurred when a rat chose an arm not baited with food and a working error occurred when a rat chose the same arm twice, regardless of whether or not it was a correct or incorrect arm.

Measurement of CBF

Hemispheric and hippocampal CBF was measured by the ^{14}C-butanol indicator fractionation technique [8, 27]. A bolus of 2.5 μg Ci ^{14}C-butanol (specific activity 0.88 mCi/mmol) was injected into the vena cava coincident with the start of continuous removal of arterial blood from the tail artery through PE-50 tubing into a 250-μl Hamilton syringe using a syringe pump set at 250 μl/min. After 10 s of blood collection, the rat was decapitated at the same moment the withdrawal pump was turned off. The brain was rapidly removed and the hippocampus dissected totally from one cerebral hemisphere. The hippocampus and the other entire cerebral hemisphere were dissolved separately and processed for scintillation counting. ABGs were measured at the time of CBF determination in all rats.

Cerebral blood flow was measured in NC treated ischemic rats immediately following ischemia and 2 h after ischemia and compared with untreated ischemic controls.

Biochemical Analysis

Adenosine triphosphate (ATP), PCr, and glucose measurements were carried out in NC-treated ischemic rats and compared with untreated ischemic and untreated unoperated rats. Measurements were performed either at the end of ischemia or after 2 h of reperfusion.

All rats were killed by in-site freeze-fixation of the brain (Ponten technique) [12, 18]. The frozen brains were dissected out of the skulls and secured by brass plates with inert brain paste and cut into 20-μm sections in a microtome-cryostat at 14° C. Samples from the strata pyramidale of the paramedian and lateral regions of the CA-1 and CA-2 sectors were isolated by hand and the samples were transferred to oil well racks for assays of ATP, PCr, and glucose. All of these procedures have been described in detail [13, 14].

Measurement of Eicosanoid Metabolites

Thromboxane B_2, the stable metabolite of thromboxane A_2, and 6-keto PGF_{1a}, a derivative of PGI_2, were measured immediately after the 30-min period of ischemia in both NC-treated rats and untreated ischemic controls. The brains were frozen (as above). The brain was homogenized, diluted with acetic acid, centrifuged, and passed through glass filter paper to remove proteinaceous debris. The remaining solution was eluted over a reverse-phase Sep-Pak C-18 column (Water Associates, Milford, MA). The column was then washed, eluted, dried, and redissolved in 1 ml radioimmunoassay.

Results

There were no mortality or physiological variable differences between the groups, except that the MABP was lower in NC-treated rats ($t=2.44$, $0.01<P<0.025$). This was anticipated.

Learning

Thirty rats were studied in four groups: untreated unoperated controls ($n=4$); NC-treated sham-operated controls ($n=6$), untreated ischemic controls ($n=10$),

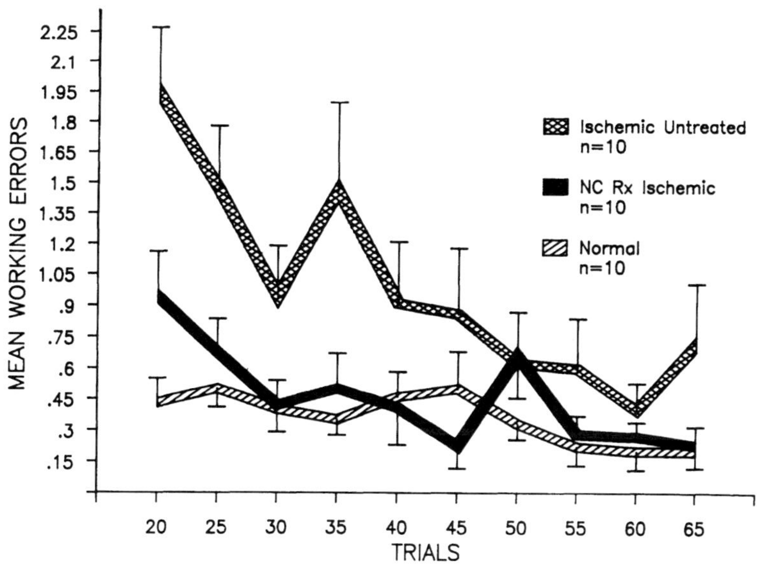

Fig. 2

and NC-treated ischemic rats ($n=10$). A two-way repeated-measure analysis of variance (ANOVA) revealed that all groups demonstrated improved performance over the testing period. There was no significant difference between NC-treated ischemic rats, NC-treated sham-operated controls and untreated unoperated controls in reference or working errors, while untreated ischemic controls made significantly more working errors than each of the other three groups ($F=7.705$, $P<0.021$) (Fig. 2). By the end of 50 trials, untreated ischemic controls were no longer abnormal, and at no time was there a difference in reference errors between untreated ischemic controls and the other groups.

CBF

Baseline hemispheric CBF in untreated unoperated controls was 118 ± 19 ml/100 g/min. There was no significant difference between baseline hemispheric hippocampal CBF. CBF was reduced to approximately 10% of baseline levels in both untreated and NC-treated rats at the end of 30 min of ischemia. CBF was still depressed 2 h after ischemia in untreated controls but was significantly higher in NC-treated rats.

ATP, PCr, and Glucose

In untreated rats, 30 min of ischemia resulted in reduction of ATP, PCr, and glucose in the paramedian CA-1. In the lateral CA-1 and CA-2, where histological damage is less severe, there was a depression of these metabolites, with significant reduction of only PCr in the lateral CA-1 and glucose in CA-2.

In the NC-treated rats a trend toward higher metabolite levels compared to untreated ischemic controls was seen immediately after ischemia but this did not reach statistical significance.

After 2 h of reperfusion, ATP had returned to normal in all regions in NC-treated rats and was significantly higher in the paramedian CA-1 than in untreated ischemic controls ($P<0.05$), Kurskal-Wallis one-way ANOVA) (Fig. 3).

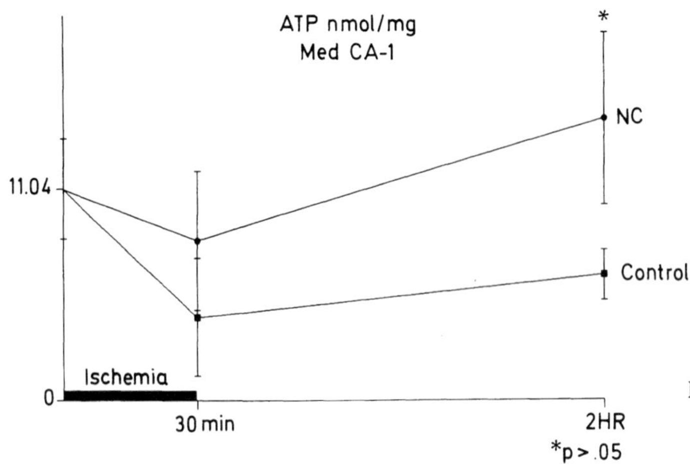

Fig. 3

PCr rose above baseline levels in all regions in NC-treated rats and was significantly higher in the paramedian and lateral CA-1 than in untreated ischemic controls ($P<0.05$, Kruskal-Wallis one-way ANOVA).

Eicosanoid Metabolites

There was no difference between NC-treated and untreated ischemic rats (Student's *t*-test) and both groups showed higher levels of both metabolites compared to baseline.

Conclusions

A number of conclusions can be drawn from the above studies:

1. NC is available to binding sites and calcium channels in neurons.
2. In the rat four-vessel ischemia model, NC improves functional outcome as measured by SEPs and learning tasks.
3. CBF is increased by NC in the reperfusion period.
4. NC causes more rapid restoration of mitochondrial function in CA-1 after 2 h of reperfusion as determined by absolute measurement of ATP and PCr.
5. NC does not improve ischemic cell damage in hippocampus as measured by light microscopy.
6. NC does not affect the production of eicosanoid metabolites immediately after ischemia.
7. NC positive effect appears to occur during the reperfusion period and not during ischemia.

Whether the drug acts primarily by improving CBF or by preventing intraneuronal calcium-activated processes cannot be conclusively answered on the basis of the results of these studies.

References

1. Barnett GH, Bose B, Little JR et al. (1986) Effects of nimodipine on acute focal cerebral ischemia. Stroke 17:884–890
2. Berger JR, Busto R, Ginsberg MD (1984) Verapamil: failure of metabolic amelioration following global forebrain ischemia in the rat. Stroke 15:1029–1032
3. Black KL, Hoff JT (1985) Leukotrienes increase blood-brain barrier permeability following intraparenchymal injections in rats. Ann Neurol 18:349–351
4. Deshpande JK, Siesjo BK, Wieloch T (1987) Calcium accumulation and neuronal damage in the rat hippocampus following cerebral ischemia. J Cereb Blood Flow Metab 7:89–95
5. Farber JL, Chien KR, Mittnacht S (1981) The pathogenesis of irreversible cell injury in ischemia. Am J Phys 102:271–281
6. Gelmers HJ, Gorter K, deWeerdt CT, Wiezer HJA (1988) A controlled trial of nimodipine in acute ischemic stroke. N Engl J Med 318:203–207

7. Gotoh O, Mohamed A, McCulloch J et al. (1986) Nimodipine and the haemodynamic and histopathologic consequences of MCA occlusion in the rat. J Cereb Blood Flow Metab 6:321–331
8. Grotta JC, Ostrow P, Fraifeld E et al. (1985) Fibrinogen blood viscosity and cerebral ischemia. Stroke 16:192–198
9. Grotta JC, Ostrow P, Fraifeld E et al. (1985) Reduction of fibrinogen and viscosity by plasma exchange dose not significantly alter ischemic injury. Stroke 16:192–198
10. Grotta JC, Spydell J, Pettigrew LC et al. (1986) The effect of nicardipine on neuronal function following ischemia. Stroke 17:213–219
11. Hakim AM (1986) Cerebral acidosis in focal ischemia. II. Nimodipine and verapamil normalize cerebral pH following MCA occlusion in the rat. J Cereb Blood Flow Metab 6:676–690
12. Lowry OH, Passonneau JV, Hasselberger FX (1964) Effect of ischemia on known substrates and cofactors of the glycolytic pathway in brain. J Biol Chem 239:18–42
13. Lowry OH, Passonneau JV (1972) A flexible system of enzymatic analyses. Academic, New York
14. McCandless DW, Feussner G, Lust WD et al. (1979) Metabolite levels in brain following experimental seizures. J Neurochem 32:743–753
15. McCord JM (1985) Oxygen-derived free radicals in post-ischemic tissue injury. N Engl J Med 312:159–163
16. Milde LN, Milde JH, Michenfelder JD (1986) Delayed treatment with nimodipine improves CBF after complete cerebral ischemia in the dog. J Cereb Blood Flow Metab 6:332–337
17. Mohamed AA, Gotoh O, Graham DI et al. (1985) Effect of pretreatment with the calcium antagonist on local CBF and histopathology after MCA occlusion. Ann Neurol 18:705–711
18. Ponten CL, Ratcheson RA, Salford LG, Siesjo BK (1973) Optimal freezing conditions for cerebral metabolites in rats. J Neurochem 21:1127–1138
19. Pulsinelli WA, Brierly (1979) A new model of bilateral hemispheric ischemia in the unanesthetized rat. Stroke 10:267–272
20. Raichle ME (1983) The pathophysiology of brain ischemia. Ann Neurol 12:2–10
21. Sakabe T, Nagai I, Ishikawa T et al. (1986) Nicardipine increases CBF but does not improve neurologic recovery in a canine model of complete cerebral ischemia. J Cereb Blood Flow Metab 6:684–690
22. Sauter A, Rudin M (1987) Calcium antagonists reduce the extent of infarction in rat MCA occlusion model as determined by quantitative MRI. Stroke 17:210–216
23. Siesjo BK, Bendek G, Koide T et al. (1985) Influence of acidosis on lipid peroxidation in brain tissues in vivo. J Cereb Blood Flow Metab 5:253–258
24. Steen PA, Newberg LA, Milde JH, Michenfelder JD (1983) Nimodipine improves CBF and neurologic recovery after complete cerebral ischemia in the dog. J Cereb Blood Flow Metab 3:38–43
25. Vanhoutte PM (1982) Effects of calcium entry blockers on tissue hypoxia. J Cereb Blood Flow Metab 2:542–544
26. Van Neuten JM, Vanhoutte PM (1980) Improvement of tissue perfusion with inhibitors of calcium ion influx. Biochem Pharmacol 29:479–481
27. Van Uitert RL, Levy DE (1978) Regional brain blood flow in the conscious gerbil. Stroke 9:67–72
28. Vibulsresth S, Dietrick D, Busto R, Ginsberg MD (1987) Failure of nimodipine to prevent ischemic neuronal damage in rats. Stroke 18:210–216
29. Vorstrup S, Andersen A, Blegvad N, Paulson OB (1986) Calcium antagonist (P4-108-068) treatment may further decrease flow in ischemic areas in acute stroke. J Cereb Blood Flow Metab 6:222–229
30. Wieloch T, Harris RJ, Siesjö BK (1982) Brain metabolism and ischemia: mechanism of cell damage and principles of protection. J Cereb Blood Flow Metab 2:55–59

Calcium Antagonists in Acute Cerebral Damage: Therapy or Risk for the Brain?

M. R. Gaab, H. G. Höllerhage, W. P. Sollmann, and E. Rickels

Introduction

Calcium antagonists, which were primarily used in the field of cardiology [8], are increasingly being investigated in a large variety of diseases. As is shown by this symposium, cerebral pathology seems to have become the second main area of clinical use.

Besides chronic disorders like cerebral sclerosis, migraine, and epilepsy, acute and severe brain damage is also suggested to benefit from calcium entry blockers. After subarachnoid hemorrhage (SAH), therapy with nimodipine is now clinically established. In supplementing the less conclusive study from Allen et al. [1], the Canadian multicenter trial [13] and the British cooperative study (Teasdale et al., this volume) show a convincing improvement in mortality and neurological outcome with oral nimodipine. With this therapy we also found an increase in CBF without signs of "intracerebral steal" and without significant side-effects [3, 6]. However, intravenous administration of nimodipine as well as other calcium antagonists has not been proven to yield any benefit in SAH. On the contrary, in initial rCBF investigations using stable xenon CT we found signs of intracerebral steal after 10 μg/kg nimodipine given over 20 min. With nimodipine given intravenously we also found a tendency to arterial hypotension, especially at doses above 1 mg/h, with very individual characteristics (5%–10% of patients [3, 6]).

These clinical observations alone suffice to demonstrate the risk of side-effects not only in patients with SAH but also in those with other kinds of acute cerebral damage where the use of nimodipine is suggested, e.g., stroke [8], cerebral hypoxia [14], and head injury [12]. In these patients not only the disease endangers the balance between cerebral blood flow, intracranial pressure, blood-brain barrier function, and edema formation. The interference of calcium antagonists with these factors may also be unfavorable: cerebral autoregulation, blood pressure control by peripheral vascular resistance, and the K^+ clearance function of the glia are all at risk with calcium antagonists. We therefore tried to evaluate the risk in clinical and experimental investigations with nimodipine.

Neurosurgical Department, Medizinische Hochschule (MHH), Konstanty-Gutschow-Str. 8, D-3000 Hannover 61, FRG.

Cerebral Ischemia and Calcium
Ed. by A. Hartmann and W. Kuschinsky

Investigations and Results

Clinical Observations

The relaxing effect of nimodipine on the smooth muscle of intracranial vessels involves the risk of autoregulation failure and consequent intracranial pressure (ICP) increase due to a rise in cerebral blood volume (CBV). Uncontrolled dilation of extracranial vessels may interfere not only with the systemic arterial pressure (SAP) and the cerebral perfusion pressure (CPP) but also with the intrapulmonary vessels which control the right-to-left intrapulmonary shunting. These two factors were clinically considered.

ICP, SAP, and CPP

In 12 patients with severe head injury but without intracranial space-occupying lesions, as confirmed by CT, nimodipine was infused intravenously in increasing doses [3]. ICP (GAELTEC microtransducer, epidurally [7]) and SAP (femoral catheter [7]) were continuously recorded. The data were evaluated on-line by computer (Neuromonitor [7]), including the analysis of intracranial pulse amplitudes (PAs).

After sampling the resting pressures for 4 h, nimodipine infusion was started at a rate of 0.5 mg/h for 2 h, followed by rates of 1 mg/h and 2 mg/h for 2 h each, if the increase in dose was tolerated in terms of the ICP and CPP.

In all patients, a significant drop in SAP of 12 ± 5 mmHg was seen with 2 mg/h, whereas the small SAP decreases with 0.5 and 1 mg/h were insignificant. A moderate tendency to a rise in ICP was seen only with 2 mg/h (+3.5 mmHg [3]), which was not significant ($P = 0.2$). The drop in CPP with 2 mg/h, however, was clearly significant (-14 ± 8 mmHg).

In the individual patient, however, much more dramatic effects can be seen than would be expected from these statistics: Whereas in the ten patients without intracranial hypertension (resting ICP below 15 mmHg) only the drop in SAP, i.e., no increase in ICP, was noted, both patients with intracranial hypertension (mean resting ICP > 25 mmHg, tendency towards pressure waves) were severely disturbed [3]. In both patients, a steep increase in ICP combined with a drop in SAP was seen, resulting in a fall in CPP to 42 mmHg (1 mg/h) and even 20 mmHg (2 mg, [3])! In both patients, the therapy with 2 mg/h had immediately to be terminated. Thereafter ICP, SAP, and CPP did continuously normalize within 4 h in both patients [3].

That the rise in ICP was caused by cerebral vasodilatation resulting in an increase in CBV is suggested by the analysis of the intracranial PAs [3]: The mean resting PA of 4.6 ± 0.8 mmHg increased to 5.6 ± 0.8 mmHg with 2 mg/h. This increase in PA of 1 mmHg corresponds to > 20%!

Systemic Side-effects

Information on systemic side-effects was derived from a running randomized study performed in our department on intravenous nimodipine after SAH. Whereas no comments on neurological outcome in these patients can be made here, we constantly found systemic side-effects using the Swan-Ganz catheter during intensive care [9]. These side-effects have already been described in the literature [2, 9]:

1. A drop in SAP, especially with doses of >15 μg/kg/h, with a very individual reaction and course. In about 5% of the patients, the drop in SAP does not stabilize and requires a dose reduction.
2. An increase in cardiac output of about 15% (compensatory for the decrease in peripheral vascular resistance).
3. A permanent (!) increase in intrapulmonary shunting (Q_s/Q_t) of more than 10% (Fig. 1). That this adverse increase in intrapulmonary shunting is caused by the nimodipine infusion is shown by the rapid normalization of the pathological shunt fraction after the end of the infusion ("on/off effect," Fig. 1), which confirms the initial observation from [9].

Animal Experiments

The clinical increase in ICP as well as the rise in CBF [3, 6] was suspected to be attributable to a disturbance of cerebral autoregulation, as suggested by Symon et al. [17]. We therefore investigated the effects of nimodipine on autoregulation, blood-brain barrier function, brain edema formation, and cerebral ion homeostasis in rat experiments [4, 5, 10, 11, 15].

Cerebral Autoregulation

Two randomized groups of ten rats each were treated with nimodipine (12–14 μg/kg/h) or with the same volume of the alcoholic nimodipine vehicle via the femoral vein. In anesthesia with ketamine (100 mg/kg) and xylazine (20 mg/kg) intraperitoneally, the cortical cerebral blood flow (lCBF) was measured on the right hemisphere after trepanation using H_2 clearance with a multiwire surface electrode [5]. lCBF was calculated from the clearance after breathing 8% H_2, excluding the first half-time from evaluation [10]. The SAP was monitored in addition (femoral catheter).

With repetitive measurements of lCBF, variations in SAP were induced by infusion of 6% hydroxyethyl starch with 0.6% norfenefrine (Novadral, pressure increase) and by subsequent hemorrhage (pressure decrease).

In the controls, a typical autoregulation curve of the lCBF was found (lower limit ca. 70 mmHg, upper limit ca. 140 mmHg, Fig. 2). At this plateau the lCBF was regulated at a nearly constant level. In the nimodipine group, however, the autoregulation was almost abolished (Fig. 2): a slight decrease but no real plateau was seen in the range where autoregulation would be expected. The lCBF was

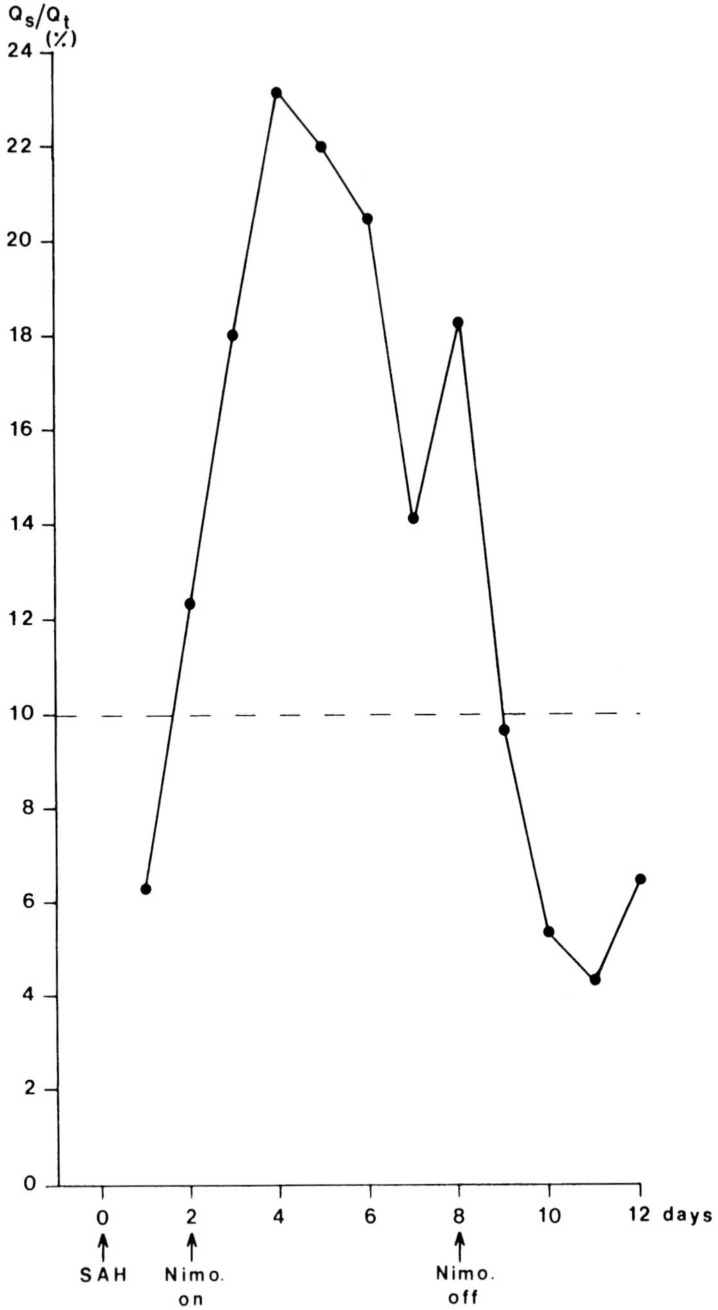

Fig. 1. The course of intrapulmonary shunting in a patient with SAH shows a clear increase during i.v. nimodipine treatment

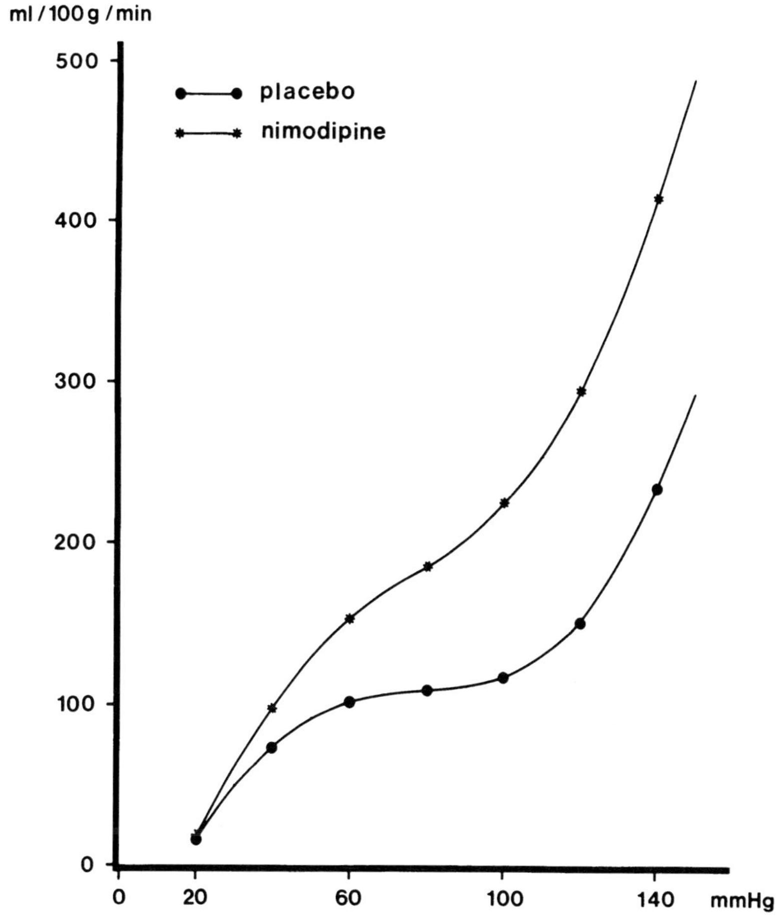

Fig. 2. Nimodipine and autoregulation. The autoregulation is impaired; instead of the normal plateau a steep lCBF/blood pressure correlation is found

markedly increased by nimodipine at all blood pressure levels. Interestingly, in three of the ten rats treated with nimodipine a massive SAH was observed at mean SAP levels of about 150 mmHg. In contrast, only one SAH was seen in the controls at SAP > 200 mmHg!

Blood-Brain Barrier (BBB) Function

Again two groups of ten rats were randomized and treated with the same protocol. Twenty minutes after infusion of nimodipine/vehicle, 15 mg/kg body wt. 2.5% Evans blue was injected. The SAP was then elevated to 180 mmHg systolic by infusion of 0.6% norfenefrine and was maintained at this level for 6 min. The animals were then sacrificed and perfused with Ringer's solution; the brains were removed, examined for Evans blue extravasation (Table 1), and frozen in nitrogen. Ten-micron frozen sections were examined for Evans blue fluorescence

Table 1. Investigation of blood-brain barrier by Evans blue

Evans blue staining	+	–	Total
Nimodipine	7	3	10
Placebo (vehicle)	1	9	10
All animals	8	12	20

$\chi^2 = 7.13$, $P < 0.01$.

[4]. Only in one control animal was some Evans blue extravasation found; this was limited to a patchy staining of the cerebellar hemispheres. In seven nimodipine-treated animals, however, diffuse Evans blue staining was observed (significant, Table 1). In these animals, fluorescence microscopy revealed that the dye had left the intravascular space, and the entire neuropil was diffusely stained.

Edema Formation

Again, groups of ten rats were randomized and treated with nimodipine (1 mg i.p.) or with vehicle. In addition, one group of nimodipine-treated animals received 3 mg norfenefrine to antagonize the constant drop in SAP seen with nimodipine alone [4, 11]. Fifteen minutes before this therapy, a cold brain injury was produced in each rat by placing a cooling stamp on the right skull [4, 11]. Blood pressures were measured for 2 h, and 24 h after injury the brains were removed and investigated for edema parameters (contents of water, sodium, and potassium [4, 11]).

In animals treated with vehicle, no decrease in SAP was seen after injury. In animals receiving nimodipine intravenously, however, a drop in SAP to 80 mmHg (60%) was observed. In these animals, the water content of the right hemisphere was less than in the vehicle animals. However, in spite of only mildly elevated water content in the injured hemisphere, the sodium uptake in the cold-injured tissue was not reduced compared to the controls – it was even higher [4, 11]. The suspicion that the reduced water uptake after nimodipine alone does not indicate a better preservation of the BBB but only a reduced water filtration due to lack of "transmural pressure" (SAP!) is confirmed by the data from animals treated with nimodipine and norfenefrine. As the drop in SAP is antagonized at least in part in these animals, much more brain edema develops: water and sodium content in the right hemisphere are significantly increased compared to untreated as well as to vehicle animals [4, 11]!

→

Fig. 3 a–c. Nimodipine in rats under the conditions of temporary and permanent anoxia. Laser Doppler lCBF decreased with nimodipine, with no posthypoxic hyperperfusion (**a**); extracellular K^+ instability (**b**) prevented any delay in the terminal Ca^{2+} decrease (**c**). Scale spread at first hypoxia – 90 s between the two *arrows*

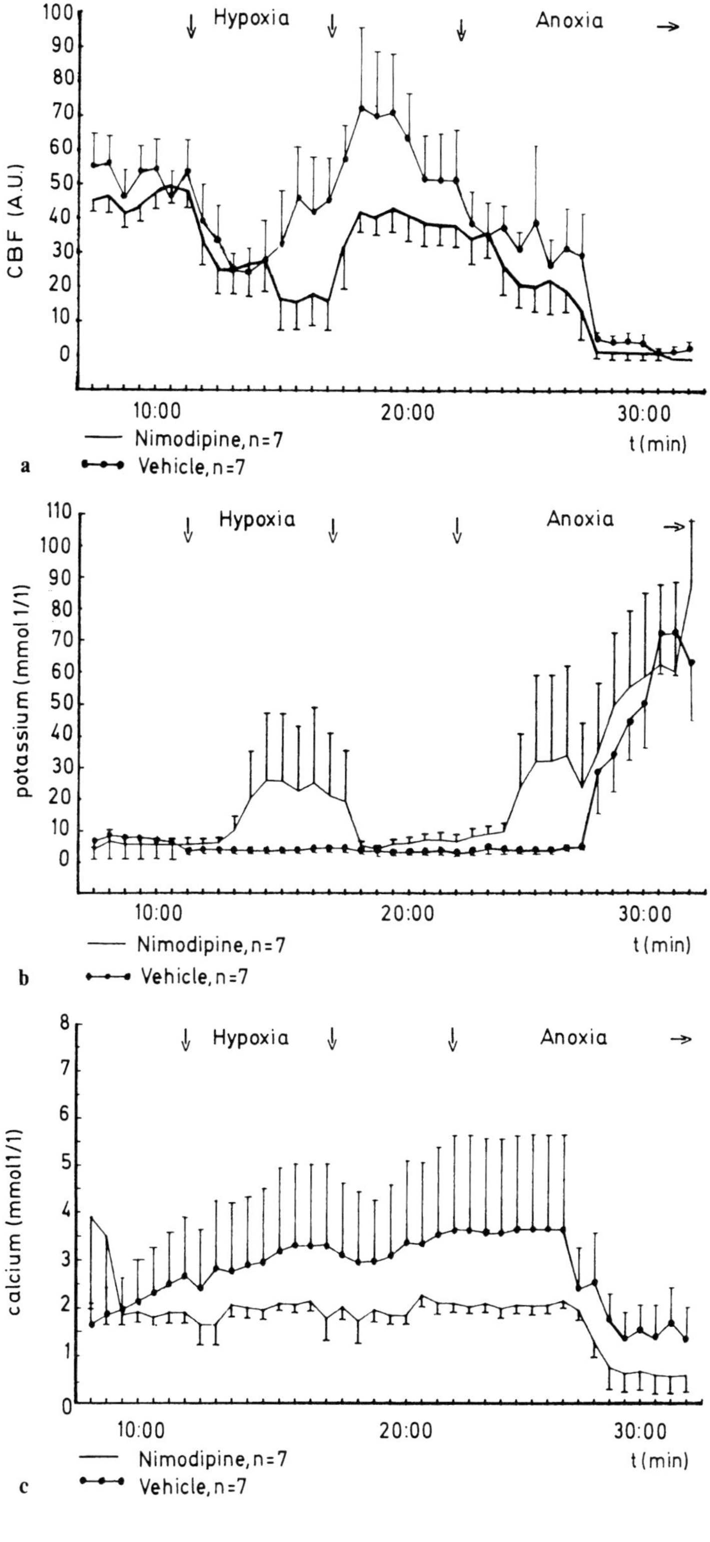

Hypoxia
Anoxia
CBF (A.U.)
10:00
20:00
30:00
t (min)
Nimodipine, n=7
Vehicle, n=7
a
Hypoxia
Anoxia
potassium (mmol 1/1)
10:00
20:00
30:00
t (min)
Nimodipine, n=7
Vehicle, n=7
b
Hypoxia
Anoxia
calcium (mmol 1/1)
10:00
20:00
30:00
t (min)
Nimodipine, n=7
Vehicle, n=7
c

Ion Homeostasis

Randomized groups of animals were treated with nimodipine (12–14 μg/kg/h) or by vehicle intravenously. On the free brain surface, relative lCBF (laser Doppler), cortical PO_2 (surface electrode [4, 5]), and extracellular K^+ and Ca^{2+} (ion-selective microelectrodes [15]) were continuously measured and evaluated by computer [15]. The EEG was registered over both hemispheres and evaluated by FFT; SAP and central venous pressures were measured in addition.

After 15 min rest, a 90-s hypoxia was achieved by breathing pure N_2, followed by a 15-min recovery period; a final anoxia by breathing N_2 was then induced (Fig. 3). Surprisingly the laser Doppler "lCBF" (Fig. 3a) was not elevated but decreased by nimodipine during temporary and permanent anoxia (Fig. 3a); this could be explained by the much steeper drop in SAP with hypoxia in these animals. Whereas a clear posthypoxic hyperperfusion was observed in the vehicle animals, no hyperperfusion was found with nimodipine infusion (Fig. 3a).

Regarding the extracellular Ca^{2+}, the terminal drop was *not* delayed by nimodipine; a striking finding was the increase in extracellular Ca^{2+} before the terminal breakdown (Fig. 3c). The extracellular K^+ was significantly *elevated* with nimodipine (Fig. 3b). Whereas the initial 90-s hypoxia period was well tolerated by the control animals without any extracellular spread in K^+, we found a significant K^+ increase with nimodipine as early as 40–50 s after the onset of hypoxia (Fig. 3b). This instability of extracellular cerebral K^+ introduced by nimodipine in hypoxia corresponds to the findings of Symon [16, 17] in experimental ischemia.

Discussion

Our clinical observations and our experiments are restricted to nimodipine as a calcium antagonist. However, the combination of possible therapeutic activities with considerable (?) side-effects suggests that evaluation of other calcium antagonists in acute cerebral damage must be cautious.

No doubt remains regarding the therapeutic or – more likely – the prophylactic effect of nimodipine given orally after SAH. This is now the therapeutic standard in the prevention of delayed neurological deficit after SAH [1, 13; Teasdale et al., this volume], and the risks of oral administration seem to be acceptable. However, with intravenous infusion and perhaps with higher enteral dosages, the side-effects may override any benefit and could become a real risk for the patient, especially in the unstable conditions of vital functions after severe acute brain damage. No doubt exists on the risk of a dramatic drop in blood pressure with a too rapid infusion. In addition to the data presented, drops in systolic arterial pressure to below 90 mmHg, especially with doses >2 mg/h, were observed in SAH patients. This is already a risk for the "normal brain"; in patients with acute brain damage, however, the combination of a drop in SAP with an increase in ICP, resulting in an overproportional decrease in CPP, may be fatal. It is also obvious that a combination of nimodipine therapy with hypervolemia and artificial hypertension, which was suggested in patients with symptomatic cerebral vasospasm after SAH, may be dangerous: The impairment of cerebral autoregulation and the risk of BBB damage at a lower upper SAP limit clearly speaks against a combination of nimodipine, hypervolemia, and hypertension.

An elevation of the SAP is often required in the ICU using a higher intravenous dose of nimodipine; however, with preexisting BBB damage and brain swelling, the edema formation can be enhanced by the combination of nimodipine and hypertension. The increase in extracellular K^+ in our hypoxia experiments, which is confirmed by Symon's observations in ischemia [16, 17], may also indicate a considerable risk for clinical treatment of acute brain damage by means of calcium antagonists. The increase in the Ca^{2+} break-in threshold achieved by nimodipine is antagonized by this instability of extracellular potassium, and could prevent any protective net effect on ion disturbances in the injured brain. However, results in experimental SAH are different; the protection of K^+ spread in such experiments [16] may indicate a protective effect under the special conditions of SAH.

Nevertheless, the demonstration of considerable risk factors, at least with the dihydropyridine nimodipine, suggests that a reserved and cautious attitude is appropriate toward the use of calcium antagonists in acute brain damage, and that at least well-monitored further investigations are needed. Not only the risk of specific intracranial side-effects like intracranial hypertension and impairment of autoregulation and BBB function, but also extracranial reactions must be considered. In patients with acute brain damage at the margin of decompensation, a drop in blood pressure and especially the considerable increase in intrapulmonary shunting may at least be inconvenient.

Summary

Cerebral disorders are a new major area of therapy with calcium antagonists. In acute brain damage requiring intensive care, however, not only the desired therapeutic effects but also the risk of side-effects must be considered. In clinical and experimental investigations with nimodipine in high parenteral doses, we found a significant drop in arterial and cerebral perfusion pressure with the risk of severe ICP/CPP decompensation in the individual patient. This may be attributed to an impairment of cerebral autoregulation by nimodipine. A further consequence of this unphysiological vasodilatation is a lowering of the blood-brain barrier threshold and an increase in edema formation, as seen in experiments with cold brain injury. In addition to these intracranial side-effects, extracranial disturbances must also be considered, such as an increase of intrapulmonary shunting. These data suggest that precautionary further evaluation of calcium antagonists for the treatment of acute brain damage is necessary.

References

1. Allen GS, Ahn S, Preziosi TJ et al. (1983) Cerebral arterial spasm – a controlled trial on nimodipine in patients with subarachnoid hemorrhage. N Engl J Med 308:619–624
2. Casthely PA, Rabinowitz L, Litwak B et al. (1984) Intrapulmonary shunting during deliberate hypotension with nifedipine, diltiazem and labetalol. Anesthesiology 61:A489
3. Gaab MR, Haubitz I, Brawanski A et al. (1985) Acute effects of nimodipine on the cerebral blood flow and intracranial pressure. Neurochirurgia 28 (Suppl 1):93–99
4. Gaab MR, Höllerhage HG, Walter GF et al. (1988) Brain edema, autoregulation and calcium antagonism. An experimental study with nimodipine. In: Long DM et al. (eds) Brain edema. Baltimore (in press)
5. Gaab MR, Höllerhage HG, Zumkeller M et al. (1987) The effect of the Ca-antagonist nimodipine on cerebral blood flow autoregulation. J Cereb Blood Flow Metab 7 (Suppl 1):170
6. Gaab MR, Rode CP, Schakel EH et al. (1985) Zum Einfluß des Ca-Antagonisten Nimodipin auf die globale und regionale Hirndurchblutung. Klin Wochenschr 63:8–15
7. Gaab MR, Ungersböck K, Hufenbeck B (1986) Evaluation of ICP by computerized bedside monitoring: methods and clinical significance. Neurol Res 8:44–52
8. Gelmers HJ, Gorter K, De Weerdt CJ et al. (1988) A controlled trial of nimodipine in acute ischemic stroke. N Engl J Med 318:203–207
9. Guggenberger H, Kottler B, Heuser D (1986) Cardio-respiratorische Problematik beim Einsatz eines Calciumantagonisten zur Vasospasmusprophylaxe bei Subarachnoidalblutung. Anaesthesist 35:429–432
10. Höllerhage HG, Gaab MR, Zumkeller M et al. (1988) The influence of nimodipine on cerebral blood flow autoregulation and blood-brain barrier. J Neurosurg 69:919–922
11. Hocheder M (1982) Beziehungen zwischen Hirnödem und Symstemblutdruck bei Therapie mit dem Kalziumantagonisten Nimodipine. M. D. Thesis, Julius-Maximilian-University, Würzburg (Germany)
12. Kostron H, Rumpf E, Stampfl G et al. (1985) Treatment of cerebral vasospasm following severe head injury with the calcium influx blocker nimodipine. Neurochirurgia 28 (Suppl 1):103–109
13. Petruk KC et al. (1988) Nimodipine treatment in poor-grade aneurysm patients. Results of a multicenter double-blind placebo-controlled trial. J Neurosurg 68:505–517
14. Roine RO, Kaste M, Kinnunen A et al. (1987) Safety and efficacy of nimodipine in resuscitation of patients outside hospital. Br Med J 294:20
15. Sollmann WP, Gaab MR, Rickels E (1988) Der Einfluß von Calciumantagonisten auf die zerebrale Kationenhomöostase und Sauerstoffspannung unter Hypoxie. 39. Jahrestagung Dtsch Ges Neurochirurgie, Köln, 8–11. 5
16. Symon L (1988) Calcium antagonists and brain ischemia – experimental observations (this volume)
17. Symon L, Harris RJ, Branston NM (1982) Calcium ions and calcium antagonists in ischemia. Acta Neurochir 63:267–275
18. Zwieten PA (1985) Calciumantagonisten-Terminologie, Klassifizierung und Vergleich. Arzneimittelforschung/Drug Res 35:298–301

Subarachnoid Hemorrhage

Calcium Channel Blockade in Cerebral Vasospasm

L. Brandt, H. Säveland, B. Ljunggren, and B. Romner

Introduction

In aneurysmal subarachnoid hemorrhage (SAH), aside from the direct effect of the initial bleed, most of the death and disability can be traced to rebleeding and/or ischemic deterioration ("cerebral vasospasm") [13]. Consequently, early aneurysm operation has been increasingly proposed to eliminate rebleeds and possibly to decrease ischemic deterioration of delayed onset by intraoperative evacuation of blood-contaminated cerebrospinal fluid (CSF) and clots [16, 21]. During recent decades a large number of studies have been undertaken in attempts to elucidate the pathophysiology of cerebral vasospasm and delayed ischemic dysfunction (DID) after SAH. Such studies have, however, "produced more questions than answers" [26].

As early as 1839, John Hunter considered constriction of arteries to be one of the physiological mechanisms responsible for the control of hemorrhage. From later work it is well known that mechanical, chemical, and electrical stimulation may induce arterial constriction [12]. The significance of various mechanisms possibly involved in cerebral vasospasm and DID remains controversial. The relationship between the amount and distribution of subarachnoid blood and the development of cerebral vasospasm was reported by Fisher and co-workers, who found that when subarachnoid blood was not detected or was distributed diffusely on computerized tomography (CT), severe vasospasm was almost never encountered [9]. In the presence of subarachnoid blood clots larger than 5 × 5 mm or layers of blood 1 mm or more thick in fissures and vertical cisterns, severe spasm ensued almost invariably. There was also a close correlation between the site of the major subarachnoid blood clots and the location of severe angiographic vasospasm [9].

Transmembrane influx of extracellular calcium through specific calcium channels is associated with the excitation-contraction coupling of vascular smooth muscle and plays an important role in cerebral ischemia. Consequently, it has been proposed that drugs affecting the influx of calcium to the vascular smooth muscle cell should be used in the period following aneurysmal SAH in order to counteract ischemia of delayed onset [1, 5, 6]. Experimental studies have revealed a pronounced cerebrovascular dilatatory effect of calcium channel blockers of the dihydropyridine group [1, 2, 6–8]. The mechanisms behind a selective cerebrovascular effect of calcium antagonists have not been established,

Department of Neurosurgery, University Hospital of Lund, S-221 85 Lund, Sweden.

Cerebral Ischemia and Calcium
Ed. by A. Hartmann and W. Kuschinsky

but may be related to a greater dependence on extracellular calcium for contractile activation of cerebral than of peripheral vessels [20]. A selective action on cerebral vessels may contribute to a beneficial effect of calcium antagonists in several cerebral disorders, particularly in the syndrome of cerebral vasospasm after SAH but also in cerebral ischemia associated with acute vessel occlusion where the clinical use of calcium antagonists so far has been promising.

Beside the well documented vascular actions in vitro as well as in vivo, the beneficial effects of nimodipine in ischemia may point to the possibility that in addition to its cerebrovascular site of action, nimodipine may also have direct anti-ischemic neuronal effects. Nimodipine binding receptors have been demonstrated in different species, including man [23]. The nimodipine receptor was found to be highly specific and stereoselective for the dihydropyridines with calcium antagonistic action. Studies on the transfer kinetics of nimodipine revealed that the influx constant to the brain was much higher compared to nifedipine and also that the regional distributions differed markedly [14]. Nimodipine exhibited a distinctly preferential distribution to the gray matter and concentrated in different structures of the brain that are known to have high concentrations of binding sites for dihydropyridines. From these studies it was concluded that high transfer of nimodipine across the blood-brain barrier as well as the specific region of distribution may explain the preferential cerebral therapeutic effectiveness of the drug.

Due to the immediate effects of the first bleed, only every second patient suffering from rupture of an intracranial aneurysm will "recover" in the initial stage in a good or fairly good condition. While awaiting surgery, many of these patients die or become disabled from recurrent hemorrhage or from secondary ischemic deterioration or both. Until recently, timing of operation for ruptured intracranial aneurysms was considered controversial not only in those patients who had survived the first bleed in a poor condition but also in those who were in a favorable condition following the first ictus. It appears that following the introduction of microneurosurgery with the possibility of achieving a good neurological recovery in approximately 75% of grade I–III patients, the only logical approach to the timing of aneurysm repair is to perform the procedure in the acute stage [19]. At the first international workshop on intracranial aneurysms in Tokyo in April 1986 it was quite evident that practically all neurosurgeons present were in favor of acute stage surgery in good grade patients [15].

In a study of 137 patients in neurological grades I–III (Hunt and Hess grading system) who were operated within 72 h after the bleed and who had not received any specific anti-ischemic treatment, the incidence of DID with fixed neurological deficits was 13% [17].

In 1983, Allen and co-workers reported the results of oral nimodipine treatment in 125 neurologically normal patients enrolled within 96 h after an aneurysmal SAH. Severe neurological deficits occurred in 2% of the nimodipine-treated patients versus 13% of the placebo group [3]. The following year the results of the first open trials conducted with the intravenous form of nimodipine combined with early aneurysm surgery were reported [4, 18]. In both studies, comprising altogether 120 patients treated with early surgery and intravenous nimodipine (2 mg/h), the incidence of DID leading to permanent neurological deficits was below 5%.

Subsequently, Philippon and co-workers showed a significant reduction in poor outcomes from vasospasm alone in a group of patients who had been treated with 60 mg nimodipine every 4 h for 3 weeks. Severe neurological outcomes including death from spasm alone occurred in 26% of placebo-treated patients as compared to 6% of nimodipine-treated patients [24].

In 1988 the results of the first randomized double-blind, placebo-controlled study of poor grade aneurysm patients were presented. DID with permanent deficits from vasospasm alone were significantly less frequent in the nimodipine group (7%) than in the placebo-treated patients (27%). There was no difference in the incidence or severity of late angiographic vasospasm, confirming results previously reported and suggesting that nimodipine is not effective in preventing the delayed type of large vessel narrowing [28].

In another 1988 double-blind, placebo-controlled study, it was demonstrated that intravenous nimodipine treatment was associated with a significant decrease in mortality and a better neurological outcome at 3 months. The combination of early aneurysm operation and intravenous nimodipine resulted in the most favorable neurological outcome [22].

Clinical Material

At our department 150 consecutive patients with a ruptured intracranial supratentorial aneurysm were subjected to operation within 72 h after the hemorrhage combined with intravenous nimodipine treatment during the acute stage. After clipping of the ruptured aneurysm an intravenous nimodipine infusion was started at a dose of 2 mg/h $\approx$ 0.5 μg/kg body weight per minute. In addition, the exposed arterial segments were irrigated with a 2.5×10^{-5} M nimodipine solution. The intravenous infusion was continued for at least 7 days postoperatively by the use of a low dose infusion pump.

One hundred and twenty-six patients were in a good or fairly good clinical grade (I–III according to the grading system of Hunt and Hess) preoperatively, while the remaining 24 patients were in a poor grade. The most common aneurysm location was in the anterior cerebral artery (ACA) territory (61 individuals) followed by the internal carotid artery (ICA) (47 individuals); the aneurysm originated from the middle cerebral artery (MCA) in 29 cases.

Among the patients preoperatively in grades I–III, 21 had arterial hypertension (AH) prior to the bleed. In a series of 137 historical controls not subjected to nimodipine treatment, 31 had AH.

Results

Preoperative grade and final outcome for the whole series of nimodipine-treated patients subjected to early operation are given in Table 1. Overall there was a 75% success rate.

The incidence of DIDs leading to permanent sequelae was 4% (6/150) in the total series. In a previous series not subjected to nimodipine treatment the corre-

Table 1. Preoperative condition and outcome in 150 patients subjected to early aneurysm operation (within 72 h post-SAH) and intravenous nimodipine

Preop. grade	Good recovery	Fair outcome	Poor outcome	Dead	Total	Success rate [%]
I	20		1		21	95
II	64	6	3	1	74	86
III	20	5	2	4	31	65
IV	7	3	2	1	13	54
V	2	2	6	1	11	18
Total	113	16	14	7 (5%)	150	75

Table 2. Delayed ischemic deficit

Aneurysm location	Historical controls	Nimodipine treated
ICA	7/47 (15%)	0/41 (0%)
MCA	5/29 (17%)	1/48 (2%)
ACA	6/61 (10%)	5/61 (8%)
Total	18/137 (13%)	6/150 (4%)

Table 3. Hypertension vs normotension: unfavorable outcomes

	Without nimodipine	With nimodipine
Normotensive	25/106 (24%)	17/98 (17%)
Hypertensive	17/31 (55%)	4/21 (19%)

sponding incidence of DID was 13% (18/137). Of the six patients in the nimodipine series with permanent deficits ascribed to vasospasm, five had aneurysms in the ACA territory; in one patient the aneurysm originated from the MCA while none of the 41 patients with ICA aneurysms developed DIDs (Table 2).

Of the patients with arterial hypertension, 19% (4/21) had an unfavorable outcome. The corresponding figure for the normotensive individuals was 17% (17/98). In the series of historical controls the corresponding figures were 55% and 24% respectively (Table 3). (All patients in grades I–III preoperatively.)

Among the hypertensive individuals in the nimodipine-treated series 1 of 21 patients (5%) developed delayed ischemia with permanent deficits compared to 11 of 31 (35%) in the non-nimodipine-treated historical controls.

Discussion

The incidence of DID with fixed neurological deficits in grade I–III patients subjected to aneurysm operation in the acute stage varies between 13% and 20% or more in the literature [17, 27, 29]. In the present study, ischemic dysfunction of delayed onset with permanent damage occurred in 4%. This markedly lowered incidence of DID supports the statement that "nimodipine significantly reduces the occurrence of severe neurologic deficits from spasm" [3]. The patients in the present series who developed fixed neurological deficits of delayed onset all had complicated surgery, significant intracerebral hematoma, advanced age, and/or very massive blood contamination on preoperative CT, and all had aneurysms in the ACA complex except for one with an MCA aneurysm. It is generally accepted that patients with complicating factors as mentioned run a higher risk of developing late cerebral dysfunction.

The most striking beneficial effect of nimodipine was seen in patients with AH prior to the bleed. One explanation for this may be an increased vessel sensitivity for nimodipine in hypertensive individuals. This is also supported by experimental data on cerebral arteries from hypertensive individuals, where the sensitivity to nimodipine seems to be tenfold [25]. Another explanation may be that the direct neuronal protective effect of nimodipine is accentuated in hypertensive individuals.

Spontaneously hypertensive experimental animals are more vulnerable than normotensive animals to bilateral ligation of the common carotid artery, which may indicate a reduced capacity for compensatory vasodilatation [10]. Furthermore, the cerebrovascular resistance during maximum vasodilatation is considerably higher in hypertensive rats than in normotensive rats [11]. This may suggest that hypertensive patients are more vulnerable to severe arterial narrowing, especially when this narrowing affects the penetrating arteries. The accumulating experimental data of a direct anti-ischemic neuronal effect of nimodipine suggest that this protective effect would be expected to be more pronounced in hypertensive individuals.

In conclusion, the present study offers further support to the concept that preventive nimodipine treatment combined with early aneurysm surgery reduces the incidence of delayed ischemic cerebral dysfunction to a minimum.

References

1. Allen GS, Bahr AL (1979) Cerebral arterial spasm: Part 10. Reversal of acute and chronic spasm in dogs with orally administered nifedipine. Neurosurgery 4:43–47
2. Allen GS, Banghart SB (1979) Cerebral arterial spasm: In vitro effects of nifedipine on serotonin-, phenylephrine-, and potassium-induced contractions of canine basilar and femoral arteries. Neurosurgery 4:37–42
3. Allen GS, Ahn HS, Preziosi TJ et al. (1983) Cerebral arterial spasm – a controlled trial of nimodipine in patients with subarachnoid hemorrhage. N Engl J Med 308:619–624
4. Auer LM (1984) Acute operation and preventive nimodipine improve outcome in patients with ruptured cerebral aneurysms. Neurosurgery 15:57–66
5. Brandt L, Andersson KE, Bengtsson B, Edvinsson L, Ljunggren B, McKenzie ET (1979) Effects of nifedipine on pial arteriolar caliber: an in vivo study. Surg Neurol 12:349–352
6. Brandt L, Andersson KE, Edvinsson L, Ljunggren B (1981) Effects of extracellular calcium and of calcium antagonists on the contractile responses of isolated human pial and mesenteric arteries. J Cereb Blood Flow Metab 1:339–347

7. Brandt L, Andersson KE, Edvinsson L, Ljunggren B (1981) Effects of extracellular calcium and calcium antagonists on the contractile responses of isolated human pial and mesenteric arteries. J Cereb Blood Flow Metab 1:339–347
8. Brandt L, Ljunggren B, Andersson KE, MacKenzie E, Tamura ET, Teasdale G (1982) Effects on feline cortical pial microvasculature of topical application of a calcium antagonist (nifedipine) under normal conditions and in focal ischemia. J Cereb Blood Flow Metab 3:44–50
9. Fisher CM, Kistler JP, Davis JM (1980) Relation of cerebral vasospasm to subarachnoid hemorrhage visualized by computerized tomographic scanning. Neurosurgery 6:1–9
10. Fujishima M, Ogata J, Sugi T et al. (1976) Mortality and cerebral metabolism after bilateral carotid artery ligation in normotensive and spontaneously hypertensive rats. J Neurol Neurosurg Psychiatry 39:212–217
11. Johansson B, Nilsson B (1979) Cerebral vasomotor reactivity in normotensive and spontaneously hypertensive rats. Stroke 10:572–576
12. Kapp J, Mahaley MS Jr, Odom GL (1968) Cerebral arterial spasm. Part 1: Evaluation of experimental variables affecting the diameter of the exposed basilar artery. J Neurosurg 29:331–338
13. Kassell NF, Torner JC (1984) The international co-operative study on timing of aneurysm surgery – an up-date. Stroke 15:566–570
14. Van den Kerckhoff W, Drewes LR (1985) Transfer of the Ca-antagonists nifedipine and nimodipine across the blood-brain barrier and their regional distribution in vivo. J Cereb Blood Flow Metab 5:459–460
15. Kikuchi H, Fukushima T, Watanabe K (eds) (1986) Intracranial aneurysms. Surgical timing and techniques. Proceedings of the First International Workshop on Intracranial Aneurysms (IWIA). Nishimura, Japan, pp 1–376
16. Ljunggren B, Brandt L, Kågström E, Sundbärg G (1981) Results of early operations for ruptured aneurysms. J Neurosurg 54:473–479
17. Ljunggren B, Säveland H, Brandt L (1983) Causes of unfavorable outcome after early aneurysm operation. Neurosurgery 13:629–633
18. Ljunggren B, Brandt L, Säveland H, Nilsson PE, Cronquist S, Andersson KE, Vinge E (1984) Outcome in 60 consecutive patients treated with early aneurysm operation and intravenous nimodipine. J Neurosurg 61:864–873
19. Ljunggren B, Brandt L, Säveland H et al. (1987) Management of ruptured intracranial aneurysm – a review. Br J Neurosurg 1:9–32
20. McCalden TA, Bevan JA (1981) Sources of activator calcium in rabbit basilar artery. Am J Physiol 241:H129–H133
21. Mizukami M, Kawase T et al. (1982) Prevention of vasospasm by early operation with removal of subarachnoid blood. Neurosurgery 10:301–307
22. Öhman J (1987) Prevention of ischemic dysfunction in patients with aneurysmal subarachnoid hemorrhage. Presented at 2nd World Congress of Neurosciences; Session 132: Mechanism and prevention of cerebral vasospasm. Budapest, Hungary
23. Peroutka SJ, Allen GS (1983) Calcium channel antagonist binding sites labelled by ^{3}H-nimodipine in human brain. J Neurosurg 59:933–937
24. Philippon J, Grob R, Dagreou F, Guggeiari M, Rivierez M, Viars P (1986) Prevention of vasospasm in subarachnoid hemorrhage. A controlled study with nimodipine. Acta Neurochir (Wien) 82:110–114
25. Ryman T (1988) Unpublished data
26. Sundt TM Jr, Davis DH (1980) Reactions of cerebrovascular smooth muscle to blood and ischemia: primary versus secondary vasospasm. In: Wilkins RM (eds) Cerebral arterial spasm. Williams and Wilkins, Baltimore, pp 244–250
27. Taneda M (1982) Effect of early operation for ruptured aneurysms in prevention of the late ischemic symptoms. J Neurosurg 57:622–628
28. Petruk K, West M, Mohr G et al. (1988) Nimodipine treatment in poor-grade aneurysm patients. Results of a multicenter double-blind placebo-controlled trial. J Neurosurg 68:505–517
29. Yamamoto I, Hara M, Ogura K et al. (1983) Early operation for ruptured intracranial aneurysms: comparative study with computed tomography. Neurosurgery 12:169–174

Cerebral Ischemia Following Subarachnoid Hemorrhage – The Efficacy of Nimodipine as Therapy

B. WEIR, F. ESPINOSA, M. NOSKO, J. LEWIS, L. DISNEY, C. KRUEGER, D. COOK, and M. GRACE

The rupture of a cerebral aneurysm commonly results in the deposition of a large volume of subarachnoid clot which adheres to the adventitia of the basal conducting arteries. As the erythrocytes lyse over a period of several days, spasmogenic substances are released which permeate the arterial vessel wall and result in a slowly developing and sustained vasoconstriction. The likelihood of a degree of vasoconstriction developing which is observable angiographically is proportional to the volume of blood in the subarachnoid space. Severe diffuse angiographic vasospasm is commonly associated with delayed onset ischemic neurological deficit. Whether the latter develops or not is a function of many factors, including the anatomy of the cerebral vasculature, the nature of collateral vessels, associated atherosclerosis, the age of the cerebral cortex, cardiac output, arterial blood pressure, and intracranial pressure.

Experimental Studies

At the University of Alberta a series of experiments have been performed on the cynomolgus monkey using the direct application of autologous blood clot to microsurgically exposed basal arteries. This model results in radiological, neurological, and pathological change which is identical to that which occurs in humans following subarachnoid hemorrhage.

In the first series, using microsurgical techniques, 33 monkeys had a frontotemporal craniectomy and unilateral opening of the subarachnoid cisterns. Cerebrospinal fluid was drained and fresh clot (averaging about 7 ml) was placed directly on the adventitia of the larger arteries of the anterior circulation unilaterally. Thirty monkeys studied for 7–14 days after subarachnoid clot application were allocated to two equal treatment groups: one received placebo and the other nimodipine administered by nasogastric tube, 1 mg/kg every 8 h. Before and after clot application neurological status, cerebral blood flow, computerized tomography, and angiographic vessel caliber were monitored. In the placebo group, delayed ischemic neurological deficit developed in one monkey 4 days following clot application. The deficit persisted to sacrifice at day 14. No such deficit occurred in the nimodipine group of animals. The effect of nimodipine on vessel caliber at the dosage employed was equivocal. Significant vasospasm (a reduction of 31%

Departments of Surgery and Pharmacology, The University of Alberta, Edmonton, Alberta T6G 2B7, Canada.

Cerebral Ischemia and Calcium
Ed. by A. Hartmann and W. Kuschinsky

–100%) developed in 87% (26/30) of the animals. Vasospasm was slightly more common in the placebo group (Fig. 1): in this group on days 7 and 14 the incidence of vasospasm was significantly higher than in the nimodipine group but the average percentage reduction in vessel caliber of the maximally constricted vessel in each monkey was not significantly different between the two groups (Fig. 2) [1].

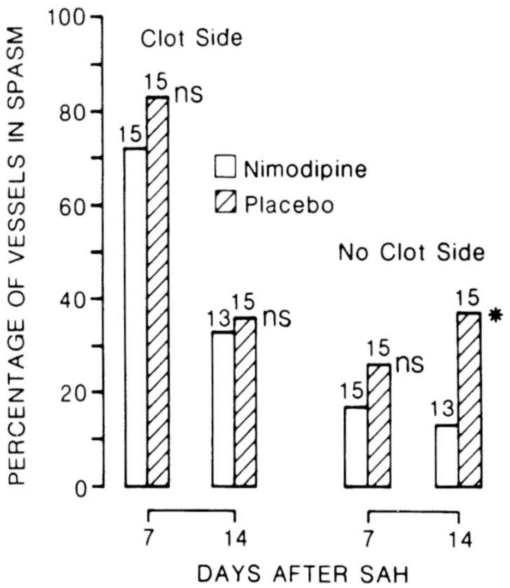

Fig. 1. Incidence of vasospasm (more than 10% reduction in vessel caliber) based on the four arterial sites measured on each side in 58 angiograms of the 30 monkeys. Vasospasm was significantly more common in the non-clot side on day 14 in the placebo group, but there was no significant difference on the clot side at any time. Figures above the columns indicate the number of animals studied at those times. *SAH*, subarachnoid hemorrhage; * $p<0.05$; *ns*, not significant. With permission, Journal of Neurosurgery

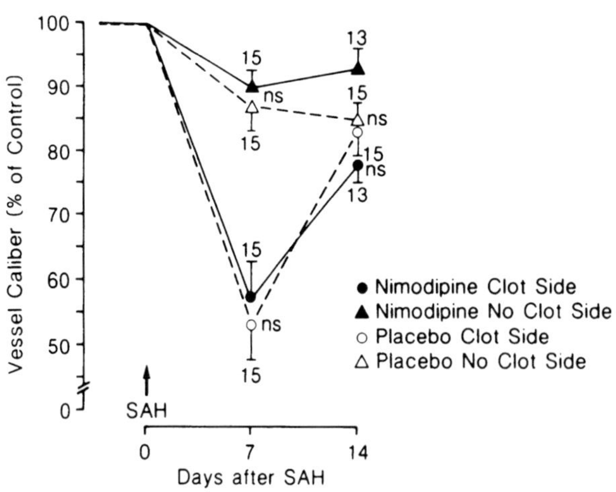

Fig. 2. Time course of vasospasm and mean (±SE) percentage change in vessel caliber based on measurements of the vessel maximally constricted in each monkey. In both groups, vasospasm was significantly greater ($p<0.001$) on the clot side on day 7. Between groups, there was no significant difference in degree of vasospasm. Figures above and below standard error bars indicate number of animals studied at those times. *SAH*, subarachnoid hemorrhage; *ns*, not significant. With permission, Journal of Neurosurgery

Twelve of the 30 monkeys studied from both groups underwent cerebral fixation on day 14 by intraarterial perfusion of glutaraldehyde at controlled pressure. The basal arteries were studied by light microscopy and scanning and transmission electron microscopy. Cerebral angiography on day 7 showed that the vasospasm was significantly more common and more severe on the clot side compared to the control or nonclot side. Angiographic vasospasm was less severe on day 14 immediately prior to sacrifice than it had been on day 7. On scanning electron microscopy, 80% of the 20 middle cerebral artery specimens that had been in spasm on day 7 angiograms showed marked corrugation; in some instances the endothelium had a fish scale appearance. All of the middle cerebral arteries examined from the clot side by transmission electron microscopy which had been in spasm on day 7 showed marked changes such as endothelial swelling, subendothelial proliferation, corrugation of the elastic lamina, and myonecrosis. With few exceptions none of the basilar arteries or middle cerebral arteries from the nonclot side showed these abnormalities. The pathological findings of vessel in spasm were considered to be slightly less severe in the nimodipine group (1 mg/kg/8 h) but the medication clearly did not prevent such abnormalities from occurring. Since considerable changes were present in the vessel wall even after radiologically visible vasospasm had abated, we concluded that vasospasm is due to long-lasting smooth muscle constriction and not the vessel wall thickening caused by a cellular or subcellular infiltrate [2].

Since it was possible that this initial study had used too low a dose of nimodipine to prevent angiographically visible vasospasm, a subsequent series of experiments was conducted to eliminate this possibility. The potential efficacy of the calcium channel blocker nimodipine in the prevention of chronic cerebral vasospasm and delayed ischemia after subarachnoid hemorrhage in monkeys was examined in another blind, randomized, placebo-controlled trial. An average of 6.4 ml autologous hematoma was placed directly against the major anterior cerebral vessels in the right basal subarachnoid space of twenty four monkeys. The animals were randomized into one of four groups and were again treated with oral administration for 7 days with nimodipine (3, 6, or 12 mg/kg) or placebo. Two other monkeys underwent the surgical procedure without clot placement. Drug administration was started between 14 and 20 h following clot placement. Parameters monitored before and following subarachnoid hemorrhage included neurological status, angiographic vessel caliber, and cerebral blood flow. Significant vasospasm (25%–100% reduction of vessel caliber) was present on day 7 on the clot side in 83% of animals. There was no significant difference in the incidence of vasospasm among the four groups. Similarly, there was no significant difference in the mean percentage vessel caliber reduction after subarachnoid hemorrhage among the four treatment groups (Fig. 3). There was no vasospasm on day 7 in the sham-operated animals. One animal receiving high dose nimodipine (12 mg/kg/q 8 h) developed a delayed ischemic neurological deficit on day 5 following the clot application. A second animal in the 12 mg/kg/q 8 h group developed transient neurological deficit which lasted between days 4 and 7 following the subarachnoid hemorrhage. Scanning electron microscopy demonstrated the morphology of severe vasospasm (Fig. 4) [3].

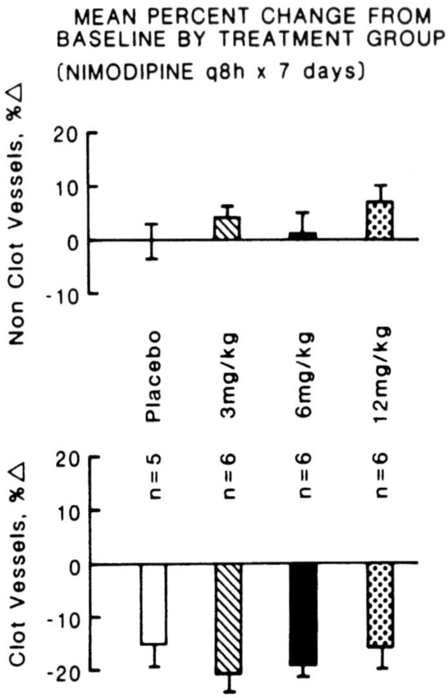

Fig. 3. Grouped mean percentage change in vessel caliber from base line value of major vessels (intradural internal carotid artery, middle cerebral artery, and anterior cerebral artery) by side and by treatment group. With permission, Journal of Neurosurgery

The effect of nimodipine on the in vitro reactivity of cerebral vessels after subarachnoid hemorrhage was then studied. With the use of this primate model of chronic cerebral vasospasm, 12 female cynomolgus monkeys had induction of subarachnoid hemorrhage by the direct placement of an average of 6.4 ml autologous hematoma against the major anterior cerebral vessels in the right basal subarachnoid space (day 0). The animals were then randomized into one of four treatment groups within 14–20 h after clot placement (3, 6, or 12 mg/kg/q 8 h orally) or placebo. Sacrifice was performed on day 7. The right and left middle cerebral arteries were immediately resected and placed in oxygenated Krebs' solution. Ring preparations from the arteries were employed in organ baths and dose-response curves to varying concentrations of norepinephrine, 5-hydroxytryptamine, and potassium chloride were obtained. There was a highly significant reduction in the response of the middle cerebral artery tissue from the clot side relative to the nonclot side to all three agonists (Fig. 5). The clot side contractility was not influenced by nimodipine treatment at any of the four doses tested (Fig. 6). The nonclot side arteries of the 12 mg/kg treatment group demonstrated significantly enhanced reactivity for all three agonists (Fig. 7). Oral treatment with high dose nimodipine enhanced the reactivity of normal cerebral vessels to the agonists tested but it did not affect the reactivity of arteries in chronic spasm at any of the four doses tested [4].

The cardiopulmonary effects of nimodipine in chronic vasospasm were also examined. Thirty cynomolgus monkeys had placement of blood clot as indicated above on day 0. They underwent a 7-day, randomized, blind trial comparing vari-

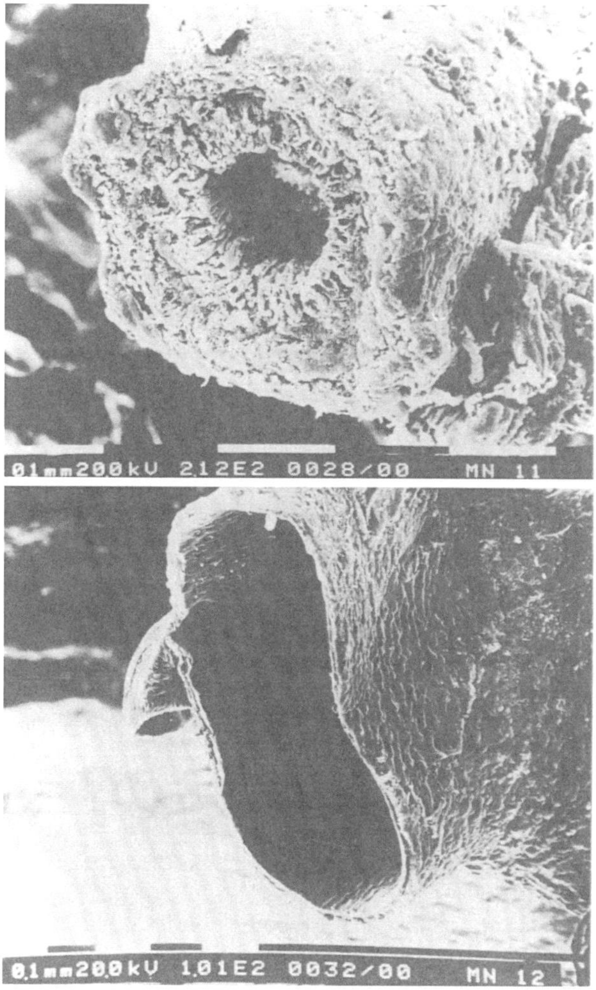

Fig. 4. Representative scanning electron micrographs of the middle cerebral arteries of a monkey treated with nimodipine, 6 mg/kg p.o. q 8 h. *Top*: right middle cerebral artery (clot side) (× 212). *Bottom*: left (non-clot side) (× 101). Note the relative magnifications, which further emphasize the degree of constriction present. With permission, Neurosurgery

ous doses of nimodipine administered by nasogastric tube compared to placebo, sham, and no treatment groups. The sham group underwent craniectomy without clot placement so that a total of 36 animals in all were operated upon. Differences in cardiopulmonary indices between day 0 and day 7 were compared within and between groups (Fig. 8). No significant differences were obtained in the sham and no treatment groups. The nimodipine 6 and 12 mg/kg groups showed significant decreases in blood pressure. Systemic vascular resistance was increased in the placebo and 3 mg/kg groups and decreased in the 12 mg/kg group. Stroke index was increased in the 12 mg/kg group. Cardiac index and stroke index correlated positively with nimodipine dosage. There were no pronounced changes in pulmonary artery wedge pressure, central venous pressure, alveolar-arterial oxygen pressure difference, arteriovenous oxygen content difference, and percentage of shunting [5].

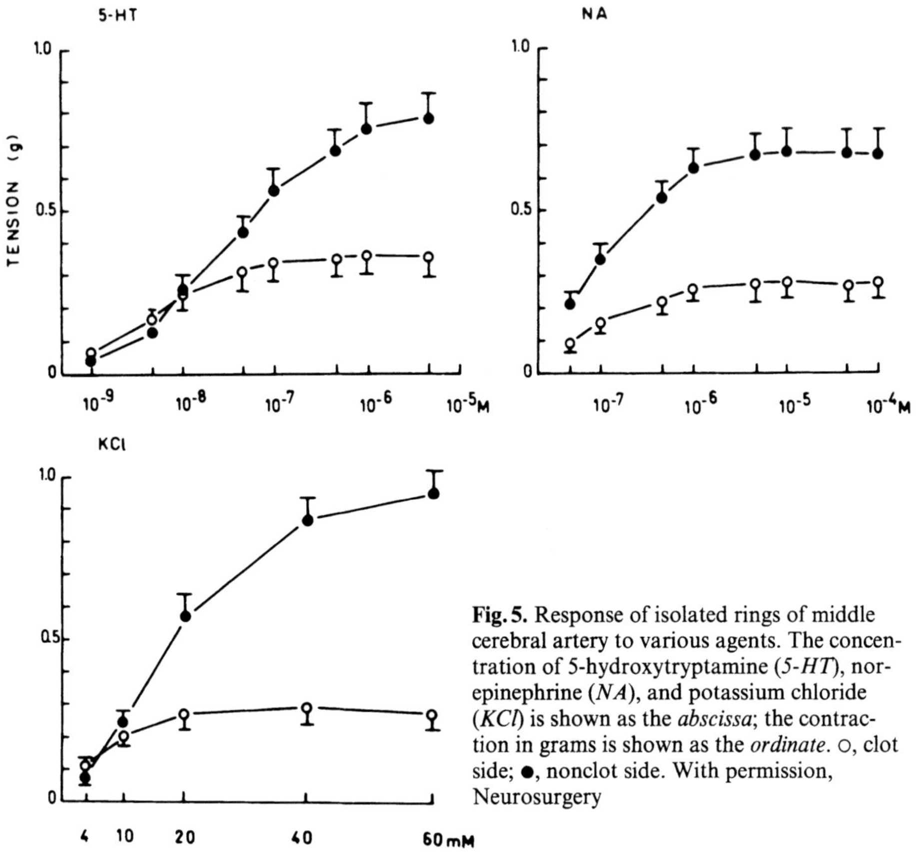

Fig. 5. Response of isolated rings of middle cerebral artery to various agents. The concentration of 5-hydroxytryptamine (*5-HT*), norepinephrine (*NA*), and potassium chloride (*KCl*) is shown as the *abscissa*; the contraction in grams is shown as the *ordinate*. ○, clot side; ●, nonclot side. With permission, Neurosurgery

On the basis of the foregoing studies, it was concluded that cerebral vascular spasm of radiologically visible arteries in a cynomolgus monkey does not appear to be modified by nimodipine. It was considered possible that cerebral arteries from this species are unusually resistant to the action of calcium antagonist. To test this, parallel studies on the in vitro response of cerebral arteries from monkey, dog, and man were carried out. Rings of basilar or middle cerebral artery were tested with potassium chloride (Fig. 9), 5-hydroxytryptamine (Fig. 10), norepinephrine (Fig. 11), prostaglandin $F_{2\alpha}$ (Fig. 12), and hemoglobin (Fig. 13). The responses were then reexamined in the presence of various concentrations of nimodipine. There was a significant variation among the species in sensitivity to nimodipine, with the vessels from the monkey being more resistant to nimodipine than those from other species. As anticipated, there was a considerable difference in the ability of nimodipine to block the difference agonists. Responses to potassium chloride were blocked at low concentrations of nimodipine in all species, whereas norepinephrine and 5-hydroxytryptamine were more resistant. In all the species tested, hemoglobin and prostaglandin $F_{2\alpha}$ were antagonized poorly even by high concentrations of nimodipine. It was concluded that if these agonists play a major role in the development of vasospasm and subsequent delayed ischemia, then cal-

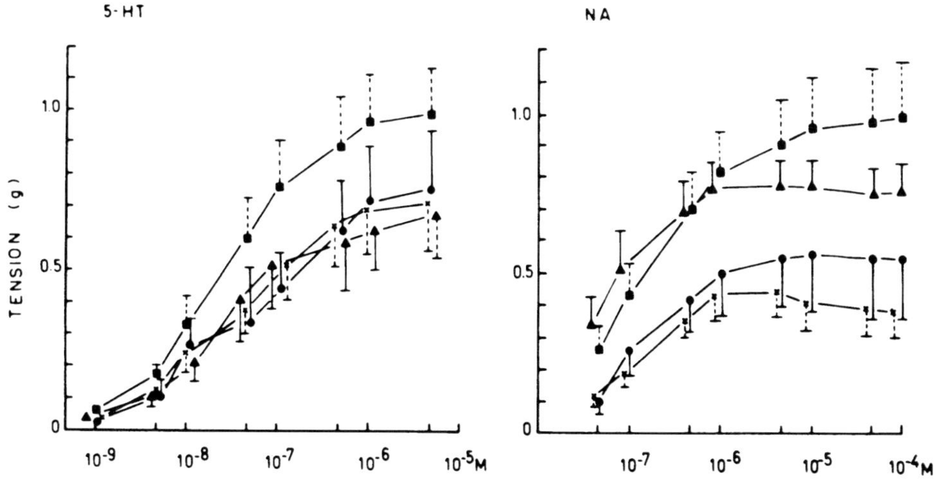

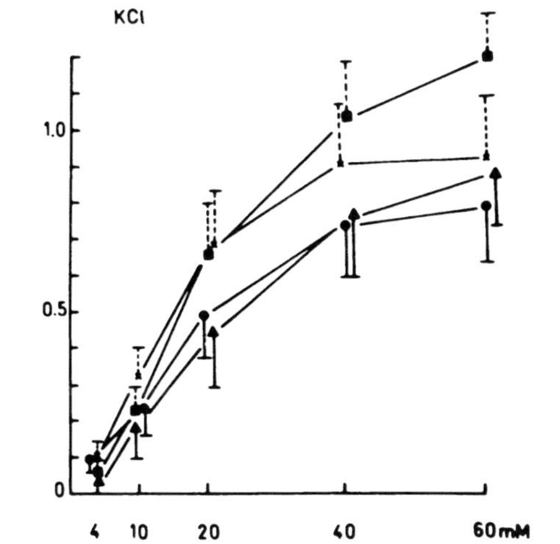

Fig. 6. Response of isolated rings of middle cerebral artery to various agents. The concentration of 5-hydroxytryptamine (*5-HT*), norepinephrine (*NA*), and potassium chloride (*KCl*) is shown as the *abscissa*; the contraction in grams is shown as the *ordinate*. The data in this figure were obtained from arteries in vasospasm (clot side). ●, placebo; ×, after treatment for 7 days with nimodipine, 3 mg/kg p.o. q 8 h; ▲, after treatment for 7 days with nimodipine, 6 mg/kg p.o. q 8 h; ■, after treatment for 7 days with nimodipine, 12 mg/kg p.o. q 8 h. With permission, Neurosurgery

cium antagonists exert a beneficial effect by some mechanism other than dilatation of spastic arteries [6].

The effects of intravenous nimodipine on the cerebral blood flow and cardiopulmonary indices in the cynomolgus monkey were examined acutely (about 2 h). Six animals received continuous infusion of nimodipine 2 μg/kg/min, 5 μg/kg/min, or placebo. Cerebral blood flow was determined by intraarterial injection of ^{133}Xe and calculated by the initial slope index technique. Cardiopulmonary indices were measured by thermodilution Swan-Ganz flow-directed catheter. No significant changes were noted in the placebo group in any variable. The cerebral blood flow gradually increased to 22% above baseline in the low dose nimodipine group and 34% in the high dose group (Fig. 14). There were no significant changes in cardiovascular parameters; however, trends were noted in the 5 μg/kg/min group, with heart rate and cardiac index initially increasing while systemic vascular resistance and mean arterial blood pressure fell. Systemic vascular resistance remained decreased, cardiac index increased, while heart rate and mean ar-

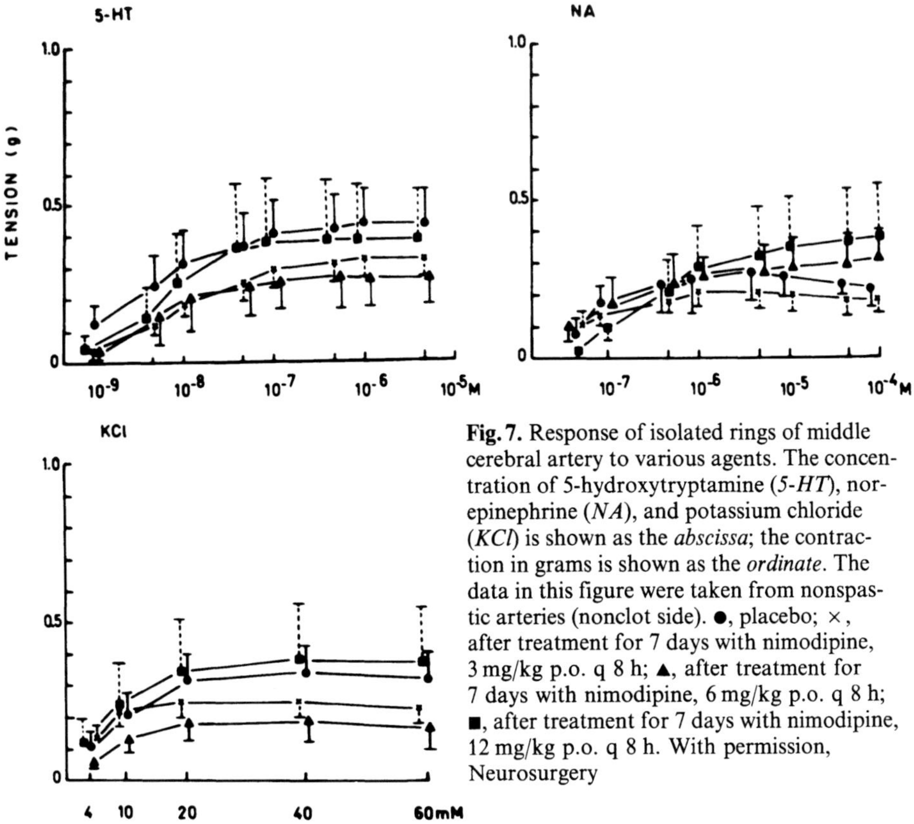

Fig. 7. Response of isolated rings of middle cerebral artery to various agents. The concentration of 5-hydroxytryptamine (*5-HT*), norepinephrine (*NA*), and potassium chloride (*KCl*) is shown as the *abscissa*; the contraction in grams is shown as the *ordinate*. The data in this figure were taken from nonspastic arteries (nonclot side). ●, placebo; ×, after treatment for 7 days with nimodipine, 3 mg/kg p.o. q 8 h; ▲, after treatment for 7 days with nimodipine, 6 mg/kg p.o. q 8 h; ■, after treatment for 7 days with nimodipine, 12 mg/kg p.o. q 8 h. With permission, Neurosurgery

terial blood pressure returned to baseline values. Continuous intravenous infusion of nimodipine increased cerebral blood flow in the cynomolgus monkey and simultaneously affected systemic vascular resistance although physiological reflexes were able to maintain heart rate and mean arterial blood pressure at baseline levels in the intact animal.

In humans there is no doubt that nimodipine at certain dose levels can cause hypotension (Fig. 15).

In a further test of the hypothesis that an insufficient dosage of nimodipine had been administered to the spastic arteries using our high dose oral regimen, a further series of experiments were carried out with the intrathecal administration of nimodipine in the primate model of chronic cerebral vasospasm. The safety, prevention, and treatment of chronic vasospasm by repeated administration of intrathecally applied nimodipine was evaluated. Twenty-four female cynomolgus monkeys were randomized into three groups of eight: sham, clot, and clot plus intrathecal nimodipine. All animals underwent a subarachnoid hemorrhage induction following baseline angiography. The nimodipine was administered postoperatively by percutaneous injection into an Ommaya reservoir of 1 ml (0.2 mg) q 8 h for 6 days. The Ommaya reservoir was connected to a catheter with multiple pores which was placed into the region of the autologous blood clot.

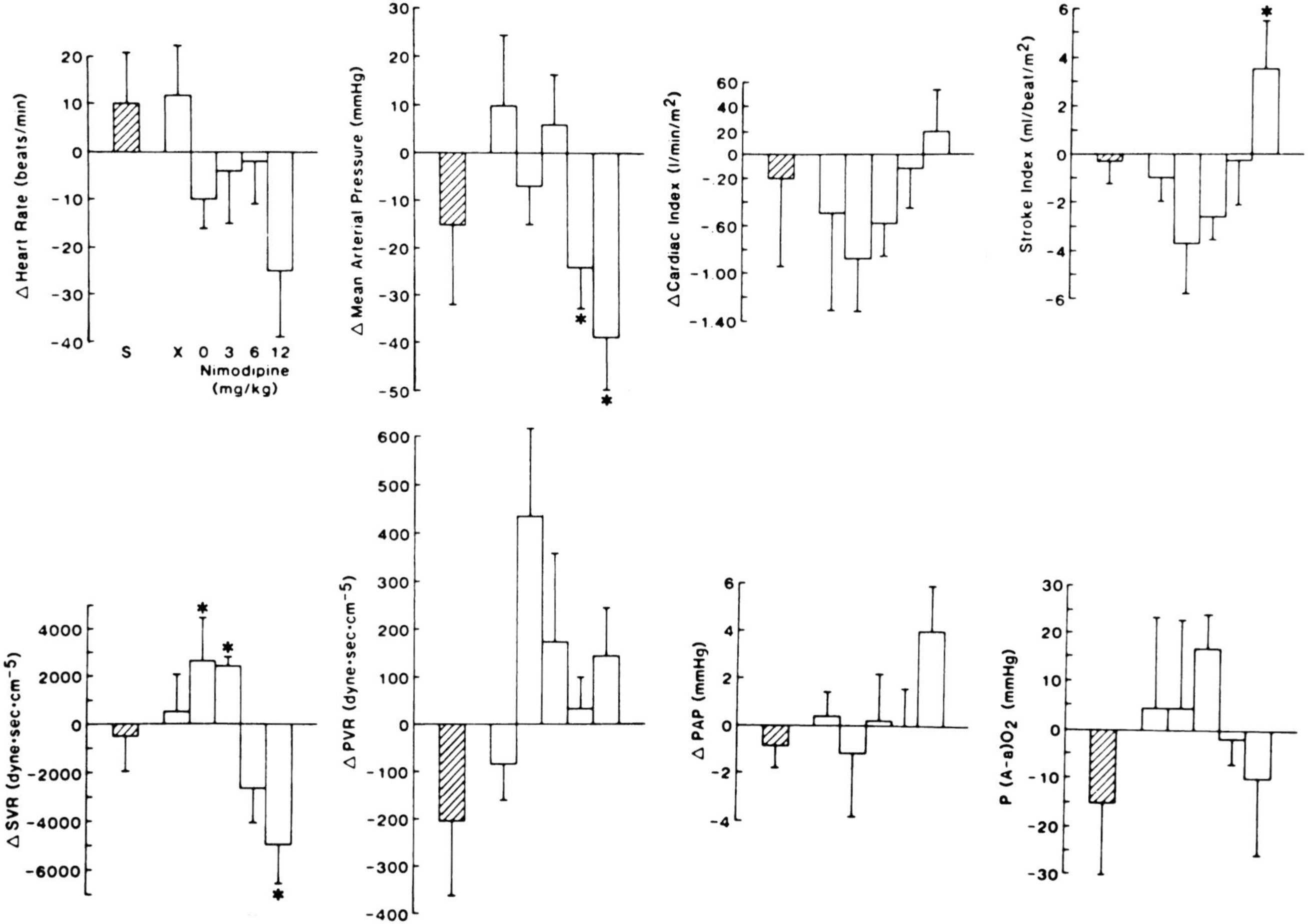

Fig. 8. The cardiopulmonary effects of nimodipine. Shown are day 7 (post-SAH) minus base line values (mean ± standard error of the mean) for each treatment group. *S*, sham group; *X*, no treatment group; *SVR*, systemic vascular resistance; *PVR*, pulmonary vascular resistance; *PAP*, pulmonary artery pressure. Nimodipine, 0 mg/kg, $n=5$; all other groups, $n=6$. *Asterisk* indicates significant difference from the placebo group ($P<0.05$). With permission, Neurosurgery

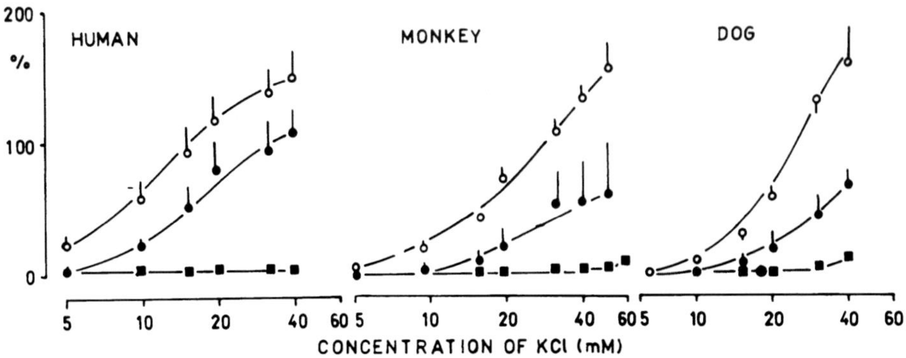

Fig. 9. Effects of different concentrations of nimodipine on the response to potassium chloride (*KCl*) of specimens from the middle cerebral arteries of human, monkey, and dog. The percent maximum initial response is given as the ordinate, and logarithm of molar concentration of KCl as the abscissa. *Open circles*: time-control specimens without nimodipine; *closed circles*: specimens exposed to 10^{-9} M nimodipine; *squares*: specimens exposed to 10^{-8} M nimodipine. *Bars* represent standard errors for a mean of eight specimens. With permission, Journal of Neurosurgery

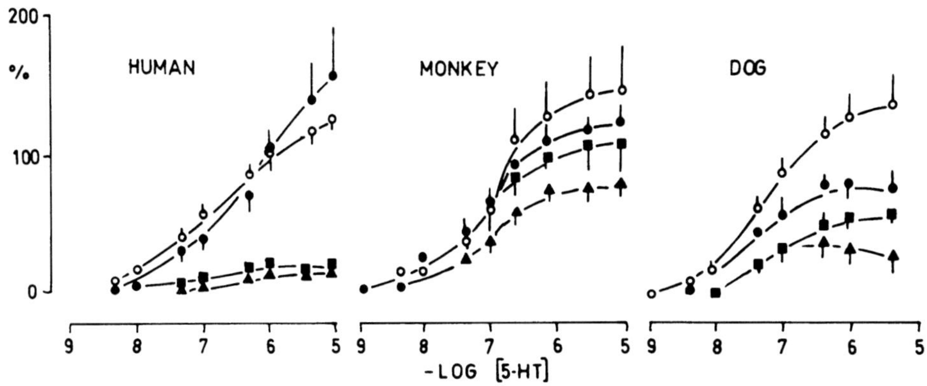

Fig. 10. Effects of different concentrations of nimodipine on the response to 5-hydroxytryptamine (*5-HT*) of specimens from the middle cerebral arteries of human, monkey, and dog. The percent maximum initial response is given as the ordinate, and logarithm of molar concentration of 5-HT as the abscissa. *Open circles*: time-control specimens without nimodipine; *closed circles*: specimens exposed to 10^{-9} M nimodipine; *squares*: specimens exposed to 10^{-8} M nimodipine; *triangles*: specimens exposed to 10^{-7} M nimodipine. *Bars* represent standard errors for a mean of eight specimens. With permission, Journal of Neurosurgery

Angiography was performed on day 7. Intrathecally applied nimodipine was not effective in preventing angiographic vasospasm (Fig. 16). It also did not seem to decrease the degree of pathological change in the vessels when compared to controls. One animal in the nimodipine group died 2 h after an intrathecal injection secondary to a respiratory arrest. Transient sedation and hypoventilation were common. No adverse pathological effects were noted. Intrathecal nimodipine in the usual dose did not produce a significant dilatation of vessels in moderate or severe spasm when a final intrathecal injection was given and angiography repeated 2 h thereafter (Fig. 17). Only one animal in eight showed modest diffuse

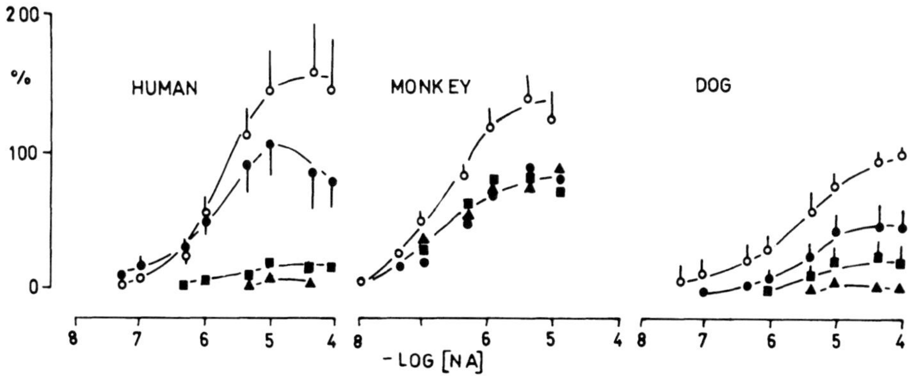

Fig. 11. Effects of different concentrations of nimodipine on the response to noradrenaline (*NA*) of specimens from the middle cerebral arteries of human, monkey, and dog. The percent maximum initial response is given as the ordinate, and logarithm of molar concentration of NA as the abscissa. *Open circles*: time-control specimens without nimodipine; *closed circles*: specimens exposed to 10^{-9} M nimodipine; *squares*: specimens exposed to 10^{-8} M nimodipine; *triangles*: specimens exposed to 10^{-7} M nimodipine. *Bars* represent standard errors for a mean of eight specimens. With permission, Journal of Neurosurgery

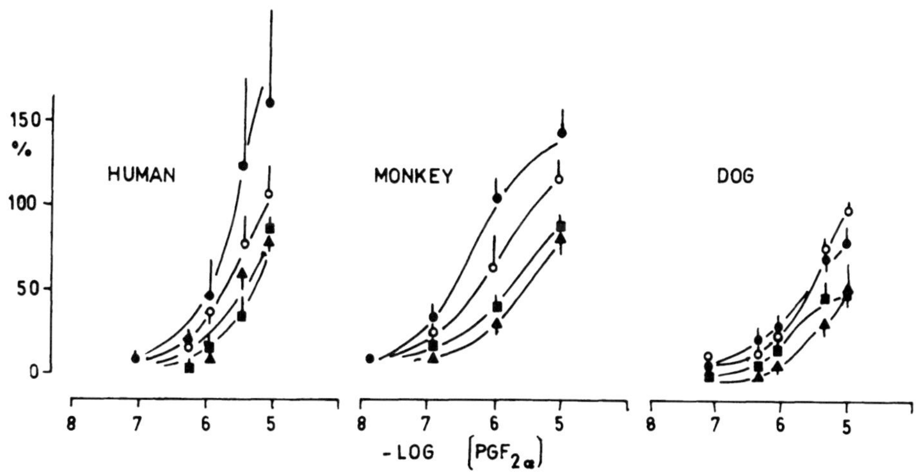

Fig. 12. Effects of different concentrations of nimodipine on the response to prostaglandin $F_{2\alpha}$ ($PGF_{2\alpha}$) of specimens from the middle cerebral arteries of human, monkey, and dog. The percent maximum initial response is given as the ordinate, and logarithm of molar concentration of $PGF_{2\alpha}$ as the abscissa. *Open circles*: time-control specimens without nimodipine; *closed circles*: specimens exposed to 10^{-9} M nimodipine; *squares*: specimens exposed to 10^{-8} M nimodipine; *triangles*: specimens exposed to 10^{-7} M nimodipine. *Bars* represent standard errors for a mean of eight specimens. With permission, Journal of Neurosurgery

dilatation of vasospastic cerebral vessels after an intrathecal nimodipine injection. Three other animals in the group developed dilatation of the basilar artery only. This vessel had been in mild spasm. In summary it seemed that a large dose of nimodipine is capable of causing inconstant dilatation of a few radiologically observable vessels which are not initially in severe spasm (Fig. 18) [7].

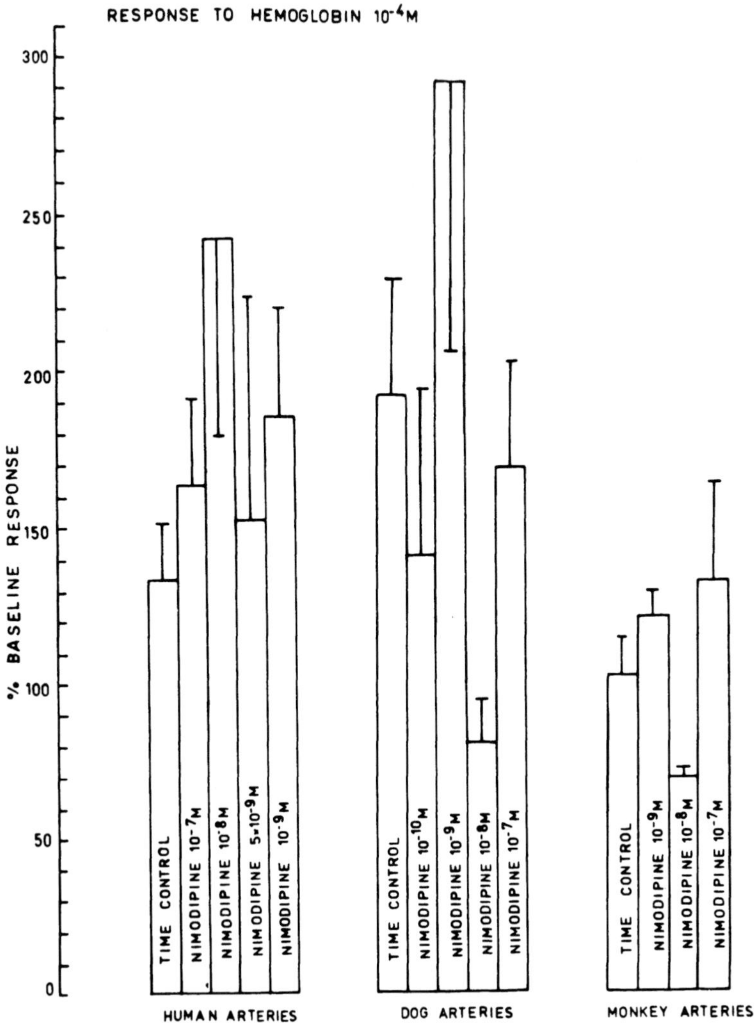

Fig. 13. Effects of different concentrations of nimodipine on the response to 10^{-4} M hemoglobin of specimens from the middle cerebral arteries of human, monkey, and dog. Percentage initial response to 10^{-4} M is given as the ordinate. *Bars* represent standard errors for a mean of eight specimens. With permission, Journal of Neurosurgery

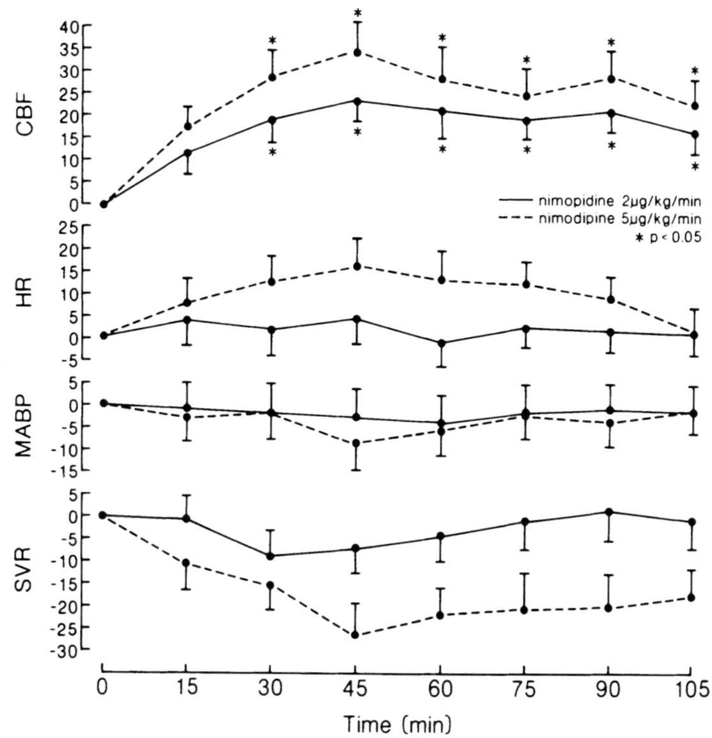

Fig. 14. Effect of continuous nimodipine infusion in cerebral blood flow (*CBF*), heart rate (*HR*), mean arterial blood pressure (*MABP*), and systemic vascular resistance (*SVR*). Time as *abscissa* and percentage change from baseline as *ordinate* (mean ± SD, $n = 6$). CBF was measured by intraarterial injection ^{133}Xe and calculated by the initial slope index method in the cynomolgus monkey

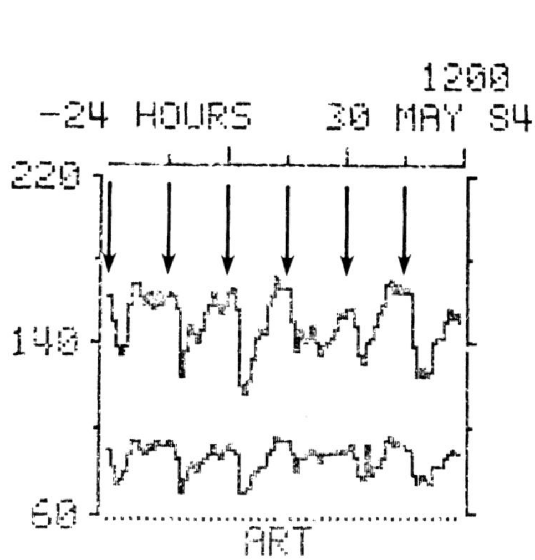

Fig. 15. An 80-kg male was admitted on day 0 as grade 2 with a ruptured posterior communicating aneurysm. He was operated on the same day and received nimodipine 120 mg by mouth a 4 h for 21 days. He was discharged on day 12 from hospital. The record demonstrates the systemic hypotensive action of this dose of the drug

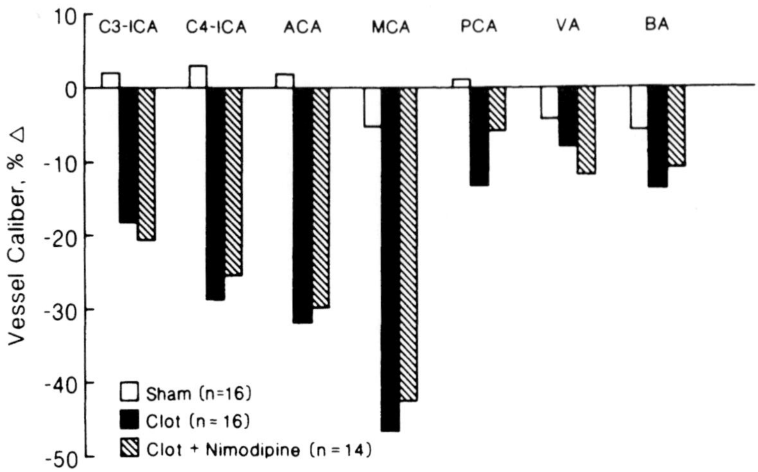

Fig. 16. Percentage of change in vessel caliber from base line of the cerebral arteries. The right- and left-sided data have been combined for bilateral vessels. *C3-ICA*, cavernous internal carotid artery; *C4-ICA*, supraclinoid internal carotid artery; *ACA*, anterior cerebral artery; *MCA*, middle cerebral artery; *PCA*, proximal pericallosal artery; *VA*, vertebral artery; *BA*, basilar artery. With permission, Neurosurgery

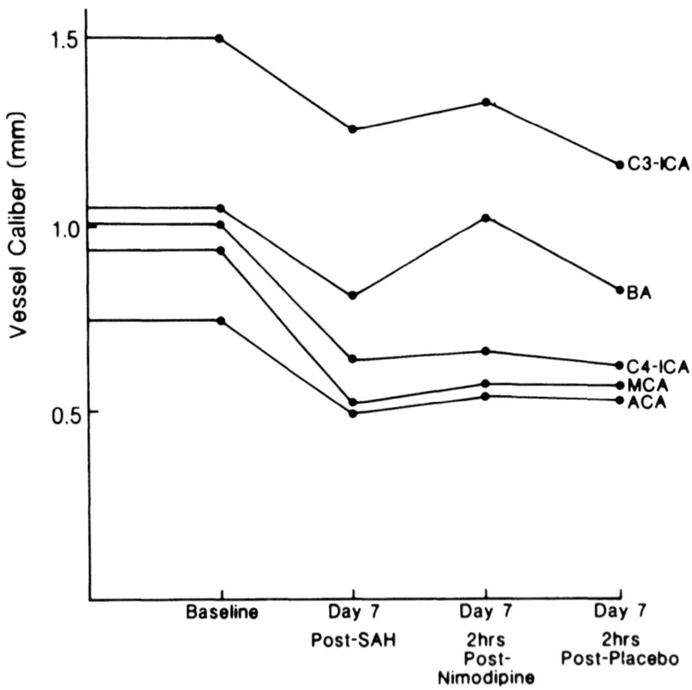

Fig. 17. Mean vessel caliber change of cerebral arteries from base line ($n = 30$) to day 7 post-SAH induction ($n = 30$) and 2 h after an intrathecal injection of nimodipine ($n = 16$) or placebo solvent ($n = 14$). Measurements of base line and day 7 angiograms for clot and clot plus nimodipine groups have been combined. Right- and left-sided data have also been combined. *C3-ICA*, cavernous internal carotid artery; *BA*, basilar artery; *C4-ICA*, supraclinoid internal carotid artery; *MCA*, middle cerebral artery; *ACA*, anterior cerebral artery. With permission, Neurosurgery

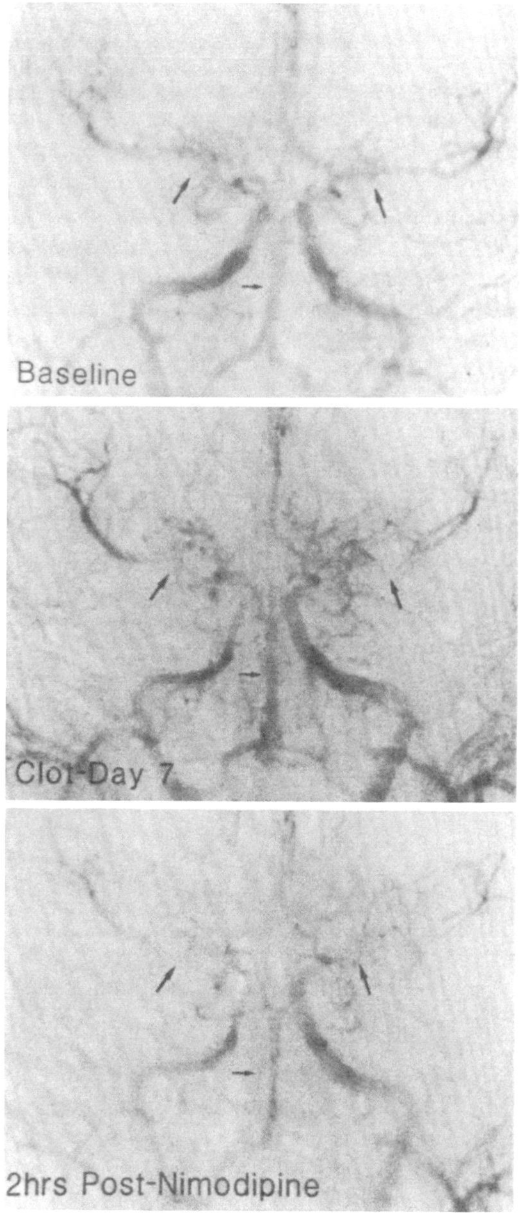

Fig. 18. Base line angiogram of a clot group monkey (*top*), day 7 post-SAH angiogram (*middle*), and day 7 post-intrathecal nimodipine injection angiogram (*bottom*). Severe vasospasm is present at day 7 post-SAH induction with no dilation of vessels occurring after an intrathecal nimodipine injection. Seven of eight animals showed this response to an intrathecal nimodipine injection. *Large arrows*, middle cerebral arteries; *small arrows*, basilar arteries. With permission, Neurosurgery

Clinical Studies

Neurosurgeons at the University of Alberta have participated in three prospective, double-blind trials of nimodipine. In the first study nimodipine was used in good grade patients who were treated between 1979 and 1982. Study medication was begun within 96 h of the ictus. The initial dose of nimodipine was 0.7 mg/kg adjusted to the nearest full 10-mg capsule. Thereafter 0.35 mg/kg were given q 4 h for 21 days. The endpoints were the development of a neurological deficit from cerebral arterial spasm and the severity of the deficit at the end of the 21-day treatment. One hundred and twenty-five neurologically normal patients with ruptured aneurysms took part in this prospective, double-blind, randomized, and placebo-controlled trial. A deficit from cerebral vasospasm that persisted and was severe or caused death by the end of the 21-day treatment occurred in 8 of 60 patients on placebo and only 1 of 56 given nimodipine. Analysis of the amount of basal subarachnoid blood on pre-entry CT scans in patients developing deficits from vasospasm showed that an increase in subarachnoid blood was not associated with a worse neurological outcome among patients who received nimodipine as opposed to the situation for those given placebo. No significant side-effects to nimodipine were encountered [8].

A subsequent multicentered trial was undertaken in an attempt to demonstrate a dose-response curve to nimodipine. This prospective, double-blind, randomized trial was conducted on 214 patients from many North American centers. The experimental design was essentially the same as the first study with different dose levels. The patients were neurologically normal following their subarachnoid hemorrhage from aneurysm. A severe, persistent deficit from vasospasm occurred in 2.3% of 170 patients who did not violate protocol. The dosages employed were 30, 60, or 90 mg/q 4 h for 21 days. At the two higher doses, there was a significant decrease in the overall incidence of all deficits from vasospasm. There was a decrease in blood pressure in 10% of patients at the two higher doses although hypotension was not prolonged or severe [9].

Our latest clinical participation was in a multicentered, randomized, placebo-controlled, double-blind trial of nimodipine in poor grade aneurysm patients carried out at 17 Canadian hospitals. Of 188 patients in the trial, 32 were excluded for protocol violations and two for statistical considerations, which left 154 patients for valid outcome analysis. Treatment with nimodipine was associated with a significantly better neurological outcome. Twenty-one (29.2%) of 72 nimodipine-treated patients had a good outcome at 3 months after subarachnoid hemorrhage compared to eight (9.8%) of 82 placebo-treated patients (Fig. 19). Delayed ischemic deficits from vasospasm alone were significantly less frequent in the nimodipine group with permanent deficits occurring in five nimodipine-treated patients (6.9%) and 22 placebo-treated patients (26.8%) (Fig. 20). The better 3-month outcome rate and reduction in delayed ischemic deficits from vasospasm alone occurred in both grade 3 and grade 4 patients with no difference between nimodipine- and placebo-treated patients being found in the grade 5 cases. Repeat angiography after day 4 was carried out in 124 patients. There was no significant difference in the incidence of moderate or severe diffuse spasm,

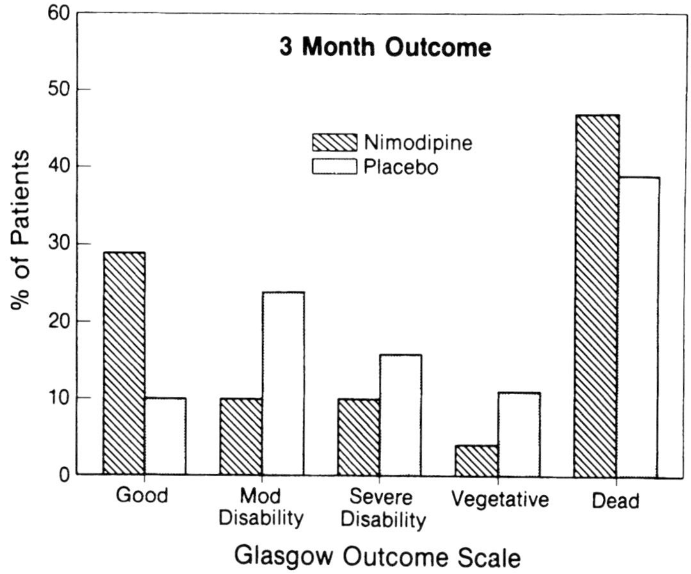

Fig. 19. Comparison of outcome based on the Glasgow Outcome Scale at 3 months after subarachnoid hemorrhage for 72 nimodipine-treated patients and 82 placebo-treated patients. With permission, Journal of Neurosurgery

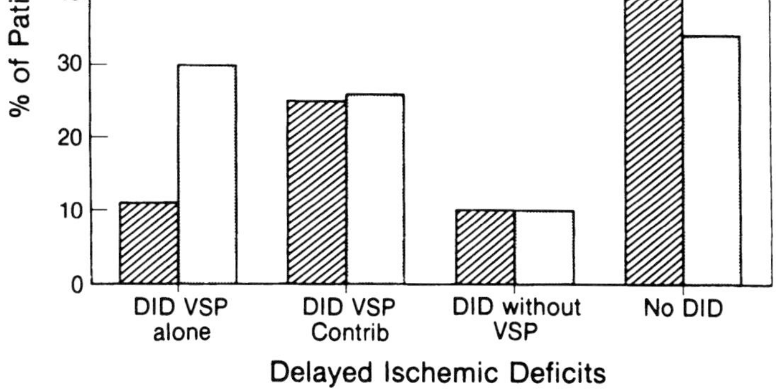

Fig. 20. Incidence of permanent delayed ischemic deficits (*DID*) and their etiology in the nimodipine and placebo groups. *VSP*, vasospasm. With permission, Journal of Neurosurgery

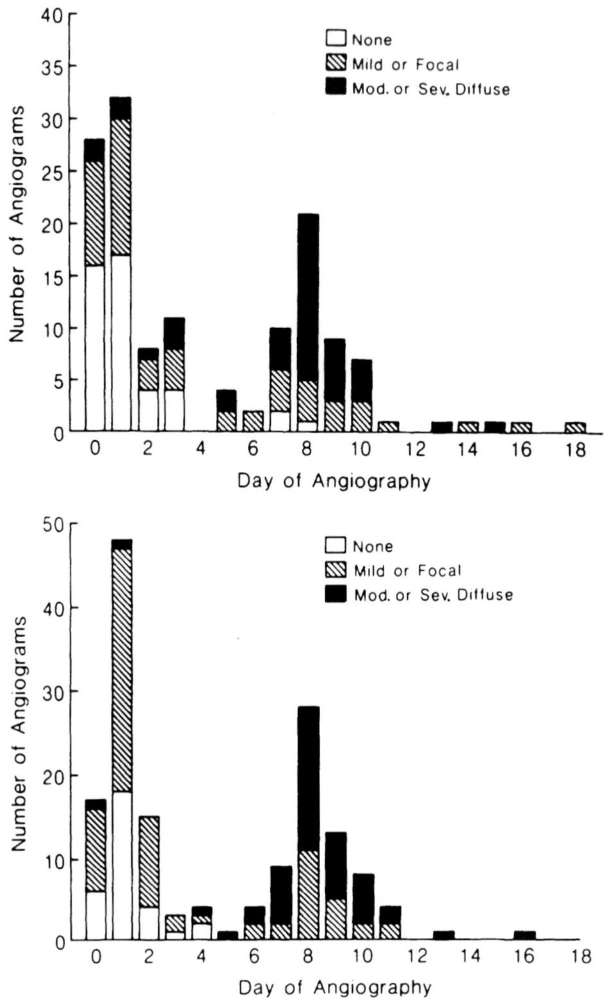

Fig. 21. Degree of angiographic spasm correlated with day of angiography. *Top*: In the nimodipine group, 139 angiograms were performed on 75 patients, including three patients with follow-up angiography performed before medication error. *Bottom*: In the placebo group, 159 angiograms were performed on 85 patients, including three patients with follow-up angiography performed before medication error. With permission, Journal of Neurosurgery

which was seen in 64.3% of nimodipine-treated patients and 66.2% of placebo-treated patients (Fig. 21). Fewer infarcts at 3 months post-SAH attributable to vasospasm were seen in nimodipine cases and such infarcts were smaller, but these differences did not achieve significance (Fig. 22). Nimodipine treatment in poor grade patients with subarachnoid hemorrhage therefore resulted in an increase in the number of good outcomes and a reduction in the incidence of delayed neurological deterioration due to vasospasm. This improvement was not brought about at the cost of an increased number of severely disabled or vegetative patients. The therapeutic effect appears to be from a mechanism other than the prevention of large vessel diffuse vasospasm as visualized on angiography [10].

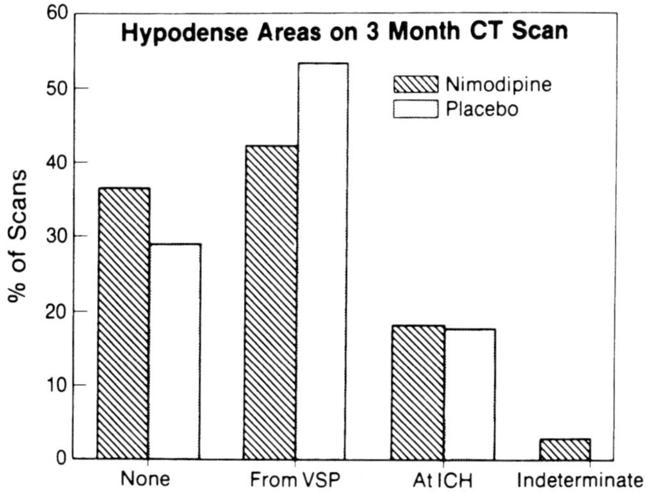

Fig. 22. Incidence and etiology of hypodense areas seen on computerized tomography (CT) at 3 months after subarachnoid hemorrhage that were consistent with infarction. *VSP*, vasospasm; *ICH*, intracerebral hemorrhage. With permission, Journal of Neurosurgery

Summary

The result of these and many studies from other centers appear to establish that nimodipine improves outcome following subarachnoid hemorrhage but the mechanism by which it does so remains in doubt. While it was originally employed in the hope that it would prevent or reverse radiological vasospasm, this does not seem to be the case. Perhaps it operates by causing dilatation of the vascular bed at a level beyond that which can be visualized at angiography. It may also have a direct protective effect on neurons and glia in the face of ischemia. The cardiovascular changes resulting from its use may on balance be beneficial or result in therapeutic corrections (such as increased fluid administration to counteract hypotension) which are subsequently therapeutic.

The search must therefore go on for agents which can prevent or reverse cerebral vasospasm, which remains an undoubted cause of delayed ischemic neurological deficit of significant proportions. Notwithstanding this, the introduction of nimodipine appears to have been a significant therapeutic advance.

References

1. Espinosa F, Weir B, Overton T, Castor W, Grace M, Boisvert D (1984) A randomized placebo-controlled double-blind trial of nimodipine after SAH in monkeys. Part 1: Clinical and radiological findings. J Neurosurg 60:1167–1175
2. Espinosa F, Weir B, Shnitka T, Overton T, Boisvert D (1984) A randomized placebo-controlled double-blind trial of nimodipine after SAH in monkeys. Part 2: Pathological findings. J Neurosurg 60:1176–1185
3. Nosko M, Weir B, Krueger C, Cook D, Norris S, Overton T, Boisvert D (1985) Nimodipine and chronic vasospasm in monkeys: Part 1. Clinical and radiological findings. Neurosurgery 16:129–136

4. Krueger C, Weir B, Nosko M, Cook D, Norris S (1985) Nimodipine and chronic vasospasm in monkeys: Part 2. Pharmacological studies of vessels in spasm. Neurosurgery 16:137–140
5. Nosko M, Norris S, Weir B, King EG, Grace M (1986) Nimodipine and chronic vasospasm in monkeys: Part 3. Cardiopulmonary effects. Neurosurgery 18:261–265
6. Nosko M, Krueger C, Weir B, Cook D (1986) Effects of nimodipine on in vitro contractility of cerebral arteries of dog, monkey and man. J Neurosurg 65:376–381
7. Lewis PJ, Weir BKA, Nosko MG, Tanabe T, Grace MG (1988) Intrathecal nimodipine therapy in a primate model of chronic cerebral vasospasm. Neurosurgery 22:492–500
8. Allen GS, Ahn HS. Preziosi TJ et al. (1983) Cerebral arterial spasm – a controlled trial of nimodipine in patients with subarachnoid hemorrhage. N Engl J Med 308:619–624
9. Allen G (1987) Personal communication. In: Weir B (ed) Aneurysms affecting the nervous system. Williams & Wilkins, Baltimore, p 671
10. Petruk KC, West M, Mohr G et al. (1988) Nimodipine treatment in poor-grade aneurysm patients. Results of a multicenter double-blind placebo-controlled trial. J Neurosurg 68:505–157

Treatment of Symptomatic Cerebral Vasospasm After Subarachnoid Hemorrhage by Intraarterial Perfusion with Nimodipine

D.-K. BÖKER and R. WÜLLENWEBER

Cerebral vasospasm after spontaneous subarachnoid hemorrhage (SAH) is often associated with an unsatisfactory outcome. But one has to realize that in a proportion of cases angiographically proven severe spasm is not symptomatic whereas only slight spasms may be associated with severe neurological deficits. Thus "angiographic spasm" is not identical with the syndrome of "clinical angiospasm". The pathogenesis of cerebral vasospasm is not yet clearly understood. Among several other factors calcium plays a role in the formation of vasospasm in that it initiates contraction of the smooth muscle cell of the vessel wall. Furthermore an intracellular overload with calcium causes damage to glial as well as to neuronal cells [6, 10]. Thus the use of calcium antagonists could be beneficial in the prevention or treatment of cerebral vasospasms [12] and in protecting the brain against intracellular calcium overloading. Clinical and experimental experiences with the calcium antagonist nimodipine after intravenous and/or oral administration indicate that the incidence of neurological deficits attributed to vasospasm may be reduced [1, 3–5, 8, 11, 13–15]. Until now there has been little experience with intraarterial administration of nimodipine in the treatment of symptomatic vasospasm. Thus Grotenhuis et al. [7] reported on a limited series of six cases in which slow bolus injection of the calcium antagonist failed to exhibit a convincing effect. We have tried a longer lasting perfusion of the cerebral arteries immediately after angiographic demonstration. In the following the technique and results in this still very limited series are reported.

Case Reports

Case 1

A 45-year-old female suffered spontaneous SAH 3 days before transfer to our hospital. A few hours beforehand she had developed a severe right-sided hemiparesis and had become drowsy. Cerebral angiography was performed immediately after admission and an anterior communicating artery aneurysm was found. Furthermore there were severe spasms of the supraclinoidal segment of the left-sided internal carotid artery and of the left anterior and middle cerebral arteries in the A_1 and M_1 sections, respectively.

Department of Neurosurgery, University of Bonn, Sigmund-Freud-Str. 25, D-5300 Bonn 1, FRG.

Cerebral Ischemia and Calcium
Ed. by A. Hartmann and W. Kuschinsky

Because the clinical deterioration had occurred only a few hours beforehand and because of the severity of the spasms we decided to try an intraarterial perfusion of the left internal carotid artery with nimodipine. Through the 7-Fr angiography catheter a 3-Fr catheter was introduced into the left internal carotid artery and the 7-Fr catheter was drawn back with its tip in the thoracic aorta, where it remained during the whole procedure. Nimodipine was administered via the 3-Fr catheter directly into the left internal carotid artery at a dosage of 0.2 mg/h for 90 min. This dosage was chosen as about 10% of the blood volume passes one carotid artery and 0.2 mg/h corresponds to 10% of the dosage normally administered intravenously.

Accidentally the 3-Fr catheter slipped back into the common carotid artery so that perfusion of the external carotid artery occurred as well. As a clinical correlate a marked flush of the left half of the patient's face was seen. Nevertheless, during the perfusion with nimodipine a slight improvement of the hemiparesis was observed. We could not register any side-effects of the intraarterial treatment. Intravenous administration of the calcium antagonist at 2 mg/h for 7 days was started immediately after intraarterial treatment. At the end of treatment neurological records of the patient were completely normal. For technical reasons we could not perform control angiography after the perfusion, but the clinical improvement of the patient as early as during the intraarterial administration of nimodipine made us feel that this form of treatment could be helpful.

Case 2

A 46-year-old woman experienced spontaneous SAH 9 days before she was transferred to our hospital. Neurological examination upon admittance disclosed a somnolent but cooperative patient with a moderate right-sided hemiparesis. There were no sensory deficits nor were there reflex differences. As in case 1, the paresis had developed only a few hours previously. Cerebral angiography was performed the same day and revealed an aneurysm of the anterior communicating artery. However, there were also marked spasms of the major vessels bilaterally (Figs. 1 A, 2 A). The angiogram of the vertebrobasilar system showed retrograde filling of both anterior cerebral arteries via an anastomosis to the posterior pericallosal arteries, thus suggesting that the spasms might have developed earlier than the clinical deterioration indicated.

As the neurological deficit had occurred only a few hours beforehand and because of the severity of the spasms we decided to try intraarterial nimodipine treatment. It followed the same protocol as described in the preceding case. Immediately afterwards we performed control angiography, which showed almost complete disappearance of the spasms (Figs. 1 B, 2 B). However, the aneurysm had enlarged in diameter as well, so the patient was operated on immediately. Postoperatively intravenous administration of nimodipine was continued at 2 mg/h for 8 days. The neurological deficits disappeared totally within the first postoperative week. Control angiography 3 weeks after surgery proved correct clipping of the aneurysm. No cerebral angiospasm was detectable, and the patient was discharged without any neurological deficit.

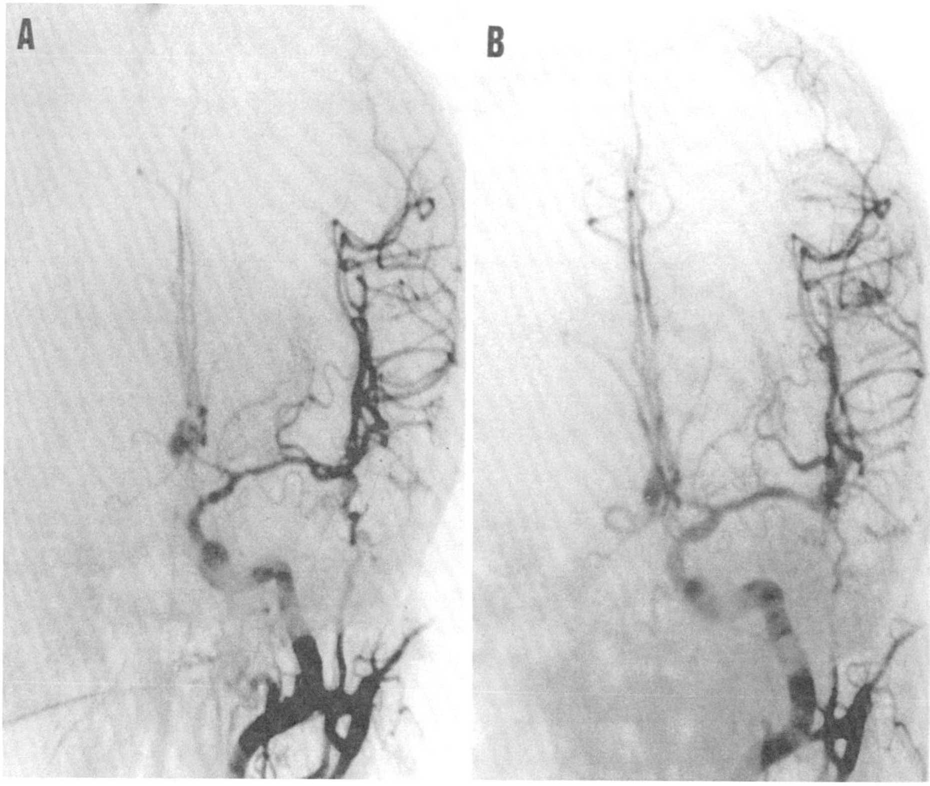

Fig. 1 A, B. Case 2, 46-year-old female. Left carotid angiogram, anteroposterior view. Aneurysm of anterior communicating artery. **A** Angiospasm of left supraclinoidal carotid artery and the anterior and middle cerebral arteries. **B** After 90 min of intracranial nimodipine angiospasm is markedly reduced

Our experience in this case with augmentation of the aneurysm during intraarterial nimodipine administration led us to the conclusion that this treatment might better be restricted to postoperative symptomatic vasospasm.

Case 3

A 42-year-old man underwent surgery for a right-sided supraclinoidal carotid aneurysm on the 3rd day after spontaneous SAH. Pre- and immediately postoperatively no neurological deficits were recorded. On day 11 after surgery he developed a left-sided hemiparesis which became worse although treatment with oral nimodipine at 180 mg/day was started. Two days later the paresis had become almost complete. CT scan did not reveal any sign of cerebral infarction, but angiographically severe spasm of the right internal carotid artery and anterior cerebral artery was demonstrated (Fig. 3 A). We began treatment with initial intravenous administration of nimodipine at 2 mg/h for 30 min. A control angiogram showed

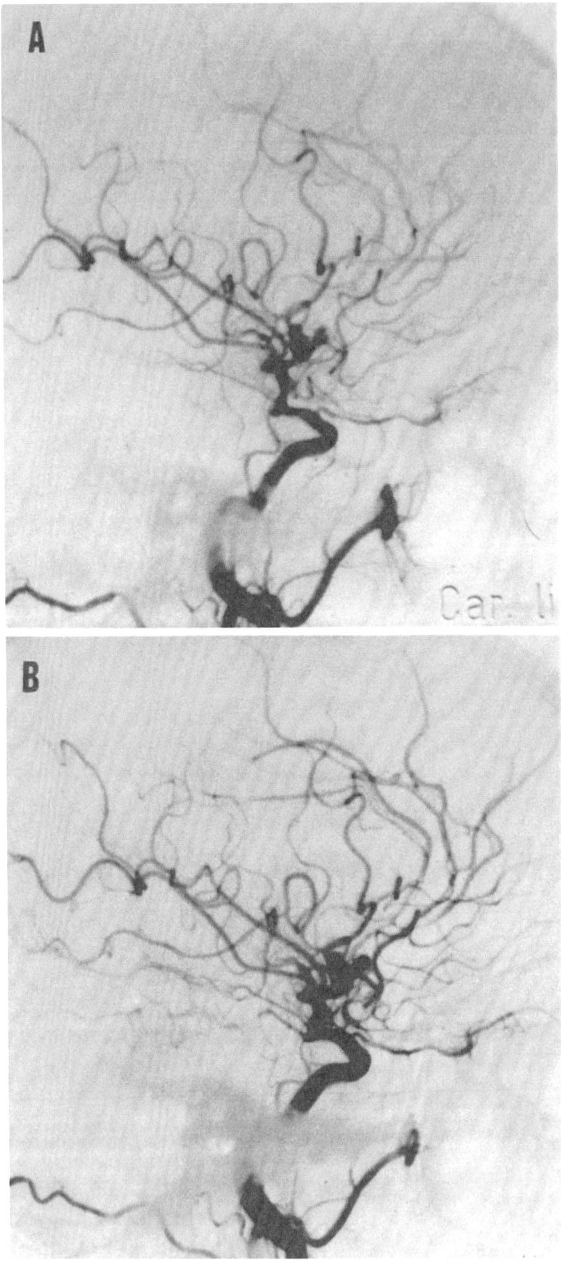

Fig. 2 A, B. Same case as in Fig. 1; left carotid arteriogram, lateral view. **A** Angiogram before treatment. **B** After 90 min there is a marked increase in the diameter of the supraclinoid carotid, and no spasm is detectable; however, there is also an increase in the size of the aneurysm

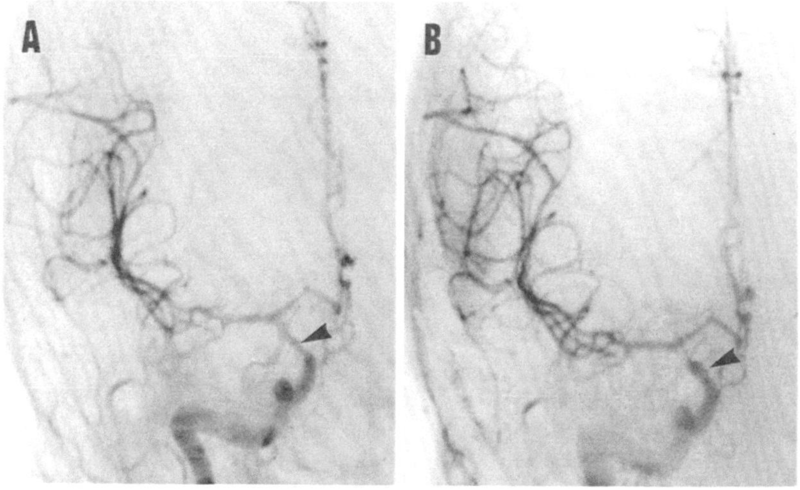

Fig. 3 A, B. Case 3, 40-year-old man. Right carotid angiogram after clipping of a supraclinoidal carotid aneurysm. **A** Marked spasm of the carotid and anterior cerebral arteries. **B** After intraarterial perfusion with nimodipine, partial resolution of the spasm is seen, with persistence in the vicinity of the clip

unaltered spasms as compared to the pretreatment films. Therefore we began perfusion of the right internal carotid artery according to the schedule described in case 1. After 90 min control angiography was again performed, which showed that the angiospasms had resolved to a large extent (Fig. 3 B). Only in the direct vicinity of the clip was there further constriction of the carotid artery. Treatment was continued by intravenous administration of nimodipine for 13 days at 2 mg/h and oral administration at 240 mg/day for 5 days. The neurological records improved and the patient was discharged with only a slight hemiparesis.

Case 4

A 32-year-old woman was operated on for an aneurysm of a small branch of the left middle cerebral artery on the 3rd day after spontaneous SAH. Intravenous nimodipine administration at 2 mg/h was started the day before operation to prevent angiospasm and was continued until the fifth postoperative day. Oral administration of nimodipine at 240 mg/day followed. Neurological records of the patient were noted pre- and postoperatively. At the time of dose reduction the patient began to complain of headache, but the neurological records were normal later.

On day 10 after surgery control angiography was performed, showing complete clipping of the aneurysm but also marked spasm of the left supraclinoidal carotid artery and the left A_1 and M_1 segments (Figs. 4 A, 5 A). We immediately started intraarterial perfusion with nimodipine according to the schedule described in case 1. Angiography after 90 min showed partial but incomplete reso-

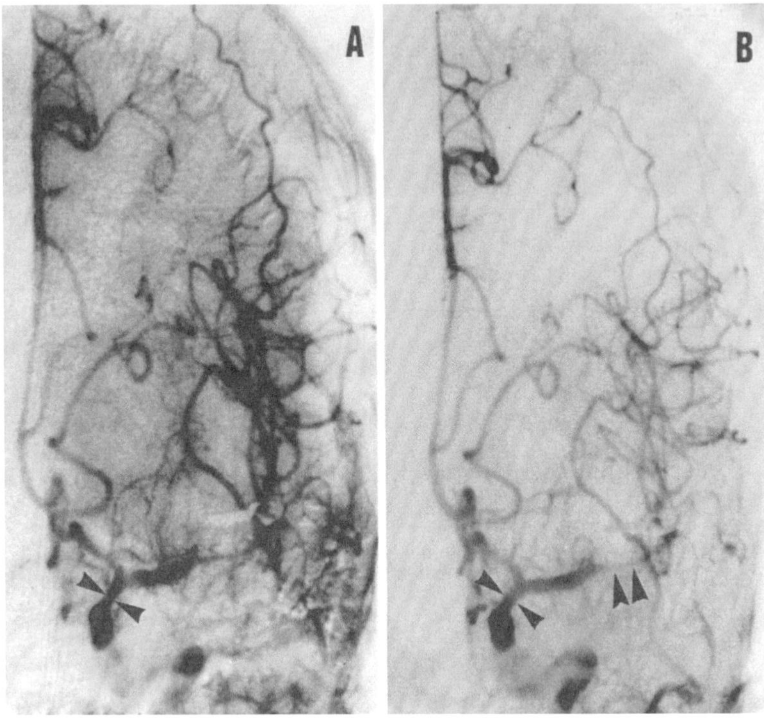

Fig. 4 A, B. Case 4, 32-year-old woman. Postoperative left carotid angiogram after clipping of an aneurysm of the left middle cerebral artery; anteroposterior view. **A** Spasm in the supraclinoidal carotid and distal M_1 segment. **B** After 90 min of arterial perfusion with nimodipine, partial resolution of the spasm is seen, with persistence in the M_1 segment

lution of the spasm (Figs. 4 B, 5 B). Spasm persisted at that time mainly in the distal M_1 segment on the left. Headache had disappeared immediately after intraarterial perfusion. Intravenous nimodipine administration at 2 mg/h was continued for 5 days and oral drug administration for another 7 days. The patient remained free of signs and symptoms and was discharged without any complaints.

Discussion

The use of calcium antagonists may provide an approach to the treatment or, what would be even more effective, to the prevention of cerebral angiospasm after SAH. The routine treatment will be oral or intravenous drug administration. Topical application of calcium antagonists during surgery often results in a marked increase in the diameter of spastic vessels [2], but little is known about the duration of the effect. In the first two cases described above we decided to try intraarterial treatment because of the severity of the spasms and because the neu-

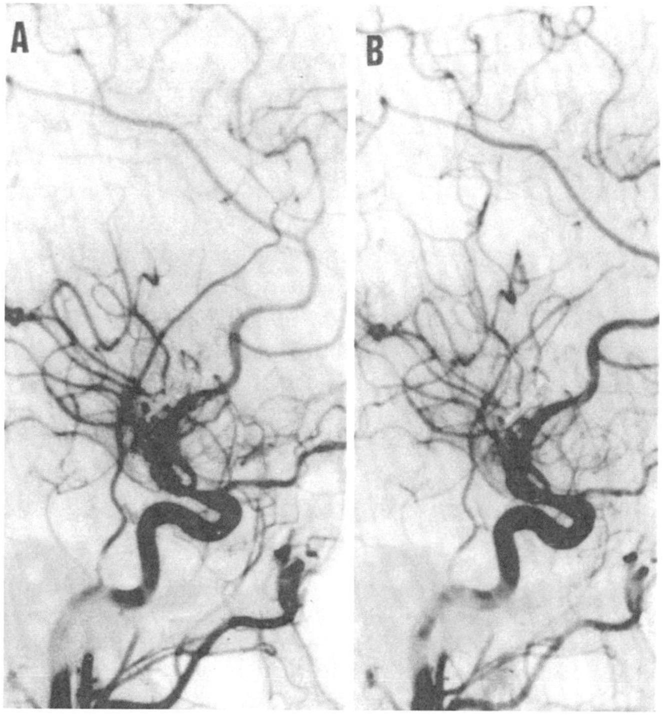

Fig. 5 A, B. Same case as in Fig. 4, lateral view. Angiogram **A** before and **B** after intraarterial nimodipine treatment. Note resolution of spasm in the supraclinoidal carotid segment

rological deficits had developed only a few hours previously. In the second case angiographic criteria gave rise to the idea that the spasms might have existed, albeit asymptomatically, for somewhat longer. As we knew that intraarterial slow bolus injections of nimodipine had had no convincing effect, we decided to try the treatment protocol described above. In our still very limited series the effect of the treatment was rather impressive. Clinical improvement was noted in all cases, and in three of them at least partial resolution of the spasms could be demonstrated. In the other case control angiography could not be performed.

Nevertheless, we will have to assume that in preoperative cases this treatment might also be of high risk: In our second case not only did the spasms disappear but the aneurysm also increased considerably in size. Thus intraarterial perfusion is probably better restricted to postoperative symptomatic angiospasm. The third and fourth cases were postoperative ones. In the third case it is remarkable that angiospasm became symptomatic as late as 11 days postoperatively. Oral and intravenous treatment were initially without effect, but the resolution of the spasms after intraarterial perfusion was almost complete after 90 min. The constriction of the carotid artery in the vicinity of the clip might be due to persisting spasms but also to the clip being positioned too close to the carotid wall.

In our fourth case headache occurred at the time of drug reduction. Angiographically spasms were shown which became better after intraarterial perfusion.

As headache disappeared at this time, we think that it was due to the spasm which developed at the time of drug reduction.

In every case the clinical benefit of intraarterial nimodipine treatment could be preserved by immediately subsequent intravenous administration. Thus we feel it advisable to follow arterial perfusion for several days by continuous intravenous administration of nimodipine.

Intraarterial perfusion with nimodipine in the treatment of symptomatic cerebral angiospasm is easy to perform in combination with angiography.

In our small series there were no side-effects attributable to the technique. Further experience will need to be acquired before a definitive judgement can be made on the real value of this protocol, but we feel that it might provide a useful tool in the treatment of symptomatic cerebral angiospasm after SAH. Because of the risk of enlargement of the aneurysm, this form of treatment should be restricted to postoperative cases.

Summary

Cerebral angiospasm after subarachnoid hemorrhage is often associated with an unsatisfactory outcome. Calcium antagonists may provide a therapeutic approach to this problem in that they may contribute to the prevention of spasm formation or the treatment of symptomatic spasm. The authors report four patients with symptomatic angiospasm in whom direct perfusion of the carotid artery with nimodipine was performed. In every case clinical improvement was noted, and in three of the four resolution of the spasms could be demonstrated angiographically.

There were no side-effects due to the perfusion technique. However, as in one case besides resolution of the spasm the aneurysm enlarged as well, this form of treatment is probably better restricted to postoperative cases, in which it also seems to be effective.

References

1. Allen GS, Ahn HS, Preziosi TF et al. (1983) Cerebral arterial spasm – a controlled trial of nimodipine in patients with subarachnoid hemorrhage. N Engl J Med 308:619–624
2. Auer LM, Ho H, Suzuki A, Ohta H (1982) Prevention of symptomatic vasospasm by topically applied nimodipine. Acta Neurochir 63:297–302
3. Auer LM (1983) Acute surgery of cerebral aneurysm and prevention of symptomatic vasospasm. Acta Neurochir 69:273–281
4. Auer LM (1984) Acute operation and preventive nimodipine improve outcome in patients with ruptured cerebral aneurysms. Neurosurgery 15:57–66
5. Böker DK, Kazner E, Sprung C et al. (1985) Clinical experience in the prevention of ischemic neurological deficits after subarachnoid hemorrhage with nimodipine. In: Auer L (ed) Timing of aneurysm surgery. Walter de Gruyter, Berlin, pp 535–544
6. Brown AW (1977) Structural abnormalities in neurones. J Clin Pathol 30(Suppl 11):155–169

7. Grotenhuis JA, Bettag W, Fiebach BJO et al. (1984) Intracarotid slow bolus injection of nimodipine during angiography for treatment of cerebral vasospasm after SAH. A preliminary report. J Neurosurg 61:231–240
8. Grotenhuis JA, Bettag W (1986) Prevention of symptomatic vasospasm after SAH by constant venous infusion of nimodipine. Neurol Res 8:243–249
9. Grotenhuis JA, Personal communication
10. Hubschmann OR, Nathanson DC (1985) The role of calcium and cellular membrane dysfunction in experimental trauma and subarachnoid hemorrhage. J Neurosurg 62:698–703
11. Kazner E, Sprung C, Adelt D et al. (1985) Clinical experience with nimodipine in the prophylaxis of neurological deficits after subarachnoid hemorrhage. Neurochirurgia 28:110–113
12. Kazda S, Towart R (1982) Nimodipine: a new calcium antagonistic drug with a preferential cerebrovascular action. Acta Neurochir 63:259–265
13. Koos WT, Perneczky A, Auer LM et al. (1985) Nimodipine treatment of ischemic neurological deficits due to cerebral vasospasm after subarachnoid hemorrhage. Neurochirurgia 28:114–117
14. Tanaka K, Gotoh F, Nuramatsu P et al. (1982) Effect of nimodipine, a calcium antagonist, on cerebral vasospasm after subarachnoid hemorrhage in cats. Drug Res 32:1529–1534
15. Voldby B, Buhl M, Petersen OF et al. (1981) Modification of cerebral vasospasm by intraventricular administration of a calcium antagonist in the dog. Acta Neurochir 59:287

Effect of Nimodipine on CO_2 Reactivity of Patients with Normal and Spastic Basal Cerebral Arteries: A Transcranial Doppler Ultrasound Study

R. W. SEILER, A. C. NIRKKO, and M. WEBER

The dihydropyridine derivative nimodipine is used with increasing frequency to combat potential ischemic complications due to vasospasm after subarachnoid hemorrhage (SAH). Experimental and clinical data [3, 4] have shown an increasing dilatatory effect with decreasing vessel caliber, so that the anti-ischemic effect of the drug may be mediated more by dilatation of the resistance arterioles than by dilatation of the spastic basal cerebral arteries. The resistance arteries and arterioles are the site of the vasoreactivity to CO_2 and blood pressure changes. Therefore, the question arises of whether the cerebrovascular reactivity is preserved by the use of nimodipine. The aim of the present study was to evaluate the effect of nimodipine on the CO_2 reactivity of patients with normal and spastic basal cerebral arteries and to investigate the effect of vasospasm on CO_2 reactivity.

Material and Methods

Method

Transcranial Doppler represents a noninvasive mean of assessing velocity in the basal cerebral arteries [1, 5]. Because the diameter of the large basal arteries can be assumed to remain constant during hypo- and hypercapnia, changes in cerebral blood flow (CBF) are proportional to changes in velocities in the basal cerebral arteries. In our previous CO_2 reactivity study, the response curve for the middle cerebral artery (MCA) velocity strongly resembled the CBF response curve to CO_2 obtained from experimental and clinical studies [10]. The same method described in detail previously was used in the present study. The patient was breathing first in the individual normocapnic range (end-tidal pCO_2 usually between 35 and 40 mmHg). Thereafter, he was instructed to hyperventilate in order to reach a hypocapnic steady state (end-tidal pCO_2 usually between 20 and 25 mmHg). In a third step the respiration mask was connected to air mixed with 5% CO_2 to induce moderate hypercapnia. CO_2 was measured with an infrared end-tidal CO_2 analyzer (capnometer 47210 A) under continuous spirometer control (Megamed 018 G). After a steady state of pCO_2 had been reached, the mean MCA velocity was recorded at a depth of 4.5–5.5 cm for 3–5 min with our laboratory range-gated 2-MHz Doppler instrument [1].

Department of Neurosurgery, University Hospital, Inselspital, CH-3010 Bern, Switzerland.

Cerebral Ischemia and Calcium
Ed. by A. Hartmann and W. Kuschinsky

Normal Arteries

Six patients without SAH (two with asymptomatic aneurysms, four with unruptured arteriovenous malformations) served as controls for the evaluation of normal arteries. The CO_2 reactivity of the MCA opposite to the lesion was determined. A baseline CO_2 reactivity was obtained in each patient first. The test was repeated under constant intravenous infusion of nimodipine 2 mg/h (given prophylactically before operation or embolization).

Spastic Arteries

Four patients with ruptured aneurysms in clinical grades I–III operated on within 1–3 days after the SAH were chosen for the evaluation of spastic arteries. A first CO_2 reactivity test with constant intravenous infusion of nimodipine 2 mg/h was done 7–10 days post-SAH, when the MCA velocities were increased to 50%–100% of the initial value as a consequence of cerebral vasospasm [2, 5]. The test was repeated 9–10 days post-SAH without nimodipine (discontinued for at least 48 h), when the MCA velocities were still increased to 50%–100% of the initial value. A third test was done 6–8 weeks later after full recovery from the SAH (controls).

Statistical Analysis

The MCA velocity value of every patient was plotted as a function of the end-tidal pCO_2. The best-fit curve for these plots was the exponential function: $y = a \times e^{bx}$, where y is velocity (cm/s), x is the end-tidal pCO_2 (mmHg), a is a theoretical quantity and represents velocity at a pCO_2 of 0 mmHg, and b corresponds to the definition of reactivity as relative velocity change divided by the pCO_2 change. Because of the variations in MCA velocity from one subject to another [1], we have chosen the MCA velocity at a pCO_2 of 35 mmHg as 100% to compare the individual measurements. If the normocapnic end-tidal pCO_2 of the investigated patient was not at 35 mmHg, the MCA velocity was calculated for this value using the individual regression curve. Then we plotted MCA velocity as a percentage of the value at a pCO_2 of 35 mmHg as a function of the end-tidal pCO_2 (mmHg) with a best-fit reactivity curve for each group for comparison. The statistical significance of any difference between groups was computed using the Mann-Whitney-Wilcoxon test on the individual reactivity values.

Results

Effect of Nimodipine on CO_2 Reactivity of Patients with Normal Arteries

Figure 1 shows the results of the CO_2 reactivity tests of the control patients done without nimodipine compared to a constant intravenous infusion of nimodipine

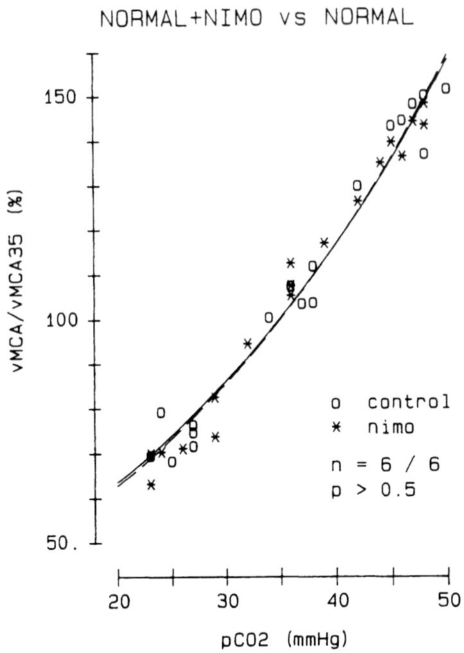

Fig. 1. MCA velocities as percentage of the velocity value at an end-tidal pCO_2 of 35 mmHg in the six patients without SAH, with and without nimodipine

2 mg/h. MCA velocities are plotted at hypno-, normo-, and hypercapnia as percentage of the velocity value at an end-tidal pCO_2 of 35 mmHg. There is no statistically significant difference between the two curves ($P > 0.5$).

Effect of Nimodipine on CO_2 Reactivity of Patients with Spastic Basal Arteries

Figure 2 demonstrates the results of the CO_2 reactivity tests of the four patients with ruptured aneurysms performed during the 2nd week post-SAH under a constant intravenous infusion of nimodipine 2 mg/h and without nimodipine. MCA velocities are plotted as in Fig. 1. There is no statistically significant difference between the two curves ($P > 0.5$).

Effect of Vasospasm on CO_2 Reactivity

Figure 3 shows the results of the CO_2 reactivity tests of the four patients with ruptured aneurysms performed during the 2nd week post-SAH without nimodipine and repeated 6–8 weeks later after full recovery. MCA velocities are plotted as in Fig. 1. The CO_2 reactivity of the patients with spastic basal arteries is significantly reduced compared to their control values after full recovery ($P < 0.05$).

Figure 4 shows the vascular reactivity, defined as relative change in velocity per change in pCO_2 as a function of the MCA velocity at a pCO_2 of 35 mmHg.

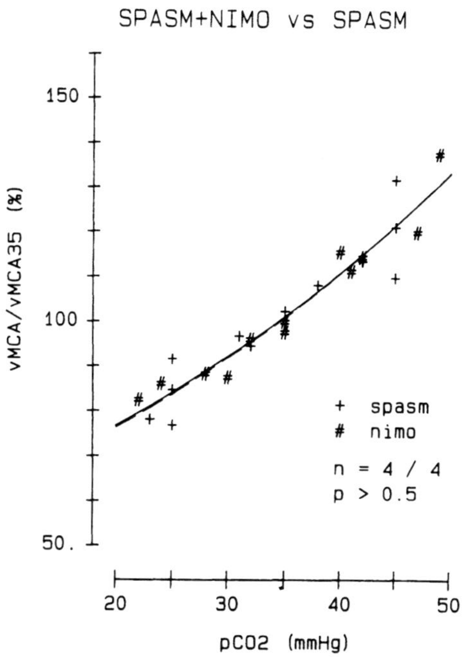

Fig. 2. MCA velocities as percentage of the velocity value at an end-tidal pCO_2 of 35 mmHg in the four SAH patients with vasospasms examined with and without nimodipine

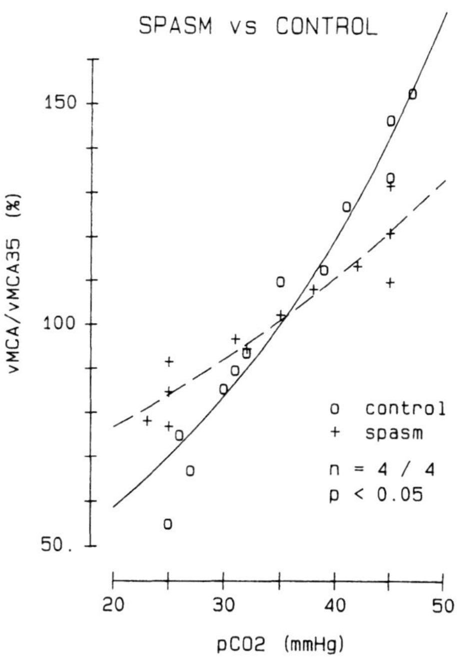

Fig. 3. MCA velocities as percentage of the velocity value at an end-tidal pCO_2 of 35 mmHg in the four patients examined in the 2nd week post-SAH without nimodipine and after full recovery 6–8 weeks later

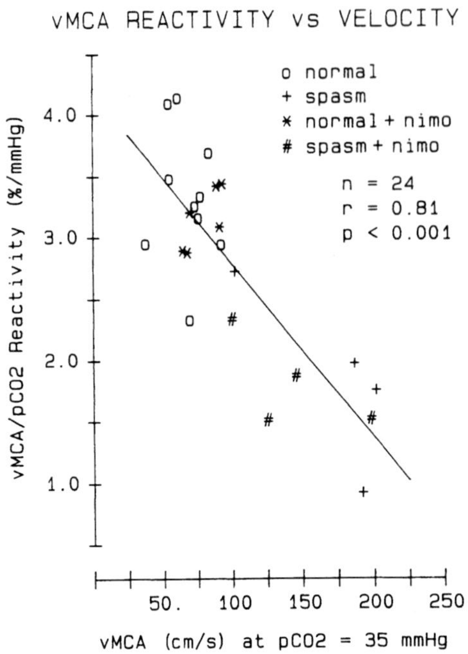

Fig. 4. CO_2 reactivity as a function of the MCA velocity at a pCO_2 of 35 mmHg in all 24 performed CO_2 reactivity tests

The values for the normal arteries and controls are within the range of 3.5% ±0.5 (mean ± SD) found in our previous study with normal subjects [10], whereas there is a progressive reduction of the vascular reactivity with increasing MCA velocities ($r = -0.81$, $P < 0.001$).

Discussion

In our human study, nimodipine reduced neither the CO_2 reactivity of patients with normal nor that of patients with spastic basal cerebral arteries. These findings are in disagreement with experimental animal investigations [6, 7, 11]. Harris et al. [6] found a severely impaired response to arterial pCO_2 changes in baboons with a dose of 0.6 µg/kg/min. Evaluating the same animal, McCalden et al. [11] described a cerebral reactivity to altered pCO_2 with a dose of 1 µg/kg/min. Haws and Heistad [7] examined the effect of nimodipine on constrictor response of pial arteries in cats. Low doses of nimodipine (0.1–0.25 µg/kg/min) decreased the vasoconstrictor response to sympathetic nerve stimulation, acute hypertension, and hypocapnia while higher doses (0.5–1.0) virtually abolished the response. The nimodipine dose used in our study was 2 mg/h i.v., corresponding to 0.5 µg/kg/min, which is comparable to the doses used in the experimental studies. In the latter investigations, anesthetized animals were examined, whereas we studied awake humans. Differences in the experimental design or species differences may explain the contradictory results.

The patients without SAH or after recovery from the bleeding showed a CO_2 reactivity of $3.5 \pm 0.5\%$ change in MCA velocity per mmHg change in end-tidal pCO_2 (Fig. 4), like the 31 normal subjects of our previous study [10]. In the 2nd week after the SAH, the vascular reactivity was significantly reduced with a close correlation between the degree of vasospasm and the degree of reduction of vasoreactivity. This is in accordance with several reports in which rCBF methods were used for the evaluation of CO_2 reactivity [8, 9, 12].

In conclusion, no significant effect of nimodipine on CO_2 vasoreactivity could be demonstrated in any of the test periods. A progressive reduction of the vascular reactivity with increasing vasospasm was found.

Summary

In six patients with unruptured aneurysms or arteriovenous malformations, the CO_2 reactivity of the middle cerebral artery opposite to the lesion was tested with and without nimodipine 2 mg/h i.v. preoperatively. Four patients with ruptured aneurysms in clinical grades I–III were operated on within 1–3 days after the hemorrhage and treated with intravenous nimodipine 2 mg/h. The CO_2 reactivity was tested during the 2nd week after the hemorrhage, when the middle cerebral artery velocity was increased by at least 50% of the initial value. Nimodipine was then discontinued and 48 h later, when the middle cerebral artery velocity was still in the same range, CO_2 reactivity was tested again. Two months later, after full recovery and normalization of the velocities, a third CO_2 reactivity test was performed in each patient as baseline control. No significant effect of nimodipine on vasoreactivity could be demonstrated in any of the test periods. A progressive reduction of the vascular reactivity with increasing vasospasm was found.

References

1. Aaslid R, Markwalder TM, Nornes H (1982) Noninvasive transcranial Doppler ultrasound recording of flow velocity in basal cerebral arteries. J Neurosurg 57:769–774
2. Aaslid R, Huber P, Nornes H (1984) Evaluation of cerebrovascular spasm with transcranial Doppler ultrasound. J Neurosurg 60:37–41
3. Auer LM (1981) Pial arterial vasodilatation by intravenous nimodipine in cats. Drug Res 31:1423–1425
4. Auer LM, Oberbauer RW, Schalk HV (1983) Human pial vascular reactions to intravenous nimodipine infusion during EC-IC bypass surgery. Stroke 14:210–213
5. Grolimund P, Seiler RW, Aaslid R, Huber P, Zurbrügg H (1987) Evaluation of cerebrovascular disease by combined extracranial and transcranial Doppler sonography. Stroke 6:1018–1024
6. Harris RJ, Branston NM, Symon L, Bayhan M, Watson A (1982) The effects of a calcium antagonist, nimodipine, upon physiological responses of the cerebral vasculature and its possible influence upon focal cerebral ischaemia. Stroke 6:759–766
7. Haws CW, Heistad DD (1984) Effects of nimodipine on cerebral vasoconstrictor responses. J Physiol 247:170–176

8. Heilbrun MP, Olesen J, Lassen NA (1972) Regional cerebral blood flow studies in subarachnoid hemorrhage. J Neurosurg 37:36–44
9. Ishii R (1979) Regional cerebral blood flow in patients with ruptured intracranial aneurysms. J Neurosurg 50:587–594
10. Markwalder TM, Grolimund P, Seiler RW, Roth F, Aaslid R (1984) Dependency of blood flow velocity in the middle cerebral artery on end-tidal carbon dioxide partial pressure – a transcranial ultrasound Doppler study. J Cereb Blood Flow Metab 4:368–372
11. McCalden TA, Nath RG, Thiele K (1984) The effects of a calcium antagonist (nimodipine) on basal cerebral blood flow and reactivity to various agonists. Stroke 3:527–530
12. Voldby B, Enevoldsen EM, Jensen FT (1985) Cerebrovascular reactivity in patients with ruptured intracranial aneurysms. J Neurosurg 62:59–67

Treatment of Subarachnoid Hemorrhage with Oral Nimodipine: Preliminary Report of the British Aneurysm Trial

G. M. TEASDALE [1], J. D. PICKARD [2], G. MURRAY [3], R. ILLINGWORTH [4], M. D. M. SHAW [5], P. FOY [5], P. HUMPHREY [5], D. M. LANG [1], R. NELSON [2], P. RICHARDS [4], J. SINAR [1], and A. SKENE [6]

Introduction

A patient who survives an initial subarachnoid haemorrhage may then die or be left disabled as a result of subsequent complications; prominent amongst the causes of such poor outcomes is cerebral ischaemia. Ischaemia may be the result of a number of factors: cerebrovascular spasm, impaired cerebrovascular reactivity, and changes in the structure of the cerebral vessel, hypotension, hypovolaemia, raised intracranial pressure, and hydrocephalus. Calcium antagonist drugs have several properties that may prevent or minimise cerebral ischaemic damage and experimental studies have shown a clear benefit from pretreatment of focal ischaemia (see McCulloch, Teasdale and Graham, and Harper in this volume).

Although subarachnoid haemorrhage provides the clearest clinical opportunity for pretreatment of ischaemia [1, 2, 4–6, 8, 9], conclusive evidence of benefit in patients has been lacking; in particular, calcium antagonists do not appear to reduce the incidence or severity of angiographic cerebral vasospasm. We therefore have conducted a prospective randomised, double-blind placebo-controlled trial, entering a sufficient number of patients for rigorous statistical analysis [8]. The aim was to establish the effect of nimodipine on the incidence of delayed cerebral ischaemic events and, in addition, on outcome following subarachnoid haemorrhage.

Methods and Patients

The trial was restricted to patients admitted to one of the participating centres within 96 h of the presenting haemorrhage. The diagnosis of subarachnoid haemorrhage was confirmed by CT in the majority of cases. Angiography was not re-

[1] Institute of Neurological Sciences, Southern General Hospital, Glasgow G51 4TF, U.K.
[2] Wessex Neurological Centre, Southampton General Hospital, Southampton SO9 4XY, U.K.
[3] Medical Statistics Unit, Department of Surgery, University of Glasgow, Glasgow G11 6NT, U.K.
[4] Department of Neurosurgery, Charing Cross Hospital, London W6 8RF, U.K.
[5] Regional Neurosurgical Unit, Walton Hospital, Liverpool, U.K.
[6] Department of Statistics, University of Nottingham, Nottingham, U.K.

Cerebral Ischemia and Calcium
Ed. by A. Hartmann and W. Kuschinsky

quired before entry to the study. Reasons for exclusion included a coma-producing subarachnoid haemorrhage in the week preceding the recent bleed.

Patients were treated with either oral nimodipine (60 mg 4 hourly) or placebo. Medication was continued for 21 days in survivors. Nimodipine or placebo was given as fast release tablets either swallowed with water or, in patients unable to swallow, crushed and administered through a nasogastric tube. The design specified a minimum 540 patients in order that the trial would have an 80% chance, with an alpha of 0.05, of detecting a 50% reduction in the occurrence of cerebral infarction if the incidence in the control group were 15%.

Patients were monitored clinically throughout their time in hospital and graded daily by a scale constructed from the Glasgow Coma Scale, features of headache and neck stiffness, and focal neurological deficit; this scale is similar to that recently approved by the World Federation of Neurosurgical Societies [10]. Minimum criteria for detection of significant deterioration were agreed in advance and all episodes were reviewed at meetings of all investigators and classified as either rebleeding, cerebral ischaemia or other. Rebleeding and infarction were divided into definite or probable on the basis of confirmatory evidence from scanning, lumbar puncture, operative findings, or post-mortem in fatal cases.

Outcome was assessed 12 weeks or more after entry according to the five-point Glasgow Outcome Scale [3] by a clinician not responsible for the patient's early management.

Results

Patients

During the 2 years and 3 months of the study a total of 554 patients were recruited: 199 in Glasgow, 151 in Southampton, 103 in Liverpool and 101 in London. Two hundred and seventy-six patients were randomised to active treatment and 278 to placebo. The two groups were very comparable in demographic data and indices of initial severity, summarised in Table 1. Differences between the two groups were small and none were statistically important.

Events During the Trial

Overall the incidence of presumed cerebral infarction was 27.6%, while that of rebleeding was 11.3%. Table 2 shows that the principal difference between the two groups was the significantly lower rate of cerebral infarction, either presumed or definite, in the patients treated with nimodipine. There was a trend to a lower occurrence of rebleeding in patients receiving nimodipine but this failed to reach significance statistically. Beneficial effects following nimodipine were observed even when patients were subgrouped according to clinical and CT features on admission. Comparison of daily systolic arterial pressures showed a mild hypotensive effect in patients taking nimodipine. Adverse reactions occurred infrequently

Table 1. Features on entry

	Nimodipine	Placebo
Total patients	278	276
Male	114 (41%)	107 (39%)
Female	164	169
Mean age (yrs)	46	48
Hypertension history	33 (12%)	23 (8%)
Fully conscious (GCS 15)	176 (63%)	171 (62%)
CT scan + ve SAH	177 (64%)	168 (61%)
Aneurysm identified	187 (67%)	181 (66%)

Table 2. Infarction and rebleeds in the two patient groups

	Nimodipine	Placebo
Infarction – definite	39	61
– probable	22	31
– total	61 (22%)	92 (33%)
Rebleeds – total	25 (9%)	38 (14%)

Table 3. Outcome at 3 months in the two patient groups

	Nimodipine	Placebo
Death	43 (16%)	60 (22%)
Vegetative	1	2
Severe disability	11 (4%)	29 (11%)
Moderate disability	24 (9%)	16 (6%)
Good recovery	199 (72%)	169 (61%)
	278	276

and included hypotension in one patient in the nimodipine-treated group and two patients in the controls. Overall eight patients discontinued nimodipine treatment and three discontinued placebo.

Outcome

Overall 146 patients had a poor outcome defined as death or vegetative or severely disabled survival at 3 months.

Significantly fewer poor outcomes occurred in the nimodipine-treated patients (20%) than in the controls (33%) ($P<0.001$). Table 3 shows that there was a reduction in deaths and disabled survivals and an increase in patients making a good recovery. The principal reason for the improvement in outcome was a reduction in deaths and disabled cases assessed as reflecting episodes of delayed ischaemia: 20 (7%) in the treated group and 34 (12%) in the controls. Other causes of death or disability were similar in the two groups; in particular, recovery in patients who had sustained a severe initial bleed did not seem to be influenced.

Discussion

The results of the study demonstrate convincingly that oral nimodipine treatment of patients with a subarachnoid haemorrhage in the previous 4 days reduces the occurrence of delayed cerebral ischaemia and reduces the risk of death or severely disabled survival [8]. In nimodipine-treated patients the occurrence of infarction was reduced by 34% (confidence intervals 13%–50%, $P<0.003$) and the risk of poor outcome was reduced by 40% (95% confidence intervals 20%–55%, $P<0.001$). The treatment was well tolerated and there were few adverse reactions ascribed to the treatment.

Whether or not these results can be extrapolated to the treatment of patients not fulfilling the criteria for this study requires judicious judgment, in particular for those admitted more than 96 h after the bleed. We specified this time limit because our hypothesis, based on previous studies, was that nimodipine should be instituted soon after the bleed in order to have a prophylactic effect upon subsequent ischaemia. In the event there was no evidence that the interval between the bleed and admission to the study influenced the findings, and, within these limitations, the delay in the institution of treatment did not preclude benefit. Although it is generally accepted that the peak incidence of delayed ischaemia is around 7–10 days following the initial bleed, the need for early and prolonged treatment was highlighted by the occurrence of infarction as soon as 2 days after the initial bleed, and as late as 41 days. Prolonged treatment may be valuable when there is a long delay between bleed and operation.

The use of oral treatment allowed nimodipine to be given for the full 21-day schedule. Some other studies have employed intravenous administration, which requires a central venous line. This is a considerable practical disadvantage and has usually led to a 2-week period of treatment.

The mechanism of action of nimodipine remains a matter for speculation. The results of the present study are consistent with its having a prophylactic effect rather than minimising damage after an ischaemic insult has occurred. It is also clear that, despite nimodipine treatment, cerebral ischaemia still developed in one in five cases. Nevertheless, we conclude that oral nimodipine, begun within 4 days and continued for 21 days, is beneficial in patients with a subarachnoid haemorrhage.

References

1. Allen GS, Ahn HS, Preziosi TJ et al. (1983) Cerebral arterial spasm – a controlled trial of nimodipine in patients with subarachnoid haemorrhage. N Engl J Med 308:619–624
2. Auer LM (1984) Acute operation and preventive nimodipine improve outcome in patients with ruptured aneurysms. Neurosurgery 15:57–66
3. Jennett B, Bond M (1975) Assessment of outcome after severe brain damage. A practical scale. Lancet I:480–484
4. Ljunggren B, Brandt L, Saveland H et al. (1984) Outcome in 60 consecutive patients treated with early aneurysm operation and intravenous nimodipine. J Neurosurg 61:864–873
5. Mee E, Dorrance D, Lowe D, Neil-Dwyer G (1988) Controlled study of nimodipine in aneurysm patients treated early after subarachnoid haemorrhage. Neurosurgery 22:484–491
6. Petruik KC, West M, Mohr G et al. (1988) Oral nimodipine treatment in poor grade aneurysm patients. Results of a multi-centre double blind placebo controlled trial. J Neurosurg 68:505–517
7. Philippon J, Grob R, Dagreou F et al. (1986) Prevention of vasospasm in subarachnoid haemorrhage. A controlled study with nimodipine. Acta Neurchir 82:110–114
8. Pickard JD, Murray GD, Illingworth R et al. (1989) Effect of oral nimodipine on cerebral infarelian and outcome after subarachnoid haemorrhage. British aneurysm trial. Brit Med J 298:636–642
9. Saveland H, Ljunggren B, Brandt L et al. (1986) Delayed ischemic deterioration in patients with early aneurysm operation and intravenous nimodipine. Neurosurgery 18:146–150
10. Teasdale GM (1988) A universal subarachnoid haemorrhage scale: report of a committee of the World Federation of Neurosurgical Societies. J Neurol Neurosurg Psychiatry 51:1457

Various

Impaired Calcium Homeostasis and Calcium Antagonists in Muscle Diseases

S. ZIERZ

Biochemical Principles of Calcium Homeostasis in Muscle

The electrical excitation of the muscle plasma membrane following a nerve impulse spreads into the T-tubule system. The electrical signal is then transferred to the sarcoplasmic reticulum, and large amounts of Ca^{2+} stored in its lumen are released into the cytosol. In the resting state at low cytosolic Ca^{2+} concentrations (about 10^{-7} M), tropomyosin sterically blocks the interaction of myosin heads with actin. When the cytosolic Ca^{2+} level is raised up to 10^{-5} M following the electrical excitation, free Ca^{2+} interacts with troponin, resulting in a conformational change of the tropomyosin, allowing the myosin heads to interact with actin. Subsequently, the Ca^{2+} concentrations of the myofibrillar compartment are rapidly restored to the resting level by uptake of Ca^{2+} into the sarcoplasmic reticulum through the Ca^{2+}-Mg^{2+} ATPase located in the sarcoplasmic reticulum [19].

The homeostasis of intracellular Ca^{2+} concentration is regulated by several Ca^{2+} binding proteins that belong to two groups: soluble nonmembranous high affinity Ca^{2+} binding proteins and membrane-intrinsic proteins [4]. Nonmembranous Ca^{2+} binding proteins such as troponin C, calmodulin, parvalbumin, and several other proteins have only a limited capacity of buffering Ca^{2+} because of their limited total amount. Their major function is in processing of the Ca^{2+} signal through Ca^{2+} induced conformational changes. Membrane-intrinsic Ca^{2+} binding proteins are Ca^{2+} transporting systems with high Ca^{2+} buffering capacity located in the plasma membrane, the sarcoplasmic reticulum, mitochondria, and other membranous organelles. The relative contribution of the various Ca^{2+} modulated proteins and transport systems to intracellular Ca^{2+} homeostasis differs markedly in different cell types. In contrast to nervous tissue and heart muscle, the Ca^{2+} transporting systems of the plasma membrane are not of primary importance for the acute regulation of intracellular Ca^{2+} concentration in skeletal muscle. Most of the Ca^{2+} required for contraction in skeletal muscle originates from the sarcoplasmic reticulum, and the reuptake of Ca^{2+} into the sarcoplasmic reticulum is essential for relaxation. The exact mechanisms leading to the Ca^{2+} release and regulating the reuptake are unknown. The main target of the Ca^{2+} signal in muscle is troponin C. The physiological significance of other

Neurologische Universitätsklinik Bonn, Sigmund-Freud-Str. 25, D-5300 Bonn, FRG.

Cerebral Ischemia and Calcium
Ed. by A. Hartmann and W. Kuschinsky

Ca^{2+} binding proteins in skeletal muscle such as calmodulin, parvalbumin, and calsequestrin is still to be established. The physiological role of mitochondrial Ca^{2+} transport remains a mystery. Although the Ca^{2+} loading capacity of mitochondria by far exceeds the total Ca^{2+} content of the cell, mitochondria are now considered to have no major share in Ca^{2+} regulation. They may, however, serve as a last line of defence in various pathological conditions accompanied by a large influx of Ca^{2+} into the cytosol.

Brody Disease

In 1969 Brody [3] first described this unusual disorder of skeletal muscle function characterized by painless contractures occurring during exercise. There was no permanent weakness of muscle but a progressive slowing of relaxation with vigorous exercise, and after repetitive maximal contraction the muscle became paralyzed in a state of marked shortening that was fully reversible after 5–15 s of rest. There was no percussion myotonia and electromyography showed no myotonic discharges but electrical silence during the periods of contracture. Ischemic exercise resulted in a normal rise of blood lactate. Muscle biopsy revealed no morphological abnormalities and phosphorylase activity and glycogen were normal. Microsomal Mg^{2+}-ATPase was within the normal range. Microsomal Ca^{2+} uptake, however, was markedly reduced and it was suggested that the reduced Ca^{2+} uptake by the sarcoplasmic reticulum caused the impaired muscle relaxation in vivo.

In 1986 Karpati et al. [14] reported four male patients from two families with symptoms similar to those of Brody's case. Electron microscopic examination of the muscle showed no significant abnormalities of the sarcoplasmic reticulum and the triadic junctions. Extensive biochemical studies on muscle biopsies revealed an extremely low rate of ATP-dependent Ca^{2+} uptake in a particulate membrane fraction. Gel electrophoresis showed a marked reduction of a 100-kD phosphoprotein, corresponding to the Ca^{2+}-ATPase of the sarcoplasmic reticulum. Immunoblots showed an almost nondetectable concentration of immunoreactive Ca^{2+}-ATPase of the sarcoplasmic reticulum. Immunohistochemical techniques revealed that this deficiency of immunoreactive Ca^{2+}-ATPase was limited to type 2 muscle fibers. Thus, a deficiency of the Ca^{2+}-transport ATPase of the sarcoplasmic reticulum seems to be the primary genetic defect in Brody disease and might explain the clinical symptoms in this disorder. Progressive impairment of muscle relaxation would occur because of a cumulative increase of cytosolic Ca^{2+}. The fact that the deficiency was only expressed in fast-twitch fibers might explain why impaired relaxation was only noted after phasic exercise in contrast to tonic activity such as maintaining posture, where no dysfunction was noted. Since after rest the patients could relax their muscles and start exercise again, it should be presumed that during rest the accumulated cytosolic Ca^{2+} is eventually restored to normal or at least near normal levels. The mechanism, however, by which this is achieved is not clear. Other Ca^{2+} transport systems might be compensatorily involved.

Recently, Danon and co-workers reported four members of a single family with sarcoplasmic Ca^{2+}-ATPase deficiency in type 2 muscle fibers [7]. In contrast to the earlier reported cases, in which a recessively transmitted disease was suggested, the four new cases appeared to follow an X-chromosomal or autosomal dominant pattern of inheritance.

Diagnostic possibilities suggested by exercise-induced muscle stiffness include myotonic disorders and defects of glucose metabolism in muscle. The aggravation of symptoms upon continuous exercise and the electrical silence during contracture distinguish Brody disease from myotonia congenita and myotonic dystrophy. Unlike McArdle disease and other disorders of glucose metabolism, the exercise-induced muscle stiffness in Brody disease is not painful.

One of Karpati's patients [14] was treated with dantrolene sodium (25 mg) or nifedipine (20 mg) four times a day but muscular function did not improve.

Malignant Hyperthermia

Malignant hyperthermia (MH) is an inherited muscle disorder that becomes clinically manifest under general anesthesia [12]. Agents which precipitate attacks of MH include halothane and succinylcholine. Episodes of MH are characterized by an increase of body temperature, hypermetabolism, and stiffness of skeletal muscle. A certain inbreed of swine represents an experimental model for MH because these animals develop episodes of MH under general anesthesia and under several forms of stress, especially before slaughter.

Although the primary molecular defect in MH is unknown, there is increasing evidence for a defective regulation of intracellular free Ca^{2+} in skeletal muscle of patients and animals susceptible to MH. Using Ca^{2+}-sensitive microelectrodes it was found that in human intercostal muscle free cytosolic Ca^{2+} at rest was 3.5-fold higher in patients with previous attacks of MH than in normal controls [17]. Intracellular ionized Ca^{2+} was measured by the same method in vivo in skeletal muscle fibers of MH-susceptible swines before and during hyperthermic episodes, showing an almost 20-fold increase of Ca^{2+} during the episode. Administration of dantrolene during the episode reduced the increased Ca^{2+} below the resting level [18]. In contrast, it was recently shown with the specific fluorescent Ca^{2+} indicator fura-2, that resting cytoplasmic Ca^{2+} concentrations were not significantly different in muscle of normal and of MH-susceptible pigs [13]. Halothane produced an increase in cytosolic Ca^{2+} with associated contracture in muscle of MH-susceptible pigs but not in normal muscle. The rise in intracellular free Ca^{2+} in skeletal muscle might result in the abnormal contraction and could activate phosphorylase kinase, leading to accelerated glycogen breakdown. Compensatory overloading of mitochondria with Ca^{2+} may also lead to uncoupling of oxidative phosphorylation and excessive heat production.

The sarcoplasmic reticulum is the focus of current theories of malfunction of Ca^{2+} regulation during MH episodes. This is supported by the fact that the muscle relaxant dantrolene has a specific therapeutic effect in both porcine and human MH. Dantrolene inhibits various manifestations of MH in vitro as well as in vivo, apparently by decreasing the release of Ca^{2+} from the sarcoplasmic reticulum. The mechanism, however, by which in normal muscle the wave of depolarization from the T-tubule is transmitted to the sarcoplasmic reticulum or to the triad is not yet understood. Dantrolene acts in the same area of the muscle. Its specific action, probably closely related to the mechanism of MH, might be used to define the lesion in MH.

Based on controlled swine studies, Ca^{2+} antagonists such as verapamil are not effective in treating attacks of MH [10].

Duchenne Muscular Dystrophy

Duchenne muscular dystrophy (DMD) is an X-linked disease characterized by progressive muscle wasting. The primary defective gene product of this disease, however, has not been identified so far and the pathogenic mechanism of the disease is still obscure.

Mokri and Engel [20] found delta-shaped lesions in nonnecrotic muscle fibers from patients with DMD, and electron microscopic studies demonstrated defects of the plasma membrane while the basal lamina was preserved. The structural defects of the sarcolemma were confirmed by experiments showing an ingress of extracellular fluid containing horseradish peroxidase into the lesions. Thus, the membrane defects might explain the leak of intracellular enzymes in DMD. Defects of the sarcolemma in DMD might cause an influx of Ca^{2+}-rich extracellular fluid into the muscle fiber. Subsequent histochemical studies demonstrated intracellular Ca^{2+} accumulations not only in hypercontracted fibers but also in nonnecrotic muscle fibers, whereas Ca^{2+} accumulation was low or absent in necrotic fibers. In nonnecrotic fibers these Ca^{2+} deposits were localized in delta-shaped regions resembling those fibers permeable for peroxidase [2]. Using the novel probe, quin-2, an internally trapped fluorescent Ca^{2+} indicator, intracellular free Ca^{2+} concentrations were found to be elevated in cultured skeletal muscle cells from patients with DMD at rest and after acetylcholine stimulation [21]. By contrast, Ca^{2+} concentrations in skin fibroblasts, lymphocytes, and transformed lymphoblasts from patients with DMD did not differ from controls, which suggests that there is no generalized membrane defect in DMD [15, 21].

Thus, intracellular Ca^{2+} accumulation in muscle fibers seems to be a significant biochemical abnormality and pathogenic factor in DMD. However, it has not yet been established if this accumulation is caused primarily by a leaky membrane or by a defect of the intracellular Ca^{2+} regulation by the sarcoplasmic reticulum, the mitochondria, or other Ca^{2+} binding proteins resulting secondarily in membrane lysis.

Continued Ca^{2+} accumulation in the muscle fiber may result in one or more of the following: (a) Ca^{2+} overloading of the mitochondria, causing reduced ATP synthesis, (b) stimulation of membrane-bound phospholipases, leading to a breakdown of membrane phospholipids, and (c) stimulation of lysosomal or Ca^{2+}-activated neutral proteases resulting in an increased protein degradation [reviewed in 27].

Based on the fact that Ca^{2+} antagonists block Ca^{2+} influx into the cell through voltage-gated Ca^{2+} channels and based on the observation of intracellular Ca^{2+} accumulation in muscle fibers of patients with DMD, the idea to use Ca^{2+} antagonists for therapy in DMD originated. Several clinical trials with Ca^{2+} antagonists have subsequently been performed [reviewed in 27]. Neither verapamil [9, 26], diltiazem [1], nifedipine [22], nor flunarizine [8] had a significant beneficial effect on the clinical course of the disease. If the abnormal accumulation of Ca^{2+} in DMD is due to pathological structural defects of the sarcolemma, regardless of whether these defects are primary or secondary, it is very unlikely that drugs blocking the Ca^{2+} influx through physiological channels would be effective in DMD.

Exertional Muscle Pain Syndrome of Unknown Cause

Recurrent myalgia and muscle cramps induced by moderate exertion are symptoms of a wide range of neuromuscular diseases such as inflammatory myopathies, early motor neuron diseases, peripheral neuropathies, Becker muscular dystrophy, certain drug-induced myopathies, endocrine myopathies, and particularly metabolic myopathies. Those metabolic myopathies with exercise-induced myalgias rather than permanent muscle weakness as a prominent feature include defects of glycogen and glucose metabolism (e.g., phosphorylase deficiency, phosphofructokinase deficiency, phosphoglyceratekinase deficiency, phosphoglyceratemutase deficiency), abnormalities of lipid metabolism (e.g., carnitine palmitoyltransferase "deficiency" [28]), defects of respiratory chain enzymes, and myoadenylate deaminase deficiency. Other causes of exercise-induced myalgia are muscle ischemia resulting from occlusion of the iliac or femoropopliteal arteries (claudicatio intermittens) and ischemia of the spinal cord or cauda equina (claudicatio spinalis). In many cases, however, detailed investigation, including electrophysiological, morphological, histochemical, and biochemical studies of the muscle, fail to identify a cause. If there are no signs for a psychogenic cause of the symptoms, the syndrome is referred to as exertional muscle pain syndrome (EMPS) of unknown cause.

Walton [25] first reported on the beneficial effect of the Ca^{2+} antagonist verapamil in two patients with EMPS of unknown cause. Extensive investigation had failed to identify a metabolic defect, and neither patient had responded to analgesics, various anticonvulsants, beta-blockers, or vasodilators. Both patients showed rapid and dramatic relief of symptoms upon treatment with verapamil 60 mg 3–4 times daily. Withdrawal of the medication after several months resulted in a return of the symptoms, which resolved again on reintroduction of the drug.

Subsequently, a double-blind, placebo-controlled crossover study of verapamil was performed on 11 patients with exertional muscle pain [16]. Three patients had McArdle disease and eight patients had EMPS of unknown cause. No patient responded to placebo and none of the patients with McArdle disease responded to verapamil, but two patients with EMPS of unknown cause experienced a marked improvement of symptoms upon verapamil. The eight patients with unclassified EMPS, however, apparently represented a heterogeneous group of disorders with different nonspecific electromyographic and histological abnormalities. Moreover, biochemical investigations were incomplete or missing in muscle of most of the patients.

In another study three patients with EMPS of unknown cause were reported, who favorably responded to a therapy with the Ca^{2+} antagonist nifedipine [24]. The daily dose of nifedipine ranged from two times 10 mg to 40 mg. All patients had previously been treated with a variety of other drugs without beneficial effect.

Exertional muscle pain syndrome of unknown cause is a diagnosis of exclusion rather than one based on positive evidence and might comprise different muscle diseases due to so far unidentified defects. The mechanism by which Ca^{2+} antagonists are beneficial in some of the cases is not known. Exertional muscle pain might arise from three circumstances: (a) as a result of ischemia, (b) as a result of mechanical damage to the muscle fiber, and (c) as a result of metabolic abnormalities that interfere with muscle energy metabolism. Ischemia and metabolic disorders presumably cause pain as a result of accumulation of metabolites that stimulate nociceptors. Thus, the beneficial effect of verapamil and nifedipine might have been due to an improvement of the microcirculation in the muscle secondary to vasodilatation resulting from the action on smooth muscle. Several patients, however, had previously failed to respond to other vasodilators. It is also possible that verapamil has a local anesthetic effect at the surface membrane level of the muscle. In contrast to heart muscle, Ca^{2+} transport through the plasma membrane plays no significant role in the acute regulation of intracellular Ca^{2+} concentration and excitation-contraction coupling in skeletal muscle. The possibility, however, remains that in skeletal muscle Ca^{2+} antagonists might influence Ca^{2+}-dependent metabolic effects such as activation of phosphorylase and Ca^{2+} uptake by mitochondria.

Other Muscle Diseases

Abnormalities of intracellular Ca^{2+} regulation have been reported in several other muscle disorders and have been reviewed in [5]: One patient with Luft syndrome, a mitochondrial myopathy caused by "loose coupling" of the mitochondria, was extensively studied by DiMauro and co-workers. Isolated muscle mitochondria of this patient showed normal Ca^{2+} uptake, but the accumulated Ca^{2+} could not be retained and was rapidly released. Studies on muscle mitochondria from patients with myotonic dystrophy by Carafoli and co-workers revealed a normal or slightly reduced uptake of Ca^{2+}, but the ability to retain the accumulated Ca^{2+} was impaired in one of the cases studied. The mitochondrial Ca^{2+} content was markedly increased in muscle of patients with myotonic dystrophy compared to controls. An earlier study [23] showed a decreased ability of total Ca^{2+} uptake in isolated microsomal fractions in patients with myotonic dystrophy that paralleled the degree of histological damage in muscle. The pathophysiological significance of these observations remains obscure. It has been suggested that these findings reflect an excessive loading of mitochondria due to an influx of Ca^{2+} from the extracellular space exceeding the buffering capacity of the mitochondria.

Therapeutic trials with Ca^{2+} antagonists have occasionally been reported in myotonic disorders. Following a therapy with verapamil, five patients with paramyotonia congenita, myotonic dystrophy, myotonia congenita, and Schwartz-Jampel syndrome showed reduced muscle stiffness on exposure to cold [6]. Subsequently, a single-blind crossover trial with nifedipine on ten patients with myotonic dystrophy failed to show convincing beneficial effects on myotonia, grip strength, or muscle fatigability [11].

Summary

Calcium plays a vital role in the regulation of muscle contraction. Several muscle diseases are thought to be caused by an impaired intracellular Ca^{2+} homeostasis affecting either the mechanism of contraction or the structural integrity of the muscle. In most of these cases, however, the exact biochemical mechanism of the Ca^{2+} derangement has not been identified.

The present review focuses on those muscle diseases in which (a) impaired intracellular Ca^{2+} homeostasis seems to be a pathogenetic factor and (b) therapeutic trials with Ca^{2+} antagonists have been reported.

References

1. Bertorini TE, Palmieri GMA, Griffin JW et al. (1988) Effect of chronic treatment with the calcium antagonist diltiazem in Duchenne muscular dystrophy. Neurology 38:609–613
2. Bodensteiner JB, Engel AG (1978) Intracellular calcium accumulation in Duchenne dystrophy and other myopathies. Neurology 28:439–446
3. Brody IA (1969) Muscle contracture induced by exercise. N Engl J Med 281:187–192
4. Carafoli E (1987) Intracellular calcium homeostasis. Ann Rev Biochem 56:395–433
5. Carafoli E, Roman I (1980) Mitochondria and disease. Mol Aspects Med 3:295–429
6. Cook JD, Henderson-Tilton AC, Daigh JD, Oliver YP (1984) Beneficial response to a Ca^{++} channel antagonist in myotonic syndromes. Neurology 34 (Suppl 1):193–194
7. Danon MJ, Karpati G, Charuk J, Holland P (1988) Sarcoplasmic reticulum adenosine triphosphatase deficiency with probable autosomal dominant inheritance. Neurology 38:812–814
8. Dick DJ, Gardner-Medwin D, Gates PG, Gibson M, Simpson JM, Walls TJ (1986) A trial of flunarizine in the treatment of Duchenne muscular dystrophy. Muscle Nerve 9:349–354
9. Emery AEH, Skinner R, Howden LC, Matthews MB (1982) Verapamil in Duchenne muscular dystrophy. Lancet I:559
10. Gallant EM, Foldes FF, Rempel WE, Gronert GA (1985) Verapamil is not a therapeutic adjunct to dantrolene in porcine malignant hyperthermia. Anesth Analg 64:601–606
11. Grant R, Sutton DL, Behan PO, Ballantyne JP (1987) Nifedipine in the treatment of myotonia in myotonic dystrophy. J Neurol Neurosurg Psychiatry 50:199–206
12. Gronert GA, Mott J, Lee J (1988) Aetiology of malignant hyperthermia. Br J Anaesth 60:253–267

13. Iaizzo PA, Klein W, Lehmann-Horn F (1988) Fura-2 detected myoplasmic calcium and its correlation with contracture force in skeletal muscle from normal and malignant hyperthermia susceptible pigs. Pflügers Arch 411:648–653
14. Karpati G, Charuk J, Carpenter S, Jablecki C, Holland P (1986) Myopathy caused by a deficiency of Ca^{++} ATPase in sarcoplasmic reticulum (Brody's disease). Ann Neurol 20:38–49
15. Klip A, Elder B, Ruiz-Funes HP, Buchwald M, Grinstein S (1985) The free cytoplasmic Ca^{++} levels in Duchenne muscular dystrophy lymphocytes. Muscle Nerve 8:317–320
16. Lane RJM, Turnbull DM, Welch JL, Walton J (1986) A double-blind, placebo-controlled, crossover study of verapamil in exertional muscle pain. Muscle Nerve 9:635–641
17. Lopez JR, Alamo L, Caputo C, Wikinski J, Ledezma D (1985) Intracellular ionized calcium concentration in muscles from humans with malignant hyperthermia. Muscle Nerve 8:355–358
18. Lopez JR, Allen PD, Alamo L, Jones D, Sreter FA (1988) Myoplasmic free Ca^{++} during a malignant hyperthermia episode in swine. Muscle Nerve 11:82–88
19. Martonosi AN (1984) Mechanisms of Ca^{++} release from sarcoplasmic reticulum of skeletal muscle. Physiol Rev 64:1240–1320
20. Mokri B, Engel AG (1975) Duchenne dystrophy: electron microscopic findings pointing to a basic or early abnormality in the plasma membrane of the muscle fiber. Neurology 25:1111–1120
21. Mongini T, Ghigo D, Doriguzzi C, Bussolino F, Pescarmona G, Pollo B, Schiffer D, Bosia A (1988) Free cytoplasmic Ca^{++} at rest and after cholinergic stimulus is increased in cultured muscle cells from Duchenne muscular dystrophy patients. Neurology 38:476–480
22. Moxley RT, Brooke MH, Fenichel GM et al. (1987) Clinical investigation in Duchenne dystrophy. VI. Double-blind controlled trial of nifedipine. Muscle Nerve 10:22–33
23. Mussini I, DiMauro S, Angelini C (1970) Early ultrastructural and biochemical changes in muscle in dystrophia myotonica. J Neurol Sci 10:585–604
24. Sufit RL, Peters HA (1984) Nifedipine relieves exercise-exacerbated myalgias. Muscle Nerve 7:647–649
25. Walton J (1981) Diffuse exercise-induced muscle pain of undetermined cause relieved by verapamil. Lancet I:993
26. Zahlmann F, Perloff JK, Durant NN, Campion DS (1983) Acute respiratory failure following intravenous verapamil in Duchenne muscular dystrophy. Am Heart J 3:104–105
27. Zierz S (1988) Calcium and calcium antagonists in Duchenne muscular dystrophy. In: Bartko D, Turcani P, Stern G (eds) New trends in clinical neuropharmacology. John Libbey, London (Current problems in neurology, vol 7, pp 39–43)
28. Zierz S, Engel AG (1985) Regulatory properties of a mutant carnitine palmitoyltransferase in human skeletal muscle. Eur J Biochem 149:207–214

Subject Index

GPSR Compliance

The European Union's (EU) General Product Safety Regulation (GPSR) is a set of rules that requires consumer products to be safe and our obligations to ensure this.

If you have any concerns about our products, you can contact us on ProductSafety@springernature.com

In case Publisher is established outside the EU, the EU authorized representative is:

Springer Nature Customer Service Center GmbH
Europaplatz 3
69115 Heidelberg, Germany

FSC
www.fsc.org
MIX
Papier | Fördert
gute Waldnutzung
FSC® C083411

Kolibri 360 GmbH
Ferdinand-Jühlke-Straße 7
99095 Erfurt, Deutschland
produktsicherheit@kolibri360.de